Coastal Engineering: Sustainability and New Technologies

Coastal Engineering: Sustainability and New Technologies

Editors

M. Dolores Esteban
José-Santos López-Gutiérrez
Vicente Negro
M. Graça Neves

MDPI

Basel • Beijing • Wuhan • Barcelona • Belgrade • Novi Sad • Cluj • Manchester

Editors

M. Dolores Esteban
ETS. Ingenieros de Caminos, Canales y Puertos
Universidad Politécnica de Madrid
Madrid
Spain

José-Santos López-Gutiérrez
ETS. Ingenieros de Caminos, Canales y Puertos
Universidad Politécnica de Madrid
Madrid
Spain

Vicente Negro
ETS. Ingenieros de Caminos, Canales y Puertos
Universidad Politécnica de Madrid
Madrid
Spain

M. Graça Neves
CERIS—Civil Engineering Research and Innovation for Sustainability
Department of Civil Engineering
NOVA School of Science and Technology
Caparica
Portugal

Editorial Office
MDPI
St. Alban-Anlage 66
4052 Basel, Switzerland

This is a reprint of articles from the Special Issue published online in the open access journal *Journal of Marine Science and Engineering* (ISSN 2077-1312) (available at: www.mdpi.com/journal/jmse/special_issues/bz_coastal_engineering).

For citation purposes, cite each article independently as indicated on the article page online and as indicated below:

Lastname, A.A.; Lastname, B.B. Article Title. *Journal Name* **Year**, *Volume Number*, Page Range.

ISBN 978-3-0365-8591-8 (Hbk)
ISBN 978-3-0365-8590-1 (PDF)
doi.org/10.3390/books978-3-0365-8590-1

Contents

Preface

Coastal engineering is a discipline in continuous evolution, where it is essential to seek a balance between the natural character of the coastal zone and the integration, to a greater or lesser extent, of human activities in that space.

The present Special Issue had the objective of gathering scientific articles dealing with any aspect of coastal engineering related to sustainability or the use of new technologies. Manuscripts could be focused on the past, the present, or what is expected in the future (short, medium, and long term). The manuscripts should have in some way shown concepts related to sustainability or the use of new technologies.

Examples of topics that were welcomed in the Special Issue were environmental aspects related to maritime works (coastal, port, and offshore activities), coastal evolution, climate change, sea level rise and its influence in the design, construction, and/or maintenance of maritime infrastructures, sustainability aspects in the sea, marine renewable energies, including offshore wind, and novel aspects such as the use of new materials, the calculation of the carbon footprint, the use of neural networks, the use of BIM, etc.

This list is an example of the topics that were intended to be covered in the Special Issue, where innovative papers on coastal engineering were welcomed.

M. Dolores Esteban, José-Santos López-Gutiérrez, Vicente Negro, and M. Graça Neves
Editors

Journal of
Marine Science and Engineering

MDPI

Editorial

Coastal Engineering: Sustainability and New Technologies

M. Dolores Esteban [1,*], José-Santos López-Gutiérrez [2], Vicente Negro [2] and M. Graça Neves [3]

[1] Environment, Coast and Ocean Research Laboratory (ECOREL-UPM), Universidad Politécnica de Madrid, 28040 Madrid, Spain
[2] Departamento Ingeniería Civil: Hidráulica, Energía y Medio Ambiente, Universidad Politécnica de Madrid (UPM), 28040 Madrid, Spain; josesantos.lopez@upm.es (J.-S.L.-G.); vicente.negro@upm.es (V.N.)
[3] CERIS—Civil Engineering Research and Innovation for Sustainability, Department of Civil Engineering, NOVA School of Science and Technology, 2829-516 Caparica, Portugal; mg.neves@fct.unl.pt
* Correspondence: mariadolores.esteban@upm.es

Citation: Esteban, M.D.; López-Gutiérrez, J.-S.; Negro, V.; Neves, M.G. Coastal Engineering: Sustainability and New Technologies. *J. Mar. Sci. Eng.* **2023**, *11*, 1562. https://doi.org/10.3390/jmse11081562

Received: 17 July 2023
Revised: 31 July 2023
Accepted: 1 August 2023
Published: 7 August 2023

Coastal engineering is a constantly evolving discipline, in which it is essential to seek a balance between the natural character of the coastal zone and the integration, to a greater or lesser extent, of human activities in that space. Between 2020 and 2023, thirty scientific articles addressing any aspect of coastal engineering related to the sustainability or application of novel technologies were published in a Special Issue. A book, entitled "Coastal Engineering: Sustainability and New Technologies", compiled these papers and presented contributions addressing environmental concerns related to the following topics: maritime works (coastal, port, and offshore activities); coastal evolution; climate change; sea level rise and its influence on the design, construction, and/or maintenance of maritime infrastructures; sustainability in the sea; marine renewable energies, including offshore wind, and novel aspects such as the application of new materials; the calculation of the carbon footprint; the use of neural networks; and the use of BIM, etc. A brief overview of the work collected is provided in this paper.

New technologies applied to different marine and coastal structures, from port infrastructures to offshore wind platforms, moored vessels to fish vessels on the sea, were presented in 18 papers.

The application of the building information modeling (BIM) method to manage port infrastructure maintenance is very scarce and is currently in its infancy. Valdepeñas et al. [1] presented a new infrastructure conservation management method that combines traditional methods with new technology in an attempt to address the gaps in the implementation of BIM in port maintenance. To demonstrate that coastal bathymetry can be estimated using panorama video by employing video stitching and wave speed inversion algorithms applied to imagery from two UAVs transiting along the coast, Fan et al. [2] extended video stitching to nearshore bathymetry for videos that were captured for the same coastal field simultaneously by two unmanned aerial vehicles (UAVs). With regard to the requirements of port handlings, and the ability of terminals to respond to emergency needs, Li et al. [3] presented twin yard crane scheduling; they considered the dynamic cut-off time and the non-crossing constraints of yard cranes in order to enhance the flexibility of container yard handling. Machine learning was used by Alvarellos et al. [4] to predict moored ship movement. With the movement data of 46 ships recorded in the Outer Port of Punta Langosteira (A Coruña, Spain) from 2015 until 2020, they created neural networks and gradient-boosting models that predict the six degrees of freedom of a moored vessel using ocean meteorological data and ship characteristics. For the same Port, Sande et al. [5] presented and analyzed field data obtained during the winters of 2017 and 2018 pertaining to the wave pressure in the crown wall structure, together with metocean variables measured on a buoy located very near to the breakwater. The data were compared with the results of the application of state-of-the-art equations for the calculation of pressures on crown walls. Park et al. [6] conducted marine experiments using V-pass and AIS terminals in order to determine the location of fishing vessels and thus improve the prompt rescue of fishing vessels in the occurrence of accident. For the oscillating water column (OWC), one of the most important wave energy converters, Juan et al. [7] presented a review of studies addressing the ways in which wave and geometric characteristics affect the performance of an OWC and those addressing the relationships between these characteristics and the hydrodynamic performance of an OWC. Floating offshore wind platforms were analyzed in three papers. For floating offshore wind platforms built using concrete, Cordal-Iglesias et al. [8] developed a beta version of a software called Arcwind in order to establish a framework for the development of an economic analysis tool. Vázquez et al. [9] presented a systematic analysis of prevalent corrosion prediction models and a tool developed for estimating diameter loss during the reinforcement of concrete structures in offshore wind farms. Encalada-Dávila et al. [10] proposed a design for a laboratory down-scaled vibration test bench for offshore jacket-type foundations in order to collect data and devise predictive maintenance strategies. The vibration test was based on a lab-scale wind turbine jacket foundation associated primarily with an offshore environment. For offshore steel structures subject to harmful vibrations, Zhang et al. [11] proposed a compensating measure containing the ocean environment in the adaptive control scheme, and presented numerical experiments conducted on a platform model with varying parameters in order to test the performance of the proposed adaptive controller. Piles are widely utilized in the foundation of streetlamp posts, coastal trestle bridges, and offshore wind power poles since they can bear vertical and lateral loads. Therefore, determining the ultimate lateral capacity of the wing piles and the extent of the plastic zone of the surrounding soil under lateral loading for piles is crucial. Wang et al. [12] provided a theoretical basis for the reinforcement of the soil around multi-wing piles in Dapeng Bay.

Numerical modelling solving Reynold Average Navier–Stokes equations, otherwise known as RANS models, has been extensively applied in coastal engineering around the world, but the models still require improvement and verification for some applications. In this book, five papers present the development and/or validation of various RANS models. Peng et al. [13] presented the simulation of a regular wave transformation on the slope of breakwater with an artificial block using DualSphysics. Brito et al. [14], applying the same model, employed DualSphysics to study the power absorption of a constrained wave energy hyperbaric converter (WEHC) under full-scale sea wave conditions. Didier et al. [15] presented methodologies using the FLUENT® numerical model to simulate interactions between waves and impermeable and porous costal structures, involving wave breaking and overtopping. Zhu et al. [16] applied Openfoam to the study of the effects of a solitary wave, simulating extreme waves generated by hurricanes or tsunamis, on the multiple semi-circle porous medium breakwater, which is used in marine engineering protection. The 3D hydrodynamic model Delft3D-FLOW was implemented and validated by Sánchez et al. [17] in order to analyze the impacts of tidal stream energy exploitation on the hydrodynamics of the Ria de Ortigueira, in the westernmost area of the Galician Rias Altas (NW Spain).

Physical modelling was used by Adibhusana et al. [18] in order to study the overtopping flow velocity (OFV) and the overtopping layer thickness (OTL), two important parameters in breakwater design. In this study, the overtopping of regular waves over a composite breakwater with tetrapod armor units was tested. A digital-image-based velocimetry method and bubble image velocimetry (BIV) were employed to measure the OFV and the corresponding image was digitized in order to obtain the OLT.

Sustainability was discussed in 12 papers.

Since many of the world's offshore platforms are approaching or have exceeded their design life, many researchers are endeavoring to study the repair and reinforcement of damaged components in offshore platforms. Zhang et al. [19] proposed a new type of wedge gripping for the grouting clamp, the most repaired component of offshore platforms. Ergodan et al. [20] investigated how preexisting biofouling and corrosion products on the vertical uncoated steel surfaces of offshore wind monopile foundations in seawater at Port Canaveral, FL, extending from the intertidal zone to the buried zone, affected the cathodic protection requirements when impressed current cathodic protection (ICCP) was applied under tidal conditions.

The accuracy of the calculation of forces and moments caused by wave action over the crown wall structures of breakwater is crucial to their design, especially the sea level rise caused by climate change and the possible damage of the armor are considered, since both aspects are not conventionally considered in most original design studies. In the context of climate change, it is vital to estimate the possible changes in security factors caused by both of these aspects, comparing the results with those of the original design. Pereira et al. [21] compared the forces and moments obtained for different equations in different scenarios, including the current and considering the sea level rise and armor damage, when applied to Ericeira breakwater.

A large number of cables and pipelines are distributed on the bottom of the Yellow River Delta, which is an important area for offshore oil and gas production; thus, soil resistance is a critical factor in their stability. Yang et al. [22] classified and calculated the variation in the soil layers using the cone penetration test (CPT) in the Yellow River Delta, where the impact from the river and the dynamics of the ocean tides make the soil composition and distribution in this area very complex.

The International Maritime Organization, with a new convention in MARPOL 73/78, announced that, throughout the world's oceans, the sulfur content in marine fuels should not exceed 0.5% of the mass. In this context, three research papers are presented in this book. The first, by Sultanbekov et al. [23], addresses the problematic active sedimentation of residual and mixed fuels caused by the loss of stability, and its relevance during storage and transportation, especially in marine contexts. Smyshlyaeva et al. [24] assessed and compared the effects these two heavy oil residues on the sedimentation stability of residual marine fuels: the vacuum residue (VacRes) and the visbreaking residue (VisRes). The third addresses the corrosion of reinforcement induced by chloride ion. Ju et al. [25] presented mesoscopic numerical models for concrete, considering various coarse aggregate contents and grading conditions, in order to further investigate and quantify the influence of comprehensive concrete material factors on chloride transport characteristics.

Maritime transport and ports are significant sources of pollution. Herrero et al. [26] presented the potential reductions in CO_2 emissions calculated for the last 11 years (2011–2021) close to urban areas in the Port of Santander (Spain). They distinguished yearly emissions per type of vessel and provided real information to port authorities so as to prioritize the installation of an On-Shore Power Supply (OPS), Cold Ironing (CI) or Alternative Marine Power (AMP) for the operation of piers/terminals in order to optimize investments and outcomes.

Regarding the sustainability of civil engineering works, including those associated with the maritime industry, Ortega et al. [27] presented a novel building material: mortars prepared using sustainable cements and lightweight aggregates comprising natural cork and expanded clay.

The development of sustainable and nature-based coastal protection solutions is critical for many coastlines vulnerable to erosion. D'Hurlaborde et al. [28] described the development of innovative artificial substrate screening methodologies using an optimized recirculating aquaculture system (RAS).

To capitalize on marine resources and minimize the exploitation of fresh water, some seawater is added during the process of preparing microbial-induced calcium carbonate precipitation (MICP) cementitious calcareous sand materials. Wang et al. [29] proposed the application of MICP technology and conducted a series of solidification tests in natural seawater and freshwater environments in order to determine the mechanical behavior of calcareous sand treated using MICP in a seawater environment.

Aquaculture is a sector with significant potential for growth, is a food source of great nutritional value, and provides substantial socioeconomic benefits. Recent research has demonstrated the potential development of systems devoted to aquaculture organisms using groundwater, as they are fundamental to preserving its quality and quantity. Ortiz et al. [30] proposed finding the resistivity of aquifer water and its relationship with the saturated formation of the coastal area of Guasave, Sinaloa, as well as the relationship between the resistivity of the aquifer water and the concentration of groundwater ions necessary for the survival of shrimp; this enabled the researchers to infer the groundwater quality required for shrimp farming and to select the ideal groundwater extraction sites for aquaculture.

Author Contributions: M.D.E., J.-S.L.-G., V.N. and M.G.N. jointly developed the concept and co-wrote this editorial. All authors have read and agreed to the published version of the manuscript.

Acknowledgments: The authors wish to thank all contributors to this Special Issue.

Conflicts of Interest: The authors declare no conflict of interest.

References

1. Valdepeñas, P.; Esteban Pérez, M.; Henche, C.; Rodríguez-Escribano, R.; Fernández, G.; López-Gutiérrez, J. Application of the BIM Method in the Management of the Maintenance in Port Infrastructures. *J. Mar. Sci. Eng.* **2020**, *8*, 981. [CrossRef]
2. Fan, J.; Pei, H.; Lian, Z. Surveying of Nearshore Bathymetry Using UAVs Video Stitching. *J. Mar. Sci. Eng.* **2023**, *11*, 770. [CrossRef]
3. Li, J.; Yang, J.; Xu, B.; Yin, W.; Yang, Y.; Wu, J.; Zhou, Y.; Shen, Y. A Flexible Scheduling for Twin Yard Cranes at Container Terminals Considering Dynamic Cut-Off Time. *J. Mar. Sci. Eng.* **2022**, *10*, 675. [CrossRef]
4. Alvarellos, A.; Figuero, A.; Carro, H.; Costas, R.; Sande, J.; Guerra, A.; Peña, E.; Rabuñal, J. Machine Learning Based Moored Ship Movement Prediction. *J. Mar. Sci. Eng.* **2021**, *9*, 800. [CrossRef]
5. Sande, J.; Neves, M.; López-Gutiérrez, J.; Esteban, M.; Figuero, A.; Negro, V. Field Campaign on Pressure on the Crown Wall at the Outer Port of Punta Langosteira Breakwater. *J. Mar. Sci. Eng.* **2022**, *10*, 1377. [CrossRef]
6. Park, C.; Jung, B.; Choi, W. Investigating the Reliability of the Location Transmitted by V-Pass Terminals: Prompt Rescue of Fishing Vessels. *J. Mar. Sci. Eng.* **2023**, *11*, 1023. [CrossRef]
7. Juan, N.P.; Negro, V.V.; Esteban, M.D.; López-Gutiérrez, J.-S. Review of the Influence of Oceanographic and Geometric Parameters on Oscillating Water Columns. *J. Mar. Sci. Eng.* **2022**, *10*, 226. [CrossRef]
8. Cordal-Iglesias, D.; Filgueira-Vizoso, A.; Baita-Saavedra, E.; Graña-López, M.; Castro-Santos, L. Framework for Development of an Economic Analysis Tool for Floating Concrete Offshore Wind Platforms. *J. Mar. Sci. Eng.* **2020**, *8*, 958. [CrossRef]
9. Vázquez, K.; Rodríguez, R.; Esteban, M. Corrosion Prediction Models in the Reinforcement of Concrete Structures of Offshore Wind Farms. *J. Mar. Sci. Eng.* **2022**, *10*, 185. [CrossRef]
10. Encalada-Dávila, Á.; Pardo, L.; Vidal, Y.; Terán, E.; Tutivén, C. Conceptual Design of a Vibration Test System Based on a Wave Generator Channel for Lab-Scale Offshore Wind Turbine Jacket Foundations. *J. Mar. Sci. Eng.* **2022**, *10*, 1247. [CrossRef]
11. Zhang, Y.; Ma, H.; Xu, J.; Su, H.; Zhang, J. Model Reference Adaptive Vibration Control of an Offshore Platform Considering Marine Environment Approximation. *J. Mar. Sci. Eng.* **2023**, *11*, 138. [CrossRef]
12. Wang, H.; Fu, D.; Yan, T.; Pan, D.; Liu, W.; Ma, L. Bearing Characteristics of Multi-Wing Pile Foundations under Lateral Loads in Dapeng Bay Silty Clay. *J. Mar. Sci. Eng.* **2022**, *10*, 1391. [CrossRef]
13. Peng, C.; Wang, H.; Zhang, H.; Chen, H. Parametric Design and Numerical Investigation of Hydrodynamic Characteristics of a New Type of Armour Block TB-CUBE Based on SPH Method. *J. Mar. Sci. Eng.* **2022**, *10*, 1116. [CrossRef]
14. Brito, M.; Bernardo, F.; Neves, M.; Neves, D.; Crespo, A.; Domínguez, J. Numerical Model of Constrained Wave Energy Hyperbaric Converter under Full-Scale Sea Wave Conditions. *J. Mar. Sci. Eng.* **2022**, *10*, 1489. [CrossRef]
15. Didier, E.; Teixeira, P. Validation and Comparisons of Methodologies Implemented in a RANS-VoF Numerical Model for Applications to Coastal Structures. *J. Mar. Sci. Eng.* **2022**, *10*, 1298. [CrossRef]
16. Zhu, K.; Jiang, R.; Sun, Z.; Qin, H.; Cheng, Z.; Wang, Y.; Zhao, E. Numerical Study on the Effects of the Multiple Porous Medium Breakwaters on the Propagation of the Solitary Wave. *J. Mar. Sci. Eng.* **2023**, *11*, 565. [CrossRef]
17. Sánchez, M.; Fouz, D.; López, I.; Carballo, R.; Iglesias, G. Effects of Tidal Stream Energy Exploitation on Estuarine Circulation and Its Seasonal Variability. *J. Mar. Sci. Eng.* **2022**, *10*, 1545. [CrossRef]
18. Adibhusana, M.; Lee, J.; Kim, Y.; Ryu, Y. Study of Overtopping Flow Velocity and Overtopping Layer Thickness on Composite Breakwater under Regular Wave. *J. Mar. Sci. Eng.* **2023**, *11*, 823. [CrossRef]
19. Zhang, B.; Zhang, Q.; Wang, T.; Wang, Z. Research on the Bearing Capacity of a Damaged Jacket Repaired by a Grouting Clamp Based on a Type of Wedge Gripping. *J. Mar. Sci. Eng.* **2020**, *8*, 973. [CrossRef]
20. Erdogan, C.; Swain, G. The Effects of Biofouling and Corrosion Products on Impressed Current Cathodic Protection System Design for Offshore Monopile Foundations. *J. Mar. Sci. Eng.* **2022**, *10*, 1670. [CrossRef]
21. Pereira, F.; Neves, M.; López-Gutiérrez, J.; Esteban, M.; Negro, V. Comparison of Existing Equations for the Design of Crown Walls: Application to the Case Study of Ericeira Breakwater (Portugal). *J. Mar. Sci. Eng.* **2021**, *9*, 285. [CrossRef]
22. Yang, Z.; Liu, X.; Guo, L.; Cui, Y.; Su, X.; Ling, X. Soil Classification and Site Variability Analysis Based on CPT—A Case Study in the Yellow River Subaquatic Delta, China. *J. Mar. Sci. Eng.* **2021**, *9*, 431. [CrossRef]

23. Sultanbekov, R.; Islamov, S.; Mardashov, D.; Beloglazov, I.; Hemmingsen, T. Research of the Influence of Marine Residual Fuel Composition on Sedimentation Due to Incompatibility. *J. Mar. Sci. Eng.* **2021**, *9*, 1067. [CrossRef]
24. Smyshlyaeva, K.; Rudko, V.; Povarov, V.; Shaidulina, A.; Efimov, I.; Gabdulkhakov, R.; Pyagay, I.; Speight, J. Influence of Asphaltenes on the Low-Sulphur Residual Marine Fuels' Stability. *J. Mar. Sci. Eng.* **2021**, *9*, 1235. [CrossRef]
25. Ju, X.; Wu, L.; Liu, M.; Jiang, H.; Zhang, W. Modelling of Chloride Concentration Profiles in Concrete by the Consideration of Concrete Material Factors under Marine Tidal Environment. *J. Mar. Sci. Eng.* **2022**, *10*, 917. [CrossRef]
26. Herrero, A.; Ortega Piris, A.; Diaz-Ruiz-Navamuel, E.; Gutierrez, M.; Lopez-Diaz, A. Influence of the Implantation of the Onshore Power Supply (OPS) System in Spanish Medium-Sized Ports on the Reduction in CO_2 Emissions: The Case of the Port of Santander (Spain). *J. Mar. Sci. Eng.* **2022**, *10*, 1446. [CrossRef]
27. Ortega, J.; Branco, F.; Pereira, L.; Marques, L. Chloride Ingress Resistance, Microstructure and Mechanical Properties of Lightweight Mortars with Natural Cork and Expanded Clay Prepared Using Sustainable Blended Cements. *J. Mar. Sci. Eng.* **2022**, *10*, 1174. [CrossRef]
28. D'Hurlaborde, A.; Semeraro, A.; Sterckx, T.; Van Hoey, G. Optimized Screening Methods for Investigation of the Larval Settlement of *Lanice conchilega* on Artificial Substrates. *J. Mar. Sci. Eng.* **2022**, *10*, 1443. [CrossRef]
29. Wang, Z.; Zhao, X.; Chen, X.; Cao, P.; Cao, L.; Chen, W. Mechanical Properties and Constitutive Model of Calcareous Sand Strengthened by MICP. *J. Mar. Sci. Eng.* **2023**, *11*, 819. [CrossRef]
30. Espinoza Ortiz, M.; Apún Molina, J.; Peinado Guevara, H.; Herrera Barrientos, J.; Belmonte Jiménez, S.; Ladrón de Guevara Torres, M.; Delgado Rodríguez, O. Evaluation of Groundwater in the Coastal Portion of Guasave, Sinaloa for White Shrimp Farming (*Penaeus vannamei*) through VES, Chemical Composition, and Survival Tests. *J. Mar. Sci. Eng.* **2021**, *9*, 276. [CrossRef]

Journal of
Marine Science and Engineering

MDPI

Article

Investigating the Reliability of the Location Transmitted by V-Pass Terminals: Prompt Rescue of Fishing Vessels

Cheor-Hong Park, Bong-Kyu Jung * and Won-Sam Choi

Department of Marine Police System, Gyeongsang National University, Tongyeong 53064, Republic of Korea; jikocom@hanmail.net (C.-H.P.); taihu98@naver.com (W.-S.C.)
* Correspondence: bkjung@gnu.ac.kr; Tel.: +82-55-772-9185

Abstract: Fishing boats are equipped to quickly rescue ships and save lives in the event of an incident at sea; therefore, determination of the incident location is imperative for a swift response. According to the 2021 marine accidents by ship use statistics, fishing and non-fishing boats accounted for 64.6 and 35.4%, respectively, of the total 3053 ships involved in an accident. In addition to V-pass and the Automatic Identification System (AIS), several other types of terminals exist; however, approximately 91% of all registered fishing boats use V-pass terminals. Therefore, it is essential to know the exact location of fishing boats. However, little research has been conducted on V-pass equipment. Therefore, in this study, marine experiments were conducted using V-pass and AIS terminals that are mainly used in fishing boats. To determine the exact location of an incident and rapidly respond and rescue, this study compared the data saved in the terminals with the data received by the Vessel Traffic Service center. In the event of a maritime incident, the radio shadow areas and causes of the error in the location transmitter must be investigated to quickly rescue the fishing boat and to determine the root cause of the incident, respectively.

Keywords: V-pass system; automatic identification system; global positioning system; marine accident; ship collision reproduction system

Citation: Park, C.-H.; Jung, B.-K.; Choi, W.-S. Investigating the Reliability of the Location Transmitted by V-Pass Terminals: Prompt Rescue of Fishing Vessels. *J. Mar. Sci. Eng.* **2023**, *11*, 1023. https://doi.org/10.3390/jmse11051023

Academic Editors: M. Dolores Esteban, José-Santos López-Gutiérrez, Vicente Negro and M. Graça Neves

Received: 19 April 2023
Revised: 5 May 2023
Accepted: 9 May 2023
Published: 11 May 2023

1. Introduction

According to the 2021 marine accidents by ship use statistics, fishing and non-fishing boats accounted for 64.6 and 35.4%, respectively, of the total 3053 ships involved in an accident, with 416 out of 512 casualties occurring on fishing boats [1].

To ensure the prompt rescue of individuals in fishing boat accidents, fishing boat location transmitters must be installed and operated in accordance with the "Fishing Boat Act," "Fishing Boat Act Enforcement Regulations," and "Notice on the Installation Standards and Operation of V-pass Devices" No. 2018-6 issued by the Korea Coast Guard.

As reported in the literature [2,3], 91.2% of the 61,717 registered fishing boats, that is, 56,279 fishing boats, operate V-pass terminals, of which 29,880 boats exceed 2 t, whereas 47% of the boats are under 2 t. This onboard location-tracking infrastructure currently does not comply with the requirement that ships with automatic identification system (AIS) devices be installed under the Ship Safety Act, which is applicable for passenger ships, oil tankers, and most other ship types with a minimum gross tonnage of 2, 50, and 300 t, respectively.

As revealed by the ship tracking information received by the ship detection system (Vessel Traffic Service; VTS) of the Marine Traffic Control Center, the final location data transmitted by fishing boats, which is used for investigating the incident factors, is stored 4–6 min before the incident.

The V-pass terminal is a system that records navigation data for periods of 4 and 6 min and subsequently transmits the data cumulatively. However, the navigation information between 30 s and 6 min is not saved depending on the time of the incident. Subsequently, the tracking information saved in the V-pass terminal installed in the stricken ship is directly extracted and used for incident analysis.

Notably, no prior research has been conducted to validate the location accuracy of V-pass terminals, and the current research on the tracking data and system is limited, except for certain studies reported in the literature [4–12].

This study compared the data saved in the V-pass terminal with the data received at the VTS through a sea test of the V-pass and AIS terminals installed on a test ship, and examined the transmission cycle and save interval. The location of AIS terminals was compared with that of numerous previous studies and the location of V-pass terminals, of which there is limited existing research data, to determine the frequency of location errors.

Section 2 presents the experimental details and compares the location-tracking capabilities of the V-pass and AIS terminals. Section 3 presents the results of this study obtained through the aforementioned test methodology. Section 4 proposes plans for the improvement of the identified problems based on the study findings. Section 5 presents the conclusions of this study and provides scope for future research.

2. Methods

Test Details and Procedure

The sea test was conducted on 9 June 2022 using V-pass (Samyung ENC, SPA-900) and AIS (Samyung ENC, AIS-50N, class B) terminals installed on the marine survey ship, Chambada, of Gyeongsang National University, and the experimental track connected Tongyeong port and Bijin-do island in Gyeongsangnam-do.

The V-pass terminal installs a location-transmitting device on a fishing boat to establish a rapid response system for marine accidents and has an SOS function. The main function of the terminal is to transmit and receive the location of boats for rapid response to incidents and emergency situations. When the SOS button on the terminal is pressed or the transmission/reception antenna is separated, an SOS notification is automatically transmitted to the Korea Coast Guard situation room.

Additionally, it is linked to the automatic ship departure and entry system; therefore, it is possible to report departure and entry without visiting the police box or branch office. The marine survey ship, Chambada, has a gross tonnage of 36 t and is 23.14 m in length, which are similar to those of inshore fishing boats [13]. Therefore, the experimental environment was similar to the transmission conditions of a location transmitter installed on an actual fishing boat. The maritime weather during the test was relatively clear with wind speeds of 5–8 m/s, and maximum and average wave heights of 1.5 and 0.6 m, respectively.

As per the test procedure, the tracking save intervals for the V-pass and AIS terminals were set to 5 s, which saved the data received by the Tongyoung coastal VTS and the data transmitted from the terminal while sailing and turning in the sea area near Bijin-do. Each track was analyzed by comparing the positions indicated on the electronic charts of the ship collision reproduction system [14–16]. The ship collision reproduction system is a system used by the Korea Coast Guard to reproduce real-time ship movements by entering the trajectory data of the ship to analyze ship collisions. It was created through the National Research and Development Project of the Korea Coast Guard from 2014 to 2017. In this study, V-pass and AIS data obtained through sea trials were input into the ship collision reproduction system and compared by displaying them on the electronic chart.

As illustrated in Figure 1, the V-pass and AIS antennae were installed at the same position above the wheelhouse of the test ship to minimize errors owing to the antenna position.

Figure 1. Automatic Identification System (AIS) and V-pass antennas installed at the same place on the ship.

Figure 2 presents the geographical positions and tracks of Gyeongsangnam-do, Tongyoung City Bijin-do, and Ogok-do, where the sea trials were conducted, on an electronic chart.

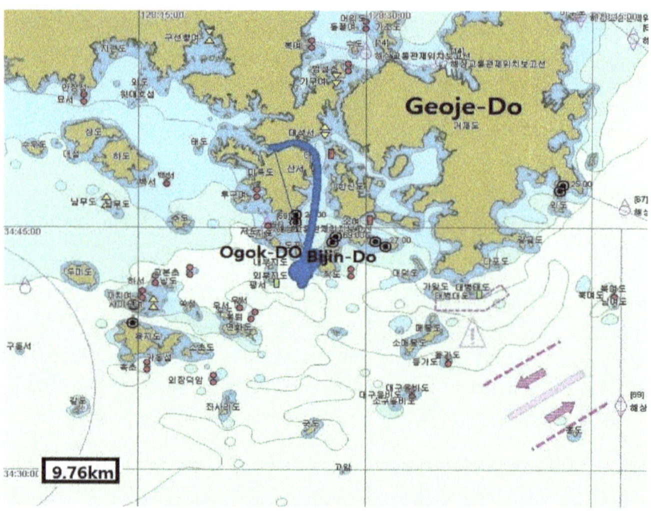

Figure 2. Sea trial location (between Bijin-do and Ogok-do).

3. Results

3.1. Comparison of the V-Pass and AIS Data Received by VTS

The data transmitted by the V-pass terminal (time zone, location, and received by VTS according to time) were compared to that transmitted by the AIS terminal. The number in the rectangles in the lower left corner of the electronic charts represents the distance scale for the horizontal length of the rectangle.

Figure 3 displays the V-pass and AIS data received by VTS from approximately 11:46 (KST) to approximately 15:53 on an electronic chart. In the test, which lasted approximately 4 h, the VTS received 269 and 223 instances of V-pass and AIS data, respectively.

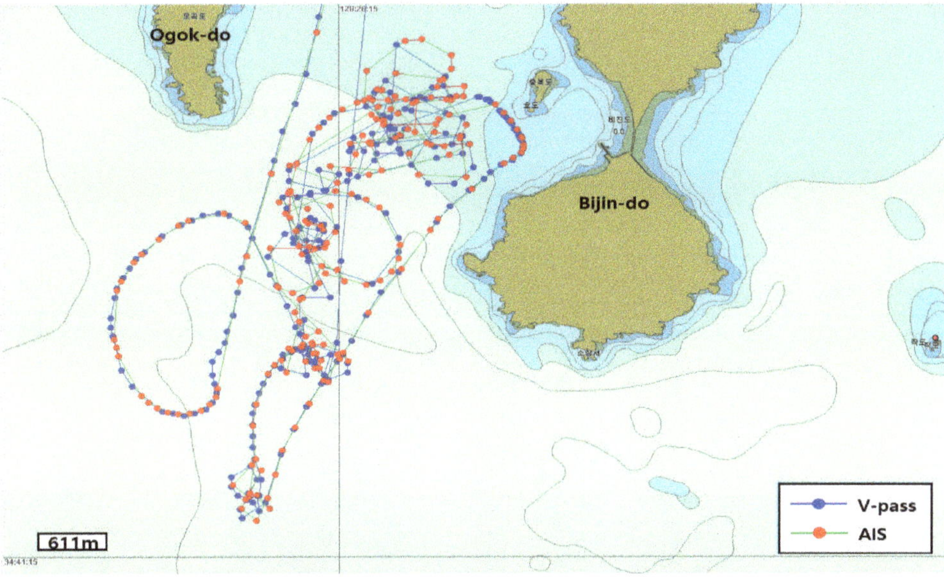

Figure 3. Comparison of the AIS and V-pass data.

The state of navigation, anchor, and changing speed in the V-pass data transmission cycle clearly did not change. These data were set to transmit in regular intervals of 30 s. Furthermore, in a particular period, the data interval was 1–7 min 30 s owing to missing data.

The reporting interval criteria for the AIS data was set to those specified in ITU-RM.1371-4 (International Telecommunication Union), which is dependent on the type of AIS class and the dynamic conditions of the ship. Class A (as mentioned in Table 1) is mainly used for international sailing vessels. The criteria in class A, such as the ship speed and course change, are more subdivided than those in class B (as mentioned in Table 2), and the reporting interval is shorter.

Table 1. AIS class A nominal reporting interval (ITU-R M.1371-4).

Dynamic Condition of the Ship	Nominal Reporting Interval
Ship at anchor or moored and not moving faster than 3 knots	3 min
Ship at anchor or moored and moving faster than 3 knots	10 s
Ship moving at 0–14 knots	10 s
Ship moving at 0–14 knots and changing course	3.3 s
Ship moving at 14–23 knots	6 s
Ship moving at 14–23 knots and changing course	2 s
Ship moving at >23 knots	2 s
Ship moving at >23 knots and changing course	2 s

Table 2. AIS class B nominal reporting interval (ITU-R M.1371-4).

Dynamic Condition of the Ship	Nominal Reporting Interval
Ship not moving faster than 2 knots	3 min
Ship moving at 2–14 knots	30 s
Ship moving at 14–23 knots	15 s
Ship moving at >23 knots	5 s

Because domestic fishing boats, excluding deep-sea fishing boats, mainly use class-B terminals, the experimental data was compared using the criteria in Table 2. The class-B terminal has a reporting interval of 30 s when the dynamic conditions of the test ship are 2–14 knots; the speed of the experimental boat was 1–14 knots. Therefore, a reporting interval of 30 s or 3 min is required depending on the speed condition.

The data was transmitted every 30 s, and the transmission data in certain sections was missing, similar to that observed for the V-pass data. In a particular period, the data was not transmitted for 1–7 min. The transmission characteristics did not deviate from the technical requirements stipulated in the Regulations on Installation Standards and Operation of Transmitting Devices, that is, every transmission was performed at intervals of 10 min or less in this experiment. Figure 4 shows three and fourteen packets of V-pass and AIS data, respectively, transmitted during a linear cruising section lasting approximately 9 min, which is a considerable difference in the number of data packets received by VTS. The AIS data was relatively regularly transmitted at intervals of 30 s–1 min, whereas the V-pass data was not transmitted for up to 7 min 30 s and the intervals between the transmission points were irregular.

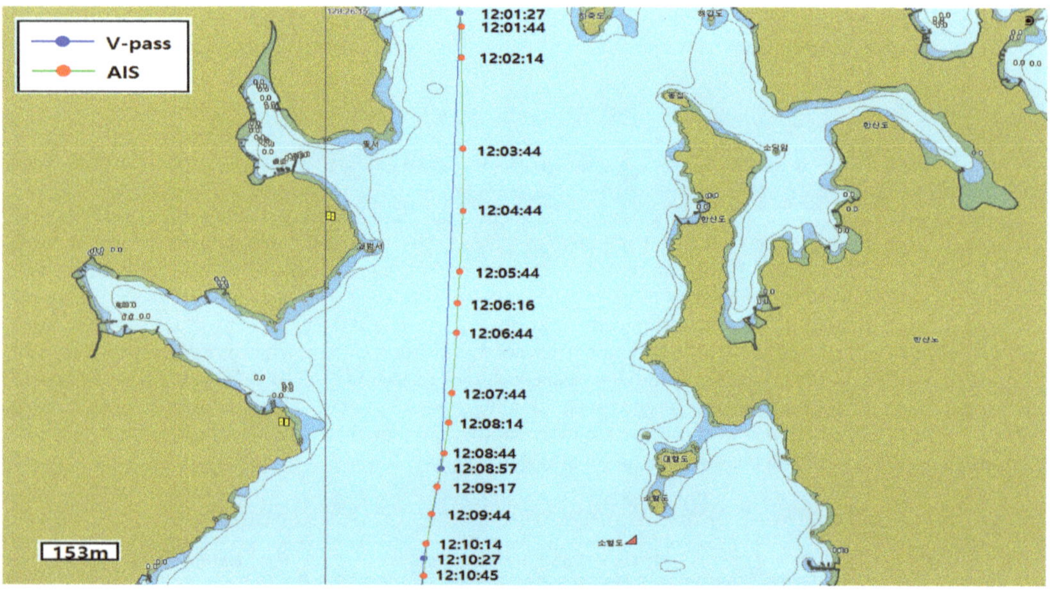

Figure 4. Linear route trajectory section where more AIS data was saved than V-pass data.

Non-storage or non-transmission of trajectory data for tens of seconds to several minutes in a linear section without speed or course changes is not problematic as the location or direction of progress over time can be predicted.

Figure 5 shows 16 and 19 V-pass and AIS data packets, respectively, for a period of approximately 20 min in a specific section of a non-linear track. Although no significant differences were observed, the AIS data was regularly received within 1 min 30 s of a turn, except for the initial track corresponding to a period of 5 min 30 s when drifting to navigation, whereas the V-pass data was received within 30 s of a turn. Clearly, the V-pass data was irregularly transmitted up to 3 min 30 s.

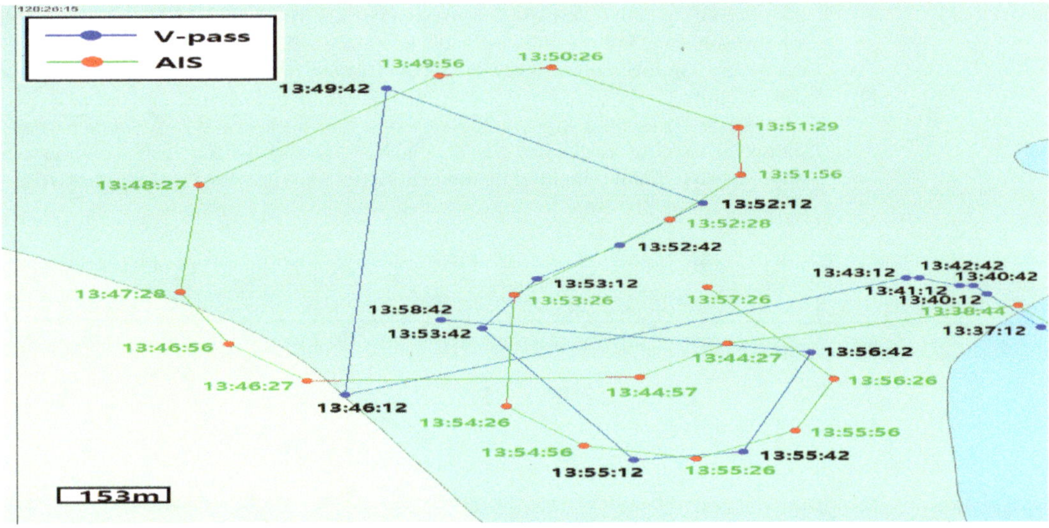

Figure 5. Sections where more AIS data was transmitted than V-pass data in a non-linear track.

If the trajectory data in a non-linear section is not transmitted for tens of seconds to several minutes, as shown in Figure 5, the location of the ship over time cannot be determined. Moreover, its direction of travel and turning circle cannot be determined.

Figure 6 presents a comparison of the trajectories in a specific section of a linear track lasting for approximately 19 min. The number of V-pass data packets (23) received by the VTS was greater than the number of AIS data packets (9). Thus, the V-pass data received was >2.5-fold greater than the AIS data.

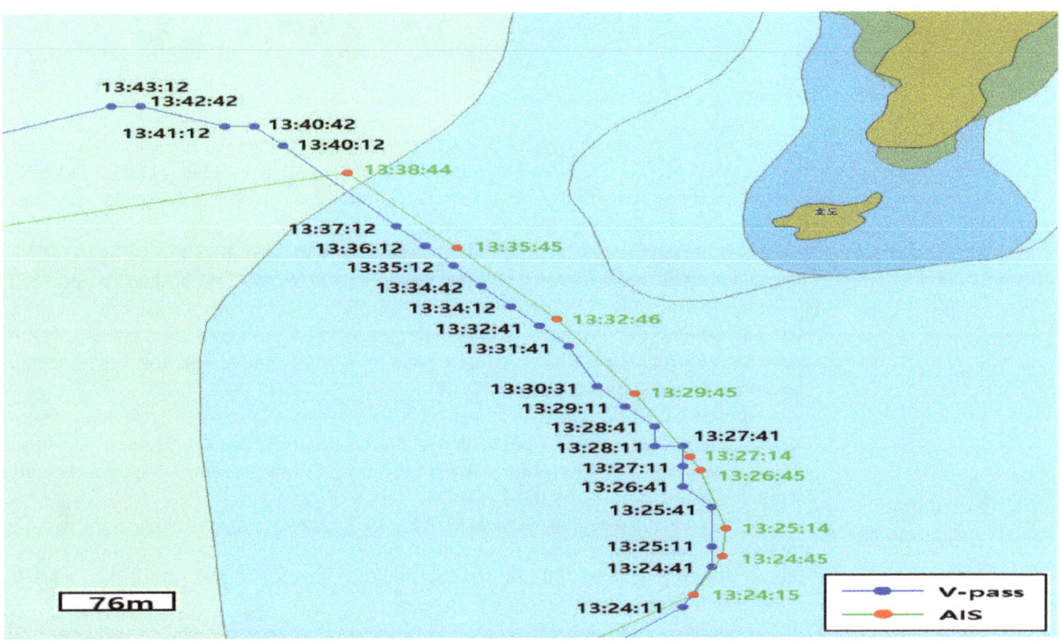

Figure 6. Linear route where more V-pass data was saved than AIS data.

According to the AIS class-B nominal reporting interval (ITU-RM.1371-4) in Table 2, when a ship does not move faster than 2 knots, the AIS terminal must transmit location data every 3 min. Simultaneously, the V-pass terminal transmitted the location data every 30 s to 1 min.

Figure 7 presents a large number of V-pass data packets and 26 AIS data packets. In particular, the data received by VTS during the 5 knots low-speed turning section lasting approximately 25 min significantly differed. Although V-pass data was regularly received, AIS data was not received for approximately 4 min in certain sections.

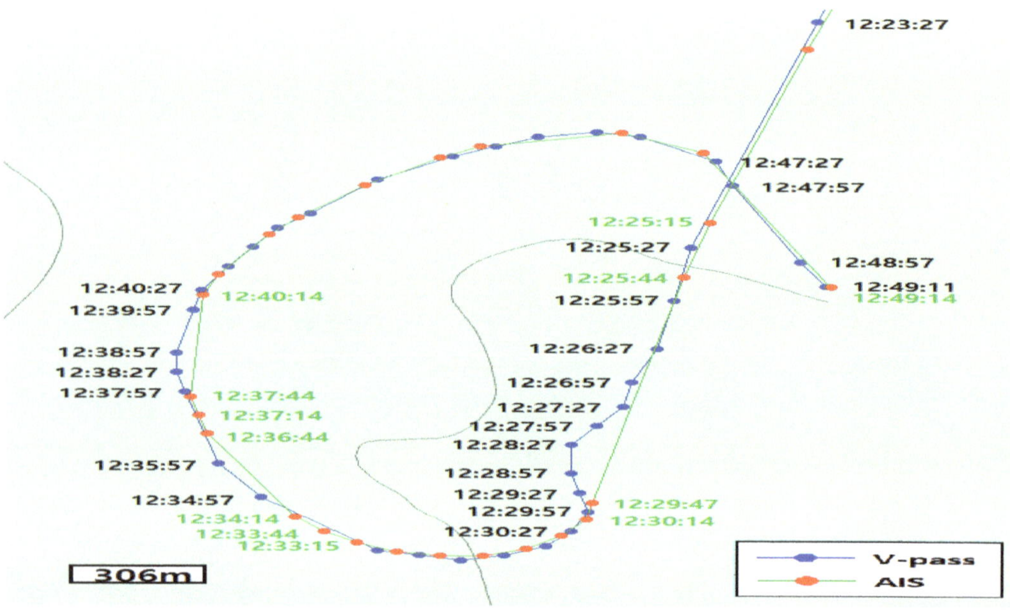

Figure 7. Curved sections where more V-pass data was saved than AIS data.

As shown in Figure 7, the AIS data did not comply with the AIS class-B nominal reporting interval (ITU-RM.1371-4) for approximately 4 min (from 12:25:44 (KST) to 12:29:47), even when the speed was approximately 4–5 knots. However, the V-pass data was simultaneously transmitted every 30 s to 1 min.

Considering the results presented in Figures 4–7, the data transmitted by the AIS and V-pass terminals and received by VTS featured several V-pass data packets corresponding to the same time period in certain sections. Moreover, the number of intervals in the AIS data packets (red dots) was relatively greater than that of the V-pass data packets. Notably, a greater volume of data received in a given period of time allows for a more detailed determination of ship movement.

Figures 4, 6 and 7 represent linear tracks or low-speed turns with relatively minimal course or speed changes. In particular, certain data in the middle of the track was not received or saved. Had this not occurred, the location of the missing time period could be easily estimated as a nearby track point.

In the case of the results illustrated in Figure 5, if certain data packets in the middle of the track are not received or saved during turning or diagonal movements, limited tracking data in the form of dots will not provide the exact turning radius of the ship; therefore, estimating the trajectory becomes challenging.

Therefore, when sailing in a rapid and continuously changing course and speed, such as when a fishing vessel is operating or turning, each terminal should have a data-save interval to track the actual course of the vessel in the data received by VTS. Therefore, a database of turn tracks must be accumulated by comparing the data received by VTS during turns set in unit intervals of 5 s through voyage experiments.

3.2. Comparison of the Data Saved in the V-Pass Terminal and the Data Received by VTS

Figure 8 is a screen-grab showing the tracking data of the ship saved every 5 s in the V-pass terminal and displayed on the electronic chart of the ship collision reproduction system. The red dots, which were transmitted by the ship every 30 s and received by VTS, are hardly visible amongst the blue dots, which were stored every 5 s. After receiving the global positioning system (GPS) signal, the difference between the data received by the VTS system and the data stored on the terminal using radio waves were comprehensively analyzed.

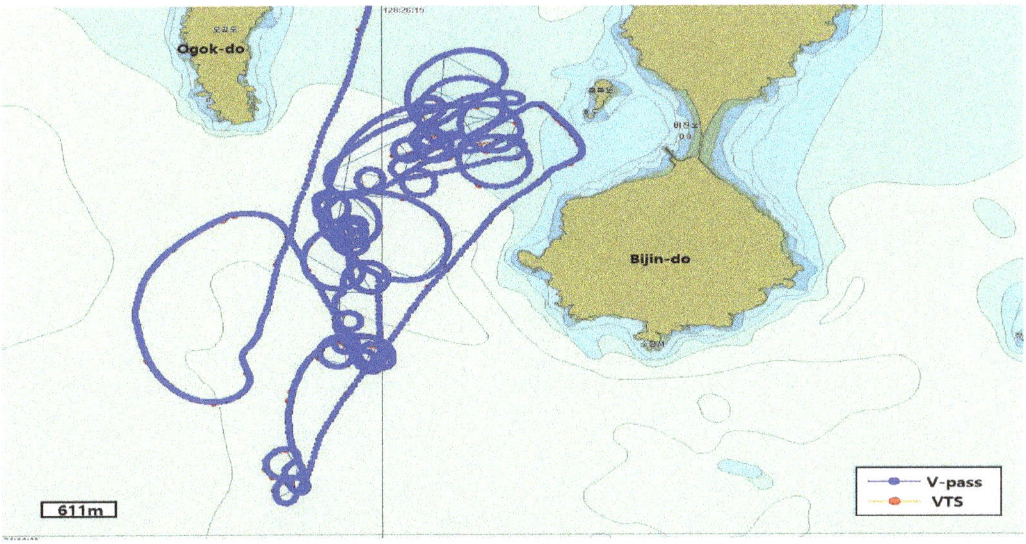

Figure 8. V-pass data rendered on the electronic chart of the ship collision reproduction system.

Figure 9 shows the Tongyoung coastal VTS radar screen, where the real-time tracking of the test ship was saved and replayed. The radar image shows the movement of the experimental ship in real time on the chart along with other nearby ships, but the stored trajectory is not as clear as that in the electronic chart of the ship collision reproduction system in Figure 8. However, it is possible to extract the trajectory of a specific ship from the stored radar image.

The blue dots and lines in Figure 10 represent the transmission data saved by the V-pass terminal in 5 s intervals, where the circular track that lasted for 4 min (14:09:42 to 14:13:42 (KST)) is represented by 47 dots. However, VTS only received five data packets during the same 4 min, which is represented with red dots and a yellow line in the figure.

Assuming that the ship was involved in an incident at approximately 14:09:42 (KST) to 14:11:12, as shown in Figure 10, VTS will send a rescue team to point B at 14:09:42 (KST), approximately 320 m from the accident point A. Although this experiment represents an extreme case, the actual situation could be more severe (as shown in Figure 3), that is, when the position of the vessel is not received by VTS for 5–7 min 30 s depending on the speed.

Similarly, the blue dots and line in Figure 11 represent the tracking data saved by the V-pass terminal in 5 s intervals. However, VTS only received four data packets within the same time period, which are marked with red dots and a yellow line. If an incident occurs at point A on the track that the fishing boat with the V-pass terminal is navigating, VTS issues a notification to the Korea Coast Guard Situation Room for point B, which is 183 m from point A, the final received location of the distressed fishing vessel, thereby triggering a nearby patrol ship.

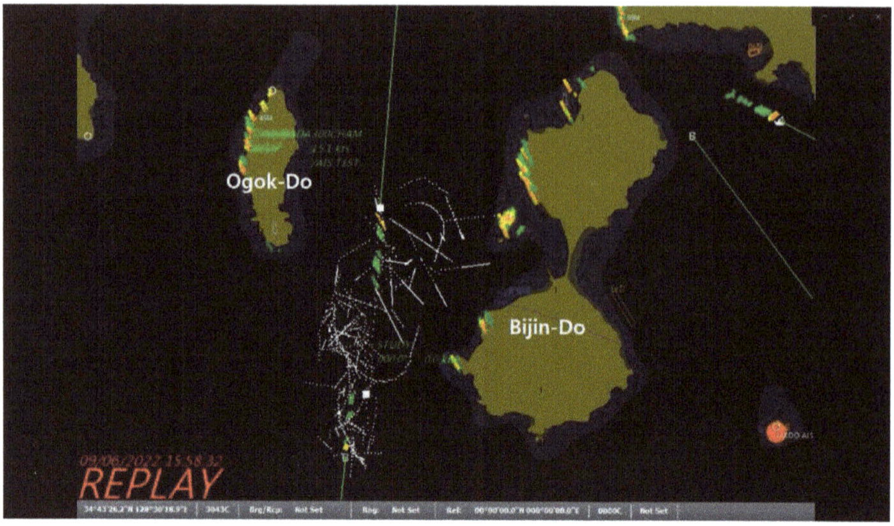

Figure 9. Screen-grab of the Tongyong coastal Vessel Traffic Service (VTS) radar screen showing the experimental track.

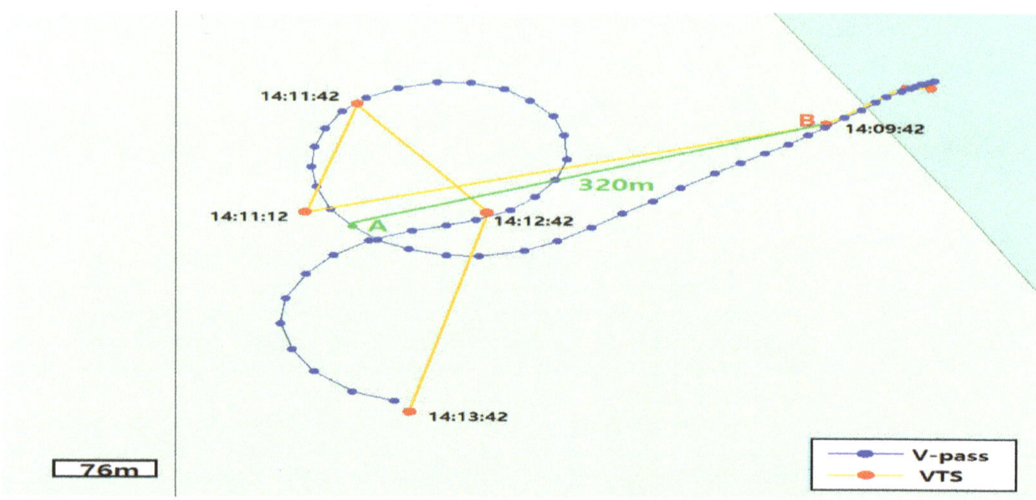

Figure 10. Comparison of the V-pass data and the five data packets received by VTS.

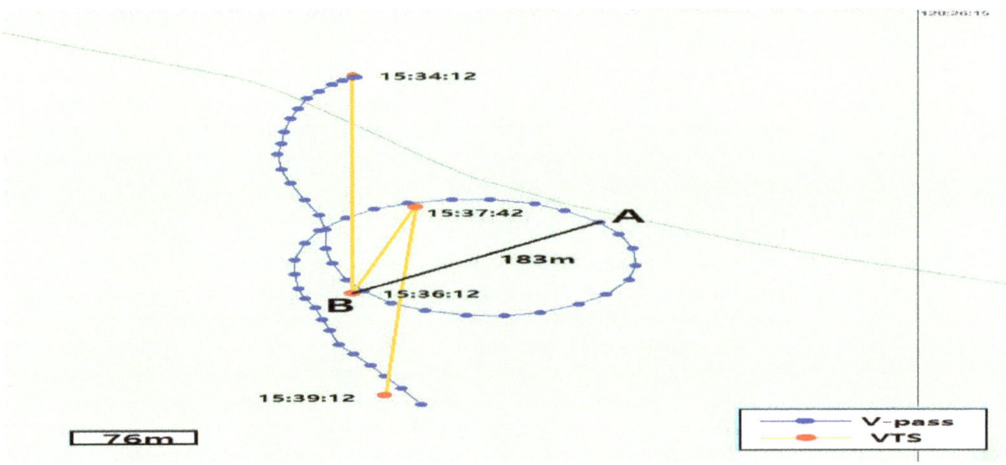

Figure 11. Comparison of the stored V-pass data and the four data packets received by VTS.

4. Discussion

This section proposes plans to improve the problems identified from the study findings. If no other fishing boats are present in the vicinity (see Figures 10 and 11), radar or other navigation equipment can be used to determine the location. However, if several fishing boats are present in the vicinity and the weather is poor, then correction cannot be determined, and the surrounding fishing boats must be directly sequentially checked. When attempting to visually identify the incident, delayed rescue operations could cause more fatalities. In particular, if the fishing boat suffers an incident and starts to sink, locating the sinking fishing boat, even with a minor positional error, will be challenging, considering that its hull will be partially visible on the water's surface.

Blank tracking sections may occur due to various factors such as model variations and manufacturing characteristics of the V-pass and AIS terminals, terminal characteristics, antenna locations and conditions, radio shadow areas, number of ships passing in the same time zone, and communication conditions. However, the data is saved in the terminal every 30 s, and this is not an erroneous feature of the terminal but that of the process through which data is transmitted to the terminal via the antenna or that of the process through which data is received by VTS through wireless communication.

As reported in the literature [17], the AIS terminal features a built-in receiver autonomous integrity monitoring system, which verifies the accuracy and integrity of the received GPS signal. The AIS message features a built-in integrity test that runs continuously at regular intervals, significantly attenuating the integrity of the AIS or causing it to stop working. Thus, a system should be designed that activates an alarm if a failure or malfunction is detected.

Moreover, considering the published research [18,19], the AIS class-B terminals currently installed and used on fishing boats can interoperate with class-A terminals; nevertheless, it is within the range that does not overload the network used for class-A communication. AIS class-B terminals are designed to transmit and receive information only in slots that are vacant within the network, and the transmission cycle of AIS class-B terminals for ship stations is already known, assuming that it is not used for transmission when the AIS networks are heavily loaded. Therefore, the dynamic information for each AIS terminal manufacturer and tracking information cannot be stored by adjusting the update period, and the cause of an incident cannot be analyzed if there is any missing information. To address these challenges, basic requirements for dynamic-information storage should be standardized for each manufacturer. Two VHF channels (CH 87 and CH 88) are currently used to selectively receive 2250 units of information per min. Consequently, system improvements are required, such as the addition of a channel that can receive more information.

As previously mentioned, research on the positional accuracy of AIS data has been conducted [20–29]; however, most of these studies focused on only a single aspect, that is, validating the positional accuracy of the V-pass terminal. As a future research direction, further research and experimentation focusing on this aspect should be conducted under unfamiliar circumstances.

5. Conclusions

A test to validate the positional error was performed in this study by comparing the data transmitted by the V-pass terminal, which is a location transmitter installed on most fishing vessels, with the location transmitted by the AIS terminal of the ship.

The experimental V-pass and AIS tracking data packets collected for approximately 4 h showed that V-pass (269 data packets) collected 46 more data packets than AIS (223).

The Tongyoung coastal VTS, which is the V-pass land-based reception station, is located in the vicinity of the test region. The regional characteristics of Tongyoung feature several islands and radio shadow areas. For optimal transmission, 480 data packets should be received in approximately 4 h. However, only 269 V-pass data packets were received, and 211 data packets, corresponding to approximately 44% of the total transmitted data, were missing. Moreover, the AIS data was missing 237 data packets, corresponding to 54% of the total transmitted data. However, this result does not conclusively indicate that the AIS position is invariably more accurate than that of the V-pass position.

If detailed navigational data can be secured within the same time, it is possible to verify the exact hourly movement of the ship and swiftly respond in the event of a marine incident. However, there is no system to receive all the information, regardless of how regularly the ship transmits its location. This can be attributed to various reasons, such as the number of ships simultaneously operating in the same area, terminal type, radio interference, limitations of the communication system, and distance from the land relay station.

The ship location transmitter is a device that allows ships to exchange their location information; therefore, this device can prevent marine incidents such as collisions. The characteristics of the test vessel used in this study were similar to those of most fishing boats, and the location transmitter was installed using similar equipment. The scope of application of the V-pass terminals, as compared to the ship location transmitter, was limited because 91% of registered fishing boats have V-pass terminals installed and operating as fishing boat location transmitters; therefore, mutual information cannot be exchanged with automatic ship-identification devices installed on general vessels, such as merchant ships, resulting in incidents involving ship collisions. The ultimate function of the location transmission device, that is, the prevention of incidents, cannot be completely realized. Moreover, these systems face reliability issues. Therefore, such system limitations can result in incidents.

For the safe operation of fishing boats [30–32], which are the ultimate motivations behind installing fishing boat location-transmitting devices, the current location of the fishing boat must be determined to establish an emergency response in the event of a marine incident and to identify the exact cause of the incident. Clarifying the root cause of such incidents can prevent the reoccurrence of similar incidents. To enhance vessel and crew safety, the V-pass system must be improved through experiments and research focusing on fishing boat location transmitters.

Author Contributions: Validation, C.-H.P. and B.-K.J.; investigation, W.-S.C.; resources, C.-H.P.; data curation, B.-K.J.; writing—original draft preparation, C.-H.P.; supervision, B.-K.J. All authors have read and agreed to the published version of the manuscript.

Funding: This research received no external funding.

Conflicts of Interest: The authors declare no conflict of interest.

References

1. National Statistical Portal. Marine Accident Statistics. Available online: https://kosis.kr (accessed on 17 April 2022).
2. Korea Coast Guard. *Statistical Yearbook of Marine Accidents*; Korea Coast Guard: Incheon, Repubilc of Korea, 2022.
3. Korea Coast Guard. *Marine Accident Statistics*; Korea Coast Guard: Incheon, Repubilc of Korea, 2021.
4. Korea Information and Communication Technology Association. Wireless Data Communication Protocol V-Pass System. 2019. Available online: http://koreaict.kr/eng/ (accessed on 17 April 2022).
5. Oh, J.H.; Kim, K.I.; Jeon, J.S.; Park, S.Y. A study on the risk analysis based on the trajectories of fishing vessel. *J. Korean Inst. Navig. Port Res.* **2014**, *6a*, 323–325.
6. Park, J.H.; Jung, H.G.; Yang, C.S. Application of V-pass using HMM fishing boat activity prediction technique. *J. Korean Soc. Coast. Disaster Prev.* **2021**, *8*, 221–227. [CrossRef]
7. *TTAK. KO-06.0281*; Wireless Data Communication Protocol for Maritime Security Safety Net. Telecommunications Technology Association: Seongnam, Republic of Korea, 2014. Available online: https://www.tta.or.kr/tta/ttaSearchView.do?key=77&rep=1&searchStandardNo=TTAK.KO-06.0281/R1&searchCate=TTAS (accessed on 17 April 2022).
8. Choe, J.U.; Park, J.H.; Kim, H.J. A Basic Study on AIS-Based Navigation Data Analysis for Remote Situation Recognition of Autonomous Ship. *J. Korean Inst. Navig. Port Res.* **2020**, *11a*, 52–53.
9. Jung, C.H.; Choi, W.K.; Park, S.H. A Study on the Improvement of AIS Equipmentthrough the Users Survey. *J. Korean Marit. Police* **2016**, *6*, 117–132.
10. *TTAK. KO-06.0281*; Wireless Data Communication Protocol for V-Pass System. Telecommunications Technology Association: Seongnam, Republic of Korea, 2019. Available online: https://www.tta.or.kr/tta/ttaSearchView.do?key=77&rep=1&searchStandardNo=TTAK.KO-06.0281/R1&searchCate=TTAS (accessed on 17 April 2022).
11. Han, H.R. A SpatioTemporal Variation Pattern Analysis of Fishing Activity in the Jeju Sea of Korea Using V-Pass Data. Ph.D. Thesis, Department of Spatial Information Engineering, The Graduate School Pukyong National University, Busan, Republic of Korea, 2021.
12. Han, J.R.; Kim, T.H.; Choi, E.Y.; Choi, H.W. A study on the mapping of fishing activity using V-pass data—Focusing on the Southeast Sea of Korea. *J. Korea Geogr. Inf. Soc.* **2021**, *24*, 112–125.
13. Kim, K.U.; Lee, W.J. A study on the advanced schemes on the welfare accommodation of fishing crew. *J. Korea Ship Saf. Technol. Auth.* **2012**, *34*, 2–17.
14. Park, C.H.; Jung, B.K.; Lee, N.W. A study on the application of the navigation analysis system for the proof of ship crimes. *Korean Assoc. Marit. Police Sci.* **2022**, *12*, 85–104. [CrossRef]
15. Korea Research Institute of Ships & Ocean Engineering. *Development of Replay System of Ships' Collision Accident*; Research Service Report; Korea Research Institute of Ships & Ocean Engineering: Daejeon, Republic of Korea, 2014; Available online: https://www.kriso.re.kr/ (accessed on 17 April 2022).
16. Ship Marine Plant Laboratory. *Development of Marine Specialized Ship Collision Reproduction System*; Research Service Report; Ship Marine Plant Laboratory: Daejeon, Republic of Korea, 2014; Available online: https://www.kriso.re.kr/ (accessed on 17 April 2022).
17. Kim, J.W.; Jeong, M. Basic Study on Improving the Reliability of AIS data: Focused on Vessel Traffic Service Operators. *J. Korea Marit. Police Assoc.* **2021**, *11*, 49–68.
18. An, J.O. *A Study on the Utilization of Marine Safety Radio Facilities and the Efficiency of Frequency Utilization*; Final Research Report; National Radio Research Institute: Naju-si, Republic of Korea, 2013; Available online: https://www.rra.go.kr/ (accessed on 17 April 2022).
19. *TTAK. KO-06.0281*; Radio Data Communication Protocol Maritime Security Network. Telecommunications Technology Association: Seongnam, Republic of Korea, 2016. Available online: https://www.tta.or.kr/tta/ttaSearchView.do?key=77&rep=1&searchStandardNo=TTAK.KO-06.0281/R2&searchCate=TTAS (accessed on 17 April 2022).
20. Kim, B.O. Message error probability analysis by AIS slot interference. In Proceedings of the Autumn Academic Conference of the Korean Society of Navigation and Harbour. *J.Korean Institute Navig. Port Res.* **2010**, *10a*, 164–166.
21. Kim, D.Y.; Hong, T.H.; Jeong, J.S.; Lee, S.J. Building an algorithm for compensating AIS error data. *J. Korean Inst. Intell. Syst.* **2014**, *24*, 181–203.
22. Jeong, J.S.; Yang, W.J. A study on the enhancement of utilization of automatic identification system. *J. Korean. Soc. Mar. Environ. Saf.* **2003**, *9*, 15–21.
23. Kim, D.W.; Ha, M.J. A Study on the Collection and Utilization of Collected Information through V-pass System, Focusing on Infringement of Fundamental Rights and Legislative Solution. *J. Korean Assoc. Marit. Police Sci.* **2021**, *9*, 310–315. [CrossRef]
24. Kim, K.I.; Jung, J.S.; Park, G.G. A Study on the Estimation of Center of Turning Circle of Anchoring Vessel using Automatic Identification System Data in VTS. *J. Navig. Port Res.* **2013**, *8*, 337–343. [CrossRef]
25. Seo, K.Y.; Hong, T.H.; Park, G.G.; Choi, C.S. Analysis of Operational State and Radio Environmont of AIS. *J. Korean Soc. Marit. Inf. Commun.* **2005**, *9*, 177–180.
26. Chao, C. A Study on Development of Expert System for Collision Avoidance and Navigation Based on AIS. Ph.D. Thesis, Department of Ship Operation Systems Engineering Graduate School of Korea Maritime and Ocean University, Busan, Republic of Korea, 2009.

27. Last, P.; Bahlke, C.; Hering-Bertram, M.; Linsen, L. Comprehensive analysis of Automatic Identification System (AIS) data in regard to vessel movement prediction. *J. Navig.* **2014**, *67*, 791–809. [CrossRef]
28. Lee, S.J.; Jeong, J.S.; Kim, M.Y.; Park, G.K. A study on real-time message analysis for AIS VDL load management. *J. Korean Inst. Intell. Syst.* **2013**, *23*, 236–261.
29. Hu, Q.; Jiang, Y.; Zhang, J.; Sun, X.; Zhang, S. Development of an automatic identification system autonomous positioning system. *Sensors* **2015**, *15*, 28574–28591. [CrossRef] [PubMed]
30. Wang, H.; Liu, Z.; Wang, H.; Graham, T.; Wang, J. An analysis of factors affecting the severity of marine accidents. *Reliab. Eng. Syst. Saf.* **2021**, *210*, 07513. [CrossRef]
31. Wang, X.; Xia, G.; Zhao, J.; Yang, Z.; Loughney, S.; Fang, S.; Zhang, S.; Xing, Y.; Liu, Z. A novel method for the risk assessment of human evacuation from cruise ships in maritime transportation. *Reliab. Eng. Syst. Saf.* **2023**, *230*, 108887. [CrossRef]
32. Fang, S.; Liu, Z.; Wang, X.; Wang, J.; Yang, Z. Simulation of evacuation in an inclined passenger vessel based on an improved social force model. *Saf. Sci.* **2022**, *148*, 105675. [CrossRef]

Journal of
Marine Science and Engineering

MDPI

Article

Study of Overtopping Flow Velocity and Overtopping Layer Thickness on Composite Breakwater under Regular Wave

Made Narayana Adibhusana [1], Jong-In Lee [1], Young-Taek Kim [2] and Yonguk Ryu [1,*]

[1] Department of Civil Engineering, Chonnam National University, Gwangju 61186, Republic of Korea; mnadibhusana@jnu.ac.kr (M.N.A.); jilee@jnu.ac.kr (J.-I.L.)

[2] Department of Hydro Science and Engineering Research, Korea of Civil Engineering and Building Technology, Goyang 10223, Republic of Korea; ytkim@kict.re.kr

* Correspondence: yuryu@jnu.ac.kr

Abstract: Overtopping flow velocity (OFV) and overtopping layer thickness (OTL) are essential parameters in breakwater design. Several empirical equations to predict these parameters are available in many works of literature, but most of the equations were derived based on impermeable structures such as sea dikes. In this study, we experimented with overtopping waves over a composite breakwater with tetrapod armor units. In the experiments, wave overtopping was generated from regular waves. We used a digital image-based velocimetry method, bubble image velocimetry (BIV), to measure the OFV and digitize the corresponding image to obtain the OLT. The patterns of OFV and OTL with respect to time steps, wave conditions, and corresponding events were provided and discussed. The application of the widely used empirical equations for sea dike to breakwater was also tested by calibrating the coefficients. New empirical coefficients and roughness factors were suggested to reduce the difference between predicted and measured OFV and OLT on breakwater through the bootstrap resampling technique. This study provides modified empirical equations on wave overtopping, which is further applicable to breakwater design.

Keywords: wave overtopping; regular wave; breakwater; bubble image velocimetry; overtopping flow velocity; overtopping layer thickness

Citation: Adibhusana, M.N.; Lee, J.-I.; Kim, Y.-T.; Ryu, Y. Study of Overtopping Flow Velocity and Overtopping Layer Thickness on Composite Breakwater under Regular Wave. *J. Mar. Sci. Eng.* **2023**, *11*, 823. https://doi.org/10.3390/jmse11040823

Academic Editor: M. Dolores Esteban

Received: 6 March 2023
Revised: 9 April 2023
Accepted: 10 April 2023
Published: 13 April 2023

1. Introduction

Wave overtopping occurs when the wave run-up is higher than the crest freeboard. Wave overtopping transfers the wave energy to the crest and rear side of the coastal defense structure. It affects the stability of the structure's crest and rip rap structure [1]. Moreover, the wave overtopping flow poses a threat to the safety of pedestrians since the crest of the structures is often used as a recreation place. As reported in EurOtop [2], in 2015, 11 people died due to wave overtopping or wave action on a coastal structure in UK. In addition, global warming leads to sea level rise, and stronger wave storms may exacerbate the wave overtopping hazard. The recent studies by Gao et al. [3,4] also show that long-period coastal waves have significant adverse effects on wave overtopping. It is not feasible to avoid wave overtopping due to the cost of an uneconomical high coastal defense structure. Therefore, it is essential to estimate wave overtopping flow behavior accurately.

Wave overtopping flows can be characterized by wave overtopping flow velocity (OFV) and wave overtopping layer thickness (OLT). These parameters are considered important when pedestrian safety on a coastal defense structure is a priority. There is extensive literature focused on OFV and OLT on sea dike crests, e.g., van Gent [5], Schüttrumpf [6], and van der Meer et al. [7]. These studies have proposed empirical design formulas to predict OFV and OLT, which are collected in the EurOtop [2] manual. Several studies also have investigated the tolerable limits for OFV and OLT to ensure pedestrian safety under overtopping flow on vertical wall structures. For example, Bae et al. [8] and Cao et al. [9].

studied the tolerable limits for these parameters on vertical wall structures. However, a few studies have focused on OFV and OLT on armored breakwater crests [10–12].

A breakwater is typically constructed in areas where water waves are severe and at high risk of wave overtopping. Existing empirical formulas used for predicting OFV and OLT on sea dikes can serve as the basis for predictions on breakwater crests. In the empirical design equations, there are reduction factors and empirical coefficients that depend on hydraulic and geometrical conditions. These variables need to be calibrated with experimental data to improve estimation accuracy [13,14]. An example of this approach is presented in Mares-Nassare et al. [10], where they adopted and calibrated existing empirical formulas on sea dikes proposed by Schüttrumpf and van Gent [15] and EurOtop [2]. They also introduced a new formula for estimating OLT on rubble mound breakwaters. OFV was estimated from the measured OLT at the middle of the crest. As suggested by Pepi et al. [14], further research is still required for different hydraulic and structure geometries.

In this study, we investigated the OFV and OLT on a composite breakwater through physical experimentation. We adopted and calibrated the existing prediction formula for the sea dike and proposed new empirical coefficients and reduction factors to extend the application of the empirical equation for OFV and OLT estimation on the breakwater. Unlike common measurement techniques for OFV and OLT, such as micro propeller and wave gauge, this study used a digital imaging technique to measure the wave overtopping flow parameters and understand the flow behavior from spatial distribution as well as temporal change.

The paper is organized as follows. Section 2 presents a literature review of previous studies on OFV and OLT. In Section 3, we describe the physical experiment setup and the measurement technique. Section 4 provides a comparison of the measured OFV and OLT with the estimation using the existing empirical equations as well as relationships of the parameters. Additionally, we provide flow patterns and the application of the new empirical coefficients and reduction factors in different wave conditions and structure geometries in this section. The interpretation of the results is discussed in Section 5. Finally, we draw a conclusion in Section 6.

2. Literature Review

Wave overtopping occurs when the wave run-up height, R_u, exceeds the structure crest freeboard, R_c. The wave run-up height is a vertical difference between the highest point of wave run-up and the mean water level (MWL). Wave run-up height can be calculated for the situation where the crest freeboard is high enough to prevent wave overtopping. During wave overtopping, the crest freeboard height is less than the wave run-up height. In that condition, the seaward slope is virtually extended to allow considering a fictitious wave run-up level, as shown in Figure 1a. This fictitious wave run-up on a sea dike structure was investigated by van Gent [16] through a prototype measurement for physical model testing and numerical modeling, and Equation (1) was proposed to estimate the fictitious wave run-up height:

$$\begin{aligned} \frac{R_u}{H_s} &= c_0\xi \text{ if } \xi \le p, \\ \frac{R_u}{H_s} &= c_1 - \frac{c_2}{\xi} \text{ if } \xi \ge p, \end{aligned} \tag{1}$$

where c_0 = 1.35, c_1 = 4.7, c_2 is given by Equation (2), p is given by Equation (3), H_s is the significant wave height of the incident wave height at the toe of the structure, and ξ is the surf similarity or Iribarren number given by Equation (4).

$$c_2 = 0.25\frac{c_1^2}{c_0}, \tag{2}$$

$$p = 0.5\frac{c_1}{c_0}, \tag{3}$$

$$\xi = \frac{\tan\alpha}{\sqrt{2\pi H_s/T^2}} \tag{4}$$

where T is the spectral wave period and α is the seaward structure slope.

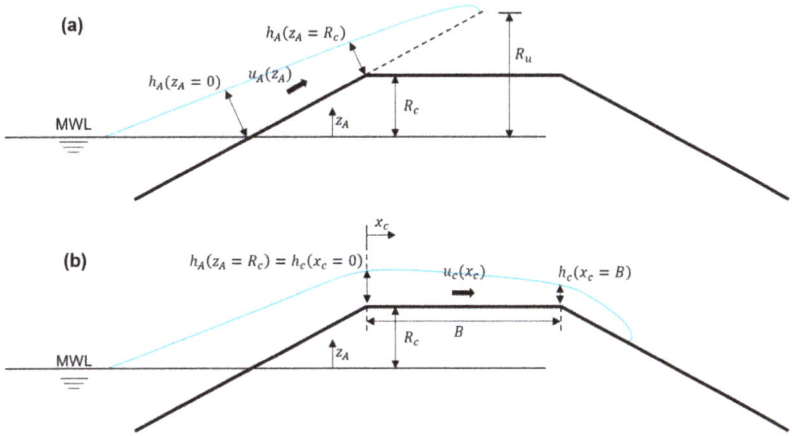

Figure 1. Definition sketch of fictitious wave run-up (**a**) and wave overtopping (**b**). The wave run-up and wave overtopping parameters are based on Schüttrumpf and van Gent [15].

Van Gent [5] and Schüttrumpf et al. [6] performed an experiment focusing on the measurement of OFV and OLT on sea dikes. van Gent [5] carried out a small-scale experiment using a sea dike with a single seaward slope $V/H = 1/4$, and Schüttrumpf et al. [6] used a dike with three different seaward slopes, $V/H = 1/3$, $1/4$, and $1/6$. In addition, Schüttrumpf et al. [6] also conducted a large-scale experiment to confirm the small-scale experiment result. In van Gent [5], the wave run-up parameters were estimated based on a fictitious wave run-up height calculated using Equations (1)–(4), while, in Schüttrumpf et al. [6], the wave run-up parameters were derived from measured wave run-up height. van Gent [5] and Schuttrumpf et al. [6] combined the findings in Schüttrumpf and van Gent [15] and proposed Equations (5) and (6) to estimate the wave run-up parameters on the seaward slope of the sea dike.

$$u_A = c_{A,u}(\sqrt{g(R_u - z_A)}), \tag{5}$$

$$h_A = c_{A,h}(R_u - z_A), \tag{6}$$

where u_A is the wave run-up velocity; h_A is the wave run-up layer thickness; z_A is the elevation from the MWL; and $c_{A,u}$ and $c_{A,h}$ are the empirical coefficients given in Table 1. The transition line between the seaward slope and crest is the initial condition for the wave overtopping flow on the crest. The overtopping flow parameters at this point can be estimated using Equations (5) and (6), with $z_A = R_c$ (Figure 1b). Schüttrumpf and van Gent [15] also proposed a method to estimate the wave overtopping flow parameters, OFV ($u_c(x_c)$) and OLT ($h_c(x_c)$), along the dike crest using Equations (7) and (8):

$$\frac{u_c(x_c)}{u_A(R_c)} = \exp(-c_{c,u}\frac{x_c\mu}{h_c(x_c)}), \tag{7}$$

$$\frac{h_c(x_c)}{h_A(R_c)} = \exp(-c_{c,h}\frac{x_c}{B}), \tag{8}$$

where x_c is the distance from the intersection of the crest and seaward slope; B is the crest width; μ is the friction coefficient; and $c_{c,u}$ and $c_{c,h}$ are the empirical coefficients given in Table 1.

Table 1. Range of the applicability and empirical coefficients from the previous studies.

	van Gent [5]	Schüttrumpf et al. [6]	van der Meer et al. [7]	EurOtop [2]	Mares-Nasarre et al. [10]
Structure	Sea dike	Sea dike	Sea dike	Sea dike	Breakwater
Slope (V/H)	1/4	1/3, 1/4, 1/6	1/3	1/3, 1/4, 1/6	2/3
R_c/H_s	0.7–2.2	0.0–2.5	0.7–2.9	-	0.34–1.75
$c_{A,u}$	1.30	1.37	$0.35 \cot \alpha$	1.4, 1.5	-
$c_{A,h}$	0.15	0.33	0.19	0.20, 0.30	0.52
$c_{c,u}$	0.50	0.50	-	-	-
$c_{c,h}$	0.40	0.89	0.13	-	-

As seen in Table 1, the empirical coefficient for wave run-up layer thickness, $c_{A,h}$, given by Schüttrumpf et al. [6], is about 2 times larger (i.e., 2.2) than that by van Gent [5]. They proposed a different empirical coefficient based on their own experimental results. This led to different results on the estimation of OLT, as shown in Mares-Nasarre et al. [10]. According to Schüttrumpf and van Gent [15], the discrepancy between these empirical coefficients was due to the different dike geometries and instruments they used. Bosman et al. [17] investigated the discrepancy of these empirical coefficients through a physical experiment. Two different dike geometries, $V/H = 1/4$ and $V/H = 1/6$ were used, with one wave condition for both dikes. They found that the seaward slope of the structure influences these empirical coefficients. Later, Lorke et al. [18] conducted a physical experiment that measured the OFV and OLT at the seaward crest edge and landward crest edge using wave gauges and micro propellers. The authors also proposed new empirical coefficients based on the seaward slope of the structure.

Van der Meer et al. [7] conducted a physical test on a dike with a slope $V/H = 1/3$ and measured the OFV and OLT at the seaward crest edge and landward crest edge. In their study, they combined their experimental results with the observation from van Gent [5] and Schüttrumpf et al. [6]. Based on this newly combined data, van der Meer et al. [7] proposed Equation (9) to estimate u_A at the seaward crest edge with a slightly different empirical coefficient, $c_{A,u}$, as shown in Table 1. The u_c along the dike crest is then estimated as the decay function given by Equation (10):

$$u_A(z_A = R_c) = c_{A,u}(\sqrt{g(R_u - z_A)}), \tag{9}$$

$$\frac{u_c(x_c)}{u_A(R_c)} = \exp(-1.4\frac{x_c}{L}), \tag{10}$$

where L is the wavelength based on the spectral wave period. The empirical coefficient in Equation (9) is $c_{A,u} = 0.35 \cot \alpha$, meaning that the slope angles of the structure is taken into account in the prediction. van der Meer et al. [7] also proposed Equation (11) to estimate h_c along the sea dike crest:

$$h_c(x_c) = c_{c,h}(R_u - R_c). \tag{11}$$

Based on their analysis, OLT decreases directly behind the seaward crest edge and then remains almost constant along the crest. The wave run-up layer thickness at the seaward edge, h_A, is 50% larger than h_c gave in Equation (11).

Based on the aforementioned studies, it can be concluded that the wave overtopping flow parameters along the crest can be estimated based on the wave run-up parameters at the seaward crest together with the empirical coefficient. However, since the seaward slope of the sea dike is impermeable and smooth, the estimation of a fictitious wave run-up height formula using Equations (1)–(4) is not applicable for rough slopes such as breakwater structures. EurOtop [2] provided an empirical formula to estimate the fictitious wave run-up on the armored front slope of a structure such as a breakwater using Equation (12):

$$\frac{R_u}{H} = 1.65 \cdot \gamma_f \cdot \gamma_\beta \cdot \gamma_b \cdot \xi, \tag{12a}$$

with the maximum value of

$$\frac{R_u}{H} = 1 \cdot \gamma_{fsurging} \cdot \gamma_\beta \cdot \left(4.0 - \frac{1.5}{\sqrt{\gamma_b \cdot \xi}}\right), \quad (12b)$$

where γ_f is the influence of roughness of the slope; γ_β is the influence of oblique wave attack; γ_b is the influence of berm; $\gamma_{fsurging}$ is a coefficient of γ_f when ξ >1.8:

$$\gamma_{fsurging} = \gamma_f + (\xi - 1.8)\frac{1-\gamma_f}{8.2}. \quad (13)$$

In the case of an impermeable and smooth seaward slope structure, where the wave attack is perpendicular to the structure without a berm, the roughness factor, γ_f, the influence of oblique wave attack, γ_β, and the influence of the berm, γ_b, are equal to 1. The γ_f value for different types of armor units can be found in [2]. For 2 layered tetrapods on a rubble mound breakwater with a permeable core, they derived $\gamma_f = 0.42$.

EurOtop [2] adopted Equations (5) and (6) to estimate the wave run-up parameters, wave run-up velocity, u_A, and wave run-up layer thickness, h_A. As shown in Table 1, EurOtop [2] specifies the empirical coefficient $c_{A,u} = 1.4$ for the slope 1/3 and 1/4 and $c_{A,u}$ = 1.5 for the slope of 1/6. In the case of the seaward structure slope between these values, EurOtop [2] suggested applying interpolation to obtain the empirical coefficient. Similarly, EurOtop [2] also provided the empirical coefficient $c_{A,h} = 0.20$ for the of slope 1/3 and 1/4 and $c_{A,h} = 0.30$ for the slope of 1/6 and suggested an interpolation method to obtain an empirical coefficient between these slopes. The OFV along the crest was then estimated using Equation (10). According to EurOtop [2], the OLT along the crest was 2/3 of that at the seaward crest edge ($h_c(x_c{>>}0)) = 2/3 \cdot h_A(R_c)$).

Recently Mares-Nasarre et al. [10] conducted an experiment on a mound breakwater with the seaward slope $V/H = 2/3$, focusing on OFV and OLT at the middle of the crest. Mares-Nasarre et al. [10] used Equation (12) proposed by EurOtop [2] to estimate the fictitious wave run-up height on three different armor units: 1-layer cubipod, 2-layers rock, and 2-layers cube. The wave run-up layer thickness at the seaward crest edge ($h_A(z_A = R_c)$) and at the middle of the crest ($h_c(x_c = B/2)$) was estimated using Equations (6) and (8). They calibrated the roughness factor, γ_f, and empirical coefficient, $c_{A,h}$, with their experiment data following procedures given in Molines and Medina [13] and proposed a new empirical coefficient, as shown in Table 1.

According to Molines and Medina [13] the roughness factor, γ_f, is a fitting parameter and needs to be calibrated based on the database. Pepi et al. [14] proposed a new formula to calculate the roughness factor for 2-layers rock mound breakwater with the seaward slope $V/H = 1/2$. Calibration of γ_f, as well as the empirical coefficient, reduced the difference between measured and estimated wave overtopping parameters [10,13,14]. This indicates that using the available empirical equation with a calibrated empirical coefficient and roughness factor has a potential application in the estimation of wave overtopping flow parameters on different coastal defense structures. Hence, in this paper, we extend the application of the empirical equation. We adopted Equation (12) to estimate the fictitious wave run-up height on the breakwater with 2-layers tetrapods and a permeable core. Then the wave run-up parameters were estimated using Equations (5) and (6). Finally, the OFV was estimated using Equation (10), and the OLT was estimated as $h_c(x_c{>>}0)) = 2/3 \cdot h_A(R_c)$ [2]. The roughness factor and empirical coefficient were calibrated following the method presented by Molines and Medina [13] and Mares-Nasarre et al. [10].

3. Materials and Methods

3.1. Experiment Setup

The experiment was performed in a two-dimensional wave flume, as shown in Figure 2. The wave flume is 56 m long, 1 m wide, and 2 m high with a transparent side wall. The wave flume is equipped with a piston-type wave maker located 5 m from the end of the flume and can produce regular and irregular waves. The 1:37.5 sloping beach starts 19 m from the wave maker, and the horizontal beach with the layer of horsehair is at the other end of the tank to absorb wave energy and reduce reflection. The wave flume was split into two sections (0.6 m to 0.4 m). The wide section was used to place the breakwater model, and the narrow section was used to measure the incident wave without being affected by the reflection from the structure model.

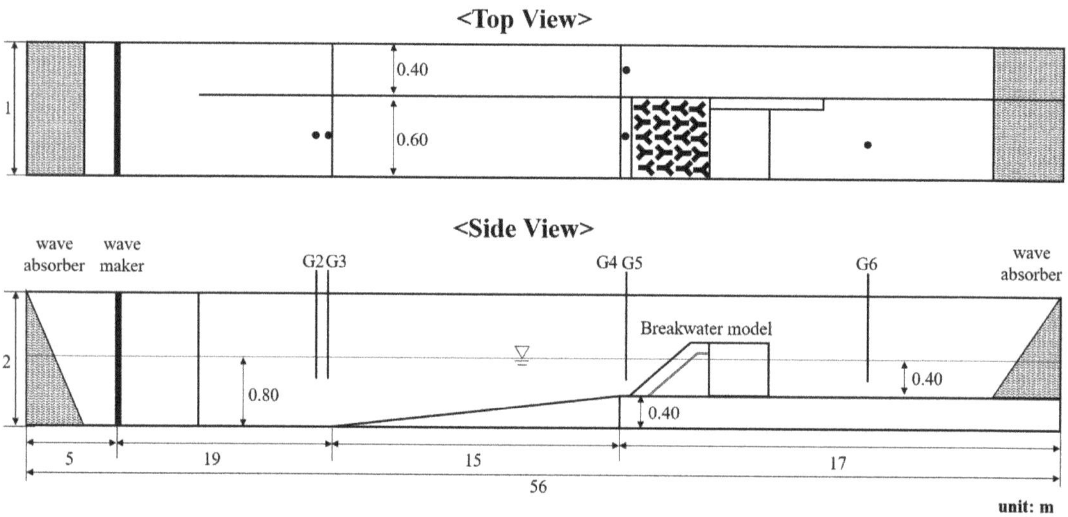

Figure 2. Dimensions of the wave flume.

The model of the composite-type breakwater (scale 1/40) was located 34.5 m away from the wave maker. The model structure consists of a caisson in the middle, armor layers on the front side, and no structure on the rear side (Figure 3). The caisson model structure has a height of 0.52 m, a length of 0.55 m and a width of 0.33 m. The water depth at the structure was kept constant at $d = 0.40$ m, and the freeboard (R_c) was 0.12 m. On the front side, the armor layers consist of three layers. First, the outer layer with the slope $V/H = 1{:}1.5$ was constructed with two layers of tetrapod (TTP) armor units. The TTP has a mass of $M = 307$ g and a layer thickness of 0.107 m. The second layer is the filter layer, made from natural rocks with a diameter $D_{n50} = 0.08$ m, a mass $M = 85$ g, and a layer thickness of 0.03 m. The third layer is the core layer made from quarry rock with a diameter $D_{n50} = 0.012$ m and a mass $M = 10$–12 g. The armor layers are the same height as the caisson part. On the crest, the TTP layer has a width of 0.22 m, and the total width of the structure is 0.55 m. In order to be representative of the general situation, the design of the structure was intended to be as simple as possible in terms of geometrical configuration. The shape of the structure, especially the armored front section, has been checked after each test to verify the constancy of the geometrical parameters.

The water surface elevation was measured using five capacitance-type wave gauges. Two gauges (named G2 and G3) were installed in the wide section 19 m downstream from the wave maker to monitor the generated waves. In the wide and narrow sections of the flume, two wave gauges (named G4 and G5) were installed 34 m downstream from the wave maker to determine the waves with and without the structure at the breakwater toe. The fifth gauge, G6, was installed behind the breakwater to monitor the transferred wave.

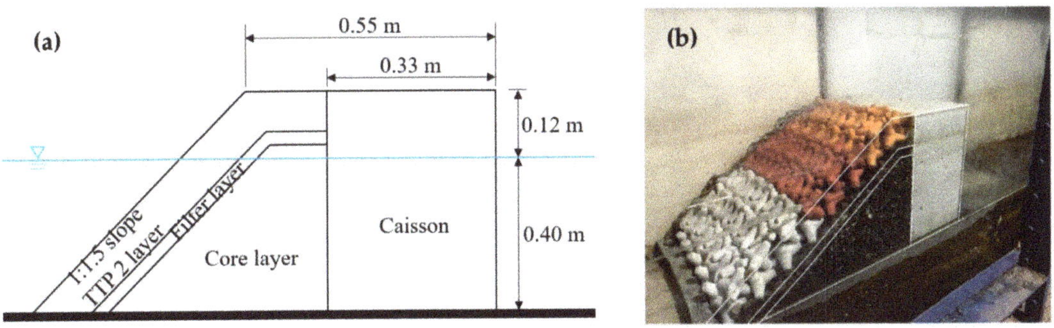

Figure 3. Composite breakwater model: (**a**) the sketch of the breakwater model and (**b**) the picture of the breakwater.

Table 2 presents the test conditions of the current experiment. The regular wave was used in this experiment since the direct relationship between wave condition and wave overtopping flow parameters can be analyzed. The regular waves have a wave period (T) of 1.5–3.0 s and a wave height (H) of 0.12–0.20 m. These wave components represent the wave conditions, including storm conditions. The water wave conditions at the breakwater toe are determined by the relative water depth, d/L, where L is the wavelength from the dispersion relationship. The relative water depth conditions were in the range of 0.028–0.114, indicating shallow to intermediate water depth. The waves were non-breaking waves, as the wave steepness, H/L, was in the range of 0.008–0.057. The Iribarren's number or the breaker parameter in this study, $\xi = \tan\alpha/\sqrt{H/L}$, was in the range of 2.795–7.216, showing a collapsing to surging wave. The relative crest freeboard, R_c/H, was in the range of 0.6–1, and the wave overtopping was observed in all relative crest freeboard conditions.

Table 2. Test conditions.

Description	Parameter	Ranges
Wave period	T	1.5–3.0 s
Wave height	H	0.12–0.20 m
Relative water depth	d/L	0.028–0.114
Wave steepness	H/L	0.008–0.057
Iribarren's number	ξ	2.795–7.216
Relative crest freeboard	R_c/H	0.6–1

In the experiment, some of the wave energy is reflected by the breakwater. These reflected waves are re-reflected towards the breakwater and lead to increasing wave height. Therefore, the measurements are only valid after the unstable part of the initial wave and before the leading edge of the re-reflected wave reaches the breakwater. In this study, the valid measurement time was defined as the arrival time between the first fully developed wave and the first re-reflected wave at the breakwater toe. By limiting the measurement time of the wave overtopping parameters, OFV and OLT, the reflection would not affect the result. From the comparison of the free surface elevations from G4 and G5, available wave components close to the target wave were selected for the analysis.

3.2. Overtopping Flow Velocity and Layer Thickness Measurement

When the wave flows overtop the breakwater structure, the flows usually contain a lot of bubbles. These bubbles make the existing velocity measurement equipment less accurate. In this study, a flow visualization technique called bubble image velocimetry (BIV) was applied to estimate the velocity of the overtopping flows. The BIV technique was first introduced by Ryu et al. [19]. This technique has the same principle as particle image velocimetry (PIV) with a flow visualization and digital image analysis to estimate the flow velocity. However, the PIV technique does not work well in multi-phase flows, such as overtopping flows, where the flows contain a lot of bubbles, due to the air–water interface, which will scatter the laser light. On the other hand, the BIV technique uses these bubbles within the flows as a tracer. These bubble textures in overtopping flow are visualized by means of a shadowgraphy technique and then captured by a high-speed camera. The measurement location in this technique is determined by adjusting the depth of field (DOF). The texture of the bubbles will appear sharp within the DOF and blurry on the outside (Figure 4a). A pair of overtopping flow images were captured by a high-speed camera and then analyzed using a cross-correlation method similar to the PIV technique in order to calculate the velocity.

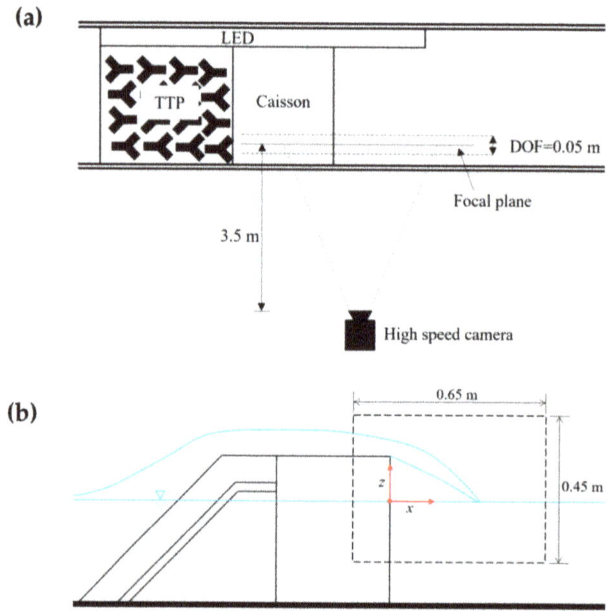

Figure 4. Scheme of the BIV setup: (**a**) Camera position and (**b**) the field of view (FOV).

Figure 4 shows the BIV technique set up and the camera field of view (FOV). A Photron high-speed camera system, FASTCAM Mini UX50, was used in this study. This camera has a CMOS image sensor with a maximum resolution of 1280 × 1024 pixels, a 12-bit dynamic range, a maximum frame rate of 2000 frames per second (fps), a maximum shutter speed of 1/256,000 s, and an internal memory capacity of 8 GB. The camera was equipped with AF Micro Nikkor 60 mm lens. In this study, the camera was placed 3.5 m from the measurement plane, and the FOV covered the rear part of the breakwater model where the overtopping flow falls and plunges into the rear water surface. The camera resolution was set at 1280 × 720 pixels, 8-bit dynamic range, and the lens aperture was set at the f-number of 2.8, yielding the 0.584 ×0.584 mm^2 spatial resolution. After several preliminary tests, the camera framing rate was set at 500 fps, and the interval between two consecutive images (Δt) was 2 ms.

The open-source software called PIVlab developed by Thielicke and Stamhuis [20] was used to analyze BIV images and calculate the velocities. Raw images were firstly inverted so that high-intensity (brightness) represents bubbles. A multi-pass algorithm with an initial window size of 32 × 32 pixels and a final window size of 16 × 16 pixels was employed with a 50% overlap between adjacent windows. Note that the estimated mean maximum bubble diameter of 12 pixels is about 3/4 of the final window size. A cross-correlation was then applied to compute velocity vectors. A median filter was used to remove spurious vectors and boundary values, and then interpolation was applied to fill the removed bad vectors. Figure 5 shows the snapshot of the velocity map of the overtopping flow passing the rear edge of the breakwater crest captured using the setup in Figure 4.

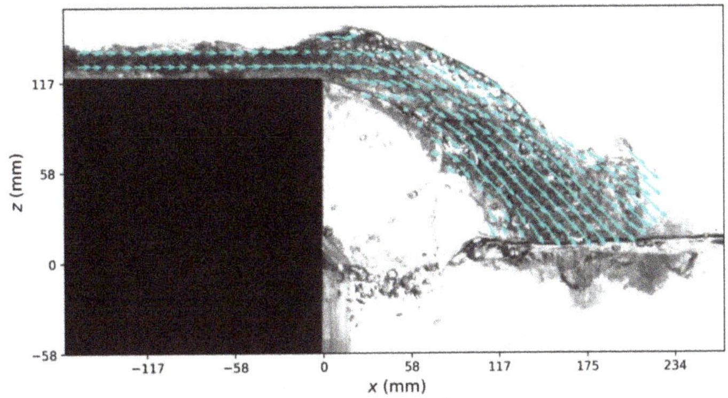

Figure 5. Snapshot of the velocity map of wave overtopping flow plunging on the rear side of the breakwater.

In this study, the OFV is the horizontal velocity at the rear edge of the breakwater crest obtained from the velocity map and is denoted as u_m. The OLT, h_m, was obtained from the digital image at the same location as the u_m by digitizing the water surface. Figure 6 shows the example of time series u_m and h_m from wave conditions $T = 2.75$ s and $H = 16$ cm. Both u_m and h_m have a pattern with a sudden increase to the maximum value at the beginning stage and a gradual decrease after that. The maximum u_m value occurred first, followed by the maximum value of h_m. This indicates that the front body of wave overtopping flows possess the largest momentum with a huge velocity and a thick layer. The fluctuation of u_m between 0.75 and 1.25 s is due to thin layer thickness and fewer air bubbles, which is considered to be an error. Nevertheless, for the analysis of u_m and h_m in the next section, the maximum values were selected.

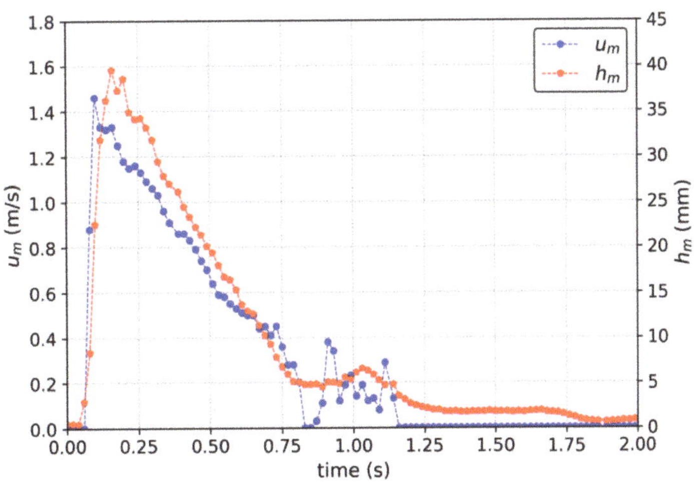

Figure 6. OFV (u_m) and OLT (h_m) at the rear edge breakwater crest. The wave condition was $T = 2.75$ s and $H = 16$ cm.

3.3. Evaluation Metric

In this study, the measurement data and the estimation using Equations (5), (6), (10) and (12) were compared. The fitting of the estimation data was analyzed using the relative root mean squared (rRMSE) given by Equation (14) and the coefficient of determination (R^2) given by Equation (15):

$$\text{rRMSE} = \frac{\text{RMSE}}{\text{MEAN}} = \frac{\sqrt{\frac{\sum_{i=1}^{N}(x_i - y_i)^2}{N}}}{\overline{x}}, \tag{14}$$

$$R^2 = 1 - \frac{\sum_{i=1}^{N}(x_i - y_i)^2}{\sum_{i=1}^{N}(x_i - \overline{x})^2}, \tag{15}$$

where x_i is the measured data; $\overline{x}$ is the average of the measured data; y_i is the estimated data; and N is the total amount of data. In addition, the correlation coefficient (r) was calculated using Equation (16), with $\overline{y}$ as the average estimated data.

$$r = \frac{\sum_{i=1}^{N}(x_i - \overline{x})(y_i - \overline{y})}{\sqrt{\sum_{i=1}^{N}(x_i - \overline{x})^2 \sum_{i=1}^{N}(y_i - \overline{y})^2}}, \tag{16}$$

4. Results

4.1. Overtopping Flow Velocity

4.1.1. Estimation of the Overtopping Flow Velocity

In this study, Equation (12) given by EurOtop [2] was used to estimate the fictitious wave run-up height. The wave run-up velocity at the seaward crest edge was estimated using Equation (5) given by Schüttrumpf and van Gent [15]. Finally, the OFV at the landward crest edge was estimated using Equation (10) given by van der Meer et al. [7]. Figure 7 shows the comparison between the measured OFV at the landward crest edge, u_m, and estimated OFV at the same point, u_e. The empirical coefficient and roughness factor used in the estimation were selected as $c_{A,u} = 1.3$, given by van Gent [5], and $\gamma_f = 0.38$, given by EurOtop [2]. The OFV estimations using the other empirical coefficients available in the literature and the roughness factor, $\gamma_f = 0.38$, are shown in Table 3. As we can see in Table 3, the best estimation was obtained when using the empirical coefficient given by van Gent [5] with rRMSE = 0.134 with $R^2 = 0.702$. The other estimation gives a similar R^2 value because the scattered pattern of the data does not change dramatically with the change of the empirical coefficient, $c_{A,u}$.

Table 3. Quantitative results on the estimation of OFV using different empirical coefficients available in the previous studies.

Previous Studies	$c_{A,u}$	rRMSE
EurOtop [2]	1.50	0.273
	1.40	0.198
Schüttrumpf et al. [6]	1.37	0.177
van Gent [5]	1.30	0.134

According to Molines and Medina [13], the roughness factor, γ_f, is a fitting parameter where its value is different depending on the formula and database. In addition, the empirical coefficients were obtained from the experiment where the structure was a sea dike with a different seaward slope. Therefore, the empirical coefficient, $c_{A,u}$, and roughness factor, γ_f, need to be calibrated with the experiment data as well.

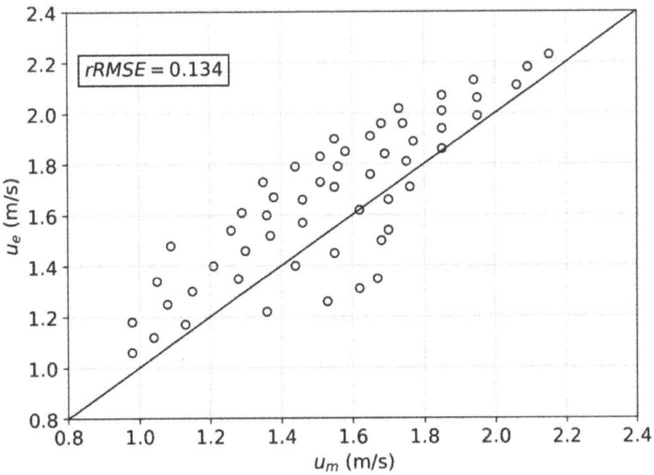

Figure 7. Comparisons of the measured and estimated OFV for $c_{A,u}$ = 1.3 and γ_f = 0.38 with rRMSE = 0.134 and $R^2 = 0.702$.

Using the empirical equation given in EurOtop [2] and the experimental data obtained in this study, an rRMSE can be used to estimate the optimum $c_{A,u}$ and γ_f, which minimizes the prediction error. However, no uncertainty information regarding the $c_{A,u}$ and γ_f are obtained. To overcome the uncertainty of the estimation of the $c_{A,u}$ and γ_f, a bootstrap resample technique was used in this study [10,13]. In the bootstrap resampling technique, a new dataset called a bootstrapped dataset was created using a resample with a replacement from the original dataset; therefore, it had the same number of samples as the original dataset. Then the mean value of bootstrapped dataset was calculated. This process was repeated several times, and the distribution of mean values was obtained.

Since there were two variables in this study, the resampling procedure shown in Mares-Nassare et al. [10] was followed. The procedure has two levels of bootstrap resampling. Firstly, 1000 resamplings were performed by optimizing both the roughness factors and the empirical coefficients. Thus, 1000 mean values of roughness factors and empirical coefficients that minimize the rRMSE were obtained, and the mean values were used to statistically characterize the parameters using percentiles of 5%, 50%, and 95%. Figure 8a shows the histogram of the empirical coefficient, $c_{A,u}$, obtained from the first level bootstrap. The P5%, P50%, and P95% were 1.18, 1.21, and 1.24, respectively.

On the second level of bootstrap resampling, the empirical coefficient value was fixed at the 50% percentile ($c_{A,u}$ = 1.21), and 1000 resamples were performed by varying the roughness factor, γ_f. The optimum roughness factors can be obtained for the empirical equation using the 50% percentile for the empirical coefficients and the existing database. Using the obtained 1000 values of each roughness factor, they were statistically characterized using the referred percentiles. Figure 8b shows the histogram from 1000 roughness factor values obtained from the second level bootstrap. The P5%, P50%, and P95% were 0.32, 0.35, and 0.40, respectively.

The new empirical coefficient, $c_{A,u}$, and roughness factor, γ_f, were used to estimate the OFV. Figure 9 shows the measured OFV at the rear edge of the breakwater crest, $u_m(x_c = B)$, as compared with the estimation given by the combination of Equations (5), (10), and (12) using the 50% percentile for the empirical coefficient and roughness factor. The rRMSE used to measure the goodness of fit is 0.112 and the coefficient of determination, R^2, is 0.755. By using the 50% percentile of the empirical coefficient and roughness factor, the estimation of OFV gives a better result.

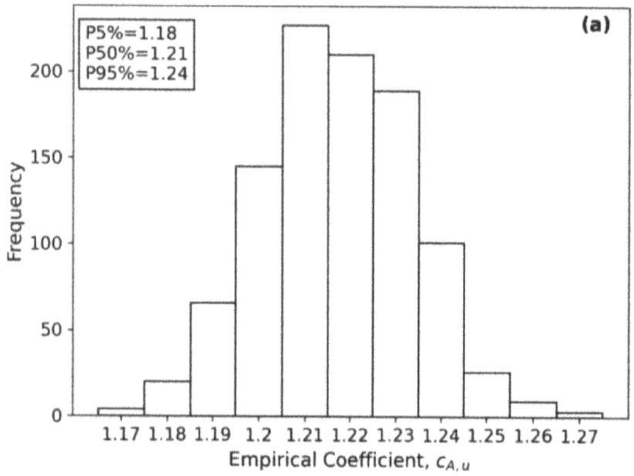

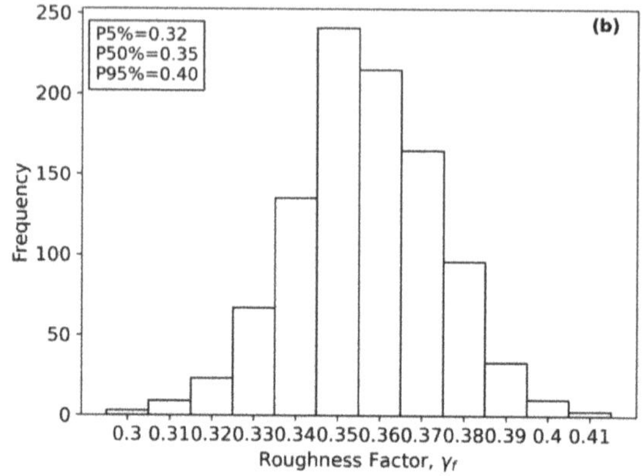

Figure 8. Histogram plot of the empirical coefficient, $c_{A,u}$, and the roughness factor, γ_f, obtained from the bootstrap resampling: (**a**) the first level bootstrap and (**b**) the second level bootstrap.

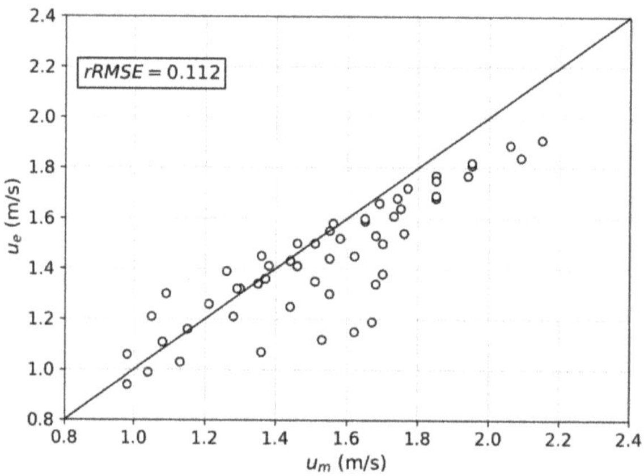

Figure 9. Comparison of the measured OFV and the estimation using $c_{A,u} = 1.21$ and $\gamma_f = 0.35$.

4.1.2. Overtopping Flow Velocity Estimation in the Different Wave Conditions

The new empirical equation with $c_{A,u} = 1.21$ and $\gamma_f = 0.35$ obtained in this study was used to estimate the OFV at the rear edge of the breakwater crest using Equations (5), (10), and (12) on the different wave conditions. The estimated OFV, u_e, was normalized with the measured OFV, u_m, and plotted against the dimensionless crest freeboard R_c/H to show agreements depending on the wave height in Figure 10. It should be noted here that R_c was constant in this study; thus, the change in the dimensionless crest freeboard value was due to the change in the wave height H. As shown in Figure 10, it is observed that u_e for the smaller wave height (i.e., the larger R_c/H) overestimates the OFV and underestimates that for the larger wave height (i.e., the smaller R_c/H). The pattern showing the sloped distribution in the plot is clear in the relatively short wave period and is seen up to $T = 2.5$ s. In the longer wave period (e.g., $T = 2.75$ s and 3 s), the estimation appears to be relatively uniform, where u_e underestimates the OFV in all the wave heights.

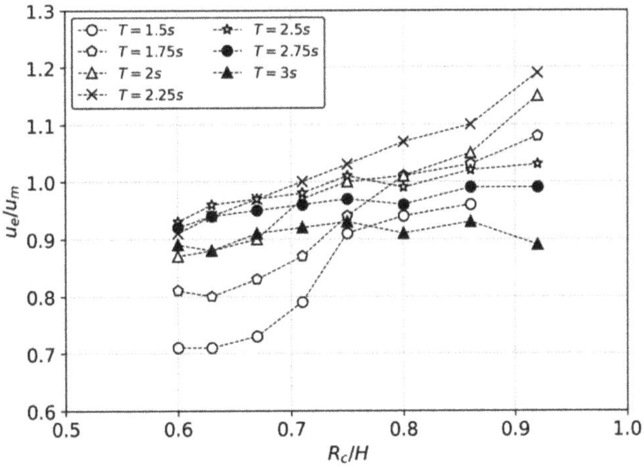

Figure 10. Estimation of the OFV relative to the measurements in the different wave conditions.

4.2. Overtopping Layer Thickness

4.2.1. Estimation of the Overtopping Layer Thickness

The measured OLT at the landward crest edge, h_m, was compared with the estimated OLT at the landward crest edge, h_e. Similar to the previous subsection for the OFV, the wave run-up height was estimated using Equation (12) by EurOtop [2]. The wave run-up layer thickness at the seaward crest edge was estimated using Equation (6) by Schüttrumpf and van Gent [15]. Finally, the OLT at the landward crest edge was estimated following EurOtop [2], where the OLT along the crest was 2/3 of that at the seaward crest edge. The empirical coefficient, $c_{A,h}$, given in the literature, and the roughness factor, $\gamma_f = 0.38$, provided by EurOtop [2], were used to estimate the OLT. Table 4 summarizes the estimation result of OLT. The best estimation was obtained using the empirical coefficient provided by [2] ($c_{A,h}$ = 0.2) with the rRMSE of 0.423 and $R^2 = 0.341$. Identical to the empirical coefficient in the OFV estimation, the empirical coefficient, $c_{A,h}$, was also derived from the experiment where the coastal structure was a sea dike.

Table 4. Quantitative results on the estimation of OLT at the rear edge breakwater crest using the empirical coefficients and roughness factor available in the previous studies.

Author	$c_{A,h}$	rRMSE
EurOtop [2]	0.20	0.423
	0.30	0.500
van der Meer et al. [7]	0.13	0.469
Schüttrumpf et al. [6]	0.15	0.429
van Gent [5]	0.33	0.546

In this study, the observation data were used to obtain the optimal $c_{A,h}$ value. The bootstrap resampling technique was applied to obtain the optimum value of $c_{A,h}$ as performed for OFV. One thousand resamples of the empirical coefficient, $c_{A,h}$, which minimizes the prediction error, were obtained. The samples were used to statistically characterize the percentiles. Figure 11 shows the histogram of 1000 resamples of $c_{A,h}$ obtained from the bootstrap resampling. The P5%, P50%, and P95% were 0.20, 0.21, and 0.22, respectively.

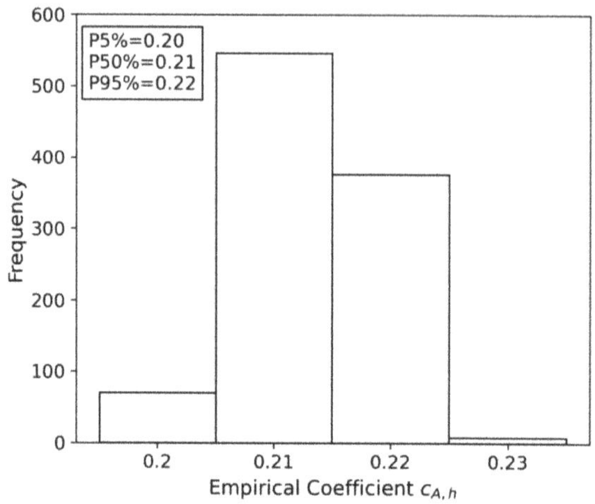

Figure 11. Histogram of the empirical coefficient, $c_{A,h}$, obtained from the bootstrap resampling.

Since the OLT and the OFV correspond to the same wave overtopping, the 50% percentile of roughness factor, γ_f, obtained from the previous section was used together with the 50% percentile of the empirical coefficient, $c_{A,h}$, to estimate the OLT. Figure 12 shows the measured and estimated OLTs using $c_{A,h} = 0.21$ and $\gamma_f = 0.35$. The goodness of fit for the estimation was rRMSE = 0.393 and $R^2 = 0.348$, which was an improvement compared with the estimation using the empirical coefficient and roughness factor available in the literature.

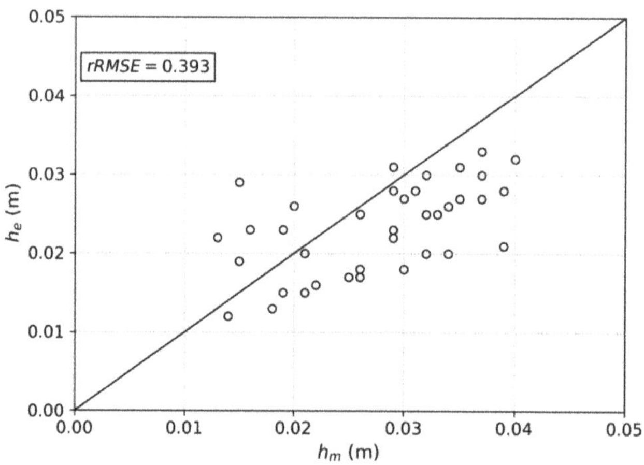

Figure 12. Comparison between the measured and estimated OLT.

4.2.2. Overtopping Layer Thickness Estimation in the Different Wave Conditions

The new empirical coefficient of $c_{A,h} = 0.21$ obtained in this study was used to estimate the OLT using Equations (6) and (12). Since the OLT and OFV correspond to the same wave overtopping, the same roughness factor, $\gamma_f = 0.35$, was used in OLT estimation. The estimated OLT, h_e, was normalized with the measured OLT, h_m, and was plotted against the dimensionless crest freeboard R_c/H to show agreements depending on the wave height in Figure 13. As shown in the figure, h_e underestimates OLT, appearing to be uniform in the relatively larger wave height (i.e., the smaller R_c/H). The pattern changes to increase as the wave height decreases (i.e., increasing R_c/H), showing overestimation.

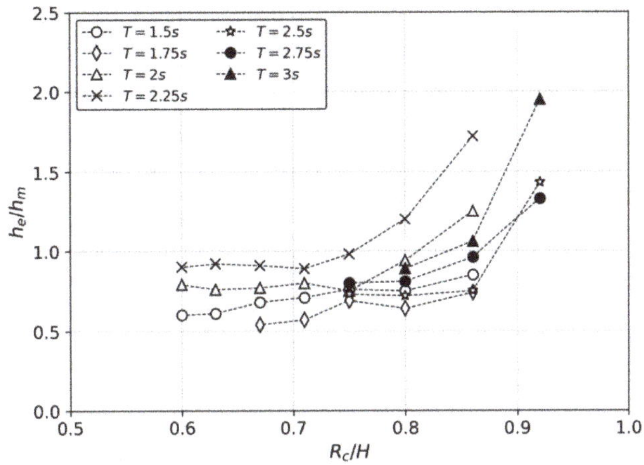

Figure 13. Estimation of the OLT relative to the measurements in the different wave conditions.

4.3. Relationship between the Overtopping Layer Thickness and the Overtopping Flow Velocity

In this study, the OFV and OLT were measured for the same wave overtopping event. Some previous studies [10,21] used the statistics of OLT to estimate the OFV. In this section, the relationship between the OFV and OLT was analyzed following the studies. Figure 14 plots the measured OFV (u_m) and OLT (h_m) for each wave condition. Contrary to the previous studies by Mares-Nasarre et al. [10] and Hughes et al. [21], where there was no clear relationship between OFV and OLT corresponding to the same overtopping event, in this study, a positive relationship was observed, as shown in the figure with the correlation coefficient, $r = 0.744$.

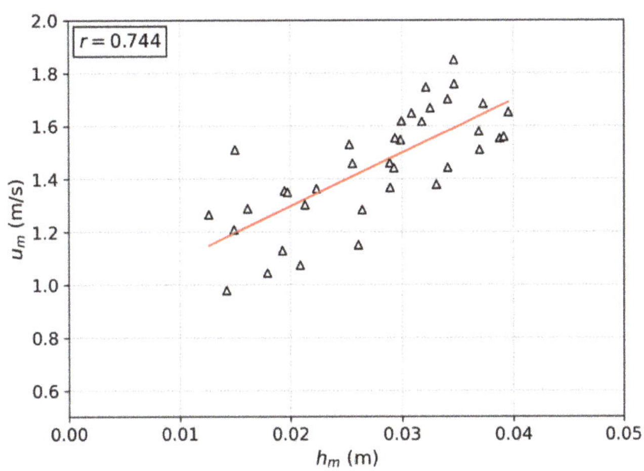

Figure 14. Comparison of the measured OFV and OLT at the rear edge breakwater crest corresponding to the same wave overtopping event.

5. Discussion

OFV and OLT on the composite breakwater can be predicted with the existing equations in the literature. The fictitious wave run-up height, R_u, needs to be estimated firstly by wave conditions at the structure toe. EurOtop [2] provides an empirical equation to estimate R_u on a permeable slope structure using Equation (12) where the roughness factor, γ_f, in this equation needs to be calibrated with experimental data to obtain the optimum estimation. With the estimated R_u, the wave run-up flow velocity and wave run-up layer thickness at the seaward crest edge of the breakwater can then be estimated using Equations (5) and (6) given by Schüttrumpf and van Gent [15]. The empirical coefficients, $c_{A,u}$ and $c_{A,h}$, in these two equations also need to be calibrated with experiment data. The roughness factor, γ_f, obtained in this study using a two-level bootstrap resampling is not significantly different from that provided by EurOtop [2]. The optimum estimation in this study is determined using the 50% percentile of the roughness factor as $\gamma_f = 0.35$. This value is slightly smaller than the roughness factor value by EurOtop [2], $\gamma_f = 0.38$, for two-layered tetrapod armors. The similarity of the roughness factor values is likely due to both experiments with the same front slope ($V/H = 1/1.5$). The empirical coefficient, $c_{A,u} = 1.21$, for the OFV in this study has a smaller value compared to $c_{A,u} = 1.30$ from van Gent [5] conducting the reliable experiments. In the OLT, the optimum empirical coefficient, $c_{A,h}$, is determined using the 50% percentile as $c_{A,h} = 0.21$. The $c_{A,h}$ value is close to $c_{A,h} = 0.20$ by EurOtop [2]. The difference in the empirical coefficient for the OFV ($c_{A,u}$) is likely explained by the different structure geometry of the experiments, as discussed in Schüttrumpf and van Gent [15]. As shown in this study, using the empirical coefficients, $c_{A,u}$ and $c_{A,h}$, and roughness factor, γ_f, calibrated with the experiments can improve the estimation of OFV and OLT on composite breakwaters.

On the other hand, the estimated OFV, u_e, fluctuates in the wave period of T = 1.5–2.25 s, where the u_e is underestimated in the relatively large wave height and overestimated in the relatively small wave height. In the longer wave period of T = 2.5–3.0 s, u_e remains relatively constant. The patterns are likely explained by the different water wave conditions as presented in Table 2. Based on the linear wave theory, the shorter wave period used in this study is mostly of the intermediate water depth condition at the structure location ($0.064 < d/L < 0.114$) and the longer wave period is close to the shallow water depth condition ($0.028 < d/L < 0.041$). The overtopping flows of the short wave period are relatively sensitive to the interaction between the structure front and water waves, and are also subject to wave breaking as the wave height increases. The Iribarren's number for the short wave period is in the range of 2.79–3.47, indicating the collapsing breaking type, which leads to wave breaking, causing nonlinearity and complicated interaction. In the OLT, a relatively uniform estimation is observed for the relatively large wave heights, but it mostly underestimates the OLT. As the wave height decreases, the estimated OLT (h_e) increases like the cases of OFV. Flows during the small wave heights are affected by most factors over the structure model and environmental conditions because the flows have relatively little momentum. The application of newly determined empirical coefficients shows the possibility of existing equations with optimized coefficients for various types of coastal structures. However, the discrepancy observed in relatively short and long waves means that a better empirical equation is needed to predict overtopping flows more accurately.

In this study, the OFV and OLT were measured from the same wave overtopping event. The maximum OFV occurred in the beginning stage first, and then the maximum OLT followed. There was little time difference between the maximum values, and they showed a similar temporal distribution pattern, which indicates the largest momentum of the flow occurs in the wavefront. From coupling OFV and OLT for the same event, there was a positive linear relationship between these two variables. This result differs from the result obtained by Mares-Nasarre et al. [10] and Hughes et al. [21], where there is no clear relationship between OFV and OLT. This difference may be due to the different methods used in the experiments. In this study, the spatial and temporal investigation using imaging techniques provided detailed information about the behavior of the overtopping flow properties, such as flow velocity and layer thickness. The OFV is determined from one of the vertically distributed horizontal velocity components, unlike the previous studies employing a point velocimeter. Moreover, the aeration of the overtopping flow is likely to lead to the inaccurate acquisition of the free surface by a capacitance-type wave gauge. From the flow images, the aeration was observed in the wave front where the max OLT occurred, which implies the possibility of errors in catching maximum values. From the positive relationship between OFV and OLT in this study, the rapidly increasing rate of momentum is expected relative to OFV and OLT.

6. Conclusions

The wave overtopping occurs when the wave run-up, R_u, is higher than the crest freeboard of a coastal structure, R_c. In such conditions, the water wave affects not only the structure's front slope but also the structure's crest and rear side. In addition, wave overtopping also poses a threat to the safety of people and vehicles on the structure crest. However, most of the previous studies have focused on wave overtopping flow parameters on impermeable structures such as sea dikes, with only a few studies on breakwaters. To address this gap, this study investigated OFV and OLT on composite breakwaters through physical experimentation. In this study, 55 physical tests for the composite breakwater with the tetrapod armors were performed on the two-dimensional wave flume. The OFV at the rear edge of the breakwater crest was measured using a digital imaging technique called bubble image velocimetry (BIV), which provides reliable measurements without disturbing the overtopping flow. The OLT was measured by digitizing the digital images obtained from the BIV technique.

The measured OFV and OLT were compared with the empirical equations provided in EurOtop [2], Schüttrumpf and van Gent [15], and van der Meer et al. [7]. The roughness factor, γ_f, in Equation (12) and the empirical coefficients, $c_{A,u}$ and $c_{A,h}$, in Equations (5) and (6) were calibrated with the experiment data. The new roughness factor for the two-layered tetrapods, $\gamma_f = 0.35$, and empirical coefficients, $c_{A,u} = 1.21$ and $c_{A,h} = 0.21$, were obtained through the bootstrap resampling technique. The rRMSE for the estimation of OFV was 0.112 and that of the OLT was 0.393. The new empirical coefficients and roughness factor obtained in this study were used to estimate the OFV and OLT in various wave conditions, which shows promising results for relevant coefficient determination. However, the agreements between the estimations and the measurements are not consistent over all wave conditions and structure geometries. In particular, the estimation for short-wave conditions showing discrepancies needs to be approached through another empirical formula. The results of this study are applicable to composite breakwater structures with armor slope $V/H = 1{:}1.5$ and relative crest freeboard $0.6 < R_c/H < 1$ in non-breaking wave condition $0.008 < H/L_o < 0.057$. More advanced empirical approaches are needed and will be further tested for irregular wave conditions as well.

From the temporal change of OFV and OLT, both parameters show the maximum values with little time difference in the wavefront, having a pattern of sudden increase and gradual decrease. Therefore, the largest momentum of the overtopping flow is expected to occur in the beginning stage of the flow. There is also a positive relationship between OFV and OLT for the same overtopping event, which implies that the momentum of the flow and overtopping volume would increase rapidly.

Author Contributions: Conceptualization, J.-I.L. and Y.R.; methodology, M.N.A. and Y.R.; experiment, M.N.A. and Y.-T.K.; validation, M.N.A.; formal analysis and investigation, M.N.A. and Y.R.; writing—original draft preparation, M.N.A.; writing—review and editing, Y.R.; supervision, Y.R. All authors have read and agreed to the published version of the manuscript.

Funding: This research was supported by the Basic Science Research Program through the National Research Foundation of Korea (NRF) funded by the Ministry of Education (NRF-No. 2022R1I1A3073069) and the Korea Institute of Energy Technology Evaluation and Planning (Project No. 20224B10200040).

Institutional Review Board Statement: Not applicable.

Informed Consent Statement: Not applicable.

Data Availability Statement: Not applicable.

Acknowledgments: This research was supported by the Basic Science Research Program through the National Research Foundation of Korea (NRF) funded by the Ministry of Education (NRF-No. 2022R1I1A3073069) and the Korea Institute of Energy Technology Evaluation and Planning (Project No. 20224B10200040).

Conflicts of Interest: The authors declare no conflict of interest. The funders had no role in the design of the study; in the collection, analyses, or interpretation of data; in the writing of the manuscript; or in the decision to publish the results.

References

1. Argente, G.; Gómez-Martín, M.E.; Medina, J.R. Hydraulic Stability of the Armor Layer of Overtopped Breakwaters. *J. Mar. Sci. Eng.* **2018**, *6*, 143. [CrossRef]
2. Van der Meer, J.W.; Allsop, N.W.H.; Bruce, T.; De Rouck, J.; Kortenhaus, A.; Pullen, T.; Schüttrumpf, H.; Troch, P.; Zanuttigh, B. *EurOtop-Manual on Wave Overtopping of Sea Defences and Related Structures: An Overtopping Manual Largely Based on European Research, but for Worldwide Application*; TU Delft: Delft, The Netherlands, 2018.
3. Gao, J.; Ma, X.; Zang, J.; Dong, G.; Ma, X.; Zhu, Y.; Zhou, L. Numerical Investigation of Harbor Oscillations Induced by Focused Transient Wave Groups. *Coast. Eng.* **2020**, *158*, 103670. [CrossRef]
4. Gao, J.; Ma, X.; Dong, G.; Chen, H.; Liu, Q.; Zang, J. Investigation on the Effects of Bragg Reflection on Harbor Oscillations. *Coast. Eng.* **2021**, *170*, 103977. [CrossRef]
5. Van Gent, M.R.A. Wave Overtopping Events at Dikes. In *Coastal Engineering 2002*; World Scientific: Singapore, 2002; pp. 2203–2215, ISBN 978-981-238-238-2.
6. Schüttrumpf, H.; Möller, J.; Oumeraci, H. Overtopping Flow Parameters on The Inner Slope of Seadikes. In *Coastal Engineering 2002*; World Scientific Publishing Company: Singapore, 2002; pp. 2116–2127.
7. Van der Meer, J.W.; Hardeman, B.; Steendam, G.J.; Schuttrumpf, H.; Verheij, H. Flow Depths and Velocities at Crest and Landward Slope of a Dike, in Theory and with The Wave Overtopping Simulator. *Coast. Eng. Proc.* **2011**, *1*, structures.10. [CrossRef]
8. Bae, H.U.; Yun, K.M.; Yoon, J.Y.; Lim, N.H. Human Stability with Respect to Overtopping Flow on the Breakwater. *Int. J. Appl. Eng. Res.* **2016**, *11*, 111–119.
9. Cao, D.; Yuan, J.; Chen, H.; Zhao, K.; Liu, P.L.-F. Wave Overtopping Flow Striking a Human Body on the Crest of an Impermeable Sloped Seawall. Part I: Physical Modeling. *Coast. Eng.* **2021**, *167*, 103891. [CrossRef]
10. Mares-Nasarre, P.; Argente, G.; Gómez-Martín, M.E.; Medina, J.R. Overtopping Layer Thickness and Overtopping Flow Velocity on Mound Breakwaters. *Coast. Eng.* **2019**, *154*, 103561. [CrossRef]
11. Mares-Nasarre, P.; Gómez-Martín, M.E.; Medina, J.R. Influence of Mild Bottom Slopes on the Overtopping Flow over Mound Breakwaters under Depth-Limited Breaking Wave Conditions. *J. Mar. Sci. Eng.* **2020**, *8*, 3. [CrossRef]
12. Mares-Nasarre, P.; Molines, J.; Gómez-Martín, M.E.; Medina, J.R. Explicit Neural Network-Derived Formula for Overtopping Flow on Mound Breakwaters in Depth-Limited Breaking Wave Conditions. *Coast. Eng.* **2021**, *164*, 103810. [CrossRef]
13. Molines, J.; Medina, J.R. Calibration of Overtopping Roughness Factors for Concrete Armor Units in Non-Breaking Conditions Using the CLASH Database. *Coast. Eng.* **2015**, *96*, 62–70. [CrossRef]
14. Pepi, Y.; Romano, A.; Franco, L. Wave Overtopping at Rubble Mound Breakwaters: A New Method to Estimate Roughness Factor for Rock Armours under Non-Breaking Waves. *Coast. Eng.* **2022**, *178*, 104197. [CrossRef]
15. Schüttrumpf, H.; van Gent, M.R.A. Wave Overtopping at Seadikes. In *Coastal Structures 2003*; American Society of Civil Engineers: Reston, VA, USA, 2003; pp. 431–443.
16. Van Gent, M.R.A. Wave Runup on Dikes with Shallow Foreshores. *J. Waterw. Port Coast. Ocean Eng.* **2001**, *127*, 254–262. [CrossRef]
17. Bosman, G.; van der Meer, J.; Hoffmans, G.; Schüttrumpf, H.; Verhagen, H.J. Individual Overtopping Events at Dikes. In *Coastal Engineering 2008*; World Scientific: Singapore, 2009; pp. 2944–2956. [CrossRef]
18. Lorke, S.; Scheres, B.; Schuttrumpf, H.; Bornschein, A.; Pohl, R. Physical Model Tests on Wave Overtopping and Flow Processes on Dike Crests Influenced by Wave-Current Interaction. *Coast. Eng. Proc.* **2012**, *1*, waves.34. [CrossRef]
19. Ryu, Y.; Chang, K.-A.; Lim, H.-J. Use of Bubble Image Velocimetry for Measurement of Plunging Wave Impinging on Structure and Associated Greenwater. *Meas. Sci. Technol.* **2005**, *16*, 1945. [CrossRef]
20. Thielicke, W.; Stamhuis, E.J. PIVlab—Towards User-Friendly, Affordable and Accurate Digital Particle Image Velocimetry in MATLAB. *J. Open Res. Softw.* **2014**, *2*, 30. [CrossRef]
21. Hughes, S.A.; Thornton, C.I.; Van der Meer, J.W.; Scholl, B.N. Improvements in Describing Wave Overtopping Processes. *Coast. Eng. Proc.* **2012**, *1*, waves.35. [CrossRef]

Journal of
Marine Science and Engineering

MDPI

Article

Mechanical Properties and Constitutive Model of Calcareous Sand Strengthened by MICP

Ziyu Wang [1,2], Xiangyu Zhao [1,3,*], Xin Chen [3], Peng Cao [4], Liang Cao [4] and Wenjing Chen [1,3]

1 Yazhou Bay Innovation Institute, Hainan Tropical Ocean University, Sanya 572025, China
2 College of Marine Science and Technology, Hainan Tropical Ocean University, Sanya 572022, China
3 College of Ecology Environment, Hainan Tropical Ocean University, Sanya 572022, China
4 Faculty of Architecture, Civil and Transportation Engineering, Beijing University of Technology, Beijing 100124, China
* Correspondence: zhaoxy@hntouedu.ntesmail.com

Abstract: To improve the mechanical properties of calcareous sand, it is proposed that microbial induced calcium carbonate precipitation (MICP) technology be used. A series of solidification tests were conducted in natural seawater and freshwater environments. The standard stress path static triaxial apparatus was used to conduct shear tests on calcareous sand and solids under varying reinforcement conditions. The composite power-exponential (CPE) model is proposed to describe the stress–strain relationship curve of the solid, and the method for determining model parameters is presented. The experimental results showed that the strength of calcareous sand with solids increased with the increase in number of reinforcement times for both test environments. Owing to the high salinity of seawater, which inhibits the activity of urease in bacterial solutions, the reinforcement strength in the seawater environment was generally lower than that in the freshwater environment. The compactness had an evident effect on the strength of the added solids. With the increase in compactness, the strength of the sample also increased, but the rate of increase was reduced. The simulation results showed that the established constitutive model can accurately describe the stress–strain relationship of microbial-reinforced calcareous sand and verified the applicability of the model.

Keywords: calcareous sand; mineralization reaction; MICP; shear test; constitutive model

Citation: Wang, Z.; Zhao, X.; Chen, X.; Cao, P.; Cao, L.; Chen, W. Mechanical Properties and Constitutive Model of Calcareous Sand Strengthened by MICP. *J. Mar. Sci. Eng.* **2023**, *11*, 819. https://doi.org/10.3390/jmse11040819

Academic Editors: José-Santos López-Gutiérrez, M. Dolores Esteban, Vicente Negro and M. Graça Neves

Received: 2 March 2023
Revised: 31 March 2023
Accepted: 4 April 2023
Published: 12 April 2023

1. Introduction

Calcareous sand is a special geotechnical medium with high calcium carbonate content of marine biogenic origin, and it is widely distributed on the continental shelf and coast in low latitude areas [1–3]. Owing to specific marine biological and chemical causes, calcareous sand has multiple edges and corners, is easily breakable, and has a high void ratio and large specific gravity [4,5]. Microorganisms induce calcium carbonate precipitation to produce urease, based on bacterial metabolism, promoting the hydrolysis of urea to carbonate ions and ammonium ions. The carbonate ion combines with the calcium ion in the environment to form calcium carbonate deposits in situ. Through particle encapsulation and inter-particle bonding, loose sand particles are cemented into a whole with a certain strength. The physical and mechanical properties of calcareous sand columns strengthened by MICP have been greatly improved. These improvements have made the columns environment friendly, efficient, and sustainable, which is in line with the requirements of island and reef engineering construction [6–9].

Xiao, Y. investigated the bearing performance of precast concrete piles embedded within calcareous sands with bio-grouting at the pile toe. The total bearing capacity of the precast concrete pile with a bio-grouted toe was 4.4 times as large as that without bio-grout [10]. Zamani, A. found that the application of MICP can lead to a decrease in soil water conductivity, and the formation of the calcium carbonate precipitate can reduce soil pore volume while improving its strength and stiffness [11]. Karimian, A. studied the effect of MICP reinforcement on the mechanical properties and microstructure of loose sand and silt, and found that there was no significant difference in the amount of calcium carbonate precipitation between loose sand and silt, and the soil strength was significantly improved [12]. Kannan, K. found that the biological stimulation method was not effective in two selected marine clays, and the undrained shear strength of MICP-treated marine clays significantly increased under the ductile limit water content [13]. Tian, Z. found that MICP can effectively improve the settlement and consolidation uniformity of marine clay during electroosmotic consolidation [14]. Islam, M.T. studied the impact of clay content on the treatment effect of MICP by selecting soils with different clay content, and found that using biologically stimulated MICP can increase the strength of cohesive soil, while the amount of calcite precipitation increases with the increase in clay content [15].

Montoya, B. found that the shear strength and stiffness of MICP-treated soil significantly increased with the increase in MICP cementation [16].

To capitalize on marine resources and reduce the use of fresh water, some seawater is added during the preparation process for MICP cementitious calcareous sand materials. The high pH value of seawater and the presence of various ionic components in seawater have a direct impact on the activity of bacteria and the mineral composition of sediment. It is, hence, necessary to study the MICP method for cementitious calcareous sand in a seawater environment. Peng Jie [17] studied the solidification effect of MICP reinforcement of coral sand in a simulated seawater environment and found that the seawater environment inhibits the production of calcium carbonate in the MICP reaction, thus inhibiting the reinforcement effect. Dong Bowen [18] found that natural seawater causes the growth of microorganisms to lag, and the effect of using seawater MICP to strengthen calcareous sand is not different from that under freshwater conditions, which verifies the feasibility of MICP reinforcement technology under seawater conditions. Gao Xinyu [19] found that the grouting round is an important influencing factor in the grouting reinforcement of sand columns in seawater environments through technical research on the reinforcement of coral sand by microbial-induced struvite under a high salt environment. With the increase in the number of grouting rounds, the unconfined compressive strength of a sand column also increases. Cheng, L. [20] considered the calcium ion in seawater as the only calcium source and found that MICP, after repeated seawater treatment, can significantly improve the strength of the sand column. At the same time, some researchers proposed that using MICP technology to modify calcareous sand with fiber reinforcement can effectively improve the toughness and ductility of the sample [21–26]. Wang, Y. conducted a rainfall erosion model test on MICP-reinforced calcareous sand, revealing the enhanced strength and erosion resistance of coastal coral sand after biological stimulation MICP treatment [27].Through simulated coastal erosion experiments, it was found that soil erosion decreased exponentially with the increase in calcium carbonate content. Under mild to moderate wave and dune slope conditions, EICP and MICP can significantly reduce dune erosion [28,29]. Hai-Lei Kou [30] and others found that the sand slope strengthened by MICP after adding recycled coconut silk (RSC) could better resist wave erosion and improve the erosion resistance of the simulated coastal slope. Xiao Yao [31] conducted a multi-gradient artificial acclimation culture test with *Bacillus pasteurii* in an artificial seawater environment and found that a seawater environment inhibits the formation of bacteria and the synthesis of urease. Moreover, the gradient acclimation could cause the bacteria to gradually adapt to the seawater environment, and the reinforcement effect of the bacterial solution after the gradient acclimation was better than that in fresh water. Yang Simeng [32] and other researchers found that when the amount of urea is three times the content of calcium ions in concentrated seawater, the efficiency of using calcium ions in the cement solution to produce precipitation can be improved.

The above research shows that using MICP to reinforce calcareous sand in a seawater environment is feasible, and the reinforcement effect is significant. The reinforcement of calcareous sand by MICP in a seawater environment is affected by water temperature, salinity, calcium source, reinforcement process and other factors. Due to previous studies focusing mainly on the reinforcement effect of calcareous sand in land environments, there are few studies on the mechanical properties of MICP-reinforced calcareous sand in seawater environments, and the main tests are unconfined compressive strength tests. Therefore, it is necessary to understand the mechanical behavior of calcareous sand treated by MICP in a seawater environment more deeply and comprehensively through triaxial experiments. Hence, in this work, for different water environment conditions, different reinforcement degree and degree of compaction, the reaction process, influence law and stress–strain relationship of consolidated sand strengthened by MICP were thoroughly studied by indoor triaxial experiments, using the CPE model to fit the relationship curve. The research results comprehensively revealed the mechanical characteristics of a microbial-reinforced calcareous sand roadbed in a seawater environment and provided a reference for the reinforcement design of a coral island reef calcareous sand roadbed in the South China Sea.

2. Materials and Methods

2.1. Reaction Principle of MICP Technology

According to the results from research conducted by Ronghui et al. [33], the MICP reaction process in a deionized water environment is as follows. Ca^{2+} in the nutrient salt first attaches to the surface of bacterial cells. After the urea is gradually hydrolyzed into an ammonium ion and carbonate ion, Ca^{2+} and $CO_3{}^{2-}$ in the cell are combined to form crystals, such as $CaCO_3$. The reaction process is shown in Figure 1, which can be simplified as follows:

$$Ca^{2+} + \mathrm{Cell} \rightarrow \mathrm{Cell} - Ca^{2+}$$

$$CO(NH_2)_2 + 2H_2O \leftrightarrow 2NH_4{}^{+} + CO_3{}^{2-}$$

$$\mathrm{Cell} - Ca^{2+} + CO_3{}^{2-} \rightarrow \mathrm{Cell} - CaCO_3$$

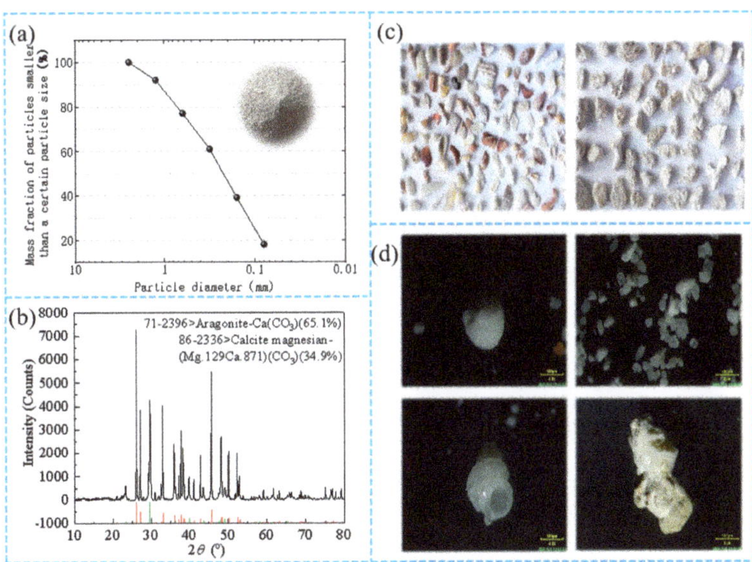

Figure 1. Gradation and composition diagram of calcareous sand. (**a**) Grain grading curve of calcareous sand sample; (**b**) X-ray fluorescence spectrum analysis diagram of calcareous sand; (**c**) images of coral reefs and shells; (**d**) calcareous sand debris under body microscope with 4 times magnification.

In the MICP mineralization reaction process in a seawater environment, Ca^{2+} in nutrient salt and a small amount of Mg^{2+} and Ba^{2+} in seawater first attach to the surface of bacterial cells. After urea is hydrolyzed into ammonium ions and carbonate ions in steps, owing to the weak alkaline environment of seawater, ions such as Ca^{2+}, Mg^{2+}, and Ba^{2+} of cells are combined with ${CO_3}^{2-}$ and OH^- to form $CaCO_3$, $Mg_5(CO_3)_4(OH)_2{\cdot}4H_2O$, $BaCO_3$, and other crystals. The reaction process can be simplified as follows:

$$Ca^{2+} + Cell \rightarrow Cell - Ca^{2+}$$

$$Mg^{2+} + Cell \rightarrow Cell - Mg^{2+}$$

$$Ba^{2+} + Cell \rightarrow Cell - Ba^{2+}$$

$$CO(NH_2)_2 + 2H_2O \leftrightarrow 2{NH_4}^+ + {CO_3}^{2-}$$

$$Cell - Ca^{2+} + {CO_3}^{2-} \rightarrow Cell - CaCO_3$$

$$Cell - Mg^{2+} + OH^- + {CO_3}^{2-} \rightarrow Cell - Mg_5(CO_3)_4(OH)_2 \cdot 4H_2O$$

$$Cell - Mg^{2+} + OH^- + {CO_3}^{2-} \rightarrow Cell - Mg_5(CO_3)_4(OH)_2 \cdot 4H_2O$$

$$Cell - Ba^{2+} + {CO_3}^{2-} \rightarrow Cell - BaCO_3$$

2.2. *Bacteria and Culture*

The bacterium used for the experiment was *Bacillus pasteurii*, which is widely distributed in natural soil and is the most widely used bacterium in the field of geotechnical engineering MICP. The bacterium can promote the hydrolysis of urea from the highly active urease produced by its metabolism, thus causing the rapid deposition of calcium carbonate. The main components and content of the bacterial culture medium were: 20 g/L of urea, 15 g/L of peptone, 5 g/L of soybean peptone, and 5 g/L of sodium chloride. The pH of the medium was adjusted to 7.3 with NaOH solution. The prepared culture medium was sterilized for 30 min in a 120 °C autoclave and then cooled to room temperature on an ultra-clean table. The bacteria were inoculated into the culture medium at a ratio of 1.0% (volume fraction) and cultured for 36 h on a constant-temperature shaking table at 30 °C at a rotational speed of 220 r/ min. Thereafter, the concentration of the bacterial solution and the urease activity of the bacterial solution were measured.

2.3. *Calcareous Sand*

The calcareous sand used in the experiment was obtained from an island and reef in the South China Sea. The maximum dry density of the sample was 1.65 g/cm^3, the minimum dry density was 1.20 g/cm ^{3}, the specific gravity was 2.73, the maximum and minimum void ratios were 1.27 and 0.63, respectively, and the particle distribution curve is shown in Figure 1a. The calcareous sand was mainly composed of coral debris and marine animals such as seashells and foraminifera. The main chemical components were calcium carbonate and magnesium carbonate. The calcium content accounted for more than 90% of the total volume, as shown in Figure 1b–d.

2.4. *Experimental Procedure*

2.4.1. Determination of Microbial Concentration and Activity

A spectrophotometer was used to determine the bacterial concentration. Briefly, 5 mL of bacterial solution was extracted and combined with 10 mL of deionized water, and then its absorbance (OD_{600}) at the wavelength of 600 nm was measured. To test the urease activity of bacterial liquid, the probe of a conductivity meter was cleaned with distilled water, then placed in a prepared mixture consisting of 5 mL of bacterial liquid combined with 1.1 mol/L urea solution, which was 9 times the volume of bacterial liquid. The conductivity change in the mixed solution was measured for 5 min. The hydrolysis ability of urease was directly proportional to the change in conductivity, and the relationship between the two can be expressed as follows:

$$\textit{Urea hydrolysis amount}\ (\text{mM/min}) = \textit{conductivity change value}\ (\text{ms/cm}) \times 11.11 \quad (1)$$

The average conductivity change value measured within 5 min multiplied by the dilution factor (10 times) was the initial enzyme activity of the bacterial solution. The absorbance (OD_{600}) of the bacterial solution measured during the experiment was 1.52, and the conductivity changed to 0.96 mmol/(L·min) within 5 min.

2.4.2. Strengthening Calcareous Sand Column with MICP

The mold selected for the test was rigid PVC round pipe, with a height of 8.8 cm and a diameter of 3.9 cm. A specified amount of calcareous sand was weighed and combined with water. A specified water content sand sample was prepared and loaded into the mold one layer at a time, and the dry density of the sand column was controlled. The cultivated bacterial solution was manually injected, and the nutrient salt was intermittently injected using the peristaltic pump grouting method to conduct the consolidation test for the calcareous sand column. The experimental steps were as follows: (1) For sample preparation, after the sample was loaded, the peristaltic pump was used for 24 h to inject natural seawater into the seawater group sample at a rate of 0.1 mL/min as the seawater curing environment. Deionized water was injected for 24 h into the freshwater group sample as the freshwater curing environment. (2) The bacterial solution and nutrient salt were injected. A glass rod was used to drain and evenly inject a mixed solution consisting of 80 mL bacterial solution and 10 mL 0.05 mol/L calcium chloride solution. After standing for 12 h, a peristaltic pump was used to inject 60 mL of nutrient salt into the sample at a rate of 0.1 mL/min. The interval between two injections of nutrients was 24 h. (3) The sample was removed and after grouting, the peristaltic pump was used to wash the sample with deionized water for 24 h. The residual ions in the sample were washed, and the continuous generation of calcium carbonate precipitation was halted. The mold was removed, and the sample was placed in a 60 °C oven for 48 h. An illustration of the experimental process is shown in Figure 2.

2.4.3. Determination of Ca^{2+} Concentration

After each grouting procedure, the effluent liquid at the bottom of the sample was collected, and the concentration of Ca^{2+} in it was measured to analyze the degree of the MICP mineralization reaction. The concentration of Ca^{2+} was measured using a PXS−270 (ion meter), Pca−1−01 (calcium ion electrode), and 232−01 (reference electrode). The ion meter was calibrated with the standard solution before each measurement. Because the lowest measurement range of the ion meter was 0.1 mol/L, the solution to be measured was diluted accordingly before measurement.

2.4.4. Determination of Carbonate Content

The main component of calcareous sand is calcium carbonate; therefore, it was not suitable to measure the quantity of raw carbonate using the acid washing method. In this study, the weighing method was used to calculate the calcium carbonate content, C, generated by the mineralization reaction, by obtaining the difference between the initial mass M_1

of the sand column and the final mass M_2 of the sand column after MICP reinforcement, cleaning and drying. The expression for C is given by

$$C = \frac{M_2 - M_1}{M_2} \times 100\%. \quad (2)$$

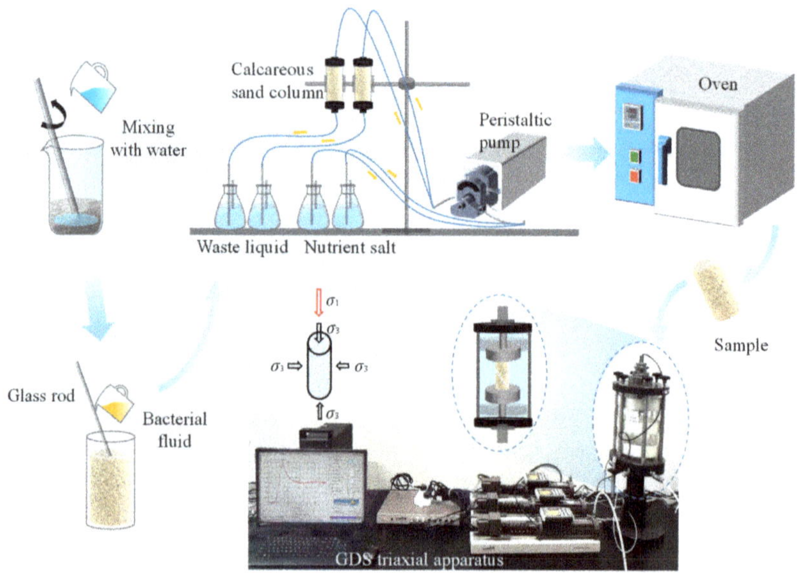

Figure 2. Schematic diagram of the reinforcement process for the MICP calcareous sand column.

2.4.5. Triaxial Compression Test and Constitutive Model Study

The standard stress path static triaxial apparatus was used to conduct loading tests on calcareous sand and solids under different confining pressures and different reinforcement conditions, and the stress–strain relationship curves of solid under different reinforcement conditions were obtained. The composite power-exponential (CPE) model was proposed to describe the triaxial-drained shear test of MICP-strengthened calcareous sand. CPE model fitting of stress–strain curves was obtained from conventional triaxial tests under different numbers of reinforcement days, compactness and confining pressure. From these, the strain-softening type curve of calcareous sand plus solids for different numbers of reinforcement days and compactness and the strain-hardening type curve of the stress–strain CPE model considering confining pressure were obtained.

3. Results

3.1. Degree of Mineralization Reaction and Amount of Carbonate Precipitation

The degree of a mineralization reaction in different water environments can be determined by measuring the concentration of calcium ion in the liquid flowing out of the sand sample after the mineralization reaction. This is achieved by collecting the effluent liquid of the sand sample after reaction with a container and measuring the concentration of Ca^{2+} in the residual liquid. It is evident from Figure 3a that the concentration of calcium ions in the effluent on the first and fourth days was very low (below 0.05 mol/L), which indicates that the mineralization reaction efficiency was high during those days, and the calcium ions in the nutrient salt were consumed to form carbonate precipitate. From the fifth day, the concentration of Ca^{2+} in the effluent increased significantly with the increase in the number of reinforcement times. When the number of reinforcement times was greater than 8, the concentration of Ca^{2+} in the effluent was between 0.8–1.0 mol/L, indicating that the mineralization reaction efficiency was low and that the amount of calcium carbonate precipitation was small. Comparing the concentration of Ca^{2+} in the outflow liquid of sand samples in both water environments, it was found that the reaction degree in both water environments was approximately equal when the number of reinforcement times was between 0 and 4. When the number of reinforcement times was greater than 5, the concentration of Ca^{2+} in the seawater environment increased marginally faster than that in the deionized water environment. This was caused by the different activity of bacteria in different water environments with time and resulted in different degrees of the mineralization reaction. With the increase in the number of reinforcement times and the decrease in bacterial activity, the mineralization reaction efficiency decreased.

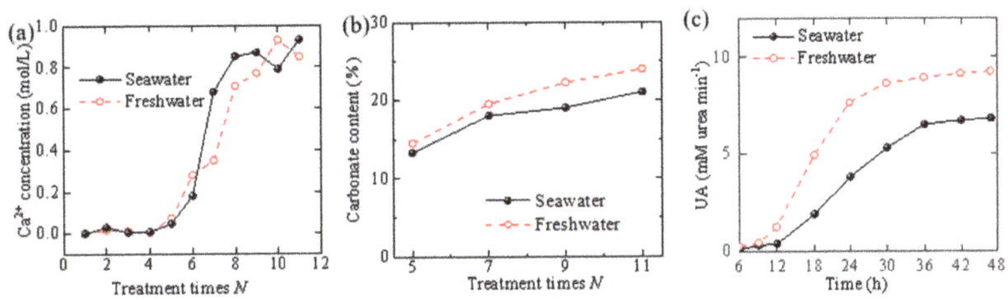

Figure 3. Variation diagram of calcium ion concentration, calcium carbonate content and urease activity. (**a**) Variation in Ca^{2+} concentration with increasing number of reinforcement days; (**b**) variation in calcium carbonate content with increasing number of reinforcement days; (**c**) urease activity with time.

It is evident in Figure 3b that for both water environments, the amount of carbonate formation in the calcareous sand column increased significantly with the increase in the number of reinforcement times, while the increase rate decreased gradually. The degree of the mineralization reaction and total amount of carbonate formation in the seawater environment differ from those in the freshwater environment. It is shown in Figure 3c that the urease activity in the seawater environment was typically lower than that in the freshwater environment. This is because the higher salinity of the seawater inhibited bacterial activity. Thus, the degree of the mineralization reaction and total amount of carbonate formation in the seawater environment were typically lower than those in the freshwater environment.

Figure 4a,b show the scanning electron microscope images of the sample in the seawater environment, and Figure 4c shows the energy spectrum images of the sediment in the seawater environment; Figure 4d,e show the scanning electron microscope images of the sample in the deionized water environment, and Figure 4f shows the energy spectrum images of the sediment in the deionized water environment. Figure 4e shows that under the scanning electron microscope with 10,000 times magnification, the calcium carbonate that precipitated in the freshwater environment formed a large number of relatively complete calcite crystals, as indicated by the purple dotted line in the figure. Figure 4f shows that a large number of precipitated calcium carbonate crystals were generated in the deionized water environment. As shown in Figure 4b, under the scanning electron microscope with 10,000 times magnification, in addition to the generation of a large number of calcite crystals in the seawater environment, a small number of dolomite crystals were also attached between the crystals. It is clear from Figure 4c that the mineral components generated in the seawater environment consisted of more magnesium, barium, and other elements than in the deionized water. This is because small amounts of magnesium ions and barium ions in the seawater make the reaction process produce not only calcium carbonate, but also a small amount of magnesium carbonate, barium carbonate and other mineral components.

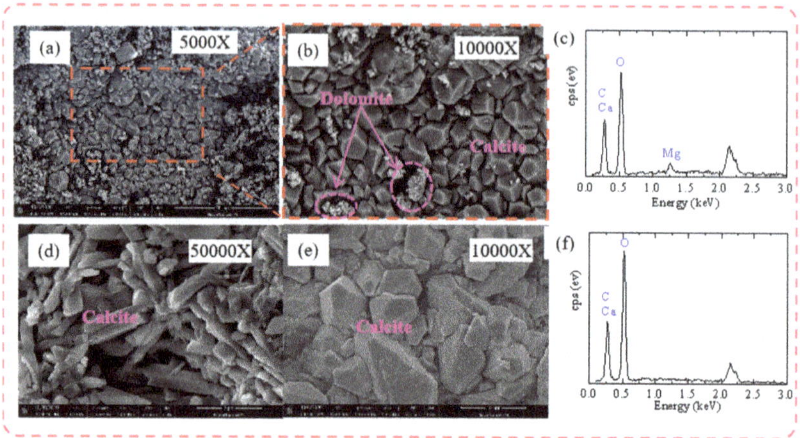

Figure 4. Scanning electron microscope images of samples and energy spectrum diagram. (**a**) Scanning electron microscope (5000 times magnification) image of seawater environment sample; (**b**) scanning electron microscope (10,000 times magnification) image of seawater environment sample; (**c**) energy spectrum of sediment in seawater environment; (**d**) scanning electron microscope (50,000 times magnification) image of deionized water environment sample; (**e**) scanning electron microscope (10,000 magnification) image of deionized water environment sample; (**f**) energy spectrum of sediment in deionized water environment.

3.2. Triaxial Compression Test

The peak strength and stress–strain relationship of MICP-strengthened specimens for different strengthening conditions were analyzed. It is evident in Figure 5a,b that the stress–strain curves of calcareous sand with respect to different reinforcement degrees in both water environments were mainly of the hardening type and softening type. When the confining pressure was 50 kPa, 100 kPa and 200 kPa, the stress–strain relationship of the reinforcement depicted a softening type, and when the confining pressure reached 400 kPa, it depicted a hardening type. It is evident from Figure 5a–c that the peak strength and residual strength of the sample increased with the increase in confining pressure in both seawater and freshwater environments.

The carbonate generated by the microbial mineralization reaction cemented the loose particles into a whole. Under the same conditions, the variation trend in the amount of carbonate generated was consistent with the peak strength, that is, the more carbonate generated, the higher the overall strength of the sample, as shown in Figure 5d. Figure 5d is a comparison diagram of the calcium carbonate content–peak strength relationships found by other research teams [18,34–37] and this study. With the increase in calcium carbonate precipitation generated by the sample, the deviatoric stress of calcareous sand plus solid increased, and the variation trend was consistent with that of existing studies. The slope of the curve was different with different reinforcement conditions.

Figure 5e,f show that the rigidity and deviatoric stress of the calcareous sand without MICP reinforcement were significantly lower than those of the sample after reinforcement, and the range of the shear band was wide. The shear-failure band of the sample with MICP reinforcement was more evident and narrower than that of the sample without MICP reinforcement.

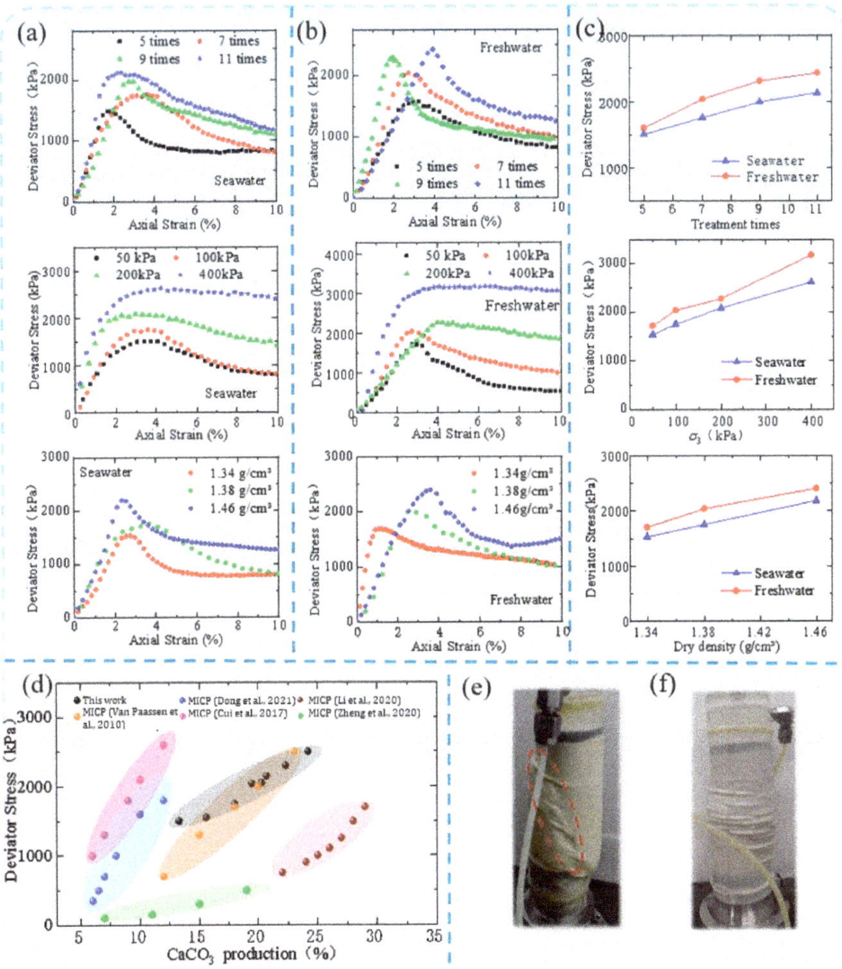

Figure 5. Stress–strain curve and failure mode diagram of sample. (**a**) Stress–strain curve of seawater environment sample; (**b**) stress–strain curve of samples in freshwater environment; (**c**) peak strength of samples under different conditions; (**d**) comparison diagram of calcium carbonate content–stress relationship [18,34–37]; (**e**) failure mode of MICP-strengthened specimen; (**f**) failure mode of the non-strengthened specimen.

3.3. Effect of Seawater on Stress–Strain Relationship

Figure 5c shows that under different conditions, the trend of shear strength variations in both seawater and freshwater environments was the same, and the shear strength in the seawater environment was less than that in the freshwater environment for the same parameter. This is because natural seawater inhibits the growth of *Bacillus pasteurii* and urease activity. These effects reduced the reaction rate and affected the reinforcement effect.

Figure 5a,b show that the samples' shear strength with respect to the number of reinforcement days in the seawater environment decreased by 7.9%, 13.6%, 14% and 12% compared with that in the freshwater environment. The strength increased with the increase in number of reinforcement days. After 11 days of reinforcement, the strength of the sample reached the maximum, 2.43 MPa and 2.13 MPa in seawater and freshwater, respectively. In both water environments, the shear strength of the sample increased with the increase in the number of reinforcement times, and the rate of increase decreased gradually. This is because with the increase in the number of nutrients injected, the amount of carbonate generated in the particle pores gradually increased, and the density and strength of the added solids increased. At the same time, as the gaps between the sand particles were gradually filled and blocked by the generated carbonate precipitation, variations in the surrounding environment of the bacteria led to the gradual reduction in bacterial activity, thus reducing the efficiency of the mineralization reaction and weakening the strength growth trend. From Figure 5c, it is evident that the strength growth rate of the specimens that were strengthened for approximately 7 days was the fastest. This is the optimum number of strengthening days to balance the strengthening effect and economic cost. The axial strain corresponding to the peak strength of the sample is defined as the failure strain. The failure strain of the sample in the seawater environment had insignificant variation compared with that in the freshwater environment. Both values were between 2% and 4%, indicating that the deformation and settlement of the calcareous sand strengthened by MICP may be easily monitored and controlled in the application of marine engineering.

The stress–strain curve of the sample under low confining pressure depicted a softening type, while that under high confining pressure depicted a hardening type and a significant increase in the peak strength. The shear strength of samples under 400 kPa confining pressure in seawater and freshwater environments increased by 75.9% and 82.3%, respectively, compared with those under 50 kPa confining pressure. The increase in strength of samples in the seawater environment was small. Therefore, in seawater or freshwater environments, the greater the confining pressure, the greater the residual strength of the sample, and the higher the deviatoric stress after reaching the peak strength. This shows that increasing the confining pressure can effectively improve the bearing capacity of calcareous sand in marine engineering.

The density had little influence on the shape of the stress–strain curve of calcareous sand samples but had a significant influence on their peak strength. In general, the higher the initial dry density of the sample, the greater its peak strength. The densities of the samples in seawater and fresh water were 1.34 g/cm^3 and 1.46 g/cm^3, respectively. Compared with the sample with a density of 1.34 g/cm^3, the strength of the sample with a density of 1.46 g/cm^3 in seawater and fresh water environments increased by 32.3% and 24.1%, respectively. Although the strength of the samples in the seawater environment was still not very large, the increase was large.

3.4. Constitutive Model

3.4.1. Model Establishment and Parameter Solution

The stress–strain relationship of consolidated calcareous sand samples under different confining pressures was characterized primarily by hardening and softening. The characteristics of the softening stress–strain curve include the following. (1) The curve passes through the coordinate origin. (2) With the gradual increase in the axial strain, the partial stress q increases monotonically and nonlinearly at first, decreases gradually after reaching the peak strength, and finally becomes stable after reaching the residual stress, that is, the first derivative of the model function approaches 0 when the strain is infinite. For the hardening type, the deviatoric stress q increases linearly with the axial strain and remains unchanged when it reaches the peak strength. Therefore, compared with the softening-type curve, the deviatoric stress of the hardening type does not decrease. In this model, the hardening type is considered as a particular form of the softening type.

In this paper, the CPE model proposed by Wang Lijin [38] was used to fit and analyze the stress–strain curve of calcareous sand. The function expression of the CPE model is

$$q = \sigma_1 - \sigma_3 = f(\varepsilon_1) = \left[(a\varepsilon_1^m - k)e^{-b\varepsilon_1^n} + k\right]p_a \tag{3}$$

where q is the deviatoric stress, σ_1 is the axial stress, σ_3 is the confining pressure, ε_1 is axial strain, p_a is the standard atmospheric pressure, and a, b, k, m, and n are test parameters. The test parameters are defined as follows:

(1) Softening-type stress–strain curve

(a) Parameter $k = q_r$

When $\varepsilon_1 \to \infty$, $f(\varepsilon_1) \to kp_a$, when the stress–strain curve shows a softening type, $k = q_r/p_a$, where q_r is the residual strength.

(b) Parameters a and m

Before the peak strength q_p, the shape of the stress–strain curve is mainly affected by $a{\varepsilon_1}^m$, thus the measured data before the peak strain ε_p conform to $q/p_a = a{\varepsilon_1}^m$, and the logarithms on both sides of the equation are obtained as follows:

$$\ln(q/p_a) = \ln a + m\ln\varepsilon_1 \tag{4}$$

It is evident that $\ln(q/p_a) - \ln\varepsilon_1$ is linear, where $\ln a$ is the intercept of the straight line on the vertical axis, and m is the slope of the straight line.

(c) Parameters b and n

After the axial strain ε_1 exceeds the peak strain ε_p, the shape of the descending section of the stress–strain curve is primarily controlled by $e^{-b{\varepsilon_1}^n}$. Therefore, the measured data of the falling section after peak strain ε_p should conform to $\ln\frac{a{\varepsilon_1}^m-k}{q/p_a-k} = b{\varepsilon_1}^n$. Let $q^* = \ln\frac{a{\varepsilon_1}^m-k}{q/p_a-k}$, taking the logarithms on both sides of the equation yields

$$\ln q^* = \ln b + n\ln\varepsilon_1 \tag{5}$$

It is evident that $\ln q^* - \ln\varepsilon_1$ also has a linear relationship, where $\ln b$ is the intercept of the straight line on the vertical axis, and n is the slope of the straight line.

(2) Hardened stress–strain curve

For the axial strain $\varepsilon_1 < 2.5\%$, the data points are fitted as above, but the parameter a is opposite. In general, the value of parameter a is approximately 0 at this time; thus, let parameter $a = 0$, then Equation (3) can be simplified as follows:

$$q = (\sigma_1 - \sigma_3) = F(\varepsilon_1) = k\left(1 - e^{-b{\varepsilon_1}^n}\right)p_a. \tag{6}$$

Equation (6) is the simplified CPE model expression of the hardening stress–strain curve.

(a) Parameter $k = q_u$ when $\varepsilon \to \infty$, $F(\varepsilon_1) \to kp_a$, and when the stress–strain curve depicts the hardening type, $k = q_u/p_a = (\sigma_1 - \sigma_3)_u/p_a$, where $q_u = (\sigma_1 - \sigma_3)_u$ is the ultimate strength.

(b) Parameter b and n

The hardening-type stress–strain curve parameter $a \approx 0$; therefore, $q^* = \ln\frac{a{\varepsilon_1}^m-k}{q/p_a-k} = \ln\frac{k}{k-q/p_a}$. Similarly, $\ln b$ is the intercept of the straight line $\ln q^* - \ln\varepsilon_1$ on the longitudinal axis, and n is the slope of the straight line.

3.4.2. Model Parameter Solution

The CPE model was applied to the data measured in the experiment to fit and analyze the calcareous sand reinforcement samples under different conditions. Figure 6a–c show the fitting curves of the stress–strain relationship of samples under different conditions. Table 1 shows the values of the CPE model parameters a, b, k, m, n and R^2. The R^2 values are all above 0.96, indicating that the CPE model can adequately simulate the MICP solidified calcareous sand sample $q - \varepsilon$ curve. Parameter a reflects the shape of the stress–strain curve before peak failure of rock and soil mass, while parameters b and n reflect the shape of the stress–strain curve after peak failure of rock and soil mass. The parameter k in the strain softening curve represents the residual strength. The final strength is depicted in the strain-hardening curve. The parameter m reflects the nature of the small-strain stage of the calcareous sand sample. When $0 < m < 1$, the curve is concave, and the greater the concave degree of the curve, the greater the m value. If $m > 1$, the curve is convex, and the larger the convex degree of the curve, the smaller the m value. It is evident in Table 1 that the m value of most calcareous sand and solids is greater than 1, which generally corresponds to the stress–strain curve of rock and soil mass with developed fissures.

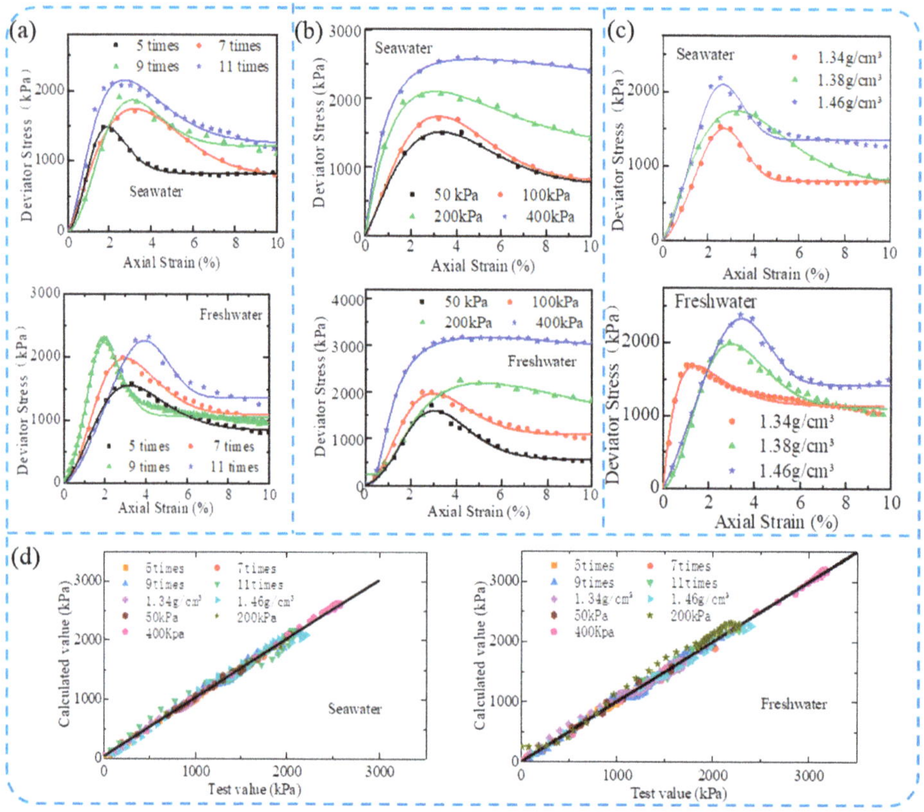

Figure 6. Fitting diagram and comparison diagram of stress–strain curve of sample. (**a**) Stress–strain fitting diagram of samples with number of reinforcement days; (**b**) stress–strain fitting diagram of samples under different confining pressures; (**c**) stress–strain fitting diagram of samples at different densities; (**d**) bias stress–fitting bias stress comparison chart.

3.4.3. Model Validation

The deviatoric stress in the actual data was considered the abscissa, and the deviatoric stress obtained by the fitting function was considered the ordinate. The comparison diagram of the actual deviatoric stress and the fitting deviatoric stress verified the fitting effect of the model on the stress–strain relationship of calcareous sand. As shown in Figure 6d, the black line in the figure is the expression of the function $y = x$. The closer in value the abscissa and ordinate are, the closer the data point is to the line, and the better the fitting effect. In contrast, if the deviation of data points from the straight line is large, the fitting effect is poor. Through comparison, it was found that the fitting difference mainly occurred at the beginning of a triaxial test, that is, at the stage of elastic deformation of calcareous sand, which indicated that the fitting effect of the CPE model on the failure form of calcareous sand at the later stage was better than that at the earlier stage, and the overall evaluation of the CPE model could adequately describe the stress–strain relationship of calcareous sand.

Table 1. Parameters of CPE model for stress–strain curve of calcareous sand column.

Water Environment	Number of Days (d)	Density (g/cm^3)	Cell Pressure (kPa)	a	b	m	n	k	R^2
Freshwater	5	1.38	100	314.6	0.69	3.2	1.2	845.9	0.992
	7			490.2	0.43	2.5	1.43	1094.5	0.985
	9			1482.3	1.27	4.63	1.42	1065.2	0.976
	11			450.5	0.004	1.41	3.68	1368.6	0.977
	7	1.34		3186.8	1.45	1.36	0.89	1135.2	0.966
		1.46		793.1	0.01	1.1	3.39	1430.1	0.995
		1.38	50	340.5	0.73	3.76	1.28	562.5	0.983
			200	479,739	7.76	−1.7	−0.65	246.1	0.999
			400	5672.3	1.42	−0.23	−1.2	42.57	0.998
Seawater	5	1.38	100	1112.8	0.98	3.12	1.4	826.3	0.996
	7			834.9	0.15	1.17	1.59	785.9	0.997
	9			303.5	0.42	2.85	1.47	1217.1	0.972
	11			1195	0.58	1.76	1.15	1255	0.975
	7	1.34		529.2	0.02	1.42	3.4	801.2	0.991
		1.46		818.9	0.05	1.37	2.86	1354.9	0.975
		1.38	50	762.2	0.13	1.1	1.59	746.1	0.995
			200	1685	0.69	1.2	0.83	1219.2	0.993
			400	76,449	3.9	−0.78	−0.35	245.39	0.998

4. Conclusions

In this paper, the effects of various factors on the physical and mechanical properties of the reinforcement of calcareous sand by microbial-induced carbonate precipitation were analyzed by varying the number of reinforcement days, sample density, confining pressure, and other conditions during the experimentation of MICP-solidified calcareous sand in seawater and freshwater environments. Using the CPE model to simulate the static test results of MICP-reinforced calcareous sand, the following interpretations were obtained:

(1) It was found that the failure strength of MICP-strengthened calcareous sand samples was affected by the degree of compaction, reinforcement effect, water environment and confining pressure. Under the same conditions, the curing effect of the freshwater environment sample was better than that of the seawater environment sample. The shear strength in the seawater environment reached more than 1.4 MPa, which meets the standard for practical application in marine engineering and shows that it is feasible to strengthen calcareous sand with MICP under seawater conditions.

(2) The strength of the calcareous sand sample strengthened by MICP increased with the increase in the number of reinforcement days, confining pressure, and density. The best number of reinforcement days was 7 days after curing, at which time the strength of the sample increased the quickest, the bearing capacity was large, and the material loss was small. With the increase in confining pressure, the failure strain of the sand column increased continuously, the hardening effect was more evident, and the residual strength also increased. These results can effectively improve the bearing capacity of calcareous sand in marine engineering. The compactness had a clear effect on the strength of the added solid. With the increase in the sample density, although the strength of the seawater environment sample was small, the increase in strength was large, indicating that the MICP curing effect can be optimized by improving the compactness of the foundation in marine engineering.

(3) In this study, the CPE model was used to fit the triaxial experimental results of calcareous sand, and the overall fitting accuracy was high. This indicates that the

model can accurately predict the stress–strain relationship during the loading process of the sample and show the process of calcareous sand sample damage, and thus can reasonably predict the application of calcareous sand in marine engineering.

5. Discussion

Although in the seawater environment, the use of microorganism-induced calcium carbonate precipitation to strengthen calcareous sand can achieve good results, there are still some unsolved technical problems that will be further studied in the follow-up work.

(1) In the process of MICP reinforcement, the by-product NH_4^+ is difficult to recover, and its large-scale application will have an impact on the environment. In order to eliminate the adverse effects of NH_4^+, the improvement of MICP technology will be the focus of future research.

Mohsenzadeh washed and recovered ammonium from soil through a two-stage treatment process and found that 86.8% of the ammonium ions could be recovered as high-purity struvite [39]. Keykha found that using zeolite suspension to pretreat sand is an effective method for preparing ammonium-free carbonate-producing bacteria. The negatively charged zeolite (natural aluminosilicate) can absorb NH_4^+ from the cement solution, bringing it to a standard level [40]. Gowthman and Yamamoto acid-dissolved bone meal (an excellent and low-cost source of calcium and inorganic phosphates) and injected it with urea and acid urease into a sand sample. It was found that the release of ammonium ions into the environment was reduced by about 50%, and the emission of toxic gaseous ammonia was reduced by about 90%, making it more environmentally friendly [41].

(2) Due to the high temperature, salinity and humidity in the coastal areas of tropical islands, coral reefs and beaches, it is also important to study the application and response of MICP-solidified calcareous sand in extreme environments.

Gowthman simulated the durability of MICP-reinforced slope soil under acid rain exposure through indoor experiments, and found that the corrosion rate mainly depends on the pH of acid rain and that lower pH conditions lead to higher corrosion rates. The intensity of acid rain also has a significant impact on the corrosion rate. The lower the transport rate, the more severe the corrosion [42]. Li found that MICP-cemented calcareous sand bodies in seawater environments have higher resistance to dry–wet cycles than those cemented in freshwater environments. Dry–wet cycles weaken the surface roughness of particles and the strength of interparticle cementation. Macroscopically, the strength and stiffness of MICP-cemented calcareous sand bodies decrease [43].

Author Contributions: Investigation, writing original draft preparation and data curation, X.Z.; writing review and editing, resources, data curation, project administration, and funding acquisition, Z.W.; formal analysis, X.C.; investigation, P.C.; visualization, L.C.; software, W.C. All authors have read and agreed to the published version of the manuscript.

Funding: This research was supported by the National Natural Science Foundation of China (Grant No. 42162024, 41602322), the Hainan Provincial Natural Science Foundation of China (Grant No. 421RC592), key research and development projects in Hainan Province (Grant No. ZDYF2017100), and Student innovation project of Yazhou Bay Innovation Institute of Hainan Tropical Ocean University (Grant No. 2022CXYXSCXXM06).

Informed Consent Statement: Written informed consent has been obtained from the patient(s) to publish this paper.

Data Availability Statement: The data used during the study are available from the first author and corresponding author by request.

Conflicts of Interest: The authors declare no conflict of interest.

References

1. Xiao, P.; Liu, H.L.; Xiao, Y.; Stuedlein, A.W.; Evans, T.M. Liquefaction resistance of bio-cemented calcareous sand. *Soil Dyn. Earthq. Eng.* **2018**, *107*, 9–19. [CrossRef]
2. Wang, X.Z.; Weng, Y.L.; Wei, H.Z.; Meng, Q.S.; Hu, M.J. Particle obstruction and crushing of dredged calcareous soil in the Nansha Islands, South China Sea. *Eng. Geol.* **2019**, *261*, 105274. [CrossRef]
3. Qin, Y.; Yao, T.; Wang, R.; Zhu, C.Q.; Meng, Q.S. Analysis of high-pressure consolidation deformation of calcareous sediment based on particle breakage. *Rock Soil Mech.* **2014**, *35*, 3123–3128. [CrossRef]
4. Shahnazari, H.; Rezvani, R. Effective parameters for the particle breakage of calcareous sands: An experimental study. *Eng. Geol.* **2013**, *159*, 98–105. [CrossRef]
5. Li, X.; Liu, J.K. One-dimensional compression feature and particle crushability behavior of dry calcareous sand considering fine-grained soil content and relative compaction. *Bull. Eng. Geol. Environ.* **2021**, *80*, 4049–4065. [CrossRef]
6. Liu, L.; Liu, H.A.; Stuedlein, A.W.; Evans, T.M.; Xiao, Y. Strength, stiffness, and microstructure characteristics of biocemented calcareous sand. *Can. Geotech. J.* **2019**, *56*, 1502–1513. [CrossRef]
7. Xiao, Y.; Stuedlein, A.W.; Xiao, Y.; Ran, J.Y.; Evans, T.M.; Cheng, L.; Liu, H.L.; van Paassen, L.A.; Chu, J. Effect of Particle Shape on Strength and Stiffness of Biocemented Glass Beads. *Am. Soc. Civ. Eng.* **2019**, *145*, 06019016. [CrossRef]
8. Bu, C.M.; Lu, X.Y.; Zhu, D.X.; Liu, L.; Sun, Y.; Wu, Q.T.; Zhang, W.T.; Wei, Q.K. Soil improvement by microbially induced calcite precipitation (MICP): A review about mineralization mechanism, factors, and soil properties. *Arab. J. Geosci.* **2022**, *15*, 863. [CrossRef]
9. Qian, C.X.; Wang, X.; Yu, X.N. Research and Application Development of Microbe Cement. *J. Mater. Eng.* **2015**, *43*, 92–103. [CrossRef]
10. Xiao, Y.; Stuedlein, A.W.; Pan, Z.Y.; Liu, H.L.; Matthew Evans, T.; He, X.; Lin, H.; Chu, J.; van Paassen, L.A. Toe-Bearing Capacity of Precast Concrete Piles through Biogrouting Improvement. *J. Geotech. Geoenviron. Eng.* **2020**, *146*, 06020026. [CrossRef]
11. Zamani, A.; Montoya, B.M.; Gabr, M.A. Investigating challenges of in situ delivery of microbial induced calcium carbonate precipitation (MICP) in fine-grain sands and silty sand. *J. Can. Geotech. J.* **2019**, *56*, 1889–1900. [CrossRef]
12. Karimian, A.; Hassanlourad, M. Mechanical behaviour of MICP-treated silty sand. *Bull. Eng. Geol. Environ.* **2022**, *81*, 285. [CrossRef]
13. Kannan, K.; Bindu, J.; Vinod, P. Engineering behaviour of MICP treated marine clays. *Mar. Georesources Geotechnol.* **2020**, *38*, 761–769. [CrossRef]
14. Tian, Z.; Tian, Z.F.; Tang, X.W.; Li, J.; Xiu, Z.L.; Xue, Z.J. Improving settlement and reinforcement uniformity of marine clay in electro-osmotic consolidation using microbially induced carbonate precipitation. *Bull. Eng. Geol. Environ.* **2021**, *80*, 6457–6471. [CrossRef]
15. Islam, M.T.; Chittoori, B.C.S.; Burbank, M. Evaluating the applicability of biostimulated calcium carbonate precipitation to stabilize clayey soils. *J. Mater. Civ. Eng.* **2020**, *32*, 04019369. [CrossRef]
16. Montoya, B.; DeJong, J.T. Stress-Strain Behavior of Sands Cemented by Microbially Induced Calcite Precipitation. *J. Geotech. Geoenviron. Eng.* **2015**, *141*, 04015019. [CrossRef]
17. Peng, J.; Tian, Y.M.; Yang, J.G. Experiments of coral sand reinforcement using MICP in seawater environment. *Adv. Sci. Technol. Water Resour.* **2019**, *39*, 58–62. [CrossRef]
18. Dong, B.W.; Liu, S.Y.; Yu, J.; Xiao, Y.; Tu, B.X. Evaluation of the effect of natural seawater strengthening calcareous sand based on MICP. *Rock Soil Mech.* **2021**, *42*, 1104–1114. [CrossRef]
19. Gao, X.Y. Study on Microbial-Induced Struvite Solidification of Coral sand in High Salt Environment. Master's Thesis, Huaqiao University, Quanzhou, China, 2020.
20. Cheng, L.; Shahin, M.A.; Cord-Ruwisch, R. Bio-cementation of sandy soil using microbially induced carbonate precipitation for marine environments. *Géotechnique* **2014**, *64*, 1010–1013. [CrossRef]
21. Yao, D.F.; Wu, J.; Wang, G.W.; Wang, P.B.; Zheng, J.J.; Yan, J.Y.; Xu, L.; Yan, Y.J. Effect of wool fiber addition on the reinforcement of loose sands by microbially induced carbonate precipitation (MICP): Mechanical property and underlying mechanism. *Acta Geotech.* **2021**, *16*, 1401–1416. [CrossRef]
22. Fang, X.W.; Yang, Y.; Chen, Z.; Liu, H.L.; Xiao, Y.; Shen, C.N. Influence of Fiber Content and Length on Engineering Properties of MICP-Treated Coral Sand. *Geomicrobiol. J.* **2020**, *37*, 582–594. [CrossRef]
23. Xiao, Y.; He, X.; Evans, T.M.; Stuedlein, A.W.; Liu, H.L. Unconfined Compressive and Splitting Tensile Strength of Basalt Fiber-Reinforced Biocemented Sand. *Am. Soc. Civ. Eng.* **2019**, *145*, 04019048. [CrossRef]
24. Xiao, Y.; Chen, H.; Stuedlein, A.W.; Evans, T.M.; Chu, J.; Cheng, L.; Jiang, N.J.; Lin, L.; Liu, H.L.; Aboel-Naga, H.M. Restraint of P-article Breakage by Biotreatment Method. *Am. Soc. Civ. Eng.* **2020**, *145*, 04020123. [CrossRef]
25. Wang, R.; Pan, X.H.; Tang, C.S.; Lu, C.; Wang, D.L.; Dong, Z.H.; Shi, B. Dynamic behaviors of MICP and fiber-treated calcareous sand under dynamic triaxial testing. *Rock Soil Mech.* **2022**, *43*, 2643–2654. [CrossRef]
26. Yin, L.Y.; Tang, C.S.; Zhang, L. Experimental Study on Mechanical Behavior of Micp-fiber Reinforce Treated Calcareous Sand. *Geol. J. China Univ.* **2021**, *27*, 679–686. [CrossRef]
27. Wang, Y.J.; Jiang, N.J.; Han, X.L.; Liu, K.W.; Du, Y.J. Biochemical, Strength and Erosional Characteristics of Coral Sand Treated by Bio-Stimulated Microbial Induced Calcite Precipitation. *Acta Geotech.* **2022**, *17*, 4217–4229. [CrossRef]

28. Dubey, A.A.; Jack, H.P.; Ravi, K.; Dhami, N.K.; Mukherjee, A. Biopolymer-biocement composite treatment for stabilisation of soil against both current and wave erosion. *Acta Geotech.* **2022**, *17*, 5391–5410. [CrossRef]
29. Liu, K.W.; Jiang, N.J.; Qin, J.D.; Wang, Y.J.; Tang, C.S.; Han, X.L. An experimental study of mitigating coastal sand dune erosion by microbial- and enzymatic-induced carbonate precipitation. *Acta Geotech.* **2021**, *16*, 467–480. [CrossRef]
30. Kou, H.L.; Liu, J.H.; Zhang, P.; Wu, C.Z.; Ni, P.P.; Wang, D. Ecofriendly improvement of coastal calcareous sandy slope using recycled shredded coconut coir (RSC) and bio-cement. *Acta Geotech.* **2022**, *17*, 5375–5389. [CrossRef]
31. Xiao, Y.; Deng, H.F.; Li, J.L.; Cheng, L.; Zhu, W.X. Study on the domestication of *Sporosarcina pasteurii* and strengthening effect of calcareous sand in seawater environment. *Rock Soil Mech.* **2022**, *43*, 395–404. [CrossRef]
32. Yang, S.M.; Peng, J.; Wen, Z.L.; Liu, Z.M.; Leng, M.; Xu, P.X. Application of concentrated seawater as calcium source solution in sand reinforcement using MICP. *Rock Soil Mech.* **2021**, *42*, 746–754. [CrossRef]
33. Rong, H.; Qian, C.X. Characterization of microbe cementitious materials. *Chin. Sci. Bull.* **2012**, *57*, 770–775. [CrossRef]
34. Li, H.; Tang, C.S.; Liu, B.; Lu, C.; Cheng, Q.; Shi, B. Mechanical behavior of MICP-cemented calcareous sand in simulated seawater environment. *Chin. J. Geotech. Eng.* **2020**, *42*, 1931–1939. [CrossRef]
35. Van Paassen, L.A.; Ghose, R.; van der Linden, T.J.M.; van der Star, W.R.L.; van Loosdrecht, M.C.M. Quantifying Biomediated Ground Improvement by Ureolysis: Large-Scale Biogrout Experiment. *J. Geotech. Geoenviron. Eng.* **2010**, *136*, 1721–1728. [CrossRef]
36. Cui, M.J.; Zheng, J.J.; Zhang, R.J.; Lai, H.J.; Zhang, J. Influence of cementation level on the strength behaviour of bio-cemented sand. *Acta Geotech.* **2017**, *12*, 971–986. [CrossRef]
37. Zheng, J.J.; Wu, C.C.; Song, Y.; Cui, M.J. Study of the strength test and strength dispersion of MICP-treated calcareous sand. *J. Harbin Eng. Univ.* **2020**, *41*, 250–256. [CrossRef]
38. Wang, L.Q.; Lu, Z.G.; Shao, S.J. Composite power-exponential nonlinear model of rock and soil mass. *Chin. J. Rock Mech. Eng.* **2017**, *36*, 1269–1278. [CrossRef]
39. Mohsenzadeh, A.; Afaki, E.; Gowthaman, S.; Nakashima, K.; Kawasaki, S.; Ebadi, T. A two-stage treatment process for the management of produced ammonium by-products in ureolytic bio-cementation process. *Int. J. Environ. Sci. Technol.* **2022**, *19*, 449–462. [CrossRef]
40. Keykha, H.A.; Mohamadzadeh, H.; Asadi, A.; Kawasaki, S. Ammonium-Free Carbonate Producing Bacteria as an Ecofriendly Soil Biostabilizer. In *Geotechnical Testing Journal*; ASTM International: West Conshohocken, PA, USA, 2018; Volume 42. [CrossRef]
41. Sivakumar, G.; Moeka, Y.; Kazunori, N.; Volodymyr, I.; Satoru, K. Calcium phosphate biocement using bone meal and acid urease: An eco-friendly approach for soil improvement. *J. Clean. Prod.* **2021**, *319*, 128782. [CrossRef]
42. Sivakumar, G.; Kazunori, N.; Satoru, K. Durability analysis of bio-cemented slope soil under the exposure of acid rain. *J. Soils Sediments* **2021**, *21*, 2831–2844. [CrossRef]
43. Li, Y.L.; Guo, Z.; Xu, Q.; Li, Y.J. Experimental study on dry and wet cycling of MICP cemented calcareous sand in seawater environment. *J. Zhejiang Univ.* **2022**, *56*, 1740–1749. [CrossRef]

Journal of
*Marine Science
and Engineering*

MDPI

Article

Surveying of Nearshore Bathymetry Using UAVs Video Stitching

Jinchang Fan [1], Hailong Pei [1,2,*] and Zengjie Lian [1]

[1] Key Laboratory of Autonomous Systems and Networked Control, Ministry of Education, Unmanned Aerial Vehicle Systems Engineering Technology Research Center of Guangdong, South China University of Technology, Guangzhou 510640, China
[2] Peng Cheng Laboratory, Shenzhen 518100, China
* Correspondence: auhlpei@scut.edu.cn

Abstract: In this paper, we extended video stitching to nearshore bathymetry for videos that were captured for the same coastal field simultaneously by two unmanned aerial vehicles (UAVs). In practice, a video captured by a single UAV often shows a limited coastal zone with a lack of a wide field of view. To solve this problem, we proposed a framework in which video stitching and bathymetric mapping were performed in sequence. Specifically, our method listed the video acquisition strategy and took two overlapping videos captured by two UAVs as inputs. Then, we adopted a unified video stitching and stabilization optimization to compute the stitching and stabilization of one of the videos separately. In this way, we can obtain the best stitching result. At the same time, background feature points identification on the shore plays the role of short-time visual odometry. Through the obtained panoramic video in Shuang Yue Bay, China, we used the temporal cross-correlation analysis based on the linear dispersion relationship to estimate the water depth. We selected the region of interest (ROI) area from the panoramic video, performed an orthorectification transformation and extracted time-stack images from it. The wave celerity was then estimated from the correlation of the signal through filtering processes. Finally, the bathymetry results were compared with the cBathy. By applying this method to two UAVs, a wider FOV was created and the surveying area was expanded, which provided effective input data for the bathymetry algorithms.

Keywords: bathymetry; video stitching; UAV; background identification; cBathy

Citation: Fan, J.; Pei, H.; Lian, Z. Surveying of Nearshore Bathymetry Using UAVs Video Stitching. *J. Mar. Sci. Eng.* **2023**, *11*, 770. https://doi.org/10.3390/jmse11040770

Academic Editors: M. Dolores Esteban, José-Santos López-Gutiérrez, Vicente Negro and M. Graça Neves

Received: 5 March 2023
Revised: 26 March 2023
Accepted: 30 March 2023
Published: 31 March 2023

1. Introduction

Coastal zone mapping plays a crucial role in oceanography, and water depth is a critical parameter that can directly reflect nearshore topography. However, the nearshore topography will be frequently changed due to wave motion, extreme typhoon weather, and human activities. It incurs tremendous problems for coastal zone management, for which we need to measure the nearshore bathymetry on a timely basis [1]. An accurate and convenient bathymetry method is urgently needed to be put forward and applied to practical work.

A convenient method is to operate an unmanned boat equipped with a sonar system weighing several hundred kilograms and then, measure depth using the acoustic principle. The disadvantage of this method is that it requires a lot of human resources and financial resources each time and cannot meet the flexible operation and low-cost needs. In addition, the airborne light detection and ranging (LiDAR) system was first developed by the United States military and provides high-precision bathymetry in cleaning waters. It is also not suitable for the above needs because of expensive costs and unreasonable spatial resolution [2]. Furthermore, satellite imagery is used for coastline topographic mapping and depth estimation [3,4] but is challenging to operate. Pressure sensor array [5] also can estimate depth continuously by measuring swell propagation velocity, and it is expensive to install these devices. Traditional methods for gathering bathymetric data include a small watercraft [6] outfitted with a real-time kinematic global positioning system (RTK-GPS) that measure depths accurately to a few millimeters in calm seas or a jetski [7] with an echo sounder attached behind it. In beaches with regular low wave times and easy access to the sea for small vessels, such a monitoring strategy can be carried out several times a month at most [8]. Consequently, finding more accessible alternatives to obtaining bathymetric data with the high spatial and temporal resolution is of significant importance.

More recently, some scholars paid attention to obtaining the information they wanted from coastal video imagery of waves based on the linear dispersion relationship [9–13], which is mathematically described by Equation (1).

$$\omega^2 = gk\tanh(kh) \tag{1}$$

where ω is the angular frequency ($2\pi/T$, T means wave period, or $2\pi f$, f means wave frequency), g is the gravitational acceleration constant. k represents wavenumber ($2\pi/L$, L means wavelength). An additional parameter is wave celerity c (L/T). Since the focus is on water depth information h, Equation (1) can be deformed to Equation (2).

$$h = \frac{c}{2\pi f}\text{arctanh}(\frac{2\pi f c}{g}) \tag{2}$$

Holman [14] creatively collected the wave propagation characteristics from coastal imagery and performed the nearshore depths. The water depth can be calculated as long as we know two of these (f, c, k, L, T).

The method based on coastal video imagery is regarded as a promising way to observe nearshore depth because of its low cost and easy operation. The critical point of this method is how to estimate the related parameters in Equation (1) or Equation (2) precisely. According to previous studies, there are two different ways to complete the process of variable estimation: the frequency method and the temporal method. The frequency method was first laid out by Holman [14] and was then refined in the cBathy algorithm [14]. Over the last few decades, many types of research [8,11,15] were inspired by cBathy in actual work. Additionally, video monitoring stations play an important role in bathymetric methods, and video-based algorithms were also extended [16–19]. Recent advancements in UAV technology and cost-cutting measures make it possible to use instruments designed for video monitoring stations at locations where measurements are needed but no video station is accessible, either due to a lack of a high vantage point or the necessity for only a single survey. The use of videos captured by UAVs to obtain bathymetry was previously investigated in recent publications [10,20–22].

Classical photogrammetry reconstructs 2-D topography from stacking aerial images and has existed as a field for quite some time. However, the method mentioned above is either based on a fixed camera for mapping or requires the UAV camera to hover at a certain height. The cBathy algorithm and the reference [22] require a certain observation time for sea level water movement, which can only provide the sea wave movement data with a fixed angle of view. That is to say, one flight can only carry out terrain inversion for a fixed area. If it is necessary to carry out bathymetric mapping for a long-distance coastline, the original method can only be used to carry out sectional sampling through separate fixed-point sampling, which greatly reduces the efficiency of mapping. Simultaneous mapping of multiple cameras is realized to obtain a wide field of view (FOV) in a single UAS [23] but it is not suitable for commercial UAVs with limited conditions, from which we were inspired and proposed another method.

The objective of this paper is to demonstrate that coastal bathymetry can be estimated from panorama video using video stitching and wave speed inversion algorithms applied to imagery from two UAVs transiting along the coast. The presented system enables the collection of mobile, short-dwell time series by background feature points identification on the shore. Additionally, the contribution of the paper can be concluded as follows.

1. The UAVs video stitching creates a wider FOV and improves the bathymetric mapping ranges;
2. The process of video stitching eliminates some of the rectification biases.

The rest of this paper is arranged as follows: in Section 2, the acquisition of video and video stitching is described. Then, pixel intensity signals are extracted from time stack imagery, and several signal processes are described in Section 3. Related parameters for bathymetry and algorithm performance are evaluated in Section 4. The discussion is provided in Section 5. Conclusions are summarized in Section 6.

2. Video Processing

2.1. Video Acquisition Scheme

According to the linear dispersion relationship, identifying relevant parameters in Equation (1) determines the final effect. After actual testing and data analysis, a suitable site needs to present a low-gradient slope in the intertidal area. The video acquisition work was selected in Shuang Yue Bay (22°35′43.1″ N, 114°52′35.3″ E), Huizhou, China, eventually, a sandy coast that faces the western Pacific Ocean. Figure 1 shows a satellite map of Shuang Yue Bay. This area has a beach coastline of about 5 km in the direction of the longshore and a remarkable horizon in the cross-shore direction to observe swells. The angle between the incoming swells' propagation direction and the coastline is almost 0. The waves on this site had perfect motion characteristics, such as clear crest lines and ideal amplitude. Thus, we chose the bay as this work's image acquisition site and obtained several UAV videos of shallow water depth areas in this place under varying conditions from January to May 2022.

Figure 1. Shuang Yue Bay satellite map and image.

We used the UAV of DJI Phantom 4 RTK, equipped with a normal wide-angle lens with 24 mm focal length, which was based on a 20 M pixel CMOS chip. The video resolution of the UAV camera was 4 K *UHD TV* (3840 × 2160). However, the feature of the camera is that it supersampled the frame in record state rather than photo state, reducing the FOV from 84° to 80.13°. To simulate small commercial UAVs, we reduced the pixel resolution to *FHD* (1920 × 1080). Due to a continuous video being required, the frame rate of the captured video was set to 30 fps. The RTK refresh rate was too low to represent the exact position of the camera. Commonly, relying on the traditional theoretical method [24], we placed several ground control points (GCPs, Figure 2) on the beach to solve the extrinsic parameter of the camera.

Figure 2. Ground control point.

The formula between camera parameters and the coordinate system can be shown in Equation (3),

$$p_{uv} = KTp_w \quad (3)$$

where K is the camera intrinsic matrix that contains focal lengths and principal point, T the camera extrinsic matrix containing rotation matrix R and translation vector t, depending on camera orientation and location, often defined as $T = \begin{bmatrix} R & t \\ 0^T & 1 \end{bmatrix}$. p_{uv} and p_w represent the coordinates of the pixel coordinate system and world coordinate system, respectively. The yaw angle of the camera should face the wave propagation to record swell motion correctly. In addition, the intersection angle between this camera's yaw angle and the wave propagation direction should not exceed 75° because an optical camera cannot capture the crest line [25]. Therefore, the yaw angle was set to approximately 30~45° relative to the region of interest to meet the above requirements. As a rule of thumb for depth inversion video collected processes, UAV should be deployed as high as possible to ensure the algorithm's field of view and better effect. Bearing this in mind and ensuring the UAV did not exceed a safe and controllable altitude, we decided to operate the UAV hovering at the height of 60~120 m. The video duration ranged from 4 to 15 min.

2.2. Image Processing

The core data required by the algorithm were to obtain the time series of the location where the water depth must be measured. Thus, these captured videos should be converted into sequential orthoimages that can conveniently extract time series precisely. It is unnecessary to preserve every frame for a high frame-rate video because of increasing computation. We are setting an appropriate frequency for down-sampling operation and retaining information of keyframes only. In this paper, the sampling frequency was set to 2 Hz, which meant that one image frame was extracted in 0.5 s from videos and combined into a complete-time sequence. The down-sampling operation would not affect the extraction of crucial information about waves because the periods of offshore waves were about 10 s.

As per the hardware structure design of the camera, there would be more or less distortion in the video and photo during recording. Camera calibration aims to obtain the intrinsic parameters and distortion coefficient for subsequent image transformation. This process refers to the method mentioned in [26].

2.3. Video Stitching and Stabilization

With a fixed angle of view, the video captured by an UAV for bathymetry was often limited by the camera's field of view (FOV). Stitching the video is a way to increase the horizontal FOV of the camera, and adding another camera appropriately reduces the influence of the rectification bias of the first camera. In this paper, bundled video path and path optimization will be used for video stitching. The details of this algorithm will be described below.

To stabilize an input video, Liu et al. [27] divided the video into grids spatially, as shown in Figure 3. A single homography $F_i(t)$ is estimated between neighboring frames in the original video, where i represents each grid at frame t. It is estimated by tracked features between adjacent frames. The camera path can be defined as the multiplication of a series of continuous homography.

$$C_i(t) = F_i(t) \cdot F_i(t-1) \cdots F_i(1), 1 \leq t \leq T, 1 \leq i \leq m^2 \tag{4}$$

where T is the number of frames in the video file and m^2 is the size of the m-by m grids. Local changes in the gird usually have a better transition to the stitching effect. The camera path can first be computed the feature trajectories by KLT tracker [28]. Given the original path $C = \{C(t)\}$, what needs to be calculated is the optimized video path and it is defined as the optimized path $P = \{P(t)\}$:

$$\Theta(P_i) = \sum_t (||P_i(t) - C_i(t)||^2 + \lambda \sum_{r \in \Omega_t} \omega_{t,r} \cdot ||P_i(t) - P_r(t)||^2) \tag{5}$$

where Ω_t denotes the neighborhood at frame t. The data term $||P_i(t) - C_i(t)||^2$ enforces the optimized path to be close to the original one, and the next term $||P_i(t) - P_r(t)||^2$ mainly stabilizes the optimized path. λ and $\omega_{t,r}$ balances the two terms. λ empirically set to 5. $\omega_{t,r}$ can be calculated by two Gaussian functions [27]:

$$\omega_{t,r} = G(||r - t||) \cdot G(||C_i(r) - C_i(t)||) \tag{6}$$

If all the grids participate in path optimization, the above Equation becomes:

$$E^{stable}(P) = \sum_t (\Theta(P_i) + \sum_t \sum_{j \in N(i)} ||P_i(t) - P_j(t)||^2) \tag{7}$$

where $P = \{P_i | 1 \leq i \leq m^2\}$, $j \in N(i)$ means the grid j is the neighbor of the gird cell i. According to the reference [29,30], E^{stable} works like a stabilizer that reduces shakiness during stitching. Additionally, P can be obtained by iteration [27,31], whose initial value can be set as $P = C$. Recall that video stitching is to create a wider FOV. P^A and P^B are the optimized paths generated by video A and video B, respectively, H is a single homography used to stitch the two videos. As for the process of stitching, there is the following optimization function that achieves stitching and stabilization at the same time:

$$E(P^A, P^B, H) = E^{stable}(P^A) + E^{stable}(P^B) + \beta \cdot E^{stitch}(P^A, P^B, H) \tag{8}$$

$$E^{stitch}(P^A, P^B, H) = \sum_t \sum_k \left|\left| P^A_{\bar{i}}(t) \cdot C^A_{\bar{i}}(t)^{-1} \cdot v^A_k(t) - H \cdot P^B_{\bar{j}}(t) \cdot C^B_{\bar{j}}(t)^{-1} \cdot v^B_k(t) \right|\right|^2 \tag{9}$$

where $v^A_k(t)$ and $v^B_k(t)$ are the kth feature point calculated by SIFT [32] at frame t of video A and video B, respectively. $\bar{i}$ and $\bar{j}$ are the grids where the feature points $v^A_k(t)$ and $v^B_k(t)$ are located. $E^{stitch}(P^A, P^B, H)$ means using SIFT to stitch frames. β was set between 10 to 30, depending on the parallax between the two videos.

In our study, the videos of two UAVs were spliced into a panoramic video. However, the above video path optimization process will more or less change the camera's optical center position and distortion degree, similar to as-projective-as-possible image warping in the method [33]. The frames should be transformed as little as possible to maintain pixel information integrity. The improvement of this paper is that the path of one of the videos is not optimized, which can also reduce the amount of calculation. Assume that the path of video A is not optimized, Equation (8) becomes:

$$E(P^A, P^B, H) = E^{stable}(P^B) + \beta \cdot E^{stitch}(P^A, P^B, H) \tag{10}$$

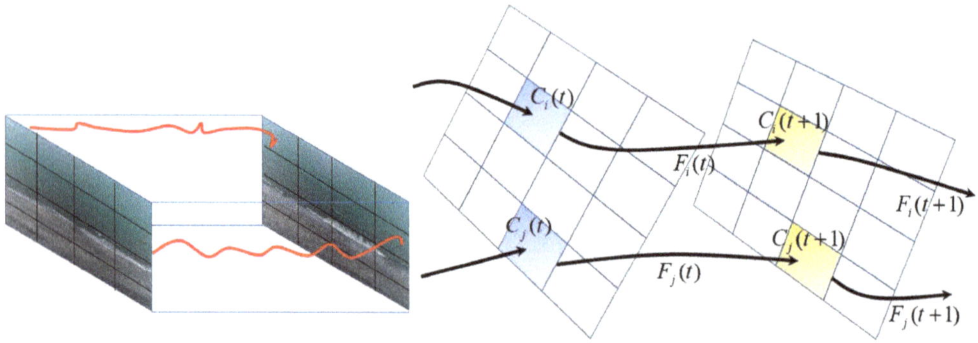

Figure 3. Each grid has its own bundled path, consisting of a sequence of homography.

The original Equation (7) optimizes the path of two videos. In this process, the optimized paths will be stabilized and remove shakiness first, then the feature matching method will be used for stitching. Note that P^A and P^B are also influenced by the stitching term. By setting $P^A(t) \equiv C^A(t)$, $P^B(t)$ tends to $P^A(t)$, i.e., the right frame B absorbs all perspective distortions (Figure 4). With known $P^A(t)$ and H, we can only calculate $P^B(t)$ by iteration. In theory, with the number of iterations ξ increasing, the camera B extrinsic matrix (in Equation (3)) is tending to the camera A extrinsic matrix:

$$\begin{cases} \lim\limits_{\xi \to \infty} R_B^{(\xi)} = R_A \\ \lim\limits_{\xi \to \infty} t_B^{(\xi)} = t_A \end{cases} \tag{11}$$

Based on the above principles, we planned two flight routes, and the distance between two UAVs was reasonably adjusted from 100 m to 130 m according to the lens parameters of the cameras. All the waypoint action setting was completed on the software DJI terra. As for other parameters such as λ and $\omega_{t,r}$ in Equation (5), β in Equations (8) and (10), we set and performed them the same as reference [30]. Table 1 shows the flight parameters of the two UAVs. Figure 5 demonstrates the flight routes and panoramic results of stitching.

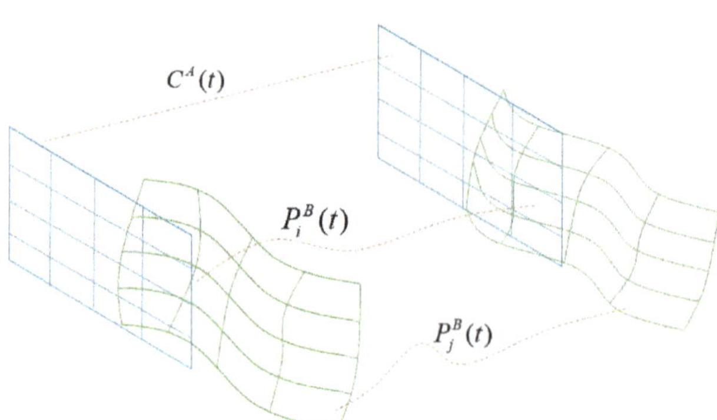

Figure 4. The process of unified stitching and stabilization.

Table 1. The flight parameters of two UAVs.

UAV Number	Height (m)	Camera Yaw (°)	Camera Pitch (°)	Linear Speed (m/s)
A	79	−148.5	−35	0.2
B	80	−148.5	−35	0.2

2.4. Orthorectification and Background Identification

An essential step in this algorithm is orthorectification to sequence the images sampled in the previous section.

The orthorectification transforms imagery into maps that remove remote sensing equipment and terrain-related geometric distortions [34]. We extracted the region of interest (ROI) frame by frame to achieve this goal. Thus, the orthorectification of each image consisted of mainly four stages.

(1) The first step is to determine the real-world coordinates of ROI by RTK-GPS;
(2) The second step is to determine the pixel resolution;
(3) The third step is to calculate the ROI pixel coordinates using GCPs;
(4) The last step is to reorganize these pixels into a complete image for the algorithm's input.

It is worth mentioning that the extrinsic parameters of each image need to be solved once because the UAV may be affected by its mechanical vibration and environmental factors such as wind-related shaking. There were six parameters required to be estimated in Equation (3), that is Equation (11), position information (x_w, y_w, z_w), and camera Euler angles (roll, pitch and yaw, shown in Figure 6) in R.

$$z_c \begin{bmatrix} u \\ v \\ 1 \end{bmatrix} = KR\begin{bmatrix} I & t \end{bmatrix} \begin{bmatrix} x_w \\ y_w \\ z_w \\ 1 \end{bmatrix} \tag{12}$$

We use black-and-white sheets as GCPs to identify them in each frame quickly. For some UAVs with simple equipment and a low price, the data of the inertial measurement unit (IMU) and GPS positioning module carried by them may not be the most accurate, which often leads to difficulties in estimating camera extrinsic parameters, especially on frames with an insufficient number of GCPs due to disappearance of GCPs during UAV movement. In the subsequent recognition, with static or simple camera motion, the pixel coordinates of GCPs are determined by color threshold classification or template matching. Considering the complex camera motion, background identification [35] is probably an effective method (Figure 7). The motion of the background points can be closely approximated to the camera motion by clustering and recognizing them on the feature stream of continuous frames. As a visual odometry, the product of the homography matrix H computed from the background points between frames is temporarily used to estimate the camera motion path M, which is similar to Equation (4).

$$\begin{cases} M(t) = H(t) \cdot H(t-1) \cdots H(1), 1 \le t \le T \\ M(t) = H(t) \cdot M(t-1) \end{cases} \tag{13}$$

where T represents the number of frames in a video file. Estimate the camera pose by the above formula, then list the values of the world coordinates and calculate the pixel coordinates to resample the frames by using Equation (12).

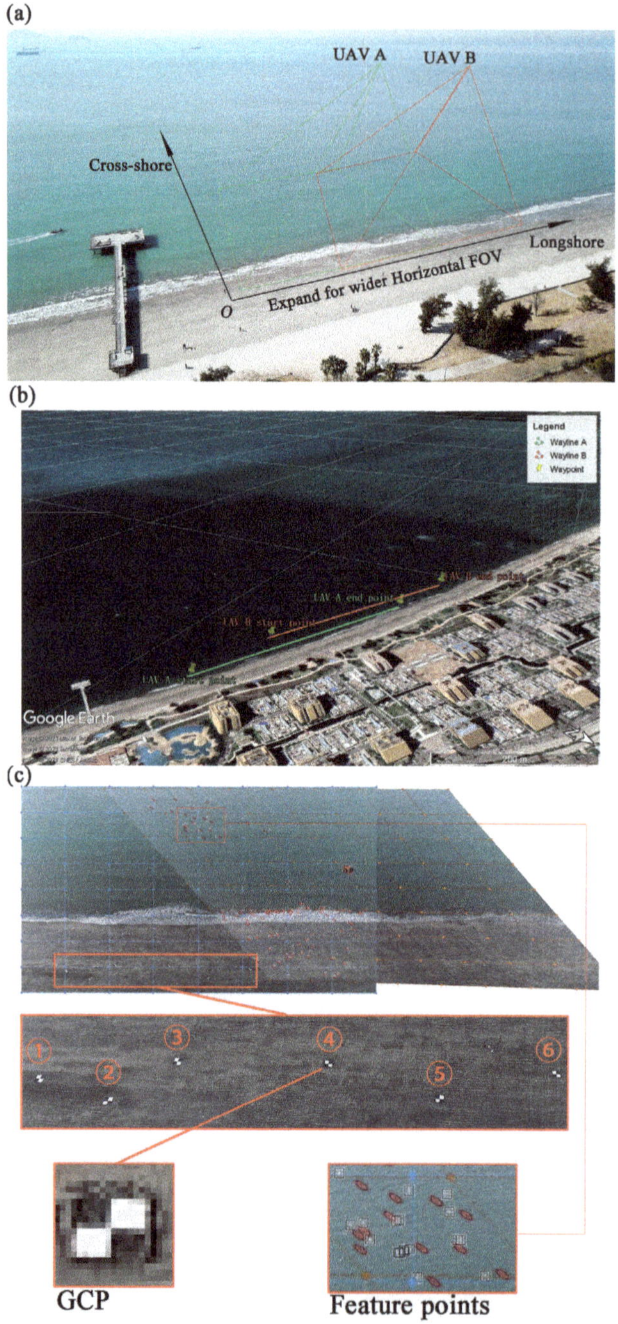

Figure 5. Schematic diagram of flight routes. (**a**) Planning two UAVs to create a wider horizontal FOV. (**b**) The flight routes of the two UAVs. (**c**) The panoramic results of stitching with grids. We placed six GCPs on the beach. The white squares and the red circles are the feature points of the left and right videos, respectively.

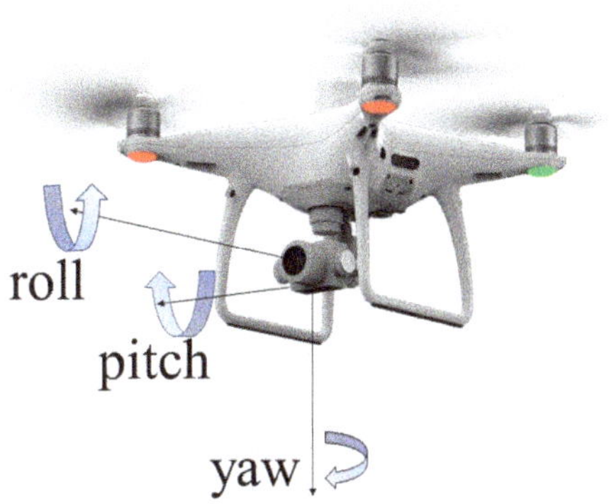

Figure 6. Camera Euler angles.

Figure 7. Background identification. The green points are marked as background points, while the red points are marked as foreground points.

All steps are shown in Figure 8 for a single UAV. We chose a GCP as the origin of the world coordinate system and determined the pixel area corresponding to the whole ROI. Table 2 demonstrates the configuration of orthoimages.

Figure 8. Schematic diagram of interest region selection. (**a**) The target area is shown in the blue line in the original scene image. (**b**) Unified pixel resolution orthoimage.

Table 2. Configuration of orthoimages of a single video.

Pixel Resolution	Cross-Shore Range	Longshore Range
0.5 m	0~200 m	0~100 m

3. Signal Extraction

3.1. Time Stack-Based Pixel Intensity Signal

In the previous section, we solved the orthorectification of down-sampling images to ensure strong consistency in pixel resolution. The motion characteristics of waves can be directly reflected by the pixel fluctuation. Hence, the pixel intensity is the unique identification of the wave signals in this work.

The pixel intensity signals can be extracted from a series of orthorectification images. Figure 9 shows the time stack image of one specified transect.

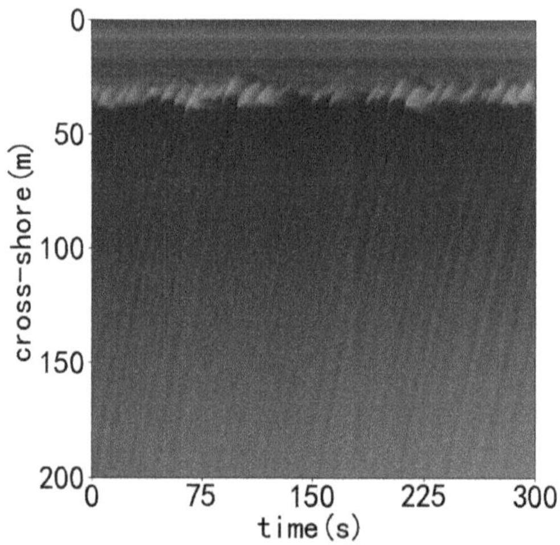

Figure 9. Time stack image in one specified transect.

3.2. Filtering Process

Ideally, a swell can be approximated as a superposition of multiple standard sine waves. However, many short wind waves unrelated to swell movement exist on the sea surface. It means that the pixel intensity signals are mixed with noise independent of water depth. Additionally, it is difficult to directly obtain the available signal from the down-sampling images under the joint action of various other uncertain factors without properly handling by analysis of raw signals. Therefore, we filtered the signal three times to obtain the signal component linked to the water depth.

The first filtering stage eliminates high-frequency noise such as short wind waves by an image filtering method. Figure 10 shows that the Gaussian low-pass filter significantly attenuated pixel intensity fluctuations. The first step of filtering separates the pixel-intensity signal from high-frequency noise. Undoubtedly, the noise filtering process enormously increased accuracy in this study.

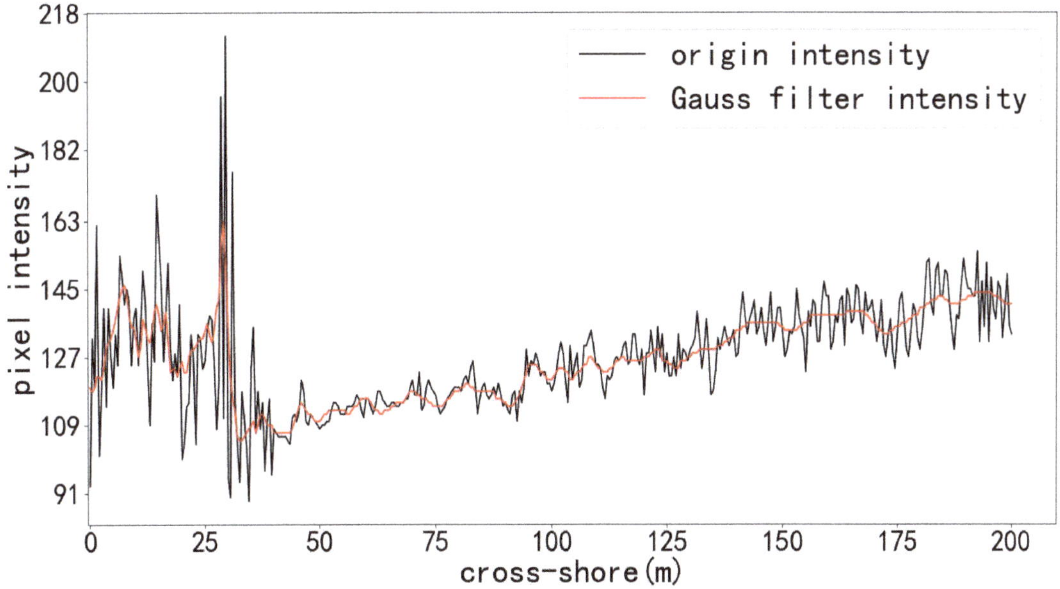

Figure 10. Comparison between Gaussian low-pass filtered signal and the original signal.

The second filtering stage desires to get rid of the influence of other irrelevant factors. A band-pass filter will cut the pixel intensity signal with a frequency range of 0.05~0.5 Hz. As shown in Figure 11, we analyzed the pixel intensity signal based on time-stack images by the fast Fourier transform. We found that the main components were concentrated in the above frequency bands. It can be considered that the frequencies outside this range were independent of swell waves [36]. These irrelevant frequency components may be generated by breaking waves and optical noise.

The last but crucial step was to purify the frequency components of time-stack signals. After a 0.05~0.5 Hz band-pass filter process, the signal will still contain a great variety of useless features. Selecting several representative frequency components can reduce the amount of computation and highlight the correlation between different signals. The principle of choosing the dominant frequency is based on the cross-correlation in a wavelength range. A correlation analysis method inspired by [14] was adopted in this work. We enumerated some possible frequency bands in advance according to frequency distribution. As stated by Equation (14), a cross-spectral matrix was computed between all pixel intensity signals within a wavelength range in a cross-shore direction for each frequency band.

$$C_{ij}(f) = \langle \hat{G}(x_i, y_i, f)^*, \hat{G}(x_i, y_i, f) \rangle \quad (14)$$

where superscript * indicates the complex conjugate. The Fourier transform of the intensity signal at one pixel would be performed in Equation (15).

$$G(x, y, f) = FT(S_{pixel}(x, y, t)) \quad (15)$$

Subsequently, we created the coherence squared for all the potential frequency bands based on the cross-spectral matrix to determine which band had the most significant impact on signals. The bands that contributed the most to the cross-spectrum will finally be retained.

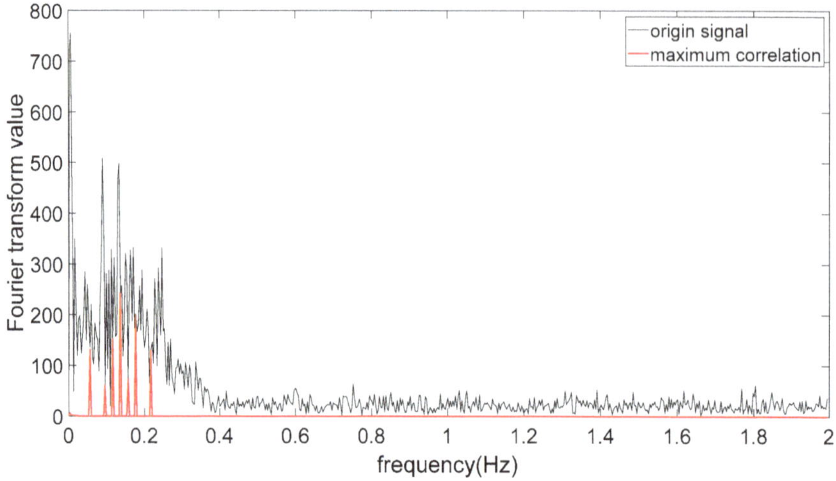

Figure 11. Frequency distribution of raw pixel intensity signal.

4. Bathymetry Results

In this paper, our bathymetric mapping method mainly referred to the estimation theory of wave celerity [5], the contributions made by Lippman and Holman in determining the dominant frequency of ocean waves, and the method of Tsukada [22]. Moreover, the method in this paper is called timeCor since this method's main signal processing flow is in the temporal domain and uses the cross-correlation analysis method between signals. Meanwhile, the bathymetry mapping results of temporal cross-correlation analysis (timeCor) will be compared with those of the cBathy algorithm [14].

4.1. Wave Celerity and Frequency Estimation

In this study, the direction of wave propagation was considered perpendicular to the coastline after orthorectification. The wave celerity was estimated within a suitable range in the cross-shore direction. Firstly, we determined the optimal range of wave celerity estimation according to whether it can reflect a complete propagation characteristic of swells. We used a time delay method to determine the range. Concretely, an empirical time lag Δt was fixed to 3 s. Then, we chose a reference pixel location i such that the pixel at 150 m offshore in Figure 12a (the red line), all neighboring pixel locations j from 1 to $i-1$ (0~149 m) participated in the calculation of the correlation coefficient according to Equation (16).

$$Cor(x_{ij}, y_{ij}) = \langle I(x_i, y_i, t), I(x_j, y_j, t+\Delta t) \rangle \quad (16)$$

As shown in Figure 12b, for each pixel in the cross-shore direction, the range between the closest point with the maximum positive correlation coefficient (the white line) and reference pixel is regarded as the suitable range for wave celerity estimation. The wave celerity of the reference point was evaluated from 134 to 150 m in Figure 12a. We calculated the cross-correlation coefficient between all pixels and the reference point in this range to determine the propagation time. Then, wave celerity was confirmed by the linear fitting method. The slope of the fitting line is regarded as wave celerity. Figure 13 illustrates the wave celerity estimation process.

According to Equation (2), the represented frequency of swell should be known before obtaining the water depth. The weighted average method Equation (17) was used to calculate the representative frequency of each wave. The variable i was configured from 1 to N, which means that the power spectrum value of each valid frequency band would be weighted once.

$$f_{rep} = \frac{\sum_{i=1}^{N} S_i f_i}{\sum_{i=1}^{N} S_i} \tag{17}$$

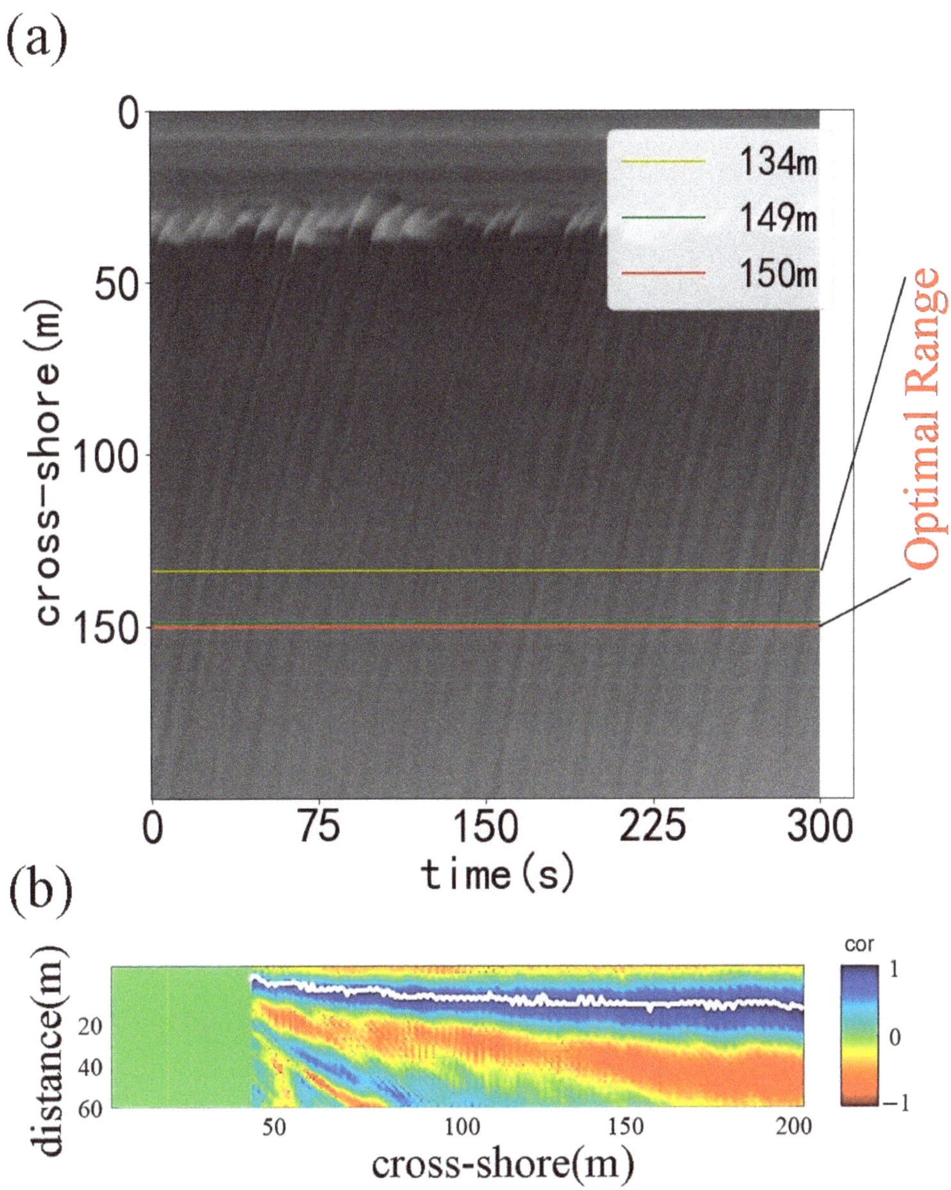

Figure 12. Optimal range estimation. (**a**) The cross-correlation calculation is carried out with this reference point (the red line) towards the shore. An optimal range is determined. (**b**) The relationship between pixel intensity cross-correlation and distance in the whole region.

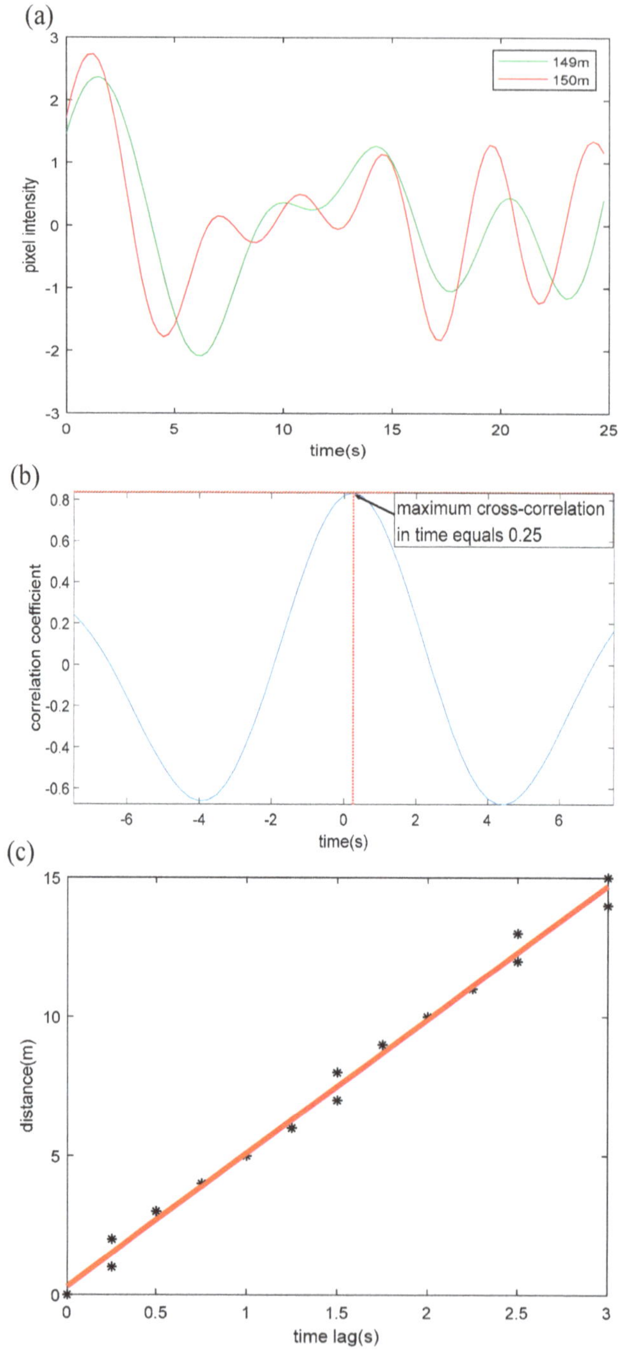

Figure 13. Wave celerity estimation. (**a**) The pixel fluctuation of adjacent pixels shows a specific correlation. (**b**) Maximum cross-correlation determines the propagation time of the wave. (**c**) The propagation time and distance between each pixel and the reference pixel in the suitable estimation range are determined and carried out the linear fitting method.

4.2. Bathymetry Result for A Single UAV

We used the unmanned remote-control boat with RTK-GPS and single beam sonar system to carry out the task of bathymetry to regard as the ground true. We chose two transects to evaluate the final effect. Figure 14 demonstrates the result for a single UAV based on Figure 8 and Table 2.

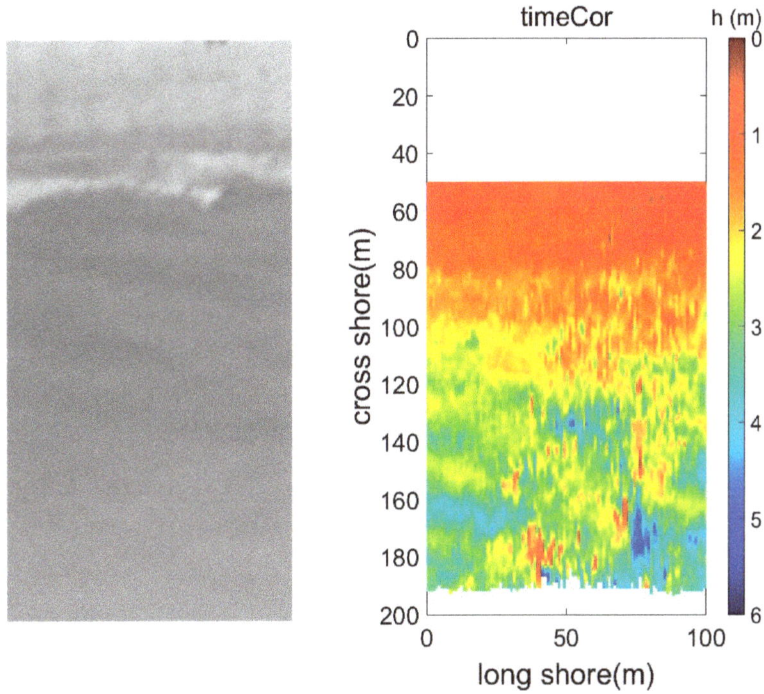

Figure 14. Bathymetry results of the target area, which longshore is 100 m and cross-shore is 200 m.

We evaluate some reference transects longshore with ground truth by boat in Figure 15. It can be seen that the error mainly comes from far offshore. The reason is that this method estimates in a one-dimensional direction, and characteristics of pixel intensity are obviously not captured at sea surface far away from shore.

In this paper, we compared the results of the cBathy algorithm and the method in Figure 16. The depth estimation area of the cBathy algorithm cannot be set prior to analysis, which will cause a significant error when the correlation of pixel signals is analyzed. An unreasonable selection of frequency bands will also affect the bathymetry. Additionally, the source of error mainly comes from the difference between each algorithm. The principle of timeCor is based on the degree of cross-correlation between signals, further analyzing the cross-power spectrum, calculating the energy distribution and correlation coefficient, and estimating the wave celerity and main frequency. The difference from cBathy is that each estimation is only performed in the one-dimensional cross-shore direction, as a result of which its result is less robust than cBathy one. Additionally, cBathy estimates in two dimensions. However, timeCor's calculation process is simple, and the whole process does not need to set too many parameters, which is suitable for temporary single estimation along the coast. A single cBathy depth estimation is often affected by external factors such as environment and acquisition, which leads to the inability to reverse the appropriate depth. The algorithm provides a Kalman filter link. If there is long-term observation data, the area that cannot be estimated at a single time can be compensated by the observation value and prediction value, but this paper is only based on a single estimation. Actually, in the waters far away from the coastal zone, the UAV camera did not obtain relatively complete pixel intensity features, or at a certain angle, the pixel feature will be more or less affected by the camera's angle of view. Hence, the comparison between the two is only to confirm the consistency of the bathymetry results rather than determine which is better.

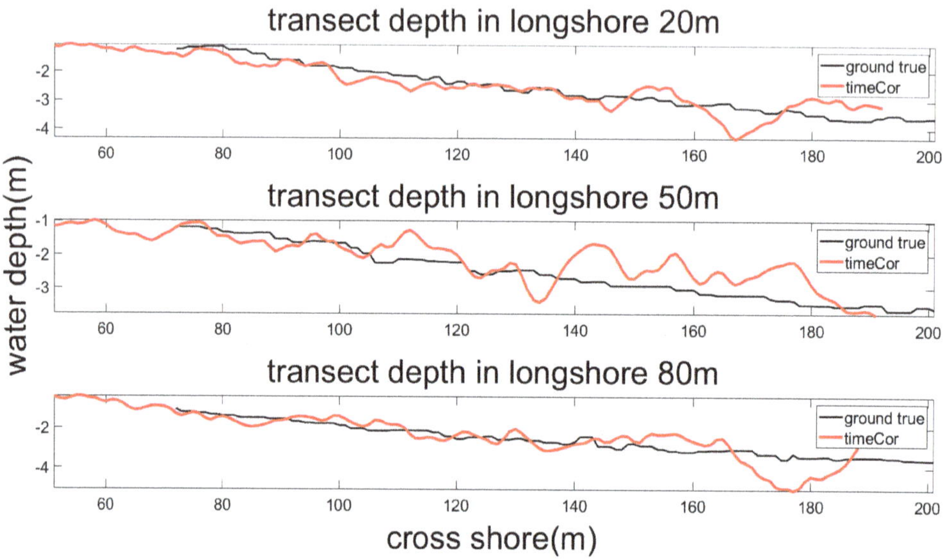

Figure 15. Comparison between ground true and the timeCor result.

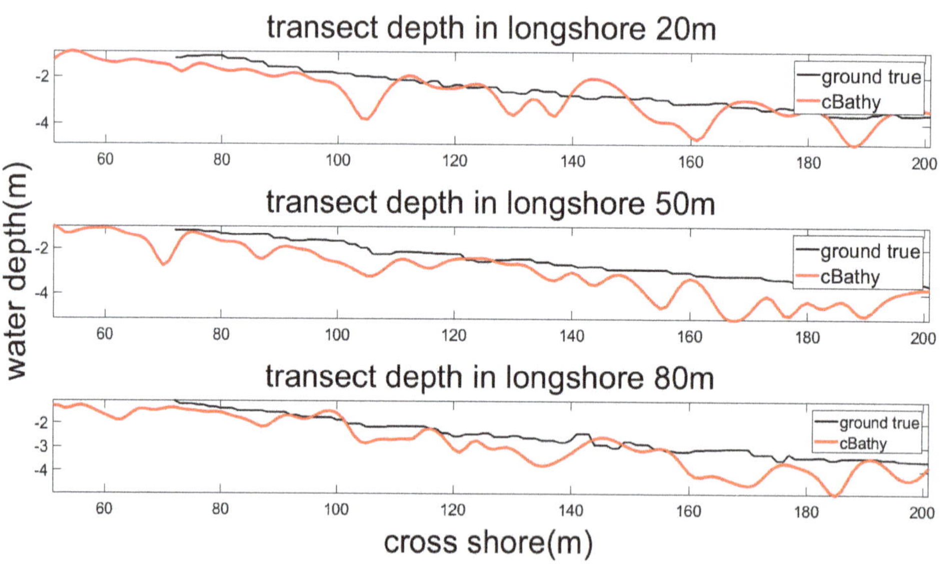

Figure 16. Comparison between ground true and the cBathy.

4.3. Stitching Result

From the previous section, we can see that the mapping view of a single UAV was limited. To achieve dual UAV bathymetry, we adjusted the placements of two UAVs. The validation of the above theory in Section 2.3 was carried out for the selected flight parameters. With the same bathymetric algorithms, we expanded the ROI area (Figure 17) and obtained considerable results, whose longshore ranged from 0 m to 200 m. In fact, our boat did not measure the real water depths over such a wider range. The purpose of this paper was more to demonstrate the effectiveness of video stitching rather than the accuracy of bathymetry algorithm (timeCor). Therefore, we considered the results of the more robust cBathy to be closer to the ground truth in this subsection. Figure 18 displays the bathymetric mapping water depths of the two algorithms. The convergence of the two algorithms was in good agreement with the experimental results due to the fact that they were both based on the water depth solved by the linear dispersion relationship, although there are some differences in the core and calculation. Moreover, we compared the depth curves of the two (Figure 19), and the mean absolute error (*MAE*) was used to describe the similarity of the depth curves. The invalid points in the curves were eliminated. The smaller the *MAE* was, the more similar the two curves are. Meanwhile, the similarity function $MAE(longshore)$ was drawn to represent the overall difference between the two results. Additionally, *MAE* is defined as:

$$MAE = \frac{1}{n}\sum|h_1(x) - h_2(x)| \tag{18}$$

where n is the number of the sum of the valid points, h means the water depth, and x represents cross-shore distance.

Figure 17. The range of ROI (200 m × 200 m) and its orthoimage.

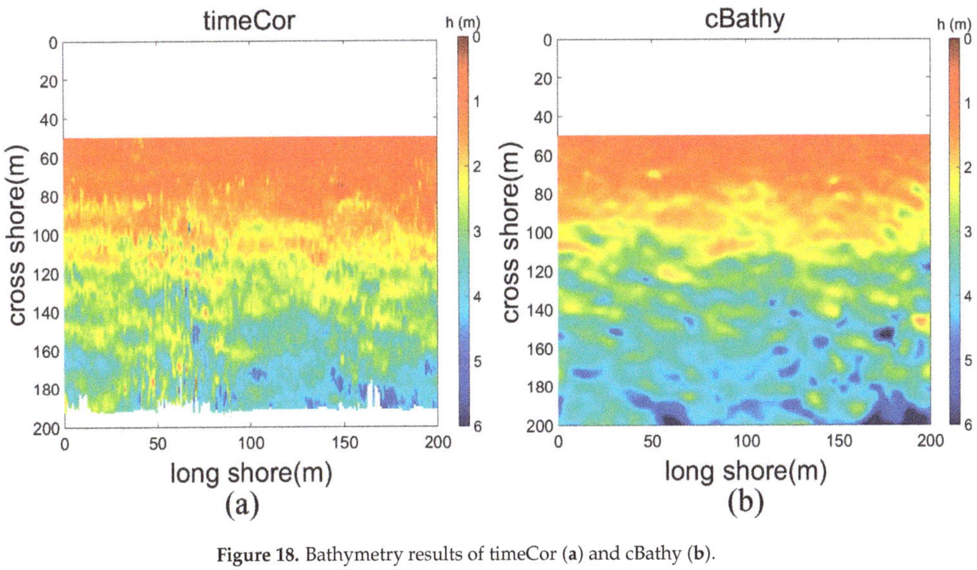

Figure 18. Bathymetry results of timeCor (**a**) and cBathy (**b**).

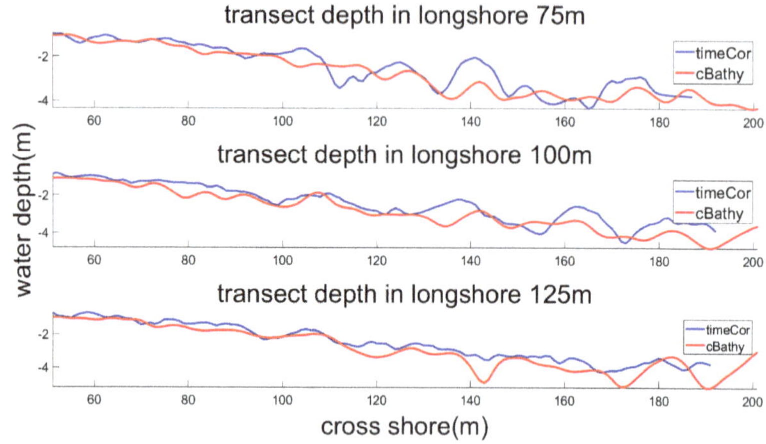

Figure 19. Comparison between timeCor and cBathy in longshore 75 m, 100 m, 125 m.

We calculated the global mean square error (MSE) and the global root mean square error (RMSE) with the two results, whose values were 0.37 m^2 and 0.60 m, respectively. It can be seen that the depth curves at about 50 m and 200 m longshore were very inconsistent in Figure 20. A persuasive conclusion supported by [22] is that there exhibits an underestimation bias in marine areas, which results from the remote distance from the camera (Figure 21). As a result of the addition of a camera during the process of video stitching, the rectification bias of the other camera would, to some extent, be compensated. For instance, the far-side swell waves often affect the results error of camera A alone, whereas camera B typically relies on its distance from camera A to compensate for the rectification bias of camera A. Thus, to further limit the influence of rectification bias while capturing videos, camera Euler angles and camera distances should be adjusted according to the field site. Meanwhile, the image blending strategy can be appropriately abandoned if the videos are severely out of synchronization. Generally speaking, our video stitching method can create a wider FOV to expand the surveying area and provide effective input data for the bathymetry algorithms.

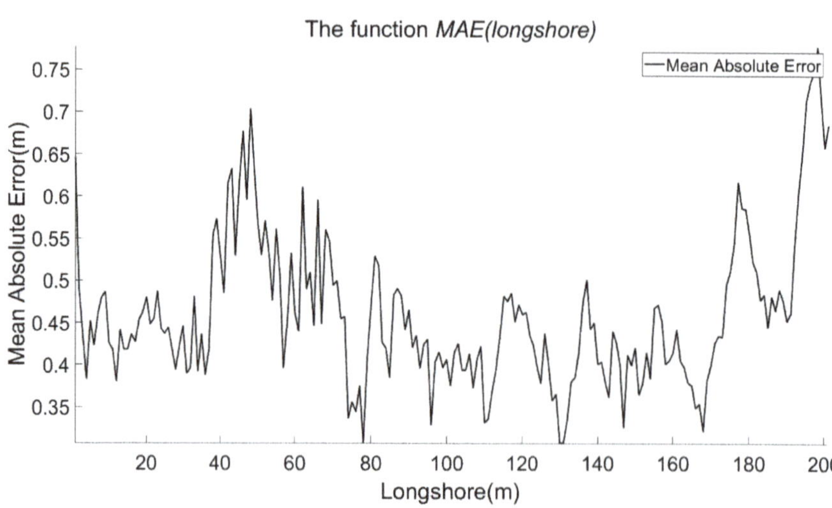

Figure 20. The diagram that MAE verses longshore.

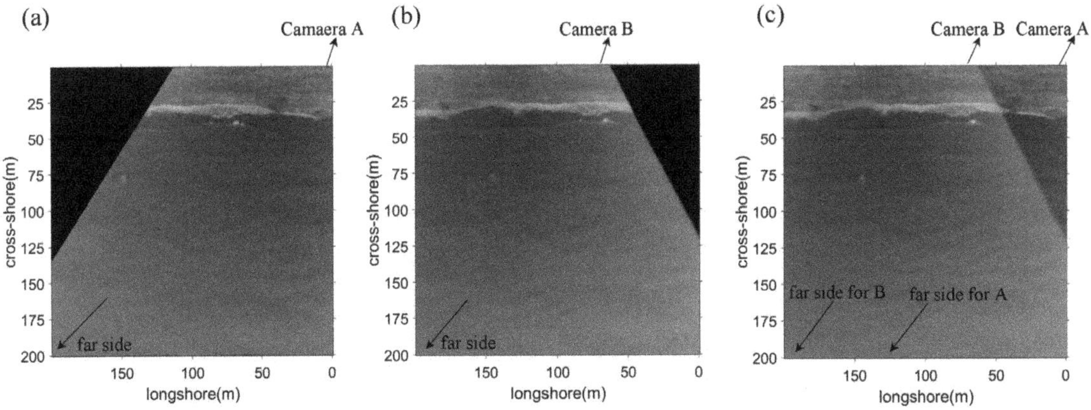

Figure 21. The far-side swell waves affect the results error of camera A alone, whereas camera B compensate for the rectification bias of camera A. (**a**) Orthoimage of camera A. (**b**) Orthoimage of camera B. (**c**) Orthoimage of panoramic image.

5. Discussion

5.1. Source of Error

Although the results in the previous section showed the present method can reasonably realize the bathymetric mapping algorithm based on video stitching, this method should be further discussed regarding the existing problems and its applicability. In addition, all of the given outcomes used a methodology that needed knowledge of GCPs to optimize the camera's exterior orientation. Improper operation of GCPs also introduces errors to experimental results, both in site layout and postprocessing.

Although the deviation of the beach affects the specific image position of the GCPs coordinates, when the frames were under orthographic rectification, the error of the pixel coordinates of each frame of GCPs would accumulate. Furthermore, in frames where GCPs are not visible, background identification plays a temporary role. Camera motion is estimated by identifying fixed feature points in the field, but the main fixed feature points were all distributed on the beach. Since it is impossible to place stationary structures on a sea surface that is filled with moving swell waves, a small mismeasurement of one-side feature points could lead to a significant bias in the far field. However, the bias, in this case, tended to be exposed only for long-term measurements. As shown in Figure 22, estimating camera motion using the image alone was equivalent to visual odometry [37], and the error accumulation was acceptable for a short-time bathymetric mapping. Otherwise, the monocular system will suffer from the scale-drift issue.

To obtain the final panoramic video, we transformed the input videos by the estimated stitching and stabilization variable in Equations (8)–(10). In the overlapping region (Figure 17), background on the shore was stitched satisfactorily, but not surprisingly, foreground objects such as swell waves had subtle ghost artifacts. As shown in Figure 23, we performed bathymetry inversion of the common field captured by the two UAVs separately, and unsurprisingly found that the two results were similar, whose RMSE was 0.2 m, MSE was 0.04 m^2 and MAE was 0.078 m. The main reason for this result was the slight parallax between the two cameras with similar camera Euler angles and relatively short distances, which can be summarized as parallax bias.

5.2. Wide Vertical FOV

An area worth discussing is the further cross-shore region, with dense information that loses a lot. Given the underestimation bias in this part of the region, it is still scarcely understood whether the method studied can solve this problem. Current experiments show that video stitching is usually based on the spatial domain. The base stitching method used in this work is SIFT-based feature matching [32]. We attempted to add a UAV in the vertical direction (in front of the first camera, Figures 24–26) to capture distant fields, which tried to expand the length of the cross-shore. Nonetheless, in the process of feature matching, there are often mismatches due to the feature points being different from each other at different angles of view at every moment. Overall, a large parallax problem exists between cameras. When iterating with Equation (15), either the value $P_i(t)$ is divergent or the stitching results perform badly. In this case, video stitching probably worked when the cameras were close enough, but the vertical fields of the cross-shore did not meet our expectations (Figure 27).

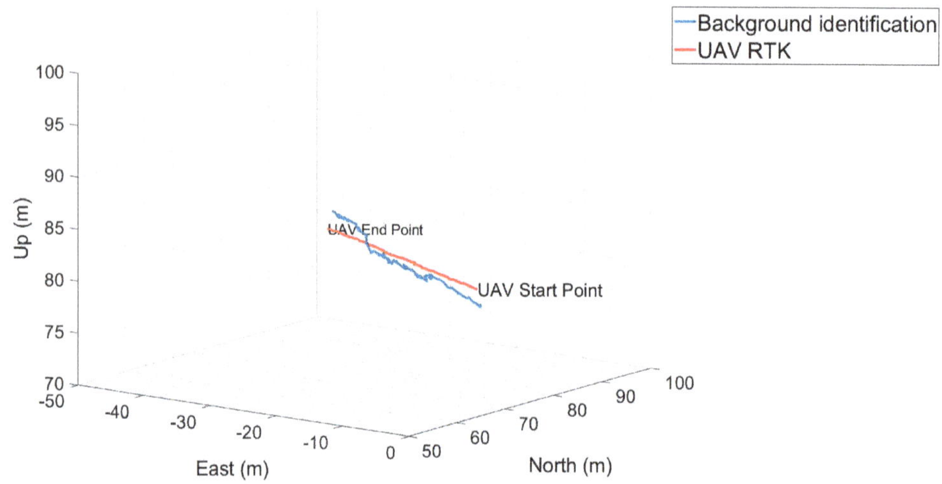

Figure 22. UAV RTK recorded the flight route, while background identification is to calculate the flight route.

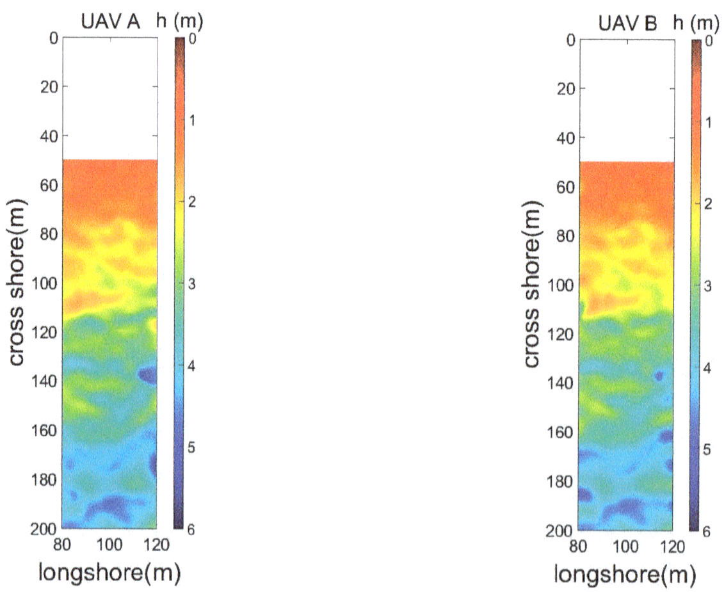

Figure 23. Bathymetry results of the common field captured by the two UAVs.

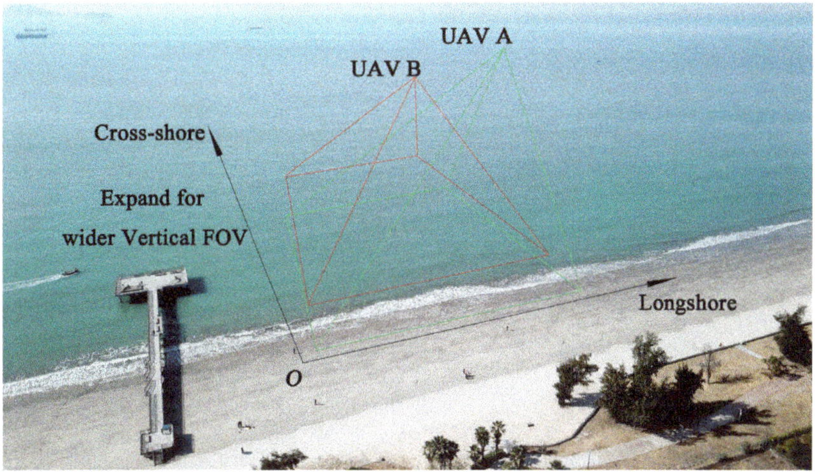

Figure 24. Two UAVs were placed to capture distant cross-shore.

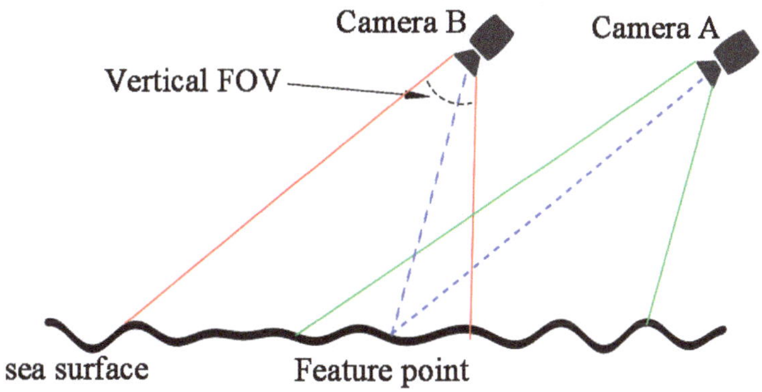

Figure 25. Inappropriate viewing angle cause strong parallax.

Figure 26. The result of a feature mismatch.

Figure 27. The green rectangle is the FOV of a single UAV. The red is the extension of the stitching.

Therefore, it is necessary to continue research related to the development of bathymetry algorithms based on optical cameras, using photogrammetric techniques based on UAV swarms. Taking into account the optical complexity of waters, one should put forward diverse ideas as well as practice experiments to verify theories.

6. Conclusions

This study developed a method of nearshore bathymetry surveying using UAVs' video stitching. By proving the effectiveness and limitations of single UAV bathymetry mapping, a second UAV was introduced to compensate for the shortcomings of the former. According to the results, this method expands the horizontal FOV of UAV mapping by planning the flight routes of two UAVs. Furthermore, while improving the efficiency of surveying, a part of the problem of how rectification bias affects the mapping results was solved. However, this work still cannot eliminate some errors for further cross-shore region, as well as parallax bias in stitching. This means that there is still work to be carried out on the research of the bathymetric mapping algorithm. Further research from the perspective of algorithm optimization will be carried out.

Author Contributions: Conceptualization, J.F. and H.P.; methodology, J.F.; software, J.F.; validation, J.F., H.P. and Z.L.; formal analysis, J.F.; investigation, J.F. and H.P.; resources, J.F., H.P. and Z.L.; data curation, J.F. and Z.L.; writing—original draft preparation, J.F.; writing—review and editing, J.F.; visualization, J.F.; supervision, J.F. and H.P.; project administration, J.F. All authors have read and agreed to the published version of the manuscript.

Funding: This research was supported in part by the Fundamental Research Funds for the Central Universities; and in part by Scientific Instruments Development Program of NSFC under Grant 61527810.

Institutional Review Board Statement: Not applicable.

Informed Consent Statement: Not applicable.

Data Availability Statement: The data that support the finding of this study are available on request from the corresponding author, H.P., upon reasonable request.

Conflicts of Interest: The authors declare no conflict of interest.

References

1. Stockdon, H.F.; Holman, R.A. Estimation of wave phase speed and nearshore bathymetry from video imagery. *J. Geophys. Res. Oceans* **2000**, *105*, 22015–22033. [CrossRef]
2. Sallenger, A.H., Jr.; Krabill, W.B.; Swift, R.N.; Brock, J.; List, J.; Hansen, M.; Holman, R.A.; Manizade, S.; Sontag, J.; Meredith, A.; et al. Evaluation of airborne topographic lidar for quantifying beach changes. *J. Coast. Res.* **2003**, *1*, 125–133.
3. Almeida, L.P.; Almar, R.; Bergsma, E.W.; Berthier, E.; Baptista, P.; Garel, E.; Dada, O.A.; Alves, B. Deriving high spatial-resolution coastal topography from sub-meter satellite stereo imagery. *Remote Sens.* **2019**, *11*, 590. [CrossRef]
4. Almar, R.; Bergsma, E.W.; Maisongrande, P.; de Almeida, L.P. Wave-derived coastal bathymetry from satellite video imagery: A showcase with Pleiades persistent mode. *Remote Sens. Environ.* **2019**, *231*, 111263. [CrossRef]
5. Almar, R.; Bonneton, P.; Senechal, N.; Roelvink, D. Wave celerity from video imaging: A new method. In *Coastal Engineering 2008*; World Sientific: Singapore, 2009; Volume 5, pp. 661–673.
6. MacMahan, J. Hydrographic surveying from personal watercraft. *J. Surv. Eng.* **2001**, *127*, 12–24. [CrossRef]
7. Dugan, J.P.; Morris, W.D.; Vierra, K.C.; Piotrowski, C.C.; Farruggia, G.J.; Campion, D.C. Jetski-based nearshore bathymetric and current survey system. *J. Coast. Res.* **2001**, *17*, 900–908.
8. Honegger, D.A.; Haller, M.C.; Holman, R.A. High-resolution bathymetry estimates via X-band marine radar: 1. beaches. *Coast. Eng.* **2019**, *149*, 39–48. [CrossRef]
9. Plant, N.G.; Holland, K.T.; Haller, M.C. Ocean wavenumber estimation from wave-resolving time series imagery. *IEEE Trans. Geosci. Remote Sens.* **2008**, *46*, 2644–2658. [CrossRef]
10. Matsuba, Y.; Sato, S. Nearshore bathymetry estimation using UAV. *Coast. Eng. J.* **2018**, *60*, 51–59. [CrossRef]
11. Simarro, G.; Calvete, D.; Luque, P.; Orfila, A.; Ribas, F. UBathy: A new approach for bathymetric inversion from video imagery. *Remote Sens.* **2019**, *11*, 2722. [CrossRef]
12. Thuan, D.H.; Almar, R.; Marchesiello, P.; Viet, N.T. Video sensing of nearshore bathymetry evolution with error estimate. *J. Mar. Sci. Eng.* **2019**, *7*, 233. [CrossRef]
13. Santos, D.; Abreu, T.; Silva, P.A.; Santos, F.; Baptista, P. Nearshore Bathymetry Retrieval from Wave-Based Inversion for Video Imagery. *Remote Sens.* **2022**, *14*, 2155. [CrossRef]
14. Holman, R.; Plant, N.; Holland, T. cBathy: A robust algorithm for estimating nearshore bathymetry. *J. Geophys. Res. Oceans* **2013**, *118*, 2595–2609. [CrossRef]
15. Bergsma, E.W.; Almar, R. Video-based depth inversion techniques, a method comparison with synthetic cases. *Coast. Eng.* **2018**, *138*, 199–209. [CrossRef]
16. Liu, H.; Arii, M.; Sato, S.; Tajima, Y. Long-term nearshore bathymetry evolution from video imagery: A case study in the Miyazaki coast. *Coast. Eng. Proc.* **2012**, *1*, 60. [CrossRef]
17. Nieto, M.; Garau, B.; Balle, S.; Simarro, G.; Zarruk, G.; Ortiz, A.; Tintoré, J.; Álvarez Ellacuría, A.; Gómez-Pujol, L.; Orfila, A. An open source, low cost video-based coastal monitoring system. *Earth Surf. Process. Landforms* **2010**, *35*, 1712–1719. [CrossRef]
18. Rodriguez-Padilla, I.; Castelle, B.; Marieu, V.; Morichon, D. Video-Based Nearshore Bathymetric Inversion on a Geologically Constrained Mesotidal Beach during Storm Events. *Remote Sens.* **2022**, *14*, 3850. [CrossRef]
19. Simarro, G.; Ribas, F.; Álvarez, A.; Guillén, J.; Chic, Ò.; Orfila, A. ULISES: An open source code for extrinsic calibrations and planview generations in coastal video monitoring systems. *J. Coast. Res.* **2017**, *33*, 1217–1227. [CrossRef]
20. Simarro, G.; Calvete, D.; Plomaritis, T.A.; Moreno-Noguer, F.; Giannoukakou-Leontsini, I.; Montes, J.; Durán, R. The influence of camera calibration on nearshore bathymetry estimation from UAV videos. *Remote Sens.* **2021**, *13*, 150. [CrossRef]
21. Sun, S.H.; Chuang, W.L.; Chang, K.A.; Kim, J.Y.; Kaihatu, J.; Huff, T.; Feagin, R. Imaging-Based Nearshore Bathymetry Measurement Using an Unmanned Aircraft System. *J. Waterw. Port Coast. Ocean Eng.* **2019**, *145*, 04019002. [CrossRef]
22. Tsukada, F.; Shimozono, T.; Matsuba, Y. UAV-based mapping of nearshore bathymetry over broad areas. *Coast. Eng. J.* **2020**, *62*, 285–298. [CrossRef]
23. Brodie, K.L.; Bruder, B.L.; Slocum, R.K.; Spore, N.J. Simultaneous mapping of coastal topography and bathymetry from a lightweight multicamera UAS. *IEEE Trans. Geosci. Remote Sens.* **2019**, *57*, 6844–6864. [CrossRef]
24. Holland, K.T.; Holman, R.A.; Lippmann, T.C.; Stanley, J.; Plant, N. Practical use of video imagery in nearshore oceanographic field studies. *IEEE J. Ocean. Eng.* **1997**, *22*, 81–92. [CrossRef]
25. Perugini, E.; Soldini, L.; Palmsten, M.L.; Calantoni, J.; Brocchini, M. Linear depth inversion sensitivity to wave viewing angle using synthetic optical video. *Coast. Eng.* **2019**, *152*, 103535. [CrossRef]
26. Kannala, J.; Brandt, S.S. A generic camera model and calibration method for conventional, wide-angle, and fish-eye lenses. *IEEE Trans. Pattern Anal. Mach. Intell.* **2006**, *28*, 1335–1340. [CrossRef]
27. Liu, S.; Yuan, L.; Tan, P.; Sun, J. Bundled camera paths for video stabilization. *ACM Trans. Graph. (TOG)* **2013**, *32*, 78. [CrossRef]
28. Baker, S.; Iain, M. Lucas-Kanade 20 years on: A unifying framework. *Int. J. Comput. Vis.* **2004**, *56*, 221–255. [CrossRef]

29. Su, T.; Nie, Y.; Zhang, Z.; Sun, H.; Li, G. Video stitching for handheld inputs via combined video stabilization. In Proceedings of the SIGGRAPH ASIA 2016 Technical Briefs, Macao, China, 5–8 December 2016; pp. 1–4.
30. Nie, Y.; Su, T.; Zhang, Z.; Sun, H.; Li, G. Dynamic video stitching via shakiness removing. *IEEE Trans. Image Process.* **2017**, *27*, 164–178. [CrossRef]
31. Guo, H.; Liu, S.; He, T.; Zhu, S.; Zeng, B.; Gabbouj, M. Joint video stitching and stabilization from moving cameras. *IEEE Trans. Image Process.* **2016**, *25*, 5491–5503. [CrossRef]
32. Lowe, D.G. Distinctive image features from scale-invariant keypoints. *Int. J. Comput. Vis.* **2004**, *60*, 91–110. [CrossRef]
33. Zaragoza, J.; Chin, T.J.; Brown, M.S.; Suter, D. As-projective-as-possible image stitching with moving DLT. In Proceedings of the IEEE Conference on Computer Vision and Pattern Recognition, Portland, OR, USA, 23–28 June 2013; pp. 2339–2346.
34. Laliberte, A.S.; Herrick, J.E.; Rango, A.; Winters, C. Acquisition, orthorectification, and object-based classification of unmanned aerial vehicle (UAV) imagery for rangeland monitoring. *Photogramm. Eng. Remote Sens.* **2010**, *76*, 661–672. [CrossRef]
35. Zhang, F.L.; Wu, X.; Zhang, H.T.; Wang, J.; Hu, S. Robust background identification for dynamic video editing. *ACM Trans. Graph.* **2016**, *35*, 12. [CrossRef]
36. Matsuba, Y.; Shimozono, T.; Tajima, Y. Observation of nearshore wave-wave interaction using UAV. *Coast. Eng. Proc.* **2018**, *36*, 12. [CrossRef]
37. Nistér, D.; Naroditsky, O.; Bergen, J. Visual odometry. In Proceedings of the 2004 IEEE Computer Society Conference on Computer Vision and Pattern Recognition, 2004. CVPR 2004, Washington, DC, USA, 27 June–2 July 2004; p. 1.

Journal of
Marine Science
and Engineering

MDPI

Article

Numerical Study on the Effects of the Multiple Porous Medium Breakwaters on the Propagation of the Solitary Wave

Kui Zhu [1], Runxiang Jiang [1], Zhaolong Sun [1], Hao Qin [2], Zeqi Cheng [2], Yang Wang [3,*] and Enjin Zhao [2]

1 College of Electrical Engineering, Naval University of Engineering, Wuhan 430033, China
2 College of Marine Science and Technology, China University of Geosciences, Wuhan 430074, China
3 Haikou Marine Geological Survey Center, China Geological Survey, Haikou 570100, China
* Correspondence: wangyang01@mail.cgs.gov.cn

Abstract: Submerged breakwater, as an important marine engineering structure, can effectively absorb wave energy and is widely used in marine engineering protection. As a new type of breakwater, porous medium breakwater has a certain influence on wave propagation. However, the influence of multiple porous medium submerged breakwaters on extreme waves remains to be studied. In the study, considering effects of extreme waves generated by hurricanes or tsunamis, the influence of the solitary wave on the multiple semi-circle porous medium breakwater is systematically investigated. According to the computational fluid dynamics theory, a numerical tank is established, in which the porous medium module is coupled. The computational capability of this model is verified first. Then, depending on the model, a series of cases are carried out to study the effects of different porous medium breakwaters on the propagation of the solitary wave. The results show that when the porosity is 0.8, the force on the submerged breakwaters is the smallest, and the water level and the velocity decrease seriously. With the increase in the diameter of the submerged breakwater, the wave profile gradually becomes flat. The higher the wave height, the more serious the wave deformation after passing through the submerged breakwater.

Keywords: hydrodynamic load; porous medium; flow field; submerged breakwater

Citation: Zhu, K.; Jiang, R.; Sun, Z.; Qin, H.; Cheng, Z.; Wang, Y.; Zhao, E. Numerical Study on the Effects of the Multiple Porous Medium Breakwaters on the Propagation of the Solitary Wave. *J. Mar. Sci. Eng.* **2023**, *11*, 565. https://doi.org/10.3390/jmse11030565

Academic Editors: M. Dolores Esteban, José-Santos López-Gutiérrez, Vicente Negro and M. Graça Neves

Received: 20 January 2023
Revised: 19 February 2023
Accepted: 23 February 2023
Published: 7 March 2023

1. Introduction

Waves are an important dynamic factor in the ocean. In order to weaken the influence of the wave on the coastal structures, a breakwater is constructed, which is a hydraulic structure to block the wave propagation and keep the water surface stable. The interaction between the wave and the breakwater has received much attention in coastal engineering and it is necessary to study the hydrodynamic characteristics around the breakwater to ensure its safe and application.

Therefore, various breakwaters are developed and investigated [1–3]. Osman et al. numerically investigated soliton wave solutions, solitary wave solutions and so on. They used a range of methods including generalized Kudryashov and Chebyshev series to obtain exact travelling wave solutions of nonlinear evolution equations [4–6]. Han and Dong investigated the interaction between solitary waves and submerged breakwaters based on smoothed particle hydrodynamics, which showed that the potential energy is transmitted more easily through a submerged breakwater than kinetic energy [7]. Nassiraei et al. investigate the long wave (tsunami) forces on caisson breakwaters using numerical modeling. Verifications of the simulation results using experimental data, analytical formulas and empirical formulas show that a numerical model is capable of modeling the aforesaid problem with enough accuracy [8]. Mahmoudof and Hajivalie numerically studied the hydraulic response of smooth submerged breakwaters (SMB) to irregular waves and proposed a comprehensive parameter to describe the responses, transmission, and reflection of SMB [9]. Zhao et al. investigated the wave interaction with a fluid-filled-membrane-submerged breakwater. A modified ghost-cell immersed boundary method was adopted. The results showed that, in lower excess internal pressure, the wave reflection, vortex shedding and energy dissipation were significant [10]. Bautista et al. studied the interaction between long water waves and two fixed submerged breakwaters of wavy surfaces, which showed that the largest values of the reflection coefficient occur for large values of the number of the undulations on the wavy surfaces [11]. Chen et al. investigated wave interaction with a twin floating breakwater through an introduced a mooring line computation δ-SPH model, which indicated that the twin floating breakwater is superior to the single pontoon floating breakwater on wave-attenuation performances [12]. Liang et al. investigated the interaction between a new box-type breakwater and the wave which showed that a larger deck width for the proposed breakwater was more conducive to dissipating the energy of longer period waves [13]. Magdalena et al. used three-layer depth-integrated equations to calculate wave amplitude reduction in a domain with n submerged porous media blocks. When one of the two layers has a greater friction coefficient, increasing its thickness will produce a greater wave attenuation [14]. Aristodemo and Filianoti performed laboratory experimental and δ-LES-SPH simulations to analyze solitary wave-induced forces at submerged rigid breakwaters. It was found that, by increasing the relative height of the breakwaters, the wave loads tended to grow, particularly for the horizontal ones [15]. Liang et al. studied the wave transmission over double submerged breakwaters using non-hydrostatic wave model SWASH. It was found that dissipation of super harmonic wave components is more obvious than that of lower harmonic wave components [16]. Lyu et al. developed an iterative multipole solution for water wave interaction with a submerged, partially perforated semi-circular breakwater based on the linear potential theory. The results showed that enlarging porosity will significantly decrease wave reflection and horizontal wave force, but has little effect on vertical wave force [17]. Liu and Wang studied the box-type floating breakwaters, extensively using the SPH. The results showed that the larger and lighter the breakwaters, the better the performance is [18]. Metallinos et al. simulated the regular and irregular wave propagation over the submerged porous breakwaters using an extended Boussinesq-type model, which showed that the porous medium break can effectively weaken wave propagation [19]. Li and Zhang investigated the effects of three-dimensional interactions of obliquely incident waves and breakwaters on wave potential energy reduction based on the energy transmission coefficient. The transmitted wave energy was reduced more significantly as the incident wave steepness increases or the submergence depth decreases, since more agitation occurs around the breakwater [20].

In addition to the investigation between the wave and the breakwater, the protection of the breakwater to the different coastal structures was also studied. Xue et al. studied the influence of the breakwater composed of three identical semicircular shells and a rectangular base on the wave loads on the coastal bridges. The results demonstrated an increased effectiveness of this breakwater to dissipate wave energy [21]. Sun et al. numerically investigated the effects of the submerged porous breakwater, which showed that the breakwater is effective in reducing the hydrodynamic loads applied to the coastal bridge decks [22]. Patil and Karmakar investigated the wave interactions with the submerged rubble mound breakwater (SRMB) and the submerged floating tunnel (SFT) using multi-domain boundary element method (MDBEM). Due to the SFT, the SFT's safety is improved [23]. He et al. established a mathematical model to solve the wave diffraction and radiation problem of floating piers with a submerged annular ripple breakwater. It was found that the wave excitation force and hydrodynamic coefficient of the floating pier change periodically with the increase in the distance from the breakwater [24]. Zhao et al. investigated the effects of the different porous medium breakwaters on the propagation and run-up of the solitary wave. With the increase in the wave height, the force difference attenuation coefficient on the breakwater decreases [25]. Guo et al. developed an analytical model for a submerged breakwater over variable bottom topography based on linear wave theory. It was found that the full-wave reflection is drastically reduced and the wave energy is dissipated [26].

Koley et al. investigated the performance of a submerged flexible porous membrane which is floating over a porous breakwater. The results demonstrated that the resonating behavior diminishes with the introduction of structural porosity in the membrane wave barrier [27]. Koraim et al. experimentally studied the hydrodynamic efficiency of a porous seawall under the regular wave, in front of which is a submerged breakwater. The submerged breakwater decreases the run-up on the seawall [28].

As a new type of marine protection structure, porous medium breakwaters have a certain influence on wave propagation. However, the influence of multiple porous medium submerged breakwater on extreme waves remains to be studied. Based on the previous study, this paper continues to improve the submerged theory and compares the different influence under different waves, which provides technical support for marine engineering construction. The rest of this paper is organized as follows. Section 2 introduces the numerical model. The verification of the model is conducted in Section 3. Section 4 discusses the effects of the multiple porous medium breakwater on the propagation of the solitary wave. Section 5 provides some concluding remarks.

2. Numerical Model

OpenFOAM (Open Field Operation and Manipulation), as an open source computational fluid dynamics software, is more and more widely used in the field of coastal engineering. The governing equations of incompressible fluid in numerical simulation include mass conservation equation and momentum equation, as follows:

$$\nabla \cdot \vec{u} = 0 \tag{1}$$

$$\frac{\partial \rho \vec{u}}{\partial t} + \nabla \rho \vec{u}\vec{u}^T = -\nabla p_r + \mathbf{g}(\mathbf{x} - \mathbf{x}_r)\nabla \rho + \mu_t \nabla \vec{u} \tag{2}$$

where ρ is the fluid density, u is velocity vector in Cartesian coordinate system, t is the time, $\nabla = (\partial/\partial x, \partial/\partial y, \partial/\partial z)$ is the differential operator; p_r is the relative pressure which is expressed as $p_r = p - \rho g x$; p is the total pressure; g is the acceleration of the gravity; $\mathbf{x} = (x, y, z)$ is the vector of spatial coordinates in Cartesian coordinate system; $\mathbf{x}_r$ is the reference position coordinate; and μ_t is the dynamic viscosity.

In the porous medium structure, the velocity through the structure decreases, and the velocity is defined as the transfer velocity, which is related to the velocity in the porous medium object with the porosity (n), as follows:

$$\vec{u} = n\vec{u}_p \tag{3}$$

where $\vec{u}_p$ is the velocity in the porous medium; Jensen (2014) [29] improved the momentum equation in the permeable layer; The continuity equation is the same as in the previous section, while the momentum equation is modified as follows:

$$(1 + C_m)\frac{1}{n}\frac{\partial \rho \vec{u}}{\partial t} + \frac{1}{n^2}\nabla \rho \vec{u}\vec{u}^T = -\nabla p_r + \mathbf{g}(\mathbf{x} - \mathbf{x}_r)\nabla \rho + \frac{1}{n}\mu_t \nabla \vec{u} - \mathbf{F}_p \tag{4}$$

where the additional quality coefficient C_m is as follows:

$$C_m = \gamma_p \frac{1-n}{n} \tag{5}$$

where γ_p is the closure coefficient which is 0.34.

The resistance formula is as follows:

$$\mathbf{F}_p/\rho = a\vec{u} + b\vec{u}\|\vec{u}\|_2 \tag{6}$$

where a and b are the resistance coefficients. In the formula proposed by Van Gent (1995) [30], the resistance coefficients a and b are calculated as follows:

$$a = \alpha \frac{(1-n)^2}{n^3} \frac{v}{{d_{50}}^2} \tag{7}$$

$$b = \beta(1 + \frac{7.5}{KC}) \frac{1-n}{n^3} \frac{1}{{d_{50}}^2} \tag{8}$$

where v is the kinematic viscosity; d_{50} is the median particle diameter of the granular material; and KC is the Keulegan-Carpenter number.

The finite volume method is adopted in the numerical solution, in which the time domain term is discretized by Euler implicit scheme, the convection term is discretized by restricted linear scheme, the diffusion term is discretized by linear scheme, and the source term and the wave-absorbing term are discretized explicitly. The numerical calculation method adopts PIMPLE algorithm, that is, a combination of the SIMPLE algorithm and the PISO algorithm. The basic idea of PIMPLE is to solve each time step with a simple steady-state algorithm (that is, to see each time step as a steady-state flow), and the step of the time step is completed by the PISO algorithm. The PIMPLE algorithm treats each time step as a steady-state flow (using Asunchi to solve the situation of large changes in two adjacent time periods), and when the steady-state solver is solved to a certain extent, the standard PISO is used as the final step of the solution. In this paper, the inlet, outlet, and bottom boundary are set as a solid wall with a non-slip boundary, and the top boundary is in direct contact with the air pressure condition. The tangential velocity is given in inflow, and the gradient of velocity components in all directions is 0 in outflow.

The model uses the VOF method to simulate the tracking of the free surface [31]. The equation is as follows:

$$\frac{\partial F}{\partial t} + \frac{1}{n}(\nabla \vec{u} F + \frac{1}{n} \nabla \vec{u}_r F(1-F)) = 0 \tag{9}$$

where $\vec{u}_r$ is the relative velocity; F is the VOF function where when it is filled with water, F is 1, and when it is filled with air, F is 0. The intermediate value will be located at or near the interface, the linear weighting method is used as follows:

$$\rho = F\rho_1 + (1-F)\rho_0 \tag{10}$$

$$\mu = F\mu_1 + (1-F)\mu_0 \tag{11}$$

where subscript 1 and 0 represent the properties of water and air, respectively.

According to the wave theory, the velocity entrance method specifies the distribution of the velocity field on the wave-making boundary at each time step. Its advantage is to strictly abide by the wave theory. If the target wave is just within the selected wave theory range, it will obtain more accurate results. However, there are two problems in this method: First, when the nonlinearity of the target wave is strong, the speed of input and extraction on the boundary in a wave period is not equal, and with the increase in calculation time, it may lead to problems of mass conservation such as wave surface uplift; secondly, the velocity entrance method cannot eliminate the secondary reflection wave. Therefore, when the velocity entrance method is adopted to simulate waves, corresponding correction measures must be taken [32]. Waves2Foam adopts the method of adding relaxation zones at the front and rear ends of the flume, which not only ensures the conservation of mass, but also eliminates the secondary reflection wave. The relaxation zone technology aims at eliminating false reflection in numerical simulation. The technique is based on the weighting between the calculated solution of the velocity field and the indicated field with the target solution. There are two forms of explicit relaxation and implicit relaxation, in which explicit/implicit refers to time integration [33]. The explicit method of relaxation zone is as follows:

$$\phi = (1 - \omega_R)\phi_t + \omega_R \phi_c \quad (12)$$

where the weighting function is ω_R as follows:

$$\omega_R = 1 - \frac{\exp \sigma^{3.5} - 1}{\exp 1 - 1} \quad (13)$$

where σ is the local coordinate in the relaxation zone.

A solitary wave, as a kind of nonlinear wave, can keep the shape and energy of waves well in long distance propagation. The wavelength tends to infinity, and the whole wave surface is higher than the still water surface. In fact, solitary waves only have one peak higher than the horizontal plane to propagate forward, which describes a moving wave with non-periodic motion [34]. The relative wave height of solitary wave is the most important factor to determine its waveform. In practical applications, solitary waves are often used to simulate extreme waves such as tsunami waves. The solitary wave used in this paper is the second-order solitary wave. The wave surface equation, water point velocity, acceleration, pressure and wave velocity are expressed as follows:

$$\frac{\eta}{h} = \varepsilon \mathrm{sech}^2 q - \frac{3}{4}\varepsilon^2 \mathrm{sech}^2 q \tanh^2 q \quad (14)$$

$$\frac{p}{\rho g h} = \frac{\eta}{h} + 1 - \frac{s}{h} - \frac{3}{4}\varepsilon^2 \mathrm{sech}^2 q [(\frac{s}{h})^2 - 1](2 - 3\mathrm{sech}^2 q) \quad (15)$$

$$\frac{u}{\sqrt{gh}} = \varepsilon \mathrm{sech}^2 q + \varepsilon^2 \mathrm{sech}^2 q [\frac{1}{4} - \mathrm{sech}^2 q - \frac{3}{4}(\frac{s}{h})^2 (2 - 3\mathrm{sech}^2 q)] \quad (16)$$

$$\frac{w}{\sqrt{gh}} = \varepsilon\sqrt{3\varepsilon}(\frac{s}{h})\mathrm{sech}^2 q \tanh q \{1 - \varepsilon[\frac{3}{8} + 2\mathrm{sech}^2 q + \frac{1}{2}(\frac{s}{h})^2 (1 - 3\mathrm{sech}^2 q)]\} \quad (17)$$

$$\frac{c}{\sqrt{gh}} = 1 + \frac{1}{2}\varepsilon - \frac{3}{20}\varepsilon^2 \quad (18)$$

$$q = \frac{\sqrt{3\varepsilon}}{2h}(1 - \frac{5}{8}\varepsilon)(x - ct) \quad (19)$$

where η represents the water level, p represents the pressure, u represents the horizontal velocity, w represents the vertical velocity, a is the wave height of solitary wave, h is the initial water depth; and the intermediate $\varepsilon = a/h$, $s = z + h$. Theoretically, the water body of the solitary wave peak is always above the still water surface and the theoretical wavelength is infinite.

3. Model Verification

This section may be divided by subheadings. It should provide a concise and precise description of the experimental results, their interpretation, as well as the experimental conclusions that can be drawn.

3.1. Solitary Wave

In order to ensure the simulation accuracy of solitary wave, the solitary wave waveform is first verified. The numerical simulation domain adopted in this study is that the length is 500 m and the height is 14 m. The water depth of the solitary wave is 8 m, and the wave height is 2 m. The wave-making boundary is located at x = 0 m. The location of the wave sensor monitor is at x = 10 m. Through the comparison between the simulation results and the theoretical wave values, as shown in Figure 1, it can be seen that the theoretical value is in good agreement with the numerical simulation results, which indicates that the established numerical flume can accurately generate solitary waves.

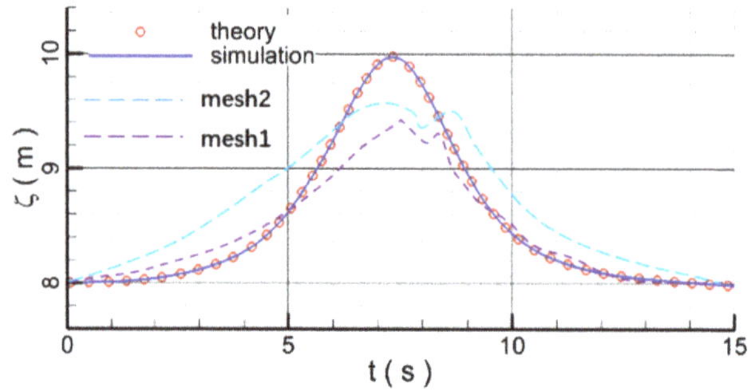

Figure 1. Solitary wave comparison between the theory values and the simulation results.

3.2. Wave Propagation over the Porous Medium Structure

Because this paper mainly discusses the propagation of solitary waves on permeable submerged breakwaters, it is necessary to verify the fluid flow ability of numerical models in permeable structures. The predicted distributions of velocity are compared with the corresponding measured data. The experiment was completed by Wu and Xiao (2013) in a glass flume [35]. The wave flume is 25 m long, 0.5 m wide and 0.6 m deep. The breakwater is composed of uniform glass balls with a diameter of 1.5 cm and a porosity of 0.52, and is installed at the bottom of the flume. The origin of the coordinate system is at the intersection of the bottom and the left side of the breakwater. The length of the simulation area is 8 m. The horizontal resolution of the grid is dx = 0.004 m, and the 40 layers are applied in the vertical direction. The distance between the entrance of the simulation domain and the breakwater is 4 m. The still water depth h is 0.106 m, and the incident solitary wave height is 0.0477 m. The Figure 2. shows the velocity and wave surface comparisons.

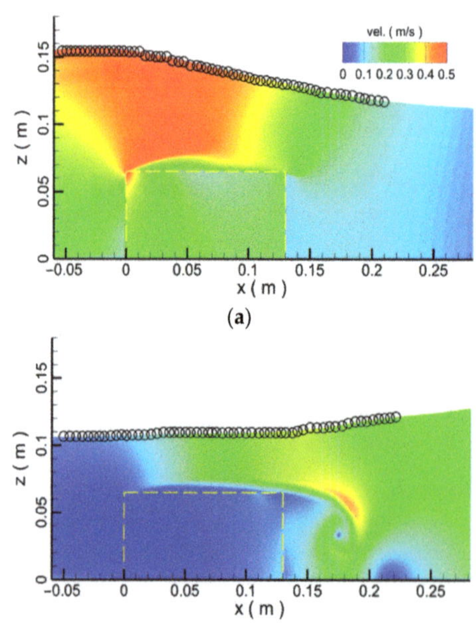

Figure 2. Flow velocity contour; (**a**) velocity contour; (**b**) water surface. (the circles represents expremental data).

The distributions of the velocity in horizontal and vertical directions recorded from −0.04 m to 0.2 m are shown in Figure 3. The figure shows that the predicted results are in good agreement with the experiment data. Overall, results clearly support the computational capability of our numerical model for describing a solitary wave propagating over the permeable breakwater.

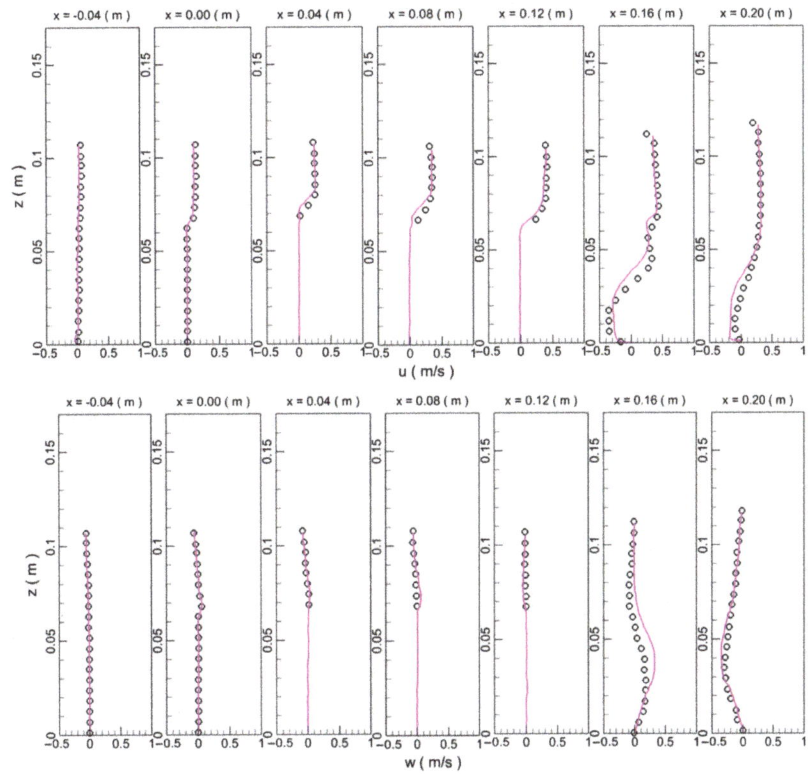

Figure 3. Comparison of velocity distribution between computation and measurement; (u) horizontal velocity; (w) vertical velocity; at t = 1.65 s.

4. Results and Discussion

Based on the verified numerical model, this section mainly discusses the influence of multiple porous semi-circular breakwaters on solitary wave propagation. The experimental layout is shown in Figure 4. The whole calculation domain is 500 m long and 16 m high. The diameter of the submerged breakwater is D. The distance between the adjacent submerged breakwaters is s. The static water depth is 8 m. The grid resolution around the submerged breakwater is 0.001 m. The center point of the first submerged breakwater is located at x = 40 m. Ten wave sensor monitors and velocity sensors are distributed above the submerged breakwater, at the positions of x = 10, 20, 30, 40, 50, 60, 70, 80, 90, and 100 m. In order to compare and analyze the force efficiently, all the forces are normalized by $\rho g v$.

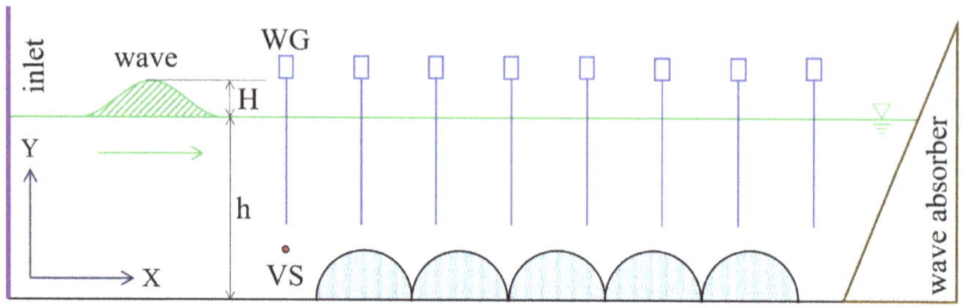

Figure 4. Computational layout for solitary wave passing over the breakwaters.

4.1. Effect of the Porosity

In this section, the influence of the breakwaters with different porosities on the propagation of the solitary wave is analyzed. Seven porosities are selected as 0.0, 0.2, 0.4, 0.5, 0.6, 0.8, and 1.0. Figure 5a shows the velocity contour of solitary waves passing over the multiple porous medium breakwaters. When the solitary wave approaches the submerged breakwater, the water level rises and the velocity increases. When the wave peak propagates to the first submerged breakwater, the velocity around the first submerged breakwater increases. When the wave crest is just above the submerged breakwater, it can be seen that the flow velocity between the water surface and the submerged breakwater are at the maximum, which indicates that the water flux per unit area increases and the overall velocity increases. In addition, the velocity inside the submerged breakwater also increases. Due to the shelter effect of submerged breakwaters, the velocity of flow between submerged breakwaters is low. When the wave peak passes over the submerged breakwater, the velocity in the gap of the submerged breakwater is larger than that in other parts, indicating that the flow in the gap is stronger due to the influence of the submerged breakwater. Figure 5b shows the vorticity contour field of solitary wave passing over the porous breakwaters. When the wave propagates to the submerged breakwater, the vortex begins to appear on the surface of the submerged breakwater due to the barrier of the submerged breakwater. When the wave peak is on the submerged breakwater, the vortex appears above each submerged breakwater, which indicates that the submerged breakwater has a certain blocking effect on the wave. The vortex gradually decreases from the foremost submerged breakwater to the rearmost submerged breakwater. After the wave crest passes over the submerged breakwater, the vortex attached to the submerged breakwater begins to grow and fall off. Due to the influence of the porous submerged breakwater, at the interface between the water body and the submerged breakwater, the vortex is flattened and presents a strip shape, and the tailed vortex presents a circle shape. Figure 6. shows the maximum velocity on the fourth and fifth breakwater at different time.

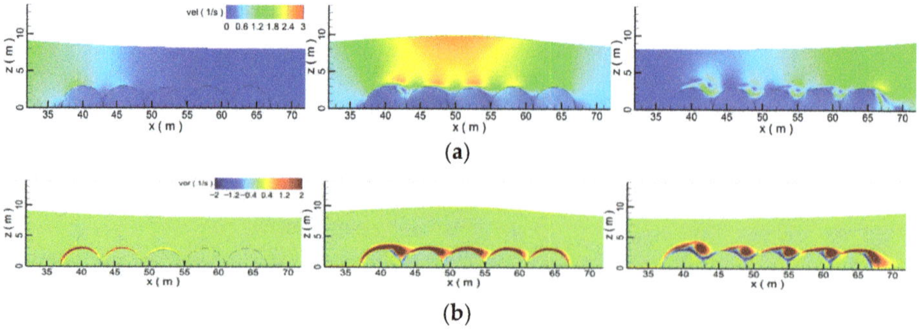

Figure 5. The flow filed around the breakwaters at different times; (**a**) velocity contour; (**b**) vorticity contour.

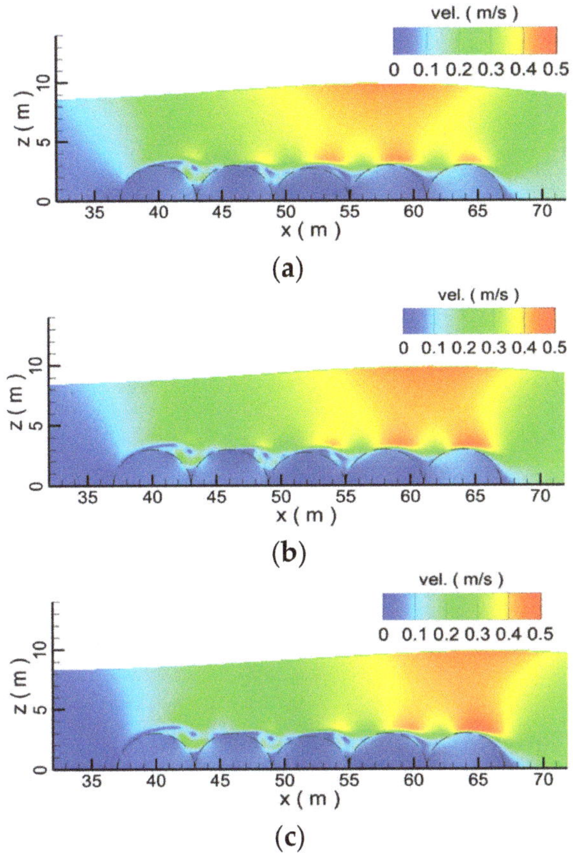

Figure 6. Maximum velocity on the fourth and fifth breakwater; (**a**) t = 12.4 s; (**b**) t = 12.8 s; (**c**) t = 13.2 s.

When the solitary wave passes over the submerged breakwater, Figure 7 shows the velocity contour under different porosities. When the porosity is 0, the submerged breakwater is solid. Since water can not pass through the submerged breakwater, the flow between the submerged breakwater and the water surfaces is the largest. Compared with the velocity above the submerged breakwater with pores, the velocity above the solid breakwater is the highest. With the increase in porosity, the flow passing through the porous submerged breakwater increases, and the velocity between the submerged breakwater and the water surfaces decreases. The velocity in the front submerged breakwater is the smallest, and the velocity in the rear submerged breakwater increases gradually, because the solitary peak has passed over the front breakwater. In addition, the various submerged breakwaters have different influences on the wave. When the porosity is 0.8, the velocity between the first and second submerged breakwaters is about 0.3 m/s, while the velocity in the fourth and fifth pores is 0.6 m/s. When the submerged breakwater does not exist, the velocity distribution of the whole flow field is uniform, and the maximum velocity is 2.0 m/s. In a word, submerged breakwaters with different porosities have different interference effects on the flow field, resulting in the change of velocity distribution in the whole flow field, thus affecting the propagation of solitary waves.

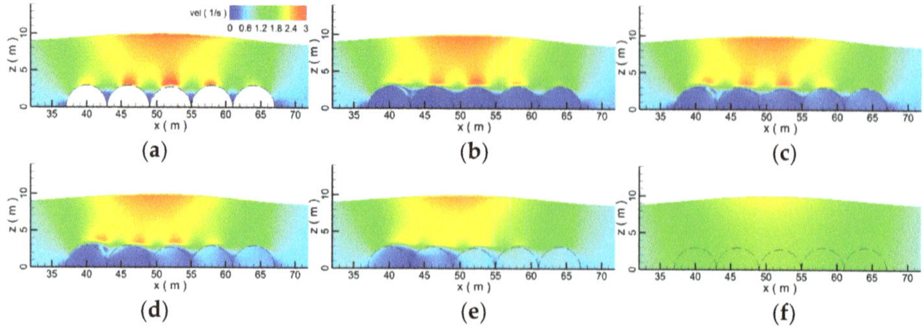

Figure 7. The velocity contour of the wave peak on the breakwaters under different porosities; (**a**) por = 0.0; (**b**) por = 0.2; (**c**) por = 0.4; (**d**) por = 0.6; (**e**) por = 0.8; (**f**) por = 1.0.

Figure 8 shows the vorticity contour around submerged breakwaters with different porosities. It can be seen that when the porosity is 0, the submerged breakwater is solid. The vorticity above the submerged breakwater is small. With the increase in porosity, the vorticity above the submerged breakwater increases. When there are pores in the submerged breakwater, the flow can pass through the submerged breakwater in the pores, and the flow interaction between the inside and outside the submerged breakwater leads to the increase in the flow intensity in the flow field. At the surface of the submerged breakwater, there are also some small vortices. When the submerged breakwater does not exist, the vortex disappears completely. In a word, when the porous submerged breakwater exists, the vortex around the submerged breakwater increases, which consumes part of the wave energy and weakens the wave propagation to some extent.

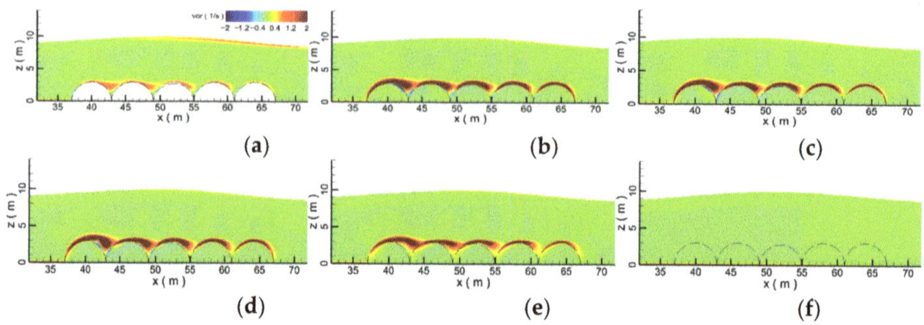

Figure 8. The vorticity contour of the wave peak on the breakwaters under different porosities; (**a**) por = 0.0; (**b**) por = 0.2; (**c**) por = 0.4; (**d**) por = 0.6; (**e**) por = 0.8; (**f**) por = 1.0.

Figure 9 shows the maximum water level and flow velocity monitored at different positions. When the waves reach the submerged breakwater, although the porosity is different, the water level recorded at x = 10.0 m is about 9.98 m, and the flow velocity is 2.05 m/s. When the water level is recorded at 20 m, it begins to change. When the porosity is 0.0 and 0.2, the water level is 9.985 m, while when the porosity is 1.0, the minimum water level is 9.95 m. When the monitoring position is at 30 m, the water level obviously rises when there is a submerged breakwater. Because of the existence of the submerged breakwater, the submerged breakwater occupies a part of the water area and the water surface rises. When the monitoring position increases from 40 m to 70 m, the wave energy is dissipated and the water level continues to decrease due to the interference of the submerged breakwater on the water flow. With the increase in porosity, when the porosity is 0.8, the water level reaches a minimum of 9.82 m. When the monitoring position is at

80 m, the wave peak passes through the submerged breakwater, and the water is suddenly released, resulting in a short rise of the water level. Then, the water level gradually drops to a gentle state. When the water passes through the submerged breakwater, the flow velocity also shows an increasing trend, because the submerged breakwater leads to the decrease in the water passing area between the water surface and the submerged breakwater, and the flow velocity increases when the flow flux is constant. When the wave peak passes through the submerged breakwater, the water passing area increases and the flow velocity decreases. In general, when the porosity is 0.8, after the waves pass through the submerged breakwater, the water level decreases significantly, and the wave velocity also decreases.

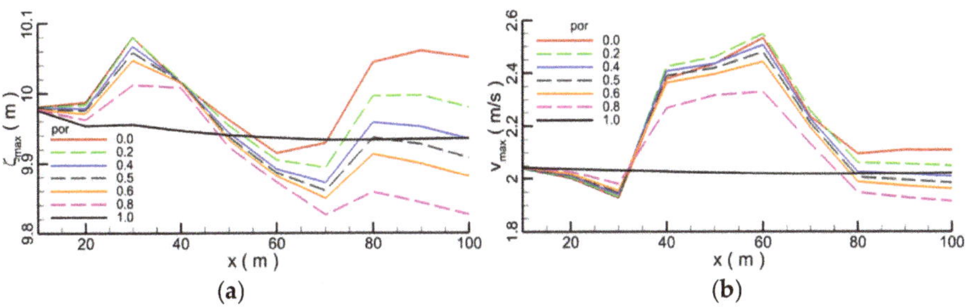

Figure 9. The maximum water level and velocity recorded under different porosities; (**a**) water level; (**b**) velocity.

Figure 10 shows the forces of different submerged breakwaters when waves pass through them. With the arrangement of the submerged breakwater from front to back, the force of the submerged breakwater gradually decreases, and the decreasing range gradually decreases, because the front submerged breakwater shields the rear submerged breakwater. With the increase in porosity, the force on the submerged breakwater gradually decreases. When the porosity increases from 0.0 to 0.8, the force on the first submerged bank decreases from 0.66 to 0.13, the force on the second submerged breakwater decreases from 0.67 to 0.14, the force on the third submerged breakwater decreases from 0.65 to 0.13, the force on the fourth submerged breakwater decreases from 0.64 to 0.12, and the force on the fifth submerged bank decreases from 0.62 to 0.11. In a word, submerged breakwaters with different porosity have different effects on wave propagation. When the porosity is 0.8, the force on the submerged breakwaters is the smallest, and the water level and velocity decreases significantly. In order to compare and analyze the force efficiently, the force are normalized by $\rho g v$, i.e., F^*.

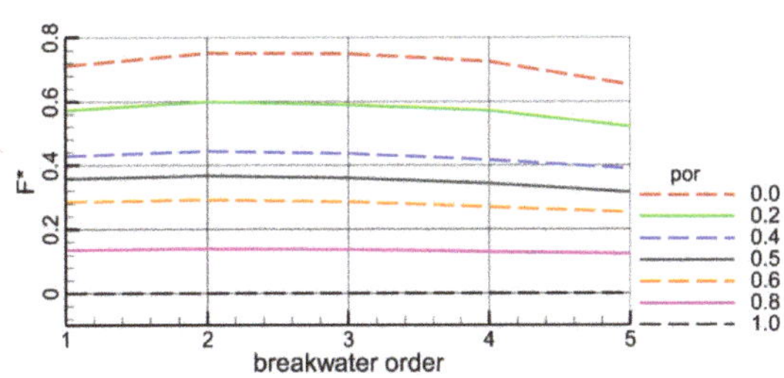

Figure 10. The maximum forces on the breakwaters with the different porosities.

4.2. Effect of the Diameter

This section discusses the influence of submerged breakwaters with different diameters on wave propagation. The porosity of the submerged breakwater is 0.5. According to the survey of the offshore waters of the South China Sea, the average radius of the diving diving dike in the coastal area is about 3 m, so 6 diameters of the submerged breakwater are selected: 1.5, 2.0, 2.5, 3.0, 3.5, and 4.0 m [36]. Figure 11 shows the velocity contour around the submerged breakwater when the peak reaches the submerged breakwater. When the diameter of the submerged breakwater is 1.5 m, the velocity inside the submerged breakwater is small, and the velocity between the submerged breakwater and the water surface is the largest with the maximum velocity of about 2.1 m/s. With the increase in the diameter, the water area above the submerged breakwater decreases, and the flow velocity increases when the flow flux is constant. When the diameter of the submerged breakwater is 2.5 m, the maximum velocity of the flow field is about 2.5 m/s, and when the diameter of the submerged breakwater is 4.0 m, the maximum velocity of the flow field is 3.0 m/s. In addition, with the increase in the diameter of the submerged breakwater, the profile of the wave gradually becomes flat.

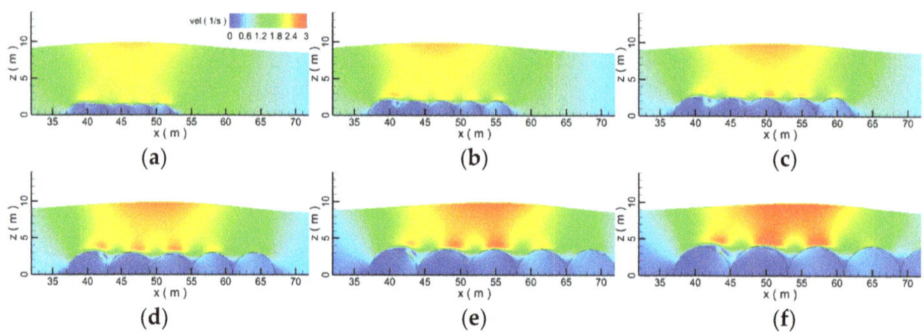

Figure 11. The velocity contour of the wave peak on the breakwaters under different diameters; (**a**) D = 1.5 m; (**b**) D = 2.0 m; (**c**) D = 2.5 m; (**d**) D = 3.0 m; (**e**) D = 3.5 m; (**f**) D = 4.0 m.

Figure 12 shows the vorticity contour around the submerged breakwater when the wave peak is above the submerged breakwater. When the diameter of the submerged breakwater is small, the vortices around the submerged breakwater are connected together to form a flat vortex. With the increase in the diameter of submerged breakwaters, the vortex in the gap between submerged breakwaters is increasing, and the vortex formed by different submerged breakwaters gradually separates, forming a convex circular vortex. When the diameter of the submerged breakwater is 3.5 m, a vortex is attached above each submerged breakwater. A pair of clockwise and counterclockwise vortices can be seen from the first submerged dike. When the diameter of submerged breakwaters is 4.0 m, compared with the vortex formed by submerged breakwaters with other diameters, the vortex reaches the maximum. It shows that the larger the diameter of submerged breakwater is, the stronger the disturbance of convection field is.

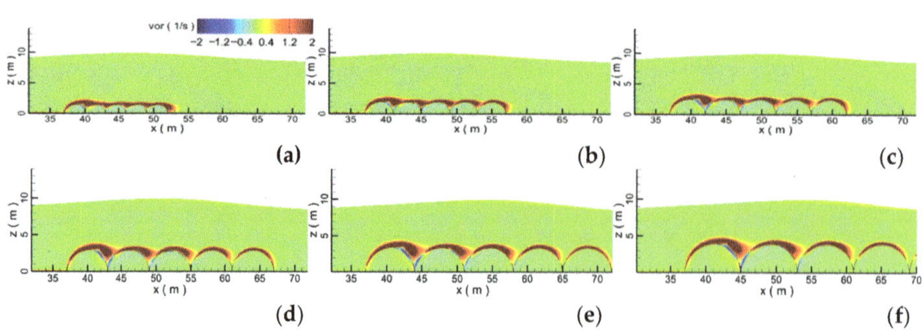

Figure 12. The vorticity contour of the wave peak on the breakwaters under different diameters; (**a**) D = 1.5 m; (**b**) D = 2.0 m; (**c**) D = 2.5 m; (**d**) D = 3.0 m; (**e**) D = 3.5 m; (**f**) D = 4.0 m.

Figure 13 shows the maximum water level and maximum flow velocities recorded at different positions. When the peak reaches the first submerged breakwater, the water area decreases and the water level rises. When the record position is at x = 30 m, the maximum water level is 10 m when the diameter of the submerged breakwater is 1.5 m, and the maximum water level is 10.11 m when the diameter of the submerged breakwater is 4.0 m. With the propagation of waves, the wave energy decreases and the maximum water level gradually decreases. When the waves pass through the submerged breakwater, the water level decreases to 9.9 m when the diameter of the submerged breakwater is 1.5 m, and 9.8 m when the diameter of the submerged breakwater is 4.0 m. When the wave reaches the submerged breakwater, the velocity is 2.04 m/s. When the wave peak passes through the submerged breakwater, the flow rate per unit area increases and the velocity increases. When the diameter of the submerged breakwater is 1.5 m, the maximum velocity is 2.2 m/s, and when the diameter of the submerged breakwater is 4.0 m, the maximum velocity is 2.65 m/s. When the peak leaves the submerged breakwater, the velocity of flow decreases rapidly. When the diameter of the submerged breakwater is 4.0 m, the velocity decreases to 1.9 m/s, and when the diameter of the submerged breakwater is 1.5 m, the velocity is 2.0 m/s. When the crest reaches the first diving dike, the water level rises due to the decrease in the area of water crossing, and the flow rate increases under the condition of a certain flow rate. As the waves propagate, the water level decreases, while the speed remains relatively high. The larger the diving dike radius, the greater the water surface rises above the first diving dike, and the water level drops most significantly after passing all the diving dikes. Therefore, the larger the radius of the diving dike, the better the wave absorption effect. In a word, the larger the diameter of submerged breakwater, the more serious the wave water level decreases, and the more severe the fluctuation of velocity, which indicates that the larger the diameter of submerged breakwater, the more significant the attenuation effect on waves.

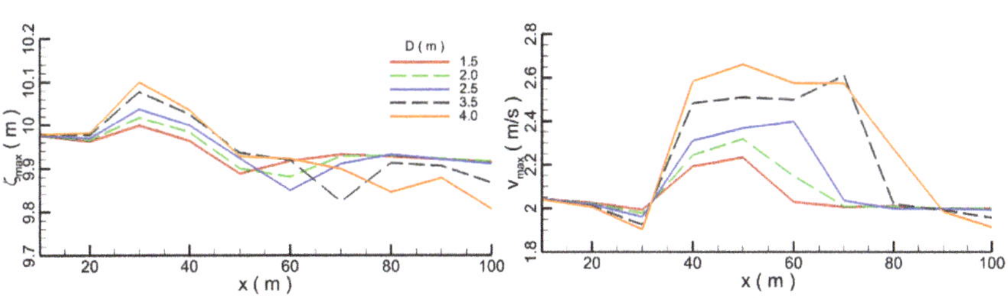

Figure 13. The maximum water level and velocity recorded under different diameters; (**a**) water level; (**b**) velocity.

Figure 14 shows the force on submerged breakwaters with different diameters. The front submerged breakwater has a certain protective effect on the rear submerged breakwater. As the submerged breakwater is arranged from front to back, the force on the submerged breakwater with the same diameter gradually decreases. When the diameter of the submerged breakwater is 4.0 m, the force on the submerged breakwater decreases from 0.48 to 0.43. The larger the diameter, the more significant the action of the water body, and the larger the force. When the diameter of the submerged breakwater increases from 1.5 m to 4.0 m, the force on the first submerged breakwater increases from 0.165 to 0.18, and the force on the fifth submerged breakwater increases from 0.16 to 0.43. In general, with the increase in the diameter of the submerged breakwater, the blocking and attenuation effect of the submerged breakwater on waves will increase, and at the same time, the force on the submerged breakwater will also increase. Therefore, the diameter of the submerged breakwater should be selected according to the actual situation, so as to ensure the wave attenuation and maintain the stability of the submerged breakwater. In order to compare and analyze the force efficiently, the force are normalized by $\rho g v$, i.e., F^*.

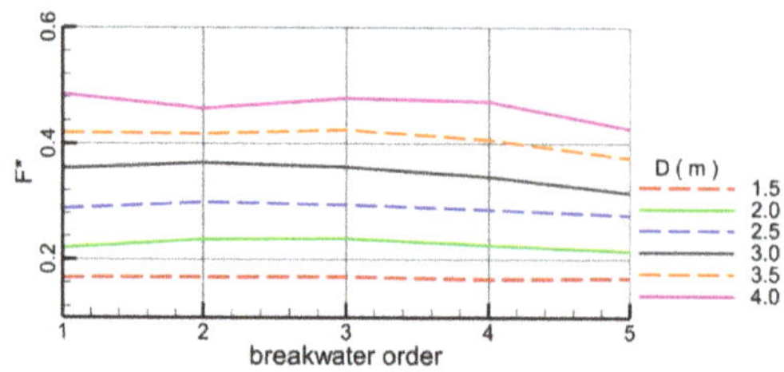

Figure 14. The maximum forces on the breakwaters with different diameters.

4.3. *Effect of the Distance*

Figure 15 shows the influence of submerged breakwaters with different distances on solitary wave propagation. When solitary waves pass through the submerged breakwater, the velocity above the submerged breakwater is larger, but the velocity in the gap between the submerged breakwaters is smaller. With the increase in the distance, the volume of water in the gap between submerged breakwaters increases, but the flow of water in the gap between submerged breakwaters is weak due to the shelter effect of submerged breakwaters. Although the distance between the submerged breakwaters has a certain influence on wave propagation, the maximum velocity around submerged breakwater in different distances is basically unchanged at about 2.7 m/s. When the distance between submerged breakwaters is 2.4 m, the velocity of water in the gap between submerged breakwaters is basically the same as that in the submerged breakwaters, which indirectly indicates that increasing the distance between submerged breakwaters appropriately is equivalent to increasing the relative length of obstacles, which is conducive to weakening the wave propagation.

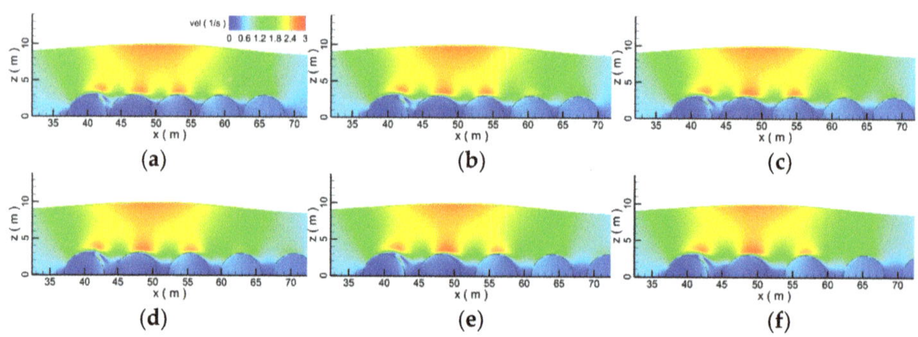

Figure 15. The velocity contour of the wave peak on the breakwaters under different distances; (**a**) s = 0.4 m; (**b**) s = 0.8 m; (**c**) s = 1.2 m; (**d**) s = 1.6 m; (**e**) s = 2.0 m; (**f**) s = 2.4 m.

Figure 16 shows the vorticity contour around the submerged breakwaters with different distances when the wave peak is above a submerged breakwater. The front submerged breakwater has the most significant blocking effect on waves, resulting in the change of water flow. Therefore, the vortex around the front submerged breakwater is the largest. As the submerged breakwaters are arranged backward, the vorticity around the submerged breakwaters gradually decreases. When the distance is 1.6 m, the vortex behind the first submerged breakwater has basically fallen off and filled the gap of the submerged breakwater. With the increase in the distance, the proportion of vortex in the gap decreases, and the vortex basically adheres to the surface of the submerged breakwater, showing the shape of tadpoles. Due to the influence of submerged breakwater seepage, a pair of clockwise and counterclockwise vortices appears on the surface of submerged breakwater. In general, with the increase in the gap between the submerged breakwaters, the disturbance to the water body is enhanced, and the vortex around the submerged breakwaters is fully developed.

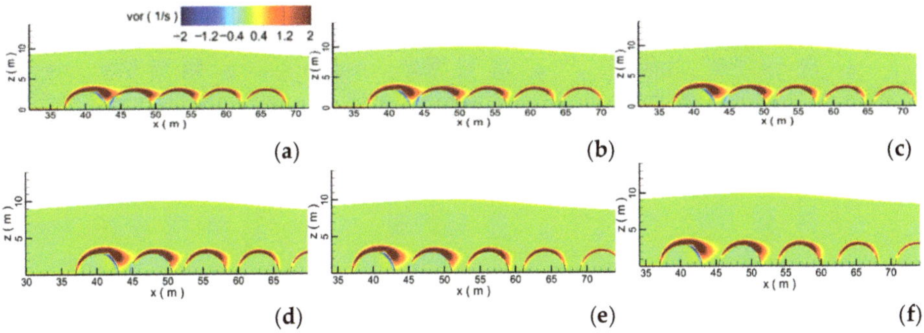

Figure 16. The velocity contour of the wave peak on the breakwaters under different distances; (**a**) s = 0.4 m; (**b**) s = 0.8 m; (**c**) s = 1.2 m; (**d**) s = 1.6 m; (**e**) s = 2.0 m; (**f**) s = 2.4 m.

Figure 17 shows the maximum water level and velocity recorded at different positions when solitary waves pass through submerged breakwaters with different distances. When the recording position is in front of 40 m, due to the existence of the submerged breakwater, the cross-section of the passing water rises, the water level uniformly rises to 10.06 m, and then begins to decline. When the recording position is 70 m, the recorded water level is the lowest. When the clearance is 0.4 m, the water level is 9.82 m, and when the distance is 2.0 m, the water level is 9.86 m. As the monitoring position moves backward and the distance increases, the water level gradually decreases. When the monitoring position is in front of 40 m, the flow rates in different gaps are basically the same. When the monitoring position is 40 m, the flow velocity reaches about 2.4 m/s. With the backward movement of the monitoring position, the monitoring flow velocity gradually decreases, and finally it basically drops to 1.92 m/s. In general, with the increase in the distance, the water level and the flow velocity decrease.

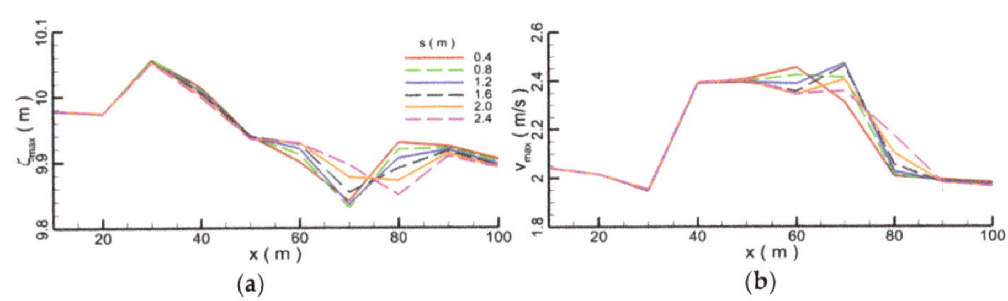

Figure 17. The maximum water level and velocity recorded with different distances; (**a**) water level; (**b**) velocity.

Figure 18 shows the maximum forces on different submerged breakwaters under different distances. It can be seen that under the same distance, the force on the second submerged breakwater is greater than that of the first submerged breakwater, because the water flow in the gap in front of the second submerged breakwater is intense, which leads to the larger force on the second submerged breakwater. When the distance is 0 m, the force on the second submerged breakwater is about 0.365, and when the distance is 2.0 m, the force on the second submerged breakwater is about 0.355. As the submerged breakwaters are arranged backward, the force on the submerged breakwaters behind them gradually decreases. When the distance is 0.0 m, the force on the fifth submerged breakwater is 0.32, and when the distance is 0.4, the force on the fifth submerged breakwater is 0.3. In general, the force on the submerged breakwater in front is greater than that of the submerged breakwater in rear, and with the increase in the gap, the force on the submerged breakwater decreases slightly. In order to compare and analyze the force efficiently, the force are normalized by $\rho g v$, i.e., F^*.

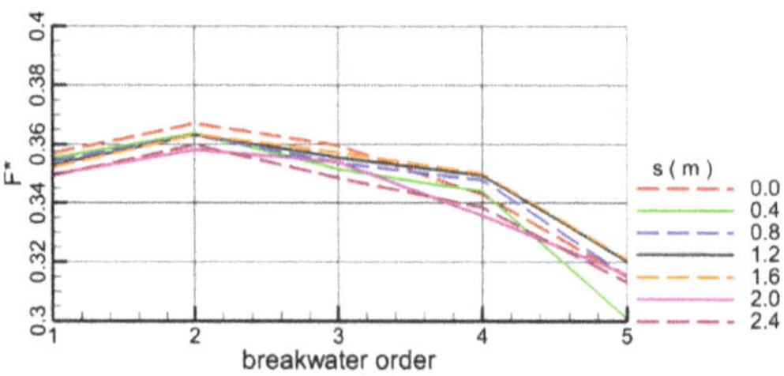

Figure 18. The maximum forces on the breakwaters with different distances.

4.4. Effect of the Wave Height

Due to the different wave heights in the sea, this section mainly discusses the influence of porous submerged breakwater on waves with different wave heights. Seven different wave heights, 1.0, 1.4, 1.8, 2.0, 2.2, 2.6, and 3.0 m, were considered in the study. Figure 19 shows the temporal evolution of the water level and velocity under the different wave heights. With the increase in wave height, the maximum water level and velocity also increase. When the wave height is 1.0 m, the recorded velocity is 0.8 m/s, and when the wave height is 3.0 m, the maximum velocity is 2.9 m/s.

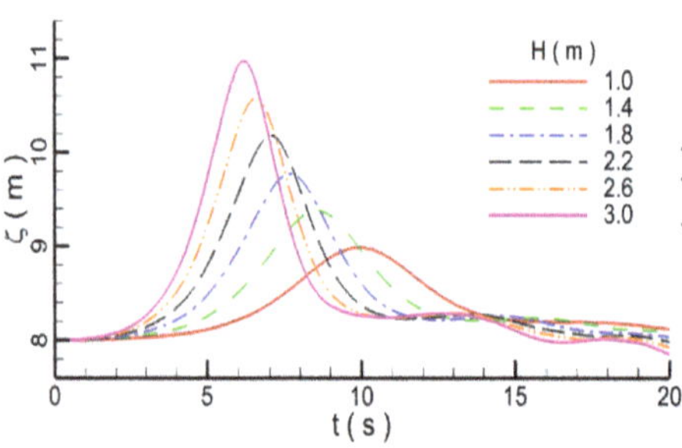

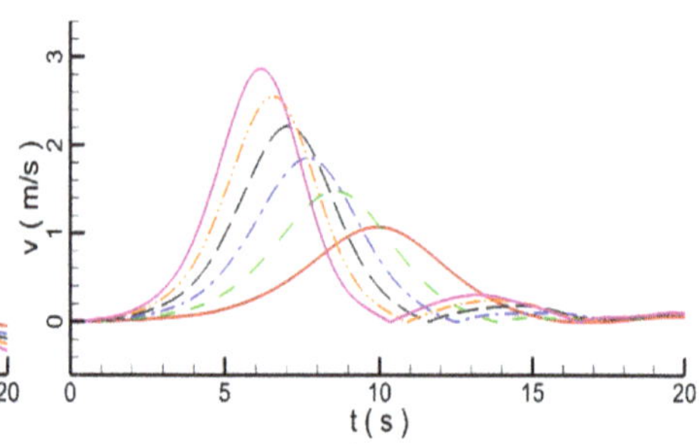

Figure 19. The temporal evolution of the water level and the velocity; (**a**) the water level; (**b**) the velocity.

Figure 20 shows the velocity contour around the submerged breakwater with different wave heights. With the increase in wave height, the velocity in the submerged breakwater increases, and the flow velocity between the submerged breakwater and the free water surface also increases. Figure 21 shows the vorticity contour around the submerged breakwater when the wave peaks with different wave heights reach the submerged breakwater. With the increase in wave height, the vorticity around the submerged breakwater increases continuously. When the wave height is 1.0 m, the vortex above the submerged breakwater is long and narrow, and there is no convex hull behind the vortex. When the wave height is 2.2 m, the convex hull formed by the vortex appears behind the submerged breakwater. The higher the wave height, the bigger the vortex behind the submerged breakwater, which indicates that the submerged breakwater has an obvious influence on waves with high wave height.

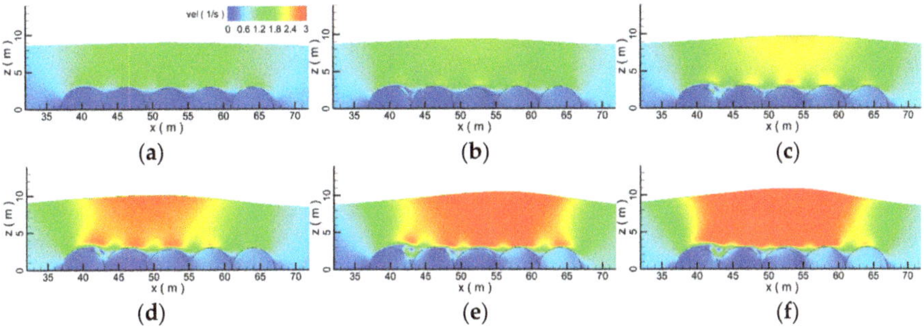

Figure 20. The velocity contour of the wave peak on the breakwaters under different wave heights; (**a**) $H = 1.0$ m; (**b**) $H = 1.4$ m; (**c**) $H = 1.8$ m; (**d**) $H = 2.2$ m; (**e**) $H = 2.6$ m; (**f**) $H = 3.0$ m.

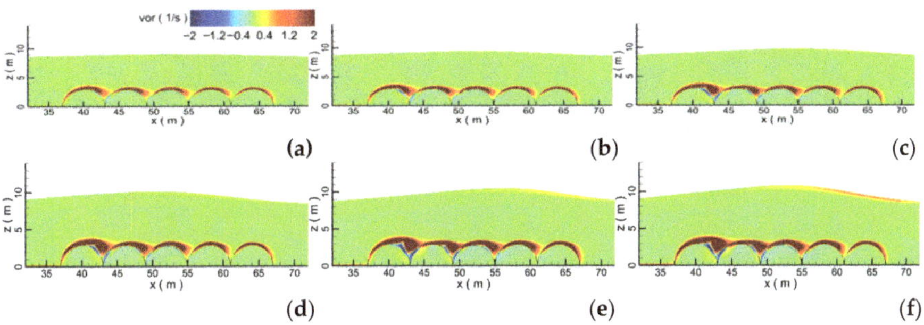

Figure 21. The vorticity contour of the wave peak on the breakwaters under different wave heights; (**a**) $H = 1.0$ m; (**b**) $H = 1.4$ m; (**c**) $H = 1.8$ m; (**d**) $H = 2.2$ m; (**e**) $H = 2.6$ m; (**f**) $H = 3.0$ m.

Figure 22 shows the maximum water levels monitored at different locations. The higher the wave, the more intense the fluctuation of water level. When the wave height is 1.0 m, the maximum water level difference monitored at different positions is 0.2 m, and when the wave height is 3.0 m, the maximum water level difference is 0.5 m, indicating that the higher the wave height, the more serious the influence of submerged breakwater on the waves. With the increase in wave height, the current velocity also increases. When the wave height is 1.0 m, the maximum velocity is 1.3 m/s and the velocity difference is 0.3 m/s. When the wave height is 3.0 m, the maximum velocity is 3.3 m and the velocity difference is 0.6 m/s. In general, with the increase in wave height, the greater the influence of submerged breakwaters on waves, the more energy consumed by wave propagation, and the more obvious the decrease in water level and velocity.

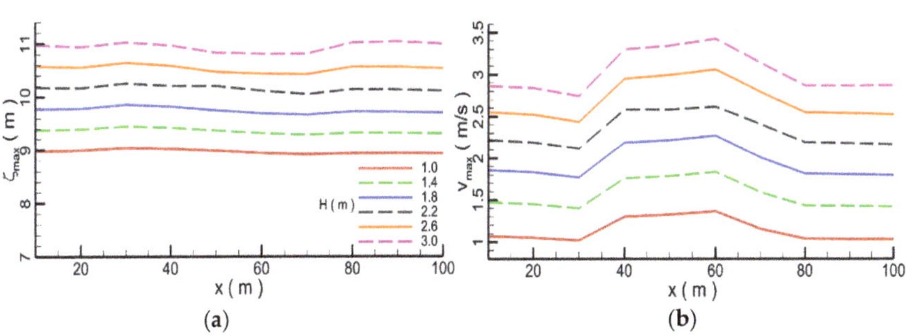

Figure 22. The maximum water level and velocity recorded with different wave heights; (**a**) water level; (**b**) velocity.

Figure 23 shows the force on the different submerged breakwaters under solitary waves with different wave heights. The force on the submerged breakwater decreases gradually from the front to the rear, because the front breakwater protects the rear submerged breakwater. When the solitary wave height is 1.0 m, the force from the first to the fifth submerged breakwater decreases from 0.2 to 0.18, and when the solitary wave height is 3.0 m, the force from the first to the fifth submerged breakwater decreases from 0.52 to 0.44. The higher the solitary wave is, the more serious the force on the submerged breakwater is, and the more the force on the submerged breakwater decreases. In general, the higher the wave height, the more obvious the blocking effect of the submerged breakwater on the waves, and the greater the force on the submerged breakwater.

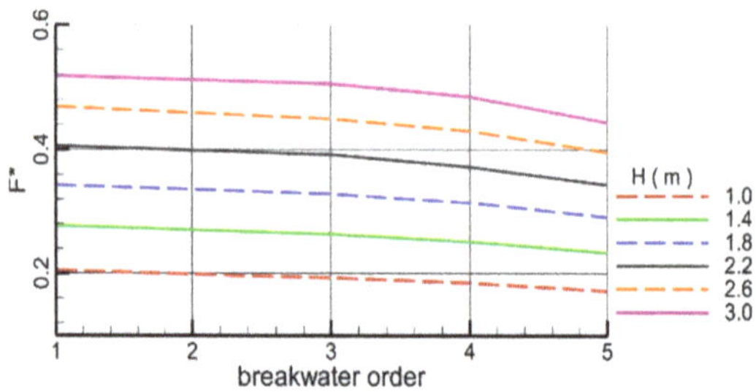

Figure 23. The maximum forces on the breakwaters with different distances.

5. Conclusions

In this paper, the influence of different submerged breakwaters on the propagation of solitary waves with different wave heights is studied by numerical simulation. The main conclusions are as follows:

(1) Due to the influence of the porous submerged breakwater, at the interface between the water body and the submerged breakwater, the vortex is squashed and presents a strip shape. Submerged breakwaters with different porosities have different interference effects on the flow field, resulting in the change of velocity distribution in the whole flow field, thus affecting the propagation of solitary waves. Submerged breakwaters with different porosities have different effects on wave propagation. When the porosity is 0.8, the force on the submerged breakwaters is the smallest, and the water level and velocity through the submerged breakwaters drop the most seriously.

(2) With the increase in the diameter of the submerged breakwater, the wave profile gradually becomes flat. When the diameter of submerged breakwaters is 4.0 m, compared with the vortex formed by submerged breakwaters with other diameters, the vortex reaches the maximum. The larger the diameter of submerged breakwater is, the stronger the disturbance of convection field is, the more serious the wave water level decreases, and the more intensive the fluctuation of the velocity, which indicates that the larger the diameter of the submerged breakwater, the more significantly the attenuation effect on the waves. With the increase in the diameter of the submerged breakwater, the blocking and attenuation effect of the submerged breakwater on the waves increases, and at the same time, the force on the submerged breakwater also increases.

(3) Appropriately increasing the distance between submerged breakwaters is equivalent to increasing the relative length of obstacles, which is conducive to weakening the wave propagation. With the increase in the distance between the submerged breakwaters, the disturbance to the water is enhanced, and the vortex around the submerged breakwaters is fully developed. With the increase in distance, the water level and flow velocity decrease. The force on the submerged breakwater in front is greater than that of the submerged breakwater in rear. With the increase in the distance, the force on the submerged breakwater decreases slightly.

(4) The higher the wave height, the more serious the wave deformation after passing through the submerged breakwater. The higher the wave height, the bigger the vortex

behind the submerged breakwater, which indicates that the submerged breakwater has serious influence on waves with high wave height. With the increase in wave height, the influence of submerged breakwaters on waves is greater, the energy consumed by wave propagation is greater, and the water level and velocity decrease more obviously. The higher the wave height, the more obvious the blocking effect of the submerged breakwater on the waves, and the greater the force on the submerged breakwater.

(5) By contrasting with a single breakwater, we obtain the conclusion that the influence of the porous breakwater on wave propagation is more significant than that of the solid breakwater. Additionally, compared with a single breakwater, the porous breakwater has a more significant impact on wave propagation.

Because the submerged breakwater can partially eliminate wave energy and the influence of water waves on the coastline, it has become a common form of breakwater structure in coastal areas. According to the above conclusions, it can be guaranteed to increase the blocking and attenuating effect on the waves under the condition that the submerged breakwater is less stressed. The challenge is that environmental loads can cause damage to the foundation around the submerged breakwater. Therefore, how to reduce the damage is one of the open questions that needs to be faced. In addition, our experiment only studied the isolated wave effects received by one form of breakwater. We will further study the influence of different forms of breakwaters by different waves, such as elliptical waves, sine waves, cosine waves, etc. in the future. It is believed that these research results can be used to better reduce the erosion damage of the coastline by waves.

Author Contributions: Conceptualization, K.Z. and R.J.; methodology, H.Q.; software, Z.C.; validation, Z.S.; investigation, Y.W.; writing—original draft preparation, E.Z. All authors have read and agreed to the published version of the manuscript.

Funding: This research was funded by National Natural Science Foundation of China (Grant No. 52001286; 52101332; 52202427), National Key Research and Development Program of China (Grant No. 2021YFC3101800), Shenzhen Science and Technology Program (Grant No. KCXFZ20211020164015024) and Shenzhen Fundamental Research Program (Grant No. JCYJ20200109110220482).

Institutional Review Board Statement: Not applicable.

Informed Consent Statement: Not applicable.

Data Availability Statement: Not applicable.

Conflicts of Interest: The authors declare no conflict of interest.

Notation List

ρ	fluid density
u	velocity vector in Cartesian coordinate system
t	time
p_r	relative pressure
p	total pressure
g	the acceleration of the gravity
$\mathbf{x}_r$	he reference position coordinate
μ_t	dynamic viscosity
u_p	velocity in the porous medium
γ_p	closure coefficient
v	kinematic viscosity
d_{50}	median particle diameter of the granular material
KC	Keulegan-Carpenter number
u_r	relative velocity
F	VOF function where when it is filled with water
σ	local coordinate in the relaxation zone
a	wave height of solitary wave
h	initial water depth

References

1. Sun, J.; Ma, Z.; Wang, D.; Dong, S.; Zhou, T. Numerical study of the run-up of a solitary wave after propagation over a saw-tooth-shaped submerged breakwater. *Int. J. Nav. Arch. Ocean Eng.* **2020**, *12*, 283–296. [CrossRef]
2. Fu, D.; Zhao, X.; Wang, S.; Yan, D. Numerical study on the wave dissipating performance of a submerged heaving plate breakwater. *Ocean Eng.* **2021**, *219*, 108310. [CrossRef]
3. Zhao, E.; Mu, L.; Hu, Z.; Wang, X.; Sun, J.; Zhang, Z. Physical and Numerical Investigations on Wave Run-Up and Dissipation under Breakwater with Fence Revetment. *J. Mar. Sci. Eng.* **2021**, *9*, 1355. [CrossRef]
4. Ali, K.K.; El Salam, M.A.A.; Mohamed, E.M.H.; Samet, B.; Kumar, S.; Osman, M.S. Numerical solution for generalized nonlinear fractional integro-differential equations with linear functional arguments using Chebyshev series. *Adv. Differ. Equ.* **2020**, *2020*, 494. [CrossRef]
5. Osman, M.S.; Rezazadeh, H.; Eslami, M.; Neirameh, A.; Mirzazadeh, M. Analytical study of solitons to Benjamin-Bona-Mahony-Peregrine equation with power law nonlinearity by using three methods. *UPB Sci. Bull. Ser. A Appl. Math. Phys.* **2018**, *80*, 267–278.
6. Akbar, M.A.; Wazwaz, A.-M.; Mahmud, F.; Baleanu, D.; Roy, R.; Barman, H.K.; Mahmoud, W.; Al Sharif, M.A.; Osman, M. Dynamical behavior of solitons of the perturbed nonlinear Schrödinger equation and microtubules through the generalized Kudryashov scheme. *Results Phys.* **2022**, *43*, 106079. [CrossRef]
7. Han, X.; Dong, S. Interaction of solitary wave with submerged breakwater by smoothed particle hydrodynamics. *Ocean Eng.* **2020**, *216*, 108108. [CrossRef]
8. Nassiraei, H.; Heidarzadeh, M.; Shafieefar, M. Numerical simulation of long waves (tsunami) forces on caisson breakwaters. *Civ. Eng.* **2016**, *32*, 3–12.
9. Mahmoudof, S.M.; Hajivalie, F. Experimental study of hydraulic response of smooth submerged breakwaters to irregular waves. *Oceanologia* **2021**, *63*, 448–462. [CrossRef]
10. Zhao, X.; Zhou, Y.; Zong, Y.; Yang, Z.; Luo, M. A CIP-based numerical simulation of wave interaction with a fluid-filled membrane submerged breakwater. *Ocean Eng.* **2022**, *260*, 111819. [CrossRef]
11. Bautista, E.; Bahena-Jimenez, S.; Quesada-Torres, A.; Méndez, F.; Arcos, E. Interaction between long water waves and two fixed submerged breakwaters of wavy surfaces. *Wave Motion* **2022**, *112*, 102926. [CrossRef]
12. Chen, Y.-K.; Liu, Y.; Meringolo, D.D.; Hu, J.-M. Study on the hydrodynamics of a twin floating breakwater by using SPH method. *Coast. Eng.* **2023**, *179*, 104230. [CrossRef]
13. Liang, J.-M.; Chen, Y.-K.; Liu, Y.; Li, A.-J. Hydrodynamic performance of a new box-type breakwater with superstructure: Experimental study and SPH simulation. *Ocean Eng.* **2022**, *266*, 112819. [CrossRef]
14. Magdalena, I.; Karima, N.; Marcela, I.; Farid, M. Generalization of a three-layer model for wave attenuation in n-block submerged porous breakwater. *Results Eng.* **2022**, *16*, 100428. [CrossRef]
15. Aristodemo, F.; Filianoti, P. On the stability of submerged rigid breakwaters against solitary waves. *Coast. Eng.* **2022**, *177*, 104196. [CrossRef]
16. Liang, B.; Wu, G.; Liu, F.; Fan, H.; Li, H. Numerical study of wave transmission over double submerged breakwaters using non-hydrostatic wave model. *Oceanologia* **2015**, *57*, 308–317. [CrossRef]
17. Lyu, Z.; Liu, Y.; Li, H.; Mori, N. Iterative multipole solution for wave interaction with submerged partially perforated semi-circular breakwater. *Appl. Ocean Res.* **2020**, *97*, 102103. [CrossRef]
18. Liu, Z.; Wang, Y. Numerical investigations and optimizations of typical submerged box-type floating breakwaters using SPH. *Ocean Eng.* **2020**, *209*, 107475. [CrossRef]
19. Metallinos, A.S.; Klonaris, G.T.; Memos, C.D.; Dimas, A.A. Hydrodynamic conditions in a submerged porous breakwater. *Ocean Eng.* **2019**, *172*, 712–725. [CrossRef]
20. Li, X.; Zhang, W. 3D numerical simulation of wave transmission for low-crested and submerged breakwaters. *Coast. Eng.* **2019**, *152*, 103517. [CrossRef]
21. Xue, S.; Xu, Y.; Xu, G.; Wang, J.; Chen, Q. A novel tri-semicircle shaped submerged breakwater for mitigating wave loads on coastal bridges part I: Efficacy. *Ocean Eng.* **2022**, *245*, 110462. [CrossRef]
22. Sun, W.; Qu, K.; Kraatz, S.; Deng, B.; Jiang, C. Numerical investigation on performance of submerged porous breakwater to mitigate hydrodynamic loads of coastal bridge deck under solitary wave. *Ocean Eng.* **2020**, *213*, 107660. [CrossRef]
23. Patil, S.B.; Karmakar, D. Hydrodynamic analysis of floating tunnel with submerged rubble mound breakwater. *Ocean Eng.* **2022**, *264*, 112460. [CrossRef]
24. He, Y.; Ji, W.; Han, B.; Yan, L.; Han, X. Influence of submerged annular rippled breakwater on hydrodynamic performance of column floating pier. *Ocean Eng.* **2022**, *247*, 110756. [CrossRef]
25. Zhao, E.; Dong, Y.; Tang, Y.; Xia, X. Performance of submerged semi-circular breakwater under solitary wave in consideration of porous media. *Ocean Eng.* **2021**, *223*, 108573. [CrossRef]
26. Guo, Y.; Mohapatra, S.; Soares, C.G. Submerged breakwater of a flexible porous membrane with a vertical flexible porous wall over variable bottom topography. *Ocean Eng.* **2021**, *243*, 109989. [CrossRef]
27. Koley, S.; Vijay, K.; Nishad, C.; Sundaravadivelu, R. Performance of a submerged flexible membrane and a breakwater in the presence of a seawall. *Appl. Ocean Res.* **2022**, *124*, 103203. [CrossRef]

28. Koraim, A.; Heikal, E.; Zaid, A.A. Hydrodynamic characteristics of porous seawall protected by submerged breakwater. *Appl. Ocean Res.* **2014**, *46*, 1–14. [CrossRef]
29. Jensen, B.; Jacobsen, N.G.; Christensen, E.D. Investigations on the porous media equations and resistance coefficients for coastal structures. *Coast. Eng.* **2014**, *84*, 56–72. [CrossRef]
30. van Gent, M.R.A. Porous Flow through Rubble-Mound Material. *J. Waterw. Port Coast. Ocean. Eng.* **1995**, *121*, 176–181. [CrossRef]
31. Zhao, E.; Qu, K.; Mu, L. Numerical study of morphological response of the sandy bed after tsunami-like wave overtopping an impermeable seawall. *Ocean Eng.* **2019**, *186*, 106076. [CrossRef]
32. Zhao, E.; Shi, B.; Qu, K.; Dong, W.; Zhang, J. Experimental and Numerical Investigation of Local Scour around Submarine Piggyback Pipeline under Steady Current. *J. Ocean Univ. China* **2018**, *17*, 244–256. [CrossRef]
33. Zhao, E.; Dong, Y.; Tang, Y.; Cui, L. Numerical study on hydrodynamic load and vibration of pipeline exerted by submarine debris flow. *Ocean Eng.* **2021**, *239*, 109754. [CrossRef]
34. Zhao, E.; Dong, Y.; Tang, Y.; Sun, J. Numerical investigation of hydrodynamic characteristics and local scour mechanism around submarine pipelines under joint effect of solitary waves and currents. *Ocean Eng.* **2021**, *222*, 108553. [CrossRef]
35. Wu, Y.-T.; Hsiao, S.-C. Propagation of solitary waves over a submerged permeable breakwater. *Coast. Eng.* **2013**, *81*, 1–18. [CrossRef]
36. Ye, J.H.; He, K.P.; Shan, J.P. Experimental Study on Stability of Revetment Breakwater Built on Reclaimed Coral Reef Islands in South China Sea under Extreme Wave Impact. *Blasting* **2019**, *36*, 11. [CrossRef]

Journal of
Marine Science and Engineering

MDPI

Article

Model Reference Adaptive Vibration Control of an Offshore Platform Considering Marine Environment Approximation

Yun Zhang, Hui Ma *, Jianliang Xu, Hao Su and Jing Zhang

Faculty of Information Science and Engineering, Ocean University of China, Qingdao 266100, China
* Correspondence: mahui@ouc.edu.cn

Abstract: Adaptive control methods are suitable for offshore steel structures subject to harmful vibrations, as they employ reference models to adapt to coastal and nearshore physics. To decrease the dependence on the accurate characteristics of the offshore platform, a compensating measure containing the ocean environment is proposed in the adaptive control scheme. With incomplete states as the driving input, external loads are approximated using a wavelet neural network frame. Numerical experiments are conducted on a platform model with varying parameters to test the performance of the proposed adaptive controller. It is shown that the adaptive weights derived from the chosen Lyapunov function are qualified both theoretically and practically. The system-output-based adaptive controller overcomes the disadvantage of state loss. The compensated disturbance environment guarantees the reliability of the restored reference system based on mismatched physics. The designed estimator as a part of the adaptive controller compensates for the deviations of the environment between the reference and the practical, resulting in a desirable reduction in the excessive vibration.

Keywords: model reference adaptive control; environmental compensation; wavelet neural network; approximation using incomplete states

Citation: Zhang, Y.; Ma, H.; Xu, J.; Su, H.; Zhang, J. Model Reference Adaptive Vibration Control of an Offshore Platform Considering Marine Environment Approximation. *J. Mar. Sci. Eng.* **2023**, *11*, 138. https://doi.org/10.3390/jmse11010138

Academic Editors: M. Dolores Esteban, José-Santos López-Gutiérrez, Vicente Negro and M. Graça Neves

Received: 30 November 2022
Revised: 29 December 2022
Accepted: 30 December 2022
Published: 7 January 2023

1. Introduction

The exploration industry in oceans has attracted rapidly growing attention. The service life of offshore steel foundations in the marine environment is determined by many factors [1]. Both floating and fixed offshore platforms are subject to undesirable vibrations led by environmental loads, such as ocean waves and sea winds. The excessive vibrations have to be mitigated to ensure the platform runs in a safe state with a relatively long lifetime [2].

The primary active device to minimize structural vibrations is the active mass damper (AMD) [3]. Regarding the massive offshore structures and deepwater drilling systems, proper control laws have been investigated in the literature on optimal control theory [4] and robust control theory [5,6]. Note that they are essentially model-based with the reliably known physics as a prior. The onshore and underwater features of the offshore platforms are highly nonlinear and time-varying, which means the dynamics should be presented in a form with varying stiffness and damping, such as a moving target [7]. In such a case, the accurate platform model that underlies the fixed feedback control gain for AMD may not be available.

To realize the vibration control goal in inaccurate models, the adaptive control theory shows its privilege in online monitoring during the control period [8]. The model-reference adaptive control (MRAC) method has wide utilization in feedback-linearisable control-affined systems with an unknown drift vector field [9]. Nowadays, the learning methods bring about a huge enhancement of the adaptive control performance [10], involving the machine learning (ML) approach and reinforcement learning (RL) approach. The soft computing methods for adaptive supplementary tools, such as the extreme learning machine (ELM), are able to replace conventional control strategies in highly nonlinear systems, such as flexible robotic grippers [11]. The control input nonlinearities for underwater vehicles are automatically compensated by the RL method [12]. The adaptive learning control policies based on perturbations of the input signals have been proved useful for both system identification and adaptive control of stochastic systems [13]. Specifically, we focus on the works about onshore and offshore adaptive structure control methods. To control the excess aerodynamic torque of a wind turbine, the RL-based method improves the adaptability of the conventional controllers against varying winds [14]. However, little research has discussed the acquisition of a reliable model with the qualified reference environment as the baseline. For the coastal environment, limited rules or experimental data sets are available for estimating ocean waves [15], while the approximation using neural networks is merely applied to the first step of predicting the wave overtopping rate for now [16]. The aforementioned data-driven methods skip the modeling part by directly identifying the time-varying parameters online or offline, while we try at least to leverage the obtained information about the offshore platform.

The identification or approximation effects using different neural networks are compared in the adaptive learning task. The radial basis function (RBF) neural network model is capable of approximating the unknown function infinitely, with a complex motion model established [17]. The approximation models constructed by methods from the RL and RBF neural networks, Gaussian process (GP), random forest (RF) and support vector regression (SVR) have the capability to fulfill the same goal about the stochastic assessment of aerodynamics and hydrodynamics [18]. We are particularly interested in the learning model based on wavelet neural networks (WNNs), as the wavelet function and wavelet transform technique are suitable for wave monitoring [19]. Noticing that the numerical wave model in the marine environment with different sea states and hydraulic power is constructed on the interaction between waves and operation systems [20], there exists the potential possibility of wave restoration using WNNs. Several works have proposed WNN-based models for wave height forecasting [21] with reliable numerical correction in predicting [22]. On the output-feedback MRAC layout, we focus on the problem to simultaneously address the state loss problem in feedback and the ocean wave compensation problem for the reference model. Motivated by the variable replacement technique complemented with learning methods to realize offshore platform vibration control [23,24], we put forward a WNN-based wave estimator to complete the state learning goal using incomplete response data.

In this paper, we design an MRAC algorithm for the offshore platform vibration controller. The time-varying dynamic model of the offshore platform is first established. Regarding the ocean environment, the time series of the acting loads is derived through reliable spectra and then restored by a WNN-based estimator. Next, the incomplete-state-based adaptive controller is proposed in the reference model with the compensated environment features, using the control input from the above reference model and the output vibration response of the target platform. The adaptive control gains are obtained from the chosen Lyapunov function. With this restored environment, the selection of the reference model has more flexibility, as shown in several cases, while the WNN-supplemented reference system has the privilege of approaching the real working environment. The contributions are listed:

1. The adaptive control method is applied to the offshore platform containing time-varying features for the purpose of vibration attenuation;
2. The reference model with an environmental compensation scheme provides the online-adjusting adaptive control force input;
3. The data-driven approximation strategy of the ocean environment is realized through a wavelet neural network.

The remainder of the paper is organized as follows. In Section 2, we provide a precise formulation of the vibration control problem for an offshore platform. The data-driven learning period is defined for marine environment restoration. In Section 3, we introduce the way to acquire a reliable reference model with the added environmental compensating neural network. In Section 4, we perform the simulation of a real offshore platform under the designed controller. An illustrative analysis is conducted on different reference models with desired vibration-suppression results provided. In Section 5, we summarize our study and discuss the potential aspects of optimization.

2. Model Description

The physical features and motion dynamics of the offshore platform and the ocean environment are introduced in this section.

2.1. Active Mass Damper System of the Offshore Platform

The active mass damper (AMD) is the control device that suppresses structural vibration. It attaches the massive damper to the controlled structure to mitigate its harmful vibration by employing active forces properly. The steel jacket-type offshore platform with an AMD is described as a single-degree-of-freedom (SDOF) system, and the dominant layout for this SDOF system is shown in Figure 1 [24].

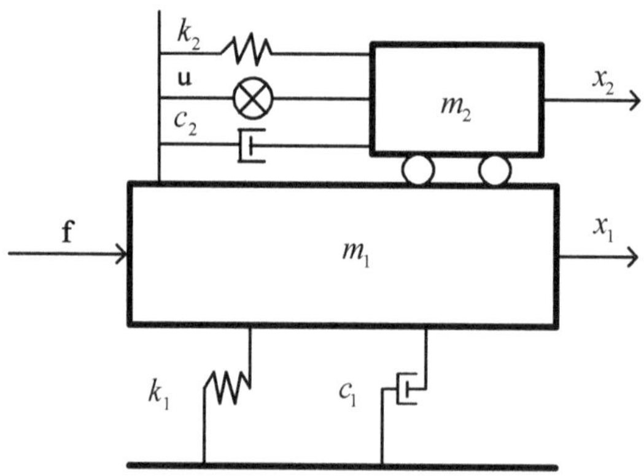

Figure 1. Simplified SDOF model of the platform with AMD on deck (adapted from [24] with permission from Elsevier, 2022).

To explain the motion, the SDOF system is defined by Newton's second law in [24,25]

$$\begin{cases} m_1(t)\ddot{x}_1(t)+c_1(t)\dot{x}_1(t)+k_1(t)x_1(t)-c_2(t)(\dot{x}_2(t)-\dot{x}_1(t))-k_2(t)(x_2(t)-x_1(t)) \\ =f(t)-u(t) \\ m_2(t)\ddot{x}_2(t)+c_2(t)(\dot{x}_2(t)-\dot{x}_1(t))+k_2(t)(x_2(t)-x_1(t))=u(t) \end{cases} \tag{1}$$

where $m_1(t)$, $m_2(t)$, $x_1(t)$, $x_2(t)$, $c_1(t)$, $c_2(t)$, $k_1(t)$ and $k_2(t)$ are the mass, displacement response, stiffness coefficients and damping coefficients of the offshore structure and AMD device, respectively; $f(t)$ is the disturbance acting on the structure, and $u(t)$ is the control force provided by the AMD device. The vibration of the platform caused by $f(t)$ should be stabilized with a properly designed $u(t)$, while perturbing $c_1(t)$, $c_2(t)$, $k_1(t)$ and $k_2(t)$ add uncertainty into the system. Using $c_1(t)$ and $k_1(t)$ as an example, the time-varying stiffness and damping are given by [24]

$$c_1(t)=2m_1(t)\xi_1(t)\iota_1(t),\ k_1(t)=m_1(t)\iota_1^2(t) \tag{2}$$

where $\xi_1(t)=\xi_1+\hat{\xi}_1\Delta\tilde{\xi}_1(t)$, ξ_1 represents the nominal value of the damping ratio; $\hat{\xi}_1$ stands for the maximum perturbation of the damping ratio; $\iota_1(t)$ represents natural frequency, $\iota_1(t)=\iota_1+\hat{\iota}_1\Delta\tilde{\iota}_1(t)$; ι_1 represents the nominal value of the damping ratio; and $\hat{\iota}_1$ stands for the maximum perturbation of the damping ratio. Here, $\xi_1(t)$ and $\iota_1(t)$ share the limitations of $\forall t\geq 0$, $|\Delta\tilde{\xi}_1(t)|\leq 1$ and $|\Delta\tilde{\iota}_1(t)|\leq 1$. Additionally, $c_2(t)$ and $k_2(t)$ share similar definitions, and we neglect the duplicates.

Inserting definition (2) into (1) yields

$$\begin{cases} \ddot{x}_1(t)+2(\zeta_1\iota_1+\xi_1\hat{\iota}_1\Delta\tilde{\iota}_1(t)+\iota_1\hat{\xi}_1\Delta\tilde{\xi}_1(t))\dot{x}_1(t)+(\iota_1^2+2\iota_1\hat{\iota}_1\Delta\tilde{\iota}_1(t))x_1(t) \\ -2m_2(t)(\xi_2\iota_2+\xi_2\hat{\iota}_2\Delta\tilde{\iota}_2(t)+\iota_2\hat{\xi}_2\Delta\tilde{\xi}_2(t))(\dot{x}_2(t)-\dot{x}_1(t))/m_1(t)+u(t)/m_1(t) \\ -m_2(t)(\iota_2^2+2\iota_2\hat{\iota}_2\Delta\tilde{\iota}_2(t))(x_2(t)-x_1(t))/m_1(t)+g_1(t)/m_1(t)-f(t)/m_1(t)=0 \\ \ddot{x}_2(t)+(\iota_2^2+2\hat{\iota}_2\iota_2\Delta\tilde{\iota}_2(t))(x_2(t)-x_1(t))-u(t)/m_2(t)+g_2(t)/m_2(t)+2(\xi_2\iota_2 \\ +\xi_2\hat{\iota}_2\Delta\tilde{\iota}_2(t)+\iota_2\hat{\xi}_2\Delta\tilde{\xi}_2(t))(\dot{x}_2(t)-\dot{x}_1(t))=0 \end{cases} \quad (3)$$

with nonlinear parts

$$\begin{cases} g_1(t)=2m_1(t)\hat{\iota}_1\hat{\xi}_1\Delta\tilde{\iota}_1(t)\Delta\tilde{\xi}_1(t)\dot{x}_1(t)-m_2(t)\hat{\iota}_2^2\Delta\tilde{\iota}_2^2(t)(x_2(t)-x_1(t)) \\ \qquad +m_1(t)\hat{\iota}_1^2\Delta\tilde{\iota}_1^2(t)x_1(t)-2m_2(t)\hat{\iota}_2\hat{\xi}_2\Delta\tilde{\iota}_2(t)\Delta\tilde{\xi}_2(t)(\dot{x}_2(t)-\dot{x}_1(t)) \\ g_2(t)=m_2(t)\hat{\iota}_2\Delta\tilde{\iota}_2^2(t)(x_2(t)-x_1(t))+2m_2(t)\hat{\iota}_2\hat{\xi}_2\Delta\tilde{\iota}_2(t)\Delta\tilde{\xi}_2(t)(\dot{x}_2(t)-\dot{x}_1(t)) \end{cases} \quad (4)$$

which are then defined as $g(t)=[g_1(t)/m_1(t)\ g_2(t)/m_2(t)]^{\mathrm{T}}$. Noting that the mass variables $m_1(t)$ and $m_2(t)$ are big enough, $g(t)$ can be deleted from the motion equation.

The SDOF system (1) has a new form as

$$\dot{x}(t)=(A(t)+M(t)F(t)N)x(t)+B(t)u(t)+D(t)f(t) \quad (5)$$

where $x(t)=[x_1(t)\ x_2(t)\ \dot{x}_1(t)\ \dot{x}_2(t)]^{\mathrm{T}}$,

$$A(t)=\begin{bmatrix} 0 & 0 & -(\iota_1^2+\iota_2^2m_2(t)/m_1(t)) & \iota_2^2 \\ 0 & 0 & \iota_2^2m_2(t)/m_1(t) & -\iota_2^2 \\ 1 & 0 & -2(\zeta_1\iota_1+\xi_2\iota_2m_2(t)/m_1(t)) & 2\xi_2\iota_2 \\ 0 & 1 & 2\xi_2\iota_2m_2(t)/m_1(t) & -2\xi_2\iota_2 \end{bmatrix}^{\mathrm{T}},$$

$$M(t)=\begin{bmatrix} 0 & 0 & \hat{\iota}_1 & 0 \\ 0 & 0 & -\hat{\iota}_2m_2(t)/m_1(t) & \hat{\iota}_2 \\ 0 & 0 & \hat{\xi}_1 & 0 \\ 0 & 0 & -\hat{\xi}_2m_2(t)/m_1(t) & \hat{\xi}_2 \end{bmatrix}^{\mathrm{T}},$$

$$N=\begin{bmatrix} -2\iota_2 & 2\iota_2 & 0 & 0 \\ 0 & -2\iota_2 & 0 & 0 \\ -2\xi_1 & 2\xi_2 & -2\iota_1 & 2\iota_2 \\ 0 & -2\xi_2 & 0 & -2\iota_2 \end{bmatrix}^{\mathrm{T}},$$

$$B(t)=\begin{bmatrix} 0 & 0 & -1/m_1(t) & 1/m_2(t) \end{bmatrix}^{\mathrm{T}},$$

$$D(t)=\begin{bmatrix} 0 & 0 & 1/m_1(t) & 0 \end{bmatrix}^{\mathrm{T}},$$

$F(t)=\mathrm{diag}\{\Delta\tilde{\iota}_1(t),\Delta\tilde{\iota}_2(t),\Delta\tilde{\xi}_1(t),\Delta\tilde{\xi}_2(t)\}$ satisfying $F^{\mathrm{T}}(t)F(t)\le I$, $x(t)\in\mathbb{R}^m$, $u(t)\in\mathbb{R}^n$, $f(t)\in\mathbb{R}^p$.

2.2. Loads Acting on the Platform

Ocean waves in conjunction with sea winds are the main excitation on the offshore platform. Wind loads usually account for around 20% of the whole external excitation, or even less in a calm wind field [24,26]. The statistical analysis of the environmental loads indicates that wind loads are usually neglected in a calm wind field. The description of waves is given by [24]

$$\begin{cases} f_j(t)=\int_0^d p_j(z,t)\varphi(z)dz \\ p_j(z,t)=C_d\rho D\sqrt{8/\pi}\sigma_{v_j}v_j(z,d,t)/2+C_m\rho\pi D^2\dot{v}_j(z,d,t)/4 \\ \qquad \triangleq \phi_{wave}(\eta(t),C_d,C_m,D,\rho) \end{cases} \quad (6)$$

where $f_j(t)$ is the wave exciting force on the platform, and d is the water depth; p_j is the wave load at the water depth z (zero at the bottom) generated from the Morison equation [24,25] via the function of water surface height $\eta(t)$, the diameter of equivalent pile cylinder D, the drag coefficient C_d, the inertia coefficient C_m and the fluid density ρ. Specifically, the velocity and acceleration of the water particle are

$$\begin{cases} v_j(z,d,t) = \eta(t)\omega_j ch(k_j z)/sh(k_j d) \\ \dot{v}_j(z,d,t) = -\eta(t) j\omega_j^2 ch(k_j z)/sh(k_j d) \end{cases} \tag{7}$$

and σ_{v_j} is the standard deviation of v_j at height z. For (7), k_j is the wave number of the jth wave component in $\omega_j^2 = g k_j \tanh(k_j d)$, g is the gravitational acceleration, $j = \sqrt{-1}$ and ω_j is the wave frequency. The wave height $\eta(t)$ can be collected from practical experiments or generated based on wave spectra.

The above descriptions of wave disturbance rely on the basic component of the wave height $\eta(t)$. With a reliable wave spectrum, such as JONSWAP [27], we can restore the essential external force time series. The disturbance item is summarized

$$f(t) = \int_0^d \phi_{wave}(\eta(t), C_d, C_m, D, \rho)\varphi(z)dz \triangleq \phi_{dis}(\eta(t)) \tag{8}$$

based on (6) and (7), where ϕ_{dis} is a description function to be identified.

2.3. Data-Dependent Approximation of the Loads

Conventional methods for model identification require the determination of proper order and the recognition of separate weight values or the total matrices. Under the data-driven concept, we skip the acquisition procedure of the order and accurate real-time values. The learning input and output have more choices of dimensions and layers as the remedy for unknown orders and parameters. The disturbance function (8) is restored by a wavelet neural network [28,29], with the output defined as (9). The input layer has p_1 nodes with an input vector x_{in}, with available candidates $x_{in}(t) = \{\eta(t), \eta(t-1), \ldots, y(t), y(t-1), \ldots, x(t), x(t-1), \ldots\}$. The hidden layer has l_1 nodes, and the output layer has n_1 nodes with an output vector y_{out}. The weights ω_{ij} between the hidden nodes and output nodes form the matrix W_{ij} and obtain the output vector of the neural network

$$y_{out} = W_{ij} h(x_{in}) \tag{9}$$

With the definition of the amplitude matrix a, transform matrix b, input weight matrix W_{jk} and output of the wavelet function f_{wav} as h, the output of the hidden layer is

$$h(x_{in}) = f_{wav}\left(\frac{W_{jk}x_{in} - b}{a}\right) \triangleq f_{wav}(\bar{x}_{in}) \tag{10}$$

We choose the Morlet wavelet function as the active function [28]

$$f_{wav}(\bar{x}_{in}) = e^{-\bar{x}_{in}^2/2} cos(1.75\bar{x}_{in}) \tag{11}$$

and its derivative is

$$f'_{wav}(\bar{x}_{in}) = -1.75 sin(1.75\bar{x}_{in}) e^{-\bar{x}_{in}^2/2} - \bar{x}_{in} cos(1.75\bar{x}_{in}) e^{-\bar{x}_{in}^2/2} \tag{12}$$

Define

$$E = \frac{1}{2}(y_{ideal} - y_{out})^{\mathrm{T}}(y_{ideal} - y_{out}) \triangleq \frac{1}{2} e^{\mathrm{T}} e \tag{13}$$

with the learning rates r_1-r_4, the updating laws for weights at the qth iteration are

$$\begin{cases} W_{ij}^{(q+1)} = W_{ij}^{(q)} + \Delta W_{ij}^{(q+1)} \\ W_{jk}^{(q+1)} = W_{jk}^{(q)} + \Delta W_{jk}^{(q+1)} \\ a^{(q+1)} = a^{(q)} + \Delta a^{(q+1)} \\ b^{(q+1)} = b^{(q)} + \Delta b^{(q+1)} \end{cases} \tag{14}$$

which are illustrated by the gradient descent method based on (13)

$$\begin{cases} \Delta W_{ij}^{(q+1)} = -r_1 \dfrac{\partial E}{\partial W_{ij}^{(q)}} \triangleq -r_1 e h(\bar{x}_{in}) \\ \Delta W_{jk}^{(q+1)} = -r_2 \dfrac{\partial E}{\partial W_{jk}^{(q)}} \triangleq -r_2 e W_{ij}^{(q)} f'_{wav}(\dfrac{W_{jk}^{(q)} \bar{x}_{in} - b^{(q)}}{a^{(q)}}) \bar{x}_{in} (a^{(q)})^{-1} \\ \Delta a^{(q+1)} = -r_3 \dfrac{\partial E}{\partial a^{(q)}} \triangleq r_3 e W_{ij}^{(q)} f'_{wav}(\dfrac{W_{jk}^{(q)} \bar{x}_{in} - b^{(q)}}{a^{(q)}})(W_{jk}^{(q)} \bar{x}_{in} - b^{(q)})(a^{(q)})^{-2} \\ \Delta b^{(q+1)} = -r_4 \dfrac{\partial E}{\partial b^{(q)}} \triangleq r_4 e W_{ij}^{(q)} f'_{wav}(\dfrac{W_{jk}^{(q)} \bar{x}_{in} - b^{(q)}}{a^{(q)}})(a^{(q)})^{-1} \end{cases} \tag{15}$$

When the neural network finishes the training on the given data set, the disturbance (8) identified by (9) after updates (14) and (15) will achieve the optimal weights W_{ij}^*, W_{jk}^*, a^* and b^*.

3. Main Results

3.1. Environmental Compensation Scheme inside the Reference System

Define an environment-compensated reference model

$$\begin{cases} \dot{x}_m(t) = A_m x_m(t) + B_m u_m(t) + D_m \hat{f}_m(t) \\ y_m(t) = C x_m(t) \end{cases} \tag{16}$$

where $x_m(t)$, $u_m(t)$, $\hat{f}_m(t)$, $y_m(t)$, A_m, B_m and D_m are the references of $x(t)$, $u(t)$, $f_m(t)$, $y(t)$, $A(t)$, $B(t)$ and $D(t)$. The reference control input can be given by

$$u_m(t) = -R_m^{-1} B_m^{\mathrm{T}} P_m x_m(t) \tag{17}$$

in which P_m satisfies

$$A_m^{\mathrm{T}} P_m + P_m A_m + Q_m = 0 \tag{18}$$

with weight matrices of the cost function satisfying $Q_m > 0$ and $R_m > 0$. The external disturbance $\hat{f}_m(t)$ is defined based on function ϕ_{dis},

$$\hat{f}_m(t) = \phi_{dis}(x_{in}(t), W_{ij}^*, W_{jk}^*, a^*, b^*) \tag{19}$$

where $\hat{f}_m(t)$ represents the item to restore the disturbance force $f(t)$, which is actually the $f_m(t)$ collected in the training set; $x_{in}(t) = \{\eta_m(t), \eta_m(t-1) \ldots, x_m(t), x_m(t-1) \ldots, y_m(t), y_m(t-1) \ldots\}$ contains the reference inputs that are chosen for different approximation models, where $\eta_m(t)$ is related to $f_m(t)$. The training of (8) approximated by (9) stops with

$$||\hat{f}_m(t) - f_m(t)|| < e_m \tag{20}$$

where e_m is a positive constant vector.

Based on the approximation model (19), the following theorem gives an adaptive compensated reference model control law with the structure assigned by Figure 2, where θ^* represents the overall optimal hyperparameters $W_{ij}^*, W_{jk}^*, a^*, b^*$ when the adaptive iteration finishes.

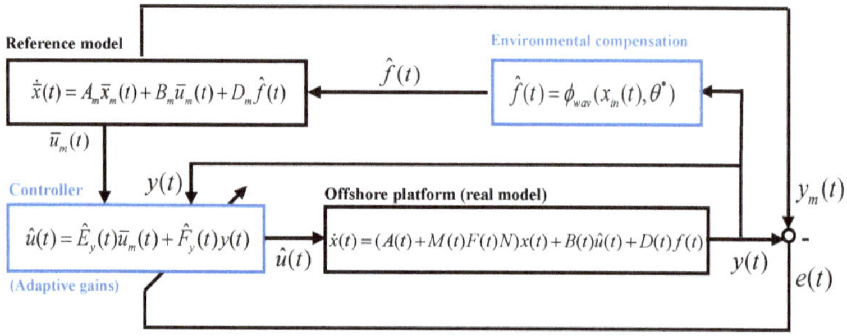

Figure 2. Model reference adaptive vibration controller with compensated environment.

3.2. Model-Based Adaptive Controller Design

Theorem 1. *For the offshore platform* (3)*, subject to random disturbance, choose a rough reference model stating*

$$\begin{cases} \dot{\bar{x}}_m(t) = A_m\bar{x}_m(t) + B_m\bar{u}_m(t) + D_m\hat{f}(t) \\ \bar{y}_m(t) = C\bar{x}_m(t) \end{cases} \tag{21}$$

where $\bar{x}_m(t)$ and $\bar{y}_m(t)$ are the reference states of $x(t)$ and $y(t)$, and $\bar{u}_m(t)$ is the reference control input

$$\bar{u}_m(t) = -(\bar{R}_m + B_m^{\mathrm{T}}\bar{P}_m B_m)^{-1} B_m^{\mathrm{T}}\bar{P}_m A_m\bar{x}_m(t) \tag{22}$$

in which $\bar{P}_m$ is derived from

$$\bar{P}_m = A_m^{\mathrm{T}}(I + B_m\bar{R}_m^{-1}B_m^{\mathrm{T}})^{-1}\bar{P}_m A_m + \bar{Q}_m \tag{23}$$

with $\bar{R}_m > 0$, $\bar{Q}_m > 0$. The environmental compensation model derived from (19) *is replaced by*

$$\hat{f}(t) = \phi_{dis}(x_{in}(t), W_{ij}^*, W_{jk}^*, a^*, b^*) \tag{24}$$

where $x_{in}(t) = \{y(t), y(t-1)\dots\}$ contains the reference state input to be chosen in different approximation models. There exists an output feedback adaptive control law

$$\hat{u}(t) = \hat{E}_y(t)\bar{u}_m(t) + \hat{F}_y(t)y(t) \tag{25}$$

which guarantees a stable system

$$\begin{cases} \dot{x}(t) = (A(t) + M(t)F(t)N)x(t) + B(t)\hat{u}(t) + D(t)f(t) \\ y(t) = Cx(t) \end{cases} \tag{26}$$

where C decides the system output $y(t)$ out of the system state $x(t)$ with adaptive weights $\hat{E}_y(t)$, and $\hat{F}_y(t)$ updating by

$$\begin{cases} \dot{\hat{E}}_y^{\mathrm{T}}(t) = \bar{P}_{yE}^{-1}\bar{u}_m(t)\bar{e}^{\mathrm{T}}(t)C\bar{P}_{ye}B_m \\ \dot{\hat{F}}_y^{\mathrm{T}}(t) = \bar{P}_{yF}^{-1}y(t)\bar{e}^{\mathrm{T}}(t)C\bar{P}_{ye}B_m \end{cases} \tag{27}$$

in which $\bar{e}(t) = \bar{y}_m(t) - y(t)$, $\bar{P}_{yF} > 0$, $\bar{P}_{yE} > 0$, $\bar{P}_{ye} > 0$, and it is the solution of

$$A_m^{\mathrm{T}}\bar{P}_{ye} + \bar{P}_{ye}A_m + \bar{Q}_e = 0 \tag{28}$$

where the quadratic weight matrix $\bar{Q}_e$ *satisfies* $\bar{Q}_e > 0$.

Proof. First, the reference control input (22) (which is another form of (17)) for (21) is derived from the quadratic cost function [23]

$$J = \bar{x}_m^T Q_m \bar{x}_m + \bar{u}_m^T R_m \bar{u}_m \tag{29}$$

It satisfies the stability conditions under the optimal theory with a maximum principle. The reference system (21), controlled by the reference control law (22), is stable.

Second, the compensated environmental disturbance requires the vibration response output of the rough reference model (16) and several features of the external disturbance in the form

$$\hat{f}_m(t) = \phi_{dis}(x_{in}(t), W_{ij}^*, W_{jk}^*, a^*, b^*) \tag{30}$$

The offline training should be performed on the historical data set with the goal (20). When the training completes, the real-time environmental loads will be approximated by the neural hyperparameters and the designated instant input. With regard to the same group of obtained neural hyperparameters, they are acquired by the training sample $x_{in}(t) = \{x_m(t), x_m(t-1)\ldots, y_m(t), y_m(t-1)\ldots\}$ and $f_m(t)$ from (16) and then applied to output $\hat{f}(t)$ using real $x_{in}(t) = \{x(t), x(t-1)\ldots, y(t), y(t-1)\ldots\}$ from (26). Note that the state $x(t)$ is not available in this situation together with $\eta(t)$ and $f(t)$. Therefore, we should use $x_{in}(t) = \{y(t), y(t-1)\ldots\}$ in both training and testing. The training period is adjusted with $x_{in}(t) = \{y_m(t), y_m(t-1)\ldots\}$. Given a non-negative constant vector e_ϵ, the approximated $\hat{f}(t)$ from (24) satisfies

$$||\hat{f}(t) - f(t)|| < e_\epsilon \tag{31}$$

which is the generalization goal of the neural network. It is achieved after numerical training [30,31].

Considering (5) and (16), the adjusting period of the adaptive weights should have the goal for the reference matrix D_m and approximated loads $\hat{f}(t)$ as

$$\lim_{t\to\infty} D_m\hat{f}(t) - D(t)f(t) \to 0 \tag{32}$$

It is noticed that real values for $D(t)$ and $f(t)$ are required for parameter correction, yet they are not given. There is no optimal value serving as the basis for approximation. As an alternative, the bound constraint of the parameters is given as follows for the proof of stability. Application with (31) and (32) added with a constant $D, D(t) \leq D$ satisfies

$$||D_m\hat{f}(t) - D(t)f(t)|| \leq ||(D_m - D)f(t) + D_m e_\epsilon|| \tag{33}$$

Define the maximum disturbance load as f_{max}, the bound constraint for (32) should be

$$||D_m - D|| * ||f(t)|| + ||D_m|| * ||e_\epsilon|| \leq ||D_m - D|| * ||f_{max}|| + ||D_m|| * ||e_\epsilon|| \tag{34}$$

Each term on the right side of the above inequality is constant, which means there is an upper bound for (34). With the proper reference matrix D_m, the adaptive law satisfies the premise (32).

Third, the error system between (3) and (26) is

$$\begin{aligned} \dot{e}(t) =& \dot{x}_m(t) - \dot{x}(t) \\ =& A_m e(t) + [A_m - A(t) - M(t)F(t)N - B(t)\hat{F}_y(t)]x(t) \\ &+ [B_m - B(t)\hat{E}_y(t)]u_m(t) \end{aligned} \tag{35}$$

To realize the ideal approximating goal (36) with a positive constant matrix e_c,

$$\lim_{t\to\infty} ||e(t)|| < e_c \tag{36}$$

the adaptive weights should converge to the optimal $\hat{E}_y^*(t)$ and $\hat{F}_y^*(t)$ with

$$\begin{cases} A_m = A(t) + M(t)F(t)N + B(t)\hat{F}_y^*(t) \\ B_m = B(t)\hat{E}_y^*(t) \end{cases} \tag{37}$$

Inserting (37) into (35) derives

$$\dot{e}(t) = A_m e(t) + B_m E_y^{*-1}(t)\tilde{F}_y(t)y(t) + B_m E_y^{*-1}(t)\tilde{E}_y(t)\bar{u}_m(t) \tag{38}$$

with the deviation between the adaptive weights and the optimal weights defined by

$$\begin{cases} \tilde{F}_y(t) = \hat{F}_y^*(t) - \hat{F}_y(t) \\ \tilde{E}_y(t) = \hat{E}_y^*(t) - \hat{E}_y(t) \end{cases} \tag{39}$$

(38) needs to satisfy the Lyapunov stability with the chosen function

$$V(t) = e^{\mathrm{T}}(t)\bar{P}_{ye}e(t) + \mathrm{tr}(\tilde{F}_y^{\mathrm{T}}(t)\bar{P}_{yF}\tilde{F}_y(t)) + \mathrm{tr}(\tilde{E}_y^{\mathrm{T}}(t)\bar{P}_{yE}\tilde{E}_y(t)) \tag{40}$$

where $\bar{P}_{ye} > 0$, $\bar{P}_{yF} > 0$, $\bar{P}_{yE} > 0$, satisfying $V(t) > 0$. Using the derivative of (40) yields

$$\begin{aligned} \dot{V}(t) =& e^{\mathrm{T}}(t)(A_m^{\mathrm{T}}\bar{P}_{ye} + \bar{P}_{ye}A_m)e(t) + 2\mathrm{tr}[\dot{\tilde{F}}_y^{\mathrm{T}}(t)\bar{P}_{yF}\tilde{F}_y(t) \\ &+ e^{\mathrm{T}}(t)\bar{P}_{ye}B_m E_y^{*-1}(t)\tilde{F}_y(t)y(t)] \\ &+ 2\mathrm{tr}[\dot{\tilde{E}}_y^{\mathrm{T}}(t)\bar{P}_{yE}\tilde{E}_y(t) + e^{\mathrm{T}}(t)\bar{P}_{ye}B_m E_y^{*-1}(t)\tilde{E}_y(t)u_m(t)] \end{aligned} \tag{41}$$

Letting the last two items on the right side of (41) be zero, we have

$$\begin{cases} \mathrm{tr}[\dot{\tilde{F}}_y^{\mathrm{T}}(t)\bar{P}_{yF}\tilde{F}_y(t) + e^{\mathrm{T}}(t)\bar{P}_{ye}B_m E_y^{*-1}(t)\tilde{F}_y(t)y(t)] = 0 \\ \mathrm{tr}[\dot{\tilde{E}}_y^{\mathrm{T}}(t)\bar{P}_{yE}\tilde{E}_y(t) + e^{\mathrm{T}}(t)\bar{P}_{ye}B_m E_y^{*-1}(t)\tilde{E}_y(t)u_m(t)] = 0 \end{cases} \tag{42}$$

Inserting (39) into (42) derives the updating rules

$$\begin{cases} \dot{\hat{F}}_y^{\mathrm{T}}(t) = -\bar{P}_{yF}^{-1}y(t)e^{\mathrm{T}}(t)\bar{P}_{ye}B_m E_y^{*-1}(t) \\ \dot{\hat{E}}_y^{\mathrm{T}}(t) = -\bar{P}_{yE}^{-1}u_m(t)e^{\mathrm{T}}(t)\bar{P}_{ye}B_m E_y^{*-1}(t) \end{cases} \tag{43}$$

Noticing that there is no optimal value $E_y^{*-1}(t)$, we set $E_y^{*-1}(t) = I$ as the substitute. Additionally, $e(t)$ is not available as we assume when using $y(t)$ (output feedback), instead of $x(t)$ (state feedback), in the control environment. In this case, we use a rough replacement in the adaptive law in

$$\begin{cases} \dot{\hat{F}}_y^{\mathrm{T}}(t) = -\bar{P}_{yF}^{-1}y(t)\bar{e}^{\mathrm{T}}(t)\bar{P}_{ye}B_m \\ \dot{\hat{E}}_y^{\mathrm{T}}(t) = -\bar{P}_{yE}^{-1}u_m(t)\bar{e}^{\mathrm{T}}(t)\bar{P}_{ye}B_m \end{cases} \tag{44}$$

and expect the adaptive rules to compensate this inaccurate setting of E_y^* and $\bar{e}(t)$. Additionally, the exchange from $e(t)$ to $\bar{e}(t)$ leads to a mismatch between the dimensions of $\bar{e}(t)$ and $\bar{P}_{ye}$. The proper weight matrix C is added in

$$\begin{cases} \dot{\hat{F}}_y^{\mathrm{T}}(t) = -\bar{P}_{yF}^{-1}y(t)\bar{e}^{\mathrm{T}}(t)C\bar{P}_{ye}B_m \\ \dot{\hat{E}}_y^{\mathrm{T}}(t) = -\bar{P}_{yE}^{-1}u_m(t)\bar{e}^{\mathrm{T}}(t)C\bar{P}_{ye}B_m \end{cases} \tag{45}$$

Inserting (39) into (45) yields the rules for adaptive weights in (27).

Finally, the non-zero part left in (41) with the quadratic weight matrix $\bar{Q}_e > 0$ satisfies (28). It is easy to obtain the qualified $\bar{P}_{ye}$, and the control performance of the rules (45) with $\bar{P}_{ye}$ replaced by $\bar{P}_m$ is given in the simulation. Therefore, (41) obtains $\dot{V}(t) < 0$, and (38) is stable. This ends the proof. □

4. Numerical Results and Discussions

We conduct several simulation experiments on the offshore platform located in Gulf of Mexico described in Table 1 [32,33]. All the numerical tests use MATLAB as the simulating tool on CPU AMD Ryzen 7 4700U. The simulation duration is several minutes, and the sampling rate is 100 hz.

Table 1. Parameters of offshore platform and sea condition.

Description	Symbol	Value	Unit
Significant wave height	H_s	5	m
Peak frequency	ω_0	0.79	rad/s
Water depth	d	218	m
Pile cylinder	D	1.83	m
Height of platform	L	249	m
First modal mass	m_1	7,825,307	kg
Nominal value of natural frequency (platform)	ι_1	2.0446	rad/s
Nominal value of damping ratio (platform)	ξ_1	2%	-
AMD device mass	m_2	78,253	kg
Nominal value of natural frequency (AMD)	ι_2	2.0074	rad/s
Nominal value of damping ratio (AMD)	ξ_2	20%	-
Drag coefficient	C_d	1.0	-
Inertia coefficient	C_m	1.5	-
Atmospheric density	$\bar{\rho}$	1.23	-
Windward resistance coefficient	C_H	1.01	-

4.1. Simulating Verification of the Offshore Platform with Active Mass Damper to Suppress Vibration

The simulation model based on (3) in the marine environment (8) is materialized with the nominal values in Table 1. The wave description (6) is established by the JONSWAP spectrum [24]. We conduct a frequency analysis on the acceleration response of the platform, which is displayed in Figure 3. The blue line represents the vibration resulting from the wave loads, and the vibration controller settings for the yellow line will be introduced in the following subsections, together with the figures displaying the total vibration responses.

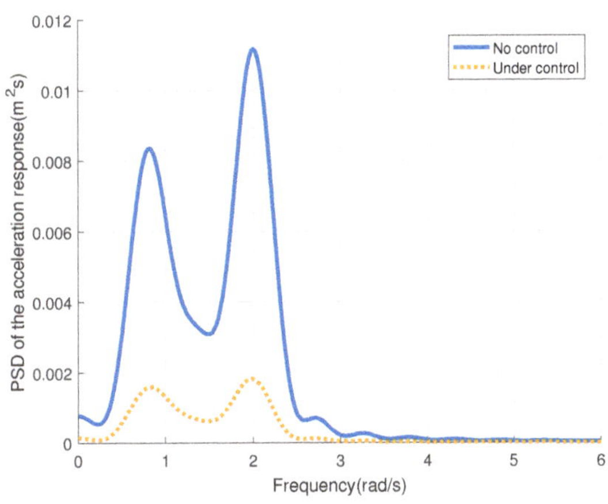

Figure 3. Power spectral density on the acceleration response of the offshore platform.

The two peaks indicate the governing vibration frequency of the platform. They are aligned with the peak frequency of the waves (0.79 rad/s) and the natural frequency of the platform (2.2 rad/s). The reduction in vibration is obvious when comparing the two lines, indicating the effectiveness of the proposed controller to be specified as follows.

4.2. Adaptive Core and Reference Input Chosen for Disturbance Approximation

In the following section, the marine environment of the reference model concerning (21), (24) and (25) is estimated by (9). Two key points of the neural network are discussed in this section: the core neural function and the input.

Firstly, we should pick up a proper core neural network for the approximator. The Morlet wavelet function (11) is used as the active function when constructing the wavelet neural network (WNN). In comparison, the most common back-propagation neural network (BPNN) is activated with the sigmoid function in the hidden layer [34]. The training data set includes the equivalent force of external loads, and the wave surface height is reconstructed at a significant height 5.0 m [24]. Considering the available data on the disturbance, the input of the neural networks in this part is designated as $\eta(t), \eta(t-1), \eta(t-2), \eta(t-3)$, and the output is $f(t)$. Their numerical values during 80s are displayed in Figure 4. The two networks have 4-7-1 nodes, 3×10^4 training samples, maximum 800 epochs and 2×10^4 testing samples. Specifically, the test set is collected when the simulated wave's significant height stays around 3.0 m, which means the fitting processes are different between the training and testing sets.

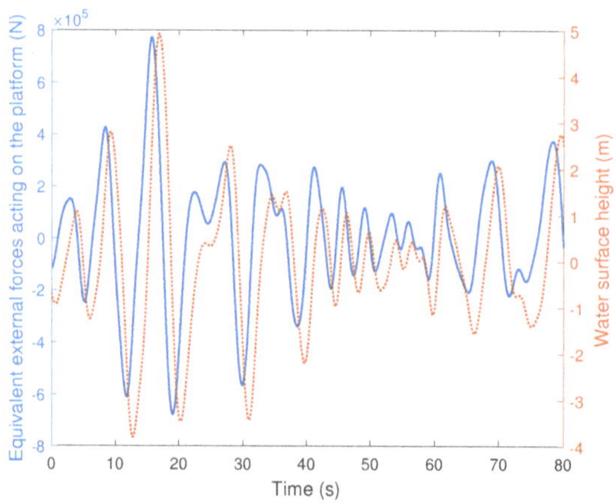

Figure 4. Ocean loads and wave height.

The proportion of the positive and negative values in the testing errors is shown in Table 2. The sum of the positive and negative proportions of each neural network should be 100% because we delete the zeros when counting the total. It is shown that the comparable BPNN has a drift in output. The testing output of BPNN is apparently biased towards the training data, which are bigger in value. The WNN keeps the same trend with the test data, with basically impartial positive and negative errors. Although the BPNN obtains smaller mean square errors in the output, as shown in Figure 5 (we display the samples between $[0.9 \times 10^4, 1.4 \times 10^4]$ for better illustration), the main object of this approximation period is to achieve a fair result during both training and testing. In that case, the WNN is more appropriate for our environmental compensation task.

Table 2. The proportion of output error from different neural networks.

	Positive Errors	Negative Errors
BPNN	80.77%	Less than 20%
WNN	61.79%	38.21%

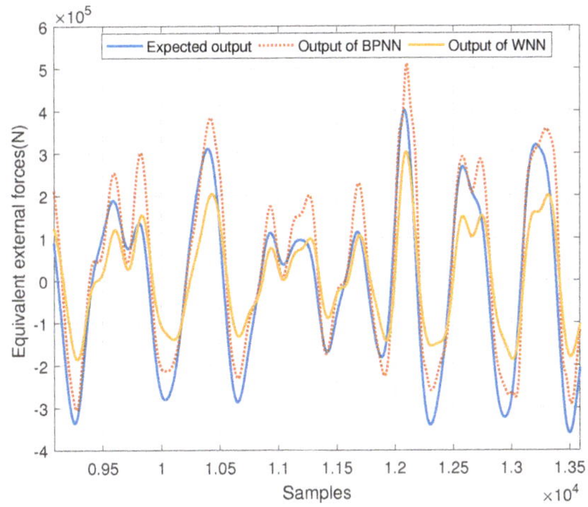

Figure 5. Comparison of the approximation models using different neural networks.

Secondly, the approximator needs suitable input data. Different types of input for the neural network are discussed in this subsection, including the system state response $x(t)$, system output response $y(t)$, wave characteristics such as wave height $\eta(t)$, or the combination of multi-class data. The most common input for identifying external wave forces should be the wave height $\eta(t)$, as mentioned above, and we display the possible input $\eta(t)$ and the defined output $f(t)$ in Figure 4. Additionally, other possible inputs are $x(t)$ and $y(t)$, and we list them in Figure 6.

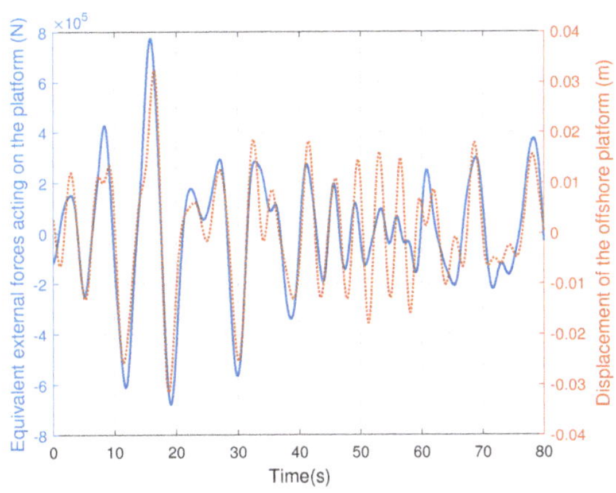

Figure 6. Ocean loads and displacement response of the platform.

We have constraints on the system (3) that no real-time disturbance data or full state data are available. Based on historical data sets including state samples and disturbance samples, we discuss the feasibility of different input. The sea surface height is theoretically more reliable as the input, as its approximation errors will not accumulate. The estimation at this time will not affect the next result, as a new $\eta(t)$ is consistently given. In comparison, when the vibration response is used as the input of the approximation model, it will change the next approximation via the feedback part of the control law $\hat{E}_y(t)\bar{u}_m(t)$. According to Figures 4 and 6, we set two different neural networks, WNN1 and WNN2. The input of WNN1 is the sea surface height $\eta(t)$, and, for WNN2, the input is the vibration response

$y(t)$. The comparison of the average error during each iteration is shown in Figure 7. It shows around 400 iterations in a possible simulation environment.

WNN1 represents the situation with qualified sensors to acquire wave features. WNN2 represents the limit that the disturbance monitoring sensors break down, and the approximation is totally dependent on the structural response of the platform itself. When the training is complete, both WNN1 and WNN2 obtain the desired estimation performance with a relatively small convergence error. Therefore, the vibration response output $y(t)$ is proved reliable in the following experiments, satisfying the constraint of the state losses ($x(t)$, $f(t)$ and $\eta(t)$) in the system.

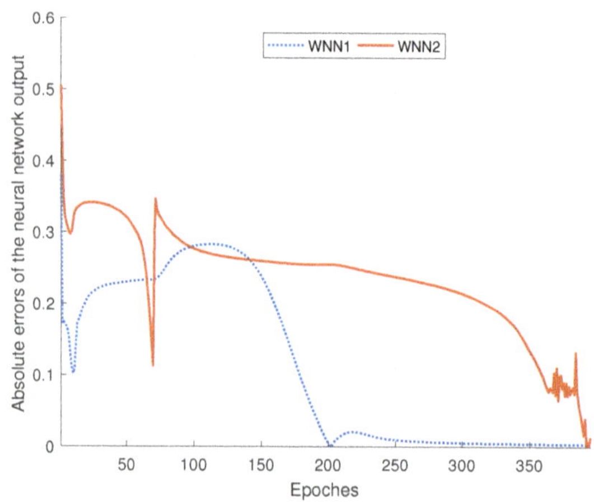

Figure 7. Comparison of the approximation models using different inputs.

4.3. Controller Test Using Different Reference Models

In this subsection, the simulation model of the offshore platform to be controlled has unknown structural parameters. The nominal physical values are shown in Table 1, while each parameter has random perturbations. The discrete nominal parameters ($t = 0$) are listed as follows.

$$\begin{cases} A(0) + M(0)F(0)N = \begin{bmatrix} 0.9998 & 0 & 0.0100 & 0 \\ 0.0002 & 0.9994 & 0 & 0.0098 \\ -0.0423 & 0.0004 & 0.9988 & 0.0002 \\ 0.0400 & -0.0441 & 0.0083 & 0.9905 \end{bmatrix} \\ B(0) = 10^{-6} \times [0, 0.0006, -0.0013, 0.1272]^{\mathrm{T}} \\ D(0) = 10^{-8} \times [0.0006, 0, 0.1277, 0.0005]^{\mathrm{T}} \end{cases} \quad (46)$$

Case 1. Control Performance with No. 1 Reference Model

In this case, (46) has 10% perturbations as $\{m_1(t) \pm 10\%$, $\iota_1(t) \pm 10\%$, $\xi_1(t) \pm 10\%$, $m_2(t) \pm 10\%$, $\iota_2(t) \pm 10\%$ and $\xi_2(t) \pm 10\%\}$. Theoretically, any other platform model can be used as a reference model under the same settings of the governing mode. In this case, we use the features from another platform as the reference model [23]. Obviously, the platform height, water depth, mass and damping characteristics of the reference and the target platform models are different. The reference model with fixed nominal values is described by the following matrices in a discrete form.

$$\begin{cases} A_m = \begin{bmatrix} 0.9998 & 0 & 0.0100 & 0 \\ 0.0003 & 0.9997 & 0 & 0.0100 \\ -0.0486 & 0.0003 & 0.9980 & 0 \\ 0.0541 & -0.0542 & 0.0046 & 0.9954 \end{bmatrix} \\ B_m = 10^{-6} \times [0, 0.0042, -0.0042, 0.8416]^{\mathrm{T}} \\ D_m = 10^{-8} \times [0.0021, 0, 0.4215, 0.0017]^{\mathrm{T}} \end{cases} \tag{47}$$

The control test is conducted in an ocean environment when the significant wave height stays at 5.0 m, with a comparable model reference adaptive controller (MRAC). We adopt the concepts [35,36] and assume that all of the information for the reference model is known. It has the adaptive law (25), in which the reference control input $\bar{u}_m(t)$ is replaced by $u_m(t)$ in (17) on the basis of (16). The disturbance item $\hat{f}(t)$ is replaced by the real value item $f(t)$, which means the limit for detection is quite strict. The wavelet neural network-based adaptive controller (WNNAC) is obtained via Theorem 1, together with an environmental compensation strategy (24), to replace (8), as shown in (21). In (29), we set $Q_m = 10^5 \times \mathrm{diag}\{1, 0, 1, 0\}$, $R_m = 10^{-4}$ for deriving (22). The other positive matrices are initially random. In short, the MRAC ideally knows all of the needed states and parameters, while the WNNAC needs approximations of certain information. The analysis of the responses is listed in Table 3 and Figures 8–10, where S_u, S_x, S_a and M_u, M_x, M_a are the standard and maximum deviations, respectively.

Table 3. Numerical analysis of the vibration response of the offshore platform.

	$S_u(10^4$ N)	S_x(m)	S_a(m/s^2)	$M_u(10^4$ N)	M_x(m)	M_a(m/s^2)
No control	-	0.0436	0.1317	-	0.1182	0.3818
MRAC	14.785	0.0571	0.2365	71.996	0.2450	0.9797
MRAC(1-300s)	5.8586	0.0237	0.0883	21.388	0.0891	0.3379
WNNAC	**1.5257**	**0.0151**	**0.0461**	**7.5821**	**0.0498**	**0.1546**

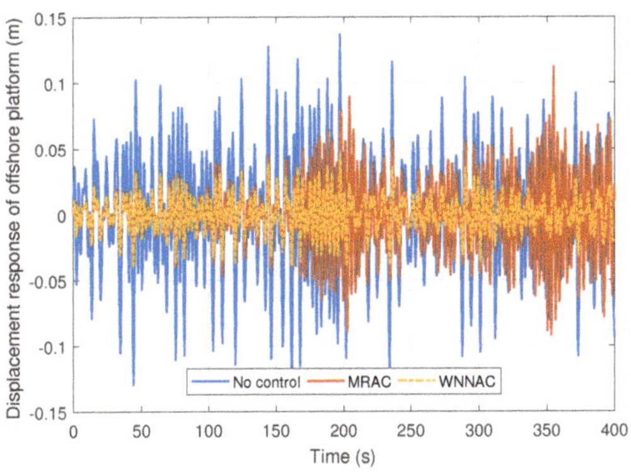

Figure 8. Displacement response of the offshore platform structure.

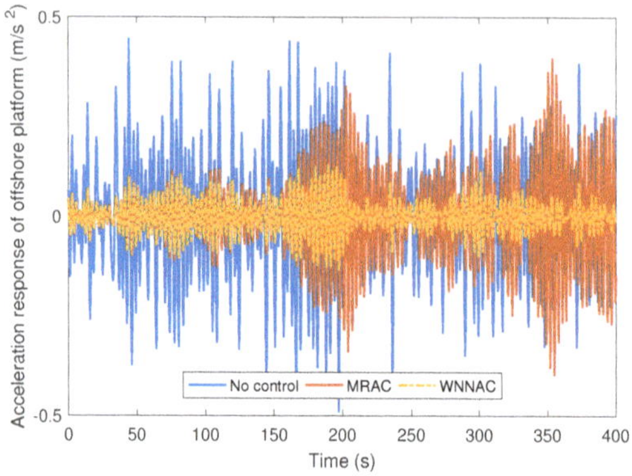

Figure 9. Acceleration response of the offshore platform structure.

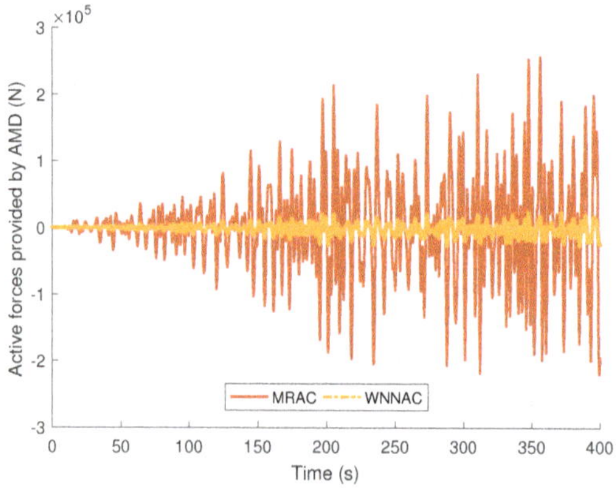

Figure 10. Control force to suppress the vibration.

Compared with the vibration response under uncontrolled conditions, the WNNAC can reduce the maximum displacement to 42.13%, the average deviation to 34.63%, the maximum acceleration deviation to 40.49% and the average deviation to 35.00%. The maximum control force provided by the WNNAC is 10.83% of that provided by the MRAC, and the average control force is 10.32%. In Figures 8 and 9, the MRAC fails in a small range of around 350 s. For the first 300 s, the WNNAC has a 64.55% reduction in maximum control compared to the MRAC with a mean reduction of 73.96%. We only show 400 s in this simulation, and the total experiment lasts 16 min, with both controllers staying stable.

Case 2. Control Performance with Different Reference Models

In order not to lose the arbitrariness, we also test the WNNAC and MRAC when the reference model has random changes based on Case 1. For example, the element in the third row and first column of A_m is 10 times the original value. B_m has random changes at each element. The reference matrices are replaced by

$$\begin{cases} A_m = \begin{bmatrix} 0.9976 & 0 & 0.0100 & 0 \\ 0.0030 & 0.9997 & 0 & 0.0100 \\ -0.4859 & 0.0003 & 0.9958 & 0 \\ 0.0531 & -0.0542 & 0.0046 & 0.9954 \end{bmatrix} \\ B_m = 10^{-6} \times [0, 0.0006, -0.0013, 0.1272]^{\mathrm{T}} \end{cases} \quad (48)$$

D_m and the other settings are the same as in (47).

The vibration responses compared with the MRAC are displayed in Figures 11–13. We can see that the WNNAC needs fewer control forces to achieve a control effect equivalent to that of the MRAC, and more details are displayed in Figure 12. The MRAC and WNNAC have basically the same reduction in the vibration responses, while the WNNAC has much smaller and more stable control forces in Figure 13.

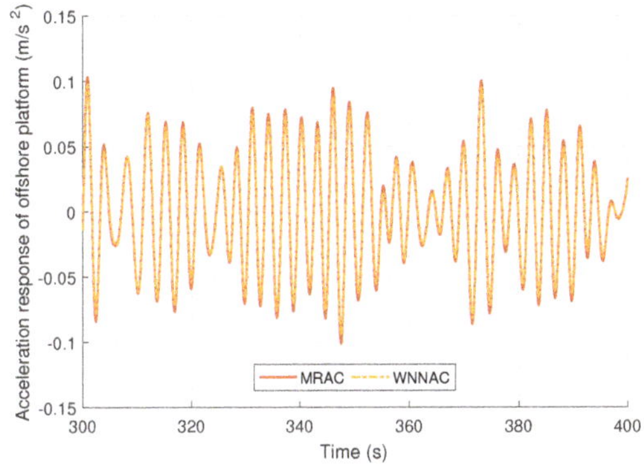

Figure 11. Acceleration response of the offshore platform structure using another reference model.

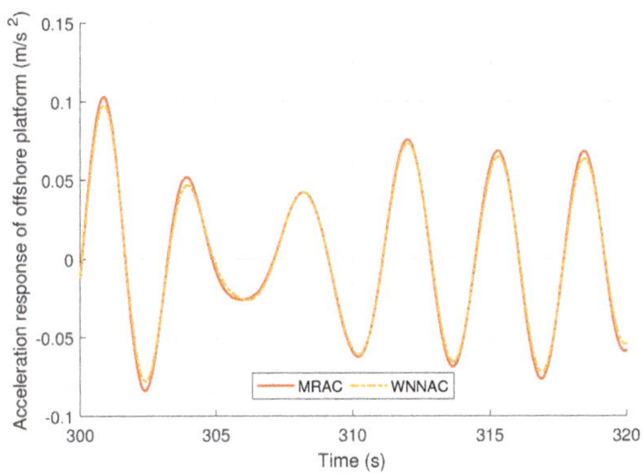

Figure 12. Acceleration response of the offshore platform structure (details).

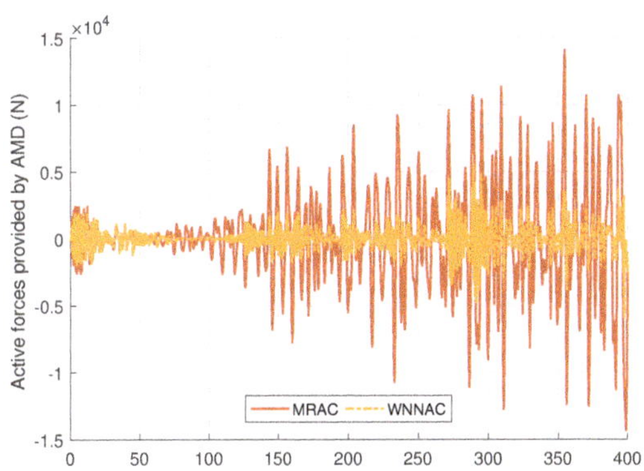

Figure 13. Control force to suppress the vibration using another reference model.

Additionally, we test the WNNAC with the nominal values of $m_1(t), \iota_1(t), \xi_1(t), m_2(t)$, $\iota_2(t)$ and $\xi_2(t)$ in the reference model and they have 0-60% random changes. The acceleration responses are shown in Figure 14 and details in Figure 15. The maximum perturbations of the reference model for each of the WNNACs are represented as 0 (blue line), 30%(red line) and 60%(yellow line). For example, the 60% perturbations are calculated by $\{m_1 \pm 60\%, \iota_1 \pm 60\%, \xi_1 \pm 60\%, m_2 \pm 60\%, \iota_2 \pm 60\%, \xi_2 \pm 60\%\}$ in (47). The purple dashed line and the green dashed line define the perturbed scope of the acceleration response. Obviously, the WNNAC constrains the vibration response to a limited extent, which means changes in the chosen reference model are acceptable.

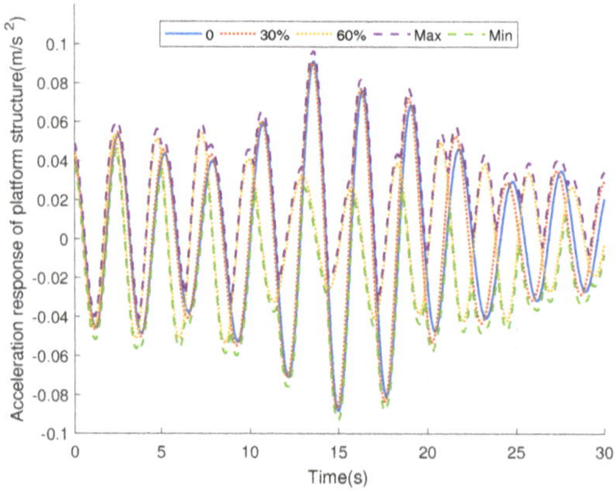

Figure 14. Control force to suppress the vibration with different perturbations in platform parameters.

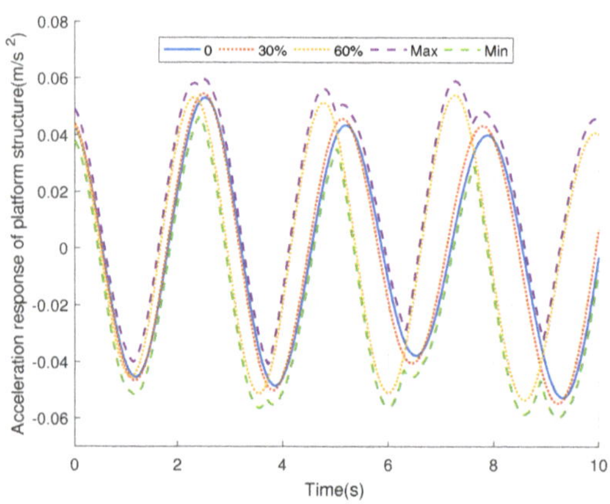

Figure 15. Control force to suppress the vibration with different perturbations in platform parameters (enlarged view of Figure 13).

5. Conclusions

The adaptive control problem for offshore platforms subject to ocean loads is discussed in this paper. The simulation tests on the numerical model of the given platform prove that the proposed environmental compensating measures reduce more than half of the control forces compared to non-adaptive controllers. Moreover, the compensated reference system is able to be transferred among different perturbing systems. The potential aspect to optimize is in the rules to follow when selecting a proper reference model, including the proper sequences of different variables. We believe that there are many other reference models that can derive a stable adaptive controller for the target platform, and to what extent our environmental approximation measure maintains effectiveness could be investigated.

Author Contributions: Conceptualization, methodology, software, validation, writing—original draft preparation, Y.Z.; resources, data curation, writing—review and editing, H.M.; visualization, supervision, project administration, J.X.; formal analysis, funding acquisition, H.S.; investigation, J.Z. All authors have read and agreed to the published version of the manuscript.

Funding: This research was funded by the Natural Science Foundation of Shandong Province (Grants No. ZR2021MD054, ZR2017QF017) and the National Science Foundation of China (Grants No. 41276085).

Institutional Review Board Statement: Not applicable.

Informed Consent Statement: Not applicable.

Data Availability Statement: Not applicable.

Acknowledgments: This research was supported by the Ocean University of China.

Conflicts of Interest: The authors declare no conflict of interest.

References

1. Erdogan, C.; Swain, G. The effects of biofouling and corrosion products on impressed current cathodic protection system design for offshore monopile foundations. *J. Mar. Sci. Eng.* **2022**, *10*, 1670. [CrossRef]
2. Hemmati, A.; Oterkus, E. Semi-active structural control of offshore wind turbines considering damage development. *J. Mar. Sci. Eng.* **2018**, *6*, 102. [CrossRef]
3. Yang, D.H.; Shin, J.H.; Lee, H.; Kim, S.K.; Kwak, M.K. Active vibration control of structure by Active Mass Damper and Multi-Modal Negative Acceleration Feedback control algorithm. *J. Sound Vib.* **2017**, *392*, 18–30. [CrossRef]
4. Zhao, Y.D.; Sun, Y.T.; Zhang, B.L.; Han, Q.L.; Zhang, X.M. Recoil control of deepwater drilling riser systems via optimal control with feedforward mechanisms. *Ocean. Eng.* **2022**, *257*, 111690. [CrossRef]
5. Zhang, W.; Zhang, B.L.; Han, Q.L.; Pang, F.B.; Sun, Y.T.; Zhang, X.M. Recoil attenuation for deepwater drilling riser systems via delayed H_∞ control. *ISA Trans.* 2022, *in press*. [CrossRef] [PubMed]
6. Zhao, Y.D.; Sun, Y.T.; Zhang, B.L.; Zhang, D. Delay-feedback-based recoil control for deepwater drilling riser systems. *Int. J. Syst. Sci.* **2022**, *53*, 2535–2548. [CrossRef]
7. Chen, Q.; Hu, Y.; Zhang, Q.; Jiang, J.; Chi, M.; Zhu, Y. Dynamic damping-based terminal sliding mode event-triggered fault-tolerant pre-compensation stochastic control for tracked ROV. *J. Mar. Sci. Eng.* **2022**, *10*, 1228. [CrossRef]
8. Wu, D.; Chen, H.; Huang, Y.; Chen, S. Online monitoring and model-free adaptive control of weld penetration in vppaw based on extreme learning machine. *IEEE Trans. Ind. Inform.* **2019**, *15*, 2732–2740. [CrossRef]
9. Calliess, J.P.; Roberts, S.J.; Rasmussen, C.E.; Maciejowski, J. Lazily Adapted Constant Kinky Inference for nonparametric regression and model-reference adaptive control. *Automatica* **2020**, *122*, 109216. [CrossRef]
10. Gaudi, J.E.; Gibson, T.E.; Annaswamy, A.M.; Bolender, M.A.; Lavretsky, E. Connections between adaptive control and optimization in machine learning. In Proceedings of the IEEE 58th Conference on Decision and Control (CDC), Nice, France, 11–13 December 2019; pp. 4563–4568.
11. Petković, D.; Danesh, A.S.; Dadkhah, M.; Misaghian, N.; Shamshirband, S.; Zalnezhad, E.; Pavlović, N.D. Adaptive control algorithm of flexible robotic gripper by extreme learning machine. *Robot. Comput.-Integr. Manuf.* **2016**, *37*, 170–178. [CrossRef]
12. Cui, R.; Yang, C.; Li, Y.; Sharma, S. Adaptive control algorithm of flexible robotic gripper by extreme learning machine. *IEEE Trans. Syst. Man Cybern. Syst.* **2017**, *47*, 1019–1029. [CrossRef]
13. Faradonbeh, M.K.S.; Tewari, A.; Michailidis, G. Input perturbations for adaptive control and learning. *Automatica* **2020**, *117*, 108950. [CrossRef]
14. TOMIN, N.; KURBATSKY, V.; GULIYEV, H. Intelligent control of a wind turbine based on reinforcement learning. In Proceedings of the 2019 16th Conference on Electrical Machines, Drives and Power Systems (ELMA), Varna, Bulgaria, 6–8 June 2019; pp. 1–6.
15. Frades, J.L.; Negro, V.; Barba, J.G.; Martín-Antón, M.; López-Gutiérrez, J.S.; Esteban, M.D.; Blasco, L.J.M. Preliminary design for wave run-up in offshore wind farms: Comparison between theoretical models and physical model tests. *Energies* **2019**, *12*, 492.
16. Oliver, J.; Esteban, M.; López-Gutiérrez, J.S.; Negro, V.; Neves, M. Optimizing wave overtopping energy converters by ANN modelling: Evaluating the overtopping rate forecasting as the first step. *Sustainability* **2021**, *13*, 1483. [CrossRef]
17. Liu, A.; Zhao, H.; Song, T.; Liu, Z.; Wang, H.; Sun, D. Adaptive control of manipulator based on neural network. *Neural Comput. Appl.* **2021**, *33*, 4077–4085. [CrossRef]
18. Richmond, M.; Sobey, A.; Pandit, R.; Kolios, A. Stochastic assessment of aerodynamics within offshore wind farms based on machine-learning. *Renew. Energy* **2020**, *161*, 650–661. [CrossRef]
19. Wu, J.; Xu, X.; Liu, C.; Deng, C.; Shao, X. Lamb wave-based damage detection of composite structures using deep convolutional neural network and continuous wavelet transform. *Compos. Struct.* **2021**, *276*, 114590. [CrossRef]
20. Brito, M.; Bernardo, F.; Neves, M.G.; Neves, D.R.C.B.; Crespo, A.J.C.; Domínguez, J.M. Numerical model of constrained wave energy hyperbaric converter under full-scale sea wave conditions. *J. Mar. Sci. Eng.* **2022**, *10*, 1489. [CrossRef]
21. Oh, J.; Suh, K.D. Real-time forecasting of wave heights using EOF-wavelet-neural network hybrid model. *Ocean Eng.* **2018**, *150*, 48–59. [CrossRef]

22. Deshmukh, A.N.; Deo, M.C.; Bhaskaran, P.K.; Nair, T.M.B.; Sandhya, K.G. Neural-network-based data assimilation to improve numerical ocean wave forecast. *IEEE J. Ocean. Eng.* **2016**, *41*, 944–953. [CrossRef]
23. Zhang, Y.; Ma, H.; Xu, J. Neural network-based fuzzy vibration controller for offshore platform with random time delay. *Ocean Eng.* **2021**, *225*, 108733. [CrossRef]
24. Ma, H.; Zhang, Y.; Wang, S.Q.; Xu, J.; Su, H. Rolling-optimized model predictive vibration controller for offshore platforms subjected to random waves and winds under uncertain sensing delay. *Ocean Eng.* **2022**, *252*, 111054. [CrossRef]
25. Chen, W.; Du, X.; Zhang, B.L.; Cai, Z.; Zheng, Z. Near-Optimal control for offshore structures with nonlinear energy sink mechanisms. *J. Mar. Sci. Eng.* **2022**, *10*, 817. [CrossRef]
26. Jahangiri, V.; Sun, C. Three-dimensional vibration control of offshore floating wind turbines using multiple tuned mass dampers. *Ocean Eng.* **2020**, *206*, 107196. [CrossRef]
27. Li, H.J.; Hu, S.L.J.; Jakubiak, C. H_2 active vibration control for offshore platform subjected to wave loading. *J. Sound Vib.* **2003**, *263*, 709–724. [CrossRef]
28. Sabir, Z.; Raja, M.A.Z.; Mahmoud, S.R.; Balubaid, M.; Algarni, A.; Alghtani, A.H.; Aly, A.A.; Le, D.N. A novel design of Morlet wavelet to solve the dynamics of nervous stomach nonlinear model. *Int. J. Comput. Intell. Syst.* **2022**, *15*, 1–15. [CrossRef]
29. Yang, C.; Yang, R.; Xu, T.; Li, Y. Computational model of enterprise cooperative technology innovation risk based on nerve network. *J. Algorithms Comput. Technol.* **2018**, *12*, 177–184. [CrossRef]
30. Chamon, L.; Ribeiro, A. Probably approximately correct constrained learning. *Adv. Neural Inf. Process. Syst.* **2020**, *33*, 16722–16735.
31. Fang, Z.; Lu, J.; Liu, F.; Zhang, G. Semi-supervised heterogeneous domain adaptation: Theory and algorithms. *IEEE Trans. Pattern Anal. Mach. Intell.* **2022**, *45*, 1087–1105. [CrossRef]
32. Ma, H.; Hu, W.; Tang, G.Y. Networked predictive vibration control for offshore platforms with random time delays, packet dropouts and disordering. *J. Sound Vib.* **2019**, *441*, 187–203. [CrossRef]
33. Ma, H.; Tang, G.Y.; Ding, X.Q. Modified-transformation-based networked controller for offshore platforms under multiple outloads. *Ocean Eng.* **2019**, *190*, 1–11. [CrossRef]
34. Chen, L.; Jagota, V.; Kumar, A. Research on optimization of scientific research performance management based on BP neural network. *Int. J. Syst. Assur. Eng. Manag.* **2021**. [CrossRef]
35. Su, G.; Wang, P.; Guo, Y.; Cheng, G.; Wang, S.; Zhao, D. Multiparameter identification of permanent magnet synchronous motor based on model reference adaptive system—Simulated annealing particle swarm optimization algorithm. *Electronics* **2022**, *11*, 159. [CrossRef]
36. Aljuboury, A.S.; Hameed, A.H.; Ajel, A.R.; Humaidi, A.J.; Alkhayyat, A.; Mhdawi, A.K.A. Robust adaptive control of knee exoskeleton-assistant system based on nonlinear disturbance observer. *Actuators* **2022**, *11*, 78. [CrossRef]

Journal of
Marine Science and Engineering

MDPI

Article

The Effects of Biofouling and Corrosion Products on Impressed Current Cathodic Protection System Design for Offshore Monopile Foundations

Caglar Erdogan and **Geoffrey Swain** *

Center for Corrosion and Biofouling Control, Florida Institute of Technology, Melbourne, FL 32901, USA
* Correspondence: swain@fit.edu

Abstract: The robustness of the cathodic protection systems utilized for offshore wind monopile foundations depends on the surface condition of the steel as well as the environmental conditions. This study investigated how preexisting biofouling and corrosion products on vertical uncoated steel surfaces extending from the intertidal zone to the buried zone affected the cathodic protection requirements when impressed current cathodic protection (ICCP) was applied under tidal conditions. The comparative results between initially clean and previously fouled and corroded panel sets showed that the fouling and corrosion products increased both the initial and mean current densities. They also altered the composition, slowed the formation, and reduced the protective properties of cathodic chalks during nine weeks of deployment in seawater at Port Canaveral, Florida.

Keywords: steel foundations; offshore wind; cathodic protection; electrodeposited films; rust; biofouling

Citation: Erdogan, C.; Swain, G. The Effects of Biofouling and Corrosion Products on Impressed Current Cathodic Protection System Design for Offshore Monopile Foundations. *J. Mar. Sci. Eng.* **2022**, *10*, 1670. https://doi.org/10.3390/jmse10111670

Academic Editor: Erkan Oterkus

Received: 3 October 2022
Accepted: 19 October 2022
Published: 5 November 2022

1. Introduction

Corrosion is one of the major factors that determine the service life of offshore wind foundations in the marine environment. Monopiles, large-diameter cylinder steel structures, are the preferred foundations in shallow water depths up to about 40 m. These dominate the existing structures in most lease areas due to costs and simplicity of design and fabrication [1,2]. Corrosion protection methods utilized for monopile foundations include corrosion allowance, coatings, and cathodic protection systems [3]. Coatings are optional for external and internal submerged surfaces [4], and these structures may not receive any cathodic protection for up to 2 years until the transition piece is installed where the sacrificial anodes are located [5]. This practice allows corrosion products to form and fouling organisms to become established on the surfaces before cathodic protection is applied. The ISO 24656:2022—Cathodic Protection of Offshore Wind Structures is the only standard that recommends immediate cathodic protection of offshore monopile foundations. Therefore, this study investigated how fouled and corroded vertical steel surfaces at different zones, including intertidal, submerged, and buried zones alter the cathodic protection design current densities under tidal cycles. The knowledge gained through this experiment may be utilized to advance the robustness and effectiveness of cathodic protection in the marine environment.

The fouling communities in marine environments are comprised of both micro- and macro-organisms. These communities may alter the corrosion dynamics at both the cathodes and anodes. The fouling established on steel surfaces may act as a barrier by limiting oxygen transfer to steel surfaces [6]. This causes the generation of both micro- and macro-galvanic cells due to the heterogeneous distribution of oxygen [6–9] and alters the conditions in the vicinity by establishing a source or a sink for chemical species [10–14]. The fouling causing localized and pitting corrosion [9,15–17] may result in local stress concentrations and reduce the fatigue life of offshore monopile foundations [4,18,19].

The accumulation of fouling on inactive anodes, for example, when they are incorporated with coating systems for corrosion prevention may alter their corrosion prevention properties. Swain et al. [20] reported that the increase in resistance and the reduction in the current output of idle Al-Zn-Hg sacrificial anodes was higher compared to working anodes due to marine growth. Rousseau et al. [21] reported that the fouling settlement reduced the effectiveness of a zinc anode utilized to keep the polarization potentials of a carbon steel structure at −1.0 V (Ag/AgCl). Hence, the anode was cleaned periodically to increase the anode current output. However, Blackwood et al. [22] reported that both Al and Zn anodes utilized to protect 316L SS performed similarly under heavy and light fouling conditions.

The biofouling community and the rust layer established on freely corroding steel surfaces cause ennoblement and influence corrosion rates [8,23–27]. The relationship between the corrosion of steel, cathodic protection and biofouling is shown in Figure 1. Guezennec et al. [11] concluded that the two eubacteria increased the cathodic current demands of X52 carbon steel samples as a result of the modification of calcium and magnesium ion balances on the surface. Eashwar et al. [28] investigated the effect of fouling on CP for mild steel, 304 stainless steel and 3004 aluminum alloy samples. They reported that the macro-fouling on steel surfaces polarized to −1.07 V (SCE) caused approximately 300 mV ennoblement and depolarized to −0.77 V and highlighted that the change in potential due to fouling could affect the CP system design and could cause underprotection. Dexter et al. [10] reported that the UNS G10180 carbon steel samples with biofilms required significantly higher cathodic corrosion currents (4.3–5.0 mA) compared to the clean samples (0.84–0.89 mA). Chen et al. [29] investigated the change in cathodic current densities when X70 carbon steel samples were immersed in a solution with and without biofilms. They concluded that the presence of biofilms decreased the potential of X70 steel and increased the current densities. Permeh et al. [30] concluded that the increased marine fouling on carbon steel surfaces increased the current densities required for cathodic polarization. Liduino et al. [31] reported an increase in cathodic current densities of AISI 1020 steel after samples were immersed in seawater for 28 days. However, the studies did not consider the effects of different corrosion zones and the effect of tides on current densities when fouling and corrosion products are present on steel surfaces.

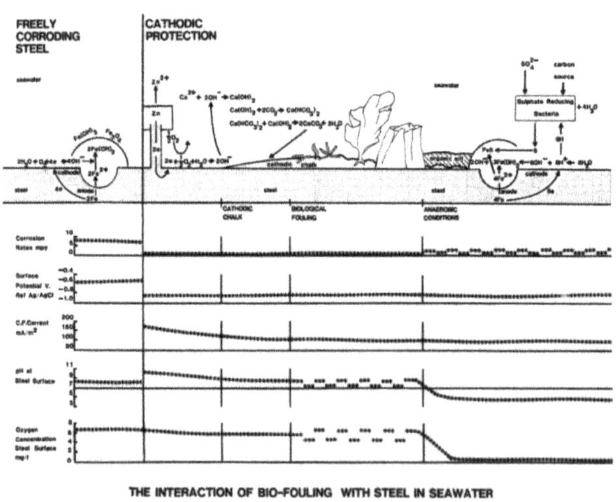

Figure 1. Interaction of biofouling with steel in seawater [32].

This study investigated the influence of biofouling and corrosion products on the cathodic current demands of previously fouled and corroded panels in seawater at Port Canaveral, FL. The steel panels were placed at the intertidal, submerged, and buried zones in seawater, thus the influence of semidiurnal tides was also recorded. The continuity between each set of panels was provided by electrical connections. The polarization potentials, current flow between panels, and total current output from each ICCP system were collected and the fouling progression was monitored. The results were compared against a clean set of panels that had not been subjected to corrosion or biofouling.

2. Materials and Methods

2.1. Test Site and Environmental Conditions

The Center's seawater test site is located at Port Canaveral, FL (28°24′30.93″ N, 80°37′39.31″ W). The data collection was performed for nine weeks from 7 October 2021 to 7 December 2021. The vertical layout of the panels was designed according to water depths and the semidiurnal tides. The predicted water depths at low and high tides were 3.70 m and 5.30 m, respectively (Figure 2) [33]. Temperature and salinity were recorded using YSI 30 conductivity meter and pH was recorded with an Onset HOBO MX2501 datalogger (Figure 3). The major ions in seawater at the test site are shown in Table 1.

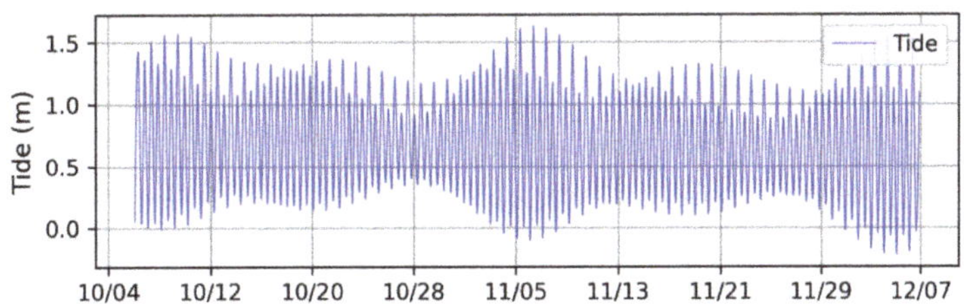

Figure 2. Tide cycle during the experiment.

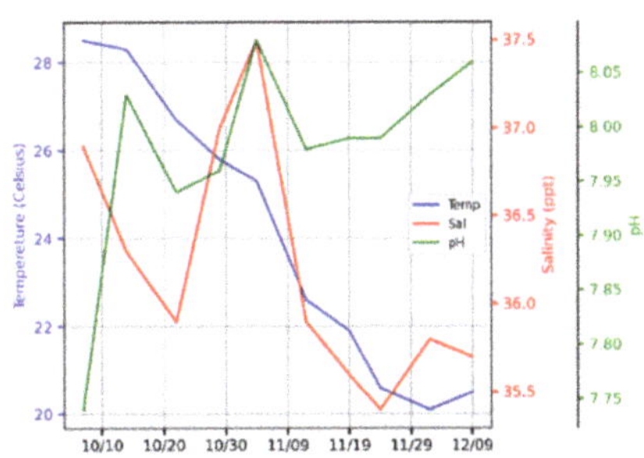

Figure 3. Temperature, salinity, and pH.

Table 1. Major ion concentrations in seawater during the experiment.

	mg/L in Seawater
Chloride [Cl^-]	19.91 ± 0.37
Sodium [Na^+]	11.06 ± 0.20
Sulfate [SO_4^{2-}]	2.79 ± 0.05
Magnesium [Mg^{2+}]	1.33 ± 0.02
Calcium [Ca^{2+}]	0.42 ± 0.01
Potassium [K^+]	0.40 ± 0.01

2.2. Materials and Experimental Setup

Two sets of A36 low carbon steel panels were used (Table 2). Each set consisted of five 300 mm by 150 mm panels with 3 mm thickness. Both sides of the steel panels were white metal sandblasted according to SSPC-SP 5 (NACE No.1) by removing corrosion products, oxide layers, grease, and oil. The surface roughness of the panels was 25.3 ± 2.2 μm (Mahr, Marsurf PS10, USA). The backs of the panels were coated with one layer of epoxy, a tie coat and a topcoat of Hempel, Hempaguard® X7, Denmark fouling control system to define the area exposed to corrosion and eliminate the weight increase due to biofouling. Two PVC frames were designed and built to allow one panel to be exposed to constant wetting and drying cycles while having a 110 cm distance between panels. Each set of panels was attached to a fixed PVC frame to have one panel (panel 1) in the intertidal zone and three panels (panels 2,3 and 4) in the submerged zone. The last panel (panel 5) was semi buried in the sediment (Figure 4). One set of panels was deployed in seawater at the test site for nine weeks prior to this experiment to allow for corrosion and the recruitment of biofouling (Figure 5).

Table 2. Composition of steel panels.

	wt%
Carbon [C]	0.25
Copper [Cu]	0.2
Iron [Fe]	rem
Manganese [Mn]	1.03
Phosphorus [P]	0.04
Silicon [Si]	0.28
Sulfur [S]	0.05

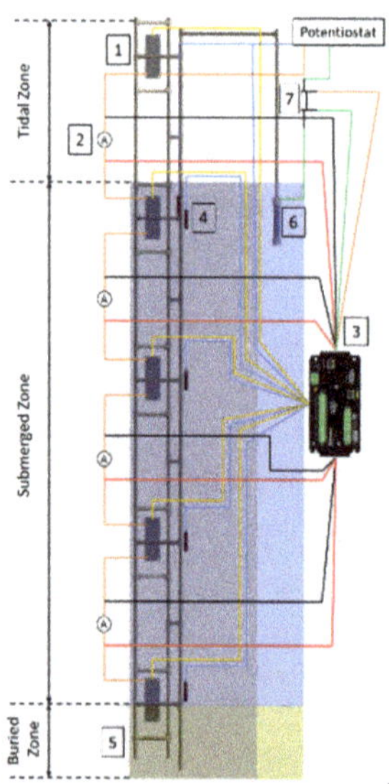

Figure 4. Experimental setup: (1) Steel Panel, (2) ZRA, (3) Datalogger, (4) Ag/AgCl Reference Electrode, (5) PVC Frame, (6) MMO Anode, (7) 1-Ohm Shunt.

On each frame, the electrical connection between panels was provided by 12 AWG wires 30 ft long. The current flow between panels was measured by using the channels of an Electrosynthesis, Model 440, USA multichannel potentiostat as zero resistance ammeters (ZRAs). Silver/silver chloride seawater reference electrodes (Ag/AgCl) were used to measure the corrosion potentials of individual panels. The reference cells were placed in perforated PVC housings which were coated with antifouling coating systems and filled with fiberglass wool to prevent influences of fouling during the experiment.

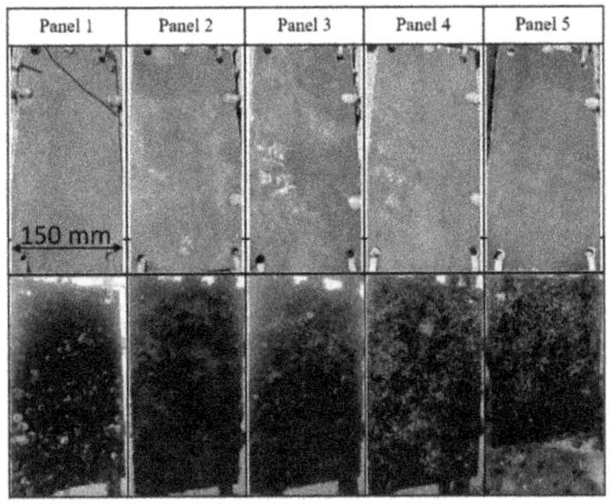

Figure 5. Clean (**top**) and previously fouled and corroded (**bottom**) panels.

The panels were cathodically protected, using ICCP, to a potential of −1.0 V (Ag/AgCl) measured at panel 2. This followed the DNV-RP-B401 recommended practice to increase the formation of cathodic chalks and reduce the current densities. The cathodic protection current was provided from a Thompson Electrochem, Ministat Precision, UK for the clean panels and a single channel of the Electrosynthesis, Model 440, USA multichannel potentiostat for the corroded and fouled panels. A 25 mm by 150 mm ribbon mesh mixed metal oxide (MMO) anode was used to apply the current to each set of steel panels. The anodes were aligned and placed 1 m away from Panel 2, the top panels in the submerged zone. The total current output from the ICCP systems for each set of panels was measured over a 1-Ohm shunt. The reference electrodes and the 1-Ohm shunts were connected to a Campbell Scientific, CR6 datalogger, USA and a Campbell Scientific, AM16/32B, USA relay multiplexer for data collection. The sampling rate for the data collection was every minute.

The panels were photographed underwater weekly for visual assessment of cathodic chalks, biofouling and corrosion products. The composition of cathodic chalks was analyzed by scanning electron microscopy (JEOL JSM-6380LV) with electron dispersive X-ray spectroscopy (EDAX Octane Elect EDS system with APEX software version 3.1). The samples were prepared with gold sputtering before the analysis. There were no replications for this pilot study in natural seawater due to the scale and the high number of sensors required.

3. Results and Discussion

The results are reported for both initially clean and previously fouled and corroded panels, and include the data from polarization potentials, electric current flows between panels, total current densities of ICCP systems, and the visual assessments of biofouling, corrosion products and cathodic chalks during and at the end of the experiment. The results for the panels on both racks are discussed in comparison to each other.

3.1. Polarization Potentials

The polarization potential selected for both sets of panels was −1.0 V (Ag/AgCl/seawater). Hartt [34] concluded that the cathodic chalks formed between the potentials −0.9 and −1.05 V (Ag/AgCl/seawater) create a less porous structure and help reduce the cathodic protection current densities. The potentials of the panels reached the design cathodic protection potentials approximately a week after immersion (Figure 6). Although the intertidal panels were polarized immediately to −1.0 V (Ag/AgCl/seawater), the polarization of clean panels took a couple of days longer compared to the previously fouled and corroded panels.

This was assumed to be due to the reduction of the total surface area by fouling and corrosion products which created a protective barrier [30]. The polarization of the bottom panel on each set was the slowest compared to the other panels due to IR drop [35].

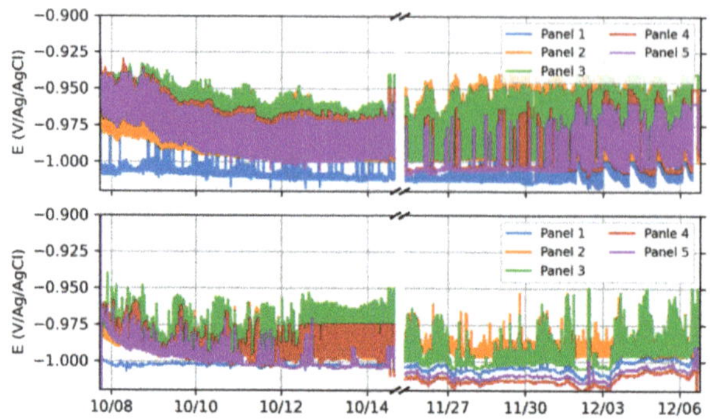

Figure 6. Polarization potentials: Clean (**top**) and previously fouled and corroded (**bottom**).

3.2. Cathodic Protection Current Densities

The cathodic protection current densities were normalized by dividing the amount of current impressed from each anode by the corresponding total steel area (m^2). The results showed that the current densities required to mitigate corrosion are highly dependent on the intertidal panels for both arrays. This was mainly due to the macro-galvanic cells formed as a result of the high partial pressure of oxygen generated during the semidiurnal tide cycles [7,8]. However, the increase in current densities was higher for the previously fouled and corroded panels when the intertidal panels were exposed to constant wetting and drying cycles [36]. The initial current density was estimated using the guidelines from the DNV recommended practice which states that it is the average current density until it attains a constant value. For the clean panel set was 175.5 mA/m^2, whereas it was 517.5 mA/m^2 for the previously fouled and corroded panels. This was due to biofouling and corrosion products causing ennoblement of the surface as well as delaying the formation of cathodic chalks that normally reduce the current densities [28,37,38].

The intertidal panels were constantly submerged during super high tides. This caused ICCP currents to converge approximatelyafter two months of immersion. However, the divergence occurred when the intertidal panels started to experience wetting and drying cycles towards the end of the experiment (Figure 7). The mean current densities for the clean and the previously fouled and corroded panel sets were around 67 mA/m^2 and 115 mA/m^2, respectively.

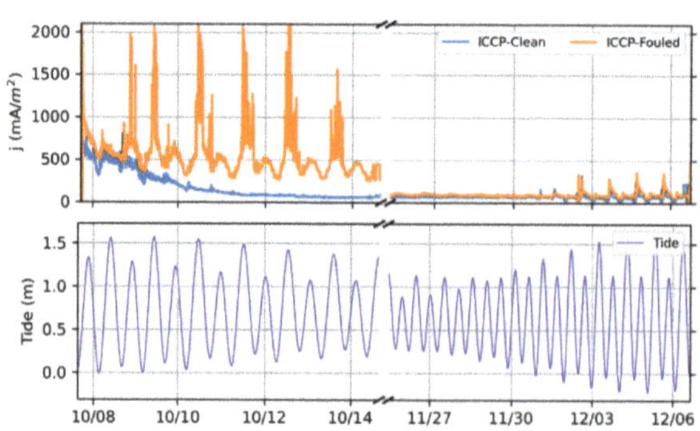

Figure 7. ICCP current density comparison: Clean vs. previously fouled and corroded (**top**) and tide (**bottom**).

3.3. Current Flow between Panels

The electric current flows measured over the zero resistance ammeters were used to measure the changes on panels that were constantly submerged and excluded the intertidal panel (panel 1). This demonstrates the differences caused by macro galvanic contributions of the steel in the intertidal zone. The current densities for the clean panels were higher in the beginning than the previously fouled and corroded panels (Figure 8). For example, the total current density after two days of immersion for the initially clean panels 2, 3, 4, and 5 was around 483 mA/m^2, whereas it was 361 mA/m^2 for the previously fouled and corroded panels. This was due to both micro- and macro-fouling creating a barrier in the vicinity of the surface and blocking oxygen transfer to the surface [6]. Additionally, the corrosion products on the steel surface comprised of magnetite (Fe_3O_4), α-FeOOH, and β-FeOOH created a barrier with high resistance and reduced the current flow into the panels [39].

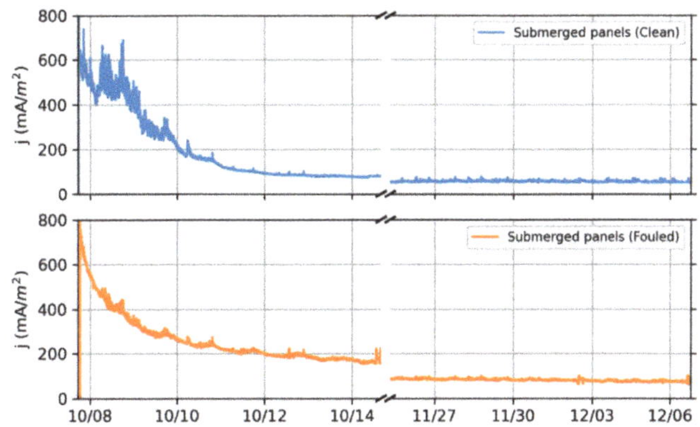

Figure 8. Current densities of constantly submerged panels: Clean (**top**) and previously fouled and corroded (**bottom**).

The current densities for the submerged clean panels decreased faster than the previously fouled and corroded panels and stabilized almost in five days of immersion (Figure 8). Although the initial current densities for previously fouled and corroded panels were low, it required almost a month for current densities to stabilize. The current densities for the fouled and corroded panels were higher than those measured for the clean panels even after the current densities were stabilized. For example, towards the end of the experiment the current density required for the clean panels was around 53 mA/m^2, whereas it was around 83 mA/m^2 for the previously fouled and corroded panels. This was due to the formation of cathodic chalks, for example., calcium carbonates ($CaCO_3$) and magnesium hydroxides ($Mg(OH)_2$) on clean panels which created a protective layer and reduced the current densities required to prevent corrosion [16,33]. However, the cathodic chalks formed on the fouled and corroded panels were less protective due to their morphology being disrupted by the biofouling and corrosion products [28,36,37].

3.4. Fouling

The fouling on panels was assessed at the end of the experiment after a light cleaning. The results showed that the density of fouling organisms on an initially clean set of panels was higher than the previously fouled and corroded one, although the latter was immersed in seawater longer [40]. This was due to the stable cathodic chalk formation on the clean panels. The fouling on panel 1 of the initially clean set was less than the rest of the panels on that array since constant wetting and drying cycles reduced the fouling settlement rate [24]. The main macrofoulers on panels were tubeworms and barnacles. Initially clean panels had some encrusting bryozoans and tunicates on constantly submerged panels. The previously fouled panels had some oysters (Figure 9). Scraping off the cathodic chalks and fouling from initially clean panels revealed that chalk formation and the settlement happened at the same time [41] and the areas under the barnacles were covered with dark corrosion products assuming that they were a thin layer of Fe(II)-based corrosion products with magnetite [42] (Figure 10).

Figure 9. Final visual assessment of the initially clean (**top**) and the previously fouled and corroded (**bottom**) panels after light cleaning.

Figure 10. Condition of the initially clean panel after scraped off.

3.5. Corrosion, Cathodic Chalks and EDX Analysis

The corrosion and cathodic chalks formed on initially clean and previously fouled and corroded panels showed visual differences. The clean panels showed no signs of corrosion except the intertidal panel since they were completely covered with cathodic chalks. The panels 2, 3, and 4 of the previously fouled and corroded array had bright orange areas after cathodic protection with an ICCP system for nine weeks (Figure 11). The buried areas for both arrays were free of corrosion and covered with a thin layer of chalk.

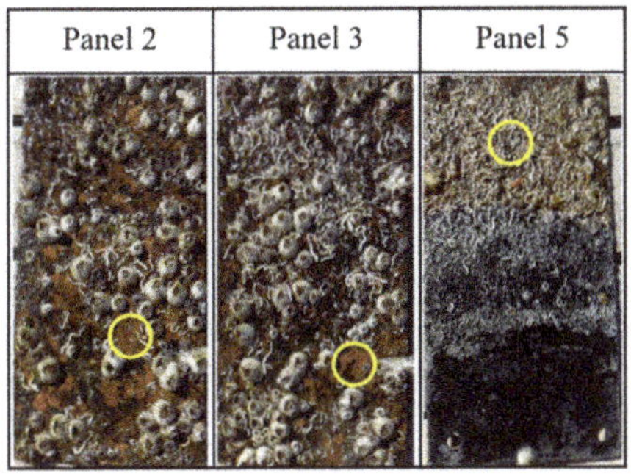

Figure 11. Areas analyzed with EDX.

The products on the bright orange areas on previously fouled and corroded panels were analyzed with EDX to determine their composition. The EDX analyses showed that the cathodic chalks formed at these areas were infused with initially present corrosion products and contained Fe element in addition to Ca and Mg. The sample taken from the buried zone of panel 5 was mainly Ca and Mg [13,27], however there was no Fe present (Figure 12). This may have been due to the iron being held in the black iron oxide films which were not removed during sampling.

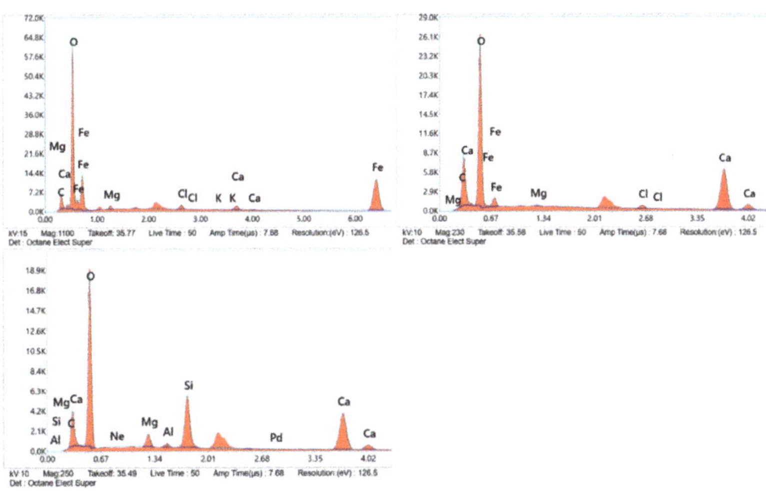

Figure 12. EDX analysis of previously fouled and corroded panels: Panel 2 (**top left**), Panel 3 (**top right**), and Panel 5 (**bottom left**).

The data collected for this study was specific to the environmental conditions at Port Canaveral, FL between 7 October 2021 and 7 December 2021. It was also specific to the condition of previously fouled and corroded panels. The amount of fouling and corrosion products was a result of the deployment in seawater at Port Canaveral, FL for nine weeks between 5 August 2021 and 7 October 2021. The ratio of fouling and corrosion products may alter the results presented in this paper. Therefore, it is recommended to deploy steel panels and perform similar experiments at locations with different environmental conditions.

4. Conclusions

The fouling and corrosion products established on offshore monopile foundations between deployment in marine environments and receiving any protection may alter the design requirements of cathodic protection systems. These systems can be optimized and designed more efficiently by understanding the influence of fouling and corrosion products under dynamic conditions due to tidal cycles.

This study demonstrated that the fouling and corrosion products on vertical uncoated steel surfaces in seawater extending from the intertidal zone to the buried zone increased the cathodic protection current densities. The initial current densities for the clean panels and the previously fouled and corroded panels were 175.5 mA/m^2 and 517.5 mA/m^2, respectively. The mean current density was 67 mA/m^2 for the clean panels and it was 115 mA/m^2 for the previously fouled and corroded panels. The current density values became closer during constant immersion due to super high tides. However, wetting and drying cycles towards the end of the experiment caused higher fluctuations of current densities for the previously fouled and corroded panels.

The initial current density for the clean panels (489 mA/m^2) below the waterline was higher than the previously fouled and corroded panels (361 mA/m^2). However, due to the immediate formation of stable cathodic chalks on the clean panels, the currents decreased faster. The current densities for both arrays became stable during the experiment, and they were lower for the clean panels (53 mA/m^2) than the previously fouled and corroded panels (83 mA/m^2).

The visual assessments showed no signs of corrosion and a more diverse fouling community on the clean panels with an ICCP system below the waterline. The fouled and corroded panels had areas that were bright orange in the submerged zone. The EDX analysis of the samples from these areas showed that the cathodic chalks included Fe as well as Ca and Mg. This demonstrated that the fouling as well as the corrosion products changed the composition of cathodic chalks.

The fouling and corrosion products on uncoated steel surfaces have been shown to increase the current demand for cathodic protection systems and alter the composition of cathodic chalks while reducing their protection properties. This study highlights the benefits of applying cathodic protection to offshore monopiles immediately after they are deployed in seawater.

Author Contributions: Conceptualization, C.E. and G.S.; Writing—Original Draft, C.E.; Formal analysis, C.E.; Investigation, C.E.; Methodology, C.E. and G.S.; Project administration, C.E.; Supervision, G.S.; Visualization, C.E.; Writing—Review and Editing, C.E. and G.S. All authors have read and agreed to the published version of the manuscript.

Funding: This research received no external funding.

Institutional Review Board Statement: Not applicable.

Informed Consent Statement: Not applicable.

Data Availability Statement: The data that support the findings of this study are available from the corresponding author upon request.

Acknowledgments: We would like to acknowledge the Office of Naval Research [grant number N00014-20-1-2243] for access to the sea water test site they support at the Center for Corrosion and Biofouling Control, Port Canaveral, Florida. We offer special thanks to Dan Kuchma from the Tufts University and the Bureau of Ocean Energy Management for their indirect support for this project.

Conflicts of Interest: The authors declare no conflict of interest.

References

1. Gupta, B.K.; Basu, D. Offshore wind turbine monopile foundations: Design perspectives. *Ocean Eng.* **2020**, *213*, 107514. [CrossRef]
2. Bisoi, S.; Haldar, S. Dynamic analysis of offshore wind turbine in clay considering soil–monopile–tower interaction. *Soil Dyn. Earthq. Eng.* **2014**, *63*, 19–35. [CrossRef]
3. Erdogan, C.; Swain, G. Conceptual Sacrificial Anode Cathodic Protection Design for offshore wind monopiles. *Ocean Eng.* **2021**, *235*, 109339. [CrossRef]
4. Price, S.J.; Figueira, R.B. Corrosion Protection Systems and Fatigue Corrosion in Offshore Wind Structures: Current Status and Future Perspectives. *Coatings* **2017**, *7*, 25. [CrossRef]

5. DNVGL. *Recommended practice–Corrosion Protection for Wind Turbines*; DNVGL-RP-0416; DNVGL: Bærum, Norway, 2016.
6. De Brito, L.V.R.; Coutinho, R.; Cavalcanti, E.H.S.; Benchimol, M. The influence of macrofouling on the corrosion behaviour of API 5L X65 carbon steel. *Biofouling* **2007**, *23*, 193–201. [CrossRef]
7. Wyatt, B.S.; Preston, J.; Jacob, W.R. Cathodic Protection of Offshore Renewable Energy Infrastructure. In Proceedings of the Eurocorr Conference, Brussels, Belgium, 7–11 September 2020.
8. Erdogan, C.; Swain, G. The effect of macro-galvanic cells on corrosion and impressed current cathodic protection for offshore monopile steel structures. *Ocean Eng.* **2022**, *265*, 112575. [CrossRef]
9. Cui, Z.; Chen, S.; Dou, Y.; Han, S.; Wang, L.; Man, C.; Wang, X.; Chen, S.; Cheng, Y.F.; Li, X. Passivation behavior and surface chemistry of 2507 super duplex stainless steel in artificial seawater: Influence of dissolved oxygen and pH. *Corros. Sci.* **2019**, *150*, 218–234. [CrossRef]
10. Dexter, S.C.; Lafontaine, J.P. Effect of Natural Marine Biofilms on Galvanic Corrosion. *Corrosion* **1998**, *54*, 851–861. [CrossRef]
11. Guezennec, J.; Edowling, N.J.; Bullen, J.; White, D.C. Relationship between bacterial colonization and cathodic current density associated with mild steel surfaces. *Biofouling* **1994**, *8*, 133–146. [CrossRef]
12. Permeh, S.; Lau, K. Identification of steel corrosion associated with sulfate-reducing bacteria by electrochemical noise technique. *Mater. Corros.* **2022**. [CrossRef]
13. Carré, C.; Zanibellato, A.; Jeannin, M.; Sabot, R.; Gunkel-Grillon, P.; Serres, A. Electrochemical calcareous deposition in seawater. A review. *Environ. Chem. Lett.* **2020**, *18*, 1193–1208. [CrossRef]
14. Little, B.J.; Blackwood, D.J.; Hinks, J.; Lauro, F.M.; Marsili, E.; Okamoto, A.; Rice, S.A.; Wade, S.A.; Flemming, H.C. Microbially Influenced Corrosion—Any Progress? *Corros. Sci.* **2020**, *170*, 108641. [CrossRef]
15. Lv, M.; Du, M. A review: Microbiologically influenced corrosion and the effect of cathodic polarization on typical bacteria. *Rev. Environ. Sci. Bio/Technol.* **2018**, *17*, 431–446. [CrossRef]
16. Thompson, A.A.; Wood, J.L.; Palombo, E.A.; Green, W.K.; Wade, S.A. From laboratory tests to field trials: A review of cathodic protection and microbially influenced corrosion. *Biofouling* **2022**, *38*, 298–320. [CrossRef]
17. Tian, H.; Xin, J.; Li, Y.; Wang, X.; Cui, Z. Combined effect of cathodic potential and sulfur species on calcareous deposition, hydrogen permeation, and hydrogen embrittlement of a low carbon bainite steel in artificial seawater. *Corros. Sci.* **2019**, *158*, 108089. [CrossRef]
18. Shojai, S.; Schaumann, P.; Braun, M.; Ehlers, S. Influence of pitting corrosion on the fatigue strength of offshore steel structures based on 3D surface scans. *Int. J. Fatigue* **2022**, *164*, 107128. [CrossRef]
19. Khodabux, W.; Liao, C.; Brennan, F. Characterisation of pitting corrosion for inner section of offshore wind foundation using laser scanning. *Ocean Eng.* **2021**, *230*, 109079. [CrossRef]
20. Swain, G.W.; Patrick-Maxwell, J. The Effect of Biofouling on the Performance of Al-Zn-Hg Sacrificial Anodes. *Corrosion* **1990**, *46*, 256–260. [CrossRef]
21. Rousseau, C.; Baraud, F.; Leleyter, L.; Jeannin, M.; Gil, O. Calcareous deposit formed under cathodic protection in the presence of natural marine sediments: A 12 month experiment. *Corros. Sci.* **2010**, *52*, 2206–2218. [CrossRef]
22. Blackwood, D.J.; Lim, C.S.; Teo, S.L. Influence of fouling on the efficiency of sacrificial anodes in providing cathodic protection in Southeast Asian tropical seawater. *Biofouling* **2010**, *26*, 779–785. [CrossRef]
23. Rajala, P.; Sohlberg, E.; Priha, O.; Tsitko, I.; Väisänen, H.; Tausa, M.; Carpén, L. Biofouling on Coated Carbon Steel in Cooling Water Cycles Using Brackish Seawater. *J. Mar. Sci. Eng.* **2016**, *4*, 74. [CrossRef]
24. Daille, L.K.; Aguirre, J.; Fischer, D.; Galarce, C.; Armijo, F.; Pizarro, G.E.; Walczak, M.; De la Iglesia, R.; Vargas, I.T. Effect of Tidal Cycles on Bacterial Biofilm Formation and Biocorrosion of Stainless Steel AISI 316L. *J. Mar. Sci. Eng.* **2020**, *8*, 124. [CrossRef]
25. Eashwar, M.; Subramanian, G.; Palanichamy, S.; Rajagopal, G.; Madhu, S.; Kamaraj, P. Cathodic behaviour of stainless steel in coastal Indian seawater: Calcareous deposits overwhelm biofilms. *Biofouling* **2009**, *25*, 191–201. [CrossRef] [PubMed]
26. Choudhary, S.; Garg, A.; Mondal, K. Relation Between Open Circuit Potential and Polarization Resistance with Rust and Corrosion Monitoring of Mild Steel. *J. Mater. Eng. Perform.* **2016**, *25*, 2969–2976. [CrossRef]
27. Wang, W.; Li, W.; Song, L.-Y.; Fan, W.-J.; Liu, X.-J.; Zheng, H. Numerical simulation and re-design optimization of impressed current cathodic protection for an offshore platform with biofouling in seawater. *Mater. Corros.* **2018**, *69*, 239–250. [CrossRef]
28. Eashwar, M.; Subramanian, G.; Chandrasekaran, P.; Manickam, S.T.; Maruthamuthu, S.; Balakrishnan, K. The interrelation of cathodic protection and marine macrofouling. *Biofouling* **1995**, *8*, 303–312. [CrossRef]
29. Chen, X.; Wang, G.; Gao, F.; Wang, Y.; He, C. Effects of sulphate-reducing bacteria on crevice corrosion in X70 pipeline steel under disbonded coatings. *Corros. Sci.* **2015**, *101*, 1–11. [CrossRef]
30. Permeh, S.; Lau, K.; Boan, M.E.; Tansel, B.; Duncan, M. Cathodic Polarization Behavior of Steel with Different Marine Fouling Morphologies on Submerged Bridge Elements with Cathodic Protection. *J. Mater. Civ. Eng.* **2020**, *32*, 04020184. [CrossRef]
31. Liduino, V.; Galvão, M.; Brasil, S.; Sérvulo, E. SRB-mediated corrosion of marine submerged AISI 1020 steel under impressed current cathodic protection. *Colloids Surf. B Biointerfaces* **2021**, *202*, 111701. [CrossRef]
32. Swain, G. *Corrosion and Types of Corrosion–Protection of Marine Materials. Center for Corrosion and Biofouling Control*; Internal Report; Florida Institute of Technology: Melbourne, FL, USA, 2022.
33. NOAA Tide Predictions—8721604 Trident Pier, Port Canaveral, FL. Available online: https://tidesandcurrents.noaa.gov/noaatidepredictions.html?id=8721604 (accessed on 7 January 2021).

34. Hartt, W. 2012 Frank Newman Speller Award: Cathodic Protection of Offshore Structures—History and Current Status. *Corrosion* **2012**, *68*, 1063–1075. [CrossRef]
35. Revie, R.W.; Uhlig, H.H. *Corrosion and Corrosion Control: An Introduction to Corrosion Science and Engineering*, 4th ed.; John Wiley & Sons, Inc.: Hoboken, NJ, USA, 2008; ISBN 9780470277270.
36. De Romero, M.; de Rincón, O.; Ocando, L. Cathodic Protection Efficiency in the Presence of SRB: State of the Art. In Proceedings of the NACE—International Corrosion Conference Series, Atlanta, GA, USA, 22–26 March 2009.
37. Dexter, S.C.; Lin, S.H. Effect of Marine Bacteria on Calcareous Deposition. *Mater. Perform.* **1991**, *30*, 16–21.
38. Dexter, S.C.; Lin, S.-H. Effect of marine biofilms on cathodic protection. *Int. Biodeterior. Biodegrad.* **1992**, *29*, 231–249. [CrossRef]
39. Wang, H.-H.; Du, M. Corrosion Behavior of a Low-Carbon Steel in Simulated Marine Splash Zone. *Acta Met. Sin. (Engl. Lett.)* **2017**, *30*, 585–593. [CrossRef]
40. Zhang, J.; Yu, Z.; Zhao, X.; Lan, X.; Wang, J.; Lv, X.; Zhang, C.; Duan, J.; Hou, B. The Interaction of Biofoulants and Calcareous Deposits on Corrosion Performance of Q235 in Seawater. *Materials* **2020**, *13*, 850. [CrossRef] [PubMed]
41. Hernandez, G.; Hartl, W.; Videla, H. Marine Biofilms and their Influence on Cathodic Protection: A Literature Survey. *Corros. Rev.* **1994**, *12*, 29–40. [CrossRef]
42. Refait, P.; Grolleau, A.-M.; Jeannin, M.; Rémazeilles, C.; Sabot, R. Corrosion of Carbon Steel in Marine Environments: Role of the Corrosion Product Layer. *Corros. Mater. Degrad.* **2020**, *1*, 198–218. [CrossRef]

Journal of
Marine Science and Engineering

MDPI

Article

Effects of Tidal Stream Energy Exploitation on Estuarine Circulation and Its Seasonal Variability

Marcos Sánchez [1], David Mateo Fouz [1], Iván López [1], Rodrigo Carballo [1] and Gregorio Iglesias [2,3,*]

1 Departamento de Enxeñaría Agroforestal, Universidade de Santiago de Compostela, EPSE, Rúa Benigno Ledo s/n, 27002 Lugo, Spain
2 MaREI, Environmental Research Institute & School of Engineering, University College Cork, P43 C573 Cork, Ireland
3 School of Engineering, University of Plymouth, Plymouth PL4 8AA, UK
* Correspondence: gregorio.iglesias@ucc.ie

Abstract: Residual flows are of major importance in coastal areas, driving environmental processes such as sediment transport or nutrient dispersion. Consequently, in those areas where a large tidal stream energy resource is available, prior to the installation of a tidal farm, it is imperative to assess how energy extraction affects the residual flows and, in particular, upwelling events. In this paper, the potential effects of different configurations of tidal stream farms on the residual circulation and its seasonality are analysed by means of a case study: Ria de Ortigueira, the westernmost of the Galician Rias Altas (NW Spain). For this purpose, a 3D numerical model was implemented and validated against field measurements. Next, a total of eight case studies, including the operation of bottom-fixed and floating converters under typical summer and winter scenarios, considering upwelling favourable winds, were studied. Overall, when a tidal farm operates, regardless of its configuration and the forcings considered, the resulting general residual flow pattern does not experience significant modifications. This pattern is characterized by a 2D circulation in the inner ria and a positive estuarine circulation in the middle and outer ria. The largest modifications of the residual flow are apparent in the vicinities of the plant, with maximum values of about 0.05 ms^{-1}. Outside this area, the alteration is lower than 0.01 ms^{-1} and virtually negligible at some distance from the farm where upwelling events develop.

Keywords: tidal stream energy exploitation; impact assessment; residual circulation; seasonal variability; upwelling; 3D numerical model

Citation: Sánchez, M.; Fouz, D.M.; López, I.; Carballo, R.; Iglesias, G. Effects of Tidal Stream Energy Exploitation on Estuarine Circulation and Its Seasonal Variability. *J. Mar. Sci. Eng.* **2022**, *10*, 1545. https://doi.org/10.3390/jmse10101545

Academic Editors: Mohamed Benbouzid and Rafael Morales

Received: 1 September 2022
Accepted: 17 October 2022
Published: 20 October 2022

1. Introduction

The interest in carbon-free energy sources has increased drastically, resulting to a large extent from the effects of climate change, which have become more evident over the last decades [1,2]. In this context, a large number of policies have been implemented within different legal frameworks in order to achieve more ambitious quotas of renewable energy production [3–8]. In this regard, the development of new renewable energies is a key to reduce fossil fuels consumption and to fight against climate change [9]. More specifically, marine renewable energies (MRE), and in particular tidal stream energy, stand out as one of the most promising alternatives to diversify the energy mix by means of zero emission technologies, principally due to its advantages in comparison with other types of renewable energies, such as its predictability, no land occupation, or a high load factor [10–12], along with it being mature enough to be ready for its commercial development [13].

Generally speaking, tidal stream exploitation is potentially of interest in coastal areas where peak velocities exceed 1.0–1.5 ms^{-1} [14], resulting from different factors such as large tidal prisms and a constricted morphology [15,16]. These factors are sometimes reinforced by baroclinic flows and river discharges [17]. The harnessing of tidal stream energy resource is carried out through tidal energy converters (TECs), which are usually classified according to their principle of operation as reciprocating [18] or rotating [19] devices. Rotating devices, also known as tidal stream turbines (TSTs), present the highest technology readiness level (TRL). Nowadays, third-generation TSTs [20] are designed to operate in shallow areas (approx. 1 m depth) with relatively low cut-in velocities (roughly 0.7 ms^{-1}) [17]. TSTs can be installed in different configurations, whose influence on the performance of a tidal farm was investigated in previous works [12]. Depending on their position in the water column, TSTs may be classified as [21]: (i) floating TSTs, i.e., devices anchored to the sea bottom floating below the sea surface, which harness the energy of the upper layers of the water column, and (ii) bottom-fixed TSTs, i.e., converters rigidly anchored to the sea bottom that harness the energy available in the bottom layers of the water column.

Suitable sites for installing a tidal farm are usually located in coastal areas driven by complex hydrodynamic patterns, such as estuaries or bays, which typically have high environmental value. This highlights the key role of an appropriate environmental impact assessment for informed decision making when installing a tidal farm [22–24]. In order to investigate the impacts of tidal stream energy exploitation over estuarine circulation, both transient and residual tidal flows were analysed in previous works by means of 3D numerical modelling [25], including a comparison between floating and bottom-fixed TSTs [26]. The results demonstrate that: (i) the impacts on the residual circulation extend their effects over wider regions than in the case of transient circulation (despite being of the same order in terms of percentage of velocity variation), and (ii) there are almost no differences regarding the impact on estuarine circulation between floating and bottom-fixed configurations, with the most significant differences occurring in the vicinity of the plants, principally due to the different vertical configuration.

For all this extensive work, in areas suitable for tidal stream energy exploitation, there are other circulation patterns to be considered, e.g., upwelling-downwelling systems, which have not been well-studied in the vicinity of a tidal farm. An example of the effects of upwelling events can be found in the Galician Rias (NW Spain)—a region with one of the largest oceanic productivities worldwide [27]. Galician Rias are flooded tectonic valleys, which are usually classified into Rias Altas and Rias Baixas according to the geographical location (north and south of Cape Fisterra, respectively). There are important tectonic and geomorphological differences between both groups [28]: the Rias Baixas are larger, deeper, and more homogeneous in their orientation (roughly NE-SW) than the Rias Altas. Upwelling events throughout the Rias Baixas have been extensively analysed in the scientific literature [29–36], and their key role in cooling and fertilizing surface waters is clearly highlighted [37]. Ekman transport is the driver of upwelling processes due to the incursion of Eastern North Atlantic Central Water (ENACW) into the Rias Baixas, which usually occurs during spring or summer in the presence of N-NE winds. As a result of the orientation of these rias (NE-SW), northerly winds cause a positive estuarine circulation, which consists in an outflow in the surface layers and inflow of cold and nutrient-laden ENACW, which ascends from the continental shelf and finally flows into the ria [38–40].

The frequency and intensity of upwelling episodes are highly influenced by the orientation of the coast, which modifies the wind velocity and direction [41,42]. Several studies [27,43] remarked that the forcing factors driving the hydrodynamics in the Rias Altas are clearly dependent on the geographic orientation of a specific coastal system, as it is much more heterogeneous than in the case of the Rias Baixas. Taking this into account, and as a result of their different coastal orientation, upwelling favourable conditions along the Rias Altas are not related to the spring-summer NE winds typical of the Rias Baixas [41]. In effect, the features of the upwelling episodes along this coastal region are partially unknown, with the specific orientation of each estuary playing a significant role in the resulting hydrographical, biogeochemical, and biological patterns during upwelling and downwelling events within Rias Altas [44].

With this in view, and in order to go further into the analysis of the impacts of a tidal farm on the residual circulation and estuarine upwelling, this work studies the effects resulting from the operation of a tidal stream plant in a specific coastal area of the Rias Altas, the Ria de Ortigueira, by means of the implementation of a 3D high-resolution hydrodynamic numerical model.

This paper is structured as follows. First, in Section 2, a brief description of the study area, the Ria de Ortigueira, is presented. Next, in Section 3, the methodology used in this work for assessing the impacts of the operation of a tidal farm on the residual circulation and estuarine upwelling is thoroughly described. Then, in Section 4, results are presented and discussed focusing on two situations: unaltered and altered conditions (i.e., without and with energy extraction, respectively). Finally, in Section 5, conclusions are drawn.

2. Study Area

The Ria de Ortigueira (Figure 1) is the westernmost of the Galician Rias Altas (NW Spain) delimited by Cape Ortegal and Pt. Estaca de Bares, presenting a total surface area of 85 km^2 and a length of 18 km along its main axis. This coastal area emerges as a promising region for tidal stream energy exploitation thanks to its complex geomorphology, with a number of constrictions, and its relatively high tidal range (maxima of 4.5 m). Previous works [16,45] show that tidal flows of approximately 2.5 ms^{-1} can be reached in the narrowing between Pt. Postiña and Pt. Cabalar, indicating the suitability of the Ria de Ortigueira as the location for a tidal stream farm.

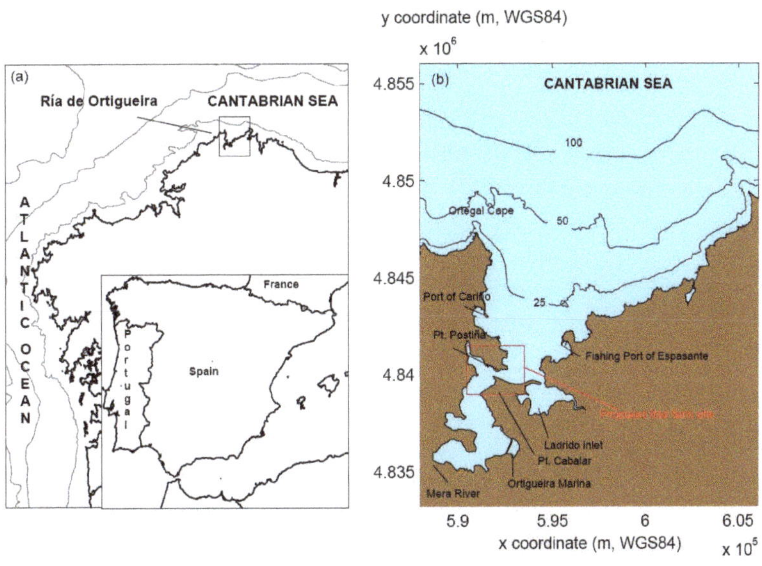

Figure 1. The Ria de Ortigueira (**b**) and its location in Galicia (NW Spain) (**a**).

Regarding the tidal regime of the Ria de Ortigueira, it is characterized by a predominance of semidiurnal components (Table 1), especially M2, as in the rest of the European Atlantic. In this regard, considering the main tidal constituents (M2, S2, K1, and O1) by means of a harmonic analysis, the tidal form factor (F) confirms the semidiurnal behaviour (F = 0.082) of this body of water [46]. In this Ria, the action of the tide is highly modified by its main geomorphological features. Ria de Ortigueira presents a gradual reduction in the width of its cross-sections from the mouth to the inner ria, ranging from more than 10 km at its ocean boundary to 150–200 m in the inner estuary. As far as the water depth is concerned, it presents an average value of 7.3 m [47] but with strong variations throughout the estuary, varying from maximums of 30–35 m at its mouth to less than 5 m throughout extensive zones in the inner ria, including large intertidal areas.

Table 1. Main tidal constituents in the Ria de Ortigueira.

Constituent	Amplitude (cm)	Phase (°)
M2	122.79	90.15
S2	42.91	121.08
N2	25.98	70.39
K2	12.03	118.76
K1	7.35	73.49
O1	6.22	324.62
P1	2.22	65.15
Q1	2.11	271.31
M4	1.45	334.80

Even though that the hydrodynamics of the Ria de Ortigueira are dominated by the tide, tides are not the only relevant forcing factor within this estuarine area, such as: (i) fluvial discharges, (ii) thermohaline conditions, and (iii) wind regime. As a result of the interaction of the different forcing factors driving the hydrodynamics of the Ria de Ortigueira, this coastal location can be classified according to the salinity structure criterion as a well-mixed estuary [16].

Regarding fluvial discharges, though there are several fluvial courses flowing into this estuary, only the River Mera presents a significant flow rate (i.e., average annual discharge of 5.45 $m^3 s^{-1}$) to be considered as a relevant hydrodynamics forcing agent in this coastal area. A seasonal analysis [17] considering mean monthly flow rates since 1970 shows that a marked intra-annual discharge variability is apparent, resulting in two different periods (autumn–winter and spring–summer) within the hydrological year. The average characteristics of fluvial discharges and thermohaline conditions of oceanic waters for these two periods are shown in Table 2.

Table 2. Average values of river discharges and thermohaline conditions during spring–summer and autumn–winter scenarios.

Scenario	River Discharge Conditions			Oceanic Conditions	
	River Flow ($m^3 s^{-1}$)	Salinity (ppt)	Temperature (°C)	Salinity (ppt)	Temperature (°C)
Spring–summer	3.07	0.051	15.60	35.68	15.51
Autumn–winter	8.48	0.051	11.10	35.48	13.17

With respect to the wind regime, pronounced patterns of seasonality are found again, as in the case of river discharges, and in line with other Galician Rias [48,49]: (i) autumn–winter scenario (prevalence of W-SW winds) and (ii) spring–summer scenario (prevalence of NE winds), hereinafter referred to for simplicity as winter and summer winds, respectively. Moreover, as mentioned in Section 1, in the Rias Altas the coastal orientation determines which wind directions produce upwelling flows. Considering the orientation of the shoreline in the Ria de Ortigueira (SSW-NNE), and despite NE winds being prevalent during spring–summer (upwelling favourable periods in other Galician Rias), SW winds are responsible for upwelling processes within this waterbody during both winter and summer scenarios. Upwelling episodes take place mainly during spring or summer; notwithstanding, they also occur during autumn or winter in both Rias Baixas [31] and Rias Altas [50].

3. Materials and Methods

3.1. Numerical Model (I): Equations

With the aim of analysing the impacts of tidal stream energy exploitation on the hydrodynamics of the Ria de Ortigueira, and more specifically on the upwelling events, the 3D hydrodynamic model Delft3D-FLOW was implemented in this estuarine area. Delft3D-

FLOW approximates by means of a finite-difference code, the Navier Stokes equations (under the shallow-water and Boussinesq assumptions) combined with the transport equation, allowed the computation of both the baroclinic and barotropic circulation [51]. In order to fully describe the resulting 3D flow structure during upwelling events [40] and how a tidal farm affects them, the numerical model was implemented in its 3D form. The model equations read:

$$\frac{\partial u}{\partial x} + \frac{\partial v}{\partial y} + \frac{\partial w}{\partial z} = Q \tag{1}$$

$$\left.\begin{aligned} \frac{Du}{Dt} &= fv - g\frac{\partial \zeta}{\partial x} - \frac{g}{\rho_0}\int_{z'=z}^{z'=\varsigma} \frac{\partial \rho}{\partial x} dz' + v_h\left(\frac{\partial^2 u}{\partial x^2} + \frac{\partial^2 u}{\partial y^2}\right) + v_v\left(\frac{\partial^2 u}{\partial z^2}\right) \\ \frac{Dv}{Dt} &= -fu - g\frac{\partial \zeta}{\partial y} - \frac{g}{\rho_0}\int_{z'=z}^{z'=\varsigma} \frac{\partial \rho}{\partial y} dz' + v_h\left(\frac{\partial^2 v}{\partial x^2} + \frac{\partial^2 v}{\partial y^2}\right) + v_v\left(\frac{\partial^2 v}{\partial z^2}\right) \end{aligned}\right\} \tag{2}$$

$$\frac{\partial p}{\partial z} = -\rho g \tag{3}$$

$$\frac{Dc}{Dt} = D_h\left(\frac{\partial^2 c}{\partial x^2} + \frac{\partial^2 c}{\partial y^2}\right) + D_v\frac{\partial^2 c}{\partial z^2} - \lambda_d c + R_s \tag{4}$$

Equation (1) represents the conservation of mass under the assumption of incompressibility (Boussinesq assumption); Equation (2) expresses the conservation of the momentum along x- and y-directions; Equation (3) represents the conservation of the momentum in the z-direction, which is simplified to the hydrostatic pressure distribution according to the shallow-water assumption; lastly, Equation (4) stands for the transport equation, which is solved for salinity and temperature constituents. In these equations, u, v, and w are the velocity components in the x-, y-, and z-directions, respectively; Q stands for the mass source intensity per unit area; t is the time; f represents the Coriolis parameter; g is the gravitational acceleration ($\approx$9.81 ms^{-2}); ζ stands for the free surface elevation relative to $z = 0$, i.e., the water level; ρ_0 and ρ represent the reference density of sea water and the density, respectively; v_h and v_v represent the horizontal and vertical eddy viscosities, respectively; D_h and D_v stand for the horizontal and vertical eddy diffusivities, respectively; c represents the thermohaline conditions (i.e., salinity or temperature); λ_d stands for decay processes; finally, R_s is the source term per unit area.

3.2. Numerical Model (II): Implementation

The numerical model covers the whole ria by means of a Cartesian grid extending towards the adjacent continental shelf up to the 150 m isobath at a distance of approx. 10 km offshore. This grid (Figure 2) is composed of 57,361 varying-size cells, with the grid size set as follows: in the inner and middle ria, a grid resolution of 50 × 50 m is adopted, progressively decreasing in the y-direction up to 50 × 150 m at the outer ria (northern boundary of the grid). Regarding the vertical discretisation, a σ-coordinate system composed of twelve layers was used [52], with the following layer thicknesses (percentages of water column) from surface to bottom: 2, 3, 5, 10, 15, 15, 15, 15, 10, 5, 3, and 2 [23,40,53].This distribution of σ-layers, which provides a higher grid resolution near the surface and seabed, ensures an adequate modelling of boundary layer flows [53].

The bathymetric data of the Ria de Ortigueira and its surroundings were obtained from nautical charts #408 and #4083 from the Navy's Hydrographic Institute (Instituto Hidrográfico de la Marina). Additionally, intertidal areas are fully described by using data from a 10 × 10 m resolution digital elevation model.

The initial conditions for the hydrodynamic numerical model are null free-surface elevation and velocity at each node of the computational grid (cold start) [54]. As far as the transport model is concerned, the initial thermohaline conditions were prescribed at the oceanic boundaries through field measurements (the reader is referred to Table 2 for a better understanding). Moreover, in order to avoid the influence of the initial conditions on the numerical results, a 30-day spin-up period was adopted to ensure their dynamical adjustment with the flow field [35,55].

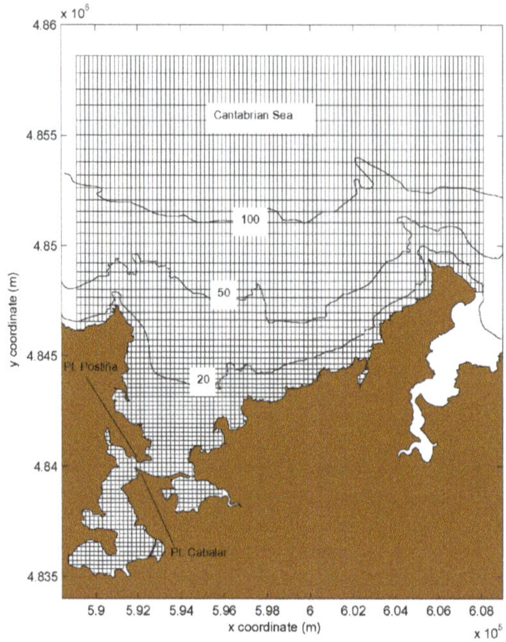

Figure 2. Numerical model grid.

Regarding the open boundary conditions, Dirichlet-type conditions were imposed along the oceanic boundaries by considering the M2 constituent and the thermohaline conditions (Table 2). At the land–water transition, null flow through the boundary and zero shear stress were imposed. At the sea bottom boundary, the shear stress, $\vec{\tau}_{b3D}$, induced by a turbulent flow was computed by means of a quadratic law:

$$\vec{\tau}_{b3D} = \frac{g\rho_0 \left|\vec{u}_b\right| \vec{u}_b}{C_{3D}^2}, \tag{5}$$

where $\vec{u}_b$ represents the magnitude of the horizontal velocity in the bottom layer, and C_{3D} represents the 3D Chézy coefficient, which is computed as:

$$C_{3D} = C_{2D} + \frac{\sqrt{g}}{\kappa} \ln\left(\frac{\Delta z_b}{2H}\right), \tag{6}$$

with k as the von Karman constant ($k = 0.41$), Δz_b the distance from the sea bottom to the nearest grid node, H the water depth, and C_{2D} the 2D Chézy coefficient, which can be expressed as:

$$C_{2D} = \frac{\sqrt[6]{H}}{n}, \tag{7}$$

where n stands for the Manning's coefficient, which is prescribed in this work as a function of the water depth [56] in order to consider the full influence of the water depth on the seabed stress [57,58].

Finally, the wind shear stress, $\vec{\tau}_s$, at the free surface is given by:

$$\vec{\tau}_s = \rho_a C_d \left|\vec{U}_{10}\right| \vec{U}_{10}, \tag{8}$$

where ρ_a represents the density of the air, $\vec{U}_{10}$ stands for the wind velocity vector at 10 m height above the sea surface, and C_d is the wind drag coefficient computed following [59,60]. Moreover, regarding the main numerical and physical parameters of the numerical model, the horizontal advection terms in Equation (2) are calculated using the cyclic method [61]; as regards the temporal discretisation, the model resorts to an alternating direction implicit (ADI) algorithm [62]; finally, the *k*-ε model was adopted for the turbulence closure [63].

The numerical model was successfully validated against field measurements obtained by means of an ADCP (Acoustic Doppler Current Profiler) and a tidal gauge deployed at two different locations in the inner and middle ria during a 19-day period. During this period, the model was forced along the oceanic boundaries with the main tidal harmonics, thermohaline conditions, wind velocity and direction and fluvial discharges of River Mera. A good agreement between computed and observed series (for both water levels and velocities) was obtained, reaching high values of the correlation coefficient R (i.e., $R \approx 0.99$ for water levels and $R > 0.82$ for flow magnitude and direction at each layer) and low values of several statistical coefficients (i.e., mean difference, *MD*, and mean square error, *MSE*). For further details of the numerical model validation, the reader is referred to previous works [12,25,26].

The implementation of the TSTs in the numerical model was conducted through the so-called momentum sink approach [22,23,64]. It consists of introducing a momentum sink term into the right-hand side of Equation (2) by means of a retarding force, which compensates the thrust force exerted by the flow on the TST [65]. Considering this, the momentum sinks can be modelled as follows:

$$\left.\begin{array}{l} M_x = \frac{F_x}{\rho} \\ M_y = \frac{F_y}{\rho} \end{array}\right\}, \tag{9}$$

with F_x and F_y as the components of the retarding force per unit volume V along x- and y-directions, respectively. These components are opposite and equal to the thrust force, and therefore, they can be expressed as:

$$\left.\begin{array}{l} M_x = -\frac{1}{2}\frac{C_T A}{V} U u \\ M_y = -\frac{1}{2}\frac{C_T A}{V} U v \end{array}\right\}, \tag{10}$$

where A represents the cross-section of a TST, U stands for the magnitude of the flow velocity, and C_T is the thrust coefficient [66]. The retarding force is considered in the vertical model layers, which correspond with the water depth at which they operate. It is important to note that the methodology applied in this study provides accurate results when analysing the modifications in the general circulation patterns considering the type of model implemented (shallow water models) and grid resolution. However, an accurate near-field analysis should resort to the implementation of high-resolution CFD (Computational Fluid Dynamics) models solving RANS (Reynolds Averaged Navier Stokes) equations.

Figure 3 shows the layout of the proposed tidal farms, which are located in the surroundings of the narrowing between Pt. Postiña and Pt. Cabalar, where the strongest tidal currents are found (see Section 1). Both plants have the same layout with a surface area of about 1500 m^2. The first plant is composed of floating TSTs occupying the upper part of the water column, and the second plant is composed of TSTs of the same characteristics but fixed to the bottom and occupying the lower part of the water column.

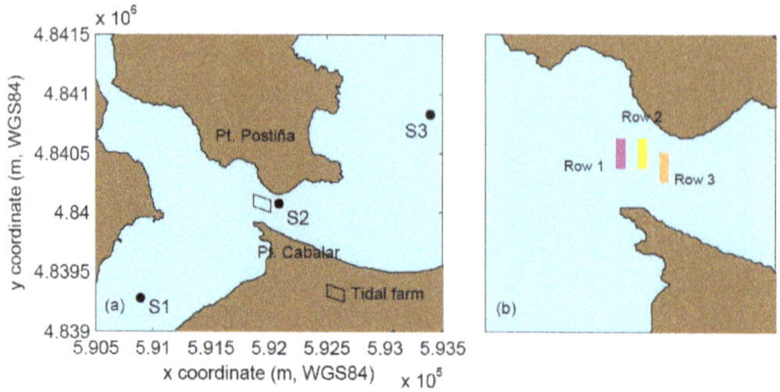

Figure 3. Location of the considered tidal stream plant, indicating the locations (S1, S2, and S3) for velocity and salinity profiles (**a**). Layout in plan view of the plants (**b**).

The TST selected is an Evopod turbine in view of its appropriateness for shallow waters areas [12,26,67], as is the case of the proposed location in the Ria de Ortigueira. Following the distance recommendations between lines of TSTs [68], 30 turbines with a sweep area of approx. 600 m^2 were distributed over three lines roughly perpendicular to the main flow direction, allowing their operation in a rather similar water depth. The technical characteristics of the Evopod turbine are provided in Table 3.

Table 3. Technical specifications of the Evopod turbine as provided by its developer (rotor diameter (D), cut-in velocity (V_{ci}), rated velocity (V_r), cut-of velocity (V_{co}), swept area (A), rated power (P_r), power coefficient in normal operation (C_{pNO}), *and* power coefficient in stall control (C_{pSC})).

D (m)	V_{ci} (ms^{-1})	V_r (ms^{-1})	V_{co} (ms^{-1})	A (m^2)	P_r (kW)	C_{pNO}	C_{pSC}
5	0.7	2.2	3.1	19.6	40	0.35	0.2

3.3. Tide-Driven Residual Circulation

The most usual procedure to compute the residual circulation induced by the tide consists of time-averaging or filtering the transient velocities computed by the numerical model [69]. There are other approaches, such as those based on the numerical solving of the so-called residual velocity evolution equations [70], which are not considered in the present application. The transient Eulerian horizontal velocity vector numerically computed (u, v) is decomposed into its periodic and residual components (u_p, v_p) and (u_r, v_r), respectively, as follows:

$$(u,v) = (u_p, v_p) + (u_r, v_r), \tag{11}$$

with the residual velocity vector computed as:

$$(u_r, v_r) = \frac{1}{T}\int_{-T/2}^{T/2}(u,v)dt, \tag{12}$$

where T represents the cut-off period of the averaging operator. A value of 12 h 25 min was adopted for this cut-off period in order to fully filter the variability of the semidiurnal tide prescribed by the dominance of the M2 tidal constituent.

3.4. Case Studies

In this work, a detailed assessment of the impacts of tidal stream energy exploitation on the residual circulation and estuarine upwelling was conducted following a seasonal approach. To this end, several case studies (CS) were undertaken with the aim of comparing the differences between altered and unaltered conditions (i.e., with and without energy extraction, respectively). Based on available data (Section 2), autumn–winter and spring–summer scenarios were defined, considering typical characteristics of their forcing factors.

First, the tide is introduced by considering its major semidiurnal component (M2). Second, fluvial discharges and thermohaline conditions correspond with those established in Table 2. Finally, the average value of SW winds is considered. The combinations of forcings leads to a total of eight different case studies, which are summarized in Table 4.

Table 4. Summary of boundary conditions and the situations simulated in the different case studies analysed.

Case Study	Tidal Plant	Wind	Fluvial and Oceanic Conditions (Table 2)
CS1	No plant	No wind	Spring–summer
CS2	No plant	No wind	Autumn–winter
CS3	No plant	SW mean summer wind (3.32 ms^{-1})	Spring–summer
CS4	No plant	SW mean winter wind (6.03 ms^{-1})	Autumn–winter
CS5	Floating plant	No wind	Autumn–winter
CS6	Bottom plant	No wind	Autumn–winter
CS7	Floating plant	SW mean winter wind (6.03 ms^{-1})	Autumn–winter
CS8	Bottom plant	SW mean winter wind (6.03 ms^{-1})	Autumn–winter

4. Results and Discussion

4.1. Unaltered Conditions

Unaltered hydrodynamic conditions (i.e., without considering energy extraction) in the Ria de Ortigueira were analysed through four case studies (CS1 to CS4), focusing this analysis on how the seasonality of river discharges and thermohaline conditions, in conjunction with the circulation induced by upwelling favourable SW winds, influence the estuarine residual circulation. With this aim, the results are provided in terms of the vectorial distribution (vector) and magnitude (colour map) of the unaltered residual circulation patterns throughout the estuary.

CS1 describes the residual circulation within the Ria de Ortigueira under summer conditions for fluvial discharges and thermohaline conditions in the absence of winds. Figure 4 shows the resulting unaltered residual circulation patterns in surface and bottom layers of the numerical grid. Two regions can be distinguished in terms of residual circulation patterns. In the inner ria (i), similar hydrodynamic patterns (i.e., a marked outflow) are present throughout the whole water column, including also a vortex rotating counterclockwise, with somewhat stronger flow velocities in surface layers (with peak velocities located in the surroundings of the narrowing between Pt. Postiña and Pt. Cabalar of about 0.20 ms^{-1} and 0.15 ms^{-1} in surface and bottom layers, respectively). On the other hand, the middle and outer ria (ii) are characterized by typical patterns of a positive estuarine circulation (i.e., inflow on the bottom layer and outflow on the surface layer) driven from river discharges flowing out in the surface layers.

Regarding CS2, the same forcing factors as CS1 were considered but now under winter conditions (Figure 5). As a result, the induced circulation is similar in both case studies, with somewhat greater velocities in the surface layer in the case of CS2, in particular in the middle and outer ria, due to the larger fluvial discharges. Therefore, it can be concluded that, in the absence of winds, the seasonality of the hydrodynamic driving agents does not lead to significant variations in the residual circulation of the Ría de Ortigueira as a consequence of the typical low riverine inputs to this estuary.

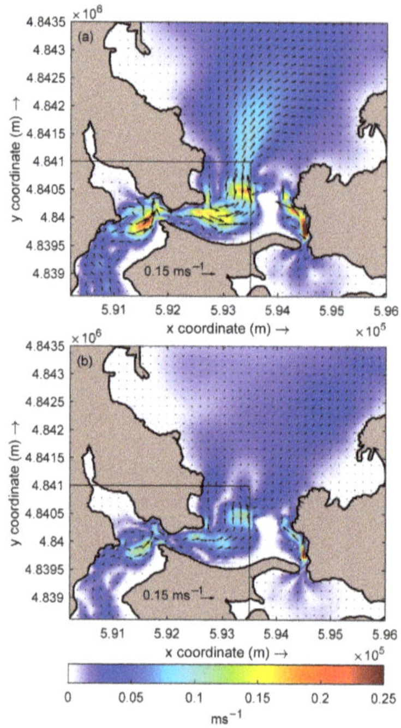

Figure 4. Residual flow pattern in Ria de Ortigueira under case study CS1 conditions. (**a**) Surface layer; (**b**) bottom layer.

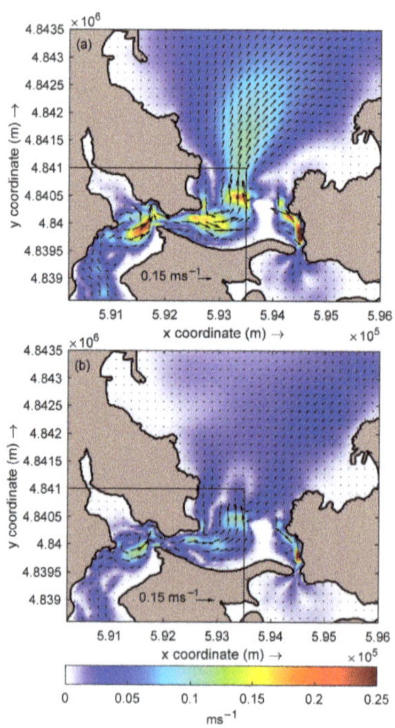

Figure 5. Residual flow pattern in Ria de Ortigueira under case study CS2 conditions. (**a**) Surface layer; (**b**) bottom layer.

The influence of river discharges driving the residual circulation in the inner part of the Ria can be further studied by analysing the vertical distribution of salinity at specific sites of interest, namely S1, S2, and S3 (Figure 3) in CS1 (spring–summer with no wind) and CS2 (autumn–winter) (Figure 6). It can be observed that this area can be considered as well-mixed throughout the whole year (autumn–winter and spring–summer scenarios), with virtual no differences in salinity throughout the vertical at S1 and S2 although with variations between seasons provoked by the different river discharges and oceanic thermohaline conditions (a greater fluvial discharge during winter reduces the salinity). This is the result of the large capacity of the tide (large tidal range in a shallow water area) to mix the overall low river discharges in this area. The capacity of the tide to mix the riverine and oceanic waters diminishes in the middle and outer ria, as the water depth and cross-sections increase, as it is apparent in S3.

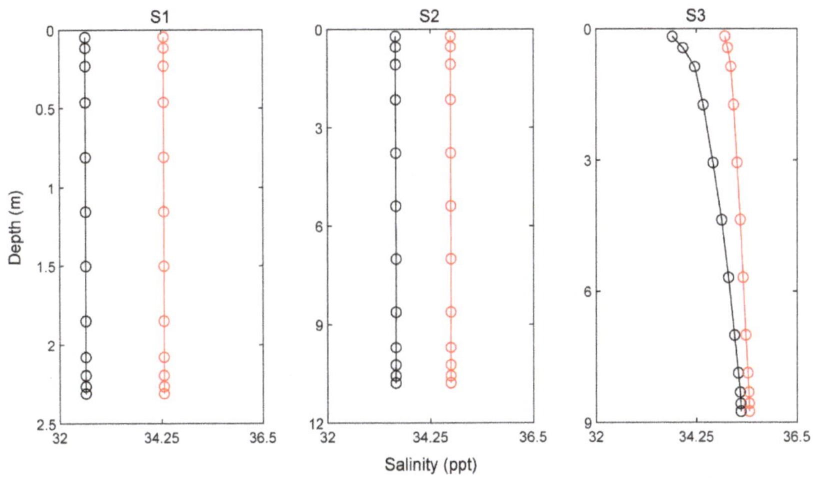

Figure 6. Salinity vertical profiles at S1, S2, and S3 during CS1 (spring–summer season) (red line) and CS2 (autumn–winter season) (black line).

The resulting low capacity of the river discharges generating residual currents is apparent in Figure 7, showing the vertical distribution of the residual circulation at S1, S2, and S3 in CS1 and CS2. The increase in discharge in the autumn–winter season is not capable of provoking significant greater residual velocities.

With respect to CS3, it corresponds to CS1 (i.e., summer conditions) but considering the action of the summer SW wind. The resulting circulation (Figure 8) could be characterized as a conventional two-layer estuarine circulation, reinforcing the residual patterns described in CS1 and CS2. In this case, the wind-induced outflow in the surface layer is strengthened, attaining values of about 0.25 ms^{-1} in the inner ria in the surroundings of the narrowing between Pt. Postiña and Pt. Cabalar and increasing the shelf-water intrusion through the bottom layers. Finally, it is worthy of mention that surface water masses tend to leave the ria next to its right margin. This pattern, which is also apparent although with less intensity in CS1 and CS2, is probably the result of wind action reinforced by the river discharges, which tend to deviate the flow to the right margin of the estuary as a consequence of the Coriolis force in the Northern Hemisphere.

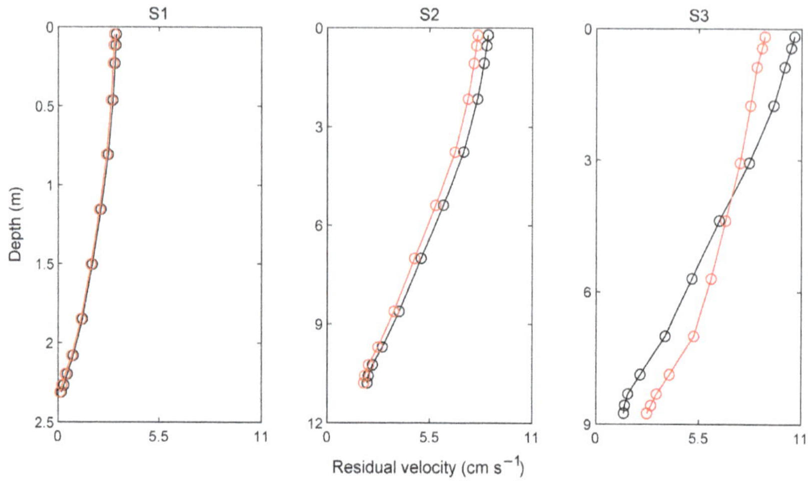

Figure 7. Vertical profiles of the magnitude of the residual circulation at S1, S2, and S3 during CS1 (spring–summer season) (red line) and CS2 (autumn–winter season) (black line).

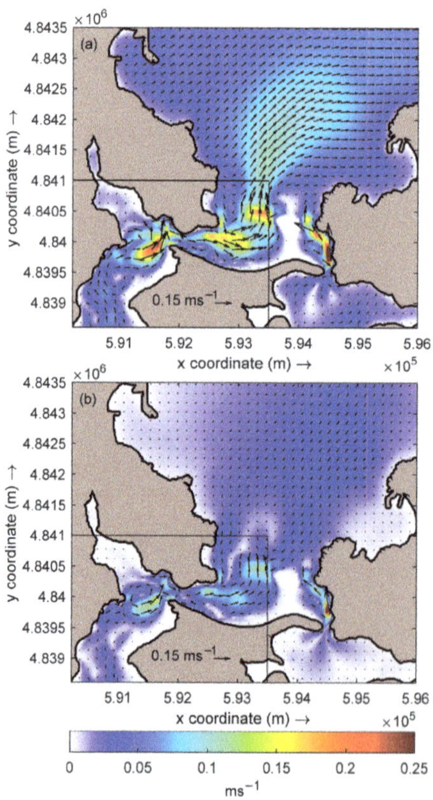

Figure 8. Residual flow pattern in Ria de Ortigueira under case study CS3 conditions. (**a**) Surface layer; (**b**) bottom layer.

As far as CS4 is concerned, it considers the same forcing factors as in the case of CS3 but under the action of a winter SW wind. The results obtained (Figure 9) are similar to those obtained in CS3 (Figure 8). However, due to the larger wind velocity and fluvial discharges under winter conditions, the residual currents induced in CS4 are stronger than in CS3, in particular in the surface layer, which is more influenced by wind action and river discharges, reaching values of up to 0.25 ms^{-1}.

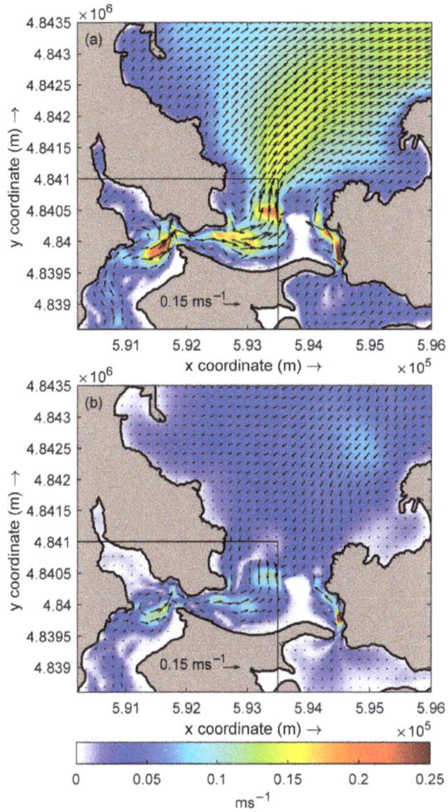

Figure 9. Residual flow pattern in Ria de Ortigueira under case study CS5 conditions. (**a**) Surface layer; (**b**) bottom layer.

Overall, the aforementioned comparison of the case studies corresponding to unaltered conditions (CS1 to CS4) shows how the typical low fluvial inputs discharging into the Ria de Ortigueira are not capable of inducing significant residual currents; on the contrary, the seasonality of the wind regime plays a key role in the residual circulation within this semi-enclosed waterbody.

4.2. Altered Conditions

The altered hydrodynamic conditions (i.e., considering energy extraction) in the Ria de Ortigueira that would result from the operation of a tidal farm were analysed through four additional case studies (CS5 to CS8), which consider different configurations (i.e., floating and bottom-fixed converters). As a consequence of the results provided by CS1 to CS4, which show the limited effects of the seasonality of thermohaline conditions and fluvial discharges on the resulting residual patterns, CS5 to CS8 only consider typical winter thermohaline and riverine conditions (summer scenarios are not retained) under no wind action and upwelling favourable winds.

With the aim of assessing the impacts of the energy extraction, the results are provided through a comparison with unaltered conditions case studies. In this way, the results provided by CS5 to CS8 (Figures 10–13) show the vectorial distribution of the altered residual circulation patterns (vectors) and the differences in the magnitude of the residual currents between altered and unaltered conditions (colormap) in the coastal area close to the location proposed for tidal energy exploitation (Figures 4, 5, 8 and 9).

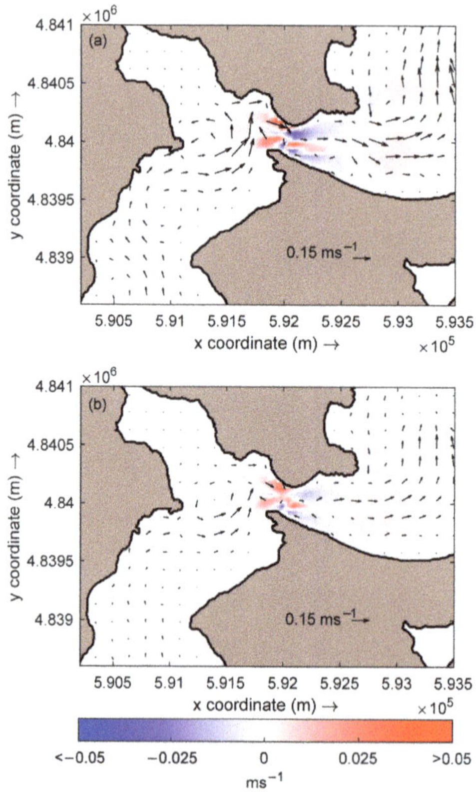

Figure 10. Residual flow pattern in Ria de Ortigueira under case study CS5 conditions (vectors); differences in the residual flow magnitude between altered flow and reference situation (CS5–CS2) (colour map). (**a**) Surface layer; (**b**) bottom layer.

CS5 computes the altered residual circulation within the study area considering winter thermohaline and riverine conditions in absence of winds. Regarding the implementation of the tidal farm and its energy extraction in the numerical model, it is conducted by considering a floating tidal farm, i.e., operating in the upper σ-layers of the grid. The results provided by CS5 (Figure 10) show that the major changes in the residual velocities are located in the constriction between Pt. Postiña and Pt. Cabalar as a consequence of the location of the proposed farm (Figure 3). Resulting from the layout of the tidal farm, which is distributed in three different lines along a W-E axis, the changes in the residual velocities along this section are spatially distributed as follows:(i) in the westernmost part of the constriction, in the surroundings of row 1, there is an apparent increase in the residual velocities towards the N and S margins of the ria for both surface and bottom layers, attaining maximum increments up to 0.05 ms^{-1} in the case of the surface layer, where the differences between unaltered and altered conditions are somewhat higher; (ii) in the central section of the constriction, in the surroundings of row 2, the situation is not homogenous, combining residual velocity increments (roughly 0.02–0.03 ms^{-1}) with weaker decrements (0.01–0.02 ms^{-1}) located W and E of row 2, respectively; finally, (iii) in the easternmost part of the constriction, in the surroundings of row 3, decrements towards the N (surface and bottom layers) and S margins (primarily in the surface layer) of the ria are apparent, along with a remarkable increase in the residual velocity within the central channel of this section (attaining maximum figures of approx. 0.05 ms^{-1}). Finally, the maximum residual velocities are of the same order than in the case of the unaltered conditions, roughly 0.25 ms^{-1}.

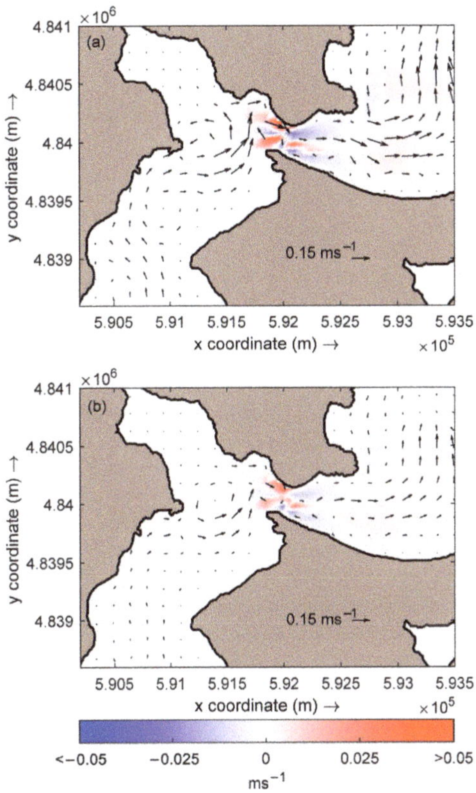

Figure 11. Residual flow pattern in Ria de Ortigueira under case study CS6 conditions (vectors); differences in the residual flow magnitude between altered flow and reference situation (CS6–CS2) (colour map). (**a**) Surface layer; (**b**) bottom layer.

In CS6, the same forcing factors and conditions as in the case of CS5 (winter thermohaline and riverine conditions in absence of winds) were input to the model but now considering a bottom-fixed tidal farm, i.e., operating in the lower σ-layers of the grid (Figure 11). The results show similar hydrodynamic patterns as in the case of the floating farm in CS5. With respect to the differences between unaltered and altered conditions, the spatial distribution of the effects of the energy extraction corresponds, in general, with those exposed in the case of CS5 (Figure 10); however, several peculiarities are now apparent: on the surface layer, the increments and decrements in the residual velocities are slightly stronger and weaker, respectively; on the contrary, on the bottom layer, the situation is roughly the opposite although with even less differences with respect to CS5. Moreover, as Figure 11 shows, there is an apparent decrement in the residual velocity (approx. 0.02 ms^{-1}) located in the surroundings of row 1, more specifically in the central channel of the ria, which is not present in the case of a floating plant.

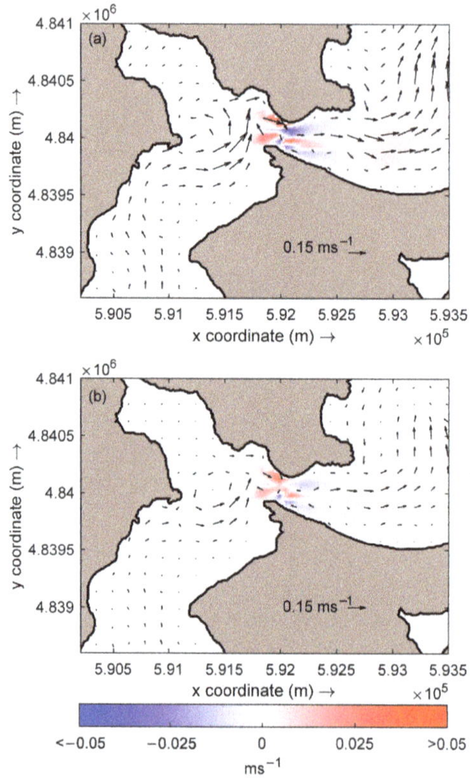

Figure 12. Residual flow pattern in Ria de Ortigueira under case study CS7 conditions (vectors); differences in the residual flow magnitude between altered flow and reference situation (CS7–CS4) (colour map). (**a**) Surface layer; (**b**) bottom layer.

As far as CS7 is concerned, winter thermohaline and riverine conditions were considered under the action of SW upwelling favourable winds. Likewise, the effects of the energy extraction were modelled by means of a floating tidal farm as in the case of CS5. The results of CS7 (Figure 12) show the influence of SW winds on the surface layers, in which greater outflow residual velocities are attained (i.e., close to the maximum figures of about 0.25 ms^{-1}) in comparison with previous altered conditions case studies in absence of winds. In the same way, bottom layers also present outflow residual flows throughout the inner ria although with weak residual velocity magnitudes. Regarding the spatial distribution of the impacts generated by the energy extraction in the vicinities of the farm, CS7 presents a similar pattern along the W-E axis, as in the case of CS5 and CS6: (i) in the surroundings of row 1, there are two major areas, located towards the N and S margin of the channel, in which the greatest increase in the residual velocities (i.e., 0.03–0.04 ms^{-1}) for both surface and bottom layers is apparent; (ii) in the surroundings of row 2, there exist several areas characterized by increments (0.02–0.03 ms^{-1}) and decrements (0.01–0.02 ms^{-1}) in the residual velocities W and E of row 2, respectively; finally, (iii) in the surroundings of row 3, the largest reductions in the residual velocities are located within two major areas towards the N (surface and bottom layer) and S (surface layer) margin of the ria; on the contrary, the maximum increments in the residual velocities (0.04 ms^{-1}) are located within the central channel of the narrowing between Pt. Postiña and Pt. Cabalar.

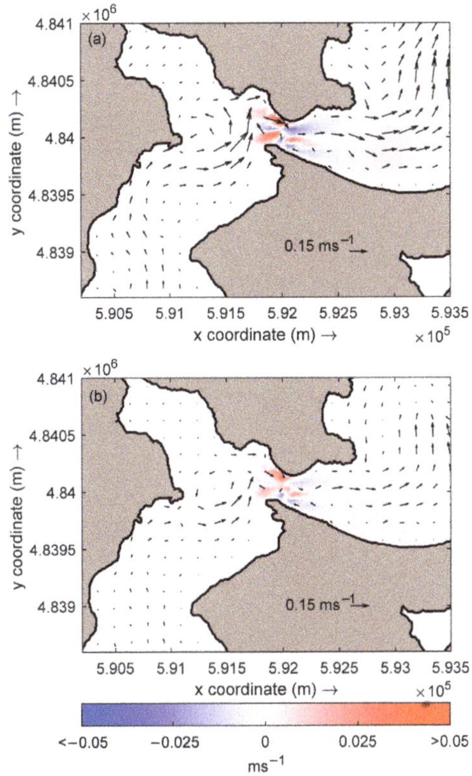

Figure 13. Residual flow pattern in Ria de Ortigueira under case study CS8 conditions (vectors); differences in the residual flow magnitude between altered flow and reference situation (CS8–CS4) (colour map). (**a**) Surface layer; (**b**) bottom layer.

Finally, CS8 computes the altered residual circulation within the inner Ria de Ortigueira produced by the operation of a bottom-fixed tidal farm under winter riverine and thermohaline conditions along with SW winds. The results obtained (Figure 13) are close, in terms of residual patterns, to those of CS6 in the absence of winds. Likewise, regarding the impacts of the energy extraction on the magnitude of the residual currents, the spatial distribution and magnitude of these impacts present, as in the case of other cases studies conducted (CS5 to CS7), a clear spatial distribution in three different areas (corresponding with the surroundings of the different lines of TSTs), which alternate increments and decrements in residual velocities. Finally, Figure 13 confirms that, in the case of a bottom-fixed tidal farm, a clear reduction in the residual velocity in the surroundings of row 1 is observed and, more specifically, in the central channel of the ria, which is not apparent in the case of a floating plant, as indicated in CS6 (Figure 11).

All in all, it can be established that the operation of the tidal farm proposed in the Ría de Ortigueira only provokes significant alterations in the residual circulation (less than 0.05 ms^{-1}) in the area occupied by the farm and in its vicinities (about 150 m distance) regardless of the tidal farm configuration and seasonal scenario considered. These modifications diminish abruptly at some distance from the plant (about 1 km), where they can be neglected. Based on previous results [26], which computed the modifications in the transient circulation in this area with the same plant configuration (both floating and bottom fixed plants), it can be established that the alteration in the magnitude of the residual flow is lower than in the transient circulation and is restricted to a smaller region. This situation contrasts with studies considering a tidal plant occupying the whole water column, in which the alteration in the residual currents is greater than that in the transient circulation [25].

5. Conclusions

In this work, a 3D numerical model was implemented in the Ria de Ortigueira, the westernmost of the Galician Rias Altas (NW Spain), in order to study the potential effects of different configurations of tidal stream farms on the residual circulation, including the estuarine upwelling and its seasonality. First, the model was successfully validated against field measurements obtained by means of an ADCP (Acoustic Doppler Current Profiler) deployed at two different locations in the inner and middle ria during a 19-day period, obtaining correlation coefficients between model and measured data very close to unity. Next, a total of eight case studies (CS) were analysed in order to obtain a detailed seasonality assessment of the impacts of tidal stream energy exploitation on the residual flow. In particular, the unaltered hydrodynamic conditions (absence of tidal farm) and altered conditions (operation of floating and bottom-fixed tidal farms) under typical summer and winter scenarios were analysed, including the absence of wind action and upwelling favourable winds.

The unaltered conditions in the absence of wind action are rather similar under winter and summer conditions. In both cases, two different regions can be distinguished in terms of residual circulation patterns throughout the ria: (i) the inner ria presents similar hydrodynamic patterns throughout the whole water column, with somewhat stronger flow velocities in surface layers, reaching peak velocities of about 0.25 ms^{-1} and 0.15 ms^{-1} in surface and bottom layers, respectively. On the other hand, the middle and outer ria (ii) are characterized by typical patterns of a positive estuarine circulation.

When the summer SW wind is considered together with summer thermohaline conditions, the resulting circulation follows a conventional two-layer estuarine circulation, reinforcing the residual patterns described above, with similar peak values. However, in the case of a typical winter SW wind, although the resulting circulation pattern is also similar, due to the larger wind velocity and fluvial discharges present, the residual currents induced are stronger, in particular in the surface layer.

Overall, when a tidal farm that partially blocks the water column (either floating or bottom-fixed) operates, the resulting modifications on the residual flow are primarily restricted to the narrowing between Pt. Postiña and Pt. Cabalar in the area occupied by the farm and in its vicinities (roughly over 100–150 m distance) regardless of the combination of tidal farm configuration and boundary conditions selected. The maximum values of the variation in the residual velocities due to the tidal farm operation are of about 0.05 ms^{-1} in the influence area of the plant. Outside of this area, the variations on the residual flow magnitude are lower than approx. 0.01 ms^{-1} and are virtually negligible at a certain distance from the farm. These results establish that the alteration in the magnitude of the residual flow caused by the operation of a tidal plant either floating or bottom-fixed is lower than alteration in the transient circulation and is restricted to a smaller region. This situation contrasts with studies considering a tidal plant that occupies the entire water column, in which the modification of the residual currents has shown to be greater than in the transient circulation.

In general, and regardless of the combination of tidal farm configuration and hydrodynamic forcing factors, the residual flow velocity increases to the W of the plant and decreases to the E of the plant both in the surface and bottom layers. In addition, in the bottom layer, the magnitude of the effects is quite similar in all the CS considered (bottom or floating plant, with and without wind action) and of less than 0.03 ms^{-1}. Although the pattern in the variations in the surface layer is also similar in the four CS, their magnitude is slightly greater when a bottom plant operates in the absence of wind, with increments in the residual flow velocities of around 0.05 ms^{-1} at the W of the plant.

This paper determined the potential impact on the residual flows in the Ría de Ortigueira caused by the operation of different tidal stream farms configurations under different typical hydrodynamic scenarios, establishing that these impacts are restricted to the surroundings of the proposed plants and virtually do not affect the upwelling events, which are of crucial importance in this coastal area. The procedure followed can be used to analyse the potential impacts on the hydrodynamics in other estuaries of interest for a tidal stream energy operation.

Author Contributions: Conceptualization, M.S., R.C. and G.I.; methodology, M.S., D.M.F., R.C. and G.I.; software, M.S., D.M.F. and R.C.; validation, M.S., I.L. and R.C.; formal analysis, M.S., I.L. and R.C.; investigation, M.S. and R.C.; resources, M.S., D.M.F., I.L. and R.C.; data curation, M.S., D.M.F., I.L. and R.C.; writing—original draft preparation, M.S. and D.M.F.; writing—review and editing, I.L., R.C. and G.I.; visualization, M.S. and I.L.; supervision, R.C. and G.I.; project administration, R.C. and G.I.; funding acquisition, I.L., R.C. and G.I. All authors have read and agreed to the published version of the manuscript.

Funding: This work was supported by the PORTOS project, which is co-financed by the Interreg Atlantic Area Programme, through the European Regional Development Fund (grant number EAPA_784/2018) and "Axudas para a consolidación e estruturación de unidades de investigación competitivas nas universidades do Sistema Universitario Galego (2020–22)" with reference number ED341B 2020/25. During this work, I. López was supported by a postdoctoral grant of the "Programa de Axudas á etapa posdoutoral da Xunta de Galicia", with reference number ED481D 2019/019.

Institutional Review Board Statement: Not applicable.

Informed Consent Statement: Not applicable.

Data Availability Statement: Not applicable.

Acknowledgments: The authors are also grateful for the support of Science Foundation Ireland and MaREI, the Marine Renewable Energy Centre of Ireland, grant SFI MAREI2_12/RC/2302/P2 Platform RA1b.

Conflicts of Interest: The authors declare no conflict of interest.

References

1. Bekun, F.V.; Alola, A.A.; Sarkodie, S.A. Toward a sustainable environment: Nexus between CO_2 emissions, resource rent, renewable and nonrenewable energy in 16-EU countries. *Sci. Total Environ.* **2019**, *657*, 1023–1029. [CrossRef]
2. Pereira, F.; Neves, M.G.; López-Gutiérrez, J.-S.; Esteban, M.D.; Negro, V. Comparison of Existing Equations for the Design of Crown Walls: Application to the Case Study of Ericeira Breakwater (Portugal). *J. Mar. Sci. Eng.* **2021**, *9*, 285. [CrossRef]
3. Fobissie, E.N. The role of environmental values and political ideology on public support for renewable energy policy in Ottawa, Canada. *Energy Policy* **2019**, *134*, 110918. [CrossRef]
4. Herath, N.; Tyner, W.E. Intended and unintended consequences of US renewable energy policies. *Renew. Sust. Energ. Rev.* **2019**, *115*, 109385. [CrossRef]
5. Nicolli, F.; Vona, F. Energy market liberalization and renewable energy policies in OECD countries. *Energy Policy* **2019**, *128*, 853–867. [CrossRef]
6. Pischke, E.C.; Solomon, B.; Wellstead, A.; Acevedo, A.; Eastmond, A.; De Oliveira, F.; Coelho, S.; Lucon, O. From Kyoto to Paris: Measuring renewable energy policy regimes in Argentina, Brazil, Canada, Mexico and the United States. *Energy Res. Soc. Sci.* **2019**, *50*, 82–91. [CrossRef]
7. Erdiwansyah; Mamat, R.; Sani, M.S.M.; Sudhakar, K. Renewable energy in Southeast Asia: Policies and recommendations. *Sci. Total Environ.* **2019**, *670*, 1095–1102. [CrossRef]
8. European Commission Proposal for a Directive Amending Directive 98/70/EC Relating to the Quality of Petrol and Diesel Fuels and Amending Council Directive 93/12/EC and Amending Directive 2009/28/EC on the Promotion of the Use of Energy from Renewable Sources [COM(2012) 595]. 2012. Available online: https://eur-lex.europa.eu/legal-content/en/ALL/?uri=CELEX%3A32015L1513 (accessed on 16 October 2022).
9. Portillo Juan, N.; Negro Valdecantos, V.; Esteban, M.D.; López Gutiérrez, J.S. Review of the Influence of Oceanographic and Geometric Parameters on Oscillating Water Columns. *J. Mar. Sci. Eng.* **2022**, *10*, 226. [CrossRef]
10. Bahaj, A.S. Generating electricity from the oceans. *Renew. Sust. Energy Rev.* **2011**, *15*, 3399–3416. [CrossRef]
11. Carballo, R.; Iglesias, G.; Castro, A. Numerical model evaluation of tidal stream energy resources in the Ría de Muros (NW Spain). *Renew. Energy* **2009**, *34*, 1517–1524. [CrossRef]
12. Sánchez, M.; Carballo, R.; Ramos, V.; Iglesias, G. Energy production from tidal currents in an estuary: A comparative study of floating and bottom-fixed turbines. *Energy* **2014**, *77*, 802–811. [CrossRef]

13. Esteban, M.D.; Espada, J.M.; Ortega, J.M.; López-Gutiérrez, J.-S.; Negro, V. What about Marine Renewable Energies in Spain? *J. Mar. Sci. Eng.* **2019**, *7*, 249. [CrossRef]
14. European Commission. *The Exploitation of Tidal Marine Currents*; EUR16683EN; EC: Brussels, Belgium, 1996.
15. Zarzuelo, C.; López-Ruiz, A.; Díez-Minguito, M.; Ortega-Sánchez, M. Tidal and subtidal hydrodynamics and energetics in a constricted estuary. *Estuar. Coast. Shelf Sci.* **2017**, *185*, 55–68. [CrossRef]
16. Iglesias, G.; Sánchez, M.; Carballo, R.; Fernández, H. The TSE index—A new tool for selecting tidal stream sites in depth-limited regions. *Renew. Energy* **2012**, *48*, 350–357. [CrossRef]
17. Fouz, D.M.; Carballo, R.; Ramos, V.; Iglesias, G. Hydrokinetic energy exploitation under combined river and tidal flow. *Renew. Energy* **2019**, *143*, 558–568. [CrossRef]
18. Liu, Z.; Qu, H.; Shi, H. Numerical study on hydrodynamic performance of a fully passive flow-driven pitching hydrofoil. *Ocean Eng.* **2019**, *177*, 70–84. [CrossRef]
19. Qian, P.; Feng, B.; Liu, H.; Tian, X.; Si, Y.; Zhang, D. Review on configuration and control methods of tidal current turbines. *Renew. Sust. Energ. Rev.* **2019**, *108*, 125–139. [CrossRef]
20. Segura, E.; Morales, R.; Somolinos, J.A.; López, A. Techno-economic challenges of tidal energy conversion systems: Current status and trends. *Renew. Sust. Energ. Rev.* **2017**, *77*, 536–550. [CrossRef]
21. Zhou, Z.; Benbouzid, M.; Charpentier, J.; Scuiller, F.; Tang, T. Developments in large marine current turbine technologies—A review. *Renew. Sust. Energy Rev.* **2017**, *71*, 852–858. [CrossRef]
22. Ramos, V.; Carballo, R.; Álvarez, M.; Sánchez, M.; Iglesias, G. Assessment of the impacts of tidal stream energy through high-resolution numerical modeling. *Energy* **2013**, *61*, 541–554. [CrossRef]
23. Ramos, V.; Carballo, R.; Sanchez, M.; Veigas, M.; Iglesias, G. Tidal stream energy impacts on estuarine circulation. *Energy Conv. Manag.* **2014**, *80*, 137–149. [CrossRef]
24. Neill, S.P.; Jordan, J.R.; Couch, S.J. Impact of tidal energy converter (TEC) arrays on the dynamics of headland sand banks. *Renew. Energy* **2012**, *37*, 387–397. [CrossRef]
25. Sánchez, M.; Carballo, R.; Ramos, V.; Iglesias, G. Tidal stream energy impact on the transient and residual flow in an estuary: A 3D analysis. *Appl. Energy* **2014**, *116*, 167–177. [CrossRef]
26. Sanchez, M.; Carballo, R.; Ramos, V.; Iglesias, G. Floating vs. bottom-fixed turbines for tidal stream energy: A comparative impact assessment. *Energy* **2014**, *72*, 691–701. [CrossRef]
27. Duarte, P.; Alvarez-Salgado, X.A.; Fernández-Reiriz, M.J.; Piedracoba, S.; Labarta, U. A modeling study on the hydrodynamics of a coastal embayment occupied by mussel farms (Ria de Ares-Betanzos, NW Iberian Peninsula). *Estuar. Coast. Shelf Sci.* **2014**, *147*, 42–55. [CrossRef]
28. Díez, J. *Las Costas*; Alianza Editorial: Madrid, Spain, 1996.
29. Alvarez, I.; Gomez-Gesteira, M.; de Castro, M.; Novoa, E.M. Ekman transport along the Galician Coast (NW, Spain) calculated from QuikSCAT winds. *J. Mar. Syst.* **2008**, *72*, 101–115. [CrossRef]
30. Gómez-Gesteira, M.; de Castro, M.; Prego, R.; Pérez-Villar, V. An Unusual Two Layered Tidal Circulation Induced by Stratification and Wind in the Ría of Pontevedra (NW Spain). *Estuar. Coast. Shelf Sci.* **2001**, *52*, 555–563. [CrossRef]
31. de Castro, M.; Gómez-Gesteira, M.; Alvarez, I.; Lorenzo, M.; Cabanas, J.M.; Prego, R.; Crespo, A.J.C. Characterization of fall-winter upwelling recurrence along the Galician western coast (NW Spain) from 2000 to 2005: Dependence on atmospheric forcing. *J. Mar. Syst.* **2008**, *72*, 145–158. [CrossRef]
32. Blanton, J.O.; Atkinson, L.P.; Castillejo, F.; Lavin, A. Coastal upwelling off the Rias Bajas, Galicia, Northwest Spain I: Hydrographic Studies. *Réun. Cons. Int. Explor. Mer.* **1984**, *183*, 79–90.
33. Villacieros-Robineau, N.; Herrera, J.L.; Castro, C.G.; Piedracoba, S.; Roson, G. Hydrodynamic characterization of the bottom boundary layer in a coastal upwelling system (Ría de Vigo, NW Spain). *Cont. Shelf Res.* **2013**, *68*, 67–79. [CrossRef]
34. Alvarez-Salgado, X.A.; Roson, G.; Perez, F.F.; Pazos, Y. Hydrographic variability off the Rias Baixas (NW Spain) during the upwelling season. *J. Geophys. Res.* **1993**, *98*, 14447–14455. [CrossRef]
35. Souto, C.; Gilcoto, M.; Fariña-Busto, L.; Pérez, F.F. Modeling the residual circulation of a coastal embayment affected by wind-driven upwelling: Circulation of the Ría de Vigo (NW Spain). *J. Geophys. Res.* **2003**, *108*, 4-1–4-18. [CrossRef]
36. Herrera, J.L.; Rosón, G.; Varela, R.A.; Piedracoba, S. Variability of the western Galician upwelling system (NW Spain) during an intensively sampled annual cycle. *An EOF analysis approach. J. Mar. Syst.* **2008**, *72*, 200–217. [CrossRef]
37. Prego, R.; Barciela, M.d.C.; Varela, M. Nutrient dynamics in the Galician coastal area (Northwestern Iberian Peninsula): Do the Rias Bajas receive more nutrient salts than the Rias Altas? *Cont. Shelf Res.* **1999**, *19*, 317–334. [CrossRef]
38. Wooster, W.S.; Bakun, A.; McLain, D.R. Seasonal upwelling cycle along the eastern boundary of the North Atlantic. *J. Mar. Res.* **1976**, *34*, 131–141.
39. Ríos, A.F.; Pérez, F.F.; Fraga, F. Water masses in the upper and middle North Atlantic Ocean east of the Azores. *Deep Sea Res. Part A Oceanogr. Res. Pap.* **1992**, *39*, 645–658. [CrossRef]
40. Iglesias, G.; Carballo, R. Seasonality of the circulation in the Ría de Muros (NW Spain). *J. Mar. Syst.* **2009**, *78*, 94–108. [CrossRef]
41. Alvarez, I.; Gomez-Gesteira, M.; de Castro, M.; Gomez-Gesteira, J.L.; Dias, J.M. Summer upwelling frequency along the western Cantabrian coast from 1967 to 2007. *J. Mar. Syst.* **2010**, *79*, 218–226. [CrossRef]
42. Torres, R.; Barton, E.D.; Miller, P.; Fanjul, E. Spatial patterns of wind and sea surface temperature in the Galician upwelling region. *J. Geophys. Res.* **2003**, *108*, 3130. [CrossRef]

43. Álvarez-Salgado, X.A.; Labarta, U.; Vinseiro, V.; Fernández-Reiriz, M.J. Environmental drivers of mussels flesh yield in a coastal upwelling system. *Ecol. Indic.* **2017**, *79*, 323–329. [CrossRef]
44. Ospina-Alvarez, N.; Prego, R.; Álvarez, I.; de Castro, M.; Álvarez-Ossorio, M.T.; Pazos, Y.; Campos, M.J.; Bernárdez, P.; Garcia-Soto, C.; Gómez-Gesteira, M.; et al. Oceanographical patterns during a summer upwelling–downwelling event in the Northern Galician Rias: Comparison with the whole Ria system (NW of Iberian Peninsula). *Cont. Shelf Res.* **2010**, *30*, 1362–1372. [CrossRef]
45. Sánchez, M.; Iglesias, G.; Carballo, R.; Fraguela, J.A. Power peaks against installed capacity in tidal stream energy. *IET Renew. Power Gener.* **2013**, *7*, 246–253. [CrossRef]
46. Pugh, D.T. *Tides, Surges, and Mean Sea-Level/a Handbook for Engineers and Scientists*; John Wiley & Sons Inc.: Hoboken, NJ, USA, 1996; p. 486.
47. Cerdeira-Arias, J.D.; Otero, J.; Barceló, E.; Río, G.d.; Freire, A.; García, M.; Nombela, M.Á.; Portilla, G.; Rodríguez, N.; Rosón, G.; et al. Hydrography of shellfish harvesting areas in the western Cantabrian coast (Rías Altas, NW Iberian Peninsula). *Reg. Stud. Mar. Sci.* **2018**, *22*, 125–135. [CrossRef]
48. Iglesias, G.; Carballo, R. Effects of high winds on the circulation of the using a mixed open boundary condition: The Ría de Muros, Spain. *Environ. Modell. Softw.* **2010**, *25*, 455–466. [CrossRef]
49. McClain, C.R.; Chao, S.Y.; Atkinson, L.P.; Blanton, J.O.; de Castillejo, F.F. Wind-driven upwelling in the vicinity of Cape Finisterre, Spain. *J. Geophys. Res.* **1986**, *91*, 8470–8486. [CrossRef]
50. Alvarez, I.; Ospina-Alvarez, N.; Pazos, Y.; de Castro, M.; Bernardez, P.; Campos, M.J.; Gomez-Gesteira, J.L.; Alvarez-Ossorio, M.T.; Varela, M.; Gomez-Gesteira, M.; et al. A winter upwelling event in the Northern Galician Rias: Frequency and oceanographic implications. *Estuar. Coast. Shelf Sci.* **2009**, *82*, 573–582. [CrossRef]
51. Deltares. *User Manual Delft3D-FLOW*; Deltares: Delft, The Netherlands, 2010.
52. Philips, N.A. A co-ordinate system having some special advantages for numerical forecasting. *J. Meteorol.* **1957**, *14*, 184–185. [CrossRef]
53. Carballo, R.; Iglesias, G.; Castro, A. Residual circulation in the Ría de Muros (NW Spain): A 3D numerical model study. *J. Mar. Syst.* **2009**, *75*, 130. [CrossRef]
54. Iglesias, G.; Carballo, R. Can the seasonality of a small river affect a large tide-dominated estuary? The case of the Ria de Viveiro, Spain. *J. Coast. Res.* **2011**, *27*, 1170–1182. [CrossRef]
55. Torres López, S.; Varela, R.A.; Delhez, E. Residual circulation and thermohaline distribution of the Ría de Vigo: A 3-D hydrodynamic model. *Sci. Mar.* **2001**, *65*, 277–289. [CrossRef]
56. Dias, J.M.; Lopes, J.F. Implementation and assessment of hydrodynamic, salt and heat transport models: The case of Ria de Aveiro Lagoon (Portugal). *Environ. Modell. Softw.* **2006**, *21*, 1–15. [CrossRef]
57. Cheng, R.T.; Casulli, V.; Gartner, J.W. Tidal, Residual, Intertidal Mudflat (TRIM) Model and its Applications to San Francisco Bay, California. *Estuar. Coast. Shelf Sci.* **1993**, *36*, 235–280. [CrossRef]
58. Cheng, R.T.; Gartner, J.W. Harmonic analysis of tides and tidal currents in South San Francisco Bay, California. *Estuar. Coast. Shelf Sci.* **1985**, *21*, 57–74. [CrossRef]
59. Smith, S.D. Wind stress and heat flux over the ocean in gale force winds. *J. Phys. Oceanogr.* **1980**, *10*, 709–726. [CrossRef]
60. Yelland, M.J.; Moat, B.I.; Taylor, P.K.; Pascal, R.W.; Hutchings, J.; Cornell, V.C. Wind stress measurements from the open ocean corrected for airflow distortion by the ship. *J. Phys. Oceanogr.* **1998**, *28*, 1511–1526. [CrossRef]
61. Stelling, G.; Leendertse, J. Approximation of convective processes by cyclic ADI methods. In Proceedings of the 2nd ASCE Conference on Estuarine and Coastal Modelling, Tampa, FL, USA, 13–15 November 1992.
62. Chung, T.J. *Computational Fluid Dynamics*; Cambridge University Press: Cambridge, UK, 2002; p. 1012.
63. Rodi, W. Examples of calculation methods for flow and mixing in stratified fluids. *J. Geophys. Res.* **1987**, *92*, 5305–5328. [CrossRef]
64. Defne, Z.; Haas, K.A.; Fritz, H.M. Numerical modeling of tidal currents and the effects of power extraction on estuarine hydrodynamics along the Georgia coast, USA. *Renew. Energy* **2011**, *36*, 3461–3471. [CrossRef]
65. Shapiro, G.I. Effect of tidal stream power generation on the region-wide circulation in a shallow area. *Ocean Sci.* **2011**, *7*, 165–174. [CrossRef]
66. Bahaj, A.S.; Molland, A.F.; Chaplin, J.R.; Batten, W.M.J. Power and thrust measurements of marine current turbines under various hydrodynamic flow conditions in a cavitation tunnel and a towing tank. *Renew. Energy* **2007**, *32*, 407–426. [CrossRef]
67. Álvarez, M.; Ramos, V.; Carballo, R.; López, I.; Fouz, D.M.; Iglesias, G. Application of Marine Spatial Planning tools for tidal stream farm micro-siting. *Ocean Coast. Manag.* **2022**, *220*, 106063. [CrossRef]
68. Myers, L.E.; Bahaj, A.S. An experimental investigation simulating flow effects in first generation marine current energy converter arrays. *Renew. Energy* **2012**, *37*, 28–36. [CrossRef]
69. Umgiesser, G. Modelling the Venice Lagoon. *Int. J. Salt Lake Res.* **1997**, *6*, 177–199. [CrossRef]
70. Salas-de-León, D.A.; Carbajal-Pérez, N.; Monreal-Gómez, M.A.; Barrientos-MacGretor, G. Residual circulation and tidal stress in the Gulf of California. *J. Geophys. Res.* **2003**, *108*, 3317. [CrossRef]

Journal of *Marine Science and Engineering*

MDPI

Article

Numerical Model of Constrained Wave Energy Hyperbaric Converter under Full-Scale Sea Wave Conditions

Moisés Brito [1,*], Francisco Bernardo [1], Maria G. Neves [2], Diogo R. C. B. Neves [3], Alejandro J. C. Crespo [4,*] and José M. Domínguez [4]

1 Research and Development Unit for Mechanical and Industrial Engineering (UNIDEMI), Department of Mechanical and Industrial Engineering, NOVA School of Science and Technology, 2829-516 Caparica, Portugal; fr.bernardo@campus.fct.unl.pt

2 CERIS—Civil Engineering Research and Innovation for Sustainability, Department of Civil Engineering, NOVA School of Science and Technology, 2829-516 Caparica, Portugal; mg.neves@fct.unl.pt

3 INEGI—Institute of Science and Innovation in Mechanical and Industrial Engineering, Campus da FEUP, Rua Dr. Roberto Frias 400, 4200-465 Porto, Portugal; dneves@inegi.up.pt

4 EPHYSLAB—Environmental Physics Laboratory, CIM-UVIGO, Campus As Lagoas, Universidade de Vigo, 32004 Ourense, Spain; jmdominguez@uvigo.es

* Correspondence: moisesbrito@fct.unl.pt (M.B.); alexbexe@uvigo.es (A.J.C.C.)

Abstract: A 2D numerical investigation of the power absorption of a constrained wave energy hyperbaric converter (WEHC) under full-scale sea wave conditions is presented. A fully non-linear numerical model DualSPHysics, based on the coupling of a smoothed particle hydrodynamics (SPH) fluid solver with a multibody dynamics solver, is used to model the interaction between wave and WEHC sub-systems. The numerical model was first validated against experimental data for a similar device, with a good accordance between PTO position and velocity. The model is then employed to study the hydrodynamics of a constrained WEHC considering several sea states, different hydraulic power take-off (PTO) damping and breakwater geometries. It is observed that the capture width ratio (CWR) is particularly sensitive to variations in the PTO damping, although the CWR absolute maximum is less sensitive considering mild variations applied to the PTO damping. Both wave height and wave period have an important effect on the CWR. The breakwater geometry is also essential for the performance of the WEHC, with a decrease in maximum CWR of about 15% for porous breakwater. These results are necessary to understand the full-scale behaviour of WEHC.

Keywords: DualSPHysics; wave energy hyperbaric converter (WEHC); non-linear numerical modelling; smoothed particle hydrodynamics (SPH); power absorption; constrained optimisation

Citation: Brito, M.; Bernardo, F.; Neves, M.G.; Neves, D.R.C.B.; Crespo, A.J.C.; Domínguez, J.M. Numerical Model of Constrained Wave Energy Hyperbaric Converter under Full-Scale Sea Wave Conditions. *J. Mar. Sci. Eng.* **2022**, *10*, 1489. https://doi.org/10.3390/jmse10101489

Academic Editor: Domenico Curto

Received: 2 September 2022
Accepted: 10 October 2022
Published: 14 October 2022

1. Introduction

A wave energy hyperbaric converter (WEHC) is a top-mounted pitching point wave energy converter (WEC) device, which mainly consists of three different sub-systems: floater, lever arm and a complex hydraulic power take-off (PTO) system [1,2]. The floater is connected by a lever arm to a hinge in the support platform where the rotational motion of the lever arm is damped by a PTO system to absorb power in operational conditions. The PTO system consists of a hydraulic pump linked to lever arm, a hyperbaric chamber and a hydropneumatic accumulator which acts as a power storage to supply a Pelton turbine coupled to a generator, converting mechanical power into electricity. The main components of the WEHC are sketched in Figure 1.

The mechanical constraints imposed by each connection among the three sub-systems, i.e., universal joint between lever arm and floater, revolute joints between lever arm and support platform and the hydraulic pump (Figure 1), have an important influence in the power absorption of WEHC. However, most of the previous studies, e.g., [2–7], have simplified the modelling of WEHC, neglecting the description of the mechanical constraints among the various sub-systems. Additionally, the PTO system is often modelled as a linear spring and damper [8]. There are two main challenges to resolve the non-linear dynamics of WEHC and to simulate sea waves in an accurate way: (i) the numerical model should predict correctly the imposed wave motion on the floater; (ii) the non-linear feedback of floater motion on wave field under several sea states and operation modes. This paper addresses the above challenges by implementing the mechanical constraints that simulate the non-linear interaction between each sub-system and understanding its effects on the power absorption under full-scale sea wave conditions. For that purpose, the objective is providing stronger evidence on the real motion of the constrained WEHC for different hydraulic PTO dampings and breakwater geometries.

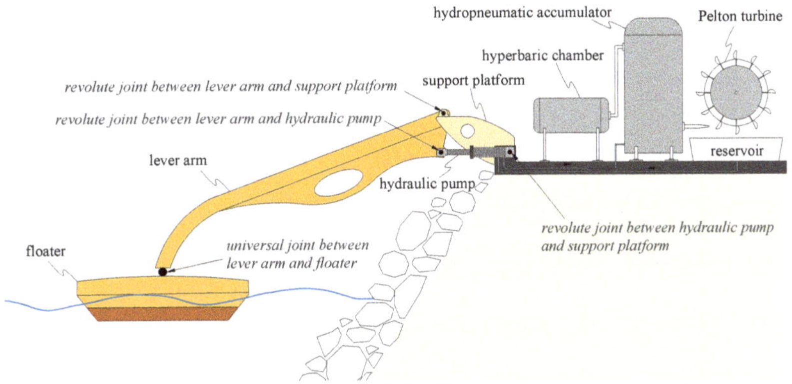

Figure 1. Schematic view of WEHC with main components (floater, lever arm, support platform and hydraulic PTO system), including the mechanical constraints among the three different sub-systems (universal joint between lever arm and floater, revolute joints between lever arm and support platform and hydraulic pump).

Efforts to model similar point-absorber-type WECs have been made by using both mesh-based and mesh-free computational fluid dynamic (CFD) methods. Mesh-based methods have been widely adopted to solve the governing equations by several codes such as COMFLOW [9], COMET [10] and AMAZON-SC [11], as well as in-house CFD codes [12–14]. Recently, [15] applied the OpenFOAM code to simulate a point-absorber-type WEC known as Wavestar, showing that the fully non-linear numerical model accurately reproduces the observed laboratory experiments of wave–floater interaction. To include PTO models, [8] developed in OpenFOAM a simulation platform for WECs by coupling a CFD-based numerical wave tank with PTO models. By comparing with other models based on the linear and partially non-linear boundary element method, their coupled model (including both the CFD and the PTO models) clearly showed that a fully non-linear numerical model is essential to resolve the wave–WEC interaction. The above listed research, among others, clearly highlighted that to accurately simulate WECs, there is a need to develop advanced models to resolve non-linear waves, physical processes of wave–WEC interaction and power absorption. However, the mesh-based method poses difficulties to include mechanical constraints. Such a method also implies solving additional equations for free-surface capturing and requires commonly expensive and complicated mesh moving algorithms to solve WECs motion. Additionally, solving mechanical constraints often requires additional complex mathematical transformations to solve non-linear complementarity problems, singularities, non-uniqueness, non-penetration and inequality constraints, which can be time-consuming when compared with solving wave–WEC interaction. Moreover, some mechanical constraints are not, in general, continuous, and such applications are not straightforward for a mesh-based method considering the ordinary differential equations that require impracticable mesh [16–18].

A promising alternative to overcome these mesh-based limitations is the mesh-free method. This method is favoured for modelling non-linear sea waves, since a mesh is not required when solving the Lagrangian form of the governing equations. Among the more up-to-date mesh-free methods, the smoothed particle hydrodynamics (SPH) has been widely adopted to study complex wave–WEC interaction. This method has been properly validated to study point-absorber-type WECs (e.g., [19–21]). SPH is able to simulate unsteady and non-linear flows, extreme deformations and complex topological evolutions, since the free surface is captured naturally [22,23]. However, given the dynamic nature of the SPH method, mechanical constraints are also difficult to implement in SPH form. Regarding this limitation, refs. [24–26] have successfully developed a coupled numerical model to simulate wave–WEC interactions. This coupling approach consists of an augmented SPH model of DualSPHysics [27] with the multibody solver of Project Chrono [28]. This is specifically designed to support the simulation of complex fluid–structure interaction considering its mechanical constraints. Hence, the SPH is used to model wave–WEC interaction, and multibody solver is used to model the flap–mechanical constraints interaction. The top-mounted pitching point absorber has also been modelled using DualSPHysics [29]. This device has a working principle similar to WEHC, however, in [29], a mechanical constraint between lever arm and floater and the effects of breakwater geometry were not considered.

Following the success of the SPH coupling approach for WEC modelling, this paper presents an analysis of the power absorption of a constrained WEHC under full-scale sea wave conditions. The details of the numerical model (coupling procedure between DualSPHysics and Project Chrono) are presented in Section 2. The numerical model validation is performed in Section 3. In Section 4, the characteristics of the capture width ratio (CWR) and PTO system are studied in both optimal and sub-optimal conditions, and the effects of wave height and wave period are analysed. Finally, Section 5 summarises the main findings.

2. Numerical Model

The coupling of an SPH fluid solver of DualSPHysics [30] with a multibody dynamics solver of Project Chrono [28] is used. This allows for efficient and accurate modelling of complex fluid–multibody interactions with any mechanical constraints, as described in [25] and [26]. The fluid–multibody interactions (e.g., non-linear waves' interaction with floaters and the interaction between floater, lever arm and PTO system) are mainly solved in three steps [26] that are described below.

In the first step, DualSPHysics solves the following continuity and Navier–Stokes equations in SPH form [31] with the density diffusion term [32]:

$$\frac{d\rho_i}{dt} = \sum_j m_j \vec{v}_{ij} \cdot \nabla_i W_{ij} + 2\delta h c_0 \sum_j \left(\rho_{ji}^T - \rho_{ij}^H\right) \frac{\vec{r}_{ij} \cdot \nabla_i W_{ij}}{\vec{r}_{ij}^2} \frac{m_j}{\rho_j} \tag{1}$$

$$\frac{d\vec{v}_i}{dt} = -\sum_j m_j \left(\frac{P_i + P_j}{\rho_i \rho_j} + \Pi_{ij}\right) \nabla_i W_{ij} + \vec{g} \tag{2}$$

where ρ is the density; t is the time; m is the mass; $\vec{v}_{ij} = \vec{v}_i - \vec{v}_j$, with $\vec{v}_i$ and $\vec{v}_j$ being, respectively, the velocity vector of the target particle i and neighbouring particle j; ∇ is the gradient operator; P is the pressure; c_0 is the reference speed of sound; g is the gravitational acceleration; ρ_{ji}^T and ρ_{ij}^H are, respectively, the total and hydrostatic part of the density according to [32]; and $W_{ij} = W(\vec{r}_{ij}, h)$ is the kernel function, which depends on the distance $\vec{r}_{ij} = \vec{r}_i - \vec{r}_j$ between i and j particles and characteristic smoothing length h. In this work, the Quintic Wendland kernel suggested by [33] is employed. δ is the delta-SPH coefficient that controls the intensity of the diffusive term. This coefficient is recommended for applications with open and moving solid boundaries, where a non-complete kernel is inevitably present, in order to reduce numerical diffusion [30]. In a strong dynamic case, this coefficient contributes with a force that may be several orders of magnitude smaller than the pressure and viscous terms, not contributing to a significant degradation of the solution. In Equation (2), Π is the viscous term described using the artificial viscosity proposed by [31].

In order to complete the set of governing equations, it is necessary to define an equation for pressure. A trivial implementation is to use an equation of state to relate the pressure to a slight variation in density, so-called Weakly Compressible SPH (WCSPH). This method requires a very small time step in order to solve the sound speed and depends on the delta-SPH term to stabilise the density field. The following Tait equation is used in DualSPHysics to determine fluid pressure based on particle density [22]:

$$P_i = \frac{c_0^2 \rho_0}{\gamma}\left[\left(\frac{\rho_i}{\rho_0}\right)^{\gamma} - 1\right] \tag{3}$$

where $\rho_0 = 1000\ \mathrm{kg/m^3}$ is the reference density, $\gamma = 7$ is the polytropic constant and c_0 is the sound speed at ρ_0, given as $c_0 = c(\rho_0) = \sqrt{\partial P/\partial \rho}|_{\rho_0}$ [30].

In DualSPHysics, a rigid-body dynamic is calculated based on the basic Newton equations, and the discretisation consists of summing the contributions from each SPH particle, as

$$M\frac{d\vec{V}}{dt} = \sum_{k\in b} m_i \frac{d\vec{v}_k}{dt} \tag{4}$$

$$I\frac{d\vec{\Omega}}{dt} = \sum_{k\in b} m_k\left(\vec{r}_k - \vec{R}_0\right) \times \frac{d\vec{v}_k}{dt} \tag{5}$$

where M is the mass and I is the inertial tensor of the rigid body b, $\vec{R}_0$ is the position of the centre of mass of a rigid body, $\vec{V}$ is the linear velocity and $\vec{\Omega}$ is the angular velocity. These vectorial quantities are computed at every time step by summing the contributions of each boundary particle k, belonging to body b.

In the second step, the linear $d\vec{V}/dt$ and angular $d\vec{\Omega}/dt$ acceleration to be applied in the centre of mass of a rigid body b are transferred to the Project Chrono library. During this time step, this library updates the motion, considering the given mechanical constraints and using the multibody dynamic model. The configuration of a rigid multibody system is described by using generalised coordinates $\vec{q} = [\vec{R}^{\mathrm{T}} + \vec{\Theta}^{\mathrm{T}}]^{\mathrm{T}}$, where $\vec{R}$ is the translational and $\vec{\Theta}$ is the rotational coordinates that define each body in the system frame [16].The dynamics of rigid bodies are characterised by a system of two differential algebraic equations that relate the time derivative of generalised coordinates and velocity through a linear transformation and the equation of the force balance that ties the inertial forces to the applied and constraint forces in the following form [18]:

$$\frac{d\vec{q}}{dt} = \vec{L}(\vec{q})\vec{V} \tag{6}$$

$$M(\vec{q})\frac{d\vec{V}}{dt} = \vec{F}_a(\vec{V},\vec{q},t) + \sum_{j\in\Phi}\left(\gamma_{j,n}\vec{D}_{j,n} + \gamma_{j,u}\vec{D}_{j,u} + \gamma_{j,w}\vec{D}_{j,w}\right) \tag{7}$$

where $\vec{V}$ is the velocity, M is the inertia matrix, $\vec{F}_a$ is the applied forces, $\vec{L}$ is the Jacobian matrix, and $\gamma_{j,n}\vec{D}_{j,n}$, $\gamma_{j,u}\vec{D}_{j,u}$ and $\gamma_{j,w}\vec{D}_{j,w}$ are the constraint forces at contact point j in normal n and tangential u and w direction [34]. Equations (6) and (7) are solved using a variable time step in Project Chrono [34–37].

In the third step, position, linear and angular velocity of the rigid body are transferred back to the DualSPHysics to update the particles that form the rigid body with the information transferred from the Project Chrono library (i.e., $\vec{V}$, $\vec{\Omega}$ and $\vec{R}_0$). The velocity of each particle of a rigid body is given by:

$$\vec{v}_i = \vec{V} + \vec{\Omega} \times \left(\vec{r}_i - \vec{R}_0\right) \quad (8)$$

The rigid bodies are also defined with a set of discrete particles [38]. The new modified Dynamic Boundary Condition (mDBC) approach is applied to describe the boundary particles [39]. This mDBC approach overcomes some of the drawbacks of the default Dynamic Boundary Condition (DBC) approach in DualSPHysics [40], resulting in less dissipation in flows near boundaries and reducing unphysical pressure and density values for the boundary and adjacent fluid particles. The motion of a floating body is described by net force on each boundary particle as the sum of the contributions of all surrounding fluid particles. The motion of a floating body is given by integrating Equation (8) in time, using the Symplectic scheme [41]. A variable time step established by the Courant–Friedrich–Lewy condition is considered [42]. This is based on the force per unit mass, viscous diffusion term and Courant–Friedrich–Lewy (CFL) condition. Lastly, DualSPHysics updates the position, velocity, density and pressure of all particles of the computational domain at the new time step.

3. Numerical Model Validation

Since no detailed data are available for WEHC, the numerical model was first validated against experimental data of [43] for a Wavestar point-absorber-type device. This selected device has a working principle similar to WEHC, based on the interaction between the three different sub-systems: floater, lever arm and hydraulic PTO system. The floater is connected by a fixed arm to a hinge in the support platform where the rotational motion is damped by a hydraulic pump to absorb power in operational conditions. In [43], the experimental tests were carried out using a 1:5-scale Wavestar model. The schematic of the computational domain with the Wavestar model is shown in Figure 2. The channel is 24 m-long and has depth in the flat-bottom section of $d = 3$ m. A passive dissipative beach located at the end of the channel with a slope of 10° is used. The wave was generated in DualSPHysics using a piston-type wavemaker with active wave absorption system (AWAS) to absorb any reflected wave energy coming from the impact with the beach and Wavestar [44]. The Wavestar model was located 14 m from the wavemaker with an initial submergence of 0.4 m. The geometry and dimensions of the Wavestar model are given in Figure 2b.

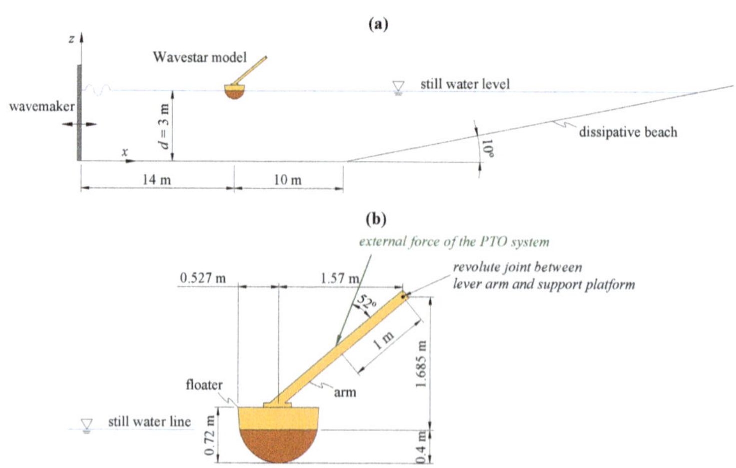

Figure 2. Schematic (side) view of the numerical set-up: (**a**) channel with the Wavestar model (not to scale); (**b**) dimensions (in m) of the Wavestar model.

In this study, four monochromatic wave conditions were considered. Table 1 presents the wave height H, wave period T and wave steepness kA, with k being the local wavenumber and A the wave amplitude of each test.

Table 1. Wave conditions considered for the validation of Wavestar.

	Test 1	**Test 2**	**Test 3**	**Test 4**
H (m)	0.1	0.15	0.25	0.25
T (s)	1.4	1.4	1.4	2.8
kA (-)	0.103	0.154	0.257	0.07

The hydraulic PTO system was modelled as an external force applied to the lever arm. This force was imposed 1 m from the revolute joint between lever arm and support platform at 52°, as shown in Figure 2b. As defined in [43], the applied PTO force is modelled as a linear spring–damper system:

$$F_{PTO}(t) = D\,\dot{z}(t) + K\,z(t) \tag{9}$$

where the first term, $D\,\dot{z}(t)$, is the linear damping force and second term, $K\,z(t)$, is the restitution force (linear spring) applied on the lever arm by the hydraulic pump, D and K being the damping and stiffness coefficients of the linear spring–damper PTO system, respectively. Similar to [43], $K = 0$, and four different damping coefficients were used $D = 0$, 50, 100 and 200 Nms.

Simulation accuracy is estimated by quantifying the discrepancy between the numerical and experimental Wavestar motion, including heaving position z and velocity $\dot{z}$. Before performing these comparisons, a convergence analysis was carried out. The numerical resolution in DualSPHysics is defined using the initial interparticle distance dp. Four different dp were considered: $dp = H/5$, $H/10$, $H/15$ and $H/20$. Similar to several applications of SPH models to simulate waves, a computational domain with $dp = H/10$ shows a good agreement between numerical results and experimental data [26]. This discretisation allows a good balance between computational cost and simulation runtime. The dp chosen in this work is $dp = 0.01$ m, which corresponds to the resolution of $dp = H/10$ for the smallest wave height tested (Test 1). With this resolution, about 624,500 particles are generated in DualSPHysics. The average time consumed was about 1 h to solve 50 s of physical time, running on an Nvidia RTX 3060.

The numerical and experimental time series of PTO position and velocity for Test 1 with different D are compared in Figure 3.

As can be observed, a good agreement was obtained between numerical results and the experimental data. However, a slight difference in both position and velocity can be found, especially for lower damping values (more amplitude of heaving motion). It is also observed that as the damping increases, the position and velocity decrease, as expected. The same behaviour was observed for other wave conditions and damping values; therefore, such results are not presented here. Figure 4 shows the time series of the PTO position and velocity for Test 2 with $D = 0$ and 200 Nms. All numerical results are consistent with the experimental data.

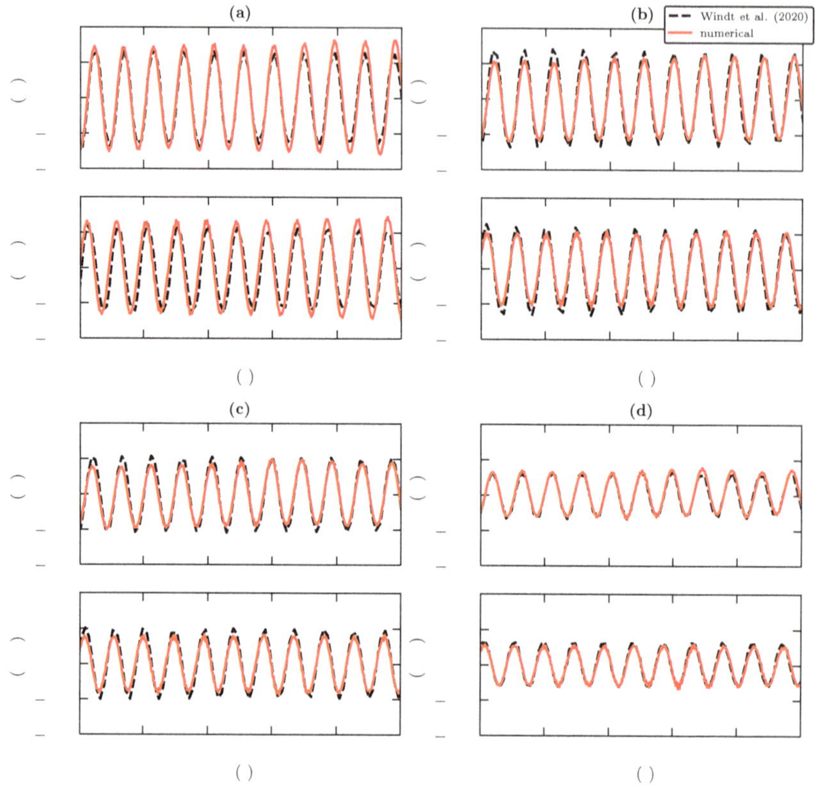

Figure 3. Time series of position (top) and velocity (bottom) of the PTO system for Test 1 with (**a**) $D = 0$, (**b**) $D = 50$ Nms, (**c**) $D = 100$ Nms and (**d**) $D = 200$ Nms [43].

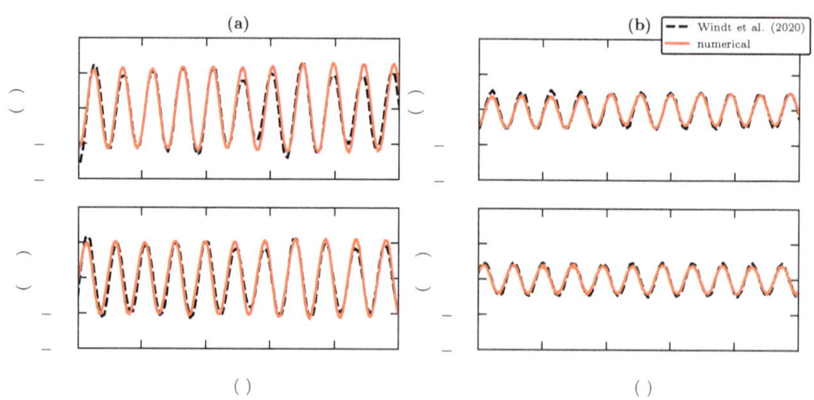

Figure 4. Time series of position (top) and velocity (bottom) of the PTO system for Test 2 with (**a**) $D = 0$ and (**b**) $D = 200$ Nms [43].

The time series for Test 3 and 4 with $D = 200$ Nms are compared in Figure 5. Similarly to Test 1 and 2, good agreement was found between the experimental data and the numerical results. Test 3 achieves slightly smaller differences with the experimental results than Test 4, even for the same wave height. This is due to a small lag in the wave period. However, the velocity appears to be slightly lagging behind when compared with the experimental velocity, with almost the same amplitude.

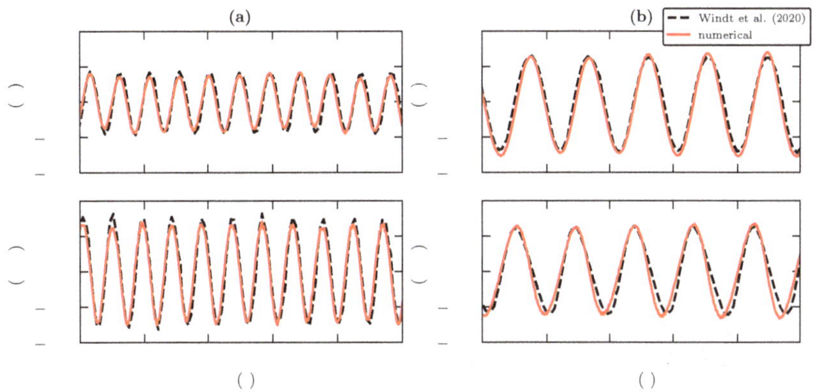

Figure 5. Time series of position (top) and velocity (bottom) of the PTO system with $D = 200$ Nms for (**a**) Test 3 and (**b**) Test 4 [43].

The time series of Wavestar motions obtained in the numerical model with different PTO values show a satisfactory agreement with the experimental data in the physical model tests. For a more quantitative comparison, the accordance between the numerical and experimental data was quantified using the average normalised root mean error (NRMSE) in amplitude. Table 2 shows NRMSE for the four tests with $D = 0$. Overall, the NRMSE is less than 10%, although in some conditions it is slightly higher, caused by the small difference in phase between the numerical results and experimental data. For all wave conditions, the NRMSE decreases with increasing D.

Table 2. Average NRMSE for the different wave states with $D = 0$.

	Test 1	Test 2	Test 3	Test 4
z (m)	10%	8%	7%	9%
$\dot{z}$ (m/s)	11%	6%	5%	13%

4. Results and Discussion

4.1. Simulation Set-Up

The power absorption of the WEHC device is investigated under full-scale sea wave conditions in the south-western coast of Portugal. The location of the western breakwater of the Port of Sines was considered for the present study. In order to correctly predict the power extracted from the waves, the first step is to correctly predict the incoming waves. For this reason, the waves chosen to simulate the device were based on the work of [45], who developed a detailed library of the most common waves and sea states along the Portuguese coast. Both regular and irregular sea states, and different operation modes and design parameters, were considered, including hydraulic PTO damping and breakwater geometries. Two different breakwater geometries were modelled: one as a straight wall with no porosity and another made by a double layer of cubic blocks with a porosity of around 40%, laying on an impermeable slope bed.

A schematic of the computational domain is shown in Figure 6. The channel is 240 m-long and has a depth in the flat-bottom region of $d = 30$ m. The WEHC device is located at the right-hand end of the channel with an initial submergence of 0.67 m. The geometry and dimensions of the device are presented in Figure 6b.

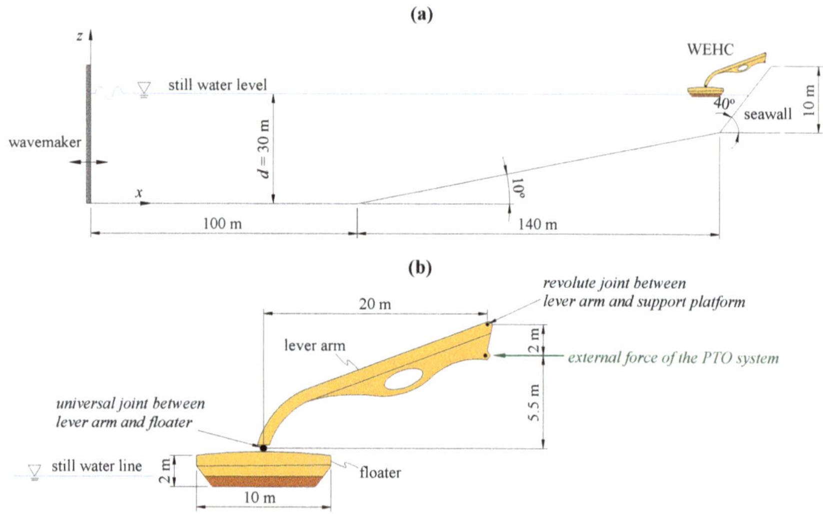

Figure 6. Schematic (side view) of the numerical set-up: (**a**) channel with the full-scale WEHC (not to scale); (**b**) dimensions (in m) of the WEHC model.

The computational domain was discretised with a $dp = 0.1$ m, i.e., $dp = H/10$ for the smallest wave height tested. The simulations were also performed on an Nvidia RTX 3060. The average computational time was about 5 h for a resolution of 564,374 particles and 300 s of physical time.

The hydraulic PTO system is modelled as a single-piston effect with the external force given as

$$F_{PTO}(t) = \begin{cases} C_{PTO} & \text{if} \quad z(t) < 0 \\ 0 & \text{if} \quad z(t) \geq 0 \end{cases} \tag{10}$$

where C_{PTO} is the constant force applied 2 m below of the revolute joint between lever arm and support platform. The friction torque on the universal joint between lever arm and floater and the revolute joint between lever arm and support platform was modelled using the Coulomb friction coefficient of 0.16 [26].

4.2. Power Absorption under Regular Sea State

The hydrodynamic response of a WEHC device under a regular sea state is analysed considering the capture width ratio (CWR), defined as the ratio between the average extracted power W_{out} and the average incoming wave power W_{inc}, which is

$$\text{CWR} = \frac{W_{out}}{W_{inc}} \tag{11}$$

The W_{out} is calculated as

$$W_{out} = \frac{1}{t_f - t_0} \int_{t_0}^{t_f} F_{PTO}(t)\dot{x}(t)dt \tag{12}$$

where $\dot{x}$ is the velocity of the point where the external force is applied (Figure 6b), and t_0 and t_f are the initial and final time, respectively.

According to linear wave theory, W_{inc} is calculated as

$$W_{inc} = \frac{1}{8}\rho g H^2 c_g \tag{13}$$

where c_g is the wave group velocity, and for this study, it is calculated as

$$c_g = \frac{\omega}{2k}\left(1 + \frac{2kd}{\sinh(2kd)}\right) \tag{14}$$

where $\omega = 2\pi f$ is the angular wave frequency, d is the water depth, f is the wave frequency and k is calculated by $\omega^2 = gk\tanh(kd)$.

Figure 7 compares the variation of CWR with C_{PTO} for six different sea states presented in Table 3. These sea states' characteristics are frequent on the south coast of Portugal [45].

Table 3. Regular wave conditions considered in the numerical tests.

	State 1	State 2	State 3	State 4	State 5	State 6
H (m)	2	2	2	2.5	2.5	2.5
T (s)	7	9	11	7	9	11
kA (-)	0.083	0.054	0.040	0.104	0.067	0.050

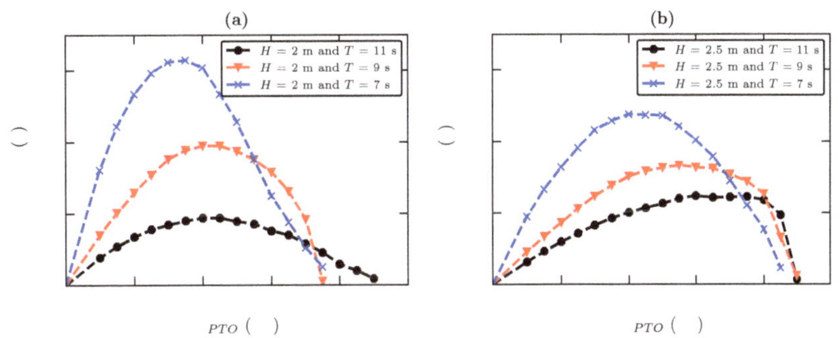

Figure 7. Influence of PTO damping on the CWR for T = 7, 9 and 11 s and (**a**) H = 2 m and (**b**) H = 2.5 m.

A large variation of CWR with PTO damping can be observed in Figure 7. These constraints, imposed by the hydraulic PTO system, have a non-negligible influence on the power absorption of WEHC. As the C_{PTO} increases, the CWR reaches its maximum. The curve shows almost similar trends for the different sea states tested, however, the maximum position is envisaged. For example, for $H = 2$ m and $T = 7$ s, the maximum CWR occurs with $C_{PTO} = 175$ kN, and for $H = 2$ m and $T = 9$ s, it occurs with $C_{PTO} = 225$ kN. A large variation of CWR is observed when the C_{PTO} increases due to the rise of the latching effect of the hydraulic PTO system, where the floater remains stationary for as long as the mass forces are unable to overcome the resisting force introduced by the PTO system. In general, the CWR decreases when the wave period increases. For $H = 2$ m and $T = 7$, the maximum CWR $\approx$ 32%, and for $H = 2$ m and $T = 11$, the maximum CWR $\approx$ 19%. A similar trend is also observed when increasing the wave height, for example, for $H = 2.5$ m and $T = 7$, the maximum CWR $\approx$ 23%. With increasing wave height, the maximum CWR occurs for a large C_{PTO}. With increasing wave period, the CWR curve becomes flatter, and the maximum occurs at a higher value of C_{PTO} for a longer period. This behaviour is caused by the non-linear waves interacting with the sea bottom that produces larger excitation forces, enlarging the bandwidth of the maximum CWR. A direct consequence of this behaviour is that it allows a simple control strategy for the pressure of the hydraulic PTO system based on wave period measurements and predictions from swell forecast models.

Figure 8 provides instantaneous snapshots of the velocity field in the vicinity of the WEHC device for $H = 2$ m and $T = 7$ s with straight breakwater. As expected, the flow field is characterised by a strong dynamical velocity field, with large acceleration near the floater. This causes differences in the turbulent field, characterised by the increase in the turbulent kinetic energy near the free surface and close to the floater.

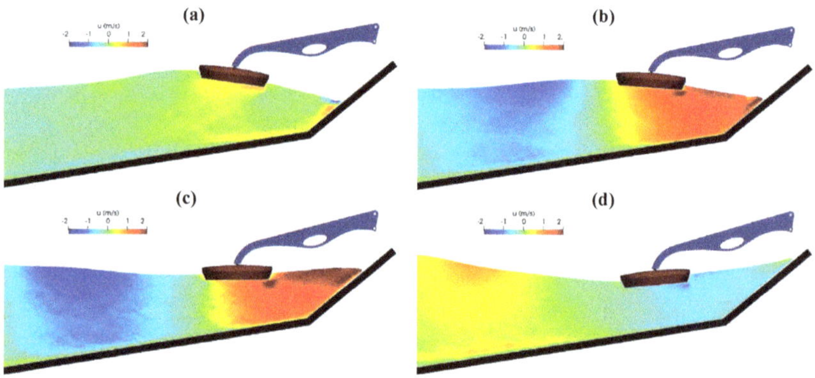

Figure 8. Snapshots of the velocity field in the vicinity of the WEHC device for $H = 2$ m and $T = 7$ s with impermeable and straight breakwater at (**a**) $t = 100$ s, (**b**) $t = 102$ s, (**c**) $t = 104$ s and (**d**) $t = 107$ s.

As seen in Figure 8, the breakwater is a straight wall with no porosity, however, the breakwaters tend to be built of shaped rocks or cement blocks. This method of construction allows for more dissipation of wave energy on the breakwater, ensuring a better protection of the onshore structures. Given the proven capability of the DualSPHysics–Project Chrono coupling for simulating fluid–structure interaction [46], one additional breakwater geometry was tested in this study in order to obtain results on the influence of more realistic breakwaters. The variation of CWR with C_{PTO} for six different sea states with $H = 2$ and 2.5 m, $T = 7$, 9 and 11 s presented in Table 3 are shown in Figure 9. Similar trends of CWR are found for different breakwater geometries, however, the maximum position is slightly envisaged. For example, for $H = 2$ m and $T = 7$ s with straight breakwater, the maximum CWR occurs with $C_{PTO} = 175$ kN, and for porous breakwater, it occurs with $C_{PTO} = 150$ kN. In general, the CWR reduces by about 5% from straight breakwater. The reflected waves are reduced due to the porosity of the breakwater. However, its effect on the excitation forces is much smaller than the incident counterpart, so it has less effect on CWR. The breakwater porosity has more effects when H increases; indeed, complex interactions arise when H increases, mainly for higher T, and the excitation forces are much affected.

In order to illustrate the influence of the breakwater porosity and geometry on the flow field, the detailed instantaneous velocity field in the vicinity of the WEHC device for $H = 2$ m and $T = 7$ s with porous breakwater is given in Figure 10.

The geometry shows the dissipation of the energy from incoming waves. Moreover, energy dissipation also causes a reduction in wave reflection, which in turn reduces the WEHC position motion, leading to a lower CWR. Even though this effect has a negative impact on energy extraction, this is closer to how breakwaters are built.

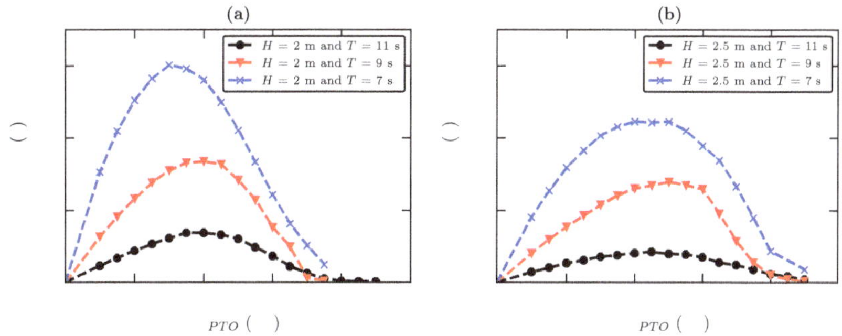

Figure 9. Influence of breakwater porosity and geometry and PTO damping on the CWR for $T = 7, 9$ and 11 s and (**a**) $H = 2$ m and (**b**) $H = 2.5$ m.

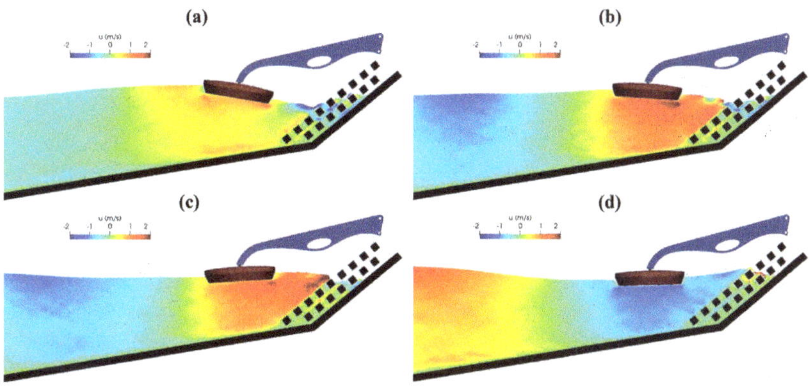

Figure 10. Snapshots of different instants of the velocity field in the vicinity of the WEHC device for $H = 2$ m and $T = 7$ s with porous breakwater at (**a**) t = 100 s, (**b**) t = 102 s, (**c**) t = 104 s and (**d**) t = 107 s.

4.3. Power Absorption under Irregular Sea State

The ability of the model to reproduce more realistic wave conditions was tested by simulating the behaviour of the WEHC device for irregular wave conditions. The following JONSWAP (Joint North Sea Wave Observation Project) spectral shape was used [47]:

$$S_\eta(f) = \alpha g^2 (2\pi)^{-4} f^{-5} \exp\left[-\frac{5}{4}\left(\frac{f_p}{f}\right)^4\right] \gamma^{\exp\left[-\frac{(f-f_p)^2}{2\sigma^2 f_p^2}\right]} \tag{15}$$

where $\alpha = 0.291$, f_p is the peak frequency, $\gamma = 3.3$ is peak enhancement factor, $\sigma = 0.07$ for $f \leq f_p$ and $\sigma = 0.09$ for $f > f_p$.

On the basis of the position time series, $x(t)$, and free surface elevation, η, the power density spectra of the WEHC motions S_x and free-surface S_η were calculated for each simulation. These power spectra were computed using Welch's technique, in which each power spectral density is estimated by dividing the data into overlapping segments, computing a modified periodogram for each segment and averaging the periodograms. The Hann window with segment length of 1024 points and overlap of 50% was used.

As can be observed in Figure 11, the spectra also properly reflect the spectral peaks, i.e., where most of the energy is concentrated and, thus, the frequency range of the greatest interest. Three sea states were considered with the significant wave height $H_s = 2.5$ m and peak period $T_p = 7, 9$ and 11 s.

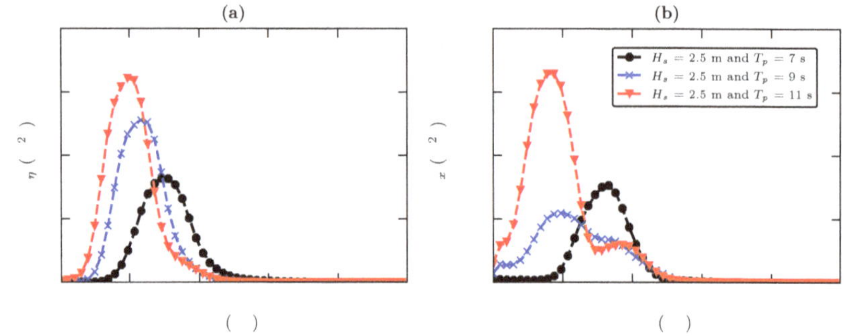

Figure 11. Power spectrum of (**a**) free-surface elevation measured at $x = 220$ m and of (**b**) horizontal position of hydraulic PTO system for $H_s = 2.5$ m and $T_p = 7, 9$ and 11 s.

The time series of each sea state for 100 s $< t <$ 300 s is shown in Figure 12. Similarly to the regular sea state, an increase in CWR with the peak period is observed. For $H_s = 2.5$ m and $T_p = 7$ s, a CWR of about 25% was found. The CWR for irregular waves is evaluated by Equation (11), where the average incident wave power was calculated as

$$W_{inc} = \rho g \int_0^\infty c_g(f) S_\eta(f) df \tag{16}$$

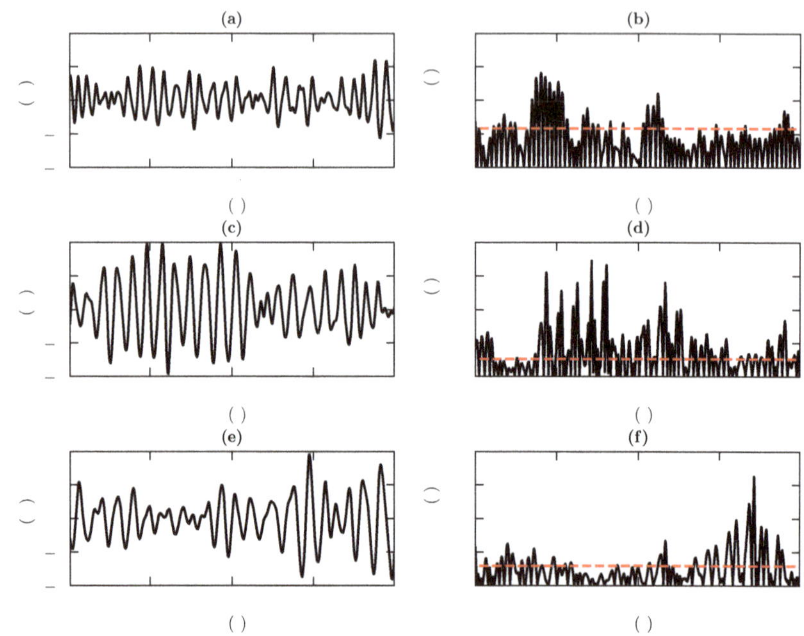

Figure 12. Detail of surface elevation and CWR for (**a**,**b**) $H_s = 2.5$ m and $T_p = 7$ s, (**c**,**d**) $H_s = 2.5$ m and $T_p = 9$ s, (**e**,**f**) $H_s = 2.5$ m and $T_p = 11$ s. In the power capture curve, the horizontal dashed red line indicates the mean power capture.

Figure 13 shows the power spectrum for three sea states considering significant wave height $H_s = 2$ m and peak period $T_p = 7, 9$ and 11 s. The spectra also properly reflect the spectral peaks.

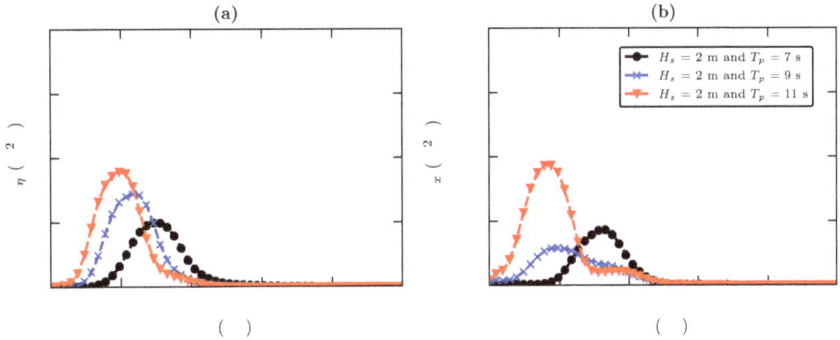

Figure 13. Power spectra of (**a**) free-surface elevation measured at $x = 220$ m and of (**b**) the horizontal position of the hydraulic PTO system for $H_s = 2$ m and $T_p = 7, 9$ and 11 s.

Figure 14 shows the time series of each sea state for 100 s $< t <$ 300 s. As for a regular sea state, an increase in CWR with the peak period is observed. For $H_s = 2$ m and $T_p = 7$ s, a CWR of about 30% was found.

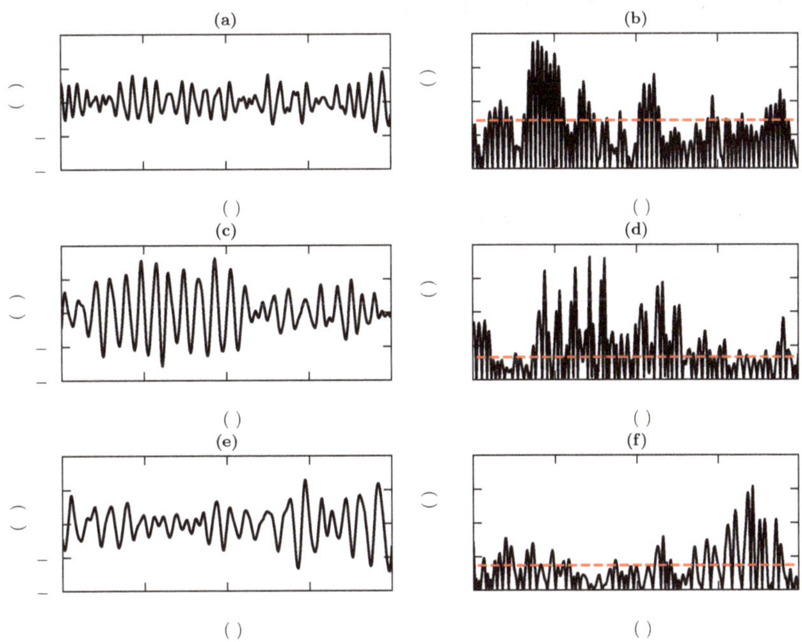

Figure 14. Detail of surface elevation and CWR for (**a**,**b**) $H_s = 2$ m and $T = 7$ s, (**c**,**d**) $H_s = 2$ m and $T = 9$ s, (**e**,**f**) $H_s = 2$ m and $T = 11$ s. In the power capture curve, the horizontal dashed red line indicates the mean power capture.

5. Conclusions

This paper presents a numerical investigation of the power absorption of a constrained wave energy hyperbaric converter (WEHC) under full-scale sea wave conditions. A fully non-linear numerical model DualSPHysics was used to model the interaction between waves and each WEHC sub-system. The numerical model was first validated against the available experimental data of a similar device, a 1:5-scale Wavestar model. A good agreement was found between the experimental data and the numerical results, with an average normalised root mean error (NRMSE) of about 10%.

The numerical model was then employed to perform a study of the non-linear hydrodynamics of a constrained WEHC under regular and irregular sea states, and for different operation modes and design parameters, including hydraulic power take-off (PTO) damping and breakwater geometry. For regular waves, it was observed that the capture width ratio (CWR) is particularly sensitive to the variations in the PTO damping, although the absolute maximum CWR is less sensitive to the slightly lower variations applied to the PTO damping. It was also found that the influence of the wave height and wave period have an important effect on the CWR. For irregular waves, a similar trend on the CWR was observed, with CWR increasing with decreasing peak period and an increase in significant wave height. In general, the efficiency of the system is very dependent on the mechanical constraints (PTO system) and on the breakwater characteristics. These results show the potential and effectiveness of this type of numerical modelling approach for understanding the non-linear dynamics of WEHC in a full-scale application.

Author Contributions: Conceptualisation, M.B. and F.B.; methodology, M.B., M.G.N. and D.R.C.B.N.; software, J.M.D. and A.J.C.C.; validation, M.B., F.B., J.M.D. and A.J.C.C.; formal analysis, M.B. and F.B.; investigation, M.B., F.B., M.G.N., D.R.C.B.N., J.M.D. and A.J.C.C.; resources, M.B.; writing—original draft preparation, M.B., F.B., M.G.N., D.R.C.B.N., J.M.D. and A.J.C.C.; writing—review and editing, M.B., F.B., M.G.N., D.R.C.B.N., J.M.D. and A.J.C.C.; visualisation, M.B., J.M.D. and A.J.C.C.; project administration, M.B.; funding acquisition, M.B. All authors have read and agreed to the published version of the manuscript.

Funding: This research was funded by Fundação para a Ciência e a Tecnologia (FCT.IP) through the grant UIDB/00667/2020 (UNIDEMI). This work was also supported by the project SURVIWEC PID2020-113245RB-I00 and by the project ED431C 2021/44 "Programa de Consolidación e Estructuración de Unidades de Investigación Competitivas". This study forms part of the Marine Science programme (ThinkInAzul) supported by Ministerio de Ciencia e Innovación and Xunta de Galicia with funding from the European Union NextGenerationEU (PRTR-C17.I1) and European Maritime and Fisheries Fund.

Acknowledgments: The authors acknowledge Fundação para a Ciência e a Tecnologia (FCT.IP) for its financial support through the grant UIDB/00667/2020 (UNIDEMI). This work was also supported by the project SURVIWEC PID2020-113245RB-I00 financed by MCIN/AEI/10.13039/501100011033 and by the project ED431C 2021/44 "Programa de Consolidación e Estructuración de Unidades de Investigación Competitivas" financed by Xunta de Galicia, Consellería de Cultura, Educación e Universidade. This study forms part of the Marine Science programme (ThinkInAzul) supported by Ministerio de Ciencia e Innovación and Xunta de Galicia with funding from the European Union NextGenerationEU (PRTR-C17.I1) and European Maritime and Fisheries Fund.

Conflicts of Interest: The authors declare no conflict of interest.

References

1. Estefen, S.F.; da Costa, P.R.; Ricarte, E.; Pinheiro, M.M. Wave Energy Hyperbaric Device for Electricity Production. In Proceedings of the Volume 5: Ocean Space Utilization, Polar and Arctic Sciences and Technology, The Robert Dean Symposium on Coastal and Ocean Engineering, Special Symposium on Offshore Renewable Energy, San Diego, CA, USA, 10–15 June 2007. [CrossRef]
2. Estefen, S.F.; Esperança, P.d.T.T.; Ricarte, E.; da Costa, P.R.; Pinheiro, M.M.; Clemente, C.H.P.; Franco, D.; Melo, E.; de Souza, J.A. Experimental and Numerical Studies of the Wave Energy Hyberbaric Device for Electricity Production. In Proceedings of the Volume 6: Nick Newman Symposium on Marine Hydrodynamics, Yoshida and Maeda Special Symposium on Ocean Space Utilization, Special Symposium on Offshore Renewable Energy, Estoril, Portugal, 15–20 June 2008. [CrossRef]
3. Garcia-Rosa, P.B.; Cunha, J.P.V.S.; Lizarralde, F.; Estefen, S.F.; Costa, P.R. Efficiency optimization in a wave energy hyperbaric converter. In Proceedings of the 2009 International Conference on Clean Electrical Power, Capri, Italy, 9–11 June 2009. [CrossRef]
4. Martínez, M.; Molina, M.; Machado, I.; Mercado, P.; Watanabe, E. Modelling and simulation of wave energy hyperbaric converter (WEHC) for applications in distributed generation. *Int. J. Hydrogen Energy* **2012**, *37*, 14945–14950. [CrossRef]
5. Machado, I.R.; Watanabe, E.H.; Garcia-Rosa, P.B. Modeling and analysis of a sea wave energy converter. In Proceedings of the 2015 IEEE 13th Brazilian Power Electronics Conference and 1st Southern Power Electronics Conference (COBEP/SPEC), Fortaleza, Brazil, 29 November–2 December 2015. [CrossRef]

6. Michele, S.; Sammarco, P.; d'Errico, M. Weakly nonlinear theory for oscillating wave surge converters in a channel. *J. Fluid Mech.* **2017**, *834*, 55–91. [CrossRef]
7. Michele, S.; Renzi, E. A second-order theory for an array of curved wave energy converters in open sea. *J. Fluids Struct.* **2019**, *88*, 315–330. [CrossRef]
8. Penalba, M.; Davidson, J.; Windt, C.; Ringwood, J.V. A high-fidelity wave-to-wire simulation platform for wave energy converters: Coupled numerical wave tank and power take-off models. *Appl. Energy* **2018**, *226*, 655–669. [CrossRef]
9. Kleefsman, K.; Fekken, G.; Veldman, A.; Iwanowski, B.; Buchner, B. A Volume-of-Fluid based simulation method for wave impact problems. *J. Comput. Phys.* **2005**, *206*, 363–393. [CrossRef]
10. Agamloh, E.B.; Wallace, A.K.; von Jouanne, A. Application of fluid–structure interaction simulation of an ocean wave energy extraction device. *Renew. Energy* **2008**, *33*, 748–757. [CrossRef]
11. Hu, Z.Z.; Causon, D.M.; Mingham, C.G.; Qian, L. Numerical simulation of floating bodies in extreme free surface waves. *Nat. Hazards Earth Syst. Sci.* **2011**, *11*, 519–527. [CrossRef]
12. Qian, L.; Causon, D.; Mingham, C.; Ingram, D. A free-surface capturing method for two fluid flows with moving bodies. *Proc. R. Soc. Math. Phys. Eng. Sci.* **2005**, *462*, 21–42. [CrossRef]
13. Bangun, E.P.; Utsunomiya, T. Evaluation of Viscous Forces Acting on A Moving Body by Navier-Stokes Solver. In Proceedings of the OCEANS 2008—MTS/IEEE Kobe Techno-Ocean, Kobe, Japan, 8–11 April 2008. [CrossRef]
14. Zhang, Y.; Zou, Q.; Greaves, D.; Reeve, D.; Hunt-Raby, A.; Graham, D.; James, P.; Lv, X. A Level Set Immersed Boundary Method for Water Entry and Exit. *Commun. Comput. Phys.* **2010**, *8*, 265–288. [CrossRef]
15. Ransley, E.; Greaves, D.; Raby, A.; Simmonds, D.; Jakobsen, M.; Kramer, M. RANS-VOF modelling of the Wavestar point absorber. *Renew. Energy* **2017**, *109*, 49–65. [CrossRef]
16. Shabana, A.A. *Dynamics of Multibody Systems*; Cambridge University Press: Cambridge, UK, 2005. [CrossRef]
17. Mazhar, H.; Heyn, T.; Pazouki, A.; Melanz, D.; Seidl, A.; Bartholomew, A.; Tasora, A.; Negrut, D. CHRONO: A parallel multi-physics library for rigid-body, flexible-body, and fluid dynamics. *Mech. Sci.* **2013**, *4*, 49–64. [CrossRef]
18. Tasora, A.; Serban, R.; Mazhar, H.; Pazouki, A.; Melanz, D.; Fleischmann, J.; Taylor, M.; Sugiyama, H.; Negrut, D. Chrono: An Open Source Multi-physics Dynamics Engine. In *Lecture Notes in Computer Science;* Springer International Publishing: Berlin/Heidelberg, Germany, 2016; pp. 19–49. [CrossRef]
19. Ropero-Giralda, P.; Crespo, A.J.; Tagliafierro, B.; Altomare, C.; Domínguez, J.M.; Gómez-Gesteira, M.; Viccione, G. Efficiency and survivability analysis of a point-absorber wave energy converter using DualSPHysics. *Renew. Energy* **2020**, *162*, 1763–1776. [CrossRef]
20. Ropero-Giralda, P.; Crespo, A.J.C.; Coe, R.G.; Tagliafierro, B.; Domínguez, J.M.; Bacelli, G.; Gómez-Gesteira, M. Modelling a Heaving Point-Absorber with a Closed-Loop Control System Using the DualSPHysics Code. *Energies* **2021**, *14*, 760. [CrossRef]
21. Tagliafierro, B.; Martínez-Estévez, I.; Domínguez, J.M.; Crespo, A.J.; Göteman, M.; Engström, J.; Gómez-Gesteira, M. A numerical study of a taut-moored point-absorber wave energy converter with a linear power take-off system under extreme wave conditions. *Appl. Energy* **2022**, *311*, 118629. [CrossRef]
22. Monaghan, J. Simulating Free Surface Flows with SPH. *J. Comput. Phys.* **1994**, *110*, 399–406. [CrossRef]
23. Colagrossi, A.; Landrini, M. Numerical simulation of interfacial flows by smoothed particle hydrodynamics. *J. Comput. Phys.* **2003**, *191*, 448–475. [CrossRef]
24. Brito, M.; García-Feal, O.; Domínguez, J.M.; Crespo, A.J.C.; Canelas, R.B.; Ferreira, R.M.L.; Neves, M.G. Coupling between DualSPHysics and Chrono-Engine: Towards large scale HPC multiphysics simulations. In Proceedings of the 11th SPHERIC International Workshop, Munich, Germany, 14–16 June 2016.
25. Canelas, R.; Brito, M.; Feal, O.; Domínguez, J.; Crespo, A. Extending DualSPHysics with a Differential Variational Inequality: Modeling fluid-mechanism interaction. *Appl. Ocean. Res.* **2018**, *76*, 88–97. [CrossRef]
26. Brito, M.; Canelas, R.; García-Feal, O.; Domínguez, J.; Crespo, A.; Ferreira, R.; Neves, M.; Teixeira, L. A numerical tool for modelling oscillating wave surge converter with nonlinear mechanical constraints. *Renew. Energy* **2020**, *146*, 2024–2043. [CrossRef]
27. Domínguez, J.M.; Fourtakas, G.; Altomare, C.; Canelas, R.B.; Tafuni, A.; García-Feal, O.; Martínez-Estévez, I.; Mokos, A.; Vacondio, R.; Crespo, A.J.C.; et al. DualSPHysics: From fluid dynamics to multiphysics problems. *Comput. Part. Mech.* **2022**, *9*, 867–895. [CrossRef]
28. Tasora, A.; Anitescu, M. A matrix-free cone complementarity approach for solving large-scale, nonsmooth, rigid body dynamics. *Comput. Methods Appl. Mech. Eng.* **2011**, *200*, 439–453. [CrossRef]
29. Zheng, X.; Chen, G.; Cao, W.; Xu, H.; Zhao, R.; Xu, Q.; Kramer, M.; Touzé, D.L.; Borthwick, A.G.; Li, Y. On the energy conversion characteristics of a top-mounted pitching absorber by using smoothed particle hydrodynamics. *Energy Convers. Manag.* **2021**, *250*, 114893. [CrossRef]
30. Crespo, A.; Domínguez, J.; Rogers, B.; Gómez-Gesteira, M.; Longshaw, S.; Canelas, R.; Vacondio, R.; Barreiro, A.; García-Feal, O. DualSPHysics: Open-source parallel CFD solver based on Smoothed Particle Hydrodynamics (SPH). *Comput. Phys. Commun.* **2015**, *187*, 204–216. [CrossRef]
31. Monaghan, J.J. Smoothed Particle Hydrodynamics. *Annu. Rev. Astron. Astrophys.* **1992**, *30*, 543–574. [CrossRef]
32. Fourtakas, G.; Dominguez, J.M.; Vacondio, R.; Rogers, B.D. Local uniform stencil (LUST) boundary condition for arbitrary 3-D boundaries in parallel smoothed particle hydrodynamics (SPH) models. *Comput. Fluids* **2019**, *190*, 346–361. [CrossRef]

33. Wendland, H. Piecewise polynomial, positive definite and compactly supported radial functions of minimal degree. *Adv. Comput. Math.* **1995**, *4*, 389–396. [CrossRef]
34. Mazhar, H.; Heyn, T.; Negrut, D.; Tasora, A. Using Nesterov's Method to Accelerate Multibody Dynamics with Friction and Contact. *ACM Trans. Graph.* **2015**, *34*, 32:1–32:14. [CrossRef]
35. Stewart, D.E. Rigid-Body Dynamics with Friction and Impact. *SIAM Rev.* **2000**, *42*, 3–39. [CrossRef]
36. Anitescu, M.; Hart, G.D. A constraint-stabilized time-stepping approach for rigid multibody dynamics with joints, contact and friction. *Int. J. Numer. Methods Eng.* **2004**, *60*, 2335–2371. [CrossRef]
37. Anitescu, M. Optimization-based simulation of nonsmooth rigid multibody dynamics. *Math. Program.* **2006**, *105*, 113–143. [CrossRef]
38. Canelas, R.B.; Domínguez, J.M.; Crespo, A.J.; Gómez-Gesteira, M.; Ferreira, R.M. A Smooth Particle Hydrodynamics discretization for the modelling of free surface flows and rigid body dynamics. *Int. J. Numer. Methods Fluids* **2015**, *78*, 581–593. [CrossRef]
39. English, A.; Domínguez, J.M.; Vacondio, R.; Crespo, A.J.C.; Stansby, P.K.; Lind, S.J.; Chiapponi, L.; Gómez-Gesteira, M. Modified dynamic boundary conditions (mDBC) for general-purpose smoothed particle hydrodynamics (SPH): Application to tank sloshing, dam break and fish pass problems. *Comput. Part. Mech.* **2022**, *9*, 911–925. [CrossRef]
40. Crespo, A.J.C.; Gómez-Gesteira, M.; Dalrymple, R.A. Boundary Conditions Generated by Dynamic Particles in SPH Methods. *Comput. Mater. Contin.* **2007**, *5*, 173–184.
41. Leimkuhler, B.J.; Reich, S.; Skeel, R.D. Integration Methods for Molecular Dynamics. In *Mathematical Approaches to Biomolecular Structure and Dynamics*; Mesirov, J.P., Schulten, K., Sumners, D.W., Eds.; Springer: New York, NY, USA, 1996; pp. 161–185. [CrossRef]
42. Monaghan, J.J.; Kos, A. Solitary Waves on a Cretan Beach. *J. Waterw. Port Coastal Ocean. Eng.* **1999**, *125*, 145–155. [CrossRef]
43. Windt, C.; Davidson, J.; Ransley, E.J.; Greaves, D.; Jakobsen, M.; Kramer, M.; Ringwood, J.V. Validation of a CFD-based numerical wave tank model for the power production assessment of the wavestar ocean wave energy converter. *Renew. Energy* **2020**, *146*, 2499–2516. [CrossRef]
44. Altomare, C.; Domínguez, J.; Crespo, A.; González-Cao, J.; Suzuki, T.; Gómez-Gesteira, M.; Troch, P. Long-crested wave generation and absorption for SPH-based DualSPHysics model. *Coast. Eng.* **2017**, *127*, 37–54. [CrossRef]
45. Pontes, M.T.; Aguiar, R.; Pires, H.O. A Nearshore Wave Energy Atlas for Portugal. *J. Offshore Mech. Arct. Eng.* **2003**, *127*, 249–255. [CrossRef]
46. Zhang, F.; Crespo, A.; Altomare, C.; Domínguez, J.; Marzeddu, A.; ping Shang, S.; Gómez-Gesteira, M. DualSPHysics: A numerical tool to simulate real breakwaters. *J. Hydrodyn.* **2018**, *30*, 95–105. [CrossRef]
47. Hasselmann, K.F.; Barnett, T.P.; Bouws, E.; Carlson, H.C.; Cartwright, D.E.; Enke, K.; Ewing, J.A.; Gienapp, H.; Hasselmann, D.E.; Kruseman, P.; et al. Measurements of wind-wave growth and swell decay during the Joint North Sea Wave Project (JONSWAP). *Deut. Hydrogr. Z.* **1973**, *8*, 1–95.

Journal of
Marine Science and Engineering

MDPI

Article

Optimized Screening Methods for Investigation of the Larval Settlement of *Lanice conchilega* on Artificial Substrates

Alice D'Hurlaborde [1,2,*], Alexia Semeraro [1], Thomas Sterckx [3] and Gert Van Hoey [1]

[1] Department Animal Science/ILVO-Marine, ILVO, Institute for Agricultural and Fisheries Research, Ankerstraat 1, 8400 Oostende, Belgium
[2] International Master of Science in Marine Biological Resources, IMBRSea, Ghent University, Krijgslaan 281/S8, 9000 Ghent, Belgium
[3] Department Benelux Division, Dredging International DEME Group, Scheldedijk 30, 2070 Zwijndrecht, Belgium
* Correspondence: alice.dhurlaborde@imbrsea.eu

Abstract: The Belgium sandy coastline is very vulnerable to erosion; therefore, development of sustainable and nature-based coastal protection solutions is important. Enhancing the settlement of the ecosystem engineer *Lanice conchilega* (Pallas, 1766) which stabilises the sediment bed, is a possible solution. In order to enhance larval settlement by artificial substrates in the field, efficient methodologies are required to screen a wide range of artificial substrates and measure how they influence currents and larval settlement. Therefore, in this study, we describe the development of innovative artificial substrate screening methodologies using an optimised recirculating aquaculture system (RAS) by: (1) analysing the capture rate of passively floating plastic particles, (2) measuring current velocity by means of an acoustic doppler velocimeter and (3) monitoring settlement of living *L. conchilega* larvae. Of the eight substrates evaluated, one was proven to significantly enhance the settlement of *L. conchilega*, namely Geotextile 3D knitted fabric with PES knit, PA spacers and wood sticks mounted at a density of 680 sticks/m^2. The results of this study show that controlled lab conditions, in conjunction with innovative methods, allowed for successful screening of a number of substrates in a short time in terms of their ability to enhance larvae settlement.

Keywords: coastal erosion; restoration; ecosystem engineer; Aulophora; recirculation aquaculture system; acoustic velocity metre; artificial substrate; geotextiles; distribution of particles

Citation: D'Hurlaborde, A.; Semeraro, A.; Sterckx, T.; Van Hoey, G. Optimized Screening Methods for Investigation of the Larval Settlement of *Lanice conchilega* on Artificial Substrates. *J. Mar. Sci. Eng.* **2022**, *10*, 1443. https://doi.org/10.3390/jmse10101443

Academic Editors: M. Dolores Esteban, Jose-Santos Lopez-Gutierrez, Vicente Negro and M. Graça Neves

Received: 9 September 2022
Accepted: 5 October 2022
Published: 7 October 2022

1. Introduction

Coastal areas represent less than 15% of Earth's land surface [1] but host 67% of the world's population [2], as well as 15 of the 20 megacities of the world [1,3]. Increasing anthropogenic activities (e.g., construction of buildings, ports and marinas) have led to intense modification of these areas over the last few decades. In Europe, the result is that 50% of the shoreline has been modified by construction [2].

Under the current climate change scenario, estimates shows that, within the next 50 years, 30% of residences situated on low coastlines and within 200 m of the sea worldwide may be severely affected by property losses due to erosion [4,5]. This number takes a different dimension in Belgium, as 85% of the coastal zone is located below 5 metres TWA of elevation, making it very vulnerable to erosion, especially with the added effect of sea-level rise and an increasing number of storms [6–8].

To face the erosion threat, societies across the world have relied on engineering and hard coastal protection solutions, such as groynes, dykes, breakwaters, jetties or sea walls [9,10]. These solutions are becoming unsuitable due to their costly and constant maintenance requirements, as well as their rigidity to adapt to the increasing erosion risk [5]. Additionally, they alter the natural adaptive capacity of the coastline to the relative sea-level rise [11]. The world needs intelligent coastal protection strategies that are sustainable, multifunctional and economically viable to help solve immediate and projected coastal erosion and flooding problems [11,12], as well as to enhance ecosystem functioning [13,14]. Mindful of the holistic approach of marine ecosystem services, new solutions based on nature-induced design have been proposed, with the creation and restoration of existing coastal ecosystems, which naturally provide coastal protection and have a capacity for self-repair and recovery [15].

The Coastbuster project allies research and industry to selectively strengthen a part of the Belgium coastline through the use of artificially enhanced biogenic reefs [16]. The common tube-dwelling polychaete *Lanice conchilega* can be considered an ecosystem engineer, colonising intertidal and subtidal sediments to depths of 1900 m [12,17–19]. The physiology, tube structure [20,21] and occurrence of *L. conchilega* aggregations [22,23], as well as their feeding habits [24], have already been described at length. Aggregations of *L. conchilega* can reach densities of thousands of individuals per m^2 [17,24] and have the ability to alter sediment properties (grain size composition or porosity), modify the hydrodynamic regime [25,26], offer refuge from predation [27], increase the stability of the habitat and oxygen supply [28] and improve the availability of attachment surfaces for larvae and small organisms [29]. Thus, the faunal community has a higher abundance and richness in areas with *L. conchilega* tubes than free bare sand [18,30].

The success of the artificial enhancement of *L. conchilega* aggregation is largely dependent on their life cycle. *L. conchilega* is an iteroparous free spawner [31]. Egg fertilisation happens within the water column and leads to trochophore larvae, which undergo a short benthic transition into the pelagic phase as aulophore larvae [32]. Finally, the aulophore larvae search for a substrate to settle on, which marks the final transition from free pelagic larvae to the sessile benthic juvenile stage [33]. As a result, the key question with respect to artificial enhancement of *L. conchilega* beds that remains partially unanswered is what kind of substrate can trigger the settlement of aulophore larvae. Selected studies have addressed this question through in situ and laboratory assessments of different types of epibenthic holdfast structures, including plastic straws [34], 3D epibenthic structure resembling macroalgae [35], metal tubes [30,36], concrete, polystyrene (PP), polyvinyl chloride (PVC), polyethene terephthalate (PET), polycarbonate (PC) [36], sticks of wood [2,36] and geotextiles [37,38].

These field studies showed that *L. conchilega* larvae can primarily use any epibenthic structure that reduces the near-bottom flow rate as a means of settlement. Enhancing larval settlement is therefore possible, although more research is needed to determine the optimal strategy and artificial substrate design. However, field trials are expensive and time-consuming, and the potential substrate candidates are numerous. Reliable estimations on settlement enhancement and reef formation would require high-frequency sampling performed with a constant methodology over short and long periods in different locations [36,38]. Laboratory conditions allow for the testing of a wide range of innovative substrate solutions in a fast and controlled manner. To optimise this testing, efficient methodologies are required to screen this wide range of artificial substrates and to measure how they influence currents and the ability to capture passively floating particles (larvae). Characterising the influence of the substrate on both hydrodynamics and capture rate is important because velocity can influence settlement [39], and the capture capacity is an essential tool for colonization of defaunated substrates [40]. Arganda-Carreras et al. [37] attempted to develop laboratory experiments to study the settlement of *L. conchilega* larvae on artificial substrates, but no significant preference was found between the tested substrates, emphasising the difficulties in screening substrates using live organisms in a laboratory context.

The objective of this study was to find innovative experimental lab conditions to compare a large range of substrates (eight were tested in this study) in terms of their ability to enhance the settlement of *L. conchilega* larvae. The first experimental method verified the ability of the substrates to trap passively floating plastic particles mimicking larvae density and size. The second experimental method consisted of measuring, with an acoustic Doppler velocimeter, the ability of a substrate to affect the near-bottom current velocity. Finally, the settlement rate of living larvae of *L. conchilega* was compared for the most promising substrates used in the previous tests.

2. Materials and Methods

2.1. Experimental Setup Design

The experiments were executed in a rearing tank (Figure 1) consisting of a two-compartment closed recirculating aquaculture system (RAS, based on the Kreisel principle [41]). This design has many advantages for rearing planktonic organisms. A laminar flow of water along one of the tank's walls allows a circular flow [42], and its shape avoids aggregation in corners. A fine net (90 µm pore size, Plansifter SEFAR NYTAL® PA, sourced from: SEFAR, Heiden, Switzerland) at the water exiting site protects the organisms from any damaging suction effect [41]. However, previous work [37] revealed that the experimental setup (RAS) needs to be optimised in order to achieve homogenisation of flow conditions. The critical aspects of the setup to be addressed were rearing tanks, inlet pipes, flowmeters and controlled water condition.

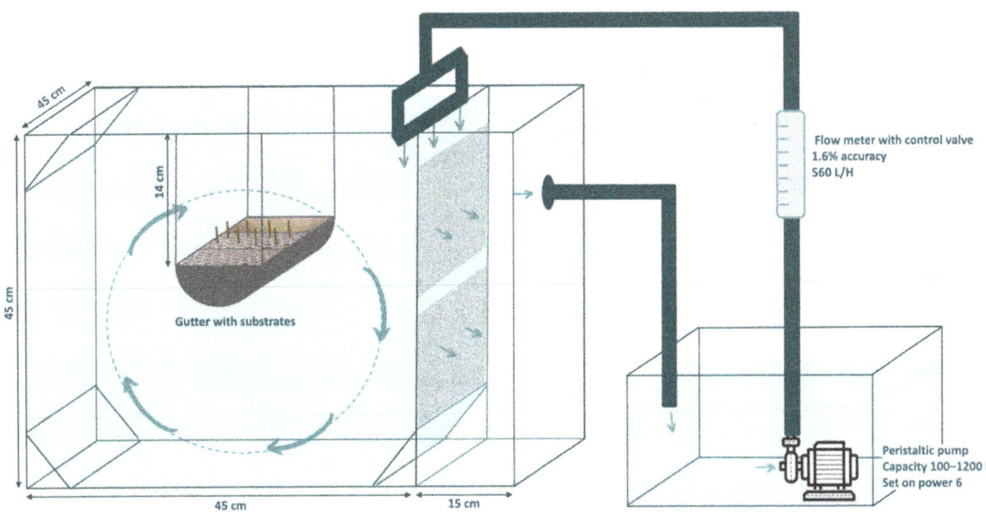

Figure 1. Illustration of the experimental design (larvae settlement experiment) and its components. The arrows represent the direction of water flow.

Observations and hydraulic calculations (pressure difference between the pipe's holes) showed that the shape of the inlet pipe used in previous experiments [37] caused variation in water flow and a lack of repeatability. The optimised water inlet system (Figure 2a) consists of two water entry points on each side of the inlet pipe with 19 3 mm holes distributed every 2 cm along the full length of the tank. The angle of the inlet pipe was carefully adjusted to achieve maximal flow homogeneity. Optimizations were based on observations and empirical trials using picture analysis [43] of the top view of the tanks containing moving plastic pellets (1 mm). The combination of a newly designed inlet pipe and its adjusted position led to suitable homogeneous water distribution in the tank (Figure 2b), providing the best experimental condition for repeatability and reproducibility within the tank and between different tanks.

In each tank, the water was pumped by a peristaltic pump (Jecod DCS–1200 pump with controller: 8 power settings, sourced from: Ocean Store, Dordrecht, Nederland) in an enclosed system. The addition of a variable-area flowmeter (Series 2000 TechFluid, 100–1000 L/H, with a reading accuracy of 1.6%, control valve, sourced from: Techfluid, Barcelona, Spain) allowed for standardization of the water injection between tanks and to monitor the flow rate during the length of the study. Optimal flow rate was identified based on the homogenous distribution of moving plastic pellets (shown by picture analysis) and based on flow stability over time. All the experiments presented in the study were performed at a flow rate of 560 L/H corresponding to the pump setting of 5 or 6.

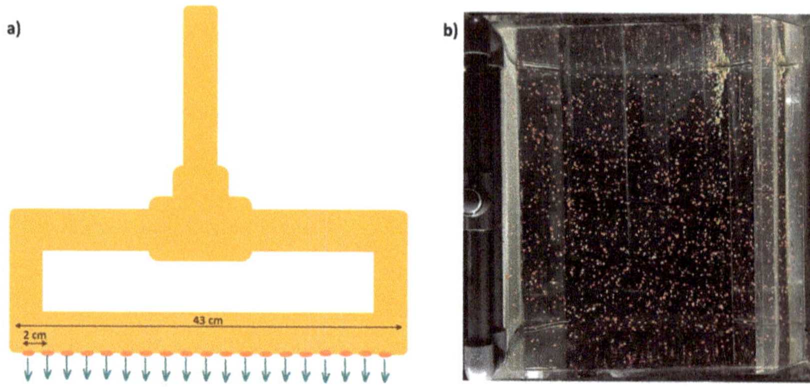

Figure 2. Optimised inlet pipe design (**a**); top view picture of the tank under optimised conditions (**b**).

The water quality was monitored during the full length of the study. The mean salinity (±1 SD) was 34 (salinity refractometer). The temperature was 14 ± 2 °C (thermometer). Continuous bubbling maintained dissolved oxygen (DO) at ~100% saturation (>8 mg/L DO). These parameters were set to match the water conditions of the coast of the North Sea, where *L. conchilega* settlement occurs [44]. The water conditions were kept constant throughout the experiments (with or without living animals).

A gutter for placing the substrates was hung inside the tank at a height of 14 cm from the top of the tank (Figure 1). In the screening tests without living larvae, the gutter was filled with shell fragments ((E) control), and only one artificial substrate was placed in the gutter at a time. In the screening test with living larvae, the gutter was divided into 3 sections with different substrates (Figure 1). The substrates had different structures, thicknesses and weaving and represented three categories: mats, sticks and a combination of mats and sticks (Figure 3). Four types of mats were tested: (A) geotextile (220 g/m^2 3D knitted fabric (10 mm thickness) based on PES knit and PA spacers), (B) non-woven geotextile Kena260 black (260 g/m^2), (C) 3 layers of non-woven geotextile Kena260 black (260 g/m^2) and (D) non-woven geotextile NW170 white (170 g/m^2). The mats were provided by Sioen industries. One type of stick was tested: (F) wooden sticks with a diameter of 5 mm and length of 5 cm (out of substrate) and positioned with a density of 680 tubes/m^2. The sizing of the wood sticks used in this experiment mimicked adult *L. conchilega* [15,45,46]. Two types of mat/stick combinations were tested: (G) substrate A with sticks and (H) substrate B with sticks.

2.2. Artificially Screening Substrates Based on Capture Rate

Artificial particles have been successfully used in the past to study settlement and flow movement [25,35]. In order to choose an artificial particle to mimic *L. conchilega* larvae, we first had to characterise the aulophore in terms of size, shape and density. The aulophore planktonic larval stage measures 2–3 mm [32,47], with a rectangular shape and an estimated density of 1.029 g/m^3. The density of the larvae was calculated according the Stroke Law:

$$v = \frac{gd^2(p-m)}{18\,\mu}$$

where v is the terminal velocity of a spherical particle; g is the gravitational acceleration—for Earth, equal to 9.80665 m/s^2; d is the particle diameter; p is the density of the particle; m is the density of the fluid; and μ is the dynamic viscosity of the fluid. The terminal velocity of the particle and the salinity and temperature of the water during the experiment were taken from an article by Bhaud et al. [47]. The density and the dynamic viscosity of the fluid were deduced from the salinity and temperature.

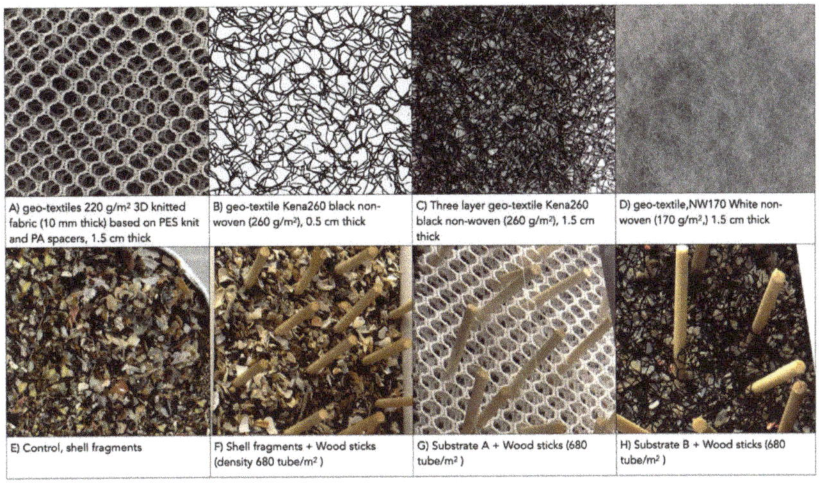

Figure 3. Pictures and descriptions of all artificial substrates tested.

The chosen artificial particles mimicking larvae were red ABS pellets, with a density of 1.05 g/m^3, a length of 4 mm and a diameter of 3 mm. They were chopped into smaller pieces around $\frac{1}{4}$ of the original size, which made them closer to the aulophore larvae population (in size and heterogeneity). In this study, pellets were quantified by volume using a 25 mL measuring cylinder.

In order to compare the capture properties of the tested substrates, the decrease in the amount of chopped plastic pellets in suspension over time was measured. A known volume of pellets (25 mL) was added to the tank in front of the inlet pipe to ensure quick dispersion. Every 5 min for a period of 20 min, all pellets still in suspension were fished out using a hand net and quantified before being reintroduced in the tank. A linear regression model was applied to the data (volume~time), and the value of the slope was used as a measurement of capture rate (R, version 4.0.2,). Five replicates were performed for each substrate candidate. This protocol was established based on trial runs (Figure 4), showing that maximal pellet capture occurs for all substrates within the first 20 min and that resuspension did not occur, regardless of the substrate. A longer measuring period would be time-consuming, providing no added value.

2.3. Screening Substrates Based on Their Ability to Affect Current Velocity

Previous studies showed that the larval settlement rate is affected by current velocity, as well as bed topography [39], and that the presence of polychaete tubes affects the surrounding current velocity [25]. In this study, we were interested in characterising the effect of different artificial substrates on the overlaying current velocity. An acoustic Doppler velocimeter (ADV) can be used to characterise current velocity in a Kreisel tank [42]. In this study, our hypothesis was that an ADV could be used to measure variation of the velocity by moving the probe over the length of the gutter. A Nortek acoustic Doppler velocity profiler (Nortek Vectrino Profiler, 15 MHz, sourced from: Nortek Scientific Acoustic Development Group Inc., Boston, MA, USA) from Flanders Hydraulics Research was used for the period of the experiment. The Vectrino Profiler sensor is a high-precision instrument that generates a short pulse of sound at a known frequency that is reflected by fine particles dissolved in water. In this study, we used artificial seeding (Polyamide (PA) beads of 60 µm) dispersed in the tank. The ADV was mounted on a tripod, the focus area was defined above the gutter in terms of height and length and always at its centre in terms of width. Each measurement lasted 1 min, with velocity profile readings taken every 0.066 s based on Nortek supplier recommendations [48]. The following data handling process was used for each test. The first step consisted of averaging the magnitude of the X, Y and Z velocity components for each depth and replicate during each measurement. The velocity magnitude (Vmag) in the 3D plane was then calculated for every depth in the sampling volume and replicated using vector addition, $V_{mag} = \sqrt{V_X^2 + V_Y^2 + V_Z^2}$. Entries with a Vmag values of less than 0.001 m/s were removed from the analysis. Velocity values below the ADV detection threshold of 0.001 m/s are either due to the signal of the bottom (null velocity) or are unreliable measurements. The first and third centimetres of the sampling volume were removed from the analysis to focus on a sampling interval of only 1 cm (highlighted in orange in Figure 5a) in order to avoid the effect of vertical velocity variability present in the Kreisel tank [42]. Finally, a quality check was performed on the data using two parameters: The small noise-to-noise ratio (SNR), with a threshold value of 15 dB; and the correlation, with a threshold value of 70% based on literature recommendations [48–50].

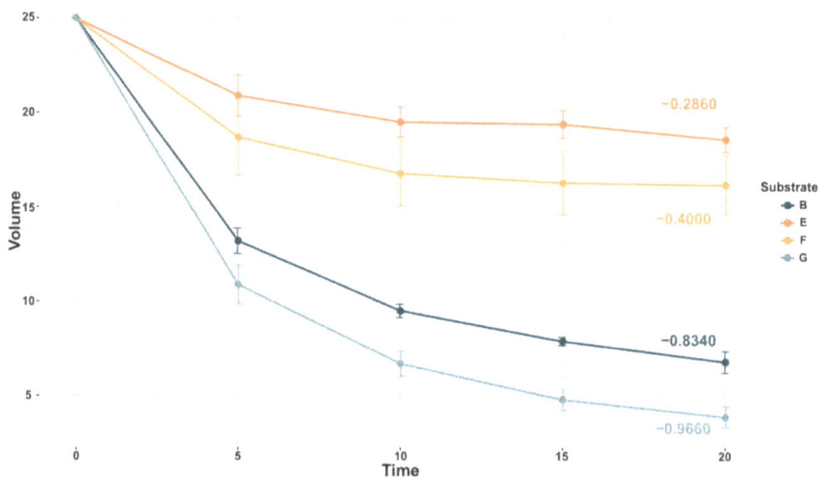

Figure 4. Regression of the floating plastic pellet volume over time due to settlement in the substrate present in the gutter. Four substrate types were tested (B: non-woven geotextile Kena260 black; E: control, shell fragments; F: shell fragments with wood sticks; G: geotextile 3D knitted fabric based on PES knit and PA spacers with wood sticks), with four replicates each. The value of the slope shown in the figure is extracted from a linear regression model.

To the best of our knowledge, this was the first time that ADV was used to measure small-current velocity above a gutter in a Kreisel tank. As a result, the optimal measurement condition had to be characterised to ensure an appropriate level of data quality. The height at which the probe was placed above the gutter was carefully identified based on data quality tests. Four height levels were tested: 101, 85, 76, 67 and 54 mm (Figure 6a) with between 3 and 5 replicates for each level (with added replicas in more sensitive positions (101 mm, 85 mm and 54 mm) due to potential bottom and surface disturbance). A quality level was attributed to each velocity measurement of the sampling volume based on the quality check criteria found in the literature (Figure 6b). Data in the category "Off limit" had a Vmag < 0.001m/s. Data in the category "Poor quality" had an SNR $\geq$ 15 or a correlation $\geq$ 70. Finally, all data deemed of good quality were labelled "Conform". The gutter was in the sampling volume of the ADV at distances of 54, 67 and 76 mm (highlighted in yellow in Figure 5a), which explains the considerable number of off-limits and poor quality values resulting from the measured null velocities (83, 70 and 50%, respectively). At distances of 101 mm and 85 mm, the quality of the data was more optimal (71 and 78% of conform data, respectively). The probe depth with the least data loss in relation to data quality was a distance of 85 mm from the gutter, which also allowed the velocity to be measured just above the substrate. Further velocity measurements in this study were performed at this depth.

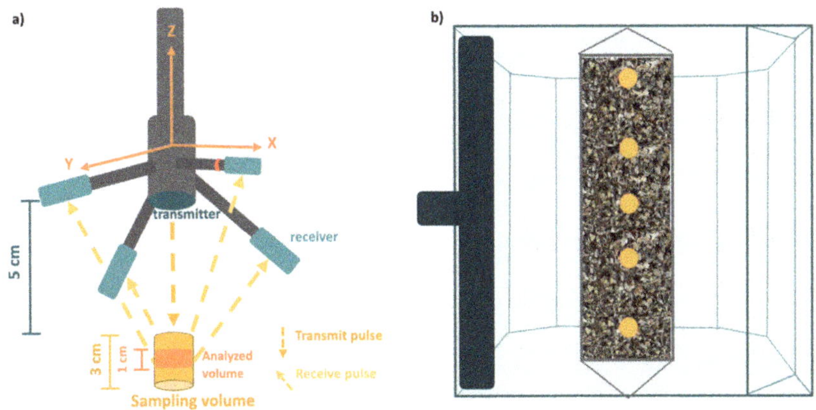

Figure 5. Screening substrates based on their ability to affect current velocity. (**a**) Illustration of Nortek Vectrino profiler probe head with the process of 3 D velocity measurement. The position of the probe sampling volume is highlighted in yellow, and the position of the analysed volume in the screening test is highlighted in orange. (**b**) Top-view illustration of the tank with the five ADV probe positions above the gutter indicted by yellow circles (5, 14, 23, 31 and 40 cm).

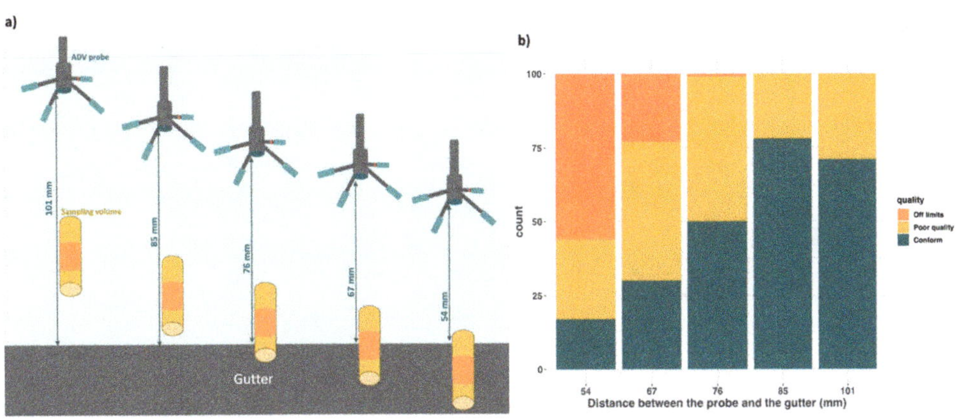

Figure 6. Probe height test. (**a**) Illustration of the probe position in relation to the gutter and the associated sampling volume (yellow). (**b**) ADV data quality, expressed in percentage, based on the probe height above the gutter.

The current velocity above the gutter was measured at a fixed height (85 mm) along the length of the gutter at five positions: 5, 14, 23, 31 and 40 cm, with 4 replicates for each position (Figure 5b). Substrate candidates were placed one at a time to cover the whole gutter surface. The average velocity for all positions was calculated for each replicate to obtain an accurate measurement of velocity, taking into account the velocity heterogeneity in the tank.

2.4. Screening Substrates Based on Larval Settlement

L. conchilega is not listed as an endangered or vulnerable species under IUCN, nor named in any international nature conservation legislation or convention. Therefore, no specific licence is required to capture the larvae. All the methodologies carried out in the field and lab work for this project were carefully regulated to ensure minimal disturbance to the animal.

The aulophore larvae were collected in the Belgian part of the North Sea (BPNS) at eight different stations on 12 April 2021 and on 30 April 2021, with the Research Vessel Simon Stevin (Vlaams Instituut voor de Zee (VLIZ)): Nieuwpoortbank 1 (51°013′12″; 20°44′24″), Nieuwpoortbank 2 (51°012′36″; 20°41′24″), Nieuwpoortbank 3 (51°010′48″; 20°37′48″), West Diep 1 (51°09′1″; 20°37′11″), West Diep 2 (51°09′18″; 20°38′36″), West Diep 3 (51°09′40″; 20°40′5″), West Diep 4 (51°010′39″; 20°39′44″) and West Diep 5 (51°011′9″; 20°39′4″). Aulophore larvae were sampled using a vertical haul WP2 plankton net with 200 µm mesh size. The sample was retrieved from the net and filtered through a 1 mm sieve to remove jellyfish and large organisms and through a 200 µm sieve to reduce the volume of water. The samples were stored in a sealed bucket filled with seawater.

Plankton samples were processed as described in [37]. The buckets with the samples were kept oxygenated until use. The content of each bucket was sorted under a magnifying glass using fractions of 200 mL. Aulophores were identified according to [32] (Figure 7a) and isolated from the sample and moved to an oxygenated 5 L beaker using a pipette.

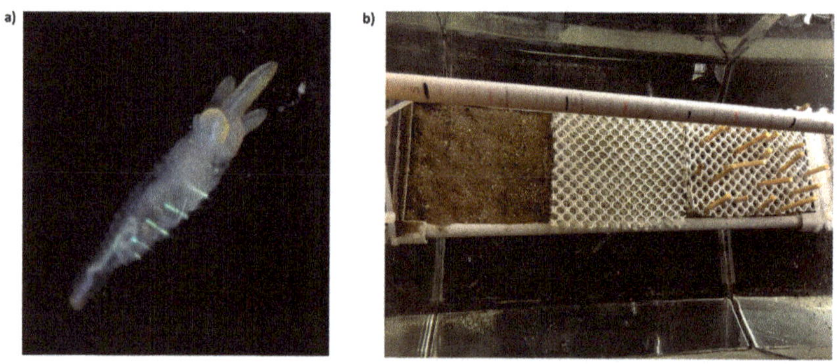

Figure 7. Larvae experiment. (**a**) L. conchilega aulophore; (**b**) organisation of the 3 substrates in the gutter.

Larvae were introduced in in front of the inlet pipe to ensure quick dispersion, and the experiment lasted 10 days. The larvae were fed three times a week with 200 mL of a mixed culture of *Nannocloropsis* and *Tetraselmis* [37,51]. The gutter was filled with fine sand, which is the natural habitat of *L. conchilega* beds [18]. In order to compare the settlement rate of larvae in each substrate, the gutter was divided into 3 sections: control (fine sand), substrate A and substrate G (Figure 7b). Three replicates were performed, and the position of the substrate in the gutter differed for each replicate. The first replicate was performed using the larvae collected on the first sampling day (100 larvae/tank), and the two other replicates were performed using larvae from the second sampling day (80 larvae/tank).

On day 10, the gutter was carefully removed from the tank, and visual observation allowed quantification of aulophores and juveniles in the substrate in the first layer of sand and in the water. The survival rate was quantified for each trial (initial number of individuals/retrieved individuals). Different initial numbers of individuals were introduced into the tanks for the first ($N = 110$) and second ($N = 85$) trials. Therefore, the distribution of individuals (aulophores and juveniles) in each of the substrates was expressed as a percentage of the retrieved population.

2.5. Statistical Tests

A one-way analysis of variance (ANOVA) was conducted to examine the effect of the substrate on each of the screening parameters (catchability, mean velocity and distribution of larvae). Residual analysis was performed to test for the assumptions of the one-way ANOVA. Outliers were assessed by box plot method normality was assessed using the Shapiro–Wilk normality test and homogeneity of variances was assessed by Levene's test. In the instance of a significant result, one-way ANOVA was followed by a Tuckey HSD test to allow for subsequent pairwise comparison tests.

An additional two-way ANOVA was conducted with a subset of the data (removing substrate D) for two screening parameters: catchability and mean velocity. The assumptions of the two-way ANOVA were met. Consequently, an analysis of simple main effects for the wooden sticks was performed with statistical significance receiving a Bonferroni adjustment. All analyses were performed using R (version 4.0.2).

3. Results

3.1. Screening of Substrate Based on Capture Rate

The benefit of using artificial substrate in the restoration of *L. conchilega* beds lies in its ability to capture pelagic larvae. Thus, the first screening phase of this study targeted the capture property of each substrate (Figure 8). The presence of an artificial substrate (A, B, C, G or H) in the gutter significantly increased the capture of the particles compared to the control with (F) or without (E) wood sticks ($p < 0.05$ for all substrates). The thickness of the geotextile appeared to be a key factor with respect to increasing the capture rate. A significant difference was found between an identical geotextile tested with different thicknesses (B and C, $p = 2.30 \times 10^{-4}$), and no significant difference was found between two different geotextiles of similar thickness (A and C, $p = 0.99$). Substrate D was removed from the analysis because it was inefficient in capturing pellets (data not shown).

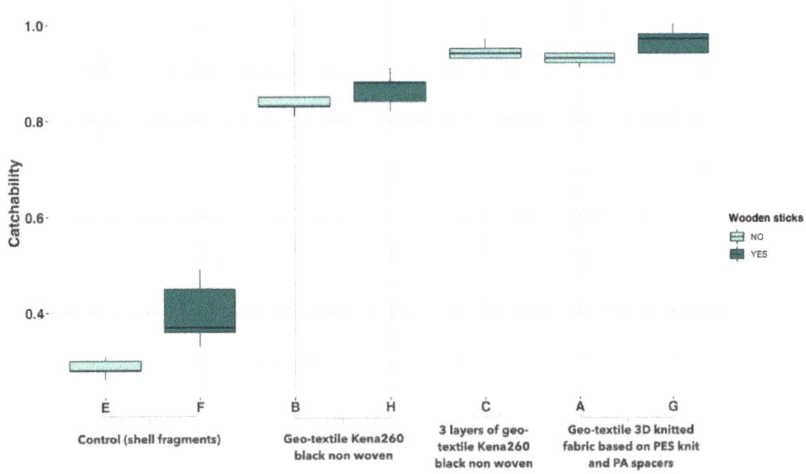

Figure 8. Box plot of the capture properties of each substrate based on plastic pellet catch rate (slope of the linear model volume of particle~time), with four replicates for each treatment. Colour represents the presence/absence of wooden sticks on the substrate.

The two-way ANOVA showed a significant effect of the presence/absence of sticks ($p = 6.58 \times 10^{-5}$) and the substrate type ($p < 2 \times 10^{-16}$) on the capture rate. A significant interaction was also shown between the sticks and the substrate ($p = 0.0254$). The analysis of simple main effect for the wooden sticks showed, when the significant threshold was $p < 0.16$, a positive effect of the presence of sticks paired with either the control ($p = 2.7 \times 10^{-5}$), geotextile 3D knitted ($p = 0.098$) or geotextile Kena260 ($p = 0.16$).

3.2. Screening of Substrates Based on Velocity

The average flow velocity overlaying the different substrates was measured (Figure 9) and showed significant differences between substrates. The average velocity above the gutter in the presence of wood sticks only (F) was significantly lower than the control (E) and all other tested substrates ($p < 4.92 \times 10^{-4}$). The presence of substrate (G) in the gutter leads to a significant increase in the measured velocity compared to the control (E) ($p < 1.37 \times 10^{-2}$). The substrate (B) also appears to increase the velocity ($p = 5.77 \times 10^{-2}$). The presence of a geotextile ((A) and (C)) does not induce a significant difference in the average velocity above the gutter in comparison to the control.

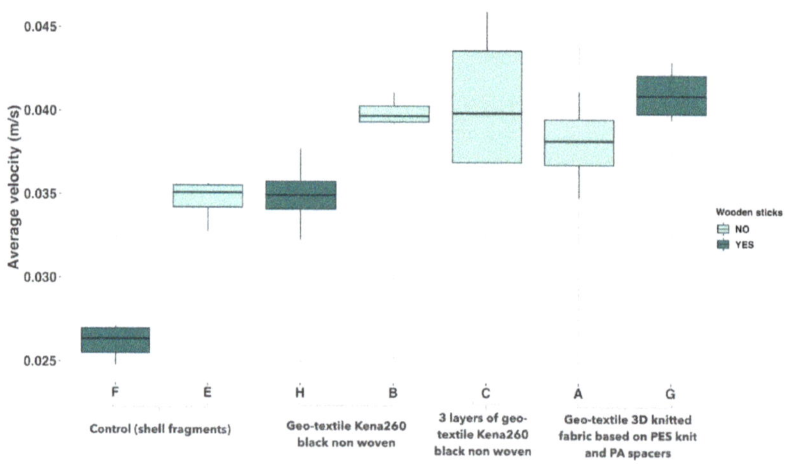

Figure 9. Box plot of the average velocity (m/s) above the gutter containing different substrate types. Four replicates were performed for each treatment, for which the average velocity is based on velocity measurement at five points along a horizontal transect above the gutter (Figure 5b). Colour represents the presence/absence of wooden sticks on the substrate.

The two-way ANOVA showed a significant effect of the presence/absence of sticks ($p = 1.07 \times 10^{-8}$) and the substrate type ($p = 1.14 \times 10^{-4}$) on the velocity. A significant interaction was also shown between the sticks and the substrate ($p = 1.23 \times 10^{-5}$). The analysis of simple main effects for the wooden sticks showed significant difference in mean velocity in both the presence and the absence of wooden sticks paired with either control ($p = 1.96 \times 10^{-6}$), geotextile 3D knitted ($p = 2.72 \times 10^{-2}$) and geotextile Kena260 ($p = 8.05 \times 10^{-4}$). The impact of wooden sticks on the velocity depended on the substrate with which it was paired; a decrease in average velocity is visible for the control and geotextile Kena260, whereas an increase is visible with geotextile 3D knitted.

3.3. Screening of Substrates Based on Larval Settlement of L. Conchilega

The first two screening phases showed that substrate G (geotextile 3D knitted fabric based on PES knit and PA spacers with added wood sticks) significantly differed from the control in terms of capture rate and average velocity; therefore, it was selected for the *L. conchilega* experiment and compared with substrate A (geotextile 3D knitted fabric based on PES knit and PA spacers without added wood sticks) and substrate E (fine sand), the control.

Living aulophore larvae were added to tanks containing a gutter with substrates G, A and E in parallel (Figure 10). After 10 days, the average mortality was 88.4% ± 1.8. Among the accounted alive individuals, a few were found outside of the substrates (residuals: juveniles = 11.9 ± 11.2; aulophores = 19.3 ± 5.1). For both life stages of *L. conchilega*, no significant difference was shown between the control and the residual ($p = 0.28$). The distribution of the juveniles was significantly higher ($p < 5.08 \times 10^{-3}$) in substrate G (58.5 ± 7.1) and substrate A (29.6 ± 6.4) compared to the control (no juveniles in natural substrate). Wooden sticks in substrate G captured almost twice as many juveniles as substrate A ($p = 5.91 \times 10^{-3}$). The same trend was observed with the aulophores, although not significant: G (18.5 ± 23.1) > A (7.4 ± 12.8) > control (2.2 ± 3.8).

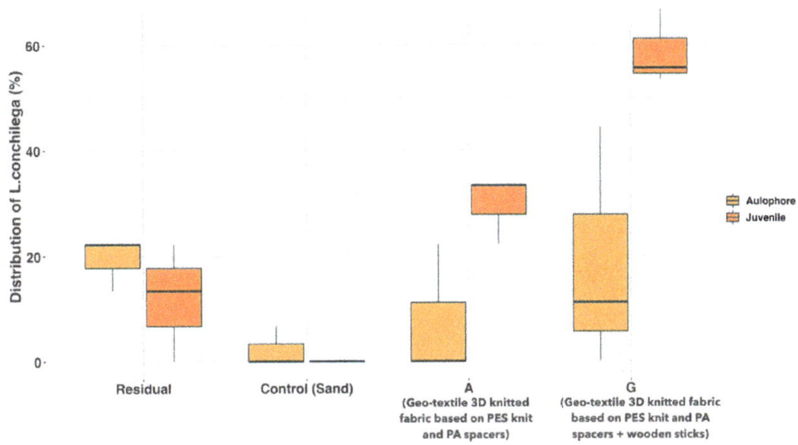

Figure 10. Box plot of distribution of *L. conchilega* juveniles (dark orange) and aulophores (light orange) in each fraction of the gutter and in the residual (water column and plastic part of the gutter). Three replicates were performed for each treatment. The distribution of individuals in each of the substrates is expressed as a percentage of the retrieved population.

4. Discussion

The objective of this study was to find innovative experimental lab conditions to compare the accuracy a wide range of substrates in terms of their ability to enhance the settlement of *L. conchilega* larvae. As field experiments are costly and time-consuming, optimal lab procedures should be developed to allow for fast and accurate screening of artificial substrates in terms of a variety of criteria, including tests with living animals. In this study, the experimental design was optimised to ensure adequate water mixing and stable flow conditions for different experiments. These conditions were not met in a previous study using the same RAS system, preventing accurate ranking of substrates [37]. The conjunction of the new design of the inlet pipe and the precise control of the water supply rate was critical in the obtention of homogeneous flow conditions in the different tanks. This design allowed for successful testing of living larvae settlement behaviour with different substrates. The optimised RAS system was also used to develop two new methods to characterise the substrates in terms of their ability to trap passively floating particles and measure their effect on overlaying current velocity.

4.1. Methodological Development

Characterising substrates in terms of their particle capture capacity is important, as they are essential tools for colonisation of defaunated substrates [40]. The method (kinetics of the number of floating particles) developed in this study allowed for accurate measurement of substrate capture properties in a relatively short amount of time, enabling successful characterisation and ranking of the tested substrates. This method could be applied to investigate the capture properties of a substrate when facing fluctuating flow rate conditions or varying particle concentrations in the water column. Improvements could still be made through the replacement of manual pellet quantification with imagery techniques, specifically multiple particle tracking (MPT) [52], which would enable a higher number of replicates and test capacity.

Characterising the influence of the substrates on hydrodynamics is important because velocity can influence settlement in two ways: The encounter rate and the attachment rate [39]. In this study, we demonstrated, for the first time, the relevance of using an ADV to measure velocity in a Kreisel tank. Nevertheless, four challenges were encountered in the use of the ADV in our system.

The first challenge concerned the density of suspended particles in the water column. Several studies demonstrate recurrent problems associated with a lack of suspended particles, leading to weak ADV signals (low correlation and low noise-to-noise ratio) [53,54]. To counteract this, in the present study, seeding was performed regularly in the main compartment in order to preserve the data quality standards.

The second challenge is associated with the probe position in terms of height. It was important to avoid impact of vertical heterogeneity above the gutter due to the circular flow pattern of the Kreisel tank [42]; secondly, the probe need to be placed as far as possible from the solid boundary layer [50]. As a result, the probe was positioned at 85 mm above the gutter to ensure optimal data quality. We tested three substrate thicknesses: 0, 5 and 15 mm. The analysed sampling volume, located 15–25 mm above the gutter, was therefore closer to the surface of the thicker substrates, which might have had an effect on the results. Further experiments should consider this parameter.

The third challenge concerned data analysis. In this study, the ADV data were converted to ASCII format, which allowed for analysis in R, which is an appropriate way to achieve the output needed in this study. Nevertheless, the use of the MAT format, the other ADV output format, should be considered if more in-depth data analyses are needed [49]. The fourth challenge concerned the measured velocity range. Our velocity measurement oscillated between 0.0011 $m.s^{-1}$ and 0.0780 $m.s^{-1}$, which is close to the ADV detection limit and low in comparison to other studies measuring flow velocity under lab conditions, in which the velocity oscillated between 0.01 $m.s^{-1}$ and 0.25 $m.s^{-1}$ [25,35,36,40]. Comparison with another method (such as laser Doppler velocimetry (LDV)) could be an interesting way to definitively validate the use of the ADV in low flow conditions [55], as the present study was conducted close to the ADV detection limit. The use of ADV to efficiently compare substrates should be further investigated, especially to define turbulent motion around the substrates. Nevertheless, the use of an ADV is a cost-efficient, fast way to get a rough insight into flow velocity changes caused by different substrates and an appropriate technique for rapid screening of multiple substrates. In conclusion, the laboratory setup and methods used in this study could be applied to other substrates, as well as different types of planktonic larvae to test capture rates and changes in flow velocities.

The aim of the present study was to test the retention efficiency of substrates with living specimens as a necessary step before recommending one particular substrate for in situ deployment. Therefore, we opted to perform tests with *L. conchilega* as a means to validate the hypothesis emerging from results obtained in the previous ranking tests. Thicker substrates with loose mesh size and 3D structure were found to be suitable candidates to enhance larval settlement. Nevertheless, working with real organisms is not straightforward, as many aspects influence success. First, the process of catching and sorting larvae did not allow for the collection of a large number of individuals and affected the choice of the number of individuals per tank. Second, *L. conchilega* larvae are very fragile organisms with a naturally high mortality rate [56], and their maintenance under lab conditions is a complex process, as illustrated by high mortality during the sorting and experimental period, as similarly reported in a previous study [37]. During our sampling, the abundance per sample was rather low compared to that reported in [37], resulting in a longer sorting time (up to 2 days). Additionally, the high mortality observed during the handling of the samples was probably due to lower oxygen levels combined with the dense presence of microalgae in the samples. Important algae blooms are frequent during the month of April in the North Sea [57], and high chlorophyll-a concentrations were measured close to the Belgian coastline on both sampling days (European Organisation for the Exploitation of Meteorological Satellites, OLCI Level 2 CHL Concentration Daily

Accumulated—Sentinel-3, on 11 April 2021 and 30 April 2021). Ideally, the sorting and start of the experiment should be done on the sampling day itself. The practical difficulties encountered during this work led to only three replicates with one performed 18 days earlier than the two others. The larvae of the second sampling campaign were, on average, larger in size, possibly indicating a difference in maturity and stage of the larvae [32], which could have an effect on their ability to settle [58]. Despite these practical challenges, a test with living animals is an important part of the substrate screening process.

4.2. Substrate Ranking

A strong understanding of the key parameters allowing for optimal settlement of larvae is essential for a successful restoration plan.

In this study, the control (shells or sand) had a pellet capture rate close to zero, with an average overlaying velocity of 3.46 cm/s, and no juveniles were found, suggesting that larval settlement on bare sand is unlikely to occur. These results are supported by the abundant literature describing the weak settlement of larvae on bare sand in comparison to artificial substrate, dead shells or adult tubes [2,30,34,36].

If the absence of a substrate is not an option, some substrates can also be easily rejected. The high density of the weave of the geotextile NW170 White non-woven (D) made it ineffective in terms of particle capture, as also observed in a previous study with living larvae [37]. Additionally, a dense mesh substrate could lead to anoxic conditions [38]. The mat density has to allow the particles/aulophore larvae to pass through. The identical capture rates of substrates A and B (with differing mesh structures: geotextile 220 g/m^2 3D knitted fabric (10 mm thick) based on PES knit and PA spacers and geotextile Kena260 black non-woven (260 g/m^2), respectively), showed that as long as the mesh size allows particles/aulophore larvae to pass through it, the substrate can be considered an artificial substrate candidate.

Our comparative study showed that one of the most discriminant criteria was the three-dimensional (3D) aspect of the substrate. This 3D aspect can be expressed on two different levels: The thickness of the geotextile and the presence of wood sticks. Our results showed that the geotextile's thickness is positively related to the pellet capture rate. Thicker geotextiles (1.5 cm), as well as multiple layers of a thin geotextile (3×0.5 cm), showed a significantly higher capture rate. In nature, adult tubes provide the 3D aspect of reefs, with a diameter of 5 mm and a length (out of sediment) that can reach 4 cm [15,45,46]. In this study, wooden sticks were used as a mimic. Our results showed that the addition of sticks increased the capture rate in all cases, confirming the results of previous studies [2,36]. As a result, elevation, created either by thickness or by sticks, seems to be a key factor. The best performance was observed with substrate G (1.5 cm thick + 4 cm wooden sticks), also achieving the highest capture rate, velocity and juvenile settlement rate. This 3D selection process is in accordance with the natural environment, as elevation relative to surrounding sediment is a characteristic of *L. conchilega* beds, and the high density of individuals influences hydrodynamics and can lead to seabed elevation from 10 to 40 cm [18,59,60].

Our comparative study showed that another discriminant criterion is the ability to create an appropriate hydrodynamic regime surrounding the area of intended settlement. The velocity measured above the substrate combined with wood sticks showed a significant reduction in the average velocity in the case of the control and the geotextile Kena260. This result is supported by the literature; the flow perturbation induced by the presence of adult tubes (or sticks) creates areas of reduced flow velocity and chaotic motion, enhancing the capture of particles [25,40]. The opposite result was observed with geotextiles, which tended to increase the overlaying flow velocity. The increase in velocity could be due to a change in the bed roughness [61]; however, further research is needed to validate this hypothesis. In the context of a turbulent system, this increase in velocity would lead to an increase in encounter rate between the pellet/larvae and the substrate [39], promoting a higher pellet/larvae capture rate. However, high velocity could also have a negative impact on the attachment rate of the pellet/larvae via mobilising forces (drag, lift and acceleration reaction [39]), which have to be considered but could not be studied in the context of this work. Along the Belgian coastline, the presence of sand mason worm aggregates is higher on the lee side of sand banks than on the exposed side [59]. The exposed side is characterised by an increase in near-bed velocities and a decrease in turbulence, while the lee side is characterised by reversed flow with high turbulence intensity [62]. These field observations substantiate our hypothesis of a higher settlement rate in turbulent systems.

In our study, 30 wooden sticks were placed in the gutter for substrates F, G and H, which represents a density of 680 tubes/m^2. The natural presence of L. conchileaga in a density superior to 500 individuals/m^2 can be considered as an aggregation [36]. Further studies are required to optimise the density of sticks and better understand the interaction between the geotextile and the sticks. Higher flow disturbance and settlement rates of larvae with higher densities (>1000 tubes/m^2) have been reported in previous studies [15,25,40]. Based on our results, we hypothesise that the presence of tube-like structures in addition to a geotextile could counterbalance the negative effect of velocity on attachment rate by locally reducing the flow velocity and increasing turbulence in a higher water layer while maintaining an optimal encounter rate.

The substrate combining the geotextile 3D knitted fabric based on PES knit and PA spacers with wood sticks (G) achieved the most promising results in all the substrate screening tests. It had the highest capture rate, with a significant increase in velocity (compared to control) and the highest presence of juveniles after 10 days. These results raise the question of the role of behaviour in the settlement of larvae. The addition of wooden sticks to the substrate considerably increased the settlement of larvae and their growth into juveniles in the experiment with living animals. Previous studies proved the hydrodynamic influence of polychaete adult tubes (or mimics, such as wooden tubes) on particle capture [25,40], with larvae observed to use the tube structure to attach and crawl down to settle in the sand. We hypothesised that substrate G would perform better because it combines the effects of a potential increase in flow velocity by the geotextile, which increases the encounter rate and the potential local reduction in velocity, and behavioural advantage due to the wood sticks, which increases the attachment rate. An additional test comparing the settlement rate of the larvae between substrate F (control + sticks) and substrate G (3D knitted mat + sticks) would further validate the positive effect of the combination of geotextile and wood sticks. In this study, one geotextile type was tested with the larvae (with and without sticks). However, geotextile Kena260 black non-woven (C) obtained a similar capture rate of pellet as the 3D kitted mat (A), with a similar effect on hydrodynamics. A comparison of the settlement rate of these two geotextiles would provide information on the possible effect of behaviour on the choice of geotextile.

5. Conclusions and Future Prospects

This work provides encouraging results with respect to the use of plastic pellet capture rate and ADV measurement as a low-cost and efficient methodology to screen artificial substrates for the enhancement of *L. conchilega* settlement. We also optimised an experimental design [37], which can be used for artificial substrate screening in a time- and cost-efficient way. We provided a list of recommendations concerning key characteristics to take into account for the development of an appropriate artificial substrate for field application. The optimal artificial substrate should have a specific structure, including a mat-type base with a loose mesh size (>5 mm diameter) and a relatively high thickness (>1.5 cm) or/and a tube-like structure rising above the mat (>5 cm length and ~5 mm diameter). This optimal artificial substrate should be able to create a dynamic flow, increasing the encounter rate between larvae and the substrate while also locally reducing the velocity in order to favour the attachment rate of the larvae on the substrate. The methodology developed in the context of this work will allow for further testing of innovative substrates in order to complete this list of characteristics to allow for the production/engineering of an adequate substrate in the form of a biodegradable material [63] that is non-destructive for the future established reef.

The efficiency of an optimal artificial substrate when facing the stress of natural hydrodynamic conditions, such as waves or tides, should be tested in a larger-scale laboratory design (e.g., flume tanks to allow for the description of turbulent currents above the substrate with varying flow velocities and wave conditions [64], as well as the optimal anchoring of the substrate. Finally, laboratory conditions would allow for the study of settlement enhancement in future ocean conditions (physical and environmental), which will make settlement even more challenging for species with planktotrophic, calcifying or weakly swimming larvae with specialised adult habitat [65]. However it is impossible to fully reproduce natural conditions in laboratory experiments, risking results with low ecological validity. Therefore, the small-scale behaviours observed and theorised under laboratory conditions must be validated within the larger, more dynamic framework of the Belgian coastal ecosystem. Under such conditions, much larger-scale hydrodynamic mechanisms strongly impact the settlement of pelagic larvae. These mechanisms can include waves, currents, fronts, coastal boundary layers and tides [64,66]. Therefore, field trials are key, as they allow for the testing of the substrates under natural conditions, with no alteration of variables and external influence on the environment. Previous field trials have elaborated on the specific techniques used to placing substrates in the field at the low waterline according to specific design to keep the substrates as close as possible to the sediment [2,34–36,38]. Preliminary trials using the same 3D geotextiles used in this study reported interesting results, opening the possibility of effectively applying this technique in the field [38].

Therefore, the research process of finding optimal substrates should follow a three-step approach. First, the capture rate and effect on current velocity of a large number of substrates can be tested with small lab experiments, such as those described in this study. In a secondary step, promising candidates can be further tested under larger-scale lab conditions to verify their performance under various hydrodynamic conditions in order to better simulate natural situations. Finally, field tests are necessary to validate the tested substrate in practice for application on an industrial scale.

Author Contributions: Conceptualization, A.D., A.S., T.S. and G.V.H.; Methodology, A.D.; Investigation, A.D.; Writing—original draft, A.D.; Writing—review & editing, A.S. and G.V.H.; Supervision, A.S. and G.V.H.; project administration, T.S. All authors have read and agreed to the published version of the manuscript.

Funding: Alice D'Hurlaborde was supported by the master programme IMBRSea (International Master's in marine biological resources). This research received funding from the Flemish innovation and entrepreneurship agency under VLAIO (Vlaams Agentschap Innoveren & Ondernemen) number HBC.2019.0037 Coastbusters 2.0.

Institutional Review Board Statement: Not applicable.

Informed Consent Statement: Not applicable.

Data Availability Statement: Not applicable.

Acknowledgments: All experiments were executed at ILVO (Flanders Research Institute for Agriculture, Fisheries and Food). We thank David Vuylsteke for his help with all technical aspects during the experiments, VLIZ (Flanders marine institute) for the use of their research vessel (the Simon Stevin) and Sioen industries for providing the textile substrates. We thank all partners for their general interest and feedback and Flanders Hydraulics for the use of the Nortek Vectrino Profiler.

Conflicts of Interest: The authors declare no conflict of interest.

References

1. In, F.; Airoldi, L.; Gibson, R.; Atkinson, R.; Gordon, J. Loss, Status and Trends for Coastal Marine Habitats of Europe. *Annu. Rev.* **2007**, *45*, 345–405. [CrossRef]
2. Pinochet, J.; Urbina, M.A.; Lagos, M.E. Marine invertebrate larvae love plastics: Habitat selection and settlement on artificial substrates. *Environ. Pollut. Barking Essex.* **2020**, *257*, 113571. [CrossRef]

3. Luijendijk, A.; Hagenaars, G.; Ranasinghe, R.; Baart, F.; Donchyts, G.; Aarninkhof, S. The State of the World's Beaches. *Sci. Rep.* **2018**, *8*, 6641. [CrossRef] [PubMed]
4. *Annual Report 2009 | UN-Habitat*; United Nations: New York, NY, USA, 2009; Available online: https://unhabitat.org/annual-report-2009 (accessed on 8 September 2022).
5. Gracia, A.; Rangel-Buitrago, N.; Oakley, J.A.; Williams, A.T. Use of ecosystems in coastal erosion management. *Ocean Coast. Manag.* **2018**, *156*, 277–289. [CrossRef]
6. Baeteman, C.; de Lannoy, W.; Paepe, R.; van Cauwenberghe, C. Vulnerability of the Belgian Coastal Lowlands to Future Sea-level Rise. *Impacts Sea-Level Rise Eur. Coast. Lowl. Inst. Br. Geogr. Spec. Publ. Ser.* **1992**, *27*, 56–71.
7. Lebbe, L.; van Meir, N.; Viaene, P. Potential Implications of Sea-Level Rise for Belgium. *J. Coast. Res.* **2008**, *242*, 358–366. [CrossRef]
8. Van Den Eynde, D.; Ponsar, S.; Luyten, P.; Ozer, J. *Analysis of Climate Changes in the Time Series of Wind Speed, Significant Wave Height and Storm Surges at the Belgian Coast*; Royal Belgian Institute of Natural Sciences, Operational Directorate Natural Environment: Brussels, Belgium, 2019; Available online: www.van-den-eynde-et-al-2019-crest-v6.pdf (accessed on 8 September 2022).
9. Charlier, R.H.; Meyer, D.; DeCroo, D. Beach protection and restoration part I: Hard structures and Beach erosion. *Int. J. Environ. Stud.* **1989**, *33*, 29–44. [CrossRef]
10. Van der Biest, K.; Verwaest, T.; Vanneuville, W.; Reyns, J.; Mostaert, F. *CLIMAR-Evaluation of Climate Change Impacts and Adaptation Responses for Marine Activities*; Flanders Hydraulics Research: Antwerp, Belgium, 2008.
11. Temmerman, S.; Meire, P.; Bouma, T.J.; Herman, P.M.J.; Ysebaert, T.; de Vriend, H.J. Ecosystem-based coastal defence in the face of global change. *Nature* **2013**, *504*, 79–83. [CrossRef]
12. Borsje, B.W.; van Wesenbeeck, B.K.; Dekker, F.; Paalvast, P.; Bouma, T.J.; van Katwijk, M.M.; de Vries, M.B. How ecological engineering can serve in coastal protection. *Ecol. Eng.* **2010**, *37*, 113–122. [CrossRef]
13. Day, J.; Psuty, N.; Perez, B. The role of pulsing events in the functioning of coastal barriers and wetlands: Implications for human impact, management and the response to sea level rise. In *Concepts and Controversies in Tidal Marsh Ecology*; Springer: Dordrecht, The Netherlands, 2002; pp. 633–659. [CrossRef]
14. Barbier, E.B.; Hacker, S.D.; Kennedy, C.; Koch, E.W.; Stier, A.C.; Silliman, B.R. The value of estuarine and coastal ecosystem services. *Ecol. Monogr.* **2011**, *81*, 169–193. [CrossRef]
15. Callaway, R. Juveniles stick to adults: Recruitment of the tube-dwelling polychaete Lanice conchilega (Pallas, 1766). *Hydrobiologia* **2003**, *503*, 121–130. [CrossRef]
16. Sterckx, T. Coastbusters: Investigation of ecosystem based coastal stabilisation solutions. In Proceedings of the 22nd World Dredging Congress, WODCON XXII, Shangai, China, 25–29 April 2019.
17. Ropert, M.; Dauvin, J.-C. Renewal and accumulation of a *Lanice conchilega* (Pallas) population in the baie des Veys, western Bay of Seine. *Oceanol. Acta* **2000**, *23*, 529–546. [CrossRef]
18. Van Hoey, G.; Guilini, K.; Rabaut, M.; Vincx, M.; Degraer, S. Ecological implications of the presence of the tube-building polychaete Lanice conchilega on soft-bottom benthic ecosystems. *Mar. Biol.* **2008**, *154*, 1009–1019. [CrossRef]
19. Borsje, B.W.; Besio, G.; Hulscher, S.J.M.H.; Blondeaux, P.; Vittori, G. *Exploring Biological Influence on Offshore Sand Wave Length*; SHOM: Leeds, UK, 2008.
20. Ziegelmeir, E. Beobachtungen über den Röhrenbau von *Lanice conchilega* (Pallas) im Experiment und am natürlichen Standort. *Helgoländer Wiss. Meeresunters.* **1952**, *4*, 107–129. [CrossRef]
21. Jones, S.E.; Jago, C.F. In situ assessment of modification of sediment properties by burrowing invertebrates. *Mar. Biol.* **1993**, *115*, 133–142. [CrossRef]
22. Carey, D.A. Sedimentological effects and palaeoecological implications of the tube-building polychaete *Lanice conchilega* Pallas. *Sedimentology* **1987**, *34*, 49–66. [CrossRef]
23. Hartmann-Schröder, G. Annelida, Borstenwürmer, Polychaeta. *Bijdr. Tot De Dierkunde.* **1992**, *61*, 219–235. [CrossRef]
24. Buhr, K.-J.; Winter, J.E. Distribution and maintenance of a lanice conchilega association in the weser estuary (FRG), with special reference to the suspension—feeding behaviour of lanice conchilega. In *Biology of Benthic Organisms: 11th European Symposium on Marine Biology, Galway, October 1976*; Elsevier: Pergamon, Greece, 1977; pp. 101–113. [CrossRef]
25. Carey, D. Particle Resuspension in the Benthic Boundary Layer Induced by Flow around Polychaete Tubes. *Can. J. Fish. Aquat. Sci.* **1983**, *40*, s301–s308. [CrossRef]
26. Heuers, J.; Jaklin, S.; Zuhlke, R.; Dittmann, S.; Gunther, C.P.; Hildenbrandt, H.; Grimm, V. A model on the distribution and abundance of the tube-building polychaete Lanice conchilega (Pallas, 1766) in the intertidal of the Wadden Sea. *Verh. Ges. Für Ökol.* **1998**, *28*, 207–215.
27. Woodin, S.A. Refuges, Disturbance, and Community Structure: A Marine Soft-Bottom Example. *Ecology* **1978**, *59*, 274–284. [CrossRef]
28. Forster, S.; Graf, G. Impact of irrigation on oxygen flux into the sediment: Intermittent pumping by Callianassa subterranea and "piston-pumping" by Lanice conchilega. *Mar. Biol.* **1995**, *123*, 335–346. [CrossRef]
29. Qian, P.-Y. Larval settlement of polychaetes. *Reprod. Strateg. Dev. Patterns Annelids* **1999**, *142*, 239–253. [CrossRef]
30. Zühlke, R.; Blome, D.; van Bernem, K.H.; Dittmann, S. Effects of the tube-building polychaeteLanice conchilega (Pallas) on benthic macrofauna and nematodes in an intertidal sandflat. *Senckenberg. Marit.* **1998**, *29*, 131–138. [CrossRef]
31. McHugh, D. A Comparative Study of Reproduction and Development in the Polychaete Family Terebellidae. *Biol. Bull.* **1993**, *185*, 153–167. [CrossRef] [PubMed]

32. Keßler, M. Die Entwicklung von Lanice concbilega [Pallas} mit besonderer eriicksichtigung der Lebensweise. *Helgoländer Wiss. Meeresunters.* **1963**, *8*, 425–476. [CrossRef]
33. Siddik, A.A.; Al-Sofyani, A.A.; Ba-Akdah, M.A.; Satheesh, S. Invertebrate recruitment on artificial substrates in the Red Sea: Role of substrate type and orientation. *J. Mar. Biol. Assoc.* **2019**, *99*, 741–750. [CrossRef]
34. Eckman, J.E. Hydrodynamic processes affecting benthic recruitment. *Limnol. Oceanogr.* **1983**, *28*, 241–257. [CrossRef]
35. Harvey, M.; Bourget, E. Recruitment of marine invertebrates onto arborescent epibenthic structures: Active and passive processes acting at different spatial scales. *Mar. Ecol. Prog. Ser.* **1997**, *153*, 203–215. [CrossRef]
36. Jaklin, S. Recruitment dynamics of North Sea macrozoobenthos in intertidal soft bottoms: Larval availability, settlement and dispersal. Ph.D. Thesis, Staats-und Universitätsbibliothek Bremen, Bremen, Germany, 2003.
37. Wyns, L.; Semeraro, A.; Sterckx, T.; Delbare, D.; van Hoey, G. Practical implementation of in vitro culture of Lanice conchilega (Polychaeta) in a coastal defence context. *Invertebr. Reprod. Dev.* **2020**, *64*, 219–236. [CrossRef]
38. Semeraro, A. *Enhancing A Lanice Reef Report of Field Trials 2017–2020*; Instituut voor Landbouwen Visserijonderzoek (ILVO): Oostende, Belgium, 2020.
39. Fingerut, J.; Hart, D.; Thomson, J. Larval settlement in benthic environments: The effects of velocity and bed element geometry. *Freshw. Biol.* **2011**, *56*, 904–915. [CrossRef]
40. Friedrichs, M.; Graf, G.; Springer, B. Skimming flow induced over a simulated polychaete tube lawn at low population densities. *Mar. Ecol. Prog. Ser.* **2000**, *192*, 219–228. [CrossRef]
41. Greve, W. The "planktonkreisel", a new device for culturing zooplankton. *Mar. Biol.* **1968**, *1*, 201–203. [CrossRef]
42. Rakow, K.C.; Graham, W.M. Orientation and swimming mechanics by the scyphomedusa Aurelia sp. in shear flow. *Limnol. Oceanogr.* **2006**, *51*, 1097–1106. [CrossRef]
43. Arganda-Carreras, I.; Kaynig, V.; Rueden, C.; Eliceiri, K.W.; Schindelin, J.; Cardona, A.; Seung, H.S. Trainable Weka Segmentation: A machine learning tool for microscopy pixel classification. *Bioinformatics* **2017**, *33*, 2424–2426. [CrossRef] [PubMed]
44. Swertz, O.C.; Colijn, F.; Hofstraat, H.W.; Althuis, B.A. Temperature, Salinity, and Fluorescence in Southern North Sea: High-Resolution Data Sampled from a Ferry. *Environ. Manag.* **1999**, *23*, 527–538. [CrossRef] [PubMed]
45. Callaway, R. Tube worms promote community change. *Mar. Ecol. Prog. Ser.* **2006**, *308*, 49–60. [CrossRef]
46. Van Hoey, G.; Vincx, M.; Degraer, S. Some recommendations for an accurate estimation of Lanice conchilega density based on tube counts. *Helgol. Mar. Res.* **2006**, *60*, 317–321. [CrossRef]
47. Bhaud, M.R.; Cazaux, C.P. Buoyancy characteristics of Lanice conchilega (Pallas) larvae (Terebellidae). Implications for settlement. *J. Exp. Mar. Biol. Ecol.* **1990**, *141*, 31–45. [CrossRef]
48. *The Comprehensive Manual for Velocimeters-Vectrino Profiler*; Nortek Manuals: Boston, MA, USA, 2018.
49. Masaló, I.; Reig, L.; Oca, J. Study of fish swimming activity using acoustical Doppler velocimetry (ADV) techniques. *Aquac. Eng.* **2008**, *38*, 43–51. [CrossRef]
50. Chanson, H. Acoustic Doppler velocimetry (ADV) in the field and in laboratory: Practical experiences. In Proceedings of the International Meeting on Measurements and Hydraulics of Sewers IMMHS'08, Summer School GEMCEA/LCPC, Bouguenais, France, 19–21 August 2008.
51. Dean, D.; Mazurkiewicz, M. Methods of culturing polychaetes. In *Culture of Marine Invertebrate Animals*; Springer: Boston, MA, USA, 1975; pp. 177–197. [CrossRef]
52. Miyake, Y.; Kimura, S.; Kawamura, T.; Horii, T.; Kurogi, H.; Kitagawa, T. Simulating larval dispersal processes for abalone using a coupled particle-tracking and hydrodynamic model: Implications for refugium design. *Mar. Ecol. Prog. Ser.* **2009**, *387*, 205–222. [CrossRef]
53. McLelland, S.J.; Nicholas, A.P. A new method for evaluating errors in high-frequency ADV measurements. *Hydrol. Process.* **2000**, *14*, 351–366. [CrossRef]
54. Blanckaert, K.; Lemmin, U. Means of noise reduction in acoustic turbulence measurement. *J. Hydraul. Res.* **2006**, *44*, 3–17. [CrossRef]
55. Voulgaris, G.; Trowbridge, J.H. Evaluation of the Acoustic Doppler Velocimeter (ADV) for Turbulence Measurements*. *J. Atmos. Ocean. Technol.* **1998**, *15*, 272–289. [CrossRef]
56. Van Hoey, G.; Vincx, M.; Degraer, S. Chapter 5: Population—dynamics of subtidal Lanice conchilega (Pallas, 1766) populations at the Belgian Continental Shelf. In *Spatio-Temporal Variability within the Macrobenthicy Abra Alba Community, with Emphasis on the Structuring Role of Lanice Conchilega*; Universiteit Gent: Ghent, Belgium, 2006; pp. 93–116.
57. Richardson, K. Algal blooms in the North Sea: The Good, the Bad and the Ugly. *Dana. Charlottenlund.* **1989**, *8*, 83–93.
58. Toonen, R.; Pawlik, J. Settlement of the gregarious tube worm Hydroides dianthus (Polychaeta: Serpulidae). II. Testing the desperate larva hypothesis. *Mar. Ecol.-Prog. Ser.* **2001**, *224*, 115–131. [CrossRef]
59. Rabaut, M.; Guilini, K.; van Hoey, G.; Vincx, M.; Degraer, S. A bio-engineered soft-bottom environment: The impact of Lanice conchilega on the benthic species-specific densities and community structure. *Estuar. Coast. Shelf Sci.* **2007**, *75*, 525–536. [CrossRef]
60. Degraer, S.; Moerkerke, G.; Rabaut, M.; Van Hoey, G.; Du Four, I.; Vincx, M.; Henriet, J.-P.; Van Lancker, V. Very-high resolution side-scan sonar mapping of biogenic reefs of the tube-worm Lanice conchilega. *Remote Sens. Environ.* **2008**, *112*, 3323–3328. [CrossRef]
61. Kashefipour, S.M.; Daryaee, M.; Ghomeshi, M. Effect of bed roughness on velocity profile and water entrainment in a sedimentary density current. *Can. J. Civ. Eng.* **2018**, *45*, 9–17. [CrossRef]

62. Kostaschuk, R. A field study of turbulence and sediment dynamics over subaqueous dunes with flow separation. *Sedimentology* **2000**, *47*, 519–531. [CrossRef]
63. Tian, K.; Bilal, M. Chapter 15-Research progress of biodegradable materials in reducing environmental pollution. In *Abatement of Environmental Pollutants*; Elsevier: Amsterdam, The Netherlands, 2020; pp. 313–330. [CrossRef]
64. Reidenbach, M.A.; Koseff, J.R.; Koehl, M.a.R. Hydrodynamic forces on larvae affect their settlement on coral reefs in turbulent, wave-driven flow. *Limnol. Oceanogr.* **2009**, *54*, 318–330. [CrossRef]
65. Bashevkin, S.M.; Dibble, C.D.; Dunn, R.P.; Hollarsmith, J.A.; Ng, G.; Satterthwaite, E.V.; Morgan, S.G. Larval dispersal in a changing ocean with an emphasis on upwelling regions. *Ecosphere* **2020**, *11*, e03015. [CrossRef]
66. Adams, T.P.; Aleynik, D.; Burrows, M. Larval dispersal of intertidal organisms and the influence of coastline geography. *Ecography* **2013**, *37*, 698–710. [CrossRef]

Journal of
Marine Science and Engineering

MDPI

Article

Influence of the Implantation of the Onshore Power Supply (OPS) System in Spanish Medium-Sized Ports on the Reduction in CO_2 Emissions: The Case of the Port of Santander (Spain)

Alvaro Herrero [1,*], Andrés Ortega Piris [1], Emma Diaz-Ruiz-Navamuel [1], Miguel A. Gutierrez [2] and Alfonso-Isidro Lopez-Diaz [3]

1 Ocean and Coastal Planning and Management R&D Group, School of Nautical Studies of Santander, University of Cantabria, C/Gamazo, 1, 39004 Santander, Spain
2 Department of Technology, Faculty of Science, Catholic University of Ávila, C/Canteros S/N, 05005 Ávila, Spain
3 Department of Electrical, Electronic, Automatic and Communications Engineering, School of Industrial and Aerospace Engineering, University of Castilla-La Mancha, Royal Arms Factory Avda, Calos III, SN, 45071 Toledo, Spain
* Correspondence: alvaro.herreromartinez@unican.es

Abstract: Society and its leaders are increasingly aware of the need to fight climate change and CO_2 emissions in the search for sustainability. Maritime transport and ports are important sources of pollution and, while industry and the rest of the large-scale emitters have achieved considerable reductions in this area, "shipping" is still not advancing at the same rate, falling behind in this race. The aim of this article is to underline the importance of an early implementation of On-Shore Power Supply (OPS), Cold Ironing (CI) or Alternative Marine Power (AMP) by making a study of potentially avoidable CO_2 emissions from vessels docked (on the basis of the EPA [Environmental Protection Agency] and ENTEC [Environmental Engineering Consultancy, an environmental and engineering consultancy in UK] methods) close to urban areas in the Port of Santander (Spain). It is the first time potential reductions have been calculated for the last 11 years (2011–2021), distinguishing yearly emissions per type of vessel and providing real information to port authorities to prioritize the installation of this technological basis for the operation of piers/terminals to optimize investments and outcomes. In this case study, results demonstrate the outcomes of ROROs, ferries, and cruises being the first target of OPS implementation (reaching total of 37.95% of total emitted tons of CO_2 during the period of study). As a clear conclusion to this paper, the reader can understand the enormous and growing potential of this technology multiplied by the continuous development, increase and implementation of green energies.

Keywords: cold ironing; sustainability; green ports; OPS; CO_2 emissions

Citation: Herrero, A.; Ortega Piris, A.; Diaz-Ruiz-Navamuel, E.; Gutierrez, M.A.; Lopez-Diaz, A.-I. Influence of the Implantation of the Onshore Power Supply (OPS) System in Spanish Medium-Sized Ports on the Reduction in CO_2 Emissions: The Case of the Port of Santander (Spain). *J. Mar. Sci. Eng.* **2022**, *10*, 1446. https://doi.org/10.3390/jmse10101446

Academic Editor: Rosemary Norman

Received: 18 August 2022
Accepted: 1 October 2022
Published: 7 October 2022

Publisher's Note: MDPI stays neutral with regard to jurisdictional claims in published maps and institutional affiliations.

1. Introduction

Since the industrial revolution, technological development and process optimization have been constants in the pursuit of companies to achieve improvements in their performance and profits. In recent years another goal has emerged, just as important as these previous ones: sustainability.

It can be stated that the first time this concept appeared as we know it today was in 1987 [1] in what is known as the Brundtland report. In it, the concept of sustainable development is defined as the process of "satisfying the needs of present generations without compromising the possibilities of those of the future to meet their own needs". In this sense, through both private and public initiative, for more than 30 years now a great effort has been made in researching and developing technologies capable of combining these objectives: the optimization of processes and the reduction of emissions.

Today there is extensive legislation, both nationally and internationally, that commits countries to the common goals of reducing emissions of polluting agents into the atmosphere. The EU has set medium- and long-term objectives [2,3] that seek to implement measures that lead member countries to respect the environment by reducing the emission of greenhouse gases: a "European Green Deal" [4]. The goal is, by 2050, to have reduced these emissions to at least 80% below 1990 levels. To do this, taking the same reference values, two milestones must be achieved: a 40% cut in emissions by 2030 and 60% by 2040.

According to Mikova [5], in addition to the legislation and efforts carried out by each country individually, collaborations between counties are essential. Similarly, synergies between the public and the private spheres are important; finally, another key factor in achieving success is the role of renewable energies and their application.

Unfortunately, despite all of the efforts made, the indicators of the European Environment Agency (EEA) [6] show that the desired results will not be achieved. According to the same source, thanks to the efforts of all parties, the EU in recent years has overseen a significant reduction in emissions. Unfortunately, however, the evidence for the transport sector shows that it has not been following this downward trend, its relative contribution to the total tons of CO_2 having increased.

"On Shore Power Supply" (OPS) consists of satisfying the energy demand of a ship from the port during its stay in dock. It is a combination of technologies that are already well-developed and mature, but which are not being fully exploited and are not yet implemented today in most of the world's ports. Most of the on-board equipment is in operation when the ship is in motion (navigating or maneuvering), while it is switched off when it is docked. At that time, the auxiliary engines of the ship itself are responsible for supplying the energy [7]. The idea of the OPS is simple: connect the ship to the land network so that it can be supplied with the necessary electrical energy while it is in port. This will allow it to stop its auxiliary engines, which are the ones that feed all the necessary services of the ship through the combustion of MGO (Marine Gas Oil), significantly reducing emissions during this period [8]. The objective is "zero emissions in port" and, in this way, by not burning diesel during the period of loading/unloading operations, emissions will be suppressed, while the noise and vibrations they produce will be eliminated.

Therefore, the reduction in emissions will depend on the origin of the energy supplied to the vessel from land. The ideal situation would be to supply energy from a renewable and clean source, minimizing that which comes from fossil fuels and optimizing the ecological performance of the installation. However, the most common situation is a mixture in which there is also a part of a fossil fuel source. It is important to control the percentages supplied. If this exceeds 80% from fossil sources, the opposite result to that pursued might be obtained and this may lead to even greater damage than that caused by the ship's engines [9]. Hence, the percentage of energy supplied from renewable sources will ultimately be responsible for optimizing emission reductions. Another advantage of the system is that emissions are relocated and moved away from urban centers and ports, emitting in already established production areas where the impact to population is marginal [9].

According to Chang [10], adopting an OPS system could reduce CO_2 emissions by 57.16% in port areas. According to Hall [11], it is estimated that the implementation of this technique in the ports of the UK could reduce CO_2, SO_2, CO and NOx emissions by around 25%, 46%, 76%, and 92% compared to what would have been emitted when using the ships' diesel auxiliary engines.

"On Shore Power Supply" (OPS) is also known as "Cold Ironing" (CI), a term which comes from the practice of steamships cooling their coal engines while moored in port. It is common practice for docked warships [12], port tugs or vessels in shipyards, but is not yet for merchant ships operating in commercial docks. The EU has established as an objective the implementation of these facilities in all European ports [13–15] by the year 2025 [16]. Government aid is essential for achieving the implementation of this initiative [17]. Without this, the ports will not invest until the ships have the technology to exploit it and vice versa.

This is the so-called 'chicken and egg dilemma' [9]. Where and how is it encouraged? That is the question that needs to be answered to achieve the objective of promoting the system both in ports and among shipowners so that, after the initial aid, the rest of the community continues along the same path [18]. The more ports that have this technology, the greater the number of vessels that will install the OPS option, achieving the desired benefits. Thus, ports such as Los Angeles or Long Beach are in the process of making the use of the OPS mandatory, leading to an optimization of the system since, if it is not used, other types of investments such as machinery renovation, incentives for voluntary speed reduction, etc., might well be found to be more convenient alternatives.

It should be noted that, in addition to the environmental, mechanical and maintenance benefits of reducing engine wear, diesel consumption, vibrations, noise, etc., the OPS has significant economic advantages [19]. Despite the large initial investment required, these advantages are more significant the higher the price of fossil fuel and the lower the price of the KWh of electricity from the supply network [16].

In view of the obvious advantages that OPS potentially offers, shown by previous investigations from other researchers, this work focuses on the maritime industry and the emissions it produces because of port activity. In it, a study is carried out to quantify the CO_2 emissions released by ships at berth and the emissions which might potentially be avoided through the implementation of the OPS system are evaluated in order, based on the results, to obtain conclusions on the importance of prompt implementation of this technology, and on objectives and priorities which can determine investments in the search for maximum optimization. This paper will calculate emissions depending on the kind of vessel and bearing in mind time alongside fuel consumed. In this paper, vessels have been classified as: Ferry, Cruise, Tanker, Chemical Tanker, Bulkcarrier, General Cargo and RORO, always counting time only while berthed and always bigger than 500 GT (Gross Tonnage). Neither time at the shipyard nor at anchorage have been considered, since objective of this paper is to classify priorities of installation of this technology on the basis of the occupation of piers/terminals, seeking to optimize investments and outcomes at the commercial port of Santander.

To achieve the proposed objective, the CO_2 emissions produced by the auxiliary engines of the ships in the port chosen for the study are calculated. Due to its volume of traffic according to yearly vessel calls, this port is a medium-sized port amongst Spanish ports [20], receiving 7 different types of vessels on a regular basis. Next, the emissions that ideally would have been produced if both the ships and the port had the necessary facilities to use the OPS system are calculated. Finally, the work concludes with the comparative analysis of the results obtained, thus quantifying the potential of the application of this measure.

2. Literature Review

There are many technologies that are currently in the process of development and numerous studies, projects and articles focused on reducing emissions in all areas of industry and, especially, in port logistics and movements [21]. These include private and industrial vehicles, freight-handling equipment, operations assistance equipment [22], barges and ocean-going vessels, etc. The maritime industry, although it is on a par with aviation and about five times lower than land transport, is one of the main sources of pollution of the atmosphere [9]. This is due to the large volume of goods transported by this mode (in 2019, despite the slowdown due to COVID-19 [23], 11.08 billion tons). For this reason, both ocean-going vessels and ports are among the main sources of gas emissions into the atmosphere, making them a key target in this fight to respect the environment and the health of all beings on the planet.

The activity of research and development of new techniques is another important field to assess. The objective is to achieve a "shipping" that respects the environment, and to achieve 'zero emissions' through technology. A great deal of research is being conducted in the field of ports in a search for the sustainable port, including initiatives, projects, and the investment of European funds (https://portosproject.eu/ (accessed on 17 August 2022)). This technological field can be divided into two: on the one hand, there are alternative energies which, although their potential is very high and promising, are still in the experimental phase. They have not been sufficiently developed for reliable exploitation: wave energy [24–29], tidal energy [30,31], and thermal gradient, etc. On the other hand, there are wind and solar energies which can be said to be mature technologies.

This manuscript focuses on the great advantages offered using OPS, but its scarce implementation is significant. According to Krämer and Czermanski [32], in 2019 only 574 ships were equipped on board with this technology, with 81% of these being container ships (466). Among other factors, it must be considered that, as Schwartz [33] points out, to install this technology on a ship it should not be more than 10 years old, since it must be considered an important investment in "retrofitting". It varies significantly depending on the type of ship. In general, installing it in an already built ship is more expensive than installing it in a newly built one. According to Yu [34], in the case of a 5000 GT (gross tonnage) ship, the final bill could amount to between 50,000 USD and 350,000 USD, while a container ship, chemical tanker or a cruise ship of about 100,000 GT may require an investment of 750,000 USD. At the same time, the port authorities must consider that they will need an estimated investment per dock of 1.5 to 2 million USD (as reported in the current IEC/ISO standard 80005), but these data could vary greatly depending on the terminal to be electrified and its requirements [16]: number of berths, distance, power to be supplied, type of vessel, etc.

The cited articles deal with the OPS, its advantages, disadvantages, and health benefits, but they are always based on the origin of the energy. This means that it must be carefully studied case by case depending on the energy supply of each port/country. As has been said before, this article will focus on the port of Santander based on the Spanish energy supply, providing real results and concrete estimations of CO_2 tons emitted from the more than 22,000 vessels' calls studied. It also opens the door to future studies, the calculation process being repeatable for other ports.

Health Background, Data and Policies

Various studies demonstrate the health risks to which populations near ports are exposed: asthma, lung cancer, heart attacks, respiratory infections, etc. [35–38]. It should be remembered that a high percentage of the world's urban centers are located in coastal areas directly affected by all these gases from the combustion of ship engines.

The emissions of greenhouse gases directly related with the maritime industry have increased from the 977 million tons in 2012 to the 1.076 million tons in 2018, an increase of 9.6%. [39] From all the gases released by engines mentioned above, this work focuses on CO_2. According to Smith [40] during the period 2007/2012, an average of 33 billion tons per year were emitted into the atmosphere. Of these, it is estimated that 3% correspond to shipping and, with respect to 2050, an increase of 250% is predicted.

As Acomi demonstrates in his study [41], there are many variables that influence these emissions from vessels (state of the engines, age, conditions, etc.), but the fuel used to power the engines will be a decisive factor both for the calculation of the tons discharged into the atmosphere and for operating costs. According to Corbett [36], the imposition of limits to reduce the speed of ships and thus emissions near ports could save 8300 lives annually and alleviate the respiratory diseases of 3 million people in the US. In fact, the port of Long Beach already in 2006 imposed a "Reduced Speed Zone (RSZ)" of 20 miles around the port, later doubling this range in 2010. In this case, governments and shipowners share the objective of reducing fuel consumption, given the significant impact on the ship's operating costs. Speed and route optimization is a major goal for emissions control [42–44].

This situation requires action to be taken by the authorities. The European Commission strongly recommends the reduction of emissions from all transport by 60%, including a reduction of between 40–50% from the maritime industry by 2050 compared to the 2005 figures. It must be stressed that in the maritime area, these measures cannot be imposed unilaterally by any of the governments or even by the European Union, since they would also affect vessels that sail under the flag of other countries over which they do not have jurisdiction outside their territorial waters [45]. It should be the IMO (International Maritime Organization) that regulates at the international level, as for example it has already done through MARPOL (International Convention for Prevention of Pollution from ships—MARitime POLlution) Annex VI, where it establishes use of fuel of a maximum of sulfur level in certain areas [14,46,47]. At the port of study, this limit insists on levels being less than 0.1% if berthed or anchored for more than 2 h. Although the European Union has shown interest in expanding these areas, it must progressively seek support and undertake measures to avoid a confrontation with the United Nations or break the Law of the Sea 1982 (UNCLOS) to which the EU is a signatory [48].

3. Methodology

3.1. Framework of Calculation

1. Scenario and reasons.
2. Authors classified vessels by type per each year.
3. Calculate (via EPA and ENTEC method) CO_2 emitted. Basis is the number of hours burning fuel and auxiliary engine power of each ship.
4. Basis is the electricity production of the grid, as authors apply % of renewable energy per each year to the total CO_2 yearly emitted, so total CO_2 tons which could have been avoided if using OPS is calculated.

3.2. Scenario—Port of Santander

The main reasons that make the Port of Santander such an interesting object of study are, on the one hand, the variety of vessels that operate there (size, types of cargo, purpose, etc.) and, on the other, the proximity to the city and population. Both factors will be considered in the conclusions to determine the docks and types of ships in which a greater benefit can be obtained after the installation of this technology.

A priori, a dock of special interest is the so-called "Muelle del Almirante" (Admiral's Dock) which welcomes cruise ships and ferries at less than 150 m from residential buildings in the city center. Given the great activity and regularity of the ships that make use of these facilities, it can be judged to be a clear candidate for the use of the OPS. The greater the number of calls and frequency of the vessels of a shipping company, the greater the return on investment and the faster the amortization will be. This recurrence will allow the implementation of a system that is 100% compatible with the regular vessels that operate in it, achieving an easy and rapid implementation, avoiding risks, reducing time, increasing safety, etc. This will be just the same way as it has been developed in other European ports such us Marseille in which is OPS fitted and which confirms, from practice, how fast and effective the repetition and specialization are, connecting and disconnecting the vessel from land easily and simply in about 15′.

On the other hand, one must not lose sight of the fact that, precisely, ferries and cruise ships are (as a rule) large vessels dedicated to the transport of people and vehicles whose great needs during periods in dock require a significant capacity to satisfy all their requirements [49]. Therefore, they always need their powerful auxiliary engines. This requirement leads to a large emission of gases from the combustion of these engines, which produces significant pollution in areas very close to the city, as well as causing high noise pollution that has already led to some port–city disputes.

3.3. Analysis of the Used Models

There are two models used for the quantification of the CO_2 emissions of vessels. [50] The first is 'Top-Down', used for the calculation of global emissions and full fleets based on their fuel consumption. The second is 'Bottom-up', based on activity and used for the calculation of specific emissions based on operating hours, installed power, etc.

This article focuses on the bottom-up model with which CO_2 emissions per year are calculated from the sum of the estimate of each of the vessels that dock and carry out operations in the dock. In this case, emissions are calculated using the EPA and ENTEC system and the results obtained are compared to evaluate possible deviations depending on the chosen method.

Both the ENTEC and EPA methods have been published by the Lloyds Register Engineering Services Data using their databases to generate the emission factors and to make these as realistic as possible in their predictions [51]. The method is based on ENTEC 2002, 2007 publications and MEPC (Marine Environment Protection Committee of the IMO) resolution 212/(63). Therefore, they are regularly used to calculate CO_2 emissions of marine origin [52].

For the study carried out in this article, a database has been generated where the port calls in the Port of Santander of ships of more than 499 GT have been registered. The sample, of 22,714 entries, only considers the hours spent in dock (18,071 calls totaling 558,626.77 h in dock) since, as seen above, the OPS can only be applied when the ship is docked. Ship emissions during navigation, anchoring, approaching port or maneuvering have not been considered in this study, the object of study is therefore the auxiliary engines.

As stated in the previous paragraph, for all the calculations carried out, the fuel used will be the MGO, whose Conversion Factor (CF) is dimensionless, according to MEPC 212/63 of 2 March [53] and the Fourth IMO GHG Study 2020 [39], which for diesel and petrol is 3206 tons of CO2 for each ton of fuel consumed. They have a standard carbon content of 0.875.

The fourth study published in July 2020 [39] introduced modifications (with respect to the third study) which are of interest and must be considered. Thus, in the case of specific fuel consumption, (SFC) (g/kWh) the emissions of the auxiliary engines go from being dependent on their power to being dependent on the age of the engine (which, as will be seen below, is also of importance in the calculation). For MGO engines built from 1984 up to 2000, SFC will be 190 g/kWh and from 2001 onwards will be 185 g/kWh.

In this sample, it is not possible to verify the year of construction of the auxiliary engines of the 18,071 vessels (min 499 GT) which, from January 2011 to December 2021, docked in the port and are part of the database. Therefore, it is the calculation of the least and most polluting situation in terms of CO2 emissions that will be analyzed and compared. For this reason, on the one hand, all the auxiliary engines will be counted as being built before the year 2000 and on the other, all with be counted as having a later date, so that the deviation of emissions between the 2 extremes will be quantified.

Considering the specific consumption of an engine (SFC) and the conversion factor (CF) of the fossil fuel used, it is possible to obtain the CO2 pollutant emission factor of an engine in g/kWh.

The pollutant emission factor (EF) is defined as the product of the specific consumption of an engine multiplied by the conversion factor of the fossil fuel used:

$$(EF) = CF \cdot SFC \quad (1)$$

The ENTEC Method is the model used by the European Commission. The formula used [54] is as follows:

$$E = t \cdot [ME \cdot LFME \cdot EF + AE \cdot LFAE \cdot EF] \quad (2)$$

In the case of our study, the emissions generated by the combustion of the main engine (ME) are not of interest, leaving the Formula (2) as follows:

$$E = t \cdot [AE \cdot LFAE \cdot CF \cdot SFC] \quad (3)$$

where:

$t(h) \rightarrow$ engine working time

AE (kW) → Auxiliary engine power

LFAE (%) → engine load—it is considered that the auxiliary engine is working at 50% of its capacity when moored

EF (g/kWh) → Emission Factor depending on fuel and engine

The EPA Method, [55] is the one accepted in the USA and is obtained in mathematical form from the tons of CO_2 emitted into the atmosphere using the following data:

1. Time that the vessel employs in the operation that is to be calculated
2. Consumption of the vessel during this period
3. A calculation of the emissions is obtained based on the fuel used

The formula used corresponds to the following equation:

$$E = EF \cdot MCR \cdot P \cdot t \quad (4)$$

where:

EF (g/kWh) → Emission Factor depending on fuel and engine

MCR (%) → Engine Work Regime. In the case of the auxiliary engines, it is considered that they are working at 100%

P (kW) → Power of auxiliary engines

t (h) → Time considered for the calculation.

Therefore, on this situation, if the methods for measuring and calculating port emissions are carefully studied, considering only those of the auxiliary engines, it can be concluded that the American and European (EPA and ENTEC) methods differ only in the percentage/load of work considered for the auxiliary engines. (EPA 100% vs. ENTEC 50%).

To apply the EPA and ENTEC methods, it is necessary to obtain the time spent in dock for each of the calls (A) and to know the power of the auxiliary engines (B) of each of the vessels.

A. The hours of stay of each call have been calculated using the registry data of the Santander Port Authority [56]. Thanks to these data, it is possible to know the exact time from which any ship docks until the time it leaves port.
B. The power of the auxiliary engines is a fundamental value. However, given the volume of the data sample used in this article, it has been decided to proceed differently depending on the type of vessel and based on information obtained from Seaweb [57]. On the one hand, due to the power needed to maintain cruise ships and ferries and to the proximity of these docks to urban centers, the values have been carefully entered scale by scale and vessel by vessel for the 2346 entries. However, for the rest of the vessels (chemical, oil, RORO (roll on-roll off), general cargo or bulk carrier), these values have been assigned by the authors based on the study of a sample. In the bibliography consulted, the question of the difficulty of assigning these values is raised due to the usual lack of information and the complicated relationship with GT, length, or other characteristics, leading some articles/authors to consider this value constant [49].

On the other hand, some dry cargo ships are fitted with cranes. It must be taken into account that, in ports such as Santander, the ships' cranes are rarely used for the loading and unloading of merchandise, since it is much faster and safer to proceed with land-based means. For this reason, it is understood as an error to consider the power of the auxiliary engines for vessels with cranes as a reference for the calculations since this really is a power that would not be used and, therefore, it would be oversizing and falsifying the result which is the object of this article. Hence, the applied values are as per Table 1.

Table 1. Auxiliary engine power values. Source: Authors.

	Chemical Tanker	Tanker	Bulkcarrier	General Cargo	RORO
GT	Pot Aux (kW)	Pot Aux (kW)	Pot Aux (kW)	Pot Aux (kW)	Pot Aux (kW)
<2000	325	350	129.13		
2001/4000	424.08	417.76	272.41	207.37	300
4001/6000	527.53	567.14	343.12	317.95	
6001/8000	541.05	704.77		370.37	
8001/10,000	544.47		384.12	395.31	354.71
10,001/12,000	709.06			451.34	
>12,000	850	1049.72			
10,001/15,000			419.2	476.73	
15,001/25,000			450.58	538.53	
25,001/35,000			563.54	735.51	
>35,001			654.16	800	
10,001/20,000					544.38
20,001/30,000					1022.96
30,001/40,000					1054.51
40,001/50,000					1224.69
50,001/60,000					1524.57
>60,001					1620.03

4. Results

The Table 2 summarizes the data obtained per group of years and type of vessel, including number of studied ships and total tons of CO_2 emitted (both via EPA and ENTEC).

Table 2. Results per type of vessel and total period of study. Source: Authors.

		2011–2013	2014–2016	2017–2019	2020–2021	TOTAL
FERRY	Number of ships	576	597	740	336	2249
	Hours alongside	2736	3817	5422	3864	15,840
Total tons of CO_2	EPA-1984/2000	3134.250	4133.770	5822.540	3880.030	16,970.590
	EPA-2000+	3051.770	4024.990	5669.310	3777.930	16,524.000
	ENTEC-1984/2000	1567.125	2066.885	2911.270	1940.015	8485.295
	ENTEC-2000+	1525.885	2012.495	2834.655	1888.965	8262.000
CRUISE	Number of ships	29	25	40	3	97
	Hours alongside	278	229	399	29	936
Total tons of CO_2	EPA-1984/2000	377.290	251.700	522.970	49.160	1201.120
	EPA-2000+	367.360	245.080	509.220	47.870	1169.530
	ENTEC-1984/2000	188.645	125.850	261.485	24.580	600.560
	ENTEC-2000+	183.680	122.540	254.610	23.935	584.765
TANKER	Number of ships	16	27	52	37	132
	Hours alongside	544	817	1965	1258	4587
Total tons of CO_2	EPA-1984/2000	178.930	216.200	739.560	464.970	1599.660
	EPA-2000+	174.220	210.510	720.100	452.730	1557.560
	ENTEC-1984/2000	89.4650	108.100	369.780	232.485	799.830
	ENTEC-2000+	87.110	105.255	360.050	226.365	778.780
CHEMICAL TANKER	Number of ships	220	148	190	86	644
	Hours alongside	4783	3677	5363	2395	16,219
Total tons of CO_2	EPA-1984/2000	1480.110	1179.100	1757.860	763.170	5180.240
	EPA-2000+	1441.150	1148.080	1711.610	743.080	5043.920
	ENTEC-1984/2000	740.055	589.550	878.930	381.585	2590.120
	ENTEC-2000+	720.575	574.040	855.805	371.540	2521.960
BULKCARRIER	Number of ships	1784	1824	1749	1235	6592
	Hours alongside	9108	81,276	92,446	62,238	32,7042
Total tons of CO_2	EPA-1984/2000	18,087.770	17,118.850	20,398.200	13,457.820	69,062.640
	EPA-2000+	17,611.780	16,668.350	19,861.400	13,103.680	67,245.210
	ENTEC-1984/2000	9043.885	8559.425	10,199.100	6728.910	34,531.320
	ENTEC-2000+	8805.890	8334.175	9930.700	6551.840	33,622.605

After calculating the CO_2 emissions of all the studied vessels that have called over the last 11 years (18,071), it is found that, despite being a small to medium-sized port in the

EU, the vessels docked have been emitting into the atmosphere for a total of 558,626.77 h in areas very close to urban centers. In addition to the constant noise caused by the operation of the auxiliary engines, it has been determined, by the EPA method, that these vessels have emitted a total of 158,227 tons into the atmosphere if considering auxiliary engines being built before the year 2000, and 154,112 tons if built at a later date. On the other hand, by the ENTEC method, 79,139 tons were emitted if considering engines being built between before the year 2000, and 77,056 tons with a later date.

In view of the results obtained, the difference in emissions between tons of CO_2 (based on the year of construction of the auxiliary engines: 1984/2000 or 2000+) is approximately 2.62%. Therefore, from now on for the rest of the operations and, given that the aim is to obtain the tons emitted, the greenest condition will continue to be accepted: all engines will be considered as being built after the year 2000.

As can be seen in the Figure 1, the evolution of the percentage of energy from renewable sources (including hydraulic, hydro-wind, wind, solar photovoltaic, solar thermal, other renewables, and renewable waste) compared to those from non-renewable sources, we arrive at the conclusion that, even in the worst scenario (year 2012), it would have been possible to reduce CO_2 emissions into the atmosphere by 30%. Moreover, the evolution and increase in renewable energies has gained pace in recent years, with this percentage becoming more and more significant (with renewable energies being even higher at certain times of the year compared to non-renewables).

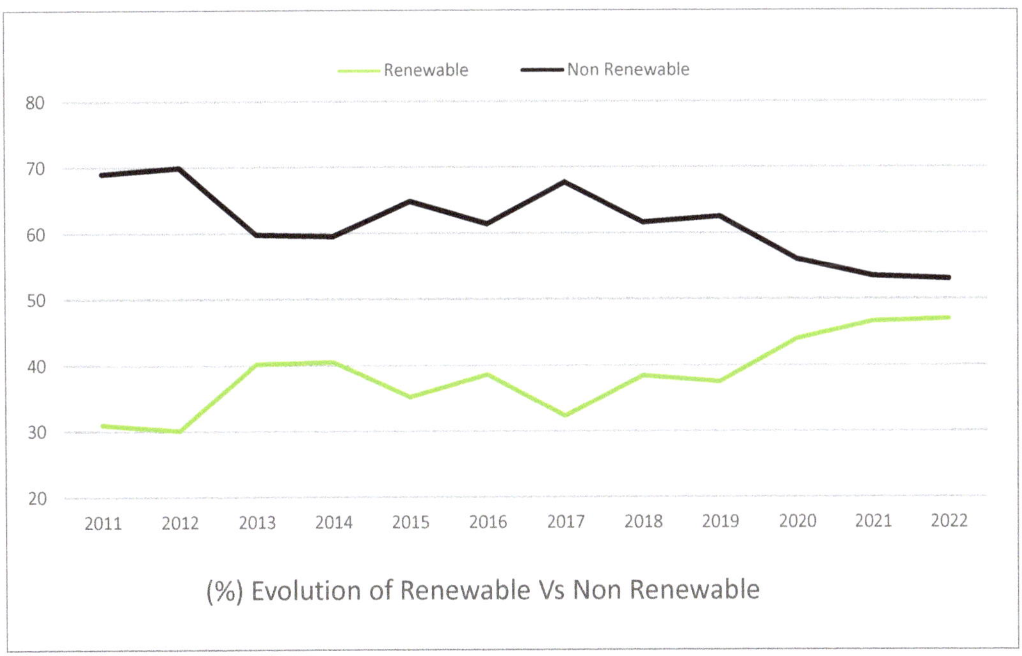

Figure 1. Graphic of the Evolution of Renewable versus Non-Renewable energies. Source: www.ree.es [58] https://www.ree.es/en/datos/generation/evolution-renewable-non-renewable (accessed on 6 January 2022).

If a direct reduction is applied in the percentage of clean energy from the Figure 1 on the tons of emitted CO_2 calculated per year on Table 2, the result will be tons that would have been avoided through the application of the OPS in the Port of Santander.

Formula as per below:

Emitted tons (Table 2) × Percentage of Renewable Energy origin (Table 3) → Potential Save of CO_2 tons (Table 4).

Table 3. Evolution of Renewable versus Non-Renewable energies. Source: www.ree.es [58] https://www.ree.es/en/datos/generation/evolution-renewable-non-renewable (accessed on 6 January 2022).

%	2011	2012	2013	2014	2015	2016	2017	2018	2019	2020	2021	2022
Renewable	31	30.1	40.2	40.5	35.2	38.6	32.3	38.4	37.5	44	46.6	47
Non-Renewable	69	69.9	59.8	59.5	64.8	61.4	67.7	61.6	62.5	56	53.4	53

Table 4. Potential saves of CO_2 per year/type of vessel/method. Source: Authors. Calculation of emissions per type of vessel and per year that could have been avoided taking the auxiliary engines to have a manufacturing date after the year 2000 (least contaminating situation) and considering the percentage of green energy produced per year.

		2011–2013	2014–2016	2017–2019	2020–2021	TOTAL
FERRY		576	597	740	336	2249
Potential save of CO_2 tons	EPA ENTEC	1021.874 510.937	1.542.659 771.3293	2083.966 1041.983	1691.343 845.6715	6339.842 3169.921
CRUISE		29	25	40	3	97
Potential save of CO_2 tons	EPA ENTEC	125.3275 62.66374	92.83066 46.41533	181.2379 90.61894	22.30742 11.15371	421.7034 210.8517
TANKER		16	27	52	37	132
Potential save of CO_2 tons	EPA ENTEC	61.34269 30.67135	79.32283 39.66142	264.673 132.3365	205.7943 102.8971	611.1328 305.5664
CHEMICAL TANKER		220	148	190	86	644
Potential save of CO_2 tons	EPA ENTEC	486.0212 243.0106	437.1079 218.5539	618.5519 309.2759	338.0673 169.0337	1879.748 939.8741
BULKCARRIER		1784	1824	1749	1235	6592
Potential save of CO_2 tons	EPA ENTEC	5992.793 2996.396	6350.675 3175.337	7171.244 3585.622	5950.531 2975.266	25,465.24 12,732.62
GENERAL CARGO		752	696	681	488	2617
Potential save of CO_2 tons	EPA ENTEC	1935.544 967.7719	2528.522 1264.261	2017.093 1008.546	1868.55 934.275	8349.708 4174.854
RORO		1243	1507	1799	1191	5740
Potential save of CO_2 tons	EPA ENTEC	2294.555 1147.278	3662.479 1831.239	4384.614 2192.307	5548.578 2774.289	15,890.23 7945.114

Therefore, it can be inferred that if the OPS had been used in the Port of Santander to supply energy to the vessels docked over the last 11 years, potentially the CO_2 emission of 589,857.604 tons into the atmosphere would have been avoided as per EPA method and 29,478.802 tons as per ENTEC method. Moreover, it should be noted that the reduction in emissions would have grown proportionally with the percentage of clean energy, leading to an increase in the optimization of this resource.

5. Discussion

If energy sources in Spain are analyzed, according to the report prepared by the consulting firm Deloitte for the Wind Energy Business Association in which the benefits provided by clean energy [59] are quantified, in 2020 wind energy accounted for 22% of the energy generated in Spain, with an upward progression which is expected to reach 50,333 MW of installed power by 2030. According to the same study, between 2011 and 2020 the emission of 260 million tons of CO_2 into the atmosphere was avoided. All these efforts, however, are still far from enough. Among other potential objectives, it is necessary to increase this capacity and produce green energy that is supplied to ships in port, increasing the differential and the advantages of using the OPS compared to conventional auxiliary engines. Even though a small percentage of energy from fossil sources must be used, through this system it is possible to reduce emissions while delocalizing pollution. These emissions, as they occur in areas far from urban centers, will have less harmful effects on the population than in the case of being emitted in ports and urban areas. Thus, through a combination of technology and mature renewable energies such as wind, solar energy etc., it is possible to supply the necessary electrical energy so that ships in port can turn off their auxiliary engines, reducing local emissions to 0.

Another factor to bear in mind is the option of producing electrical energy from nuclear power. This energy has determined production capacity and important disadvantages such us risks of leaks, wasted nuclear fuel, potential disaster, or others, but according to Eide [60] this could lead to a reduction in emissions of up to 95%.

Knowing that the total number of calls in the port of Santander were 22,714 and considering that ships during their stay in the shipyard and ships smaller than 499 Gt (reduced to 18,071 ships) were discarded from this study, the estimation of CO_2 emissions into the atmosphere can be obtained. Such a high figure for a small-to-medium-sized port suggests the great number of ports of similar or larger sizes, all of this justifying the growing interest in promoting the OPS as an optimal tool for reducing emissions in search of the sustainability proposed as long ago as 1987 but still, in 2022, far from being achieved.

Transferring the data obtained in this study to a graph (Figure 2), obviously, emissions depend fundamentally on the number of vessels docked and time spent alongside burning fuel to produce their needed energy. For the Port of Santander, having mainly RORO and bulk traffic, the result highlights how bulk carriers and RORO vessels are those that emit the most tons in total values.

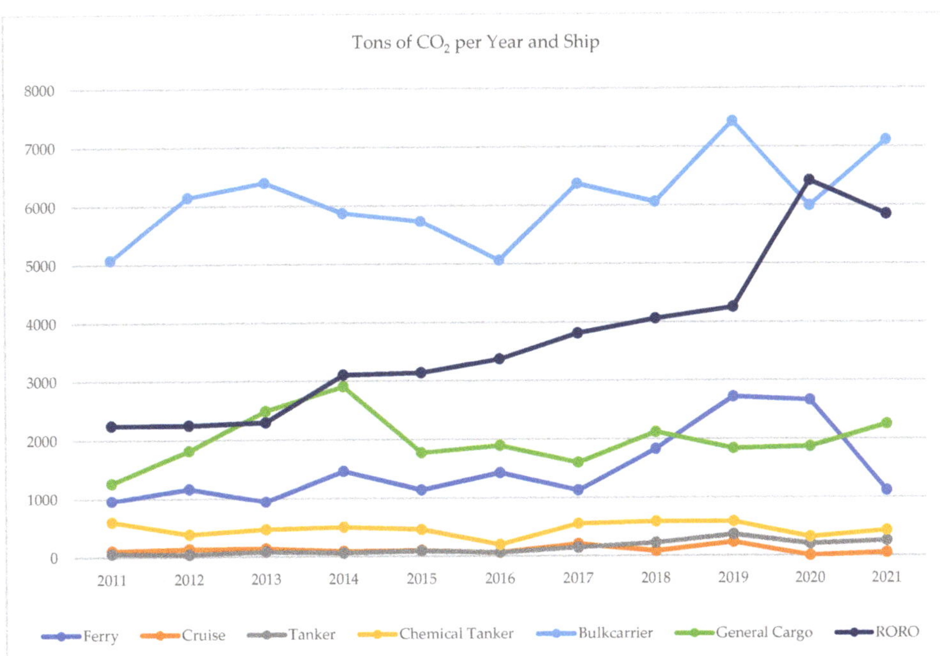

Figure 2. Emissions per year and type of vessel. Source: Authors.

The total percentage of emissions accumulated by each type of vessel over the 11 years of the study is shown in Figure 3.

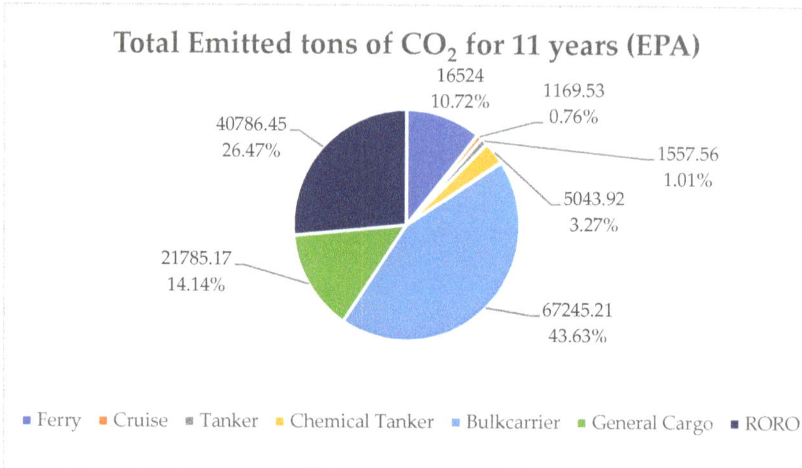

Figure 3. Tons of CO_2 emitted per type of vessel. Source: Authors.

After analyzing the data, and with a view to implementing the OPS system, the feasibility, and the possibility of improving the reduction of emissions must be considered. Ideally, all vessels and docks should be equipped with this system, but due to the costly initial investment required by the ports and shipowners, this implementation must be gradual, and a balance must be found that allows the results to be maximized.

The type of ship that accounts for 43.63% of the accumulated tons is the bulk carrier. Most of these vessels belong to the tramp market. In other words, they do not have any regular or specific routes and, in many cases, call at the port of Santander less than once a year. In addition, these ships belong to too wide a variety of shipowners, without there being any large owners that group together a high percentage of vessels of this style. Therefore, today, in the bulk of docks where these vessels operate, it is not considered feasible to get the most out of the OPS. Hence, a great deal of interest is not to be expected from ports or shipowners who will not be able to receive a return on their investment. The longer the OPS is used, the greater the benefits. It will not be an initial target for the OPS installation.

The case of ROROs, which have emitted slightly over 26.4% of the total accumulated emissions, is radically different. This type of vessel is a clear target, with great potential to be the ideal candidate for the introduction of this technology. The automotive industry is currently making great investments to reduce its carbon footprint and, as is logical, logistics is one of the processes it is looking to improve. In this sense, RORO shipowners are already making large investments in R&D, developing technologies such as rigid sails [61] or alternative fuels [62] among others. In short, these are vessels that emit large amounts of CO_2, are moored between 6 and 12 h, although sometimes for even longer and, more importantly, they are ships with a great regularity that belong to a few shipowners and could obtain a clear benefit in the reduction of consumption. Virtually all of them cover regular lines, calling regularly at the same ports and making numerous annual calls. Both ports and ships will be able to implement this technology, reducing connection/disconnection times, saving fuel, reducing noise, improving the quality of life for the inhabitants, port workers, sailors, etc.: in short, they will be able to optimize the investment.

In view of the above, general cargo vessels (which do not have the necessary regularity) will not be among the initial objectives of the OPS in a port such as Santander, but the group of ferries (which occupies the 4th position of the total accumulated emissions of CO_2) will be. From the data in this study, it can be inferred that during 2021 (still heavily affected by the COVID-19 epidemic) they have called at port 149 times, of which 142 calls have been made by just 2 ships in the continuous rotation of a regular line. Based on the data obtained, in 2021 these two ships emitted, in a total of 832 h, 921.3 tons into the atmosphere according to the EPA method. Therefore, this recurrence would allow the development of a system specifically designed for them that meets their needs and allows them to achieve the desired objectives. In addition, the dock for these operations is literally less than 150 m from inhabited buildings and from the city center, which, from a socio/sanitary point of view, multiplies the interest on the part of the authorities, passengers, and citizens of Santander, while the fuel savings will justify that of the shipowners.

As Bouman [63] points out in his article, the maximum reduction in emissions is not based on one single solution but must be a combination of different technological and operational measures, etc. Therefore, if in addition to achieving improvements and advances in the docked ships, this is combined with other measures aimed at reducing emissions and environmental improvements, such as controlled speed zones, emission limits for vehicles in port (trucks, machinery, cars, cranes . . .), technological improvements for process automation, electric/ecological vehicles [64], energy efficiency improvements in port buildings, installation of solar panels, etc., a significant reduction in CO_2 emissions would be achieved, thus achieving the targets set.

According to the data provided in this study, approximately 38.26% of the tons of the CO_2 emitted by ships docked in the port of Santander could have been avoided. This result goes in the same line as that proposed by WJ Hall [11], who concludes that the impact of the OPS on CO_2 emissions will be −37.8% in Spain. In this manuscript figures are the result for the last 11 years, but this percentage is not linear, since it varies depending on weather conditions (mainly wind and sun) and the resources and generating plants available in the system. Hence, this percentage will be increased significantly in the future as the multiple projects and investments of the green energy generation come into operation, or as new technologies that are currently in the experimental phase are developed and established. Therefore, it can be stated that the OPS will be a far more advantageous technique in the coming years than in the past, as it is logical to assume that if 37.5% of the energy produced in Spain in 2019 were renewable, that figure would be 44% in 2020, 46.6% in 2021 and forecast for 47% in 2022, and the next few years will be the period with the greatest expansion and consolidation of the OPS.

It should be noted that this improvement in the performance of the system does not entail extra investments or costly upgrades or fine-tuning of the facilities, since this optimization is achieved at the energy source, in the production mode. The OPS is simply based on the supply of electricity. Therefore, the sooner this type of facility is installed and put into service, the more years it will serve everyone and the greater the number of tons of CO_2 emissions will be avoided.

On the other hand, in the area of the port, from the point of view of health, the vessels moored where the OPS is a common or even mandatory feature would avoid the emission of 100% of the greenhouse gases (from the combustion of auxiliary engines) harmful to living beings in highly populated areas and urban centers, such as ports. In this way, the prevention of cardiorespiratory diseases, allergies and other diseases related to this type of contamination is promoted, leading to an increase in life expectancy and quality of life, a relief in the pressure on hospitals and health centers and even significant savings for the national health systems so heavily punished in recent years by the aging of populations, various pathologies, or even global epidemics such as COVID-19. It is also true, as a weakness of this technology/this paper, if nergy provided from the grid to the boats has a fuel-burning origin, emissions will only be shifted (delocalized) from the area of the port to the power plant areas (usually settled already far from urban centers where the impact to population is marginal), potentially increasing health problems for living beings in the area.

According to the estimations of Winkel [9] a saving of over 2.94 billion Euros could be made if all European ports were equipped with the OPS, while according to Vaishnav [65] the saving would be between 70 and 150 million dollars if between 1/4 and 2/3 of the vessels that call in US ports had this type of technology. If it is considered that the installation of this technology in an average port can cost 7.4 million Euros [66] and that its amortization is estimated to take 7 years (less if subsidies are received), it can be stated that there is plenty of scope for governments to aid and promote the private investment of port authorities or shipowners in return for a subsequent reduction in health spending. In the case of Spain, with an investment of around 6.2 million Euros of which 1.6 million (26%) come from European funds [67], there is a project called the 'OPS Master plan' [68] through which this type of installations is being promoted with the clear objective that all the general interest ports should be equipped with this technology by the year 2030. Among the ports participating in this project are Algeciras, Barcelona, Valencia, Bilbao, Motril, etc.

The policies of governments and international organizations have for many years been promoting and aiding the installation of green energy sources [69]. The rewards of these policies are now being felt, as there has been a considerable increase in the use of this clean energy over other sources of energy.

Most of the major companies are already aware of the importance of reducing emissions. In addition to the clear goal of reducing costs and optimizing profits, they value very highly and invest heavily in minimizing their carbon footprint, and their logistics play a fundamental role in this regard. Shipping companies, ports, carriers, operators, etc. must comply with these requirements to maintain a fruitful business relationship. The object of this study is the world of maritime transport and port activity. In this area, these initiatives are advancing little by little, timidly making their way, but there is still much to be done. Bearing in mind that the vessels studied in the Port of Santander make up only a small part of the total calls at the national level, and the dependence of countries on the maritime sector for their international trade (74% of imported/exported goods of the EU and 37% of exchanges within the EU are by sea) [70], it can be concluded that there is an enormous potential for reducing emissions that can be achieved without the need to develop new technologies, simply by applying existing ones. Even though the concept of the 'green port' is based on the combination of several different techniques, the OPS stands out as a key part in achieving significant reductions in emissions into the atmosphere (greenhouse gases) based on already developed and mature technologies with a guarantee of success. As Winkel states [9] in his study, if all European ports had these facilities, a potential reduction of 800,000 tons would be obtained. As the introduction of new technologies and mandatory measures must be done progressively, this work justifies that these investments should move forward based on the tons emitted and the regularity of the ships that call at the different ports. The characteristics of each of these vessels must be considered, regularity being a key point in decision-making. Thus, it is concluded that in the port of Santander the initial targets and the focus of attention should be on RORO ships and ferries as well as on the docks where they operate to optimize the initial investments.

Hence, given the great environmental and health benefits underlined in this study, and although there is still a long way to go before it becomes a reality, Spain already has a strategic plan published in December 2021 for a progressive installation of these dock electrification systems [71]. The port of Bilbao has a project that has an investment of 51.8 million Euros for the progressive installation of OPS in 7 docks [72], and the port of Barcelona (through the project Nexigen) will invest 110 million Euros to electrify their facilities [73]. On the other hand, the European Union has not stopped urging its members to encourage and promote the installation and use of the OPS with the hope of a prompt implementation of these systems in all the ports of the area [13,15] and of the world.

6. Conclusions

The following conclusions can be drawn from the data and results of this study:

1. The current measures are not enough to meet the emission reduction targets set by international organizations and governments. Bearing in mind that Santander is a medium-sized port and emissions could have been reduced by 38.26%, the OPS system has great reduction potential if applied to all ports in the area. It is also compatible with the introduction of other preventive measures.
2. From an economic point of view, the greater the fuel/electricity price difference, the more attractive it will be for private entities to undertake the necessary investments.
3. The OPS system is an innovative ship power supply system where its development should focus on the standardization, safety, and speed of the system. Technological advances in electricity production and the increase in green energy generation plants will directly improve the OPS' emission reduction potential without the need for new investments.
4. The reduction in emissions into the atmosphere will be greater the longer this system is in use.
5. Regularity is a key factor in its development, both on ships and in ports. Decision makers need to carefully evaluate pier priorities case by case per each port/traffic. From the results obtained in this work, it is deduced that, for the Port of Santander, the initial objective should be to implement the OPS for RORO, Ferry, and Cruise ships, which account for 37.95% of the total CO_2 emissions during the period studied.
6. OPS environmental performance varies directly depending on the source of the energy supplied.
7. It is inferred that between 2011 and 2021, the reduction in emissions in Santander could have reached an average of 38.26%. Hall [11] already in 2010 estimated potential CO_2 reduction of OPS around 37.8%. Its reducing capacity is not constant, but it is yearly increasing. It is expected that this positive trend will continue multiplying benefits during the next years due to the significant development of green energies in recent times.
8. Regardless of the source of energy, a vessel docked and connected to the OPS reduces auxiliary engines port emissions locally to 0, improving the quality of life of the port community and all those located in its area of influence. (In the case of releasing CO_2 emissions to the atmosphere, these are relocated where energy production facilities are settled).

Author Contributions: Conceptualization, A.H., A.O.P. and E.D.-R.-N.; methodology, A.H., A.O.P., E.D.-R.-N. and A.-I.L.-D.; software, M.A.G., A.H. and A.-I.L.-D.; validation, A.H., A.O.P., E.D.-R.-N., M.A.G. and A.-I.L.-D.; formal analysis, A.H., A.O.P., E.D.-R.-N. and M.A.G.; investigation, A.H., A.O.P., E.D.-R.-N., M.A.G. and A.-I.L.-D.; resources, A.H., A.O.P., M.A.G. and A.-I.L.-D.; data curation, A.H., E.D.-R.-N., M.A.G. and A.-I.L.-D.; writing—original draft preparation, A.H. and A.O.P.; writing—review and editing, A.H., A.O.P., E.D.-R.-N., M.A.G. and A.-I.L.-D.; visualization, A.H., A.O.P., E.D.-R.-N., M.A.G. and A.-I.L.-D.; supervision, A.H., A.O.P., E.D.-R.-N., M.A.G. and A.-I.L.-D.; project administration, A.H., A.O.P., E.D.-R.-N., M.A.G. and A.-I.L.-D. All authors have read and agreed to the published version of the manuscript.

Funding: This research received no external funding.

Institutional Review Board Statement: Not applicable.

Informed Consent Statement: Not applicable.

Conflicts of Interest: The authors declare no conflict of interest.

References

1. United Nations. Report of the World Commission on Environment and Development. In Proceedings of the United Nations General Assembly—42nd Session, New York, NY, USA, 20 November 1987.
2. European Commission. *A Roadmap for Moving to a Competitive Low Carbon Economy in 2050*; European Commission: Brussels, Belgium, 2011.
3. European Parliament. Deployment of Alternative Fuels Infrastructure. 2014. Available online: https://eur-lex.europa.eu/legal-content/EN/TXT/PDF/?uri=CELEX:32014L0094&from=EN (accessed on 25 May 2021).
4. European Commission. The European Green Deal—COM/2019/640 Final. 2019. Available online: https://eur-lex.europa.eu/legal-content/EN/TXT/?uri=COM%3A2019%3A640%3AFIN (accessed on 25 February 2021).
5. Mikova, N.; Eichhammer, W.; Pfluger, B. Low-carbon energy scenarios 2050 in north-west European countries: Towards a more harmonised approach to achieve the EU targets. *Energy Policy* **2019**, *130*, 448–460. [CrossRef]
6. European Environment Agengy. Greenhouse Gas Emissions from Transport in Europe. 2019. Available online: https://www.eea.europa.eu/data-and-maps/indicators/transport-emissions-of-greenhouse-gases-7/assessment (accessed on 8 January 2021).
7. Song, S.-K.; Shon, Z.-H. Current and future emission estimates of exhaust gases and particles from shipping at the largest port in Korea. *Environ. Sci. Pollut. Res.* **2014**, *21*, 6612–6622. [CrossRef]
8. Du, Y.; Chen, Q.; Quan, X.; Long, L.; Fung, R.Y.K. Berth allocation considering fuel consumption and vessel emissions. *Transp. Res. Part E Logist. Transp. Rev.* **2011**, *47*, 1021–1037. [CrossRef]
9. Winkel, R.; Weddige, U.; Johnsen, D.; Hoen, V.; Papaefthimiou, S. Shore side electricity in Europe: Potential and environmental benefits. *Energy Policy* **2016**, *88*, 584–593. [CrossRef]
10. Chang, C.-C.; Wang, C.-M. Evaluating the effects of green port policy: Case study of Kaohsiung harbor in Taiwan. *Transp. Res. Part D Transp. Environ.* **2012**, *17*, 185–189. [CrossRef]
11. Hall, W.J. Assessment of CO_2 and priority pollutant reduction by installation of shoreside power. *Resour. Conserv. Recycl* **2010**, *54*, 462–467. [CrossRef]
12. Paul, D.; Haddadian, V. Cold ironing-power system grounding and safety analysis. In Proceedings of the Fourtieth IAS Annual Meeting. Conference Record of the 2005 Industry Applications Conference, Hong Kong, China, 2–6 October 2005; Volume 2, pp. 1503–1511.
13. European Union. Promotion of Shore-Side Electricity for Use by Ships at Berth in Community Ports. 2006. Available online: https://eur-lex.europa.eu/legal-content/EN/TXT/PDF/?uri=CELEX:32006H0339&from=EN (accessed on 17 March 2021).
14. European Parliament. Council Directive 2012/33/UE. 2012. Available online: https://eur-lex.europa.eu/legal-content/ES/TXT/PDF/?uri=CELEX:32012L0033&from=IT (accessed on 23 March 2021).
15. European Union. Directive 2014/94/Eu Deployment of Alternative Fuels Insfrastructure. 2014. Available online: https://eur-lex.europa.eu/legal-content/EN/TXT/PDF/?uri=CELEX:32014L0094&from=ES (accessed on 5 April 2021).
16. Zis, T.P. Prospects of cold ironing as an emissions reduction option. *Transp. Res. Part A Policy Pract.* **2019**, *119*, 82–95. [CrossRef]
17. Molavi, A.; Lim, G.J.; Shi, J. Stimulating sustainable energy at maritime ports by hybrid economic incentives: A bilevel optimization approach. *Appl. Energy* **2020**, *272*, 115188. [CrossRef]
18. Wu, L.; Wang, S. The shore power deployment problem for maritime transportation. *Transp. Res. Part E Logist. Transp. Rev.* **2020**, *135*, 101883. [CrossRef]
19. Yiğit, K.; Kökkülünk, G.; Parlak, A.; Karakaş, A. Energy cost assessment of shoreside power supply considering the smart grid concept: A case study for a bulk carrier ship. *Marit. Policy Manag.* **2016**, *43*, 469–482. [CrossRef]
20. Goverment, S. Puertos del Estado. 2022. Available online: https://www.puertos.es/es-es/estadisticas/Paginas/estadistica_mensual.aspx (accessed on 20 September 2022).
21. Du, K.; Monios, J.; Wang, Y. Green port strategies in China. In *Green Ports*; Elsevier: Amsterdam, The Netherlands, 2019; pp. 211–229.
22. Díaz-Ruiz-Navamuel, E.; Piris, A.O.; Pérez-Labajos, C.A. Reduction in CO_2 emissions in RoRo/Pax ports equipped with automatic mooring systems. *Environ. Pollut.* **2018**, *241*, 879–886. [CrossRef]
23. United Nations. Review of Maritime Transport. 2020. Available online: https://unctad.org/webflyer/review-maritime-transport-2020 (accessed on 15 November 2021).
24. Cabral, T.; Clemente, D.; Rosa-Santos, P.; Taveira-Pinto, F.; Morais, T.; Belga, F.; Cestaro, H. Performance assessment of a hybrid wave energy converter integrated into a harbor breakwater. *Energies* **2020**, *13*, 236. [CrossRef]
25. Rosa-Santos, P.; Taveira-Pinto, F.; Clemente, D.; Cabral, T.; Fiorentin, F.; Belga, F.; Morais, T. Experimental study of a hybrid wave energy converter integrated in a harbor breakwater. *J. Mar. Sci. Eng.* **2019**, *7*, 33. [CrossRef]
26. Cascajo, R.; García, E.; Quiles, E.; Correcher, A.; Morant, F. Integration of marine wave energy converters into seaports: A case study in the Port of Valencia. *Energies* **2019**, *12*, 787. [CrossRef]
27. Murai, M.; Li, Q.; Funada, J. Study on power generation of single Point Absorber Wave Energy Converters (PA-WECs) and arrays of PA-WECs. *Renew. Energy* **2020**, *164*, 1121–1132. [CrossRef]
28. Bertram, D.V.; Tarighaleslami, A.H.; Walmsley, M.R.W.; Atkins, M.J.; Glasgow, G.D.E. A systematic approach for selecting suitable wave energy converters for potential wave energy farm sites. *Renew. Sustain. Energy Rev.* **2020**, *132*, 110011. [CrossRef]
29. de Antonio, F. Wave energy utilization: A review of the technologies. *Renew. Sustain. Energy Rev.* **2010**, *14*, 899–918.

30. Ramos, V.; Carballo, R.; Álvarez, M.; Sánchez, M.; Iglesias, G. A port towards energy self-sufficiency using tidal stream power. *Energy* **2014**, *71*, 432–444. [CrossRef]
31. Hua-Ming, W.; Xiao-Kun, Q.; Lin, C.; Lu, T.; Qiao, W. Numerical study on energy-converging efficiency of the ducts of vertical axis tidal current turbine in restricted water. *Ocean Eng.* **2020**, *210*, 107320. [CrossRef]
32. Krämer, I.; Czermański, E. Onshore power one option to reduce air emissions in ports. *Nachhalt. Manag. Forum Sustain. Manag. Forum* **2020**, *28*, 13–20. [CrossRef]
33. Schwartz, H.; Gustafsson, M.; Spohr, J. Emission abatement in shipping–is it possible to reduce carbon dioxide emissions profitably? *J. Clean. Prod.* **2020**, *254*, 120069. [CrossRef]
34. Yu, J.; Voß, S.; Tang, G. Strategy development for retrofitting ships for implementing shore side electricity. *Transp. Res. Part D Transp. Environ.* **2019**, *74*, 201–213. [CrossRef]
35. Sorte, S.; Rodrigues, V.; Borrego, C.; Monteiro, A. Impact of harbour activities on local air quality: A review. *Environ. Pollut.* **2020**, *257*, 113542. [CrossRef] [PubMed]
36. Corbett, J.J.; Winebrake, J.J.; Green, E.H.; Kasibhatla, P.; Eyring, V.; Lauer, A. Mortality from ship emissions: A global assessment. *Environ. Sci. Technol.* **2007**, *41*, 8512–8518. [CrossRef]
37. Bailey, D.; Plenys, T.; Solomon, G.M.; Campbell, T.R.; Feuer, G.R.; Masters, J.; Tonkonogy, B. *Harboring Pollution: Strategies to Clean Up US Ports*; NRDC: New York, NY, USA, 2004; pp. 1–85.
38. Quaranta, F.; Fantauzzi, M.; Coppola, T.; Battistelli, L. Analysis of the Pollution Level and Possible solutions. *J. Marit. Res.* **2012**, *9*, 81–86.
39. MEPC, IMO. Reduction of GHG Emissions from Ships. Available online: https://safety4sea.com/wp-content/uploads/2020/08/MEPC-75-7-15-Fourth-IMO-GHG-Study-2020-Final-report-Secretariat.pdf (accessed on 25 September 2021).
40. Smith, T.W.P.; Jalkanen, J.P.; Anderson, B.A.; Corbett, J.J.; Faber, J.; Hanayama, S.; O'keeffe, E.; Parker, S.; Johanasson, L.; Aldous, L. Third IMO GHG Study 2014. Available online: https://greenvoyage2050.imo.org/wp-content/uploads/2021/01/third-imo-ghg-study-2014-executive-summary-and-final-report.pdf (accessed on 2 February 2022).
41. Acomi, N.; Acomi, O.C. The influence of different types of marine fuel over the energy efficiency operational index. *Energy Procedia* **2014**, *59*, 243–248. [CrossRef]
42. Psaraftis, H.N.; Kontovas, C.A. Green maritime transportation: Speed and route optimization. In *Green Transportation Logistics*; Springer: Berlin, Germany, 2016; pp. 299–349.
43. Psaraftis, H.N.; Kontovas, C.A. Speed models for energy-efficient maritime transportation: A taxonomy and survey. *Transp. Res. Part C Emerg. Technol.* **2013**, *26*, 331–351. [CrossRef]
44. Wang, S.; Meng, Q. Sailing speed optimization for container ships in a liner shipping network. *Transp. Res. Part E Logist. Transp. Rev.* **2012**, *48*, 701–714. [CrossRef]
45. Hermeling, C.; Klement, J.H.; Koesler, S.; Köhler, J.; Klement, D. Sailing into a dilemma: An economic and legal analysis of an EU trading scheme for maritime emissions. *Transp. Res. Part A Policy Pract.* **2015**, *78*, 34–53. [CrossRef]
46. European Parliament. Council Directive 1999/32/UE. 1999. Available online: https://eur-lex.europa.eu/legal-content/EN/TXT/PDF/?uri=CELEX:31999L0032&from=ES (accessed on 14 October 2021).
47. European Parliament. Concil Directive 2005/33/UE. 2005. Available online: https://eur-lex.europa.eu/legal-content/ES/TXT/PDF/?uri=CELEX:32005L0033&from=EN (accessed on 17 August 2022).
48. Tichavska, M.; Tovar, B.; Gritsenko, D.; Johansson, L.; Jalkanen, J.P. Air emissions from ships in port: Does regulation make a difference? *Transp. Policy* **2019**, *75*, 128–140. [CrossRef]
49. Stolz, B.; Held, M.; Georges, G.; Boulouchos, K. The CO_2 reduction potential of shore-side electricity in Europe. *Appl. Energy* **2021**, *285*, 116425. [CrossRef]
50. Maragkogianni, A.; Papaefthimiou, S.; Zopounidis, C. Current methodologies for the estimation of maritime emissions. In *Mitigating Shipping Emissions in European Ports*; Springer: Berlin, Germany, 2016; pp. 25–35.
51. Dolphin, M.J.; Melcer, M. Estimation of ship dry air emissions. *Nav. Eng. J.* **2008**, *120*, 27–36. [CrossRef]
52. Piris, A.O.; Díaz-Ruiz-Navamuel, E.; Pérez-Labajos, C.A.; Chaveli, J.O. Reduction of CO_2 emissions with automatic mooring systems. The case of the port of Santander. *Atmos. Pollut. Res.* **2018**, *9*, 76–83. [CrossRef]
53. MEPC. Directrices Sobre el Método de Cálculo del Índice de Eficiencia Energética (EEDI) de Proyecto Para Buques Nuevos. MEPC. 2012. Available online: https://www.directemar.cl/directemar/site/artic/20190212/asocfile/20190212091101/mepc_212_63_.pdf (accessed on 18 August 2021).
54. Grebot, B.; Scarbrough, T.; Ritchie, A.; Mahoney, C.; Noden, R.; Sobey, M.; Whall, C. *Study to Review Assessments Undertaken of the Revised MARPOL Annex VI Regulations*; Entec UK Limited: London, UK, 2010.
55. U.S. Environmental Protection Agency. EPA. Available online: https://www.epa.gov/air-emissions-factors-and-quantification/basic-information-air-emissions-factors-and-quantification (accessed on 12 December 2021).
56. Santander Port Authority. Available online: https://www.puertosantander.es/cas/home.aspx (accessed on 10 January 2022).
57. Sea-Web AIS Life. Available online: https://maritime.ihs.com/ (accessed on 25 January 2022).
58. Goverment, S. Spanish National Grid. Available online: https://www.ree.es/es/datos/generacion/estructura-generacion-emisiones-asociadas (accessed on 12 February 2022).
59. Eólica, A.E. Asociación Empresarial Eólica. Available online: https://www.aeeolica.org/sobre-la-eolica/la-eolica-espana (accessed on 15 November 2021).

60. Eide, M.S.; Chryssakis, C.; Endresen, Ø. CO_2 Abatement Potential towards 2050 for Shipping, including Alternative Fuels. *Carbon Manag.* **2013**, *4*, 275–289. [CrossRef]
61. WalleniusWilhelmsen. WWL—Wind Powered RORO. 2022. Available online: https://www.walleniuswilhelmsen.com/news-and-insights/highlighted-topics/orcelle (accessed on 23 March 2022).
62. Hoegh. Hoegh—Zero Carbon New Building. 2022. Available online: https://www.hoeghautoliners.com/news-and-media/news-and-press-releases/hoegh-signs-contract-with-china-merchants-heavy-industry-to-build-a-series-of-its-zero-carbon-ready-aurora-class-vessels (accessed on 25 March 2022).
63. Bouman, E.A.; Lindstad, E.; Rialland, A.I.; Strømman, A.H. State-of-the-art technologies, measures, and potential for reducing GHG emissions from shipping–a review. *Transp. Res. Part D Transp. Environ.* **2017**, *52*, 408–421. [CrossRef]
64. Ecological Cars. *Estrecho Digit.* 2022. Available online: https://www-elestrechodigital-com.cdn.ampproject.org/c/s/www.elestrechodigital.com/2022/01/26/el-puerto-de-santander-sustituye-trece-vehiculos-contaminantes-por-otros-ecologicos/amp/?p=139190 (accessed on 20 July 2021).
65. Vaishnav, P.; Fischbeck, P.S.; Morgan, M.G.; Corbett, J.J. Shore power for vessels calling at US ports: Benefits and costs. *Environ. Sci. Technol.* **2016**, *50*, 1102–1110. [CrossRef]
66. Innes, A.; Monios, J. Identifying the unique challenges of installing cold ironing at small and medium ports–The case of Aberdeen. *Transp. Res. Part D Transp. Environ.* **2018**, *62*, 298–313. [CrossRef]
67. European Commission. Masterplan for OPS in Spanish Ports. Available online: https://ec.europa.eu/inea/sites/default/files/fiche_2015-eu-tm-0417-s_final.pdf (accessed on 15 August 2022).
68. Europe Facility (CEF) for Transport. OPS Masterplan for Spanish Ports. Available online: http://poweratberth.eu/?page_id=38&lang=en (accessed on 23 September 2021).
69. European Union. Green Ports. 2022. Available online: https://greencportsproject.eu/ (accessed on 15 May 2022).
70. 295 FINAL—An Engine for Growth. 2013. Available online: EUR.COM (accessed on 10 April 2021).
71. European Union. OPS Master Plan. 31 December 2021. Available online: https://poweratberth.eu/?lang=english (accessed on 11 May 2021).
72. Mercantil, E. OPS Bilbao Port. *El Merc.* 2022. Available online: https://elmercantil.com/2022/03/21/el-puerto-de-bilbao-prepara-la-electrificacion-de-sus-muelles-para-2025/ (accessed on 19 December 2021).
73. Canal, E. EL Canal MARITIMO. *Canal Marit.* 2022. Available online: https://www.diarioelcanal.com/puerto-barcelona-one-ocean-summit-sistemas-ops-2028/ (accessed on 5 January 2021).

Journal of
Marine Science and Engineering

MDPI

Article

Bearing Characteristics of Multi-Wing Pile Foundations under Lateral Loads in Dapeng Bay Silty Clay

Hao Wang [1,*], Dewei Fu [1], Tiantian Yan [1], Deng Pan [2], Weiwei Liu [3] and Liqun Ma [4]

1 School of Civil Engineering, Architecture of Henan University, Kaifeng 475004, China
2 Water Resources Survey and Monitoring Center, Henan Provincial Institute of Natural Resource Monitoring and Land Remediation, Zhengzhou 450046, China
3 Engineering Technology Research Center for Embankment Safety and Disease Control, Ministry of Water Resources, Zhengzhou 450003, China
4 The College of Geography and Environment Science, Henan University, Kaifeng 475004, China
* Correspondence: wanghao8023@henu.edu.cn

Abstract: This study provides a theoretical basis for reinforcement of the soil around multi-wing piles. Limit analysis was used to determine the ultimate lateral capacity (ULC) of three- and four-wing piles in Dapeng Bay silty clay. The effects of the pile–soil interaction coefficient α, wing width B_w, and lateral-load direction β on the ULC of the pile and the shear plastic zone range of the surrounding soil were analyzed. The normalized ULC of the three-wing pile decreased when the wing–diameter ratio increased. When B_w was 0.15 m and α was 0.4, the ULC of the four-wing pile was 19% higher than that of the three-wing pile. As β increased, the normalized ULC of the four-wing pile decreased, whereas that of the three-wing pile went through a minimum at 30°. The size of the soil shear plastic ring did not depend on α for either pile type; it increased around the three-wing (but not the four-wing) pile with changes in β. However, there was also a double plastic ring of broken soil around the four-wing pile. The four-wing pile had a more symmetrical influence on the soil around the pile than the three-wing pile.

Keywords: limit analysis method; multi-wing pile; ultimate lateral capacity; clay; lateral load

Citation: Wang, H.; Fu, D.; Yan, T.; Pan, D.; Liu, W.; Ma, L. Bearing Characteristics of Multi-Wing Pile Foundations under Lateral Loads in Dapeng Bay Silty Clay. *J. Mar. Sci. Eng.* **2022**, *10*, 1391. https://doi.org/10.3390/jmse10101391

Academic Editors: M. Dolores Esteban, José-Santos López-Gutiérrez, Vicente Negro and M. Graça Neves

Received: 1 September 2022
Accepted: 26 September 2022
Published: 29 September 2022

1. Introduction

Because piles can bear vertical and lateral loads, they are widely used in the foundation of street-lamp posts, coastal trestle bridges, and offshore wind-power poles [1–5]. The action of typhoon and tide on a bay trestle will cause the trestle to cut and move sideways. Wings can be added along the pile to improve its ultimate lateral capacity (ULC) [6]. Different forms of wing piles have been developed [7,8]. Multi-wing piles with three or four wings at the same angle are especially common. There will be some saturated clay around such winged piles in a bay trestle. Therefore, it is important to determine the ULC of the wing piles and the extent of the plastic zone of the surrounding soil under lateral loading.

Many calculation methods for studying the factors that influence wing–pile ULC have been proposed. Dührkop and Grabe [9] modified the key parameters of the p–y-curve procedure to calculate the bearing capacity of the wing pile. Zhou et al. [10,11] used complex-variable elasticity and the finite-element method to calculate the ULC of noncircular piles. Nasr [6] determined that wing–pile ULC is affected by the geometry and position of the wing plate. Moreover, when the wing–plate length is less than 0.4 times the pile length, the lateral bearing capacity of the pile increases with wing–plate length; however, when it is greater than half the pile length [7], the influence of the wing–plate length on ULC becomes insignificant. Yaghobi et al. [12] calculated that the ULC of a wing pile with a diameter of 40 mm was 29% higher than that of a regular pile with the same diameter. Moreover, three-dimensional finite element analysis shows that the length of the wing plate has more effect on ULC than does the width. Increasing the length of the wing plate improves the bearing capacity and reduces the displacement of the pile head [13,14].

The response of pile foundations under lateral loads are usually analyzed using the beam-on-nonlinear-Winkler-foundation (BNWF) model in granular soil [15,16]. Limit–analysis theory, which considers the plastic deformation of soil, is widely used to study the ultimate bearing capacity of irregular-section piles and the relationship between the pile and soil interactions. Randolph and Houlsby [17] accurately calculated the ULC of circular piles from the plasticity theory. Martin [18,19] proposed that the actual ULC of circular piles is between the upper- and lower-limit solutions by limit-analysis theory. Keawsawasvong and Ukritchon [4] found that the upper- and lower-limit solutions from limit analysis, obtained with the OPTUM G2 software program, allow the ULC of I-shaped piles in undrained clay to be calculated perfectly; load direction also affected ULC. Zhou et al. [11] determined that, as the wing plate becomes thinner, it exerts a more significant effect on the normalized ULC of an XCC pile; load direction again has some effect. Truong and Lehane [20] noted that, when a pile fails, the displacement vector of the soil around it changes accordingly. The displacement–vector range of the soil around a concave pile is larger than that around a rectangular pile with the same section area, and the vector displacement of the soil around a circular pile is smaller than that around triangular, concave, and rectangular piles.

Limit-analysis theory can also be used to study the ULC of pile groups. Zhou et al. [21] proposed a design method for XCC pile groups in undrained clay. For a triangular composite foundation [22], a wide pile spacing can realize the best combination of pile groups, thus improving ULC. Modifying the single-pile *p–y* curve was also used in the analysis of the lateral loading behavior of pile groups [23,24].

The aforementioned studies investigated the ULCs of piles with different cross-sections (round, square, and h-shaped) in undrained clay and the failure mechanism of the soil around the piles. However, the ULCs of three-wing and four-wing piles in bay trestles have not been studied. In particular, the soil plastic-zone range around a wing pile, which partly determines the soil reinforcement range, has not been investigated. Hence, this study uses the upper and lower solutions from limit analysis to study the ULC of rigid multi-wing piles in Dapeng Bay and to determine the range of the silty-clay plastic zone. Furthermore, the effects of the three main factors (the pile–soil interaction coefficient, wing–diameter ratio, and lateral-load direction) on the ULC of a multi-wing pile are analyzed, and the effects of parameter changes on the shear plastic-zone range of the soil are explored. This provides a basis for the reinforcement of the soil around multi-wing piles in bay areas.

2. Research Area and Methodology

2.1. Description of the Calculated Section of the Multi-Wing Pile

A sightseeing trestle is planned to be built in the coastal bay area of Dapeng, Yantian District, Shenzhen, near E114°23′30″, N22°26′30″. The bay is trough-shaped with exposed Quaternary deposits (Q_{me}) and Quaternary alluvial–pluvial sediments (Q_{al+pl}), with binary or multiple structures. The upper part is silty clay with an average thickness of 2.5 m, and the lower part is an interlayer of sandy soil and clay soil with an average thickness of 5 m. The maximum wind speed of typhoons in this area is greater than 40 m/s, the maximum wind speed under non-typhoon conditions is 10–29 m/s, and the main wind direction is northeast. In order to resist the horizontal loads caused by typhoons and tides, the foundation of the sightseeing trestle uses multi-wing piles. The research process for the design and selection of multi-wing piles is shown in Figure 1. The physical and mechanical characteristics of the soil around the winged pile are shown in Table 1.

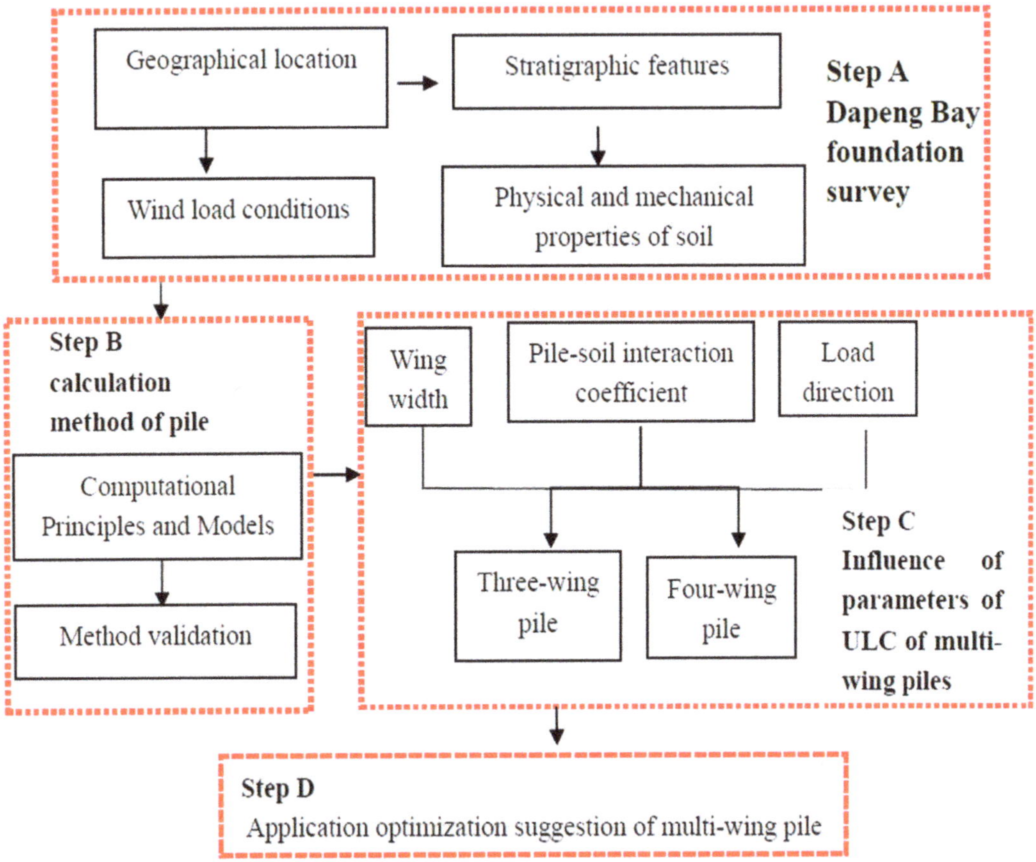

Figure 1. Flowchart for the research processes.

Table 1. Physical and mechanical properties of Dapeng Bay soil.

Soil Classification	Average Soil Depth h (m)	Saturated Density r_{sat} (g/cm^3)	Compression Modulus Es (Mpa)	Shear Strength S_u (kPa)	Liquid Index I_p	Foundation-Bearing Capacity f_a (kPa)
Silty clay	2.5	2	30	50	4.4	140
Sand and clay interlayer	5	2.13	60	-	6	200

2.2. Calculated Section of a Multi-Wing Pile

Two types of multi-wing pile have been selected for the Dapeng Bay trestle foundation design: three-wing and four-wing. The calculated section of the three-wing pile includes three wings and one circular pile (Figure 2). The angle between the wings is 120°, the width of each wing is B_w, and the diameter of the circular pile is D. The lateral load acts on the section of the pile head, and the angle between the lateral load direction and the wing is β. When $\beta = 0$, the direction of the lateral load aligns with the wing orientation; when $\beta = 60°$, the lateral load acts along the angular bisector of two wings. Similarly, the calculated section of the four-wing pile includes four wings and one circular pile (Figure 3). In the four-wing pile, the lateral load acts along the angular bisector of two wings when $\beta = 45°$. Other parameters are the same as for the three-wing pile. The value of the interaction coefficient α_i between the multi-wing pile and the soil varies from 0 to 1. The soil around the pile is silty clay with undrained shear strength S_u.

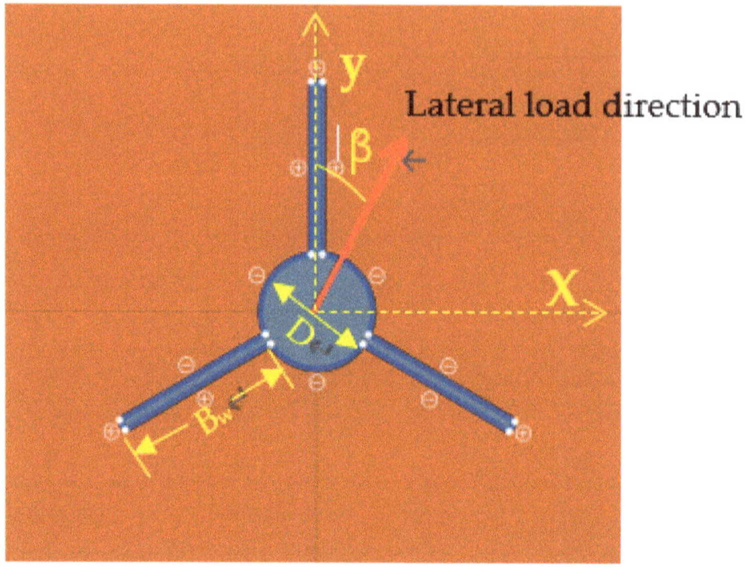

Figure 2. Calculated section of three-wing pile.

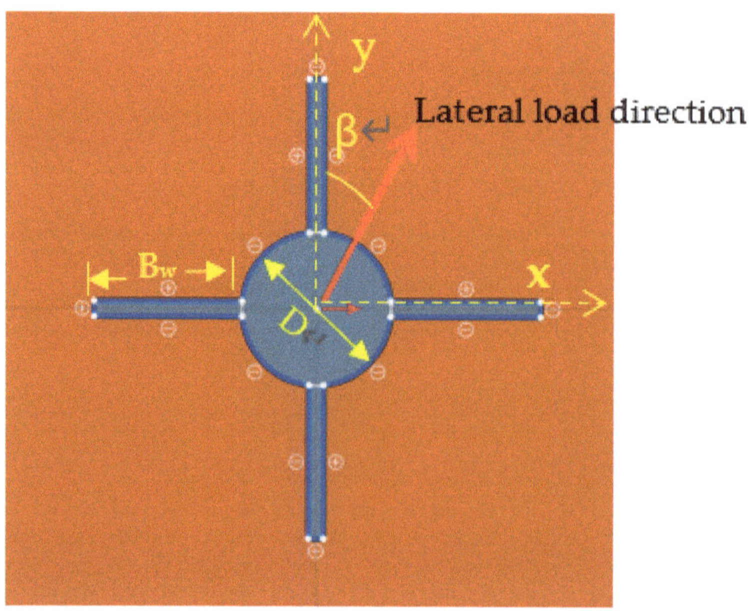

Figure 3. Calculated section of four-wing pile.

In the calculation, the multi-wing pile will be considered completely rigid. Its normalized ULC (denoted N_F) is given by [25]:

$$N_F = \frac{F}{S_u D_e} = f\,(\frac{D_e}{B_w},\ \alpha,\ \beta) \tag{1}$$

where F is the ultimate lateral load of the pile, and D_e is the equivalent diameter of the multi-wing–pile cross-section, which is related to the maximum section width of the pile [9]. The equivalent diameter of the three-wing–pile cross-section is

$$D_e = D + B_w\ (1 + \cos\ 60^\circ); \tag{2}$$

and that of the four-wing pile cross-section is

$$D_e = D + 2B_w \tag{3}$$

N_F is affected by changes in three parameters (wing–diameter ratio D_e/B_w, the lateral-load direction β, and the pile–soil interaction coefficient *a). F* can be calculated by performing finite-element limit analysis (FELA) using OPTUM G2 software [4,11]. The FEM basis of OPTUM G2 is to use the upper bound theorem and the lower bound theorem of limit analysis to solve the ultimate load of the structure [11]; the lower-bound and upper-bound solutions provide a safe estimate of the real ULC. Thereafter, by substituting Equation (2) or Equation (3) into Equation (1), the empirical formula for the normalized ultimate bearing capacity of the multi-wing pile can be obtained through parameter fitting; this is very convenient for engineering.

2.3. Verifying the Calculation Results for the Pile–Soil Interaction

Randolph and Houlsby [17] used the lower-bound theorem of limit analysis to find the ULC of a 0.5-m-diameter circular pile in saturated soft clay (Figure 4). In this study, a numerical model was established to calculate the ultimate bearing capacity of the same circular pile by conducting FELA in OPTUM G2. The results were compared with those of Randolph and Houlsby to determine the relevant calculation parameters of the pile–soil model. When considering the influence of the calculation-model boundary on ULC [26,27], the established numerical soil model considers a plane circle with diameter 8 m and an internal circular pile diameter (D) of 0.5 m. The material of the plane circle was saturated clay, and its strength (S_u) was 50 MPa. The pile was made of a rigid material; the plate material is set on the contact surface around the pile to study the change of the pile–soil interaction coefficient.

According to the difference and convergence rate of the calculation results of the calculation model after several trial calculations, the number of model elements was set to 10,000, and an adaptive grid was selected (shown in Figure 4a). The initial number of elements was 1000, and the number of adaptive iterations was three. The boundary of the soil model was fully constrained. When the interaction coefficient between the pile and soil changed from 0 to 1, two cases of N_F were calculated according to the upper- and lower-bound solutions of the limit analysis; the difference between them was small (Figure 4b). Randolph and Houlsby 's solution fell between the upper- and lower-bound solutions and was in good agreement with the latter.

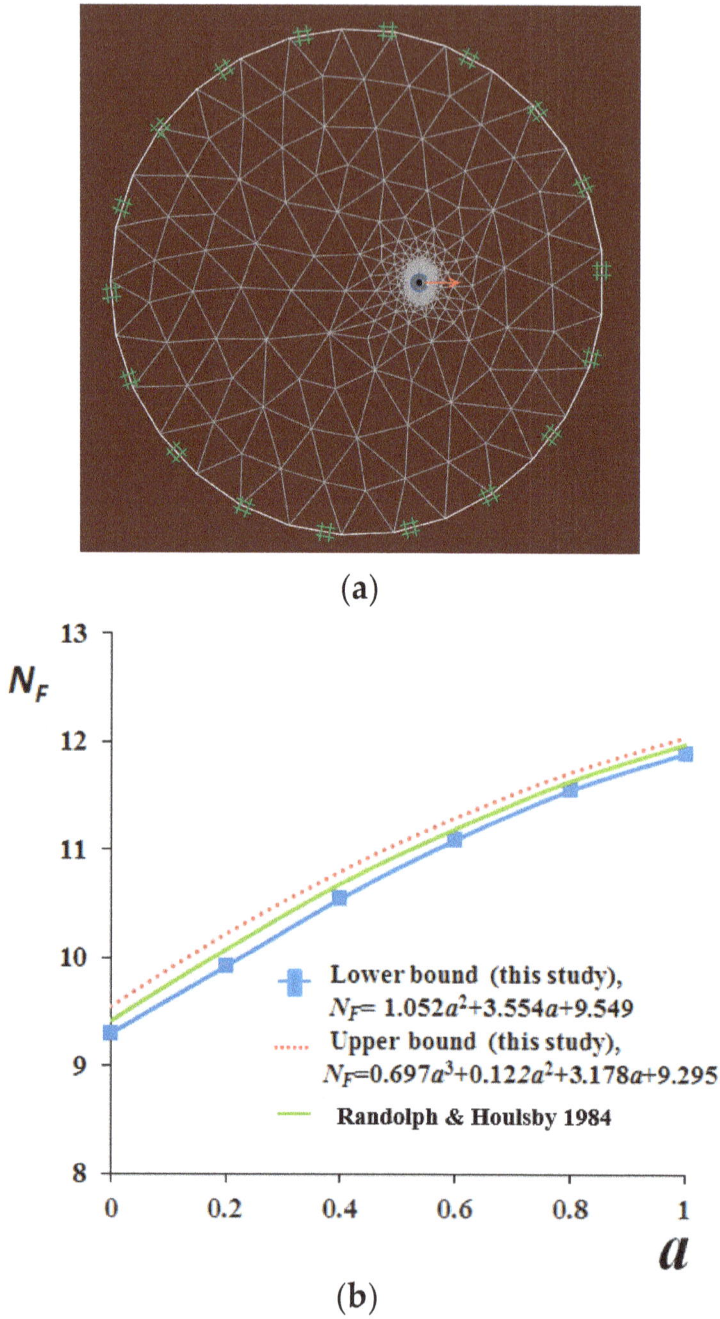

Figure 4. Calculation results for circular-pile model. (**a**) circular-pile-model self-adaptive computing grid; (**b**) a comparison of results calculated by Randolph and Houlsby (1984) with those using the method from this study. (α: pile–soil interaction coefficient; N_F: normalized ultimate lateral capacity).

3. Numerical Methods of Analysis

3.1. Three-Wing Pile

3.1.1. Effect of Wing–Diameter Ratio on Normalized Ultimate Lateral Capacity

The width of the wing plate affects the bearing capacity during experimentation on the wing pile [8]. To study the interaction between the pile ULC and the wing width B_w under the lateral load, OPTUM G2 was used to calculate the ULC of the three-wing pile. The calculation model for the three-wing pile is shown in Figure 2. The established numerical model of the soil was a plane circle with diameter 8 m and internal circular-pile diameter 0.2 m. The thickness of the wing plate was 0.02 m. The direction of the lateral load (β) was 0°. The material of the plane circle was saturated clay, and its strength (S_u) was 50 MPa. The pile body is made of steel pipe with a thickness of 10 mm, and the wing plate is a steel plate with a thickness of 8 mm. Compared with silty clay in Dapeng Bay, their strength is higher than soil, so the pile and three wings were assumed rigid. Wing widths of 0.05 m, 0.1 m, 0.15 m, 0.2 m, 0.25 m, and 0.3 m were considered. According to Equation (2), the equivalent diameters of the three-wing pile (D_e) were therefore 0.275 m, 0.35 m, 0.425 m, 0.5 m, 0.575 m, and 0.65 m. Similarly, the wing–diameter ratios (D_e/B_w) were 5.5, 3.5, 2.8, 2.5, 2.3, and 2.17.

The number of model elements was set to 10,000, and the adaptive grid was selected. The initial number of elements was 1000, and the number of adaptive iterations was three. The boundary of the soil model was fully constrained. The relationship between N_F and D_e/B_w is shown in Figure 5—to observe the influence of the change of the pile–soil interaction coefficient on the bearing capacity of multi-wing pile and provide a basis for the design of multi-wing pile under the most dangerous situation. The pile–soil interaction coefficients changed from 0 to 1, according to the limit analysis.

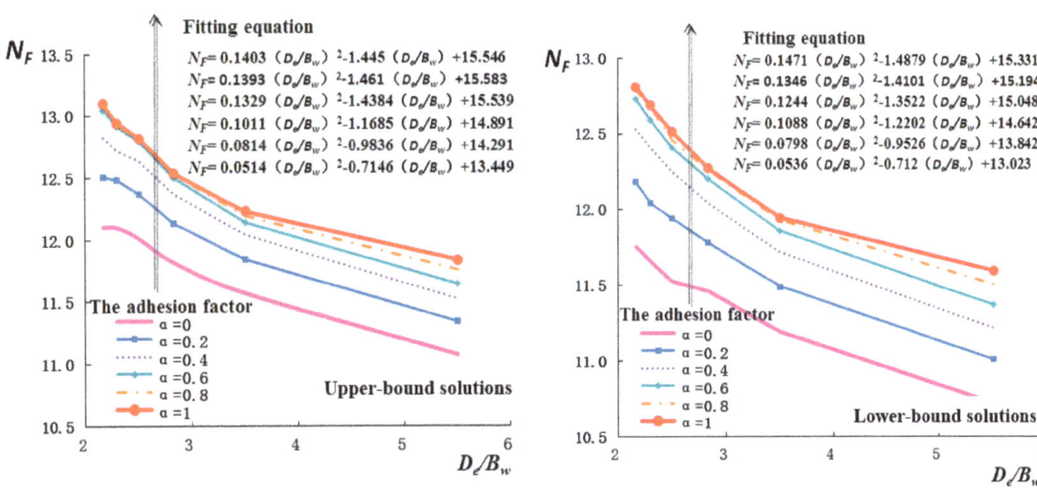

Figure 5. Relationship between the normalized ultimate lateral capacity (N_F) of the three-wing pile and the wing–diameter ratio (D_e/B_w) when lateral-load angle β = 0. (**Top**) Upper-bound and (**Bottom**) lower-bound solutions from limit analysis with different pile-soil interaction coefficients (adhesion factors).

Although the pile–soil interaction coefficients are different, the fitted-curve trends in Figure 5 are identical. When the soil loading interaction coefficient is greater than 0.4, the different fitting curves overlap. The influence of the wing–diameter ratio on the shear plastic zone of the soil is shown in Figure 6; the pile–soil interaction coefficient was set to 0.4, according to the upper-bound solution from limit analysis.

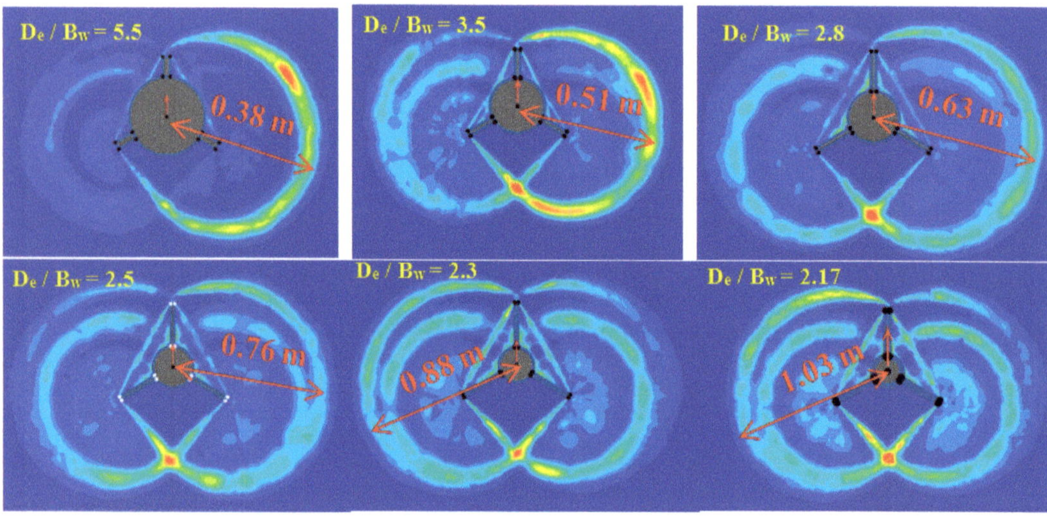

Figure 6. Influence of various wing–diameter ratios of the three-wing pile on the shear plastic zone of the soil when load angle $\beta = 0$ and pile–soil interaction coefficient $a = 0.4$. (D_e: effective diameter; B_w: wing width.).

3.1.2. Effect of Pile–Soil Interaction Coefficient on Normalized Ultimate Lateral Capacity

The pile–soil interaction coefficient affects the pile-bearing capacity and the shear plastic ring of the soil [11]. To study this, the ULC of the three-wing pile was calculated by OPTUM G2. The calculation model for the three-wing pile is shown in Figure 2. The other model parameters were the same as those previously described. The relationship between the pile–soil interaction coefficient (α) and the normalized ultimate lateral capacity of the three-wing pile (N_F) when α changed from 0 to 1, according to the solution of the limit analysis, is illustrated in Figure 7.

Although the pile–soil interaction coefficients are different, the fitted-curve trends in Figure 7 are identical. When the wing–diameter ratio was lower than 2.3, the soil had a double-circular plastic zone (see Figure 6). The influence of the pile–soil interaction coefficient on the shear plastic zone of the soil when D_e/B_w was set to 2.3, according to the upper bound solution of the limit analysis, is shown in Figure 8.

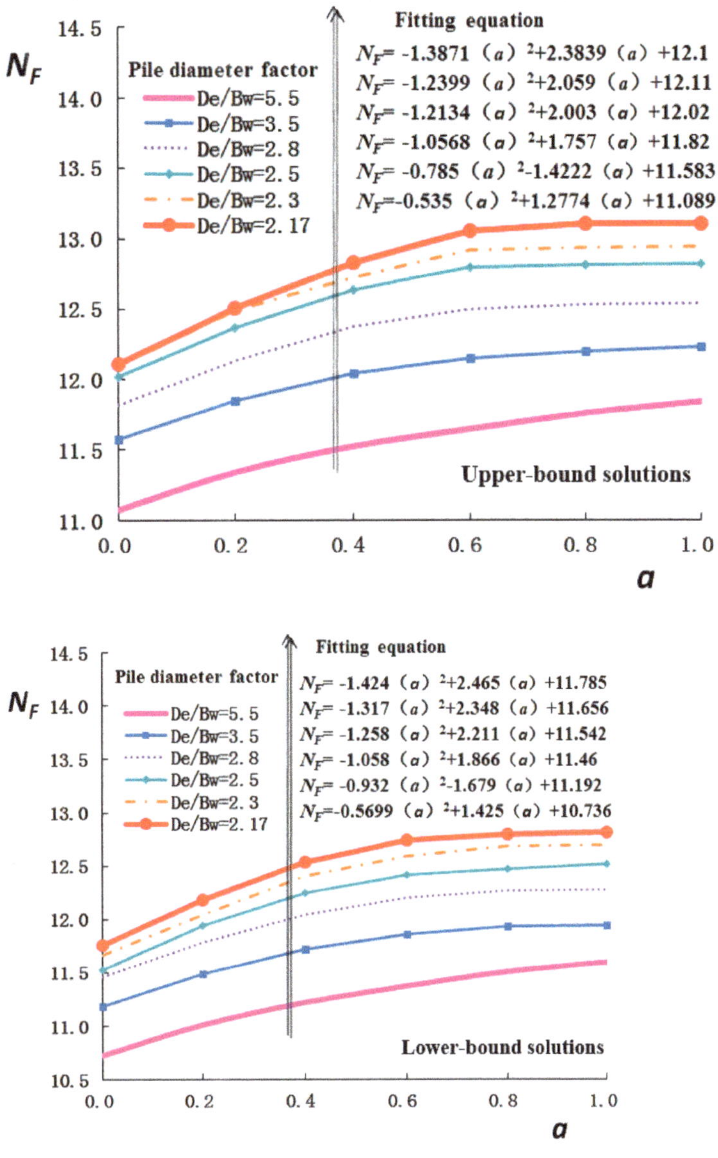

Figure 7. Relationship between the normalized ultimate lateral capacity (N_F) of the three-wing pile and the pile–soil interaction coefficient (α) when lateral-load angle $\beta = 0$. (**Top**) Upper-bound and (**Bottom**) lower-bound solutions from limit analysis with different wing-to-diameter ratios. (D_e: effective diameter; B_w: wing width.).

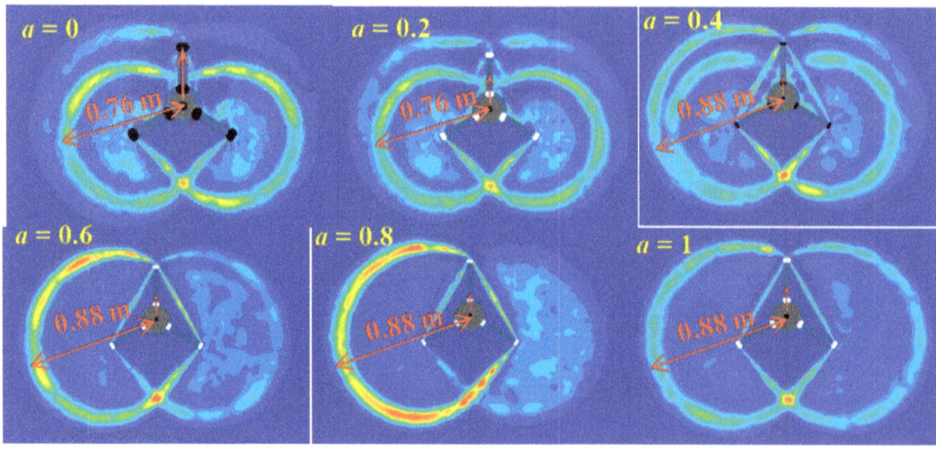

Figure 8. Influence of the different pile–soil interaction coefficients a on the shear plastic zone of the soil around the three-wing pile when lateral-load angle $\beta = 0$ and wing–diameter ratio is 2.3.

3.1.3. Effect of the Lateral-Load Direction on Normalized Ultimate Lateral Capacity

Different directions of the load affect the ultimate lateral bearing capacity of piles with different cross-sections [8]. The ULC of the three-wing pile was calculated by OPTUM G2 to elucidate the influence of the lateral load direction. The calculation model for the three-wing pile is shown in Figure 2. The wing width was set to 0.2 m. The lateral-load angle (β) ranged from 0° to 60°; it can be expressed in radians as $\beta' = \beta\pi/180°$. The wing–diameter ratio D_e/B_w was set to 2.5. The other model parameters were the same as those previously described. The relationship between β' and N_F as the pile–soil interaction coefficients changed from 0 to 1, according to limit analysis, is shown in Figure 9.

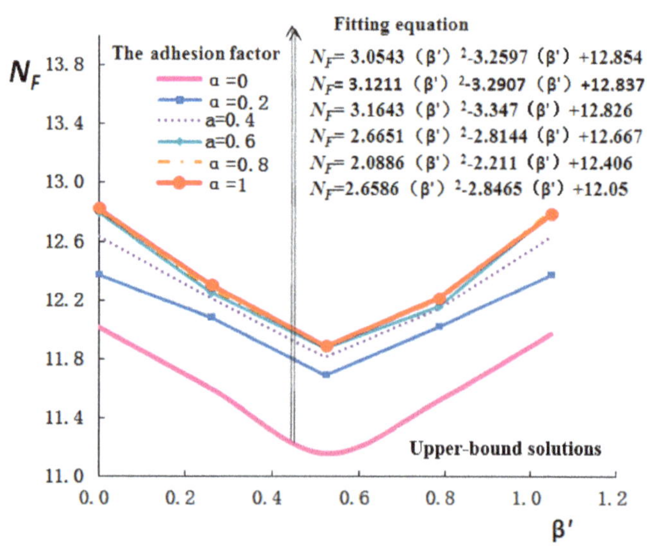

Figure 9. *Cont.*

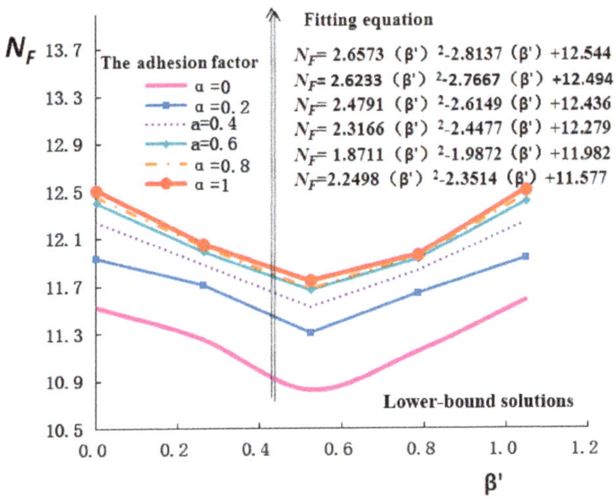

Figure 9. Relationship between the normalized ultimate lateral capacity (N_F) of the three-wing pile and the direction of the lateral load (β′) when the wing-to-diameter ratio D_e/B_w = 2.5. (**Top**) upper-bound and (**Bottom**) lower-bound solutions from limit analysis with different pile–soil interaction coefficients (adhesion factors).

Although the pile–soil interaction coefficients are different, the fitted curve trends in Figure 9 are identical. When the soil loading interaction coefficient exceeded 0.4, the range of the plastic zone of the soil in Figure 8 stopped increasing. D_e/B_w was 2.5 when the pile–soil interaction coefficient was 0.4. The influence of the pile–soil interaction coefficient on the shear plastic zone of the soil, according to the upper bound solution of limit analysis, is shown in Figure 10.

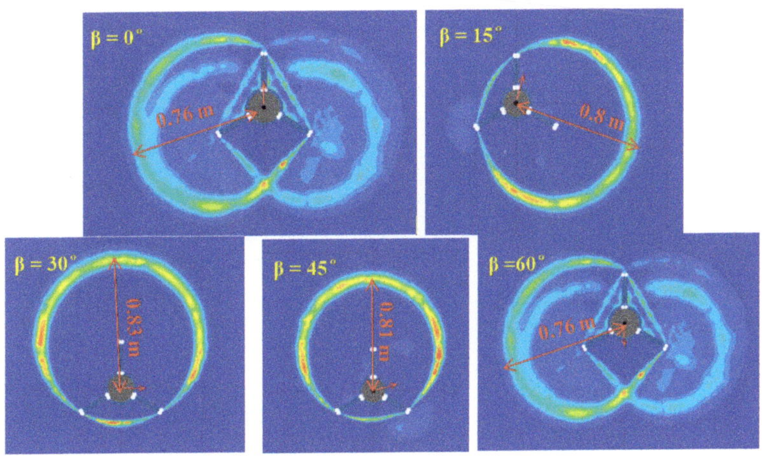

Figure 10. Effect of different lateral-load directions on the shear plastic zone of the soil around the three-wing pile when pile–soil interaction coefficient a = 0.4 and D_e/B_w = 2.5. (D_e: effective diameter; B_w: wing width.).

3.2. Four-Wing Pile

3.2.1. Effect of Wing–Diameter Ratio on Normalized Ultimate Lateral Capacity

To study the interaction between the four-wing pile and the soil under the lateral load, the ULC of a four-wing pile was calculated using OPTUM G2. In addition, the influence of the wing width on ULC was studied. The calculation model for the four-wing pile is shown in Figure 3. The established numerical model of the soil was a plane circle with diameter 8 m and an internal circular pile with diameter 0.2 m. The thickness of the wing plate was 0.02 m. The lateral load angle (β) was 0°. The material of the plane circle was saturated clay with S_u = 50 MPa. The pile and four wings were assumed rigid. The wing widths were set to 0.05 m, 0.1 m, 0.15 m, 0.2 m, 0.25 m, and 0.3 m. According to Equation (3), the equivalent diameters were 0.3 m, 0.4 m, 0.5 m, 0.6 m, 0.7 m, and 0.8 m, corresponding to wing–diameter ratios 6, 4, 3.3, 3, 2.8, and 2.7.

The calculation conditions were the same as in the three-wing case above. Figure 11 shows the relationship between D_e/B_w and N_F for the four-wing pile when the pile–soil interaction coefficients changed from 0 to 1, according to the solutions of the limit analysis.

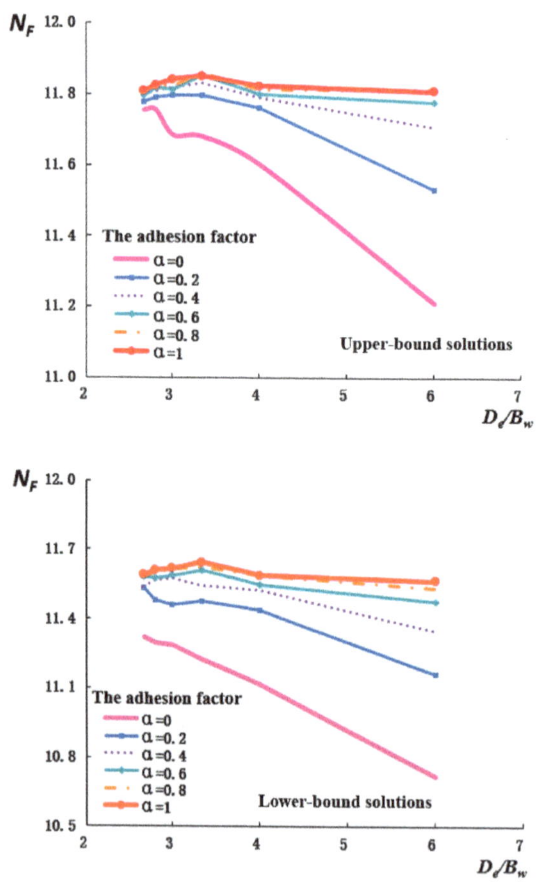

Figure 11. Relationship between the normalized ultimate lateral capacity (N_F) of the four-wing pile and the wing–diameter ratio (D_e/B_w) when lateral-load angle $\beta = 0$. (**Top**) Upper-bound and (**Bottom**) lower-bound solutions from limit analysis with different pile–soil interaction coefficients (adhesion factors).

Although the pile–soil interaction coefficients were different, the fitted curve trends in Figure 11 were identical. When the soil loading interaction coefficient was greater than 0.4, the different fitting curves overlapped. The influence of various wing widths on the shear plastic zone of the soil around the four-wing pile for a pile–soil interaction coefficient of 0.4, according to the upper bound solution of the limit analysis, is shown in Figure 12.

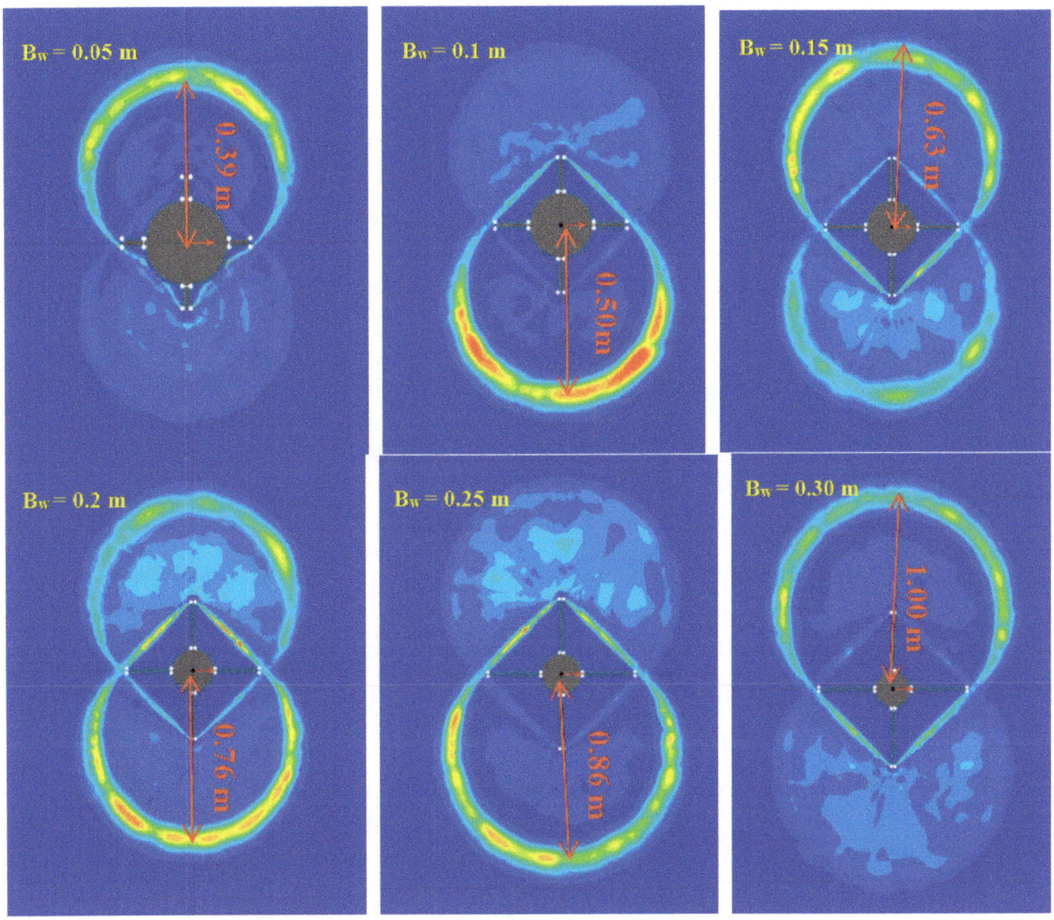

Figure 12. Influence of various wing widths B_w on the shear plastic zone of the soil around the four-wing pile when lateral-load angle β = 0 and pile–soil interaction coefficient a = 0.4.

3.2.2. Effect of Pile–Soil Interaction Coefficient on Normalized Ultimate Lateral Capacity

The influence of the pile–soil interaction coefficient on the ULC of the four-wing pile, (calculated by OPTUM G2) was studied. The calculation model for the four-wing pile is shown in Figure 3. The other model parameters were the same as those previously described. The relationship between the pile–soil interaction coefficient (a) and the normalized ultimate lateral capacity of the four-wing pile (N_F) as a changed from 0 to 1, according to limit analysis, is presented in Figure 13.

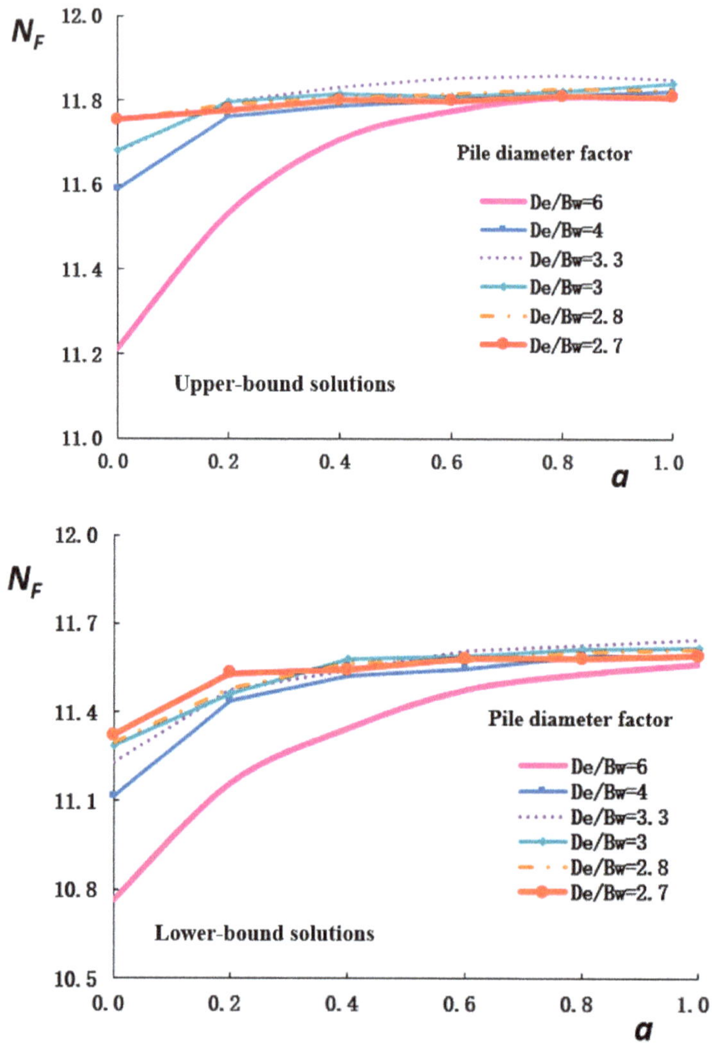

Figure 13. Relationship between the normalized ultimate lateral capacity (N_F) of the four-wing pile and the pile–soil interaction coefficient (α) when lateral-load angle $\beta = 0$. (**Top**) Upper-bound and (**Bottom**) lower-bound solutions from limit analysis with different wing-to-diameter ratios. (D_e: effective diameter; B_w: wing width.).

The changing trends of the fitted curves in Figure 13 are the same under the different wing–diameter ratios. When the wing width is more than 0.15 m, as shown in Figure 12, the soil has a double-circular plastic zone. The influence of the pile–soil interaction coefficient on the shear plastic zone of the soil around the four-wing pile, when B_w is set to 0.15, according to the upper bound solution of the limit analysis, is shown in Figure 14.

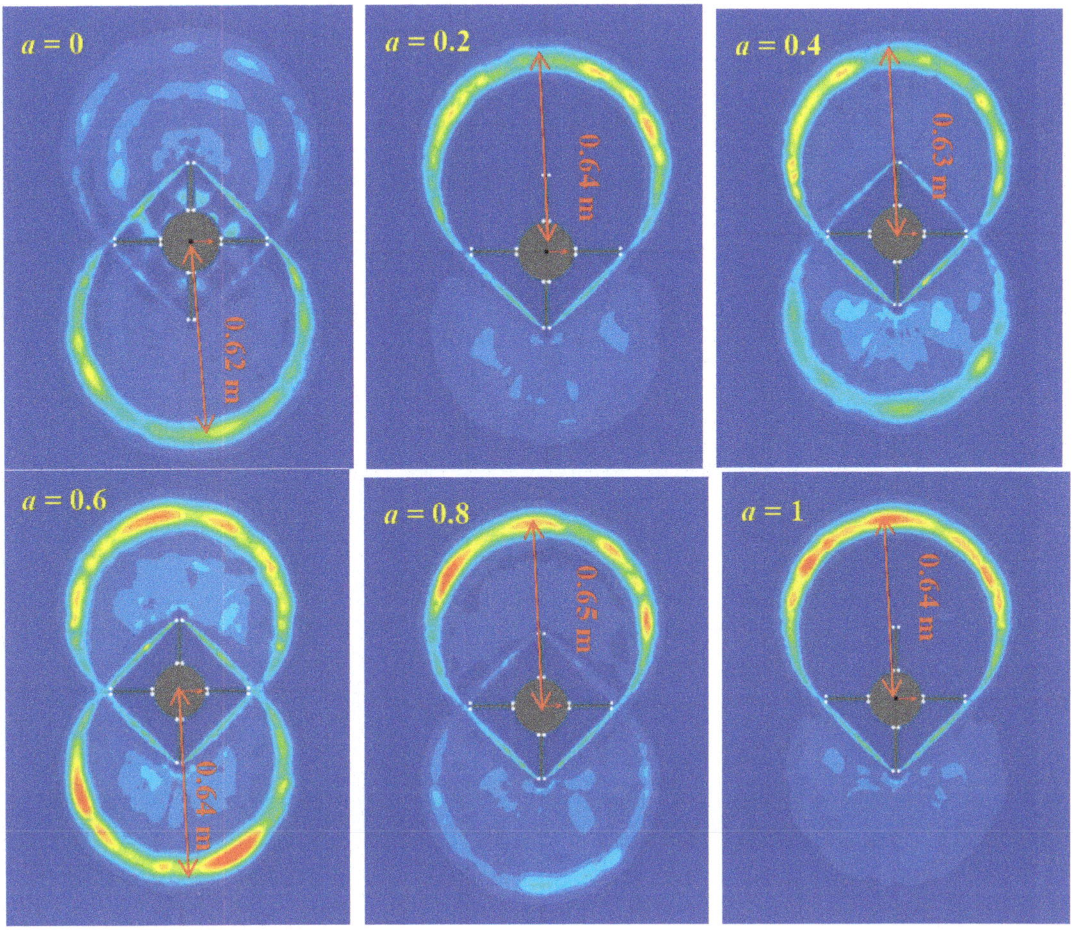

Figure 14. Influence of the different pile–soil interaction coefficients a on the shear plastic zone of the soil around the four-wing pile when lateral-load angle $\beta = 0$ and wing–diameter ratio $D_e/B_w = 3.3$.

3.2.3. Effect of Lateral-Load Direction on Normalized Ultimate Lateral Capacity

To study the influence of the direction of the lateral load, OPTUM G2 was used to calculate the ULC of the four-wing pile. The calculation model for the four-wing pile is shown in Figure 3. The wing width was set to 0.2 m. The direction of the lateral load (β) ranged from 0 ° to 45 °; the wing–diameter ratio (D_e/B_w) was 3. The other model parameters were the same as those previously described. The relationship between the direction of the lateral load (β′) and the normalized ultimate lateral capacity of the four-wing pile (N_F) as the pile–soil interaction coefficients changed from 0 to 1, according to the upper- and lower-bound solutions of the limit analysis, is shown in Figure 15.

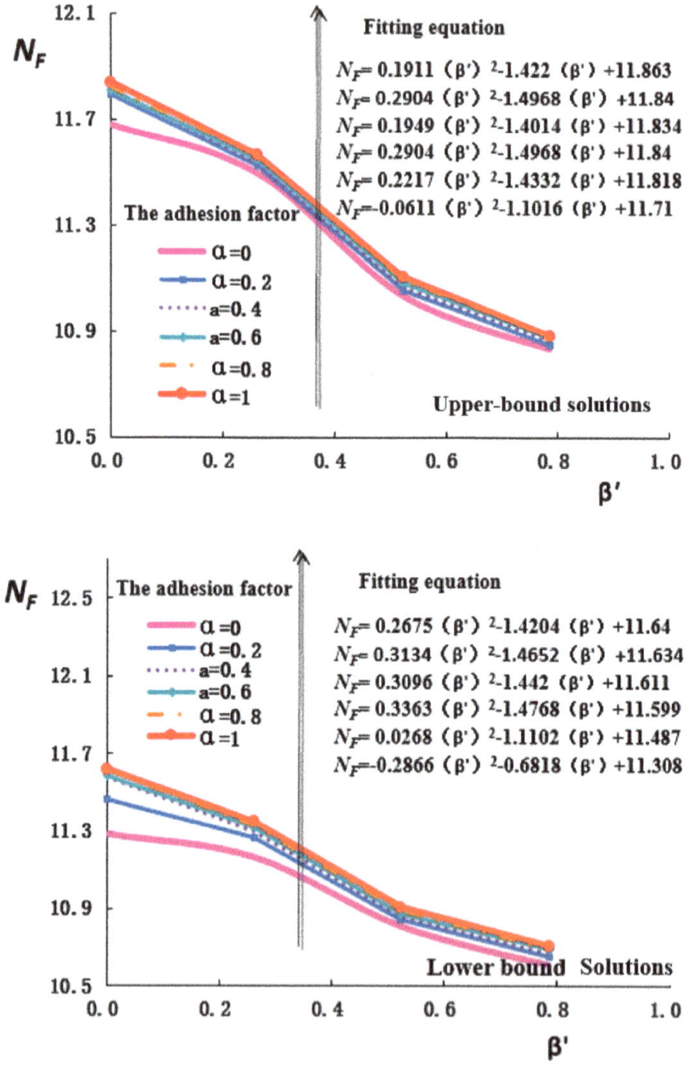

Figure 15. Relationship between the normalized ultimate lateral capacity (N_F) of the four-wing pile and the direction of the lateral load (β′) when the wing-to-diameter ratio D_e/B_w = 3. (**Top**) upper-bound and (**Bottom**) lower-bound solutions from limit analysis with different pile–soil interaction coefficients (adhesion factors).

Despite the different pile–soil interaction coefficients, the changing trends of the fitted curves in Figure 15 are identical. When the soil loading interaction coefficient was greater than 0.4, the range of the plastic zone of the soil in Figure 14 no longer increased. By setting the pile–soil interaction coefficient to 0.4, D_e/B_w was 3. The influence of the lateral-load direction on the shear plastic zone of the soil around the four-wing pile according to the upper-bound solution of the limit analysis is shown in Figure 16.

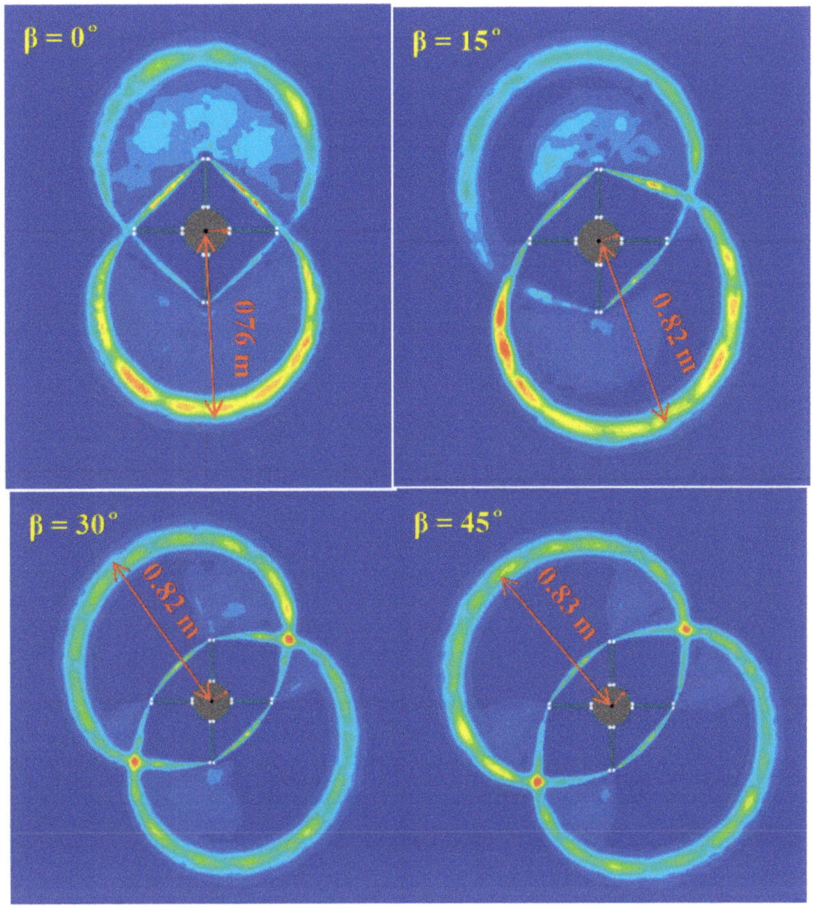

Figure 16. Effect of the different lateral-load directions on the shear plastic zone of the soil around the four-wing pile when pile–soil interaction coefficient a = 0.4 and wing–diameter ratio D_e/B_w = 3.

4. Results and Discussion

4.1. Three-Wing Pile

4.1.1. Effect of Wing–Plate Width and Pile–Soil Interaction Coefficient on ULC

Figure 5 show that the upper- and lower-bound solutions for N_F for the three-wing pile as D_e/B_w range from 2.17 to 5.5; it can be observed that N_F decreases with an increase in the wing–diameter ratio. The fitting curves are quadratic, and the correlation coefficient is greater than 0.99. When the wing–diameter ratio (D_e/B_w) is 2.5, that is, the width of the wing plate is 0.2 m, the slope of ULC curve of the three-wing pile becomes slower, so the recommended wing width of the three-wing pile is 0.2 m; the normalized ULC of the three-wing pile rises with the pile–soil interaction coefficient (α) in the different curves. However, when α is greater than 0.4, the curves overlap, and the influence of the pile–soil interaction coefficient on N_F is weakened. The parameter D_e/B_w, although small, significantly influences N_F for the three-wing pile (see Figure 5). Keawsawasvong and Ukritchon [27] discovered that, as the side length perpendicular to the direction of the force becomes smaller, the normalized ULC of an I-section pile decreases, mainly because of the increased contact area between the wing plate and soil.

Figure 6 shows that, as D_e/B_w decreases, the shear plastic ring of the broken soil gradually expands. When D_e/B_w is 5.5, the maximum distance (L) from the shear plastic ring to the three-wing pile center is 0.38 m, which is 1.9 times the pile diameter (D). The proportion of the soil plastic zone is defined as L/D. When D_e/B_w was 3.5, 2.8, 2.5, 2.3, or 2.17, L/D was 2.55, 3.15, 3.8, 4.4, or 5.15, respectively. Fitting this data to a quadratic equation relates the proportion of soil plastic zone to the wing–diameter ratio as

$$\frac{L}{D} = 0.5141\left(\frac{D_e}{B_w}\right)^2 - 4.81\frac{D_e}{B_w} + 12.834 \tag{4}$$

the correlation coefficient R is 0.96.

The shear plastic ring range of the soil around the three-wing pile is positively related to the normalized ultimate bearing capacity. As the width of the wing plate increases, the shear plastic zone of the soil changes from a single ring to multiple rings (see Figure 6). This shows that increasing the width of the wing plate will increase damage to the soil around the pile.

Figure 7 show that N_F increases as α varies from 0 to 1. The fitting curves are quadratic, and the correlation coefficient is greater than 0.99. However, when α is greater than 0.6, the increase in N_F is not marked; this is consistent with the change trend in Figure 8, which represents the range of the shear plastic zone ring of the soil. When a is greater than 0.4, the maximum distance from the plastic ring to the pile center no longer increases but remains at 0.88 m, which is more than four times the pile diameter. This is a result of the process required to ensure all the soil in the plastic ring reaches the maximum strength. Zhou et al. [11] determined that when the pile–soil interaction coefficient increases, a square plastic zone of the soil appears around an XCC pile. Keawsawasvong and Ukritchon [4] investigated ULC of H piles and determined that the area of the soil failure shear zone increases with the pile–soil interaction coefficient. The soil shear zone changes from an ellipse to a circle, and the plastic zone inside the ellipse decreases. In addition, the pile–soil interaction coefficient affects the distribution of the plastic rings on both sides of the three-wing pile (see Figure 8), which may be affected by the form of the pile section. Because the cross-section of the pile is different, the range of the load transfer from the pile to the soil is different as well.

4.1.2. Effect of Lateral-Load Direction on ULC

Figure 9 show that N_F of the three-wing pile first decreased and then increased with β', following a quadratic relationship, for all α in the range 0 to 1. However, the calculation results from the lower-bound solutions of XCC piles show that, when the interaction coefficient is greater than 0.4, the bearing capacity is reduced [11]. This difference is mainly because the ULC of the three-wing pile is restricted by the pile–soil interaction coefficient when the load direction changes.

Figure 10 shows that, as the direction of the lateral load changes from 0 to 60°, the maximum distance from the shear plastic ring to the pile center first increases and then decreases. When β is 0° or 60°, the shear plastic ring of the soil appears as a double ring around the three-wing pile. When β is 30°, the shear plastic ring of the soil appears as a single ring; the distance from the plastic ring to the pile center is the largest, and the ULC is the smallest. Therefore, the size of the shear plastic ring has less influence on the ULC of the three-wing pile than the number of plastic rings. Ukritchon and Keawsawasvong [28] determined that, when the load direction of the rectangular pile changes, the extent of the plastic zone expands towards the direction of the lateral load. However, there is a larger change in the direction of the farthest point of the shear plastic ring to the center of the three-wing pile as the load direction changes (Figure 10). This shows that the shear plastic area of the soil around the three-wing pile changes with the load direction. Therefore, the soil reinforcement should be from the pile center to the maximum radius of the shear plastic ring around the pile.

4.2. Four-Wing Pile

4.2.1. Effect of Wing-Plate Width and Pile–Soil Interaction Coefficient on ULC

Figure 11 show the upper- and lower-bound solutions for N_F for the four-wing pile as D_e/B_w ranges from 2.67 to 6. N_F increases with α for the different curves. When α is greater than 0.4, N_F first increases and then decreases with an increase in the D_e/B_w; the maximum occurs when D_e/B_w is 3.3 (see Figure 11), so the recommended width of the four-wing pile is 0.15 m. This is different from the three-wing–pile case (see Figure 5), mainly because ULC and equivalent diameter are changing during the calculation of N_F. When the wing width is equal to 0.15 m and the pile–soil interaction coefficient is equal to 0.4, the ULC of the four-wing pile is 295.79 kN, whereas that of the three-wing pile is 262.94 kN: the ULC increased by 19.9% (see Figures 5 and 11). It is recommended to use the four-wing pile if the engineering cost is sufficient in the foundations of Dapeng Bay. Murphy [29] indicated that the ULC of two-wing piles increased by 16% and 36% for two different wing lengths, compared with that of a round pile of the same diameter. This shows that both width of wing panels and their lengths affect the ULC of the wing pile.

Figure 12 shows that, as B_w of the four-wing pile decreases, the shear plastic ring of the soil gradually expands. When B_w is greater than 0.2 m, the shear plastic ring of the soil around the four-wing pile is less than that of the three-wing pile (see Figures 6 and 12). However, the maximum distance of the plastic ring to the four-wing–pile center is 1 m, whereas the maximum distance of the plastic ring to the three-wing–pile center is 1.03 m. The difference is small, which shows that increasing the number of wing plates has little effect on the range of the soil shear plastic ring. As in the calculation for the three-wing pile, L/D was fitted to a quadratic in D_e/B_w:

$$\frac{L}{D} = 0.5061\left(\frac{D_e}{B_w}\right)^2 - 5.22\frac{D_e}{B_w} + 15.07 \quad (5)$$

The correlation coefficient R was 0.96.

Figure 13 show the upper- and lower-bound solutions for N_F for the four-wing pile increase as α varied from 0 to 1. The change trend was similar to that of the three-wing pile (see Figure 7). However, when α was greater than 0.2, the increase in N_F was not prominent, which is consistent with the range of the plastic ring that belongs to the broken soil in Figure 14. Comparing three- and four-wing piles, the pile–soil interaction coefficient has relatively little effect on the maximum distance from the plastic ring to the pile center (see Figure 14).

4.2.2. Effect of Lateral-Load Direction on ULC

Figure 15 show that the upper- and lower-bound solutions for N_F decrease quadratically as β goes from 0 to 45°. When α is 0.4, according to the upper-bound solutions, the maximum N_F of the four-wing pile is 11.8179 and the minimum N_F is 10.866, a decrease of 9.9%. Under the same conditions, the maximum N_F of the three-wing pile in different directions is 6.5% lower than its minimum value (see Figure 9). This shows that the normalized ULC of the four-wing piles is more sensitive to direction changes than that of three-wing piles. Babu and Viswanadham [7] considered the difference in ULC of the four-wing pile was 5% in dense sand, as the direction of the lateral load varied between 0 and 45° because the influence of direction on wing–pile ULC depends on the type of soil.

Figure 16 shows that, when β is between 0° and 45°, the maximum distance from the shear plastic ring of soil to the pile center increases and the shear plastic ring of the soil appears as a double ring around the four-wing pile. However, the shear plastic ring of the soil mainly appears as a single ring around the three-wing piles when β is between 15° and 45° (see Figure 10). Pile-failure tests have revealed that cracks around three-wing piles are less than those of four-wing piles. This shows that four-wing piles have a more symmetrical influence on the surrounding soil than three-wing piles, and that higher reinforcement is needed for soil around the four-wing pile than around a three-wing pile.

The ULC of piles in undrained clay is also affected by factors such as the pile length, pile diameter, and soil gravity density [30–33]. The position of the wing plate along the pile body also affects the plastic zone of the soil on the side of the pile [7]. These issues need further investigation via three-dimensional limit analysis. The current research should be applicable to clay, silt, and silty clay strata deposited in still water environments such as bays, and further research is needed for coarse-grained strata deposited in dynamic water environments.

5. Conclusions

The ULCs of three- and four-wing piles with different wing widths, pile–soil interaction coefficients, and load directions were investigated in saturated silty clay. The shear plastic zone of the soil around the pile was studied to find conditions where the multi-wing pile reached its ultimate lateral capacity. Based on the upper- and lower-limit solutions, the following conclusions can be drawn:

(1) When the empirical equation of ULC of the three-wing piles that varied with each parameter was fitted (with a correlation coefficient greater than 0.99), the width of the wing plate was found to have a significant influence on ULC of the wing pile.
(2) The normalized ULC of the three-wing pile increased with the pile–soil interaction coefficient. The change trend was similar in the four-wing pile; however, when the pile–soil interaction coefficient was greater than 0.2, the increase in the normalized ULC of the four-wing pile was not prominent.
(3) For both three-wing and four-wing piles, the pile–soil interaction coefficient had relatively little effect on the maximum distance from the shear plastic ring of the soil to the pile center.
(4) When the wing width was 0.3, the maximum distance from the shear plastic ring of the soil to the three-wing pile center was 1.03 m, which is more than five times the pile diameter. However, under the same conditions, the four-wing pile did not show any increase in the shear plastic zone of the soil around the pile.
(5) There was a double shear plastic ring of soil around the four-wing pile. The four-wing piles had a more symmetrical influence on the soil around the piles than the three-wing ones. When both the three-wing pile and the four-wing pile reach their respective ultimate bearing capacities, the reinforcement needed for the soil around a four-wing pile will therefore be higher than for that around a three-wing pile.

Author Contributions: Conceptualization: H.W., D.F., T.Y., L.M., W.L. and D.P.; Methodology: D.F. and D.P.; Software: H.W.; Validation: L.M., W.L. and H.W.; Formal analysis: T.Y. and L.M.; Investigation: H.W.; Resources: H.W.; Data curation: T.Y.; Writing—Original draft preparation: H.W.; Writing—Review and editing: H.W., D.F., T.Y., L.M., W.L. and D.P.; Visualization: T.Y.; Supervision: H.W.; Project administration: L.M.; Funding acquisition: W.L. All authors have read and agreed to the published version of the manuscript.

Funding: This research was funded jointly by the China Postdoctoral Science Foundation (Grant No. 2020M672202), the Youth project of Henan Natural Science Foundation (Grant No. 212300410125), Key projects of universities in Henan Province (Grant No. 20A570002), and the Engineering Research Center for Embankment Safety and Disease Control of Ministry of Water Resources (No. LSDP202202).

Institutional Review Board Statement: Not applicable.

Informed Consent Statement: Not applicable.

Data Availability Statement: Not applicable.

Acknowledgments: The authors would like to thank Starfin Australia for leading the Winged Pile Study.

Conflicts of Interest: The authors declare no conflict of interest.

References

1. Zhou, H.; Yuan, J.; Liu, H. A general analytical solution for lateral soil response of non-circular cross-sectional pile segment. *Appl. Math. Modell.* **2019**, *71*, 601–631. [CrossRef]
2. Deb, T.K.; Singh, B. Response and capacity of monopod caisson foundation under eccentric lateral loads. *Mar. Georesour. Geotechnol.* **2018**, *36*, 452–464. [CrossRef]
3. Zhu, B.; Ren, J.; Ye, G.L. Wave-induced liquefaction of the seabed around a single pile considering pile–soil interaction. *Mar. Georesour. Geotechnol.* **2018**, *36*, 150–162. [CrossRef]
4. Keawsawasvong, S.; Ukritchon, B. Undrained lateral capacity of I-shaped concrete piles. *Songklanakarin J. Sci. Technol.* **2017**, *39*, 751–758. [CrossRef]
5. Broms, B.B. Lateral resistance of piles in cohesive soils. *J. Soil Mech. Found. Div.* **1964**, *90*, 27–63. [CrossRef]
6. Nasr, A.M.A. Experimental and theoretical studies of laterally loaded finned piles in sand. *Can. Geotech. J.* **2014**, *51*, 381–393. [CrossRef]
7. Babu, K.V.; Viswanadham, B.V.S. Numerical studies on lateral load response of fin piles. *Geomech. Geoengin.* **2018**, *14*, 85–98. [CrossRef]
8. Sakr, M.A.; Azzam, W.R.; Wahba, M.A. Model study on the performance of single-finned piles in clay under lateral load. *Arab. J. Geosci.* **2020**, *13*, 1–16. [CrossRef]
9. Dührkop, J.; Grabe, J. Laterally loaded piles with bulge. *J. Offshore Mech. Arct. Eng.* **2008**, *130*. [CrossRef]
10. Zhou, H.; Liu, H.L.; Ding, X.; Kong, G. A p–y curve model for laterally loaded XCC pile in soft clay. *Acta Geotech.* **2020**, *15*, 3229–3242. [CrossRef]
11. Zhou, H.; Liu, H.; Wang, L.; Kong, G. Finite element limit analysis of ultimate lateral pressure of XCC pile in undrained clay. *Comput. Geotech.* **2018**, *95*, 240–246. [CrossRef]
12. Yaghobi, M.H.; Hanaei, F.; Fazel Mojtahedi, S.F.F.; Rezaee, M. Numerical finite element analysis of laterally loaded fin pile in sandy soil. *Innov. Infrastruct. Solut.* **2019**, 14. [CrossRef]
13. Peng, J.-R.; Rouainia, M.; Clarke, B.G. Finite element analysis of laterally loaded fin piles. *Comput. Struct.* **2010**, *88*, 1239–1247. [CrossRef]
14. Bienen, B.; Dührkop, J.; Grabe, J.; Randolph, M.F.; White, D.J. Response of piles with wings to monotonic and cyclic lateral loading in sand. *J. Geotech. Geoenviron. Eng.* **2012**, *138*, 364–375. [CrossRef]
15. Adeel, M.B.; Jan, M.A.; Aaqib, M.; Park, D. Development of simulation basedp-multipliers for laterally loaded pile groups in granular soil using 3D nonlinear Finite Element Model. *Appl. Sci.* **2021**, *11*, 26. [CrossRef]
16. Adeel, M.B.; Aaqib, M.; Pervaiz, U.; Rehman, J.U.; Park, D. Numerical response of pile foundations in granular soils subjected to lateral load. *Geomech. Eng.* **2022**, *28*, 11–23. [CrossRef]
17. Randolph, M.F.; Houlsby, G.T. The limiting pressure on a circular pile loaded laterally in cohesive soil. *Géotechnique* **1984**, *34*, 613–623. [CrossRef]
18. Martin, C.M.; Randolph, M.F. Upper-bound analysis of lateral pile capacity in cohesive soil. *Géotechnique* **2006**, *56*, 141–145. [CrossRef]
19. Martin, C.M.; White, D.J. Limit analysis of the undrained bearing capacity of offshore pipelines. *Geogr. Tech.* **2012**, *62*, 847–863. [CrossRef]
20. Truong, P.; Lehane, B.M. Effects of pile shape and pile end condition on the lateral response of displacement piles in soft clay. *Géotechnique* **2018**, *68*, 794–804. [CrossRef]
21. Albusoda, B.S.; Al-Saadi, A.F.; Jasim, A.F. An experimental study and numerical modeling of laterally loaded regular and finned pile foundations in sandy soils. *Comput. Geotech.* **2018**, *102*, 102–110. [CrossRef]
22. Taghavi, A.; Muraleetharan, K.K. Analysis of laterally loaded pile groups in improved soft clay. *Int. J. Geomech.* **2017**, *17*, 04016098. [CrossRef]
23. Zhou, H.; Liu, H.; Li, Y.Z.; Ding, X. Limit lateral resistance of XCC pile group in undrained soil. *Acta Geotech.* **2020**, *15*, 1673–1683. [CrossRef]
24. Zhao, Z.; Kouretzis, G.; Sloan, S.; Gao, Y. Ultimate lateral resistance of tripod pile foundation in clay. *Comput. Geotech.* **2017**, *92*, 220–228. [CrossRef]
25. Butterfield, R. Dimensional analysis for geotechnical engineering. *Géotechnique* **2009**, *49*, 357–366. [CrossRef]
26. Fatahi, B.; Basack, S.; Ryan, P.; Zhou, W.H.; Khabbaz, H. Performance of laterally loaded piles considering soil and interface parameters. *Geomech. Eng.* **2014**, *7*, 495–524. [CrossRef]
27. Keawsawasvong, S.; Ukritchon, B. Ultimate lateral capacity of two dimensional plane strain rectangular pile in clay. *Geomech. Eng.* **2016**, *11*, 235–252. [CrossRef]
28. Ukritchon, B.; Keawsawasvong, S. Undrained lateral capacity of rectangular piles under a general loading direction and full flow mechanism. *KSCE J. Civ. Eng.* **2018**, *22*, 2256–2265. [CrossRef]
29. Murphy, G.; Doherty, P.; Cadogan, D.; Gavin, K.G. Field experiments on instrumented winged monopiles. *Proc. Inst. Civ. Eng. Geotech. Eng.* **2016**, *169*, 227–239. [CrossRef]

30. Bisaws, S.K.; Mukherjee, S.; Chakrabarti, S.; De, M. Experimental investigation of free head model piles under lateral load in homogenous and layered sand. *Int. J. Geotech. Eng.* **2015**, *9*, 363–378. [CrossRef]
31. Pan, J.L.; Goh, A.T.C.; Wong, K.S.; Teh, C.I. Model tests on single piles in soft clay. *Can. Geotech. J.* **2000**, *37*, 890–897. [CrossRef]
32. Zhou, H.; Yuan, J.; Liu, H.; Kong, G. Analytical model for evaluating XCC pile shaft capacity in soft soil by incorporating penetration effects. *Soils Found.* **2018**, *58*, 1093–1112. [CrossRef]
33. Martin, C.M. The use of adaptive finite element limit analysis to reveal slip-line fields. *Geotech. Lett.* **2011**, *1*, 23–29. [CrossRef]

Journal of
Marine Science and Engineering

MDPI

Article

Field Campaign on Pressure on the Crown Wall at the Outer Port of Punta Langosteira Breakwater

José Sande [1,*], Maria Graça Neves [2], José-Santos López-Gutiérrez [3,*], M. Dolores Esteban [3], Andrés Figuero [1,*] and Vicente Negro [3]

1 Water and Environmental Engineering Group (GEAMA), Universidad de A Coruña, 15071 A Coruña, Spain
2 Nova School of Science and Technology (FCT NOVA), Universidade NOVA de Lisboa, Campus de Caparica, 2829-516 Caparica, Portugal
3 Environment, Coast and Ocean Research Laboratory (ECOREL-UPM), Universidad Politécnica de Madrid, 28040 Madrid, Spain
* Correspondence: jose.sande@udc.es (J.S.); josesantos.lopez@upm.es (J.-S.L.-G.); andres.figuero@udc.es (A.F.)

Abstract: Punta Langosteira port, located in A Coruña (Spain), was monitored during the winters of 2017 and 2018, measuring wave pressure in the crown wall structure. Furthermore, the metocean variables were measured on a buoy located very close to the breakwater. This paper presents the real pressures measured at the crown wall of the breakwater during different storm events. These values are compared with the results of the application of state-of-the-art equations for the calculation of pressures on crown walls. The results obtained show the behaviour of the pressures with a crown wall fully protected by the armour, as is the case of Langosteira breakwater. Finally, several conclusions are made on the methodology for measuring the pressures using physical models and the relevance of the armour roughness.

Keywords: crown wall; rubble mound breakwater; field campaign; pressure

Citation: Sande, J.; Neves, M.G.; López-Gutiérrez, J.-S.; Esteban, M.D.; Figuero, A.; Negro, V. Field Campaign on Pressure on the Crown Wall at the Outer Port of Punta Langosteira Breakwater. *J. Mar. Sci. Eng.* **2022**, *10*, 1377. https://doi.org/10.3390/jmse10101377

Academic Editor: Theophanis V. Karambas

Received: 24 August 2022
Accepted: 22 September 2022
Published: 26 September 2022

1. Introduction

Port breakwaters are key infrastructures for maritime transport as their presence shelters ships from waves so that ships can safely carry out their operations safely in the port [1,2]. One of the most common types of breakwaters are rubble mound ones, which are an accumulation of material in an orderly way consisting of different components: armour, filters, core, berm, and crown wall. Breakwaters with crown walls require less material, and thus less occupation of the seabed and consequently a lower impact, to achieve the same overtopping rate as those without crown walls [3–5]. In fact, nowadays, it is most common to find and design seawalls with crown walls.

There are different types of crown walls, whether we consider the type of material they are made of or their shape, although they all have the same function. Crown wall design methods are not as widespread as those for other parts of rubble mound breakwaters such as the armour, filter, and so on. As the crown wall is a critical part of the breakwater, a reliable design equation is critical [6,7].

Calculating the forces on the crown walls is a complex task that is critical for an adequate design of the breakwater. There have been several failures in breakwater crown walls, so it is evident that their design must be based on experience. Moreover, the existing equations to design crown walls are based on physical models and the limitations of their use are based on the conditions in the tests used to develop them at the laboratory. So, the first step in the design of crown walls is to select the equation or equations to be used given the geometrical condition of the specific breakwater to be designed and the wave conditions of the zone.

There are several existing equations for crown wall design including Iribarren and Nogales (1964) [8], Jensen (1984) [9] and Bradbury et al. (1998) [10], Günback and Gökçe

(1984) [11], Martin et al. (1999) [12], Berenguer and Baonza (2006) [3,5], Pedersen (1996) [13,14], and Nørgaard et al. (2013) [15]. Most of these equations have been obtained as laboratory test results, each of them covering boundary conditions outside of which the equations are not reliable [16–20].

Nowadays, these formulas are starting to be applied for a great variety of wave conditions and sea levels [21], some of them out of the limit of the application of equations, which is most important considering the phenomenon of climate change [22–26].

Although the ideal is to have field trials at a 1:1 scale to verify formulas, this is not always feasible given the cost of implementation, the time it takes to analyse them to contemplate different storms, and so on. In addition, instrumentation in nature in such an aggressive environment is not always easy to install, maintain, and operate.

This leads to uncertainty when it comes to obtaining the actions to calculate the stability of the crown wall. Validating the most suitable equations with a 1:1 scale test is a must that would help to take an important step in the state-of-the-art of crown wall design. Therefore, the motivation of this research and this article is to analyse with real instrumentation results the reliability of the equation results, as well as to identify the lessons learned from laboratory tests that lead to most of this type of equation, as well as from field measurements at a 1:1 scale crown wall. The real case presented here is the Port of Langosteira, Spain, where one section of the breakwater was instrumented with eight pressure sensors, with five of them located on the front wall of the crown wall and the other three placed on top of the crown wall.

2. Methods

This manuscript aims to analyse field data obtained during eight storm events on the front face of one cross section of the crown wall at the Punta Langosteira Outer Port. Specifically, eight pressure sensors were deployed in one cross section of the main breakwater, five of which allow one to study the pressure on the crown wall front face, exposed to the waves, while the other three are to study the overtopping events on this cross section. In addition, the port has a directional wave buoy in the vicinity of the port, enabling to have simultaneous information of metocean variables.

The data presented here were obtained from the field campaign developed during the winter of 2017/2018. Eight storm events led to no null horizontal pressure registered on the eight pressure sensors. In these events, pressures in the mentioned cross section were recorded.

The methodology applied in this work has two main objectives. The first one is the analysis of the pressure data measured by the pressure sensors located in the front face of the crown wall. This information makes it possible to explain the water behaviour at the crown wall and the causes of the maximum pressure peaks. The second goal is to compare the measured pressure values with the ones resulting from the application of the different existing equations for the design of crown walls, enabling us to analyse their applicability in crown walls with similar characteristics.

Wave conditions and water levels, as well as cross-sectional characteristics, enable the estimation of maximum horizontal pressure using formulas available in the literature. Not all formulas presented in the literature estimate pressures and are valid for crown walls fully protected by the armour layer. Here, the formulas used to estimate the pressure of the front wall are as follows: (1) Günback and Gökçe (1984) [11]; (2) Martin et al. (1999) [12]; and (3) Nørgaard et al. (2013) [15]. Formulas that directly calculate forces, such as Berenguer and Baonza (2006) [3,5] or Jensen (1984) [9] and Bradbury et al. (1988) [10], were not considered.

The following chapters show the results obtained in the two methodological stages described before, as well as the conclusions obtained in the analysis of the breakwater crown wall fully protected by pieces of the protection armour.

3. Description of the Case Study

The cross section of the Punta Langosteira rubble mound breakwater monitored is described. The pressures measured in the front face of the crown wall during the winter of 2017/2018 storm events are presented and analysed here.

3.1. Rubble Mound Breakwater Description

The outer port of Punta Langosteira (Figure 1) is protected by a 3360 m main breakwater, developed in three alignments with a toe at a maximum depth of 45 m. In addition, the port is also protected by a secondary breakwater, the western breakwater, composed of Cubipod® (SATO, Madrid, Spain) [27].

Figure 1. Punta Langosteira Outer Port and the western breakwater. Location of the cross section monitored [28].

The main breakwater has a 2:1 slope, protected by a double layer of cubic blocks of 150 tons in the trunk. The crowning height of the crown wall is +25 m, and it has a zone of 20 m wide partially protected by a perimeter gallery. The upper berm of the armour layer is as high as the crest elevation of the crown wall. In other words, the crown wall is fully protected by the blocks of the armour layer. The crown wall crest level means that the overtopping occurs only for the larger wave heights.

Within the Cormoran project, the engineering firm Eptisa—Ingeniería, Instrumentación y Control (IIC) installed eight pressure sensors in one cross section of the main breakwater called Section PK 2 + 883. There, the structure is composed of a protective armour formed by two layers of cubic blocks of 150 ton placed randomly. Below this layer is a first filter formed by cubic blocks of 15 ton, which in turn rests on a 1-ton rock layer. Both the cross section and the sensors' location are shown in Figure 2. As can be seen in Figure 2, the underlayer consists of a 5-ton rock above the mean sea level and a 1-ton rock beneath it.

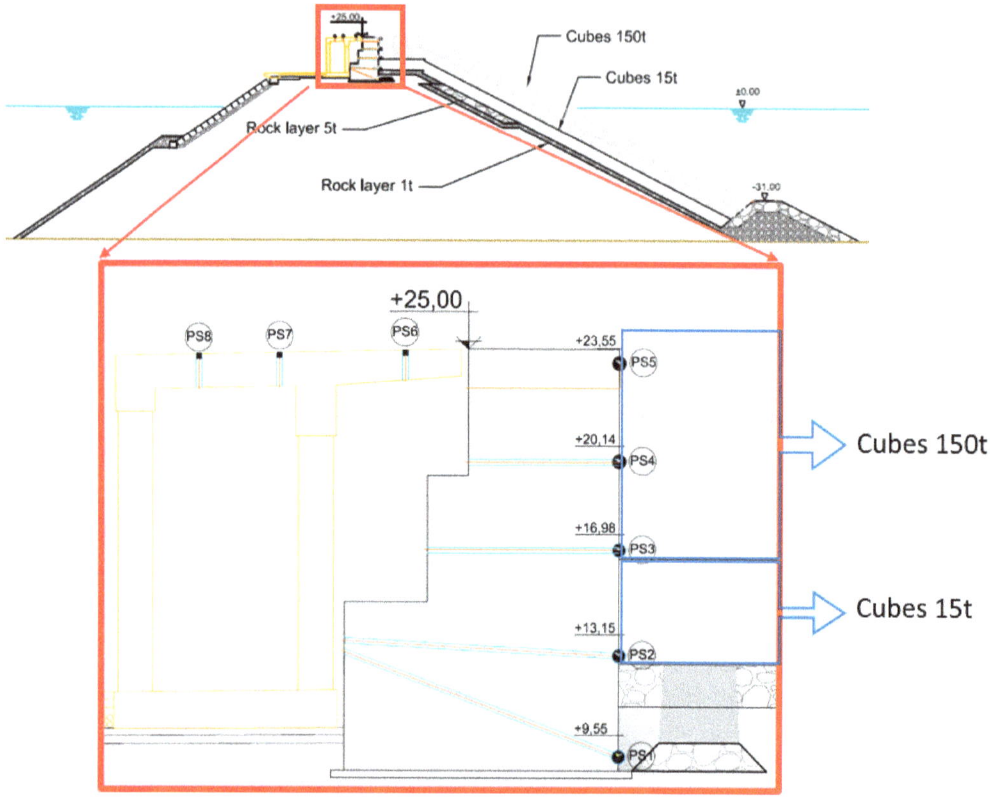

Figure 2. Schematic view of the cross section and the instrumentation.

Pressure sensors were regularly distributed along the front face of the crown wall, starting from the base (PS1) to the top of the crown wall (PS5). There is a cube placed just in front of pressure sensors PS1 and PS2. PS1 is in front of the rubble mound foundation of the cube. Pressure sensor 2 is in front of the first filter and pressure sensors PS3 to PS5 are in front of the main armour layer. All of them are more than 9.55 m above the chart datum (CD). PS6 to PS8 sensors are placed in the crown of the perimeter gallery and have helped to identify if the overtopping phenomenon occurs when the pressure was measured. These sensors give information about overtopping occurrence.

Figure 3 shows the main parameters of the monitored cross section, and Table 1 presents their values. The outer main armour layer has a 1:2 slope. The distance from the crown of the superstructure in relation to the sea level (R_c) and the distance from the superstructure foundation in relation to the sea level (F_c) depends on the water level considered.

Table 1. Values of the main parameters of the cross section.

Parameter	Value
Be	24.05 m (CD)
h	45.8 m
hw	6.0 m
B	10.0 m
Wc	24.0 m (CD)
Wf	9.05 m

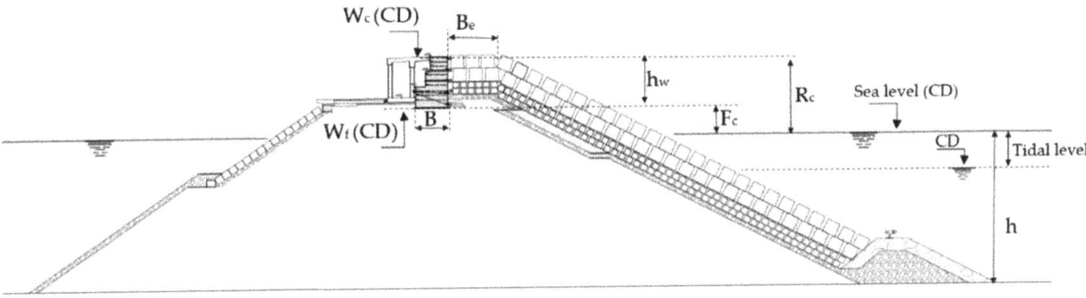

Figure 3. Main parameters of the cross section.

Finally, an important element is the crown wall configuration in the area where the sensors are located. Figure 4 shows an aerial view of the analysis area, where it can be seen how the sensors are placed just behind a block of the armour layer, which will be relevant when analysing the results and drawing conclusions.

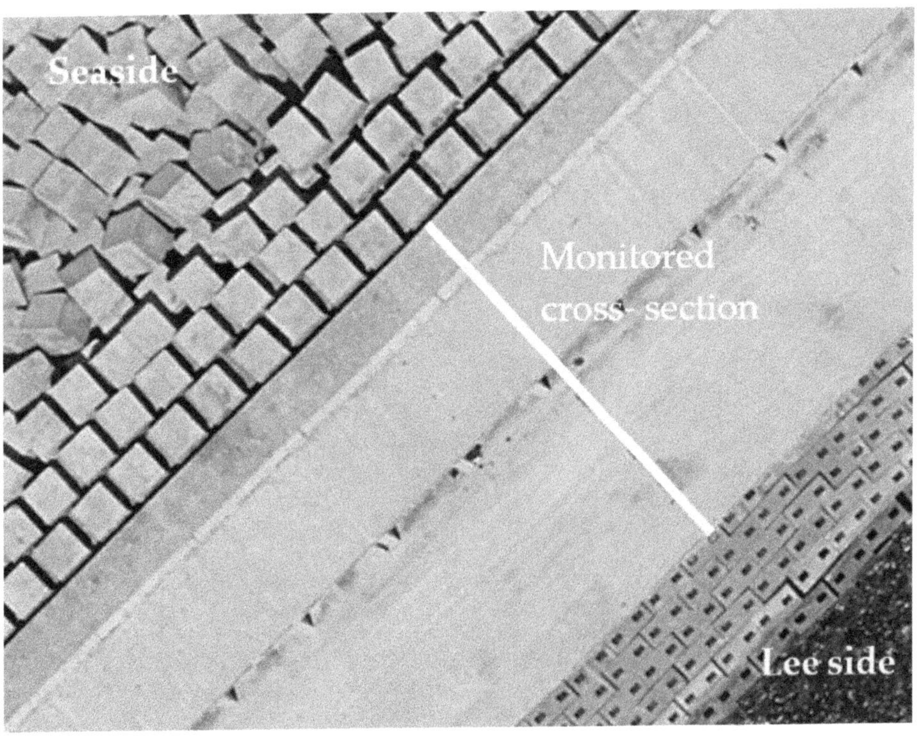

Figure 4. Aerial view of the monitored section.

3.2. Storm Events Analyzed

During the winters of 2017 and 2018, eight events occurred with measurements at pressure sensors. For these eight events, three-hourly values of significant wave height (H_s), maximum wave height (*Hmax*), peak period (*Tp*), and wave directions (*Dir*) were also available. These wave characteristics were measured at a buoy located 2300 m from the breakwater, at a depth of 60 m. The tidal level was obtained using the tide gauge located on the spur breakwater and the range measure of this variable was from −0.5 m to +4.57 m in that location.

Table 2 summarizes the main characteristics of the wave and tidal level during the eight events. As can be seen in Table 2, *Hs* is higher than 7 m and *Hmax* is higher than 9 m. Peak periods, *Tp*, vary from 14.3 s to large values of 20 s. The incident wave direction differs very little between the events, coming from the northwest and attacking the breakwater with an angle less than 20°. The sea level varies from almost high tide during storm events *c* and *d*, to almost low tide during storm events *e* to *h*.

Table 2. Wave and sea level characteristics during the eight storm events.

Storm Event	H_s (m)	*Hmax* (m)	*Tp* (s)	*Dir* (°N)	Tidal Level (m)
a	10.78	14.80	19.20	329.00	2.66
b	9.69	13.81	20.00	327.00	3.16
c	10.65	14.86	18.90	327.00	3.40
d	10.78	13.98	19.70	327.00	3.33
e	8.45	12.49	15.40	334.00	1.46
f	7.21	10.58	14.30	332.00	1.41
g	7.09	9.87	14.30	330.00	1.94
h	9.40	14.10	20.00	319.00	1.06

3.3. *Presure Measured on the Crown Wall Front Wall*

During the eight storm events presented in Table 2, time series of pressure at the pressure sensors with a frequency of 6 Hz were obtained. Figure 5 presents the time series of pressure measured (one hour in total) on the sensors during one of those storm events, storm *h*, and during the maximum pressure event registered.

It can be seen how the pressure recorded by the sensors varies depending on their position, with two different behaviours being detected for different sensors: PS1 and PS2, located in the lower area of the crown wall; and PS3, PS4, and PS5 sensors, located in the higher part of the front wall of the structure. PS1 records the highest-pressure values and for a significantly longer duration than the other sensors. As can be seen in Figure 2, all sensors are always above the sea level. Therefore, the recorded pressure data are due to the runup after wave breaking over the outer armour layer of the breakwater or due to overtopping. In addition, this cross section has several peculiarities, as can be seen in Figure 2. One of them is the presence of a cube with its rubble mound foundation on the base of the crown wall, added during construction, and placed just in front of the pressure sensors PS1 and PS2, which acts as a barrier to the flow in this area, as referred to above. The water accumulated in that area causes large pressure values at PS1, which gradually decrease over the time, depending on the permeability of the core and filter layers. This accumulation of water registered in PS1 is also registered at PS2 sensor, located above the block core level, but with a smaller effect. PS3, PS4, and PS5 sensors have always registered lower pressures than at PS1 and PS2, being more related to water reaching the crown wall, from the armour layer or from the top of the structure owing to overtopping the crown wall.

To illustrate the results obtained, Table 3 presents the maximum pressure measured by each pressure sensor and the concomitant values of the rest of the sensors during the storm of 28 October 2017 in the 19:00 sea state (storm *h* in Table 2).

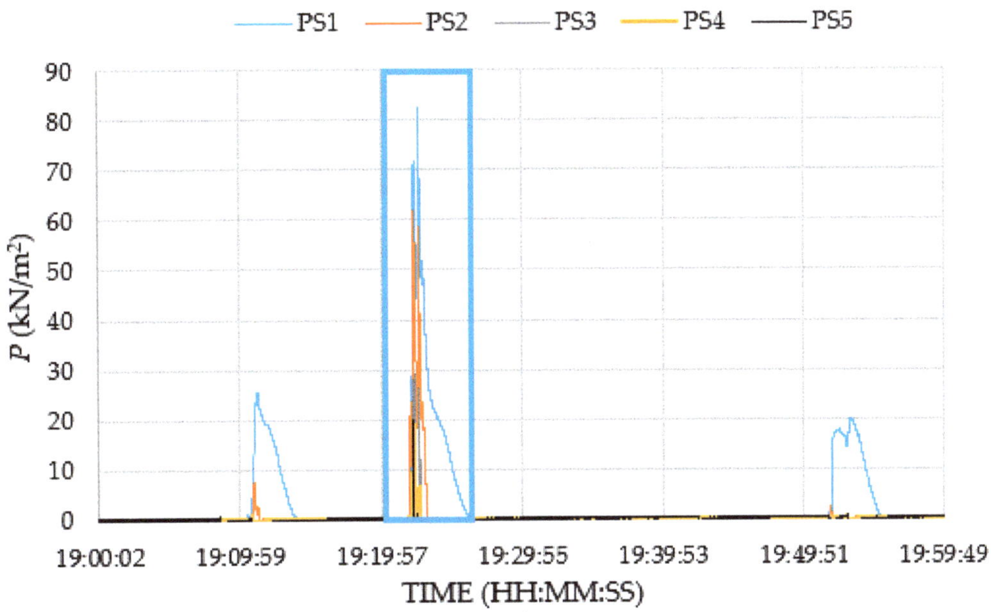

Figure 5. Time series of pressure measured at pressure sensors on 28 October 2017.

Table 3. Maximum pressure measured at the front face of the crown wall during the storm event that occurred on 28 October 2017 at the 19:00 sea state.

Time (h:min:s)	PS1 (kN/m^2)	PS2 (kN/m^2)	PS3 (kN/m^2)	PS4 (kN/m^2)	PS5 (kN/m^2)
19:22:12	35.38	39.89	12.67	7.49	**19.80**
19:22:14	58.87	36.04	20.98	**18.86**	15.75
19:22:17	70.40	53.75	**29.00**	10.71	−1.57
19:22:13	52.66	**61.64**	15.71	12.47	16.84
19:22:37	**82.59**	57.12	25.98	6.61	−1.76

As can be seen, each sensor reaches its own maximum pressure first at PS5 and at least at PS1. This seems to be caused by waves that overtopped the structure (and approach this section from the top to the bottom of the crown wall) crown wall and not by the impact of the wave impinging the structure. Moreover, as referred before, the crown wall is fully protected by the berm of the outer armour of the breakwater, so waves do not have direct impact to the crown wall. In fact, the same behaviour was found in all storm events analysed.

Table 4 summarizes the maximum pressure measured by the pressure sensors in each of the storm events divided by the *Hmax* obtained from the buoy; see Table 2.

Table 4. Maximum pressure measured at the sensors in each storm event.

Storm Event			*P*/*Hmax* (kN/m^3) at		
	PS1	PS2	PS3	PS4	PS5
a	1.28	0.20	0.12	0.04	0.14
b	2.12	0.11	0.06	0.04	0.40
c	1.13	0.15	0.12	0.09	0.29
d	4.09	2.50	0.58	0.08	0.10
e	2.17	0.28	0.06	0.04	0.10
f	0.20	0.05	0.05	0.07	0.15
g	0.72	0.08	0.04	0.05	0.10
h	5.86	4.37	2.06	1.34	1.40

As can be seen, in general, pressures are much higher in PS1 and PS2, as explained above. However, at PS3 to PS5, the pressures presented the same order of magnitude, all with values less than 0.58 kN/m^3. The only exception is storm *h*, with the highest values of $P/Hmax$ obtained in all sensors. Storm *d* presents high values at SP1, reducing at PS2 and PS3 and having the same order of magnitude of the other storms at PS4 and PS5.

Figure 6 graphically shows the maximum pressures of each sensor in the eight storms analysed. Storms *a*, *b*, *c*, *d*, and *h* have, in general, the highest recorded $P/Hmax$ values. This is because they are the events with the most demanding wave conditions. However, at PS1, the storms with the highest *Hmax*, storms *a* and c, present lower values than in other storms with lower values of *Hmax*. The difference in results could also be because the measurements have been taken in just one cross section in a breakwater of more than 3000 m in length, where the maximum pressure values could occur in another zone, especially if it depends on overtopping. It is important to note that these eight storms analysed here have coincided with overtopping events on the instrumented section verified by no null pressure at the sensors placed at the crown of the gallery of the breakwater.

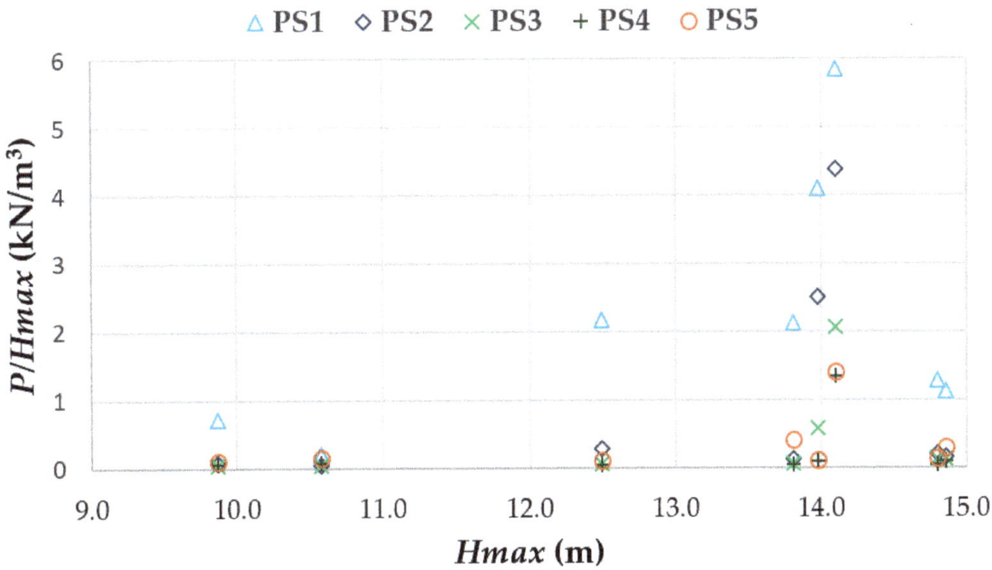

Figure 6. Maximum pressure measured on the storm events in each sensor.

Finally, in view of the analysis carried out on the measurements of the pressure sensors, as well as the conditioning factors owing to the section under analysis, it has been decided to select PS3, PS4, and PS5 sensors for the comparison of the maximum pressure measured with the resulting pressures estimated by state-of-the-art equations. In fact, the existing equations do not contemplate the effect of the placement of blocks that exist here in the core in front of the crown wall, which certainly affects the measurements on PS1 and PS2 sensors. As this structure has been found to have no influence on the measurements of the other three sensors, PS3 to PS5, the following study was proposed using only the three upper sensors.

4. Results

Pressure on the vertical face of the crown wall, calculated using formulas for the different storm events presented in Section 3.2, are presented here and compared with the pressure data obtained during the main field campaign at the main Punta Langosteira breakwater. Maximum pressures were obtained with formulas and compared with those measured for those pressure sensors selected on the vertical face of the crown wall during the field campaign.

4.1. Equation Application

The pressures on the superstructures of the rubble mound breakwaters fully protected by the armour layer are essentially due to the interaction between waves, filtered by the armour layer, and the superstructure, or due to overtopped waves reaching the superstructure.

4.2. Comparison of Pressure Data Measured and Estimated by Formulas

Günback and Gökçe (1984) [11] and Martin et al. (1999) [12] give null dynamic pressures for all storm events tested. In fact, run-up calculated by the formulas leads to non-overtopping conditions, contrary to that occurring in the field, where, for all storm events, overtopping had occurred. As the formula relates the different pressures with the run-up, R_u, and to the height of the armour berm, A_c, it leads to null dynamic pressure in the protected zone.

Martín et al. (1999) [12] is valid if the wave breaks before the superstructure of the bottom breaks ($\xi_{0p} > 3$), which occurs for all events tested when the range of $\xi_{0p,max}$ varies from 3.3 to 4.4. However, the data exceed the limits of expressions for calculating the λ parameter, representative of the decrease in dynamic pressure in the protected zone owing to the constituent materials of the main outer armour, because H/Lp is less than 0.03. So, the formula was applied outside its range of validity. Günback and Gökçe (1984) [11] do not impose validity limits on the equation. However, as referred to in Pereira et al. (2021) [21], the geometry considered in all of the formulas has the armour berm below the crest of the crown wall, and here, the crown wall is completely protected by the outer armour composed of cubes, which can lead to more overtopping conditions than with other blocks, which leads to larger wave dissipation.

Nørgaard et al.'s (2013) [15] formula was the only one that gives no null pressures for some of the storm events analysed here and, consequently, is the only one presented here. Sande et al. (2019) [27] used this formula to compare pressures measured with a physical model for this same port for the western breakwater (secondary breakwater), even with a slightly different cross section than the constructed one. They concluded that the formula represents the same behaviour of the wave forces on the crown wall, overestimating the pressures by not considering the effect of obliquity.

In 2013, Nørgaard, Lykke-Andersen, and Burcharth [15] proposed a modification to the formulation of Pedersen (1996) [13] with the aim of validating it for small depths, as it was originally limited to large and intermediate depths. The proposed modifications were based on the results of 162 tests on small-scale models and irregular wave action, considering agitation conditions at large and small depths and using the same experimental equipment used by Pedersen (1996) [13].

The first modification to the original formulation of Pedersen (1996) was made with respect to the runup only exceeded by 0.1% of the waves of the sea state considered, $R_{u,0.1\%}$ level. According to Nørgaard et al. (2013) [15], the value of $R_{u,0.1\%}$ would now be given by the following:

$$R_{u,0.1\%} = \begin{cases} 0.603\ H_{0.1\%}\ \xi_{0m} & \text{if } \xi_{0m} \leq 1.5 \\ 0.722\ H_{0.1\%}\ \xi_{0m}^{0.55} & \text{if } \xi_{m} > 1.5 \end{cases} \tag{1}$$

where $H_{0.1\%}$ is the maximum wave height only exceeded by 0.1% of the waves of the sea states considered, given by Equation (2) according to Rayleigh's distribution:

$$H_{0.1\%} = \frac{H_s}{0.538} \tag{2}$$

Knowing the value of $R_{u,0.1\%}$, and ignoring energy losses by friction, the ascending velocity of the water mass when run-up, v_0, is given by the following:

$$v_0 = \sqrt{2\,g\,(R_{u,0.1\%} - A_c)} \tag{3}$$

The dynamic pressure in the protected area of the crown wall exceeded by only 0.1% of the waves of the sea states considered, $P_{d(p),0.1\%}$, is obtained by the following equation:

$$P_{d\,(p),0.1\%} = 0.5\,\rho_w g \frac{v_0^2}{2g} = \rho_w g (R_{u,0.1\%} - A_c) \tag{4}$$

The formula is valid for $R_c/H_s < 2$, which occurs in three (storms a, c, and d) of the eight storm events analysed, although storm b has a close value, as shown in Figure 7.

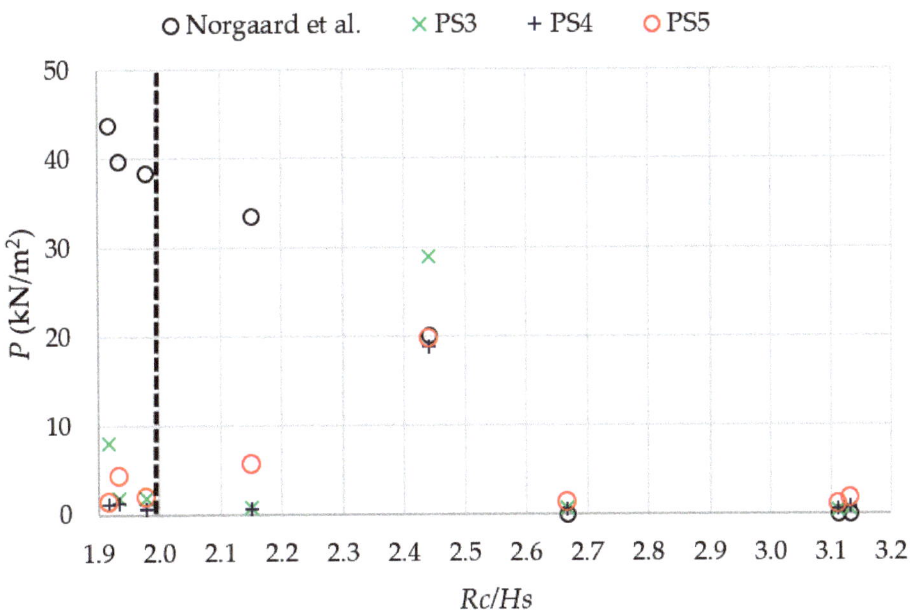

Figure 7. Horizontal pressure calculated by Nørgaard et al. (2013) [15] and measured at sensors PS3 to PS5.

Table 5 summarizes the results obtained by the formula and compared with the maximum pressure obtained in pressure sensors PS3 to PS5, as the number of pressure events is very limited, meaning $P_{0.1\%}$ cannot be calculated.

Table 5. Maximum pressure calculated by Nørgaard et al. (2013) [15] and measured in the sensors.

Storm Event	R_c/H_s (m)	$P_{\text{Norgaard et al.}}$ (kN/m^2)	Maximum Pressure (kN/m^2) at		
			PS3	PS4	PS5
a	1.98	38.51	1.83	0.59	1.99
b	2.15	33.53	0.78	0.57	5.58
c	1.93	39.80	1.82	1.26	4.31
d	1.92	43.88	8.06	1.18	1.44
e	2.67	0.00	0.69	0.49	1.24
f	3.13	0.00	0.48	0.69	1.62
g	3.11	0.00	0.40	0.48	0.97
h	2.44	20.08	29.00	18.85	19.80

Figure 7 presents the results obtained by the formula and the maximum pressure measured at pressure sensors PS3 to PS5 as a function of R_c/H_s.

The storm events presented different behaviours. For storms *a* to *d*, pressures estimated by the formula of Nørgaard et al. (2013) [15] are much higher than those measured at PS3 to PS5. For storms *e* to *g*, Nørgaard et al. (2013) [15], similar to Günback and Gökçe (1984) [11] and Martin et al. (1999) [12], give null pressures, contrary to the pressure sensors, which measured pressures up to 1.62 kN/m^2, with higher values being obtained at PS5. For storm *h*, th formula gives a maximum pressure similar to that measured at PS4 and PS5.

Storms *a* to *d* are almost within the application range of the formula for R_c/H_s, but as seen in Section 3.3, the maximum pressures occur first in PS5 and then in the lower ones, because the protection of the armour layer prevents the waves from directly impacting the sensor. Consequently, it seems that the registered pressures are mainly due to overtopping, leading to water penetrating between the blocks and the crown wall. This would explain why pressures are lower than those estimated by the equations and are higher in the pressure sensor located at a higher position and, in general, progressively reducing their values on the sensors located lower in the crown wall front face.

On the other hand, in storm events *e* to *g*, outside the validation limits of the equation, the pressures estimated by the formula are zero, corresponding to the low pressures recorded by the pressure sensors and to the less energetic weather conditions analysed for the breakwater under study.

The explanation that the authors have found for this different behaviour of the formula is the overtopping differences and three-dimensional effects due to the obliquity with which the waves arrive at the breakwater. To illustrate the referred effect of incident wave obliquity on the overtopping and measurements obtained from the sensors, Figure 8 shows overtopping events during storms *c* and *d*, both with high values of H_s and Tp and the same incident wave direction (see Table 2). In this figure, a red line has been marked at the monitored section. The image on the left shows the overtopping during storm event *c*, which occurs only at the monitored section. However, in storm *d*, overtopping occurs in several sections along the breakwater with different intensity. This explains why the pressure measured at only one monitored section could not reflect the overtopping that occurs over the levee. Consequently, it does not reflect either the pressure affected by overtopping at a crown wall or in a three-dimensional environment as in the field.

(a)  (b) 

Figure 8. Aspect of overtopping event during (**a**) storm *c* and (**b**) storm *d* (monitoring section represented by a red line).

5. Conclusions

This study shows the field results of pressures on the crown wall front face fully protected by the armour layer. The existing equations were obtained through two-dimensional physical model tests, where the sensor is usually left directly exposed to the incidence of waves to measure the pressures on the crown wall. Moreover, when a crown wall of these characteristics is designed, it is assumed that pressure is constant along the whole structure. These facts become relevant when proposing crown wall designs with this type of breakwater typology.

The results obtained raise the importance of the correct location of the sensors in both laboratory and field campaigns. In laboratory tests, it is usual that different alignments of sensors are placed to analyse the pressure generated in the whole crown wall. However, it would be interesting to analyse the influence of their placement on crown walls protected by an armour layer with different porosities. Moreover, when planning field campaigns such as the current ones, it is important to locate the sensors in such a way that they allow the maximum possible information to be obtained.

An important aspect detected in the research is that, when the aerial view of the monitored section was analysed, it was noted that the sensors are placed behind a block of the armour layer. This is critical when an instrumentation is installed in a real section.

Moreover, this case study has peculiar characteristics that affect the measurements: there is a cubic block in front of the lowest level sensor position that generates a barrier to the flow, leading the accumulation of water in that area. This sensor is where the highest-pressure values were recorded and for a significantly longer duration than the other sensors. The accumulation of water registered in this sensor is not enough to significantly affect the measurement of the other sensors located above, located at the core level, and does not seem to affect other sensors, being more related to wave action crown wall. The maximum pressure occurs firstly in the sensor located at the highest level and is progressing to the bottom; it can be due to overtopping crown wall.

Pressures measured during eight storm events were compared to the state-of-the-art equations. In all storm events, pressures are no null and overtopping occurs. The equations of Günback and Gökçe (1984) [11] and Martin et al. (1999) [12] give null pressure results in all cases tested here. Nørgaard et al.'s (2013) [15] formulas were the only ones that give no null pressures, and thus were the ones used in this research. The cases with low pressures recorded by the pressure sensors and corresponding to less energetic weather conditions were outside the range of validity of the formulas. For those cases, the formula gives null pressures. For cases where higher pressures were measured, the formula of Nørgaard et al. (2013) [15] is valid and gives higher maximum pressures than the measured ones. This behaviour is possible because the crown wall is totally protected by the armour layer, and the pressure is measured at only one monitored section, not reflecting the overtopping that occurs over the whole breakwater. The results raise the importance of the correct location of the sensors in both the laboratory and field campaigns in such a way that they allow the maximum possible information to be obtained.

Author Contributions: Conceptualization, J.S., M.G.N., J.-S.L.-G., M.D.E., A.F. and V.N.; methodology, J.S., M.G.N., J.-S.L.-G. and M.D.E., software, J.S. and M.G.N.; validation, J.S., M.G.N., J.-S.L.-G., M.D.E., A.F. and V.N.; formal analysis, J.S., M.G.N., J.-S.L.-G., M.D.E., A.F. and V.N.; investigation, J.S., M.G.N., J.-S.L.-G., M.D.E., A.F. and V.N.; writing—original draft preparation, J.S., M.G.N., J.-S.L.-G. and M.D.E.; writing—review and editing, J.S., M.G.N., J.-S.L.-G., M.D.E., A.F. and V.N.; supervision, J.S., M.G.N., J.-S.L.-G., M.D.E., A.F. and V.N. All authors have read and agreed to the published version of the manuscript.

Funding: Acknowledgement of the scientific and financial support by the Portuguese Foundation for Science and Technology (FCT) for project BSAFE4SEA—Breakwaters SAFEty control through a FORecast and decision support SystEm Analysis, Ref. PTDC/ECI-EGC/31090/2017.

Institutional Review Board Statement: Not applicable.

Informed Consent Statement: Not applicable.

Data Availability Statement: Data available upon request owing to privacy restrictions. The data are not publicly available for confidentiality reasons.

Acknowledgments: The authors are grateful to the Port Authority of A Coruña (Spain) and the Eptisa—Ingeniería, Instrumentación y Control (IIC) company for their kind collaboration and for sharing the information obtained in the Cormoran project. J.-S.L.-G. and M.D.E. are very grateful to LNEC for giving them the opportunity to perform research in the Hydraulics and Environment Department.

Conflicts of Interest: The authors declare no conflict of interest. The funders had no role in the design of the study; in the collection, analyses, or interpretation of data; in the writing of the manuscript; or in the decision to publish the results.

References

1. Li, J.; Cheng, J.; Liu, S. An experimental study on the hydrodynamic performance of the twin vertical baffles underflow breakwater. *Ocean Eng.* **2022**, *256*, 111566. [CrossRef]
2. Mahmoudof, S.M.; Eyhavand-Koohzadi, A.; Bagheri, M. Field study of wave reflection from permeable rubble mound breakwater of Chabahar Port. *Appl. Ocean Res.* **2021**, *114*, 102786. [CrossRef]
3. Berenguer, J.M.; Baonza, A. On the design of crown walls of rubble mound breakwaters. In Proceedings of the 31st PIANC Congress, Estoril, Portugal, 14–18 May 2006; pp. 608–624.

4. Lacasa, M.C.; Esteban, M.D.; López-Gutiérrez, J.S.; Negro, V.; Zang, Z. Feasibility study of the installation of wave energy converters in existing breakwaters in the north of Spain. *Appl. Sci.* **2019**, *9*, 5225. [CrossRef]
5. Han, X.; Jiang, Y.; Dong, S. Wave forces on crown wall of rubble mound breakwater under swell waves. *Ocean Eng.* **2022**, *259*, 111911. [CrossRef]
6. Negro, V.; Martín-Antón, M.; del Campo, J.M.; López-Gutiérrez, J.S.; Esteban, M.D. Crown walls in mass and reinforced concrete: The way to aesthetics in maritime works. In *Coasts, Marine Structures and Breakwaters 2017: Realising the Potential*; ICE Publishing: London, UK, 2018.
7. Martín-Antón, M.; Negro, V.; del Campo, J.M.; López-Gutiérrez, J.S.; Esteban, M.D. The Impact of Public Works in Spain: Natural, constructed and destroyed landscape. *J. Constr.* **2017**, *16*, 82–91.
8. Iribarren, R.; Nogales, C. *Obras Marítimas*; Dossat Ed: Madrid, Spain, 1964.
9. Jensen, O.J. *A Monograph on Rubble Mound Breakwaters*; Danish Hydraulic Institute: Hørsholm, Denmark, 1984.
10. Bradbury, A.P.; Allsop, N.W.H.; Stephens, R.V. *Hydraulic Performance of Breakwater Crown Walls*; Report SR146; HR Wallingford: Wallingford, UK, 1988.
11. Günbak, A.R.; Gökçe, T. Wave screen stability of rubble-mound breakwaters. In Proceedings of the International Symposium of Maritime Structures in the Mediterranean Sea, Athens, Greece, 22–25 November 1984; pp. 2.99–2.112.
12. Martín, F.L.; Losada, M.A.; Medina, R. Wave loads on rubble mound breakwater crown walls. *Coast. Eng.* **1999**, *37*, 149–174. [CrossRef]
13. Pedersen, J. Wave Forces and Overtopping on Crown Walls of Rubble Mound Breakwaters: An Experimental Study. Ph.D. Thesis, Aalborg University, Aalborg, Denmark, 1996.
14. Pedersen, J.; Burcharth, H.F. Wave force on crown walls. In Proceedings of the 23rd International Conference on Coastal Engineering, ASCE, Venice, Italy, 4–9 October 1992; pp. 1489–1502.
15. Nørgaard, J.Q.H.; Andersen, T.L.; Burcharth, H.F. Wave loads on rubble mound breakwater crown walls in deep and shallow water wave conditions. *Coast. Eng.* **2013**, *80*, 137–147. [CrossRef]
16. Negro, V.; López-Gutiérrez, J.S.; Polvorinos Flors, J.I. Comparative study of breakwater crown wall—Calculation methods. *Proc. Inst. Civ. Eng. Marit. Eng.* **2013**, *166*, 25–41.
17. Molines, J.; Herrera, M.P.; Medina, J.R. Estimation of wave forces on crown walls based on wave overtopping rates. *Coast. Eng.* **2018**, *132*, 50–62. [CrossRef]
18. CIRIA/CUR/CETMEF. Physical processes and design tools: Crown walls. In *The Rock Manual: The Use of Rock in Hydraulic Engineering*, 2nd ed.; CIRIA: London, UK, 2007.
19. United States Army Corps of Engineers (USACE). *Coastal Engineering Manual*; USACE: Washington, DC, USA, 2008.
20. Losada, M.A. Recent developments in the design of mound breakwaters. In *Handbook of Coast and Ocean Eng*; Gulf Publishing: Houston, TX, USA, 1990; pp. 939–1050.
21. Pereira, F.; Neves, M.G.; López-Gutiérrez, J.-S.; Esteban, M.D.; Negro, V. Comparison of Existing Equations for the Design of Crown Walls: Application to the Case Study of Ericeira Breakwater (Portugal). *J. Mar. Sci. Eng.* **2021**, *9*, 285. [CrossRef]
22. Webpage of Intergovernmental Panel on Climate Change (IPCC). Available online: www.ipcc.ch (accessed on 3 January 2021).
23. Diez, J.J.; Esteban, M.D.; Paz, R.; López-Gutiérrez, J.S.; Negro, V.; Monnot, J.V. Urban coastal flooding and climate change. *J. Coast. Res.* **2011**, *64*, 205–209.
24. Negro, V.; López-Gutiérrez, J.S.; Esteban, M.D.; del Campo, J.M.; Luengo, J. Action strategy for studying marine and coastal works with climate change on the horizon. *J. Coast. Res.* **2018**, *85*, 506–510. [CrossRef]
25. Negro, V.; López-Gutiérrez, J.S.; Esteban, M.D.; Matutano, C. An analysis of recent changes in Spanish Coastal Law. *J. Coast. Res.* **2014**, *70*, 448–453. [CrossRef]
26. Díaz-Sánchez, R.; López-Gutiérrez, J.S.; Lechuga, A.; Negro, V.; Esteban, M.D. Direct estimation wave setup as a medium level in swash. *J. Coast. Res.* **2013**, *65*, 201–206. [CrossRef]
27. Sande, J.; Peña, E.; Maciñeira, E.; Figuero, A. 3D physical modelling for checking Level III probabilistic design in breakwaters: A case study in Punta Langosteira Port, Spain. *Coast. Eng. J.* **2019**, *61*, 460–471. [CrossRef]
28. Webpage of Port of A Coruña. Available online: http://www.puertocoruna.com/es (accessed on 29 July 2022).

Journal of
Marine Science and Engineering

MDPI

Article

Validation and Comparisons of Methodologies Implemented in a RANS-VoF Numerical Model for Applications to Coastal Structures

Eric Didier [1,*] and Paulo R. F. Teixeira [2]

1 Hydraulics and Environment Department, National Laboratory for Civil Engineering, Av. do Brasil 101, 1700-066 Lisboa, Portugal
2 Engineering School, Federal University of Rio Grande, Av. Itália, km 8, Campus Carreiros, Rio Grande 96203-900, RS, Brazil
* Correspondence: edidier@lnec.pt

Abstract: Methodologies to be used in numerical models based on Reynolds-averaged Navier–Stokes (RANS) equations and the volume of fluid (VoF) to deal with waves over coastal structures, which involve wave breaking and overtopping and porous structures, are shown in this manuscript. Two turbulence models, k-ε NLS (non-linear Reynolds stress tensor) and k-ε SCM (stabilized closure model), that are used to avoid the growth of the eddy viscosity, are implemented in the FLUENT® numerical model. Additionally, equations of momentum and turbulence models are adapted to simulate porous media of coastal structures. Comparisons of performance of k-ε NLS, k-ε SCM and standards k-ε and k-ω SST models in several classical cases of regular and random waves on coastal structures are carried out. It was noticed that the standard k-ε turbulence model, and k-ω SST with less intensity, over-predicted eddy viscosity, caused the decay of the free surface elevation and under-estimated wave overtopping discharge. k-ε NLS and k-ε SCM turbulence models have similar performance, with slightly better results of k-ε NLS, showing good agreement with experimental ones.

Keywords: RANS-VoF; turbulence models; coastal structures; porous breakwater; wave breaking; wave overtopping

Citation: Didier, E.; Teixeira, P.R.F. Validation and Comparisons of Methodologies Implemented in a RANS-VoF Numerical Model for Applications to Coastal Structures. *J. Mar. Sci. Eng.* **2022**, *10*, 1298. https://doi.org/10.3390/jmse10091298

Academic Editors: M. Dolores Esteban, José-Santos López-Gutiérrez, Vicente Negro and M. Graça Neves

Received: 10 August 2022
Accepted: 29 August 2022
Published: 14 September 2022

1. Introduction

In the last decades, several researchers have investigated wave–structure interactions by means of numerical simulations. However, this task still imposes many difficulties due to complex phenomena which this type of case involves, such as wave reflection, wave breaking, wave run-up/down, wave–porous structure interaction and wave overtopping. Each physical process requires the different abilities of numerical models. Recently, models based on Reynolds-averaged Navier–Stokes (RANS) equations have been developed, validated and applied to this type of problem; however, they still show some issues in correctly predicting all phenomena.

Wave breaking requires an adequate turbulence closure model and the free surface tracking method. Sakai et al. [1] carried out the earliest studies involving wave breaking by using the marked-and-cell (MAC) method, developed by Harlow and Welch [2]. Later, Lemos [3] applied the volume of fluid (VoF) method, originally developed by Hirt and Nichols [4]. Lin and Liu [5] developed a RANS model by using an algebraic Reynolds stress k-ε model in combination with the VoF method, which was validated to spilling breaking waves on a sloping beach, based on experiments of Ting and Kirby [6]. The authors observed that turbulence levels near the breaking point were significantly overestimated compared with the experimental data. Bradford [7] employed the RANS-based Flow-3D model and different turbulence models to simulate the spilling and plunging breakers of Ting and Kirby [6] experiments. The author noticed that turbulent kinetic energy in the wave crest prior to breaking is overpredicted using a k-ε model, resulting in an underprediction of the breaking wave height, and suggested the use of k-ω model. Mayer and Madsen [8] modified the k-ω model, using the vorticity of the mean flow instead of the local mean velocity gradient to estimate the production of turbulent kinetic energy. Thus, the generation of the turbulent kinetic energy was overcome in the potential flow region (wave propagation zone outside the surf zone) and the wave damping over the length of the flume was avoided. Jacobsen et al. [9] implemented in the OpenFOAM model a two-phase flow solver and validated this technique in breaking waves by using the modified k-ω turbulence model proposed by Mayer and Madsen [8]. The authors showed the importance of the mesh aspect ratio on numerical results. Brown et al. [10] used different turbulence models in OpenFOAM to simulate both spilling and plunging breakers and observed the necessity of including the density explicitly in the turbulence transport equations. In addition, they concluded that the overall best model is the nonlinear k-ε model; however, the k-ω model showed improvements in relation to the solution without the turbulence model. Devolver et al. [11] used the OpenFOAM model by using both k-ω and k-ω SST turbulence models to analyze the performance of a buoyancy-modified turbulence model in wave breaking in a numerical wave flume. The authors concluded that buoyance-modified turbulence models significantly reduce the common overestimation of the turbulent kinetic energy in the flow field. Chella et al. [12,13] used the level set method and the k-ω model by means of the REEF3D model, in which a turbulence damping scheme was applied to avoid unphysical turbulence production at the interface between water and air. The authors noticed an underprediction of the turbulent kinetic energy levels in the surf zone. Larsen and Fuhrman [14] developed a formulation of the k-ω turbulence model closure to stabilize the model in nearly potential flow regions and overcome the overestimate turbulence levels in pre- and post-breaking. The authors extended the stabilized closure model formulation to other turbulence models.

The wave overtopping on coastal structures involves simultaneously wave run-up/down and wave breaking. It depends on the structure topology, geometry, material characteristics, incident wave conditions and foreshore bathymetry [15]. Numerical models must be able to simulate these complex physical processes that include turbulence, eddy vortices and strong wave–structure interactions [16]. Soliman et al. [17] investigated regular and random wave overtopping over coastal impermeable structures by using the model developed by Lin and Liu [5], which is based on RANS equations, and employing the k-ε turbulence and nonlinear Reynolds stress models, as well as the VoF method. The authors concluded that empirical equations of overtopping, used for design purposes, underestimate the amount of wave overtopping due to random waves. Tuan and Thin [18] compared both non-linear shallow water (developed by Tuan and Oumeraci [19]) and RANS models (COBRAS-UC [15]) in regular and random wave overtopping on impermeable sea dikes with low vertical crown-walls. The authors concluded that the non-linear shallow water model is efficient and reliable to estimate wave overtopping discharges at sea dikes with relatively low walls. However, the RANS-VoF model has more accuracy in high walls. Recently, de Finis et al. [20] applied the IH2VOF model to investigate the impact of waves on a storm wall for a dike-promenade structure. Several tests were carried out to study the dependence of the wave forces with the wave characteristics and structural dimensions, and a new empirical design formula was proposed.

Many coastal porous structures are applied to protect coastal infrastructures and natural beaches, including submerged breakwaters that protect coasts with minimal degradation (sometimes enrichment) of the marine environment [21]. The wave–porous structure interaction has been investigated by several researchers to understand the flow motion inside and outside of the permeable structure. The study of the action of waves on rock (rubble) mound breakwaters is difficult, because the flow within interstices of blocks, with very complex geometry, is strongly non-stationary and involves moment advection, free boundary and turbulence and air–water interaction [22]. In addition, the small scales of the porous medium require a prohibitive spatial and time discretization of numerical models.

To circumvent these difficulties, van Gent [23], Lin and Liu [5], Liu et al. [24], Nakayama and Kuwahara [25], Getachew et al. [26] and Pedras and de Lemos [27], proposed different strategies to develop momentum equations extended by Darcy and Forchheimer terms to consider the drag caused by the porous structure. Van Gent [23] developed a 1DH model based on the non-linear shallow water equations and a two vertical dimensional (2DV) model based on RANS equations to simulate the wave interaction with permeable and impermeable coastal structures. Hur [28] and Hur et al. [29] developed and applied this methodology of adding drag and inertia terms in the Navier–Stokes equations to simulate flows in porous media to 2D/3D cases.

Nowadays, the most used method applied in porous structures is the volume averaging method, in which Navier–Stokes based models are applied to volumes whose scales are larger than scales of porous structures that compose the porous media [30,31]. These models accurately represent nonlinearities, frequency dispersion and wave breaking, overcoming restrictions of simplified models. Hsu et al. [32] developed a model based on the volume-averaged Reynolds-averaged Navier–Stokes (VARANS) equations, in which the turbulence models are extended by additional terms that describe the porosity effect. Garcia et al. [33] applied this methodology to low-crested breakwaters and Lara et al. [34] analyzed cases considering random waves. Losada et al. [15] investigated the interaction of regular and random waves with a rubble mound breakwater by using the COBRAS-UC numerical model. The authors observed problems with computational efficiency, due to the mesh resolution and number of waves to be simulated. In addition, they noticed difficulties in setting empirical drag coefficients of linear and non-linear terms. Del Jesus et al. [35] reformulated VARANS equations developed by Hsu et al. [32] and developed the IH-3VOF model to simulate two-phase flows within porous media. Jensen et al. [36] revised and re-implemented VARANS equations in the OpenFOAM code. Higuera et al. [37] also implemented in this code a hybrid methodology (2D-3D) to optimize the simulation time needed to assess the three-dimensional effect wave interaction with coastal structures. Vanneste and Troch [38] validated the FLOW-3D numerical model, based on RANS equations with additional drag force term in the momentum equations [29] and the VoF method in large physical model tests on a multi-layered breakwater section.

This study aims to compare the performance of methodologies applied to wave–coastal structure interaction problems implemented in a RANS-VoF numerical model. The FLUENT® numerical model is used for this purpose, in which some methods have already been implemented by the authors to simulate wave propagation and wave interaction with coastal structures, such as wave generation, wave active absorption technique [39–41] and random wave generation by JONSWAP/TMA spectrum [41], among others. Additionally, two new major implementations are carried out in this study: the extended Darcy–Forchheimer formulation, which is used to deal with transient flows in porous medium [23]; and closure turbulence models to avoid the growth of turbulent kinetic energy and eddy viscosity and enable nearly constant wave propagation for longtime simulation in a long wave flume, using the relatively recent k-ε closure turbulence model, proposed by Larsen and Fuhrman [14], and the k-ε non-linear Reynolds stress tensor, which was developed and used by Shih et al. [42] and Lin and Liu [5]. Both implementations are fundamental in dealing with simulations of waves over impermeable and porous structures that involve wave breaking, wave run-up/down and wave overtopping. Results obtained with these two turbulence models are compared to those obtained with standard k-ε and k-ω SST turbulence models for several configurations of interaction between waves and structures.

2. Methodology

In this section, equations and methodologies of the FLUENT® model [43] used in the study cases are shown, including momentum equations, turbulence model equations, porous medium flow equations and numerical conditions. Some are implemented in the original code by means of user-defined functions (UDF).

2.1. Standard Governing Equations of the FLUENT Model

The RANS FLUENT® model employs the finite volume method to discretize the continuity and momentum equations, which, considering 2D and incompressible flows, are given by [44,45]:

$$\frac{\partial u_i}{\partial x_i} = 0 \quad (1)$$

$$\frac{\partial}{\partial t}(\rho u_i) + u_j \frac{\partial}{\partial x_j}(\rho u_i) = -\frac{\partial p}{\partial x_i} + \rho g_i + \frac{\partial \tau_{ij}}{\partial x_j} - \frac{\partial \rho\left(\overline{u_i' u_j'}\right)}{\partial x_j} + S_{Mi} \quad (2)$$

$$\tau_{ij} = \mu\left(\frac{\partial u_i}{\partial x_j} + \frac{\partial u_j}{\partial x_i}\right) \quad (3)$$

where $i, j = 1, 2$, ρ is the specific mass, g_i are components of gravity acceleration, u_i are components of velocity, p is the pressure, μ is the dynamics viscosity, τ_{ij} is the viscous stress tensor, S_{Mi} is the source term and $-\rho\left(\overline{u_i' u_j'}\right)$ is the Reynolds stress tensor, which, considering the Boussinesq hypothesis proposed in 1877, is as follows:

$$-\rho\left(\overline{u_i' u_j'}\right) = -\frac{2}{3}\rho k \delta_{ij} + \mu_t\left(\frac{\partial u_i}{\partial x_j} + \frac{\partial u_j}{\partial x_i}\right) \quad (4)$$

where μ_t is the turbulent viscosity and k is the kinetic energy of turbulence per mass.

Free surface flow motion is defined by the VoF method [4], which is based on the transport equation of the volume fraction, given by:

$$\frac{\partial f}{\partial t} + u_j \frac{\partial f}{\partial x_j} = 0 \quad (5)$$

where the volume fraction, f, is a scalar that takes values 0 in the air, 1 in the water and 0.5 in the position of the free surface.

2.2. Turbulence Models

The turbulence equations are used as a closure for the RANS equations. In this study, two standard turbulence models, k-ε and k-ω SST, available in the FLUENT® model, are employed. Equations of turbulence kinetic energy (k) and dissipation rate (ε) of the standard k-ε model are given by Harlow and Nakayama [46]:

$$\begin{aligned}\frac{\partial}{\partial t}(\rho k) + \frac{\partial}{\partial x_j}(\rho k u_j) &= \mu_t(2S_{ij}S_{ij}) - \rho\varepsilon \\ &+ \frac{\partial}{\partial x_j}\left[\left(\mu + \frac{\mu_t}{\sigma_k}\right)\frac{\partial k}{\partial x_j}\right] + S_k\end{aligned} \quad (6)$$

$$\begin{aligned}\frac{\partial}{\partial t}(\rho \varepsilon) + \frac{\partial}{\partial x_j}(\rho \varepsilon u_j) &= C_{1\varepsilon}\frac{\varepsilon}{k}\mu_t(2S_{ij}S_{ij}) - C_{2\varepsilon}\rho\frac{\varepsilon^2}{k} \\ &+ \frac{\partial}{\partial x_j}\left[\left(\mu + \frac{\mu_t}{\sigma_\varepsilon}\right)\frac{\partial \varepsilon}{\partial x_j}\right] + S_\varepsilon\end{aligned} \quad (7)$$

$$\mu_t = \rho C_\mu \frac{k^2}{\varepsilon} \quad (8)$$

where

$$S_{ij} = \frac{1}{2}\left[\frac{\partial u_j}{\partial x_j} + \frac{\partial u_j}{\partial x_i}\right] \quad (9)$$

and $\sigma_\kappa = 1.0$, $\sigma_\varepsilon = 1.3$, $C_{1\varepsilon} = 1.44$, $C_{2\varepsilon} = 1.92$, $C_\mu = 0.09$. S_κ and S_ε are the source terms.

Equations of the k-ω SST model are given by Menter [47]:

$$\begin{aligned}\frac{\partial}{\partial t}(\rho k)+\frac{\partial}{\partial x_j}(\rho k u_j) &= \frac{\partial}{\partial x_j}\left[\left(\mu+\frac{\mu_t}{\sigma_k}\right)\frac{\partial k}{\partial x_j}\right]+\mu_t(2S_{ij}S_{ij}) \\ &-\rho\beta^* f_{\beta^*}k\omega+S_k\end{aligned} \tag{10}$$

$$\begin{aligned}\frac{\partial}{\partial t}(\rho \omega)+\frac{\partial}{\partial x_j}(\rho \omega u_j) &= \frac{\partial}{\partial x_j}\left[\left(\mu+\frac{\mu_t}{\sigma_\omega}\right)\frac{\partial \omega}{\partial x_j}\right]+\frac{\alpha\alpha^*}{v_t}\mu_t(2S_{ij}S_{ij}) \\ &-\rho\beta f_\beta\omega^2+2(1-F_1)\rho\frac{1}{\omega\sigma_{\omega,2}}\frac{\partial k}{\partial x_j}\frac{\partial \omega}{\partial x_j}+S_\omega\end{aligned} \tag{11}$$

$$\mu_t=\frac{\rho k}{\omega}\frac{1}{max\left[\frac{1}{\alpha^*},\frac{SF_2}{a_1\omega}\right]} \tag{12}$$

Parameters and functions shown in Equations (10)–(12) are detailed in Appendix A and in [47].

In this study, two turbulence models, specially developed for surface gravity waves applications, are implemented in the FLUENT® model by means of UDF's. In both models, k-ε turbulence equations are modified to be applied in coastal engineering problems and to reduce or avoid the growth of the eddy viscosity and the inevitable wave decay which occurs in standard closure models, such as k-ε and k-ω models.

In the first one, k-ε NLS—non-linear Reynolds stress tensor, the standard Reynolds stress tensor shown in Equation (4)—is related to the strain rate of the mean flow by using a non-linear algebraic model, given by Shih et al. [42] and Lin and Liu [5]:

$$-\rho\left(\overline{u_i'u_j'}\right)=-\frac{2}{3}\rho k\delta_{ij}+C_d\frac{k^2}{\varepsilon}\left(\frac{\partial u_i}{\partial x_j}+\frac{\partial u_j}{\partial x_i}\right)+\rho\frac{k^3}{\varepsilon^2}D_{ij} \tag{13}$$

where D_{ij} represents the non-linear term that is:

$$D_{ij}=(C_1A_{ij}+C_2B_{ij}+C_3C_{ij}) \tag{14}$$

and A_{ij}, B_{ij} and C_{ij} are given by Lin and Liu [5] and de Finis et al. [20]:

$$A_{ij}=\left(\frac{\partial u_i}{\partial x_k}\frac{\partial u_k}{\partial x_j}+\frac{\partial u_j}{\partial x_k}\frac{\partial u_k}{\partial x_i}-\frac{2}{3}\frac{\partial u_l}{\partial x_k}\frac{\partial u_k}{\partial x_l}\delta_{ij}\right) \tag{15}$$

$$B_{ij}=\left(\frac{\partial u_i}{\partial x_k}\frac{\partial u_j}{\partial x_k}-\frac{1}{3}\frac{\partial u_l}{\partial x_k}\frac{\partial u_l}{\partial x_k}\delta_{ij}\right) \tag{16}$$

$$C_{ij}=\left(\frac{\partial u_k}{\partial x_i}\frac{\partial u_k}{\partial x_j}-\frac{1}{3}\frac{\partial u_l}{\partial x_k}\frac{\partial u_l}{\partial x_k}\delta_{ij}\right) \tag{17}$$

Coefficients C_1, C_2, C_3 and C_d are restricted to avoid nonphysical values of turbulent energy, which are [5]:

$$C_d=\frac{2}{3}\left(\frac{1}{7.4+S_{max}}\right) \tag{18}$$

$$C_1=\frac{1}{185.2+D_{max}^2} \tag{19}$$

$$C_2=\frac{1}{58.5+D_{max}^2} \tag{20}$$

$$C_3=\frac{1}{370.4+D_{max}^2} \tag{21}$$

where

$$S_{max}=\frac{k}{\varepsilon}max\left[\left|\frac{\partial u_i}{\partial x_i}\right|\right] \tag{22}$$

$$D_{max} = \frac{k}{\varepsilon}\, max\left[\left|\frac{\partial u_i}{\partial x_j}\right|\right] \tag{23}$$

In Equation (22), indexes are not summed.

Source terms (S_{Mi}) of the standard FLUENT® model for momentum equations, Equation (2), and the source term (S_k) for the turbulent kinetic energy equation, Equation (6), are implemented to consider the non-linear Reynolds stress tensor, as follows:

$$S_{Mi} = \frac{\partial}{\partial x_j}\left(\rho\frac{k^3}{\varepsilon^2}D_{ij}\right) = \frac{\partial}{\partial x_j}\left[\rho\frac{k^3}{\varepsilon^2}(C_1A_{ij} + C_2B_{ij} + C_3C_{ij})\right] \tag{24}$$

$$S_k = \rho\frac{k^3}{\varepsilon^2}D_{ij}\frac{\partial u_i}{\partial x_j} = \rho\frac{k^3}{\varepsilon^2}(C_1A_{ij} + C_2B_{ij} + C_3C_{ij})\frac{\partial u_i}{\partial x_j} \tag{25}$$

The second turbulence model implemented in the standard FLUENT® model, k-ε SCM (stabilized closure model) is based on a formulation proposed by Larsen and Fuhrman [14], in which the growth of the eddy viscosity is avoided by using the following equation:

$$\mu_t = \rho C_\mu \frac{k^2}{\tilde{\varepsilon}} \tag{26}$$

where

$$\tilde{\varepsilon} = max\left(\varepsilon,\ \ \lambda_2\frac{C_2}{C_1}\frac{p_o}{p_\Omega}\varepsilon\right) \tag{27}$$

$$p_o = 2\, S_{ij}S_{ij} \tag{28}$$

$$p_\Omega = 2\, \Omega_{ij}\Omega_{ij} \tag{29}$$

The parameter λ_2 is from 0.02 to 0.1 (in this study, $\lambda_2 = 0.05$, which is the value used by Larsen and Fuhrman [14]). S_{ij} is given by Equation (9) and Ω_{ij} is

$$\Omega_{ij} = \frac{1}{2}\left[\frac{\partial u_i}{\partial x_j} - \frac{\partial u_j}{\partial x_i}\right] \tag{30}$$

2.3. Porous Medium Adaptation

In this study, momentum equations and k-ε turbulence equations of the standard FLUENT® model are adapted to deal with flows in coastal porous structures. The original momentum equations of the standard FLUENT® model applied to porous zones are given by [43]:

$$\frac{1}{n}\frac{\partial \rho u_i^s}{\partial t} + \frac{\partial}{\partial x_j}\left(\frac{\rho u_i^s u_j^s}{n^2}\right) = -\frac{\partial p}{\partial x_i} + \rho g_i + \frac{\partial \tau_{ij}^s}{\partial x_j} - \frac{\partial \rho\left(\overline{u_i^{s'} u_j^{s'}}\right)}{\partial x_j} + S_{Mi}^s \tag{31}$$

where

$$\tau_{ij}^s = \mu\left[\frac{\partial}{\partial x_j}\left(\frac{u_i^s}{n}\right) + \frac{\partial}{\partial x_i}\left(\frac{u_j^s}{n}\right)\right] \tag{32}$$

u_i^s is the superficial velocity (mean velocity of the flow) and n is the porosity of the medium, which is the relation between the volume of fluid and the total volume, including the solid one. In Equation (31), S_{Mi}^s is the source term that represents the resistance in porous media based on the Darcy–Forchheimer formulation:

$$S_{Mi}^s = -\left(\frac{\mu}{\alpha}u_i^s\right) - C_2\frac{1}{2}\rho|u^s|u_i^s \tag{33}$$

where α and C_2 are permeability and inertial coefficients, respectively.

Coefficients of Equation (33) are determined considering the hydraulic gradient I of the extended Darcy–Forchheimer formulation used to deal with transient flows in porous medium [23]:

$$I = -\frac{1}{\rho g}\frac{\partial p}{\partial x_i} = a_p u_i^s + b_p |u^s| u_i^s + c_p \frac{\partial u_i^s}{\partial t} \tag{34}$$

where

$$c_p = \gamma \frac{(1-n)}{gn} \tag{35}$$

$\gamma = 0.34$ is the added mass empirically determined by van Gent [23]. There are some proposals to coefficients a_p and b_p, such as those of Ergun [48] and Engelund [49]; however, a more recent formulation, developed by van Gent [50], is used in this study:

$$a_p = \frac{\alpha_E \mu (1-n)^2}{\rho g D_p^2 n^3} \tag{36}$$

$$b_p = \frac{\beta_E (1-n)}{g D_p n^3} \tag{37}$$

where α_E and β_E are empirical parameters, and D_p is the particle diameter of the porous medium. In the case of coastal structures, $D_p = D_{50}$, which is expressed by

$$D_{50} = \left(\frac{M_{50}}{\rho_r}\right)^{1/3} \tag{38}$$

M_{50} is the average of the stone mass and ρ_r is the specific mass of the stone. In this study, β_E depends on the Keulegan–Carpenter number [51], as follows:

$$\beta_E = \beta_{Est}\left(1 + \frac{7.5}{KC}\right) \tag{39}$$

where KC is the Keulegan–Carpenter number, given by

$$KC = \frac{U_m T_p}{n D_{50}} \tag{40}$$

where β_{Est} is the coefficient for the stationary flow, U_m is the maximum oscillating velocity and T_p is the peak period for random incident waves or the period for regular ones.

Parameters α and C_2 are determined considering a_p and b_p, from Equation (36) and Equation (37), the hydraulic gradient I (Equation (34)) and the source term S_{Mi}^s (Equation (33)), resulting in:

$$\frac{1}{\alpha} = \frac{\alpha_E (1-n)^2}{D_p^2 n^3} \tag{41}$$

$$C_2 = \frac{2\beta_E (1-n)}{D_p n^3} \tag{42}$$

Therefore, the resistance term S_{Mi}^s is rewritten, taking into account the hydraulic gradient (Equation (34)):

$$S_{Mi}^s = -\left(\frac{\mu}{\alpha} u_i^s\right) - C_2 \frac{1}{2}\rho |u^s| u_i^s - C_A \frac{\partial \rho u_i^s}{\partial t} \tag{43}$$

The temporal term of Equation (43) is discretized by using the first order backward approximation and implemented in FLUENT® by means of UDF.

$$C_A = g c_p = \gamma \frac{(1-n)}{n} \tag{44}$$

In the standard FLUENT® model, the turbulence equations in the k-ε model are not adapted to porous zones, since it is considered that the solid of porous structure does not significantly affect the turbulence generation and dissipation rates. This assumption can be adopted when there is a high permeability of the medium and its geometric scale does not interact with the turbulence vortex scale. However, a different formulation must be used in cases of waves over porous coastal structures. In this study, the k-ε turbulence equations, developed by Nakayama and Kuwahara [25] and applied to coastal porous structures by del Jesus et al. [35] (IH3VOF) and Higuera et al. [37] (IHFOAM), are used. In this formulation, source terms S_k (Equation (6)) and S_ε (Equation (7)) are:

$$S_k = n\varepsilon_\infty \tag{45}$$

$$S_\varepsilon = nC_{2\varepsilon}\frac{\varepsilon_\infty^2}{k_\infty} \tag{46}$$

where

$$k_\infty = 3.7(1-n)n^{3/2}u_i^s u_i^s \tag{47}$$

$$\varepsilon_\infty = 39.0(1-n)^{5/2}u_i^s u_i^s \frac{1}{D_{50}} \tag{48}$$

2.4. Numerical Conditions

Cases of waves over porous coastal structures involve complex phenomena, such as wave breaking and overtopping, and flow in porous media. Therefore, specific methodologies in the RANS-VoF numerical model and numerical conditions of wave flumes and tanks must be used to simulate these cases. Considering the use of the RANS-VoF FLUENT® numerical model, the following characteristics are emphasized:

(a) Schemes. The solver scheme SIMPLEC (with standard under-relaxation factors) and the scheme PRESTO! are used for discretizing pressure. The momentum is discretized by the third-order scheme MUSCL, and the turbulence kinetic energy and dissipation rate are discretized by the second order upwind scheme. The Geo-reconstruct scheme, well adapted for modeling the complex shape of free surface flow, such as wave breaking and overtopping, is used for the VoF equation, compatible with the first order time integration scheme and variable time steps [52,53]. The implementation of the k-ε NLS and k-ε SCM turbulence models and equations for porous media of coastal structures in the FLUENT® numerical model are carried out in this study according to Sections 2.2 and 2.3.

(b) Boundary conditions. The non-slip condition is imposed on walls of the structure and the bottom of the wave flume. The atmospheric pressure is applied to the top boundary. The incident wave generation is applied to the wave-maker boundary, imposed by a UDF. Velocity component profiles, which are related to time and depth according to the wave theory, are imposed, and the corresponding free surface position is defined by the volume fraction value (0 for air and 1 for water). An active absorption technique is imposed at the wave generation [39–41] to eliminate re-reflection of regular and random waves on the flume by using the methodology proposed by Shäffer and Klopman [54], which is based on the linear shallow water theory and can be applied to a numerical wave flume [55–57]. Due to the high non-linear characteristics of the study cases, regular waves are generated by using the Fourier wave theory [58,59], with 20 terms in the series. Random waves are simulated by using 50 waves to represent the JONSWAP/TMA spectrum. The wave generation of random waves in the FLUENT® model was implemented and validated by Teixeira and Didier [41].

(c) Initial conditions. Free surface level at rest, null velocity components, hydrostatic pressure on the water and atmospheric pressure on the air are the initial conditions. In addition, initial conditions of k and ε (and ω) are imposed according to Larsen and Fuhrman [14].

(d) Spatial discretization. Computational meshes of wave flumes with coastal structures have, at least, two main zones with different characteristics: the propagation wave zone and the zone around the structure. In the former, a structured regular mesh is used, in which the free surface is well behaved, and the mesh must be refined around it. Generally, in this zone, the boundary layer on the bottom does not significantly influence the water flow and, consequently, its spatial discretization is not important and does not require a fine resolution. Mesh resolution for accurate wave propagation is defined as follows: in the horizontal direction, 70 cells per wavelength are employed; in the vertical direction, the mesh follows the rule of 20 cells per wave height, in the zone of variation of free surface flow, and it is stretched to the bottom and top of the flume [60–66]. In zones around and inside the coastal structure, the flow has a different and complex behavior and, generally, regular cells are recommended with an aspect ratio close to 1 [9–11].

(e) Time discretization. A variable time step is used for time integration, in which the maximum value is $Tp/600$ (Tp is the peak period for a random wave and the wave period for regular waves) and the minimum one is 30 times smaller than the maximum time step [52]. Six non-linear iterations per time step enable the reduction of residue by at least two orders of magnitude which are enough to obtain good accuracy in wave propagation and wave–structure interaction [52,53,60–63,65].

3. Previous Analysis of Turbulence Models and Mesh Dependency

It is well known that an accurate solution in numerical modeling is partly due to the quality of the mesh, verified by a classical mesh dependency study. However, the free surface flow modeling in cases with wave propagation involves additional degrees of difficulties. One major and known difficulty comes from the decay of the wave height for longtime modeling of wave propagation in a long numerical wave flume when conventional turbulence models are used. Larsen and Fuhrman [14] showed that only adapted turbulence models can maintain characteristics of the wave during the propagation by avoiding the growth of turbulent kinetic energy and eddy viscosity. This is achieved by new and stable closure models proposed by Larsen and Fuhrman [14]. Nevertheless, in these conditions, almost same results can be obtained by the k-ε NLS turbulence model [5,42] for wave propagation and wave overtopping, which are two cases analyzed in Section 3.1. In Section 3.2, a classical convergence study with mesh refinement is performed for the complex case of wave overtopping on an impermeable sea dike with a crown-wall using the k-ε NLS turbulence model.

3.1. Analysis of the Grow of Eddy Viscosity along a Flume in Longtime Simulation

In order to follow the analysis carried out by Larsen and Fuhrman [14], highlighting the absolute necessity of using more adapted turbulence models avoiding the growth of eddy viscosity for free surface flows modelling, four turbulence models, shown in Section 2, are tested for wave propagation over a horizontal bottom wave flume and wave overtopping on an impermeable sea dike with a crown-wall [18]. For both configurations, the incident wave is regular, with T = 1.5 s and H = 0.16 m, being considered moderately non-linear in intermediate-depth water. The horizontal wave flume length is 8 L, where L is the wave length, and the simulation time is 300 s, which corresponds to 200 waves. Mesh resolution for both cases follows the recommendations described in Sections 2.4, 3.2 and 4.2.

For analyzing the growth of eddy viscosity in space and time, the spatial average of the non-dimensional eddy viscosity $\langle\langle \nu_t \rangle\rangle/\nu$ is calculated in zones of length L along the flume: ν_t is the kinematic eddy viscosity and ν is the kinematic viscosity of the fluid. For the wave propagation case, the horizontal wave flume is divided in eight zones of length L and an active absorption is imposed at the end of the flume. Figure 1a shows a sketch of the flume and the position of the eight zones. For the wave overtopping case, a ninth zone is added, which includes the sea dikes, crown-wall and dike crest. Figure 1b shows a sketch of the wave flume for this case and the division of the flume by nine zones. An outflow condition is imposed at the end of the flume. These analyses differ slightly from Larsen and Fuhrman [14], in which the growth of eddy viscosity along the time in a computational domain of single wavelength using periodic lateral boundaries is calculated for simulating a simple progressive wave train.

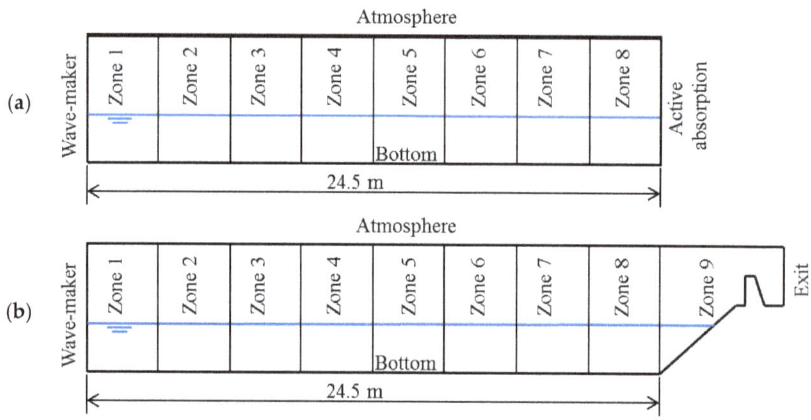

Figure 1. Sketch of the computational domain for both wave propagation (**a**) and wave overtopping cases (**b**) and the division of the flume by zones of length L for computing $\langle\langle \nu_t \rangle\rangle / \nu$.

The wave propagation over a horizontal bottom wave flume is firstly analyzed (Figure 1a). Figure 2 shows the time series of $\langle\langle \nu_t \rangle\rangle / \nu$ for three zones along the flume: zone 1, in the vicinity of the wave-maker; zone 4 and zone 7, in the middle and near the end of the flume, respectively. Figure 3 shows the time series of the free surface elevation in the middle of the wave flume, at 12.25 m from the wave-maker, for the four turbulence models (k-ε, k-ω SST, k-ε NLS and k-ε SCM). The dash-lines indicate the mean wave crest and trough obtained by the k-ε NLS turbulence model, which is considered as a reference.

Figure 2 clearly shows that the standard k-ε turbulence model results in a significant and rapid growth of eddy viscosity, which produces a non-physical decay of free surface elevation along the time, with a wave height reduction of 75% compared to the wave height obtained by the k-ε NLS turbulence model. The k-ω SST turbulence model exhibits a slightly greater growth of eddy viscosity than those of k-ε SCM and k-ε NLS ones. It results in a little decay of free surface elevation (Figure 3), which seems to indicate that the standard k-ω SST turbulence model does not include a stable closure model. The wave height reduction is 12% compared to that obtained by the k-ε NLS turbulence model. Finally, k-ε SCM and k-ε NLS turbulence models avoid unphysical behaviors of the free surface elevation and are able to simulate a nearly constant form wave propagation over long durations and along the wave flume. The value of $\langle\langle \nu_t \rangle\rangle / \nu$ is slightly smaller than 1.0 for the k-ε SCM turbulence model and around 1.7 for the k-ε NLS turbulence model on the three zones analyzed. A slight difference of the wave height, around 3%, can be noted between both models, which can be related to the methodologies of each model and minor numerical diffusion associated with numerical schemes.

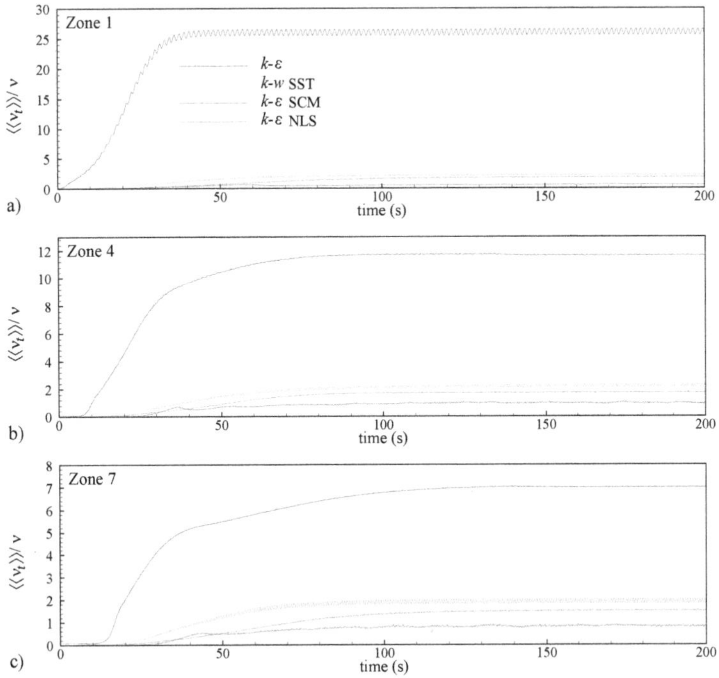

Figure 2. Time series of non-dimensional eddy viscosity (spatial average) in zone 1 (**a**), 4 (**b**), and 7 (**c**) for the wave propagation case for the four turbulence models.

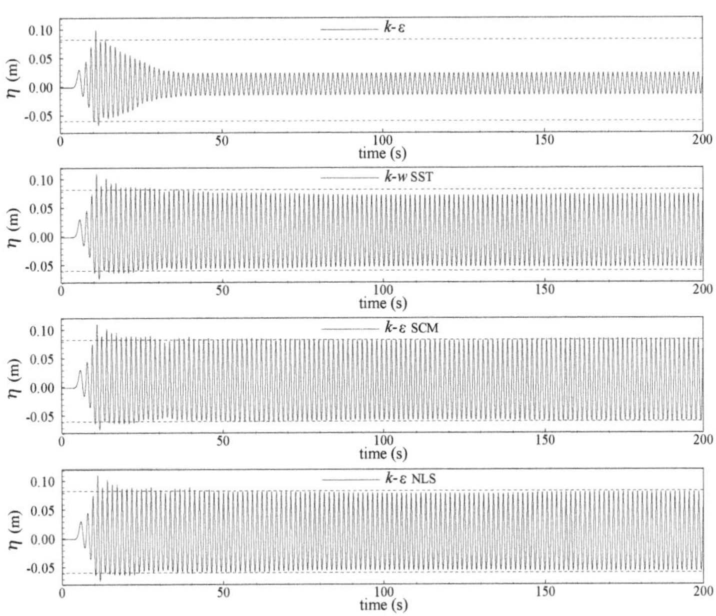

Figure 3. Time series of the free surface elevation in the middle of the wave flume for the wave propagation case for the four turbulence models—dash-lines indicate the mean wave crest and trough obtained by the *k-ε* NLS model.

The second case analyzed involves wave breaking and wave overtopping over an impermeable dike (Figure 1b). Figure 4 shows the time series of $\langle\langle \nu_t \rangle\rangle / \nu$ for the four turbulence models and three zones along the flume: zone 1, in the vicinity of the wave-maker; zone 4 and zone 7, in the middle and near the toe of the sea dike, respectively. The case corresponds to the configuration *W6S0* of Tuan and Thin [18], in which there is a vertical wall 0.06 m high on the dike crest (*W*) without a promenade (*S* is 0 cm) (see Section 4.2 for more details). Figure 5 presents the time series of the wave overtopping discharge.

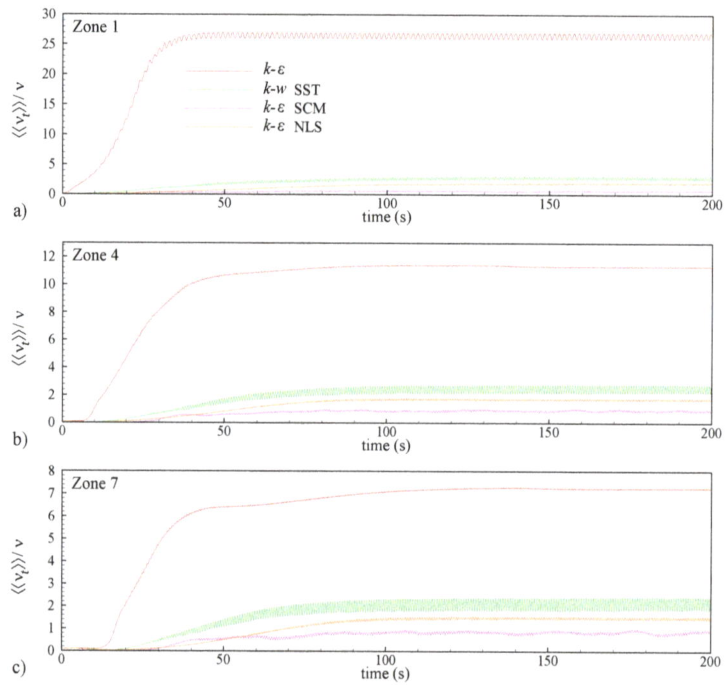

Figure 4. Time series of non-dimensional eddy viscosity (spatial average) in zone 1 (**a**), 4 (**b**), and 7 (**c**) for the wave overtopping case for the four turbulence models.

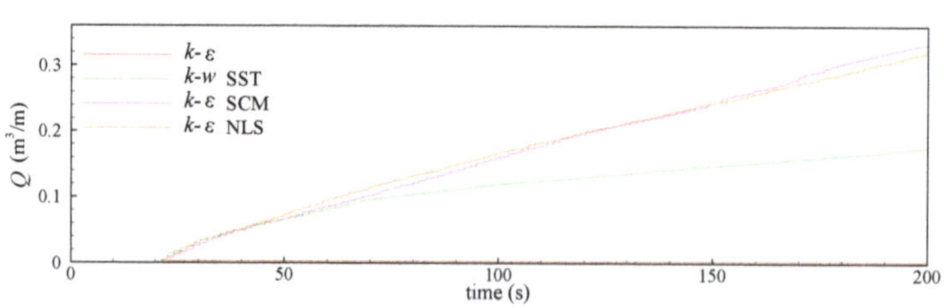

Figure 5. Time series of the wave overtopping discharge for the four turbulence models.

Figure 4 shows that the standard k-ε turbulence model results in a significant and rapid growth of eddy viscosity, similar to that observed in the wave propagation case. The k-ω SST turbulence model shows a slightly greater growth of eddy viscosity than both k-ε SCM and k-ε NLS, with a value of $\langle\langle \nu_t \rangle\rangle / \nu$ slightly smaller than 1.0 for the k-ε SCM turbulence model and around 1.7 for the k-ε NLS turbulence model. The wave overtopping discharge of the turbulence models (Figure 5) shows drastic differences. Since the significant growth of eddy viscosity using the standard k-ε turbulence model implicates a significant and unphysical decay of free surface elevation, wave overtopping discharge is completely wrong, being null in the time interval from 120 to 198 s. The other turbulence models show the same trend up to 50 s. However, k-ω SST turbulence models exhibit a decay of wave overtopping discharge after this instant, due to the growth of eddy viscosity, as demonstrated in the wave propagation case. The average wave overtopping discharge is smaller than the experimental one, 1.58 l/s/m [18], being 0.563 l/s/m for the k-ω SST turbulence model, i.e., 67% smaller than the experimental value. Only k-ε SCM and k-ε NLS turbulence models show a regular and nearly constant trend of wave overtopping discharge along the time, since both models avoid the non-physical growth of the eddy viscosity. The average wave overtopping discharges of k-ε SCM and k-ε NLS turbulence models are rather similar, at 1.727 and 1.672 l/s/m, respectively, with a difference of only 3%. Both results are slightly greater than the experimental one.

These two tests demonstrate how the standard k-ε turbulence closure model, widely used for modeling wave propagation and wave–structure interaction, can lead to severely over-predicted eddy viscosity levels, causing the decay of the free surface elevation and an under-estimated wave overtopping discharge. The k-ω turbulence model shows a correct behavior and allows accurate results for wave propagation and wave overtopping to be obtained, but only for short simulations. For long simulations, the growth of eddy viscosity, even if it remains smaller than observed in the k-ε turbulence model, leads to the decay of the free surface elevation and severe under-estimation of the wave overtopping discharge. On the contrary, two completely different models, k-ε SCM and k-ε NLS turbulence ones, based on the stabilized closure model and non-linear Reynolds stress tensor, respectively, allow avoidance of unphysical behaviors and modeling wave propagation, and a nearly constant wave overtopping discharge over long durations. Therefore, turbulence models k-ε SCM and k-ε NLS are suitable for modeling wave propagation, wave–structure interactions and wave overtopping. Nevertheless, in cases of short flumes, in which the short propagation distance limits the extent of the turbulence over-production, and short time simulations, the growth of the eddy viscosity causing the waves decay could not occur, as referred by Larsen and Fuhrman [14] and verified by the authors.

3.2. Mesh Dependency Analysis

Mesh dependency analysis is carried out for the case *W6S0* of Tuan and Thin [18], which is a representative case of the wave over coastal structure that involves wave breaking and overtopping. This case corresponds to a vertical wall on the dike crest with 0.06 m height (W) without a promenade (S is 0 cm) (see Section 4.2 for more details). The impermeable model dike is 0.70 m height, with a seaward slope 1/3 and 0.10 m of the freeboard. A regular incident wave with 2.5 s period and 0.24 m high is generated. The numerical wave flume length is reduced for the mesh dependency analysis, in which the length from the wave-maker to the toe of the seaward slope is 6.0 m, i.e., around two wave lengths. Table 1 shows characteristics of the five meshes used in this study. Meshes are unchanged in the wave propagation part of the fume, i.e., from the wave-maker to the toe of the sea dike ramp, following the recommendations described in Section 2.4. The mesh refinement is carried out only in the vicinity of the structure, from the toe of the ramp to the end of the wave flume, using an unstructured mesh with regular cells of characteristic length dl, i.e., with aspect ratio close to 1 (see Section 4.2 for more details). The cell number varies from 26,521 to 45,018 for the coarser to finer meshes (Table 1), which corresponds to a reduction of dl from 1.5 to 0.4 cm in the vicinity of the structure, i.e., the cell area is reduced by a factor of 14. The k-ε NLS turbulence model is used for the mesh dependency analysis, although the k-ε SCM turbulence model could also be used.

Table 1. Mesh characteristics in the mesh dependency analysis.

Mesh	dl ($\times 10^{-2}$ m)	Number of Cells
M-1	1.5×1.5	26,521
M-2	1.0×1.0	29,805
M-3	0.75×0.75	33,351
M-4	0.5×0.5	39,226
M-5	0.4×0.4	45,018

Figure 6 presents a time series of the wave overtopping discharge, Q, for the five meshes. Table 2 shows the average wave overtopping discharge, calculated over the time interval from 60 to 110 s, i.e., 20 wave periods, and its relative error comparing to the finer mesh M-5. It can be seen that wave overtopping discharge has the same trend for meshes M-2 to M-5, with only small differences. The coarser mesh M-1 shows a slightly different behavior. The averaged wave overtopping discharge difference between meshes M-2 to M-4 and the finer mesh M-5 is small, around or inferior to 2%, and very similar for mesh M-2 and M-4. Since mesh M-2 gives similar values to mesh M-4, mesh resolution used for mesh M-2 seems to be a good choice compared to conducted simulations, and it is adopted for simulations of Tuan and Thin [18] in Section 4.2.

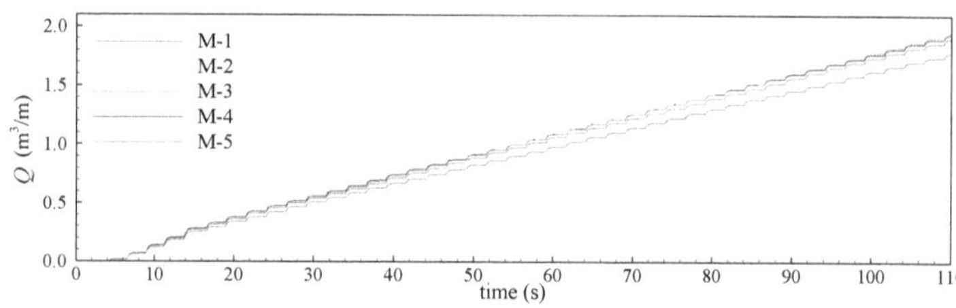

Figure 6. Time series of the wave overtopping discharge for meshes M-1 to M-5.

Table 2. Average wave overtopping discharge for meshes M-1 to M-5.

Mesh	Average Wave Overtopping Discharge (m^3/m)	Relative Error (%) to Finer Mesh M-5
M-1	16.024	6.96
M-2	16.867	2.07
M-3	17.015	1.21
M-4	16.918	1.77
M-5	17.223	-

Different to the well behaved case of the wave propagation on a flume, there is no possibility of recommending element sizes in the vicinity of coastal structures, based on dimensional characteristics of the problem, due to the complexity of involved phenomena. Therefore, in other cases presented in Section 4, element sizes are equal or close to those used by the authors in their investigations, so that comparison of results is possible.

4. Results and Discussion

Methodologies implemented in the FLUENT® numerical model are validated by comparing numerical results with experimental and numerical ones in the literature. Four configurations are tested: wave breaking over an impermeable beach of Ting and Kirby [6], wave overtopping on impermeable sea dikes with crown-walls of Tuan and Thin [18], wave interaction with a porous low-crested rubble-mound breakwater of Garcia et al. [33] and wave overtopping on a rubble mound breakwater of Losada et al. [15]. The quantitative analysis of numerical results is carried out using the *Bias* statistical parameter [67] and the

NRMSE (normalized root mean square error), which are obtained comparing numerical results with those of the reference. *Bias* is defined as follows:

$$Bias = \frac{\sum_i^n (s_i - m_i)}{n} \quad (49)$$

where s_i are values obtained by the numerical simulation, m_i are reference values and n is the sample size. *NRMSE*, as a percentage, is given by:

$$NRMSE = 100\sqrt{\sum_i^n \frac{(s_i - m_i)^2}{n}}/\overline{m} \quad (50)$$

where $\overline{m}$ is the average of the reference values.

4.1. Spilling and Plunging Wave Breakers on an Impermeable Beach

Physical experiments developed by Ting and Kirby [6] are used to validate numerical models in cases of spilling and plunging wave breaking on an impermeable bottom beach. It consists of a flume 0.4 m deep with a slope (1:35) at one end. The origin of the horizontal coordinate (x) considered by Ting and Kirby [6] is 0.7 m from the beginning of the slope. Incident waves with T = 2 s and H = 0.125 m (spilling) and T = 5 s and H = 0.128 m (plunging) are generated in the wave-maker by using the Fourier theory. A sketch of the computational domain is shown in Figure 7, in which dimensions in parenthesis represent the plunging case. The horizontal wave flume length is 20.0 m (5.2 L) and 50.0 m (4.7 L), respectively. The mesh resolution follows the recommendations described in Section 2.4 and 3.2, and an unstructured mesh with regular cells with a grid resolution close to 0.0063 × 0.0063 m (aspect ratio close to 1) is used in the slope region (Figure 8). The total number of cells is around 115,157 and 118,976 for the spilling and plunging cases, respectively.

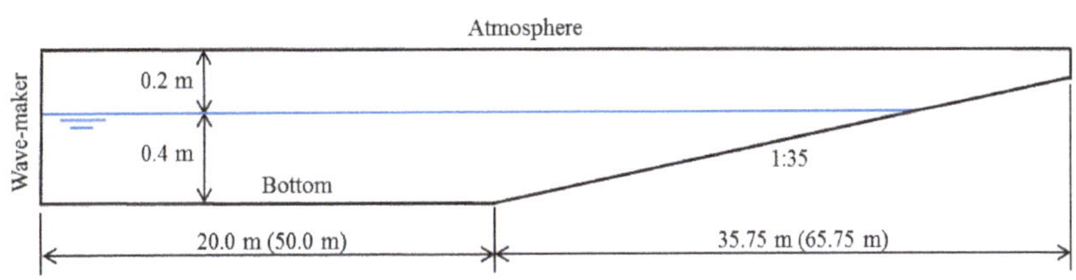

Figure 7. Sketch of the computational domain of Ting and Kirby [6] cases. Dimensions in parenthesis represent the plunging case.

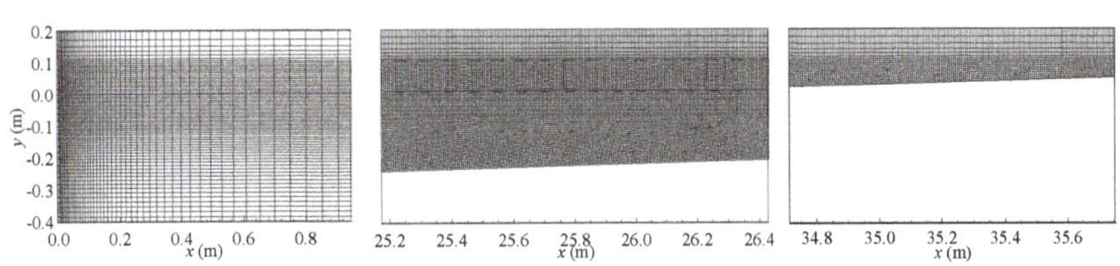

Figure 8. Details of the mesh for the spilling case (Ting and Kirby [6]) near the wave-maker, along the slope and at the end of the domain. The blue line indicates the SWL.

Figure 9 shows maximum and minimum wave height envelopes and mean water level obtained by using turbulence models k-ε, k-ω SST, k-ε NLS and k-ε SCM for the spilling case (black dashed line in Figure 9 indicates the still water level). Results are compared with experimental ones of Ting and Kirby [6] at instants from 80 to 90 s. Clearly, the standard k-ε turbulence model is highly dissipative and, consequently, it systematically underestimates the wave height envelopes and leads to unphysical behaviors. The other turbulence models have similar results, with slightly better ones reached by the *k*-ε NLS turbulence model, meanly because it accurately follows the breaking zone. Table 3 shows the *Bias* and *NRMSE* of wave height obtained by turbulence models. The lowest *Bias* is −0.01 cm of the *k*-ε SCM turbulence model, and the lowest *NRMSE* is 6.5% of the *k*-ε NLS turbulence model.

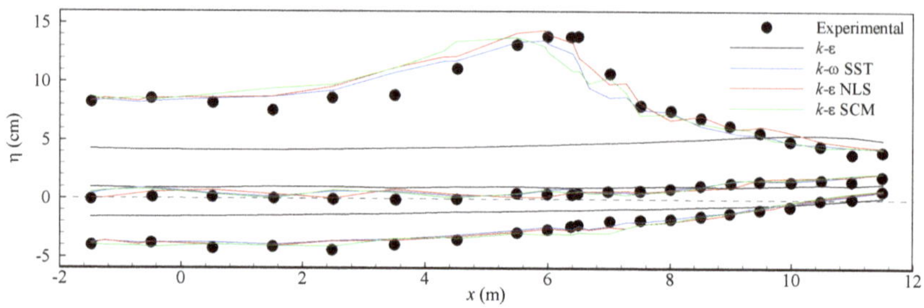

Figure 9. Maximum and minimum wave height envelopes and mean water level in the spilling case by using *k*-ε, *k*-ω SST, *k*-ε NLS and *k*-ε SCM turbulence models.

Table 3. *Bias* and *NRMSE* of wave height for spilling and plunging cases and *k*-ε, *k*-ω SST, *k*-ε NLS and *k*-ε SCM turbulence models.

	T = 2 s, H = 0.125 m		**T = 5 s, H = 0.128 m**	
	***Bias* (cm)**	***NRMSE* (%)**	***Bias* (cm)**	***NRMSE* (%)**
k-ε	−4.97	59.2	−3.58	35.4
k-ω SST	−0.38	8.4	0.05	10.6
k-ε NLS	0.25	6.5	0.80	11.9
k-ε SCM	−0.01	9.7	0.75	10.1

Figure 10 shows maximum and minimum wave height envelopes and mean water level by using turbulence models *k*-ε, *k*-ω SST, *k*-ε NLS and *k*-ε SCM at instants from 70 to 90 s for the plunging case (black dashed line in Figure 10 indicates the still water level). As the spilling case, the *k*-ε turbulence model significantly underestimates the wave height envelopes, leading to unphysical behavior, and the other turbulence models have similar results, in which the maximum wave height envelopes are overestimated before the wave breaking zone. Although the *k*-ω SST turbulence model has the lowest *Bias*, 0.05 cm, it occurred because there is a higher underestimation of the maximum wave height envelope in the wave breaking zone, which compensates the overestimation before this zone. *NRMSE* of all turbulence models are similar, from 10.1 to 11.9%, except the *k*-ε one (Table 3).

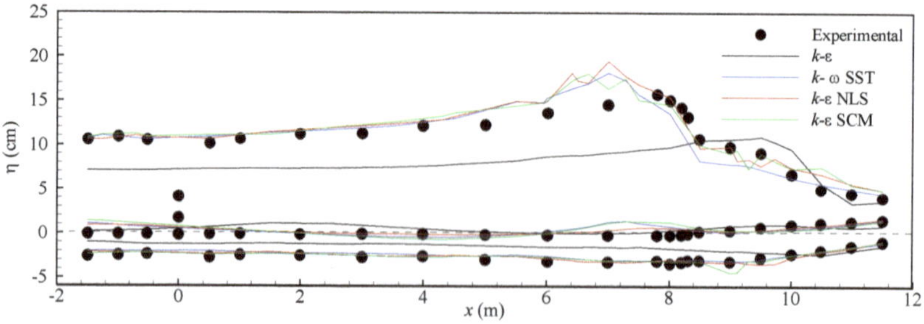

Figure 10. Maximum and minimum wave height envelopes and mean water level in the plunging case by using the *k*-ε, *k*-ω SST, *k*-ε NLS and *k*-ε SCM turbulence models.

In summary, the k-ω SST, k-ε NLS and k-ε SCM turbulence models show similar results for spilling and plunging wave breakwaters, while the k-ε turbulence model leads to unphysical behaviors.

4.2. Wave Overtopping on Sea Dikes with Crown Walls

In order to evaluate the capability of the model to simulate wave overtopping, physical experiments and numerical results obtained by Tuan and Thin [18] are used. Among the set of experiments presented by Tuan and Thin [18], the model dike was smooth and impermeable, using a height of 0.70 m and a 1/3 seaward slope (Figure 11). The vertical wall on the dike crest is 0, 4, 6 and 9 cm height (W) and the promenade width (S) is 0, 0.10 and 0.20 m. Freeboard is 0.10 and 0.15 m for regular and random wave tests, respectively. The computational domain has the same dimensions as Tuan and Thin [18], with 24.5 m from the inflow waves at the seaward boundary and the dike toe. It is important to emphasize that the physical wave flume is 45 m long, and Tuan and Thin [18] used a recorded time-series from the wave gauge at 24.5 m seaward of the structure to generate the incident waves at the boundary, whereby the free surface and the velocity components are prescribed. However, theoretical values of wave period and wave height are used in the present study to generate the incident wave in the static paddle. Fourier waves are generated for both regular wave cases (T =1.5 s, H = 0.16 m and T = 2.5 s, H = 0.24 m), and a JONSWAP-type spectrum is considered for two random wave tests (T_p = 2.2 s and H_s = 0.123 and 0.126 m). The mesh resolution follows the recommendations described in Sections 2.4 and 3.2, and an unstructured mesh with regular cells with a grid resolution close to 0.01 $\times$ 0.01 m, i.e., aspect ratio close to 1, is used in the proximity of the dike and the vertical wall (Figure 12). The total number of cells is around 70,000, varying with the length of the promenade and the wall height. The k-ε NLS and k-ε SCM turbulence models are used for this application, since both models allow accurate results of average wave overtopping discharges to be obtained, different to the standard k-ε and k-ω SST turbulence models, as verified in Section 3.1.

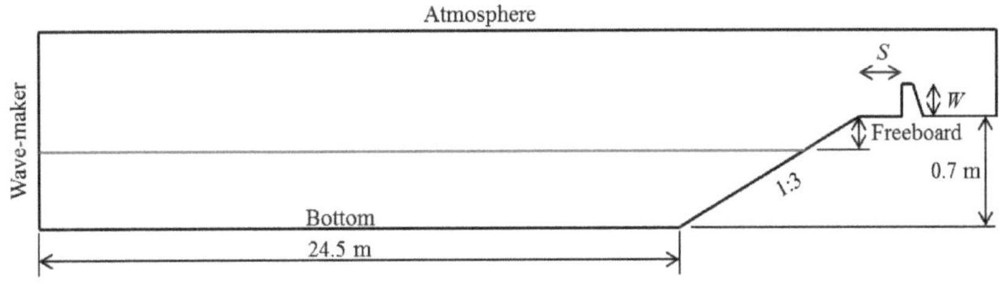

Figure 11. Sketch of the computational domain (identical to that of Tuan and Thin [18]).

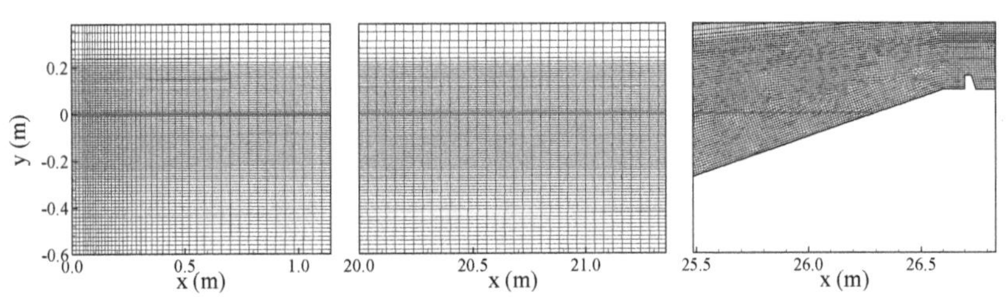

Figure 12. Details of the mesh near the wave-maker, along the flume and in the proximity of the dike and vertical wall. The blue line indicates the SWL.

Average wave overtopping discharges predicted by the numerical model compared with the experimental and numerical results of Tuan and Thin [18] are shown in Table 4 for regular (RE) and random (IR) waves. Figure 13 shows the present and Tuan and Thin

(2014) numerical results versus the experimental ones. Numerical and experimental average wave overtopping discharges in Tuan and Thin [18] are calculated from around 10 waves for regular waves and 1000 waves for random waves. In the present study, overtopping discharges for regular waves are recorded from the 30th and 50th waves, whereby the periodic stationary wave–structure interaction flow is obtained. For random waves, 1000 waves are recorded, i.e., $1000T_p$, according to Tuan and Thin [18]. The present average wave overtopping discharge is obtained from the water volume flow rate monitored after the vertical wall.

Table 4. Average wave overtopping discharge of regular (RE) and random (IR) waves: experimental and numerical from Tuan and Thin [18] and numerical predictions for *k*-ε NLS and *k*-ε SCM turbulence models.

Case		**Regular Waves**			**Average Wave Overtopping Discharge (l/s/m)**			
					Tuan and Thin [18]		**Turbulence Model**	
	***W* (cm)**	***S* (cm)**	***H* (m)**	***T* (s)**	**Exp.**	**Num.**	***k*-ε NLS**	***k*-ε SCM**
REW0S0_1	0	0	0.16	1.5	5.26	5.08	6.74	7.51
REW0S0_2	0	0	0.24	2.5	21.05	21.23	27.57	27.97
REW0S4_1	4	0	0.16	1.5	2.28	2.54	3.05	3.58
REW0S4_2	4	0	0.24	2.5	21.75	20.91	20.77	22.50
REW4S10_1	4	10	0.16	1.5	1.14	2.09	2.73	3.24
REW4S10_2	4	10	0.24	2.5	19.82	18.12	20.46	21.09
REW4S20_1	4	20	0.16	1.5	0.70	0.98	2.75	2.68
REW4S20_2	4	20	0.24	2.5	18.42	18.33	19.69	20.49
REW6S0_1	6	0	0.16	1.5	1.58	1.65	1.77	1.78
REW6S0_2	6	0	0.24	2.5	17.02	16.10	17.09	18.49
REW6S10_1	6	10	0.16	1.5	1.14	1.01	2.13	2.37
REW6S10_2	6	10	0.24	2.5	15.61	16.60	17.69	18.14
REW6S20_1	6	20	0.16	1.5	0.19	0.34	1.49	1.71
REW6S20_2	6	20	0.24	2.5	13.51	17.12	16.11	17.03
REW9S0_1	9	0	0.16	1.5	1.32	0.51	0.69	0.58
REW9S0_2	9	0	0.24	2.5	14.39	13.29	10.59	12.34
REW9S10_1	9	10	0.16	1.5	0.21	0.21	1.01	1.28
REW9S10_2	9	10	0.24	2.5	13.51	8.94	13.46	14.99
REW9S20_1	9	20	0.16	1.5	0.04	0.05	0.34	0.14
REW9S20_2	9	20	0.24	2.5	10.18	8.94	10.44	12.51
Case		**Random waves**						
	***W* (cm)**	***S* (cm)**	***H* (m)**	**T_p (s)**	**Exp.**	**Num.**	***k*-ε NLS**	***k*-ε SCM**
IRW6S10	6	10	0.123	2.2	0.224	0.281	0.145	0.155
IRW9S10	9	10	0.126	2.2	0.096	0.073	0.065	0.078

It can be seen from Table 4 and Figure 13 that both turbulence models present globally higher average wave overtopping discharges than experimental and numerical results obtained by Tuan and Thin [18], especially for small overtopping discharge. *Bias* parameters of the *k*-ε NLS turbulence model are 0.87 and 0.86 compared to experimental and numerical results of Tuan and Thin [18], respectively; conversely, those of the *k*-ε SCM turbulence model are 1.56 and 1.55, respectively. The *k*-ε SCM turbulence model exhibits an average wave overtopping discharge 6.1 and 9 times greater than experimental values for the cases REW9S10_1 and REW6S20_1, respectively, both with small wave overtopping discharge. Both cases present a larger difference compared to experimental values. Using the *k*-ε NLS turbulence model, both cases exhibit an average wave overtopping discharge 4.8 and 7.8 times greater than experimental values, respectively, slightly smaller than those obtained by using the *k*-ε SCM turbulence model. Nevertheless, the average wave overtopping discharge is well predicted in several cases using the k-ε NLS turbulence model, such as in REW6S0 and REW9S10.

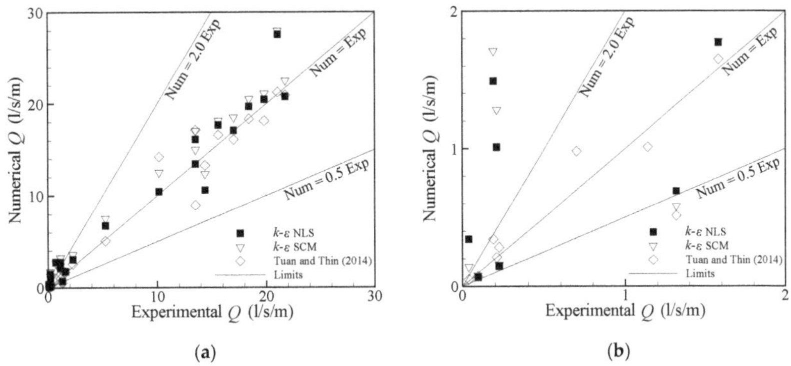

Figure 13. Numerical (k-ε NLS and k-ε SCM turbulence models and Tuan and Thin [18]) versus experimental (Tuan and Thin [18]) average wave overtopping discharge of regular and random waves: (**a**) all results and (**b**) details for small values of wave overtopping discharge.

The *NRMSE* of average wave overtopping discharges, for the set of cases presented in Table 4, is 38.6 and 34.4% for the k-ε SCM and k-ε NLS turbulence models, respectively. Results obtained using the k-ε NLS turbulence model are in better accordance with experimental data than those predicted by the k-ε SCM turbulence model. However, *NRMSE* of both models is larger than *NRMSE* of Tuan and Thin [18], which is equal to 23.2%. Nevertheless, the comparison of average wave overtopping discharges greater than 10 l/s/m shows that both turbulence models and Tuan and Thin [18] present the same order of magnitude of *NRMSE* and accuracy: *NRMSE* is 17.9% for k-ε SCM, 16.1% for k-ε NLS, and 14.5% for Tuan and Thin (2014). In cases when average wave overtopping discharges are smaller than 10 l/s/m, *NRMSE*s are quite large: 103.7% for k-ε SCM, 83.4% for k-ε NLS, and 30.4% for Tuan and Thin [18]. It is important to emphasize that the wave overtopping discharge is very sensitive to the model, experimentally and numerically, and differences found between present results and those of Tuan and Thin [18] are of the same order of magnitude as reported by Neves et al. [53].

For random waves, it can be noted that average wave overtopping discharges are small. For the case IRW9S10, k-ε SCM and NLS turbulence models and numerical results of Tuan and Thin [18] underestimate the average wave overtopping discharge, with a difference around 20% compared to experimental values, and slightly greater for k-ε NLS turbulence model, at 30%. For the case IRW6S10, k-ε SCM and NLS turbulence models underestimate the average wave overtopping discharge, with differences of 30 and 35%, respectively, compared to the experimental values, while Tuan and Thin [18] overestimates, with a difference of 25%.

Differences in average wave overtopping discharges found among turbulence models can lead to differences in the plunging breaking wave, wave dissipation after wave breaking and, consequently, in the wave overtopping. In addition, differences between k-ε SCM and NLS turbulence models used in this study and those employed by Tuan and Thin [18] could be related to different methods used for wave generation.

4.3. Wave Interaction with Porous Low Crested Rubble Mound Breakwater

Physical experiments conducted at the University of Cantabria within the framework of the European Union Design of Environmental Low Crested Coastal Defence Structures (DELOS) project are used to validate the numerical model in cases of regular and random waves impinging on low-crested rubble-mound multilayered breakwater. The case is examined by Garcia et al. [33] for regular waves and Lara et al. [34] for random waves by using the VARANS numerical model COBRAS (Cornell Breaking Waves and Structures [24,32]). Among the set of experiments, the low-crested breakwater model with 1.0 m crest width and water depth 0.4 m (freeboard with 5 cm) is tested. Figure 14 shows the sketch of the computational domain, which reproduces the general configuration of the tests. The low-crested structure is made of two porous layers of different hydraulic properties above a bottom connected to the bottom of the flume by a 1:20 slope. In the experimental test and mathematical model, the wave flume includes a flow recirculation system aimed at preventing water piling-up in the leeward side of the low-crested structure, due to the rear absorbing 1:20 slope beach at the end of the flume. In COBRAS, wave generation is carried out using a source function with a sponge-layer method used to absorb the waves that propagate in the opposite direction to the zone of interest. In the present numerical wave flume, a static wave-maker with active wave absorption, located at 4.76 m from the toe of the 1:20 bottom flume slope, generates the incident wave at the inlet flow boundary. An active wave absorption is used at the end of the flume which avoids using a flow recirculation system to maintain the mean water level in the leeward side of the low-crested structure. Fourier waves are generated for both regular incident wave cases (T =1.6 s, H = 0.07 and 0.10 m) and a TMA-type spectrum, with a γ parameter equal to 3.3 and cutoff frequencies 0.16 and 1.00 Hz [34], is considered for both random wave tests (T_p = 2.4 and 3.2 s with H_s = 0.10 m). Values of the linear, α_E, and nonlinear, β_E, coefficients governing the flow inside the porous media have been kept equal to those presented by Garcia et al. [33] for regular waves (model calibration was carried out using wave conditions: h = 0.40 m, H = 0.07 m, T = 1.6 s). For random waves, α_E is taken to be equal to 1000 for both armour layers (n = 0.53, D_{50} = 0.0387) and core (n = 0.49, D_{50} = 0.0118) and β_E equal to 0.8 and 1.2, respectively, according to Lara et al. [34]. The mesh resolution follows recommendations described in Sections 2.4 and 3.2, and an unstructured mesh with regular cells with grid resolution close to 0.0035 × 0.0035 m, i.e., aspect ratio close to 1, is used in the proximity of the low-crested structure, inside and outside the porous layers (Figure 15). The total number of cells is around 97,000. Standard k-ε, k-ε NLS and k-ε SCM turbulence models are used for this application, since they include source terms to take into account the effect of turbulence inside the porous media, which it is not the case of the k-ω SST turbulence model.

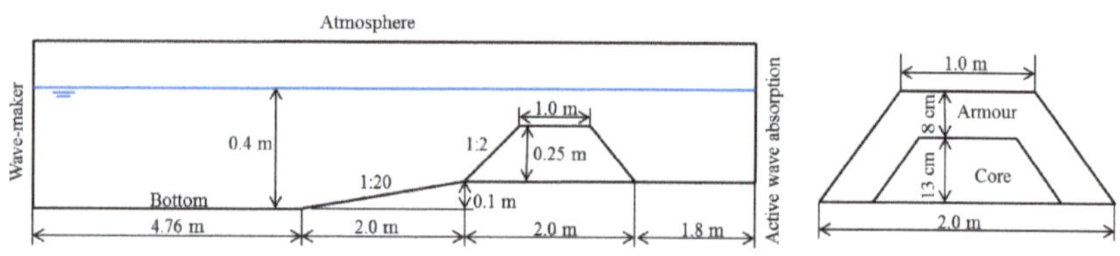

Figure 14. Sketch of the computational domain for the porous low-crested rubble-mound breakwater case.

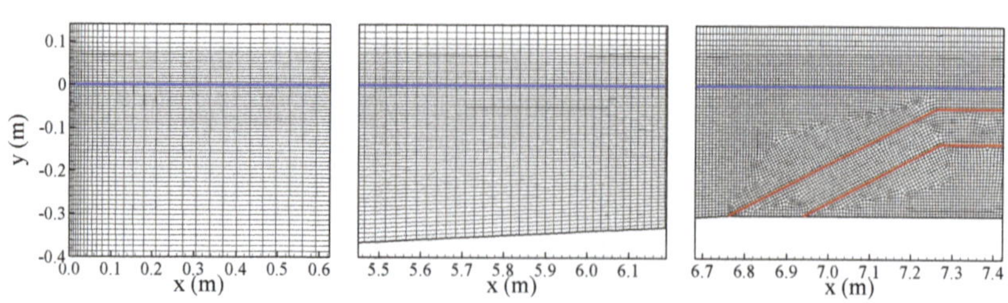

Figure 15. Details of the mesh near the wave-maker, along the flume (above the ramp) and in the proximity of the low-crested permeable breakwater. The blue line indicates the SWL and the red ones the limits of the porous layers.

Similar to Garcia et al. [33] for regular incident waves, numerical results from 10 waves, counted from 60 s of simulation, were analyzed, which ensured stabilized periodic flow conditions. For random waves, 200 waves are simulated for each test, which allows a time series long enough for analysis to be achieved, according to Lara et al. [34].

Figures 16 and 17 show results of maximum and minimum wave height envelopes and mean water level of a regular wave with a period T =1.6 s and wave height H = 0.07 and 0.10 m, respectively. Figures 18 and 19 show results of maximum and minimum wave height envelopes and mean water level of random waves with wave height H_s = 0.10 m and period T_p = 2.4 and 3.2 s, respectively. Numerical results are compared with experimental data of Lara et al. [34] in 11 free surface gauges placed before, over and after the submerged rubble-mound breakwater. Table 5 presents *Bias*, Equation (50), and *NRMSE*, Table 5 presents Bias and NRMSE of the wave height at 11 sections of measurement for regular and random waves for the three turbulence models, of the wave height at 11 sections of measurement for regular and random waves for the three turbulence models.

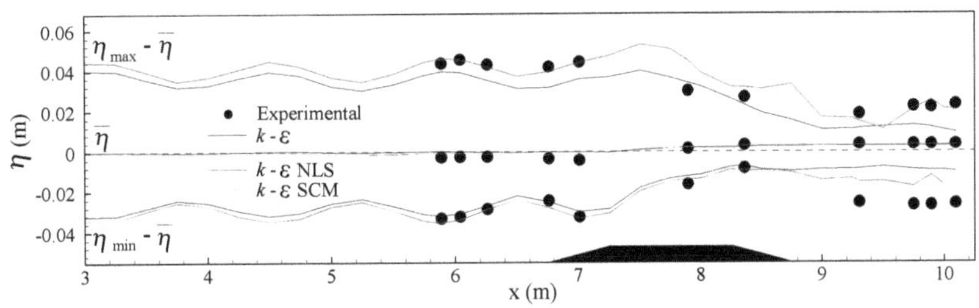

Figure 16. Maximum and minimum wave height envelopes and mean water level of *k-ε*, *k-ε* NLS and *k-ε* SCM models and the experiment (dash line indicates the still water level), T = 1.6 s, H = 0.07 m.

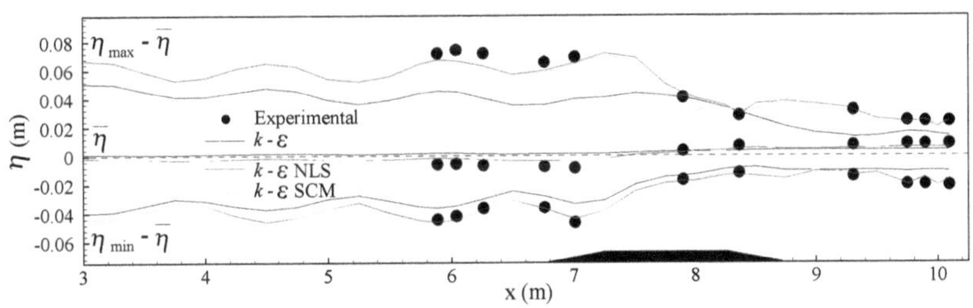

Figure 17. Maximum and minimum wave height envelopes and mean water level of *k-ε*, *k-ε* NLS and *k-ε* SCM models and the experiment (dash line indicates the still water level), T = 1.6 s, H = 0.10 m.

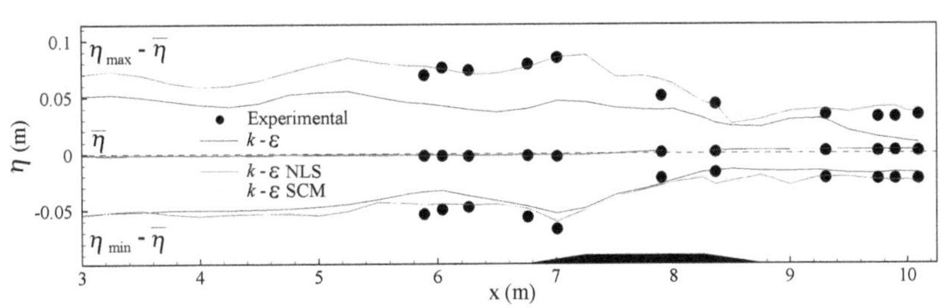

Figure 18. Maximum and minimum wave height envelopes and mean water level of *k-ε*, *k-ε* NLS and *k-ε* SCM models and the experiment (dash line indicates the still water level), T_p = 2.4 s, H_s = 0.10 m.

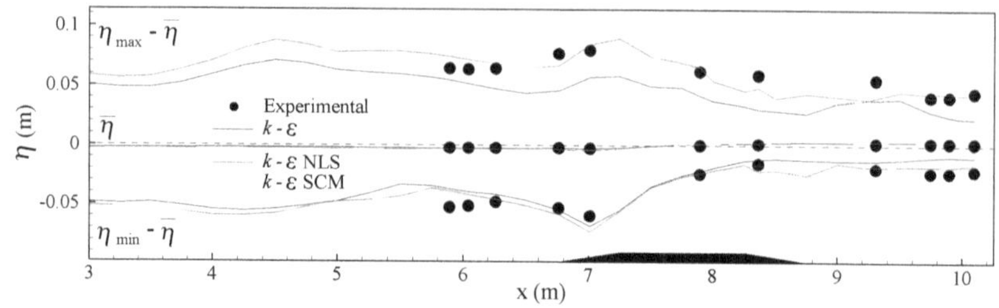

Figure 19. Maximum and minimum wave height envelopes and mean water level of k-ε, k-ε NLS and k-ε SCM models and the experiment (dash line indicates the still water level), T_p = 3.2 s, H_s = 0.10 m.

Table 5. *Bias* and *NRMSE* of wave height at the 11 sections of measurement for regular and random waves for the k-ε, k-ε NLS and k-ε SCM turbulence models.

	T = 1.6 s *H* = 0.07 m		*T* = 1.6 s *H* = 0.10 m		T_p = 2.4 s H_S = 0.10 m		T_p = 3.2 s H_S = 0.10 m	
	Bias	*NRMSE*	*Bias*	*NRMSE*	*Bias*	*NRMSE*	*Bias*	*NRMSE*
k-ε	−0.0143	30.1	−0.0246	37.1	−0.0323	39.0	−0.0609	66.0
k-ε NLS	−0.0022	15.4	−0.0028	5.8	0.0028	10.3	−0.0037	9.1
k-ε SCM	−0.0025	15.4	−0.0039	6.9	0.0026	10.2	−0.0056	13.5

Table 5 shows that *Bias* is negative for almost all cases, indicating that the numerical model underestimated the wave heights. Figures 16–19 show that the numerical model using the standard k-ε turbulence model, for the four tests, systematically underestimates the maximum and minimum wave height envelopes, the maximum wave height envelope being significantly more underestimated than the minimum one. *NRMSE* of the wave height varies from 30.1 to 66.0%. However, the numerical model using k-ε NLS and k-ε SCM turbulence models adequately reproduces the main features of the interaction of regular and random waves with submerged porous breakwater. The whole pattern of maximum and minimum wave height envelopes is accurately predicted: in the zone of the structure crest, the wave breaking, represented by the wave height damping, is correctly reproduced; in the leeward region, the transmission features due to overtopping and flow through the porous structure are also well reproduced. The mean water level variations due to wave breaking can be observed and are well predicted: the mean water level decreases at the offshore side of the breaking point and increases at the onshore side. *NRMSE* of the wave height varies from 5.8 (*T* =1.6 s, *H* = 0.10 m, using k-ε NLS turbulence model) to 18.4% (*T* = 1.6 s, *H* = 0.07 m, using k-ε turbulence model). In general, considering the four tests for each turbulence model, the average *NRMSE* is around 10 to 12%, and the best results are obtained using the k-ε NLS turbulence model. *NRMSE* on the wave height at 11 sections of measurement is 10.3 and 9.1% for k-ε NLS model, and 10.2 and 13.5% for k-ε SCM, for a random wave test with period T_p = 2.4 and 3.2 s, respectively. These values are in agreement with those obtained by Lara et al. [34] for both random waves, at 9 and 8%, respectively.

4.4. *Wave Overtopping of a Rubble Mound Breakwater*

In order to evaluate the capability of the models to simulate wave overtopping over a rubble mound breakwater, physical experiments obtained by Losada et al. [15] in the wave flume of the University of Cantabria are used. Figure 20 shows the sketch of the computational domain which reproduces the general configuration of the tests. The breakwater is built of an impermeable caisson, with 1.04 m length and 0.3 m height, installed on the rubble mound foundation. The frontal wall is located at 45 m from the wave-maker. The gravel core foundation, with a 1:2 slope, D_{50} = 0.01 m and porosity 0.48, is 0.7 m high and covered by two external layers: one layer is gravel with D_{50} = 0.035 m, and the external layer is gravel with D_{50} = 0.135 m. The porosity is 0.50 for both layers. The bottom of the flume is horizontal, the still water level is 0.8 m, and the structure freeboard is 0.2 m. In the experimental test and the COBRAS-UC numerical model [33,34], three dissipative ramps were placed at the rear end of the wave flume to absorb the transmitted waves. A source function is used in COBRAS-UC for the wave generation associated with a sponge-layer method to absorb the waves that propagate in the opposite direction to the zone of interest. In the present numerical wave flume, a static wave-maker with active wave absorption generates the incident regular wave (Fourier wave for T = 6 s, H = 0.25 m) at the inlet flow boundary and an active wave absorption is used at the end of the flume. Values of the linear, α_E, and nonlinear, β_E, coefficients governing the flow inside the porous media have been kept equal to those presented in Losada et al. [15] for regular waves (model calibration was carried out using wave conditions: h = 0.80 m, H = 0.10 m, and T = 3 and 5 s). The best parameter values are: α_E = 200 for both the armour layer and core; β_E = 0.8 for the breakwater core; β_E = 1.1 for the small gravel external layer; and β_E = 0.7 for the big gravel external layer. The mesh resolution follows the recommendations described in Sections 2.4 and 3.2, and an unstructured mesh with regular cells with grid resolution close to 0.01 × 0.01 m, i.e., aspect ratio close to 1, is used in the vicinity of the breakwater, which is the area of interest, inside and outside the porous layers (Figure 21). The total number of cells is around 77,240. The standard k-ε, k-ε NLS and k-ε SCM turbulence models are used for this application, since they include source terms that take into account the effect of turbulence inside the porous media, which is not the case of the k-ω SST turbulence model.

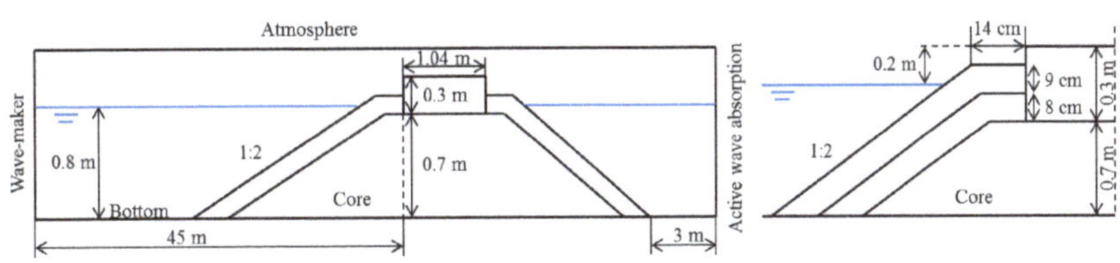

Figure 20. Sketch of the computational domain for the rubble mound breakwater case.

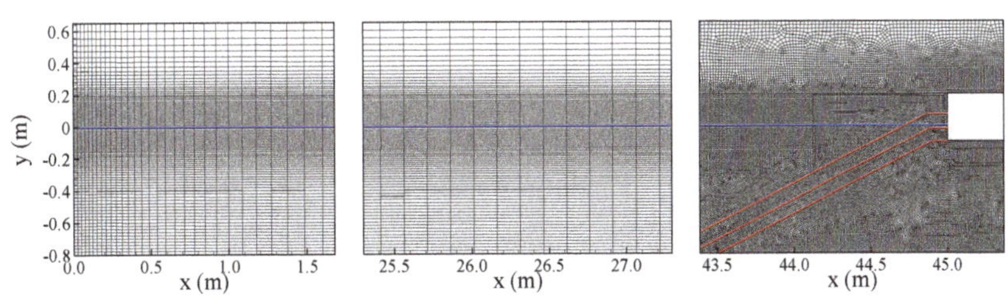

Figure 21. Details of the mesh near the wave-maker, along the flume and in the proximity of the rubble mound breakwater. The blue line indicates the SWL and the red ones the limits of the porous layers.

Figure 22 shows the free surface elevation at a wave gauge at 42.5 m from the wave-maker, i.e., located near the toe of the breakwater. Two intervals of time are presented: the first time interval, from 0 to 60 s, corresponds to the beginning of the interaction between the wave and the breakwater; and the second one, from 240 to 300 s, corresponds to the periodic stationary flow. Figure 23 shows the time series of the wave overtopping discharge. Table 6 shows the average wave overtopping discharge calculated from 138 to 300 s, i.e., considering 27 wave periods.

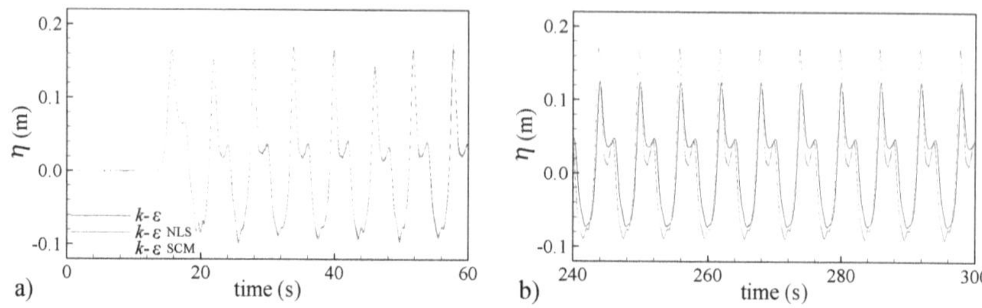

Figure 22. Time series of free surface elevation for k-ε, k-ε NLS and k-ε SCM turbulence models at the gauge x = 42.5 m (near the toe of the breakwater): (**a**) from 0 to 60 s, and (**b**) from 240 to 300 s.

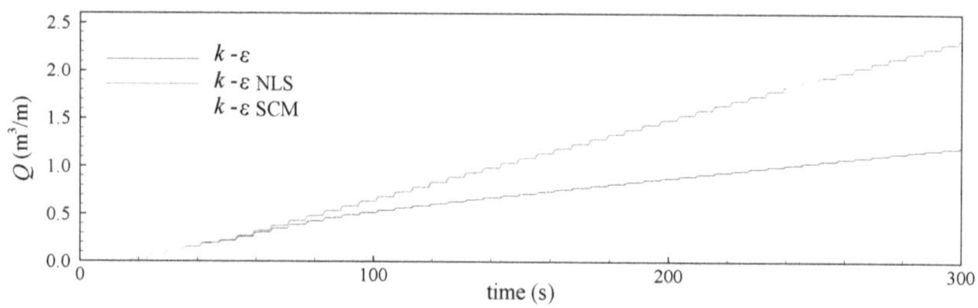

Figure 23. Time series of wave overtopping discharge for k-ε, k-ε NLS and k-ε SCM turbulence models.

Table 6. Average wave overtopping discharge for k-ε, k-ε NLS and k-ε SCM turbulence models compared to experimental and numerical values of Losada et al. [15].

	Losada et al. [15]		FLUENT		
	Exp.	**Num.**	**k-ε**	**k-ε NLS**	**k-ε SCM**
Average wave overtopping discharge ($m^3/s/m$)	0.0066	0.0063	0.0033	0.0083	0.0084
Relative error (%)	-	4.6	50.0	25.8	27.3

It can be seen that both free surface elevation and wave overtopping discharge, for the three turbulence models, are very similar until 60 s. However, wave overtopping discharge slows down using the standard k-ε turbulence model, due to the progressive growth of eddy viscosity which reduces the free surface elevation until reaching a stabilized behavior. This does not occur with both k-ε NLS and k-ε SCM turbulence models, which present the same behavior for free surface elevation and wave overtopping discharge. Both models overestimate the average wave overtopping discharge with a difference related to the experimental result of 26%, whereas the value obtained by the standard k-ε turbulence model is largely underestimated. Differences observed between numerical results and those obtained by Losada et al. [15] can be related to the wave generation and the parameter values of the porous layers, which are equal to those of Losada et al. [15], which may not be the best parameters for the present numerical model, even if the porous medium equations seem the same as the author's.

5. Conclusions

This paper showed methodologies implemented in a RANS-VoF numerical model, the FLUENT® numerical model in this case, to simulate interactions between waves and impermeable and porous costal structures, which involve wave breaking and overtopping. Performance of standard k-ε and k-ω SST turbulence models was compared with two modified turbulence models, implemented in the RANS-VoF model, which avoid the growth of the eddy viscosity: the *k*-ε NLS, developed by Shih et al. [42] and Lin and Liu [5], and *k*-ε SCM, developed by Larsen and Fuhrman [14]. Equations of momentum and turbulence models were adapted in original numerical model to take into account the porous media of coastal structures.

The performance of *k*-ε NLS, *k*-ε SCM and standard *k*-ε and *k*-ω SST models were compared in cases with wave breaking (regular waves on an impermeable beach of Ting and Kirby [6]), wave breaking and overtopping (regular and random waves over impermeable sea dikes with crow-walls, proposed by Tuan and Thin [18]), wave on a porous submerged structure (regular and random waves over a low-crested rubble breakwater of Garcia et al. [33] and Lara et al. [34]), and wave overtopping in porous structure (regular and random waves over a rubble mound breakwater of Losada et al. [15]). The main conclusions of these analyses are:

(a) Mainly for longtime simulations in long numerical wave flumes, the standard *k*-ε turbulence model severely over-predicted eddy viscosity and, consequently, caused unphysical behaviors. In addition, the decay of the free surface elevation and an under-estimated wave overtopping discharge were noticed. These observations were clearly shown through the analysis of wave envelopes for wave breaking on an impermeable beach and waves over low-crested rubble breakwater.
(b) The *k*-ω SST turbulence model also had the same tendency as the *k*-ε one, but with much less intensity.
(c) The *k*-ε NLS and *k*-ε SCM turbulence models had similar performance, avoiding unphysical behaviors and modeling wave propagation and a nearly constant wave overtopping discharge over long durations. Both turbulence models showed results with good agreement with experimental ones. The *k*-ε NLS turbulence model presented slightly better results than the *k*-ε SCM one.
(d) It is very difficult to compare the wave overtopping discharge obtained by numerical models, since it strongly depends on the simulation of phenomena that precede the overtopping, mainly in cases in which the magnitude of the discharge is small. Both wave flume length and wave generation can cause small differences in incident waves which can lead to significant differences on wave overtopping discharge. Nevertheless, the application of FLUENT® to regular and random waves over impermeable sea dikes with crow-walls showed results of average wave overtopping discharge in good accordance with experimental and numerical ones obtained by Tuan and Thin [18]. *NRMSE* of average wave overtopping discharges were 38.6 and 34.4% for the *k*-ε SCM and *k*-ε NLS turbulence models, respectively, slightly larger than Tuan and Thin [18], which is equal to 23.2%.
(e) Methodologies developed and implemented in RANS-VoF numerical models to deal with coastal porous structures had good performance, but they are hardly dependent of some empirical parameters that must be set. Regular and random waves over a low-crested rubble breakwater showed good agreement with experimental results, with a *NRMSE* varying from 5.8 to 15.4% for the *k*-ε NLS and *k*-ε SCM turbulence models, respectively, also in agreement with numerical results obtained by Garcia et al. [33]. Regular wave over a rubble mound breakwater presented a slightly larger average wave overtopping discharge than experimental and numerical results obtained by Losada et al. [15].

Although the wave generation and absorption in 3D domains are more complex, all methodologies used for 2D cases in this investigation can also be applied to the more general 3D wave tanks.

Author Contributions: Conceptualization, methodology, software, validation, investigation, writing—review, E.D. and P.R.F.T. All authors have read and agreed to the published version of the manuscript.

Funding: This research received no external funding.

Institutional Review Board Statement: Not applicable.

Informed Consent Statement: Not applicable.

Data Availability Statement: The data presented in this study are available on request from the corresponding author. The data are not publicly available due to confidential reasons.

Acknowledgments: The authors acknowledge the cooperation between the Universidade Federal do Rio Grande—FURG, Brazil, and the Laboratório Nacional de Engenharia Civil—LNEC, Portugal. The second author acknowledges the support of CNPq (project 307546/2921-7) and FAPERGS (project 21/2551-0002014-8).

Conflicts of Interest: The authors declare no conflict of interest. The funders had no role in the design of the study; in the collection, analyses, or interpretation of data; in the writing of the manuscript, or in the decision to publish the results.

Appendix A

The parameters and functions that appear in the k-ω SST model [47], given by Equations (10)–(12), are detailed in this section.

In the first term, on the RHS (right hand side) of Equation (10), which represents the effective diffusivity of k, σ_k is given by:

$$\sigma_k = \frac{1}{F_1/\sigma_{k,1} + (1 - F_1)/\sigma_{k,2}} \tag{A1}$$

where

$$F_1 = \tan\mathrm{h}\left(\Phi_1^4\right) \tag{A2}$$

$$\Phi_1 = min\left[max\left(\frac{\sqrt{k}}{0.09\omega y}, \frac{500\mu}{\rho y^2 \omega}\right), \frac{4\rho k}{\sigma_{\omega,2} D_\omega^+ y^2}\right] \tag{A3}$$

$$D_\omega^+ = max\left[2\rho\frac{1}{\sigma_{\omega,2}}\frac{1}{\omega}\frac{\partial k}{\partial x_j}\frac{\partial \omega}{\partial x_j}, 10^{-10}\right] \tag{A4}$$

y is the distance to the next surface, D_ω^+ is the positive portion of the cross-diffusion term, $\sigma_{k,1}$ = 1.176, $\sigma_{k,2}$ = 1.0 and $\sigma_{\omega 2}$ = 1.168.

The third term, on the RHS of Equation (10), represents the dissipation of k, in which β^* is

$$\beta^* = 0.09\left[\frac{4/15 + (Re_t/8)^4}{1 + (Re_t/8)^4}\right] \tag{A5}$$

where

$$Re_t = \frac{\rho k}{\mu\omega} \tag{A6}$$

and f_{β^*} is constant equal to 1. The last term on the RHS of Equation (10), S_k, is a source term.

In Equation (11), σ_ω, in the third term on the RHS, which represents the effective diffusivity of ω, is given by:

$$\sigma_\omega = \frac{1}{F_1/\sigma_{\omega,1} + (1 - F_1)/\sigma_{\omega,2}} \tag{A7}$$

where $\sigma_{\omega 1}$ = 2.0.

The second term, on the RHS of Equation (11), represents the production of ω, in which α and α^* are

$$\alpha = \frac{\alpha_\infty}{\alpha^*}\left(\frac{\alpha_0 + Re_t/R_\omega}{1 + Re_t/R_\omega}\right) \tag{A8}$$

$$\alpha^* = \alpha_\infty^*\left(\frac{\alpha_0^* + Re_t/R_k}{1 + Re_t/R_k}\right) \tag{A9}$$

where $\alpha^*_\infty = 1$, $\alpha_0 = 1/9$ and

$$\alpha_\infty = F_1\alpha_{\infty,1} + (1 - F_1)\alpha_{\infty,2} \tag{A10}$$

$$\alpha_{\infty,1} = \frac{\beta_{i,1}}{\beta^*_\infty} - \frac{\kappa^2}{\sigma_{\omega,1}\sqrt{\beta^*_\infty}} \tag{A11}$$

$$\alpha_{\infty,2} = \frac{\beta_{i,2}}{\beta^*_\infty} - \frac{\kappa^2}{\sigma_{\omega,2}\sqrt{\beta^*_\infty}} \tag{A12}$$

$R_\beta = 2.95$, $R_k = 6$, $\alpha^*_0 = 0.072/3$, $\beta_{i,1} = 0.075$, $\beta_{i,2} = 0.0828$, $\beta^*_\infty = 0.09$ and $\kappa = 0.41$.
The third term of Equation (11) represents the dissipation of ω, in which

$$\beta = F_1\beta_{i,1} + (1 - F_1)\beta_{i,2} \tag{A13}$$

and f_β is constant equal to 1. The last term RHS of Equation (11), S_ω, is a source term.
In Equation (12), $a_1 = 0.31$ and F_2 is

$$F_2 = \tan\mathrm{h}\left(\Phi_2^2\right) \tag{A14}$$

where

$$\Phi_2 = max\left[2\frac{\sqrt{k}}{0.09\omega y}, \frac{500\mu}{\rho y^2\omega}\right] \tag{A15}$$

S is the modulus of the mean rate-of-strain tensor, defined as:

$$S = \sqrt{2S_{ij}S_{ij}} \tag{A16}$$

References

1. Sakai, T.; Mizutani, T.; Tanaka, H.; Tada, Y. Vortex formation in plunging breakers. In Proceedings of the 20th Conference in Coastal Engineering, Taipei, Taiwan, 9–14 November 1986; pp. 711–723.
2. Harlow, F.H.; Welch, J.E. Numerical calculation of time-dependent viscous incompressible flow of fluid with free surface. *Phys. Fluids* **1965**, *8*, 2182–2189. [CrossRef]
3. Lemos, C.M. Wave breaking: A numerical study. In *Lecture Notes in Engineering*; Springer: Berlin/Heidelberg, Germany, 1992; Volume 71, pp. 1–185.
4. Hirt, C.W.; Nichols, B.D. Volume of fluid VOF method for the dynamics of free boundaries. *J. Comput. Phys.* **1981**, *39*, 201–225. [CrossRef]
5. Lin, P.Z.; Liu, P.L.F. A numerical study of breaking waves in the surf zone. *J. Fluid Mech.* **1998**, *359*, 239–264. [CrossRef]
6. Ting, F.C.K.; Kirby, J.T. Observation of undertow and turbulence in a laboratory surf zone. *Coast. Eng.* **1994**, *24*, 51–80. [CrossRef]
7. Bradford, S.F. Numerical simulation of surf zone dynamics. *J. Waterw. Port Coast. Ocean Eng.* **2000**, *126*, 1–13. [CrossRef]
8. Mayer, S.; Madsen, P.A. Simulations of breaking waves in the surf zone using a Navier–Stokes solver. In Proceedings of the 25th International Conference in Coastal Engineering, Sydney, Australia, 16–21 July 2000; pp. 928–941.
9. Jacobsen, N.G.; Fuhrman, D.R.; Fredsoe, J. A wave generation toolbox for the open-source CFD library: OpenFOAM(R). *Int. J. Numer. Methods Fluids* **2012**, *70*, 1073–1088. [CrossRef]
10. Brown, S.A.; Greaves, D.M.; Magar, V.; Conley, D.C. Evaluation of turbulence closure models under spilling and plunging breakers in the surf zone. *Coast. Eng.* **2016**, *114*, 177–193. [CrossRef]
11. Devolver, B.; Trouch, P.; Rauwoens, P. Performance of a buoyancy-modified k-ω and k-ω SST turbulence model for simulating wave breaking under regular waves using OpenFOAM®. *Coast. Eng.* **2018**, *138*, 49–65. [CrossRef]
12. Chella, M.A.; Bihs, H.; Myrhaug, D.; Muskulus, M. Breaking characteristics and geometric properties of spilling breakers over slopes. *Coast. Eng.* **2015**, *95*, 4–19. [CrossRef]
13. Chella, M.A.; Bihs, H.; Myrhaug, D.; Muskulus, M. Hydrodynamic characteristics and geometric properties of plunging and spilling breakers over impermeable slopes. *Ocean Model.* **2016**, *103*, 53–72. [CrossRef]
14. Larsen, B.E.; Fuhrman, D.R. On the over-production on turbulence beneath surface waves in Reynolds-averaged Navier-Stokes models. *J. Fluid Mech.* **2018**, *853*, 419–460. [CrossRef]
15. Losada, I.J.; Lara, J.L.; Guauche, R.; Gonzales-Ondina, J.M. Numerical analysis of wave overtopping of rubble mound breakwaters. *Coast. Eng.* **2008**, *55*, 47–62. [CrossRef]
16. EurOtop. Manual on Wave Overtopping of Sea Defences and Related Structures. An Overtopping Manual Largely Based on European Research, but for Worldwide Application. Available online: www.overtopping-manual.com (accessed on 21 September 2021).

17. Soliman, A.; Raslan, M.S.; Reeve, D.E. Numerical simulation of wave overtopping using two dimensional breaking wave model. *Trans. Build Environ.* **2003**, *70*, 439–447.
18. Tuan, T.Q.; Thin, N.V. Numerical study of wave overtopping on sea-dikes with crown-walls. *J. Hydro-Environ. Res.* **2014**, *8*, 367–382.
19. Tuan, T.Q.; Oumeraci, H. A numerical model of wave overtopping on seadikes. *Coast. Eng.* **2010**, *57*, 757–772. [CrossRef]
20. De Finis, S.; Romano, A.; Bellotti, G. Numerical and laboratory analysis of post-overtopping wave impacts on a storm wall for a dike-promenade structure. *Coast. Eng.* **2020**, *155*, 103598. [CrossRef]
21. Metallinos, A.S.; Klonaris, G.T.; Memos, C.D.; Dimas, A.A. Hydrodynamic conditions in a submerged porous breakwater. *Ocean Eng.* **2019**, *172*, 712–725. [CrossRef]
22. Dentale, F.; Reale, F.; Leo, A.D.; Carratelli, E.P. A CFD approach to rubble mound breakwater design. *Int. J. Nav. Archit. Ocean Eng.* **2018**, *10*, 644–650. [CrossRef]
23. Van Gent, M.R.A. Porous flow through rubbe-mound material. *J. Waterw. Port Coast. Ocean Eng.* **1995**, *121*, 176–181. [CrossRef]
24. Liu, L.-F.; Lin, P.; Sakakiyama, T. Numerical modeling of wave interaction with porous structures. *J. Waterw. Port Coast. Ocean Eng.* **1999**, *125*, 322–330. [CrossRef]
25. Nakayama, A.; Kuwahara, F. A macroscopic turbulence model for flow in a porous medium. *J. Fluids Eng.* **1999**, *121*, 427–433. [CrossRef]
26. Getachew, D.; Minkowycz, W.J.; Lage, J.L. A modified form of the k-e model for turbulent flows of an incompressible fluid in porous media. *Int. J. Heat Mass Transf.* **2000**, *43*, 909–915. [CrossRef]
27. Pedras, M.H.J.; de Lemos, M.J.S. Macroscopic turbulence modelling for incompressible flow through undeformable porous media. *Int. J. Heat Mass Transf.* **2001**, *44*, 1081–1093. [CrossRef]
28. Hur, D.-S.; Mizutani, N. Numerical estimation of the wave forces acting on a three-dimensional body on submerged breakwater. *Coast. Eng.* **2003**, *47*, 329–345. [CrossRef]
29. Hur, D.-S.; Lee, K.-H.; Yeom, G.-S. The phase difference effects on 3-D structure of wave pressure acting on a composite breakwater. *Ocean Eng.* **2008**, *35*, 1826–1841. [CrossRef]
30. Bear, J.; Bachmat, Y. *Introduction to Modeling of Transport Phenomena in Porous Media. Theory and Applications of Transport in Porous Media*; Kluwer Academic Publishers: Dordrecht, The Netherlands, 1990; Volume 4.
31. Whitaker, S. *The Method of Volume Averaging. Theory and Applications of Transport in Porous Media*; Kluwer Academic Publishers: Dordrecht, The Netherlands, 1999; Volume 13.
32. Hsu, T.-J.; Sakakiyama, T.; Liu, P.L.-F. A numerical model for wave motions and turbulence flows in front of a composite breakwater. *Coast. Eng.* **2002**, *46*, 25–50. [CrossRef]
33. Garcia, N.; Lara, J.; Losada, I. 2-D numerical analysis of near-field flow at low-crested permeable breakwaters. *Coast. Eng.* **2004**, *51*, 991–1020. [CrossRef]
34. Lara, J.L.; Garcia, N.; Losada, I.J. RANS modeling applied to random wave interaction with submerged permeable structures. *Coast. Eng.* **2006**, *53*, 395–417. [CrossRef]
35. del Jesus, M.; Lara, J.L.; Losada, I.J. Three-dimensional interaction of waves and porous structures. Part I: Numerical model formulation. *Coast. Eng.* **2012**, *64*, 57–72. [CrossRef]
36. Jensen, B.; Jacobsen, N.G.; Christensen, E.D. Investigations on the porous media equations and resistance coefficients for coastal structures. *Coast. Eng.* **2014**, *84*, 56–72. [CrossRef]
37. Higuera, P.; Lara, J.L.; Losada, I.J. Three-dimensional interaction of waves and porous coastal structures using OpenFOAM. Part I: Formulation and validation. *Coast. Eng.* **2014**, *83*, 243–258. [CrossRef]
38. Vanneste, D.; Troch, P. 2D numerical simulation of large-scale physical model tests of wave interaction with a rubble-mound breakwater. *Coast. Eng.* **2015**, *103*, 22–41. [CrossRef]
39. Didier, E.; Teixeira, P.R.F.; Neves, M.G. A 3D Numerical Wave Tank for Coastal Engineering Studies. *Defect Diffus. Forum* **2017**, *372*, 1–10.
40. Teixeira, P.R.F.; Didier, E.; Neves, M.G. A 3D RANS-VOF wave tank for oscillating water column device studies. In Proceedings of the MARINE 2017, Nantes, France, 15–17 May 2017; pp. 710–721.
41. Teixeira, P.R.F.; Didier, E. Numerical analysis of the response of an onshore oscillating water column wave energy converter to random waves. *Energy* **2021**, *220*, 119719. [CrossRef]
42. Shih, T.-H.; Zhu, J. Calculation of wall-bounded complex flows and free shear flows. *Int. J. Numer. Methods Fluids* **1996**, *23*, 1133–1144. [CrossRef]
43. ANSYS. *FLUENT—User's Guide*; ANSYS Inc.: Canonsburg, PA, USA, 2016.
44. Wilcox, D.C. *Turbulence Modeling for CFD*; DCW Industries, Inc.: La Canada, CA, USA, 1993.
45. Versteeg, H.K.; Malalasekera, W. *An Introduction to Computational Fluid Dynamics: The Finite Volume Method*; Pearson Education Limited: London, UK, 2007.
46. Harlow, F.H.; Nakayama, P. *Transport of Turbulence Energy Decay Rate*; University California Report LA-3854; Los Alamos Science Lab: Los Alamos, NM, USA, 1968.
47. Menter, F.R. Two-Equation Eddy-Viscosity Turbulence Models for Engineering Applications. *AIAA J.* **1994**, *32*, 1598–1605. [CrossRef]
48. Ergun, S. Fluid flow through packed columns. *Chem. Eng. Prog.* **1952**, *48*, 89–94.

49. Engelund, F. On the laminar and turbulent flow of ground water through homogeneous sand. *Trans. Dan. Acad. Tech. Sci.* **1953**, *3*, 356–361.
50. van Gent, M.R.A. Formulae to Describe Porous Flows. In *Communications on Hydraulic and Geotechnical Engineering;* Delft University of Technology: Delft, The Netherlands, 1992.
51. van Gent, M.R.A. Stationary and Oscillatory Flow through Coarse Porous Media. In *Communications on Hydraulic and Geotechnical Engineering;* Report No. 1993-09; Faculty of Civil Engineering, Delft University of Technology: Delft, The Netherlands, 1993.
52. Santos, G.C.; Teixeira, P.R.F.; Didier, E. Analysis of wave overtopping on an impermeable coastal structure using a RANS-VOF numerical model. In Proceedings of the 25th ABCM International Congress of Mechanical Engineering, COBEM 2019, Uberlândia, MG, Brazil, 20–25 October 2019.
53. Neves, M.G.; Didier, E.; Brito, M.; Clavero, M. Numerical and physical modelling of wave overtopping on a smooth impermeable dike with promenade under strong incident waves. *J. Mar. Sci. Eng.* **2021**, *9*, 865. [CrossRef]
54. Schäffer, H.; Klopman, G. Review of multidirectional active wave absorption methods. *J. Waterw. Port Coast. Ocean Eng.* **2000**, *126*, 88–97. [CrossRef]
55. Lara, J.L.; Ruju, A.; Losada, I.J. Reynolds averaged Navier-Stokes modelling of longwaves induced by a transient wave group on a beach. *Proc. R. Soc. A Math. Phys. Eng. Sci.* **2011**, *467*, 1215–1242.
56. Didier, E.; Neves, M.G. A Semi-Infinite Numerical Wave Flume using Smoothed Particle Hydrodynamics. *IJOPE* **2012**, *22*, 193–199.
57. Higuera, P.; Lara, J.L.; Losada, I.J. Realistic wave generation and active wave absorption for Navier-Stokes models application to OpenFOAM®. *Coast. Eng.* **2013**, *71*, 102–118. [CrossRef]
58. Fenton, J.D. A fifth-order Stokes Theory for Steady Waves. *J. Waterw. Port Coast. Ocean Eng.* **1985**, *111*, 216–234. [CrossRef]
59. Fenton, J.D. The numerical solution of steady water waves problems. *Comput. Geosci.* **1988**, *14*, 357–368. [CrossRef]
60. Barreiro, T.G. Estudo da Interacção de Uma Onda Monocromática Com um Conversor de Energia. Master's Thesis, Nova University of Lisbon, Nova School of Science and Technology, Costa de Caparica, Portugal, 2009. (In Portuguese)
61. Didier, E.; Conde, J.M.P.; Teixeira, P.R.F. Numerical simulation of an oscillation water column wave energy converter with and without damping. In Proceedings of the Fourth International Conference on Computational Methods in Marine Engineering, Lisbon, Portugal, 28–30 September 2011; pp. 206–217.
62. Conde, J.M.P.; Teixeira, P.R.F.; Didier, E. Numerical simulation of an oscillating water column wave energy converter: Comparison of two numerical codes. In Proceedings of the 21th International Offshore (Ocean) and Polar Engineering Conference-ISOPE, Maui, HI, USA, 19–24 June 2011; pp. 674–688.
63. Teixeira, P.R.F.; Davyt, D.P.; Didier, E.; Ramalhais, R. Numerical simulation of an oscillating water column device using a code based on Navier Stokes equations. *Energy* **2013**, *61*, 513–530. [CrossRef]
64. Dias, J.; Mendonça, A.; Didier, E.; Neves, M.G.; Conde, J.M.P.; Teixeira, P.R.F. Application of URANS-VOF models in hydrodynamics study of oscillating water column. In Proceedings of the SCACR2015–International Short Course/Conference on Applied Coastal Research, Florence, Italy, 28 September–1 October 2015.
65. Mendonça, A.; Dias, J.; Didier, E.; Fortes, C.J.E.M.; Neves, M.G.; Reis, M.T.; Conde, J.M.P.; Poseiro, P.; Teixeira, P.R.F. An integrated tool for modelling OWC-WECs in vertical breakwaters: Preliminary developments. *J. Hydro-Environ. Res.* **2018**, *19*, 198–213. [CrossRef]
66. Lisboa, R.C.; Teixeira, P.R.F.; Torres, F.R.; Didier, E. Numerical evaluation of the power output of an oscillating water column wave energy converter installed in the Southern Brazilian coast. *Energy* **2018**, *162*, 1115–1124. [CrossRef]
67. Willmott, C.J.; Ackleson, S.G.; Davis, R.E.; Feddema, J.J.; Klink, K.M.; Legates, D.R.; O'Donnell, J.; Rowe, C.M. Statistics for the evaluation and comparison of models. *J. Geophys. Res.* **1985**, *90*, 8995–9005. [CrossRef]

Journal of
*Marine Science
and Engineering*

MDPI

Article

Conceptual Design of a Vibration Test System Based on a Wave Generator Channel for Lab-Scale Offshore Wind Turbine Jacket Foundations

Ángel Encalada-Dávila [1,†], Lenín Pardo [2,†], Yolanda Vidal [3,4,*,†], Efraín Terán [1,†] and Christian Tutivén [1,†]

1 Mechatronics Engineering, Escuela Superior Politécnica del Litoral (ESPOL), Faculty of Mechanical Engineering and Production Science (FIMCP), Campus Gustavo Galindo, Km. 30.5 vía Perimetral, Guayaquil EC09015863, Ecuador
2 Facultad de Ingenierías, Universidad ECOTEC, Km. 13.5 Samborondón, Samborondón EC092302, Ecuador
3 Control, Data and Artificial Intelligence (CoDAlab), Department of Mathematics, Escola d'Enginyeria de Barcelona Est (EEBE), Universitat Politècnica de Catalunya (UPC), Campus Diagonal-Besos (CDB), Eduard Maristany, 16, 08019 Barcelona, Spain
4 Institute of Mathematics (IMTech), Universitat Politecnica de Catalunya (UPC), Pau Gargallo 14, 08028 Barcelona, Spain
* Correspondence: yolanda.vidal@upc.edu; Tel.: +34-934-137-309
† These authors contributed equally to this work.

Citation: Encalada-Dávila, Á.; Pardo, L.; Vidal, Y.; Terán, E.; Tutivén, C. Conceptual Design of a Vibration Test System Based on a Wave Generator Channel for Lab-Scale Offshore Wind Turbine Jacket Foundations. *J. Mar. Sci. Eng.* **2022**, *10*, 1247. https://doi.org/10.3390/jmse10091247

Academic Editors: Diego Vicinanza and Eva Loukogeorgaki

Received: 19 July 2022
Accepted: 31 August 2022
Published: 5 September 2022

Abstract: Structural health monitoring (SHM) systems are designed to continually monitor the health of structures (e.g., civil, aeronautic) by using the information collected through a distributed sensor network. However, performing tests on real structures, such as wind turbines, implies high logistic and operational costs. Therefore, there is a need for a vibration test system to evaluate designs at smaller scales in a laboratory setting in order to collect data and devise predictive maintenance strategies. In this work, the proposed vibration test system is based on a lab-scale wind turbine jacket foundation related primarily to an offshore environment. The test system comprises a scaled wave generator channel, a desktop application (WTtest) to control the channel simulations, and a data acquisition system (DAQ) to collect the information from the sensors connected to the structure. Various equipment such as accelerometers, electrodynamic shaker, and DAQ device are selected as per the design methodology. Regarding the mechanical part, each component of the channel is designed to be like the wave absorber, the mechanical multiplier, the piston-type wavemaker , and the wave generator channel. For this purpose, the finite element method is used in static and fatigue analysis to evaluate the stresses and deformations; this helps determine whether the system will work safely. Moreover, the vibration test system applies to other jacket structures as well, giving it greater utility and applicability in different research fields. In sum, the proposed system is compact and has three well-defined components that work synchronously to develop the experimental simulations.

Keywords: wind turbine; wavemaker channel; vibration system; DAQ system; accelerometer

1. Introduction

In recent years, the planet's energy crisis has prompted mankind to seek new sources of energy. The development of renewable energy sources has become key to combating the phenomenon of global warming. Among the various technological advances in the field of renewable energy sources, wind energy is already cost-effective (land-based utility-scale wind is one of the lowest-priced energy sources available today) and guarantees a clean and environmentally friendly energy source, which makes it one of the best alternatives at present [1].

Figure 1 shows a typical configuration of a wind turbine (WT). Moreover, the size of WTs has been increasing as a consequence of the desire to generate more energy and be more efficient in the process.

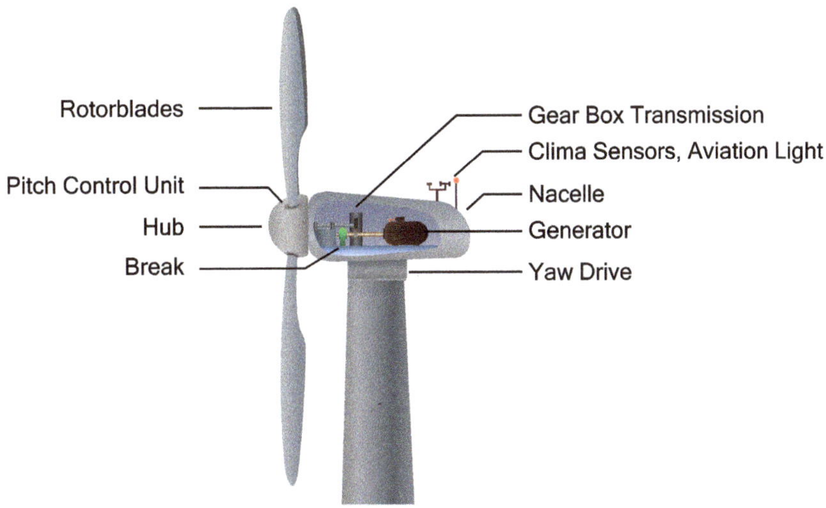

Figure 1. Main parts of a wind turbine.

Besides size, the installation environment has also been one of the factors changing over time. In the past, all WTs were placed onshore, but more recently, they have migrated to offshore environments as well. However, this strategy has both advantages and disadvantages, as new engineering challenges are hindering the process. The frequent inspection and maintenance of this type of offshore installation pose a great risk, considering the height and dimensions of the WTs and the fact that they are located on open waters. The durability of the material with which the structures are built also changes since the marine conditions are different [1]. Another major challenge is energy transport produced from the sea to the land. It is therefore necessary to improve safety conditions, minimize downtime, reduce logistics and maintenance costs, and avoid sudden or unexpected damages. To achieve it, WTs must be constantly monitored to ensure that they are in good operating condition and capable of anticipating future damages or breakdowns.

Of the many existing monitoring systems, structural health monitoring (SHM) is of vital importance since structural damage can lead to a catastrophic failure that puts the entire system at risk. The information that can be collected from an SHM system can be used to provide maintenance to WTs, without the need to be on site [2]. Through an SHM system, the unnecessary inspection times of components and time lost in change or replacement of parts can be significantly reduced. Moreover, such a system provides confidence since it is being monitored constantly. SHM systems are designed so that during the life of a particular structure, the health of the constituent materials, parts, and the overall assembly is diagnosed consistently. Structure health must remain within the specified ranges to ensure its proper functioning. Here it is noteworthy that these ranges change according to the age of the structure, environmental conditions, operating conditions, etc.

Monitoring facilitates saving or storing of the information collected as a kind of database or archive. Moreover, it provides a prognosis in terms of the evolution of a damage, the residual lifespan, the recovery process of a component, etc.

In fact, several works have been developed over the last years, where data (such as supervisory control and data acquisition or SCADA) are exploited to develop strategies based on machine learning for fault detection in WTs. For instance, Encalada-Dávila et al. [3] worked on a methodology based on ANNs (artificial neural networks) that uses SCADA data from a real wind farm to detect main bearing faults in WTs early on. Another work is that from Velandia-Cárdenas et al. [4], where a strategy for WT fault detection (related to the temperature of main gearbox) is proposed by using highly imbalanced real SCADA data. On the other hand, WT SCADA data that come from software simulation can also be employed to develop predictive maintenance strategies, such as the work done by Vidal et al. [5]. Their methodology aims to propose a data-driven multi-fault detection and classification system based on different types of faults on actuators and sensors.

SHM systems involve the integration of sensors, actuators, data transmission, processing, and so on, thus achieving their first level, which is sensing. These systems are conceived as a new way of nondestructive testing (NDT) [6].

Performing SHM tests' strategies on real WTs means incurring high logistic and operational costs, risk or difficulty, downtime, etc. In addition, several tests may be conceived as destructive since they directly affect the system. Therefore, there is a need for lab-scale test systems to evaluate the structures with different methods. In this case, this would imply using vibrational response, for instance. These systems must guarantee accuracy and realism during the simulations to obtain reliable results for analysing and processing later. Each testing workbench is designed according to the aims of the study; thus, several types of systems have been developed to study different topics, e.g., vibrational response [7,8], sensor quality [9], and WT foundation-type-based [10,11].

Now, regarding both wavemaker channels and down-scale WT foundation assessment, there are also several works to highlight. For instance, Aktaş [12] worked on a methodology to assess the wavemaker theory by using first-order and second-order wavemaker solutions. The wavemaker was piston-type within a 40 m × 1 m × 1.4 m channel (medium scale) and the structure under study was a spar-type floating WT. On the other hand, Hu et al. [13] studied the dynamic response of offshore monopile-type WT towers under wind, wave and current during installing stage, by placing a sensor network (accelerometers) on the structure's key locations. Moreover, for wave simulations, the wavemaker channel employed was 8 m × 3 m × 1.6 m and can provide a maximum current velocity of 3.0 m/s, waves height of 0.1 m, and sea wind velocity of 30 m/s. Following the same baseline, Song et al. [14] studied the dynamic response under several environmental conditions of TLP-type (tension leg platform) floating offshore WT systems. By getting support from Froude scaling, a 1:200 scale model was designed to carry out experiments in a lab-scale wave channel by means of a piston-type wavemaker. Furthermore, due to wave reflections, a wave absorber was placed on the other side of the channel to appropriately dissipate wave energy. In respect to WT foundations' scale, the structures under study can be several times larger. For instance, Zhang et al. [15] worked on a hydrodynamic model test of a NREL 5 MW monopile-type offshore WT (1:80 scale). Some tests were performed under the action of shallow water breaking waves and experimental data were collected to be further processed and analysed.

Thus, in this work, a design of a laboratory down-scaled vibration test bench for offshore jacket-type foundations is proposed. The design comprises a scaled wave generator channel, a desktop application (WTtest) to control the channel simulations, and a data acquisition system (DAQ) to collect the information from the sensors connected to the structure. Note that various equipment such as accelerometers, an electrodynamic shaker, and a DAQ device are selected as per the design methodology. Regarding the mechanical part, each component of the channel is designed like the wave absorber, the mechanical multiplier, the piston-type wavemaker, and the wave generator channel. For this purpose, the finite element method is used in static and fatigue analysis to evaluate the stresses and deformations. Finally, to simulate different wind loads, a vibration shaker is installed on the top beam that simulates the nacelle. Thus, in summary, this testing workbench addresses several key-points that add novelty and value:

- The vibration test system appropriately fits to medium/small scale laboratory environments (e.g., covering an area not greater than 5 m^2) where a test bench for lab-scale structures like the proposed one (WT jacket-type foundation) is required;
- Linear wave theory is employed as the design basis for the WGC, considering that this theory is greatly accurate for small ratios of wave height to water depth (i.e., shallow water), and wave height to wave length (deep water), which is according to these project's conditions;

- The proposed design methodology can be used as a useful guide to perform other designs based on different conditions or scales according to the engineering needs. This way, the methodology is scalable and reproducible.

The rest of the manuscript is divided into the following sections: Section 2 explains the base of the project, i.e., the lab-scale offshore WT jacket foundation; Section 3 comprehensively describes the methodology employed to carry out the phases of this work; Section 4 presents the results, and the discussion about them is drawn; Section 5 sums up the key points of the work and the scope for future ideas and research.

2. Lab-Scale Offshore Wind Turbine Jacket Foundation

An already existing and validated down-scaled WT offshore jacked-type foundation structure serves as the basis for the conceptual design of this work. In fact, on this down-scaled WT jacket-type foundation model, several studies have been developed to study damage detection and classification on collected vibration-response data.

For example, Vidal et al. [16] designed a methodology to classify several damages due to fatigue cracks localized at different levels (structure depth). Through the use of machine learning algorithms and vibrational response data, the methodology was well-applied and with accurate results. On the other hand, Feijóo et al. [17] worked on a novel strategy based on only healthy vibration-response data and by using an auto-encoder neural network to learn the normal vibrational behaviour of the structure and then to detect and localize cracks in it. Likewise, following the same study pipeline, Puruncajas et al. [18] proposed a method where CNNs (convolutional neural networks) are used to detect different sorts of structure damages by transforming vibration-response-only data into images and getting those as inputs for the algorithm. Finally, both Zugasti et al. [19] and Gómez González et al. [20] performed a similar comparative assessment of some damage detection methods on the laboratory tower (damages simulated by loosening some bolts of the different joints) by using collected vibration response data.

Figure 2 shows the structure with a 67 cm squared base on its side and roughly 108.72 kg in weight. This structure was designed for an offshore environment, as its foundation will work inside the wave generator channel (WGC).

Thus, the design of the vibration test system (VTS) does not start from zero but from an existing real down-scale model of the WT structure. The structure is 2.7 m tall and comprises the following parts:

- A top beam of 1×0.6 m with an attached modal shaker (GW-IV47 from Data Physics), which is used to simulate the nacelle mass and mainly the effects of wind excitation. By applying an electrical signal to the shaker (white noise), the vibration that is needed to excite the structure is created. The simulation of different wind speeds is also simulated with this vibration shaker, by changing the amplitude of the input electrical signal. To simulate the nacelle orientation changes, the top part of the structure can spin 0°, 30°, and 90°;
- A tower comprising three tubular sections, which are connected through bolts;
- A jacket foundation, which is based on a pyramidal structure made up of 32 steel bars (AISI 316) with different lengths, steel sheets (DC01 LFR), and a set of bolts and nuts.

Thus, using this lab-scale offshore WT jacket model as the basis, the design of the VTS is developed in the next sections.

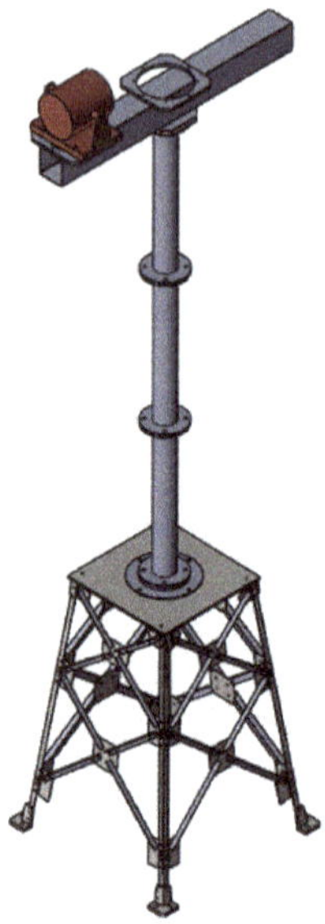

Figure 2. 3D model of the lab-scale offshore wind turbine jacket foundation.

3. Methodology

3.1. Design Methodology

To design the VTS, this work is divided into five phases:

1. Establish all the requirements of the design and then formulate a conceptual solution;
2. Select sensors (accelerometers) and a DAQ device to constitute the data acquisition system (DAS);
3. Design an application called WTtest for data collection and visualization;
4. Design a WGC and all its mechanical components, including an appropriate filter and pumping system;
5. Design the electrical and electronic connections for the whole system.

Figure 3 comprehensively explains the above-mentioned phases of the work plan. As shown, the phases are not necessarily executed sequentially but parallelized.

According to the first phase, the requirements for the project are described in Table 1. These requirements are based on the lab-scale offshore WT jacket foundation and the needs for testing at laboratory.

3.2. System Conceptual Design

The VTS design integrates a WGC that simulates the impact of waves on the WT foundation anchored inside the channel. In this regard, a set of sensors is distributed on the WT foundation for collecting data through the DAS during the tests (Figure 4 illustrates the conceptual design for the VTS). Moreover, the WTtest application configures the modal shakers and the parameters for wave generation. Further, the data collection and visualization are facilitated through the application as well.

Thus, the first part of design is the WGC, which is composed of two main parts: the complete metallic structure and the mechanism for generation of waves. Details about the first part are displayed in the subsequent sections.

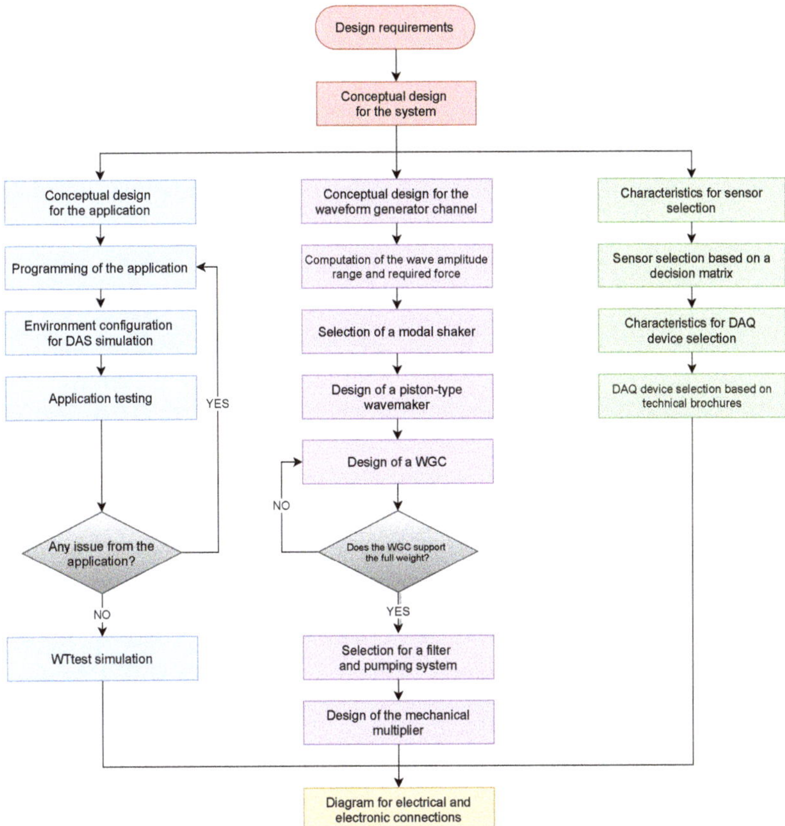

Figure 3. Flowchart of the design process for the work.

Table 1. Design requirements for the project.

Component	Requirement	Details
Waveform generator channel	Channel dimensions	3 × 0.70 × 0.95 m
	Cover material for channel	Tempered glass
	Foundation type	Jacket
	WT jacket foundation minimum mass	100 kg
WTtest application	Functions	DAS configuration, actuators' visualization and control
	Work platforms	Windows–Mac–Linux
	Licenses	Open source
	Hardware compatibility	National Instruments
Sensors and actuators	Power supply voltage	220 V, 50 Hz
	Accelerometers	8
	Accelerometer type	Triaxial
	Modal shaker type	Electromechanical

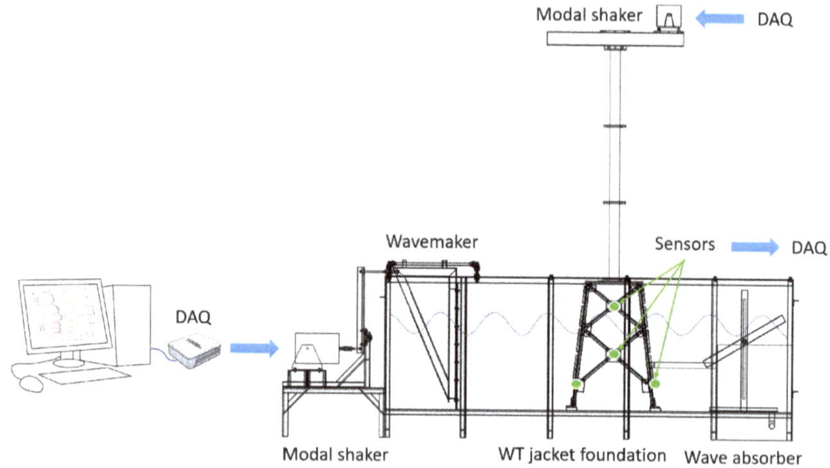

Figure 4. Conceptual design for the vibration test system (VTS) based on a wave generator channel (WGC).

3.3. *Wave Amplitude Range and Required Forces Calculation*

Reproducing waves under specific conditions in a controlled environment is a key requirement for the experimental study of offshore structures, especially offshore WTs. To generate sine waves, it is necessary to know parameters such as wave height and length as well as the maximum depth of the channel (see Section 3.3.1).

In this work, the wave shape requirements are based on the geometry of the WT jacket foundation, as shown in Figure 5.

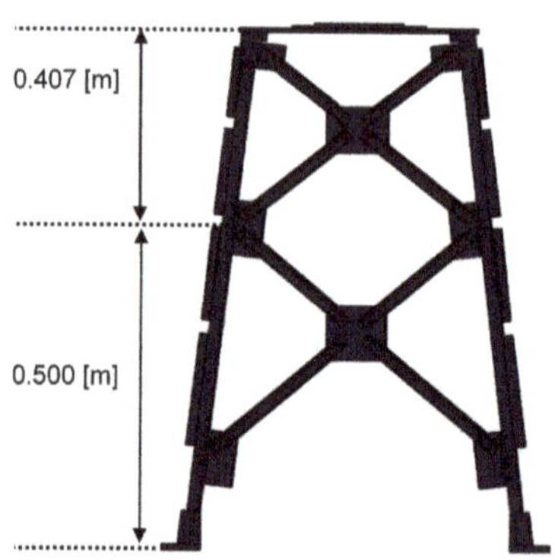

Figure 5. Base of the lab-scale wind turbine jacket foundation.

At the height of 50 cm, measured from the base, the level above which the wave height would be measured is delimited. From this level, there is a distance of 40.7 cm to reach the top of the jacket. With this distance, it is justified that the maximum wave height should be 25 cm, i.e., 12.5 cm of amplitude. In this manner, the jacket would not be completely underwater; there would be an uncovered part called the splash zone. Regarding the wave length, a typical range of lengths over which scale wave tests are carried out is selected [21], i.e., from 40 to 150 cm. Thus, the three parameters evaluated above are part of the requirements for the WGC design.

3.3.1. Wave Generation Theory

Frequency, generator displacement, and water depth in the channel are the three parameters on which the wave generation depends [22]. Figure 6 shows in detail the other characteristics of a wave, such as:

- λ: the wave length;
- H: the wave height, i.e., twice the value of the amplitude a;
- h: the channel depth;
- $\eta(x,t)$: the profile of the generated wave;
- τ: the wave period.

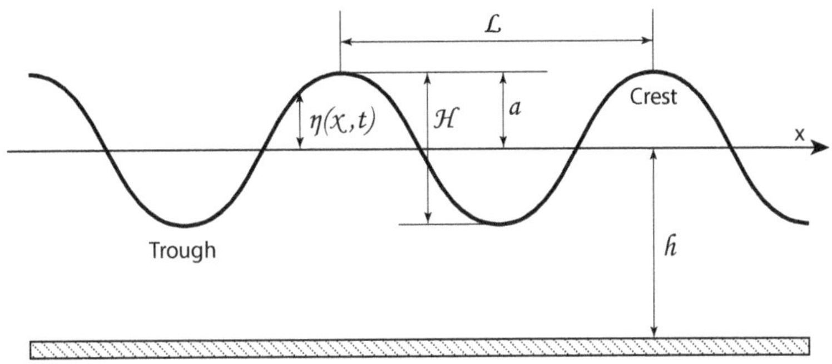

Figure 6. Main wave characteristics.

The force required to move the wave generator and the induced wave motion are variables that can be determined using linear wave theory.

In fact, in Ref. [22] there is a complete chapter (sixth) intended to explain wavemaker theory and which can be consulted to deeply understand some basis. The main reasoning in this theory is that the water displaced by the wavemaker should be equal to the crest volume of the propagating wave form. For instance, a piston-type wavemaker with a stroke S into a channel with depth h supposes a displaced volume Sh (directly related to H and k), and from that a quite interesting ratio rises, called the height-to-stroke ratio, which is finally expressed as Equation (1) indicates,

$$\frac{H}{S} = \frac{2 \cdot (\cosh 2kh\ - 1)}{\sinh 2kh\ + 2kh}. \tag{1}$$

In the "Ratio of wave height relative to stroke" section in Ref. [23], the curve of the height-to-stroke ratio versus relative channel depths is shown, where a threshold value of 2 is observed for relative depth values greater than 4. Furthermore, it is vital to clarify whether a piston-type is the appropriate wavemaker for this work.

Through successive approximations and based on geometrical design constraints, a 15 cm stroke for the piston-type wavemaker was selected. Thus, Figure 7 (computed through Equation (1)) shows the range of wave heights that could be achieved according to the selected stroke value and the above-mentioned requirements. Finally, it is possible to validate the possibility of generating 26 cm height waves with a 150 cm wave length.

3.3.2. Force Exerted by the Piston-Type Wavemaker

According to Dean and Dalrymple [22], one way to estimate the power required (P) to generate waves can be easily obtained by determining the energy flux away from the wavemaker as it is described in Equation (2),

$$P = ECn, \tag{2}$$

where Cn is the speed at which the energy is transmitted (this velocity is called the group velocity C_g), and E is the average total energy per unit cross-sectional length of the channel, measured in kg/s^2 and calculated as shown in Equation (3).

$$E = \frac{1}{8}\rho g H^2 \lambda, \tag{3}$$

where ρ is the density of water in kg/m^3, and g is the gravity acceleration ($9.81\ m/s^2$).

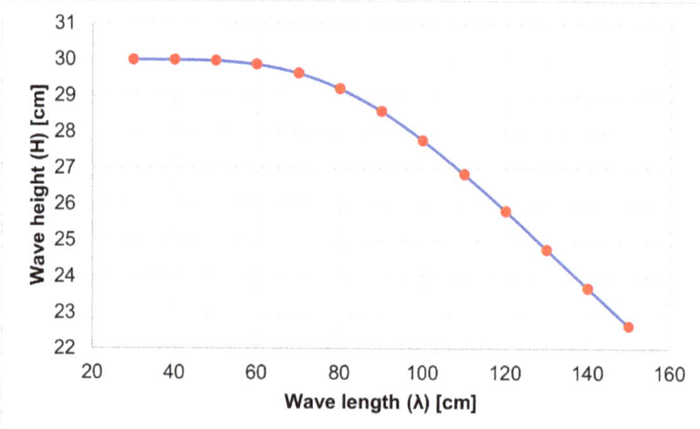

Figure 7. Wave height variation based on the defined wave length range.

The parameter C from Equation (2) is the wave velocity, measured in m/s, which for shallow water is calculated as follows in Equation (4),

$$C = \sqrt{\frac{g\lambda}{2\pi}}. \tag{4}$$

On the other hand, the parameter n from Equation (2) is a fraction relating the phase celerity C to the group velocity C_g. For shallow water, the energy and the wave profile travel with the same velocity. However, in deep water, the wave profile travels at twice the velocity of the energy, and it is calculated as follows (Equation (5)),

$$n = \frac{1}{2}\left(1 + \frac{2kh}{\sinh 2kh}\right), \tag{5}$$

where k is the wave number (the inverse of the wave length), and h is the depth of the test channel, measured in m.

In mechanics, the transmitted power is related to force and velocity. In this regard, applying and relating this relationship with Equation (2), the force exerted by the piston-type wavemaker is expressed as follows (Equation (6)):

$$F = En. \tag{6}$$

Figure 8 shows the force that the piston-type wavemaker would have to exert for different values of wave amplitude, calculated by using Equation (6). In this case, for a length of wave of 150 cm, the required force is 46.08 N.

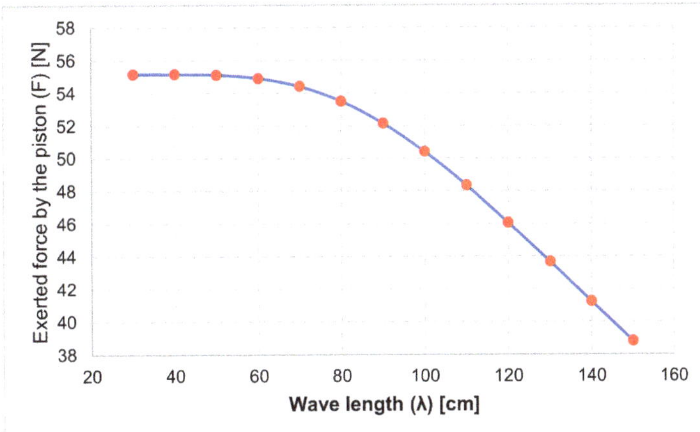

Figure 8. Force exerted by the piston-type wavemaker for the range of wave amplitudes.

3.4. Shaker Selection

The structure has an inertial shaker at the top part—the beam that represents the nacelle—which simulates the vibrations created by the wind loads on the rotor. As it can be seen in Figure 2, the shaker produces a movement along the main axis of its cylindrical body. It supposes that the force movement is also aligned with the axis of the beam (nacelle). Moreover, it must be considered that the beam can rotate like yaw rotation, so it lets modal shaker produce vibrations in any directions according the initial setting. Thus, based on the force requirement for the piston-type wavemaker—computed in the above subsection—it is within the operating range of the GW-M50 exciter, which is of DataPhysics brand.

The shaker needs an electrical amplification equipment that can enhance the signal, otherwise known as an amplified signal. In this regard, the selected amplifier is the model PA300E from DataPhysics brand, as it meets this requirement.

3.5. Piston-Type Wavemaker Design

This piston-type wavemaker is responsible for displacing the volume of water horizontally at a certain frequency, an action that generates the waves. In general, the mechanism comprises a triangular frame composed of alloyed aluminium C-profiles with an acrylic panel at the front acting as a piston-type wavemaker head. On the rear side is the connection with the vibration exciter, which facilitates the oscillatory movement of the piston-type wavemaker.

Of the basis of mechanical design, both static and fatigue analysis are quite needed. Thus, a static analysis (Von Misses stress) using the finite element method (through SolidWorks software) was performed to evaluate the behaviour of the mechanism parts against the main external force, i.e., the reaction of the displaced water volume, assuming that it directly impacts the acrylic panel. Likewise, it was important to apply a fatigue analysis, since the mechanical components are under the impact of cyclic loads while moving along water. Therefore, it is important to measure the stresses and analyse the lifespan of the components.

3.5.1. Piston-Type Wavemaker Frame Design

This structure is designed to overcome the pressure exerted by the volume of water displaced in the horizontal movement developed by the mechanism.

It consists of 50.8 × 25.4 × 3.18 mm aluminium C-profiles, forming a triangular structure that supports the acrylic panel and distributes the reaction force exerted by the volume of water, avoiding the acrylic's deformation.

The acrylic panel is fastened by HX-SHCS 0.25 hex bolts to the aluminium frame. The acrylic panel has a thickness of 9 mm, while the sides contain small clearances that allow the piston-type wavemaker to move without causing high friction on the sides of the channel.

3.5.2. Guide and Alignment System Design

The guiding system is aligned from two sides with 20 mm diameter steel shafts, which are supported by four NTN linear ball bearings (model KH2030). The bearings, being cylindrical, had to be embedded in aluminium blocks, which serve as the headers to be anchored in the aluminium frame.

The frame and the guidance alignment system, in turn, are supported by steel angles placed at the front and rear end. This enables the piston-type wavemaker to be flexible and moved vertically in case it has to be raised or lowered a few millimetres with respect to the surge channel.

3.6. Wave Generator Channel Design

The generator channel is designed to reproduce scaled marine waves using a piston-type wavemaker. It is designed using 8 mm tempered glass as side panels; this glass provides the necessary strength to support the volume of water contained inside the flume, according to a stress and strain analysis performed.

Considering that the channel is 3 m long, 70 cm wide, and 95 cm high, it is important to design a reinforced structure to provide stability and resistance in all the sections and joints. Finally, the tank structure is basically composed of 40 $\times$ 40 $\times$ 3 mm steel angles, arranged in main sections and with reinforcing angles, which are placed to distribute the forces and avoid stress concentration.

This structure supports many of the external loads of the different components of the WGC, such as the weight of water volume, the weight of the wave absorber, the weight of the WT jacket-type foundation, and the weight of the piston-type wavemaker. Only the weight associated with the WT structure is one of the largest loads for the WGC. Its weight is approximately 108.72 kg, which represents a force of 1.07 kN. As the structure has a square base, there are four support points over which this load is distributed.

With all the loads applied to the structure, Von Misses stresses and deformations were computed through a simulation on SolidWorks software. It was observed that the lateral walls experienced stresses that tend to displace them to the sides. Similar to the previous case, the critical zones are the lower part and the sidewalls of the channel. The maximum value of deformation reached is close to 0.5 mm. Therefore, it can be assured that the channel behaves in an expected way, resisting the stresses caused by the different applied loads.

3.7. Filtering and Pumping System Design

A filtering and pumping system is essential in environments where fluids, in this case water, are stored for long periods of time, as it allows for proper maintenance and cleaning. In this case, it is necessary to select a water pump with the appropriate filtration system.

For this project, it is desirable that this process is carried out within approximately 1 h. Based on the dimensions of the channel (width and length) and a height of 50 cm, the volume of water stored inside is calculated to be 1.38 m^3.

3.8. Mechanical Multiplier Design

This component connects the vibration exciter to the piston-type wavemaker and is responsible for amplifying the exciter's peak-to-peak horizontal travel (2.54 cm) enough to reach the piston-type wavemaker stroke (15 cm). As this component is a beam rotating around an axis, by law, the moment applied at both ends of the beam must be conserved. In other words, at the point of connection with the vibration exciter, the force will be six times greater than at the other extreme. In this regard, the mechanical multiplier has been designed in a 6:1 ratio.

The finite element method was also applied to this mechanism to determine the stresses and deformations that could compromise the system. It was observed that the deformations of the beam are more relevant in the upper part, since the displacement obtained is 7.7 mm. This value does not represent any type of failure since for the static analysis, it is considered that the beam is in a vertical position. Furthermore, the fatigue study is particularly significant since it gives long-term information on a component's behaviour under cyclic pressures. It was seen that with 12 cycles, the areas near the bearing would experience stress concentration, which in turn could cause an internal structural damage (physical damage). On the other hand, the areas farther away have a longer life cycle. Thus, in conclusion, the mechanical multiplier performs satisfactorily and within the expected operating limits.

3.9. DAQ Devices and Sensors Selection

The function of the accelerometers is to quantify the acceleration of the structure under test loads; the following characteristics are taken into account for their selection:

- Operating environment: Accelerometers must be capable of operating when submerged in water;
- Bandwidth: For vibration tests, an important characteristic is the bandwidth of the sensor and its resonance frequency. In a previous analysis of the lab-scale WT jacket foundation, it was found that the type of structure to be tested presented vibrations below 100 Hz [1], due to which it was chosen as the lowest frequency measured by the sensors;
- Measurement axes: Accelerometers with three measurement axes are chosen;
- Type of accelerometer: This indicates the physical principle used for the transduction of acceleration to an electrical signal. Piezoelectric and piezo-resistive (IEPE) types are identified as the most common for modal analysis, while fibre optic types are seen as high-precision technology that is reliable and efficient.

The other characteristics considered are the measurement range, sensitivity, and susceptibility to noise, which are all related to the overall quality of the sensor.

3.10. Application of Conceptual Design

Software design and implementation require models to guide the development process in a manner analogous to architectural blueprints. To create these models, visual languages such as the Unified Modelling Language (UML) are used, which promote standardization and provide a framework within the project.

UML specifies and delimits the parts of a software system through structure and behavioural diagrams; structural diagrams describe the static components, such as objects, classes, operations, relationships, and attributes, while behavioural diagrams emphasize the dynamic part that shows the state changes of the objects or the collaborations between them.

The DAQ interface application is designed by considering the following UML diagrams:

- Class diagram: Classes group the different types of objects (representations of real objects or conceptual abstractions) used in solving a problem;
- Use case diagram: It illustrates the different ways in which a user interacts with the system; each use case approximates a functionality from the point of view of an actor, which may be a person or another software system;
- Activity diagram: It shows the flow of actions to be followed in an application usage scenario.

3.11. Application Development

Commonly, the interfaces used for data acquisition are ad hoc solutions that are implemented in licensed platforms such as MATLAB™ or LabVIEW™. Python, being widely used in data science, is a widely accepted option in open-source modal analysis projects.

The application code is written in Python 3.6 and the NumPy, Pandas, and NI-DAQmx libraries due to their open-source nature and maturity in the data science field. This also helps abundant documentation to be located.

To facilitate the task of data acquisition through the application, a graphical interface is required to display in real time the input and output signals of the sensors and actuators, in addition to providing the user with a simple and intuitive means of control. This function is performed by a graphical interface comprising a set of widgets (elements that allow access to frequent functions), organized based on a layout (arrangement of elements) and styled using CSS (cascading style sheets). The graphical user interface (GUI) design is developed using the PyQt5 library and the Qt Designer tool, which creates windows by interactively positioning and adjusting widgets.

DAQ Environment Configuration

The software development process is iterative in nature and requires the use of testing and debugging tools. In the application, data acquisition simulation is used through the creation of virtual DAQ devices that are configured in the NI-DAQmx driver. Through these virtual devices, it is possible to avail all the necessary features for the operation of the system without having to deploy real hardware, speeding up the error detection and correction process.

4. Results and Discussion

In this section, the results for each part of the project are presented. Additionally, as a Supporting Information, to watch the VTS operation, please refer to the following repository: VTS Operation (https://github.com/angelswat/vibration_test_system_paper.git accessed on 16 July 2022).

4.1. Wave Generator Channel

The system for wave generation comprises several components whose mechanical design process has been explained in Section 3. Among these are the surge channel, the wave absorber, the wavemaker, and the mechanical multiplier.

For instance, to design the wave absorber, it is configured at an inclination of 30° with horizontal axis [21]. This is an optimum angle at which a higher wave absorption capacity can be obtained. Regarding the piston-type mechanism, some steel angles support the channel, the parallel axes where the wavemaker runs, and the aluminum frame with an acrylic panel.

Finally, Figure 9 shows the total assembly of the wave generator, including other elements such as the control table, the pumping and filtration system, and the installed WT support.

4.2. Accelerometer and DAQ Devices Selected

The accelerometer model selected is Dytran 3217A, with titanium alloy casing and polyurethane protection for immersion at depths up to 100 m. Here, each one has three voltage outputs with a BNC-type connector.

For the DAQ device, a solution consisting of a National Instruments compactDAQ™ 9178 chassis with eight slots for equipping input or output expansion modules and USB data connection is selected, allowing the sensor and actuator signals to be processed from the same interface. The connection of the sensors to the chassis is made through six NI-9234 vibration and sound modules with four independent analogue inputs, each one giving the maximum capacity for eight triaxial accelerometers.

The power amplifiers feeding the wave and wind vibration exciters received the control signal through a NI-9260 analogue voltage two-output module.

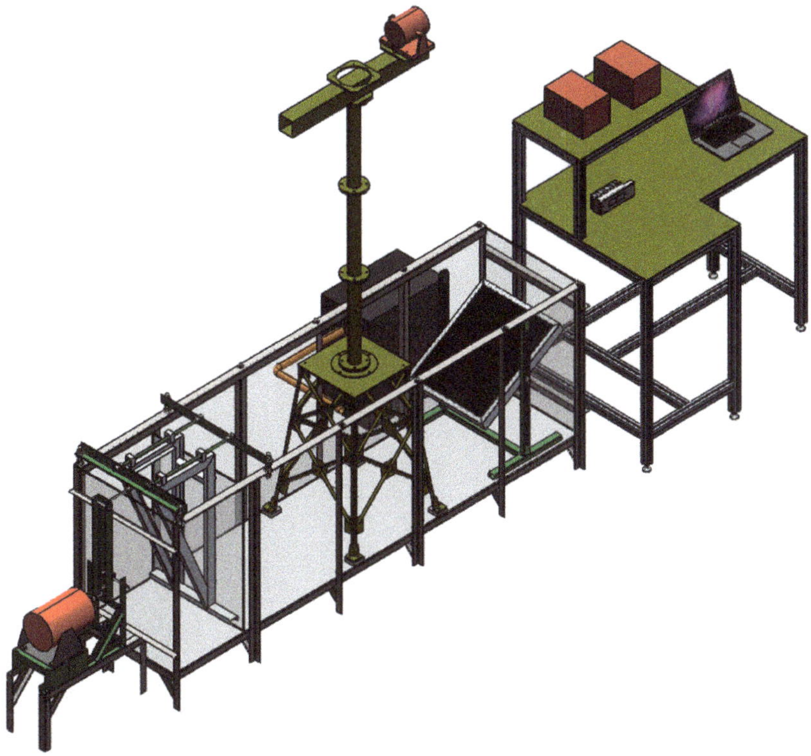

Figure 9. Complete assembly for the wave generator channel (WGC).

4.3. Data Acquisition and Control Application

The application is divided into two main windows that are accessible through tabs at the top; by default, the measurement tab containing the graphs of the signals and parameters for the test control is displayed. The four windows in the graphics area correspond to the sensors and actuators and are updated in real time as the signals get read or written by the configured DAQ device.

The time-domain response plot shows the measured acceleration in gravitational units (g) versus time in seconds (s) and is placed next to the frequency domain response plot to give a more complete view of the signals acquired by the accelerometers. To show the performance of the actuators simulating wave and wind, two more plots are placed at the bottom; the wave is represented as the height of the generated wave in centimetres and as a function of time in seconds. The wind is shown as an acceleration signal over time.

In the control area, the following options are added:

- Iteration: It indicates the number of times the data capture is repeated;
- Sampling frequency: It sets the number of samples per second that are read from the DAQ device;
- Sample: It represents the total data read from each channel configured in the DAQ device;
- Delay: It introduces a delay between each iteration for the actuators to reach a steady state or perform a variation in the test conditions.

A status bar indicating the progress (in percentage) of the capture and a log panel with important event information are also displayed. In addition, a pair of buttons to save (in comma-separated values [CSV] format) or discard the captured data are enabled at the end of the test.

In the configuration tab, the acquisition and generation options are included to choose the tasks and channels associated with the data capture. A button is also added at the top to access the task scheduler through the measurement and automation explorer of the NI-DAQmx driver (see Figure 10).

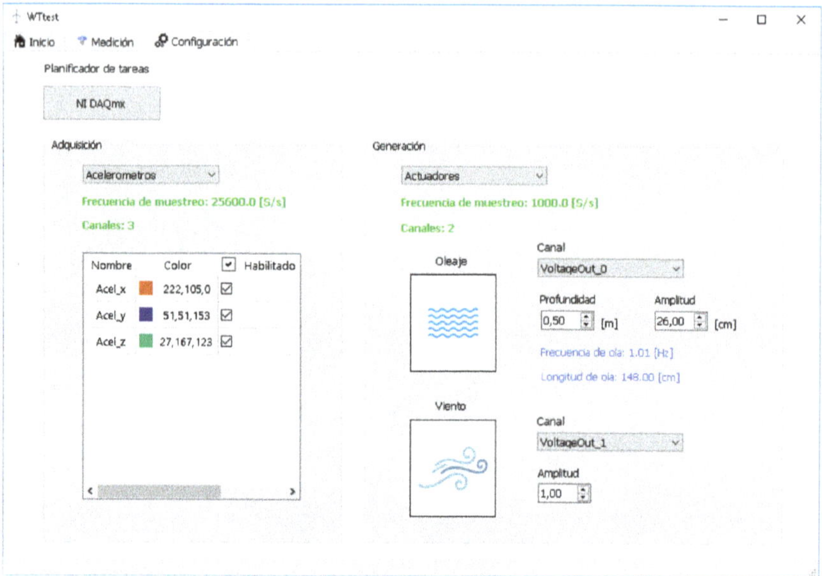

Figure 10. Configuration tab for DAQ application.

The acquisition option allows the selection of the signals and colours that represent them in the graphs, while indicating the number of channels available to read and the sampling rate in samples per second (S/s).

In the generation part, the channel selection option is added for each actuator. For the wave, the channel depth and wave height are defined as input variables; the frequency and wave length are calculated and displayed from these data.

For wind generation, only one amplitude option is included for the white Gaussian noise signal; in the acquisition part, the number of available channels and the sampling frequency are highlighted in green.

4.4. Simulation for the Application Operation

The data acquisition simulation for the application tests is executed in a Python environment using Anaconda 3 and the virtual DAQ devices configured using the NI-DAQmx driver. On the other hand, the simulated accelerations are of the sinusoidal type with a frequency of 0.25 Hz and a 5° phase shift between them.

The parameters used for the experiment are left by default at 10,000 samples with a frequency of 1000 S/s. For the actuators, a channel depth of 0.5 m and a wave height of 26 cm are set, while the amplitude of the wind generator is set to 1 (see Figure 11).

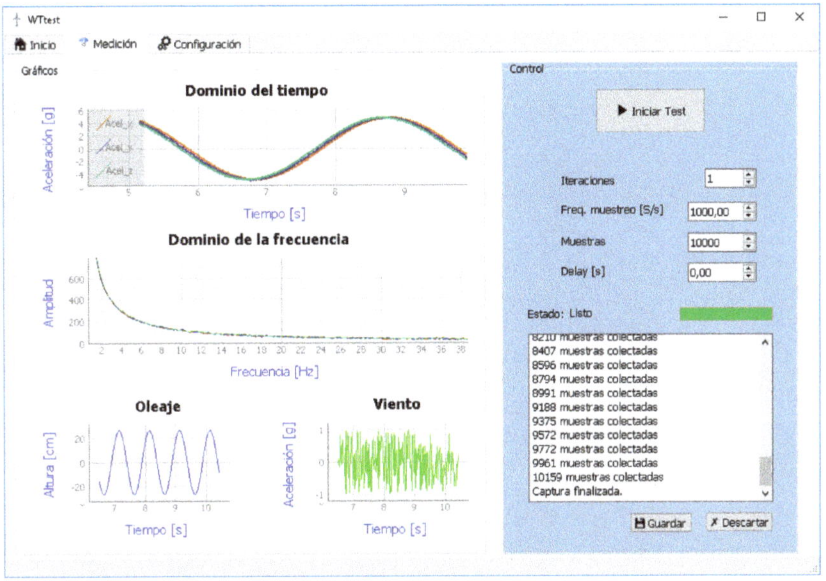

Figure 11. Simulation for the application DAQ.

4.5. Electrical and Electronic Connection Diagram

The DAQ system components are connected by following the hierarchy outlined in Figure 4 of the conceptual design. The data flow starts from the computer running the test application and the sensors/actuators and proceeds through the signal conditioning layer. This comprises the DAQ device and the power amplifiers that receive direct power. The rest of the components are indirectly powered by them (see Figure 12).

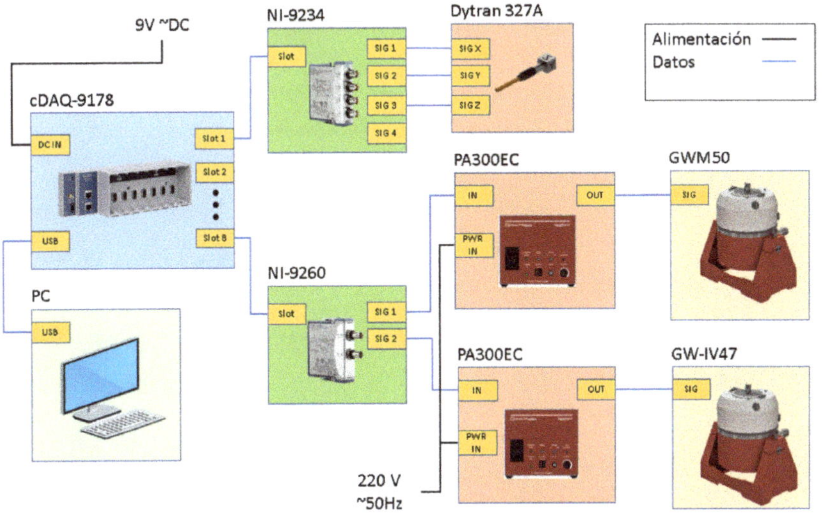

Figure 12. Electrical and electronic connection diagram.

5. Conclusions

The static and fatigue analysis of the various mechanically designed components has validated that stresses and deformations are within the permissible range, indicating that the components will work safely.

The WGC has a standard depth of 50 cm and handles a range of wave amplitudes that will be a function of the wave length, which works in the range of 30 to 150 cm.

The wave flume has a filter and pumping system that allows it to recirculate the stored water to the interior of the flume and provides it with the proper maintenance. It is worth noting that all this is done in less than an hour.

The WTtest application meets the design requirements; this means, it has a good degree of usability and allows the simulation control to be executed with the WGC.

The user interface during the simulation displays the signals acquired by the accelerometers in real time and saves the captured data as datasets in CSV format.

The cDAQ-9178 device has eight serial ports, six of which are configured as inputs for the accelerometers, while one functions as an output for the exciters. This leaves one serial port that can be used as an input or output, depending on the need.

In a nutshell, if the designed system is implemented it would be capable of producing waves and vibrations that allow the testing of WT foundations in a lab-scale environment by gathering data that would allow further analysis for failure prevention. After, this methodology can be scaled in order to be applied in real environments once all the pre-testing has been done under controlled lab-conditions.

Supplementary Materials: The following supporting information can be downloaded at: https://github.com/angelswat/vibration_test_system_paper.git, VTS operation.

Author Contributions: Conceptualization, Á.E.-D. and L.P.; methodology, L.P.; software, Á.E.-D.; validation, C.T., Y.V. and E.T.; formal analysis, Y.V.; investigation, Á.E.-D. and L.P.; writing—original draft preparation, Á.E.-D.; writing—review and editing, Y.V. and C.T.; supervision, E.T. and C.T.; funding acquisition, Y.V. All authors have read and agreed to the published version of the manuscript.

Funding: This work is partially funded by the Spanish Agencia Estatal de Investigación (AEI)—Ministerio de Economía, Industria y Competitividad (MINECO), and the Fondo Europeo de Desarrollo Regional (FEDER) through the research project DPI2017-82930-C2-1-R and by the Generalitat de Catalunya through the research project 2017 SGR 388.

Institutional Review Board Statement: Not applicable.

Informed Consent Statement: Not applicable.

Data Availability Statement: Not applicable.

Conflicts of Interest: The authors declare no conflict of interest.

References

1. Zugasti, E. Design and Validation of a Methodology for Wind Energy Structures Health Monitoring. Ph.D. Thesis, Universitat Politecnica de Catalunya, Barcelona, Spain, 2014.
2. Chen, I.W.; Wong, B.L.; Lin, Y.H.; Chau, S.W.; Huang, H.H. Design and Analysis of Jacket Substructures for Offshore Wind Turbines. *Energies* **2016**, *9*, 264. [CrossRef]
3. Encalada-Dávila, A.; Puruncajas, B.; Tutivén, C.; Vidal, Y. Wind Turbine Main Bearing Fault Prognosis Based Solely on SCADA Data. *Sensors* **2021**, *21*, 2228. [CrossRef] [PubMed]
4. Velandia-Cardenas, C.; Vidal, Y.; Pozo, F. Wind Turbine Fault Detection Using Highly Imbalanced Real SCADA Data. *Energies* **2021**, *14*, 1728. [CrossRef]
5. Vidal, Y.; Pozo, F.; Tutivén, C. Wind Turbine Multi-Fault Detection and Classification Based on SCADA Data. *Energies* **2018**, *11*, 3018. [CrossRef]
6. Tchakoua, P.; Wamkeue, R.; Tameghe, T.A.; Ekemb, G. A review of concepts and methods for wind turbines condition monitoring. In Proceedings of the 2013 World Congress on Computer and Information Technology (WCCIT), Sousse, Tunisia, 22–24 June 2013; pp. 1–9. [CrossRef]
7. Zheng, X.Y.; Li, H.; Rong, W.; Li, W. Joint earthquake and wave action on the monopile wind turbine foundation: An experimental study. *Mar. Struct.* **2015**, *44*, 125–141. [CrossRef]

8. Spanos, N.A.; Sakellariou, J.S.; Fassois, S.D. Vibration-response-only statistical time series structural health monitoring methods: A comprehensive assessment via a scale jacket structure. *Struct. Health Monit.* **2020**, *19*, 736–750. [CrossRef]
9. Mieloszyk, M.; Ostachowicz, W. An application of Structural Health Monitoring system based on FBG sensors to offshore wind turbine support structure model. *Mar. Struct.* **2017**, *51*, 65–86. [CrossRef]
10. Duan, F.; Hu, Z.; Niedzwecki, J. Model test investigation of a spar floating wind turbine. *Mar. Struct.* **2016**, *49*, 76–96. [CrossRef]
11. Wang, X.; Zeng, X.; Li, X.; Li, J. Investigation on offshore wind turbine with an innovative hybrid monopile foundation: An experimental based study. *Renew. Energy* **2019**, *132*, 129–141. [CrossRef]
12. Aktaş, K. Wave Generation and Analysis in the Laboratory Wave Channel to Conduct Experiments on the Numerically Modeled Spar Type Floating Wind Turbine. Master's Thesis, Izmir Institute of Technology, Urla, Turkey, 2020.
13. Hu, Y.; Yang, J.; Baniotopoulos, C.; Wang, X.; Deng, X. Dynamic analysis of offshore steel wind turbine towers subjected to wind, wave and current loading during construction. *Ocean. Eng.* **2020**, *216*, 108084. [CrossRef]
14. Song, J.; Lim, H.C. Study of floating wind turbine with modified tension leg platform placed in regular waves. *Energies* **2019**, *12*, 703. [CrossRef]
15. Zhang, S.; Shi, W.; Zhou, L.; Zhang, L.; Michailides, C. Model test of monopile-type offshore wind turbine under breaking waves. In Proceedings of the 30th International Ocean and Polar Engineering Conference, Shanghai, China, 11–16 October 2020.
16. Vidal, Y.; Aquino, G.; Pozo, F.; Gutiérrez-Arias, J.E.M. Structural health monitoring for jacket-type offshore wind turbines: Experimental proof of concept. *Sensors* **2020**, *20*, 1835. [CrossRef] [PubMed]
17. Feijóo, M.d.C.; Zambrano, Y.; Vidal, Y.; Tutivén, C. Unsupervised Damage Detection for Offshore Jacket Wind Turbine Foundations Based on an Autoencoder Neural Network. *Sensors* **2021**, *21*, 3333. [CrossRef] [PubMed]
18. Puruncajas, B.; Vidal, Y.; Tutivén, C. Vibration-Response-Only Structural Health Monitoring for Offshore Wind Turbine Jacket Foundations via Convolutional Neural Networks. *Sensors* **2020**, *20*, 3429. [CrossRef] [PubMed]
19. Zugasti, E.; Gómez González, A.; Anduaga, J.; Arregui, M.A.; Martínez, F. A Comparative Assessment of Two SHM Damage Detection Methods in a Laboratory Tower. *Adv. Sci. Technol.* **2013**, *83*, 232–239. [CrossRef]
20. Gómez González, A.; Zugasti, E.; Anduaga, J. Damage Identification in a Laboratory Offshore Wind Turbine Demonstrator. *Key Eng. Mater.* **2013**, *569*, 555–562. [CrossRef]
21. Castillo-Feijoó, S. Diseño y Construcción de un Tanque de Olas Para el Laboratorio de Ing. Naval. Ph.D. Thesis, Escuela Superior Politécnica del Litoral, Guayaquil, Ecuador, 2018.
22. Dean, R.G.; Dalrymple, R.A. *Water Wave Mechanics for Engineers and Scientists*; World Scientific: Singapore, 1991. [CrossRef]
23. Jang, T.; Sung, H. New nonlinear theory for a piston-type wavemaker: The classical Boussinesq equations. *Appl. Math. Model.* **2021**, *91*, 43–57. [CrossRef]

Journal of
Marine Science and Engineering

MDPI

Article

Chloride Ingress Resistance, Microstructure and Mechanical Properties of Lightweight Mortars with Natural Cork and Expanded Clay Prepared Using Sustainable Blended Cements

José Marcos Ortega [1,*], Fernando G. Branco [2], Luís Pereira [2] and Luís Marques [2]

[1] Departamento de Ingeniería Civil, Universidad de Alicante, Ap. Correos 99, 03080 Alicante, Spain
[2] ISISE, Department of Civil Engineering, University of Coimbra, R. Luís Reis Santos, 3030-788 Coimbra, Portugal
* Correspondence: jm.ortega@ua.es; Tel.: +34-96-5903-400 (ext. 2470)

Abstract: The use of lightweight aggregates in construction materials is a good solution for increasing the contribution to sustainability of civil engineering works, such as maritime ones. In this regard, the possibility of using cork granulates and expanded clay is a current research topic. The combination of eco-friendly cements with lightweight aggregates could provide solutions for developing new building materials. In this work, it has been studied mortars prepared with sustainable cements and the lightweight aggregates of natural cork and expanded clay. These cements incorporated slag, limestone and fly ash. Reference mortars with only sand as aggregate were also made. The total porosity and pore size distributions were obtained. The non-steady-state chloride migration coefficient and compressive and flexural strengths were also determined. The tests were performed at 28 days and 1 year. The differences in the total porosity between the natural cork and expanded clay series were not high, depending on the binder. Natural cork mortars showed similar or slightly higher migration coefficients than the reference and expanded clay mortars at 1 year. This adequate chloride resistance and the low mechanical strengths observed for the natural cork mortars recommend the possible use of this new aggregate in non-structural cement-based materials for civil engineering works exposed to maritime environments.

Keywords: lightweight aggregates; chloride ingress resistance; natural cork; expanded clay; slag; fly ash; limestone; sustainable cements; mechanical properties; microstructure

Citation: Ortega, J.M.; Branco, F.G.; Pereira, L.; Marques, L. Chloride Ingress Resistance, Microstructure and Mechanical Properties of Lightweight Mortars with Natural Cork and Expanded Clay Prepared Using Sustainable Blended Cements. *J. Mar. Sci. Eng.* **2022**, *10*, 1174. https://doi.org/10.3390/jmse10091174

Academic Editor: Vincenzo Crupi

Received: 17 July 2022
Accepted: 18 August 2022
Published: 23 August 2022

1. Introduction

The reduction in the environmental impacts produced by the construction industry is currently a major subject worldwide [1–3]. Therefore, the development of strategies for improving the sustainability of buildings and other civil engineering works, such as those related to maritime and coastal engineering, nowadays constitutes a relevant topic of research [4–6]. Among these strategies, two of the most popular are the use of alternative aggregates [7,8], such as lightweight or recycled ones, as well as widening the possible applications of sustainable cements [9,10].

In relation to cements with low carbon footprints, they generally incorporate supplementary cementitious materials (SCMs) as clinker replacement [11–14]. Most of these materials improve the performance of mortars and concretes, such as their durability [15,16] and mechanical properties [17,18]. Therefore, the assessment of their performance for specific civil engineering applications [19,20], or combined with other ecofriendly building materials, still constitutes a relevant field of investigation [21–23]. Regarding the standardized SCMs, fly ash [9,12,24] and ground granulated blast-furnace slag [25,26] are two of the most popular. They are active additions, and so their components can develop hydraulic and/or pozzolanic activity [9,12,27], given as products additional CSH phases, which can improve the microstructure and performance of cementitious materials [25,28].

Furthermore, there are also inert additions, such as limestone, whose main effect in these materials is as filler [29–31].

The incorporation of lightweight aggregates in building materials is also a good solution for increasing the contribution to sustainability of the construction sector. These aggregates reduce the density of materials, resulting in a lessening of the self-weight of construction elements, and they may be suitable, for example, for use in the rehabilitation and retrofitting of civil engineering structures or buildings [32,33]. Moreover, they also improve the thermal resistance of these elements [32,33]. In order to provide alternatives to the most commonly used lightweight aggregates, such as expanded clay, at present, the possibility of using cork granulates has been assessed by several works [32–34]. In this regard, to combine eco-friendly cements, which incorporate supplementary cementitious materials with cork lightweight aggregates, could be a relevant topic of study for providing new solutions for sustainable building materials.

The study of the microstructure of cementitious materials is of major importance, and especially when novel components are used, because the microstructure of these materials has a direct influence on their durability and mechanical behavior [35,36]. Regarding the durability, chloride ions are aggressive agents that can cause embedded steel corrosion in reinforced concrete elements [37,38], reducing their service life. Then, the characterization of the chloride ingress behavior of cement-based materials would be interesting in order to evaluate their ability for specific real applications, and in particular for those related to marine environments [39,40]. Their mechanical properties are also relevant because most of the fundamental parameters established by the standards and codes for specific civil engineering works are related to these properties [41].

Thus, the purpose of this research is to analyze the effects in the long term of combining sustainable cements and new cork lightweight aggregates in the pore structure, chloride ingress resistance and the mechanical strengths of mortars. Fly ash, limestone and ground granulated blast-furnace slag additions were incorporated as partial clinker substitutions in the cements used. The new lightweight aggregate studied was natural cork, whose effects, combined with eco-friendly cements, have hardly been studied, being this the main novelty of the research included here. It has been also tested mortars prepared with only sand as aggregate and with expanded clay in order to compare their performance to that noted for mortars containing natural cork.

2. Materials and Methods

2.1. Materials and Sample Preparation

Mortars that were made using different binders and aggregates were tested in this work. For a better view of the meanings and compositions of the mortar series prepared, they are compiled in Table 1.

In relation to the binders, four different binders were studied. Firstly, an ordinary Portland cement, CEM I 42.5 R (European standard EN 197-1 [42]), was used, indicated as "1" in the designation of mortars. Furthermore, mortars produced with three binders containing additions were prepared. One of them was a commercialized cement containing limestone: CEM II/B-L 32.5 N [42]. The content of the limestone of this cement was in the range from 21% to 35% in weight as clinker replacement, and it was named "2" in the designation of mortars. Another binder was a CEM III/B 32.5 N/SR [42] commercial cement, which incorporated 70% in weight of blast-furnace slag as clinker substitution. It was indicated as "3" in the designation of mortars. Lastly, a binder with fly ash was also used, which consisted of 80% in weight of CEM I 42.5 R [42] cement, and 20% in weight of this addition. This binder would verify the requirements of a standardized cement type, CEM II/A-V, following the prescriptions of the standard EN 197-1 [42]. This last binder was named as "4" for the mortar designation.

Table 1. Components (binder and aggregates) and designation of the mortar series tested.

Designation	Binder		Aggregate (Percentage in Volume)		
	Cement Type [42]	Addition (Percentage in Weight)	Sand	Natural Cork	Expanded Clay
1REF	CEM I 42.5 R	-	100%	-	-
2REF	CEM II/B-L 32.5 N	Limestone (21–35%)	100%	-	-
3REF	CEM III/B 32.5 N/SR	Blast-furnace slag (70%)	100%	-	-
4REF	Equiv. to CEM II/A-V	Fly ash (20%)	100%	-	-
1NCK	CEM I 42.5 R	-	50%	50%	-
2NCK	CEM II/B-L 32.5 N	Limestone (21–35%)	50%	50%	-
3NCK	CEM III/B 32.5 N/SR	Blast-furnace slag (70%)	50%	50%	-
4NCK	Equiv. to CEM II/A-V	Fly ash (20%)	50%	50%	-
1ECL	CEM I 42.5 R	-	50%	-	50%
2ECL	CEM II/B-L 32.5 N	Limestone (21–35%)	50%	-	50%
3ECL	CEM III/B 32.5 N/SR	Blast-furnace slag (70%)	50%	-	50%
4ECL	Equiv. to CEM II/A-V	Fly ash (20%)	50%	-	50%

The studied mortars were made using a water to binder ratio of 0.5 (in weight), and with an aggregate to cement ratio of 3:1 (in weight). For each binder, a reference series was made, using only sand as aggregate. In these reference series, the abbreviation REF was added to the designation of the mortar. In addition, several mortars were prepared replacing 50% of the sand volume by alternative lightweight aggregates, which were natural cork and expanded clay. In the designation of the series, the acronyms NCK and ECL were used, respectively, for naming the abovementioned lightweight aggregates.

The lightweight aggregates used were characterized in order to determine their bulk density and granular dimensions [32,33,43]. Natural cork granulates had a diameter of 4–5 mm and a bulk density of 70 kg/m^3, and they were manufactured by Corticeira Amorim, Portugal. The expanded clay aggregate showed dimensions of 0/2 mm and a bulk density of 490 kg/m^3, and it was produced by Argex. The particle size distributions of the aggregates used in this work are represented in Figure 1.

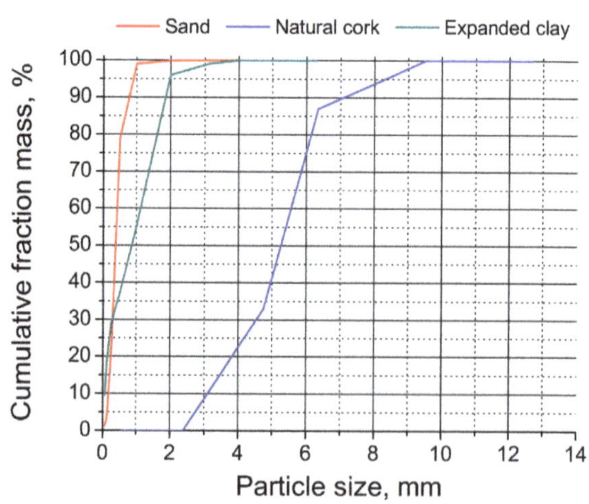

Figure 1. Particle size distribution of the aggregates used.

Prismatic and cylindrical samples were prepared. The dimensions of the first type were 4 cm × 4 cm × 16 cm, while those of the cylindrical specimens were 5 cm in height and 10 cm in diameter. All of them were demolded after the first 24 h since the setting. At that moment, the samples were kept immersed in tap water at a controlled temperature (20 ± 2 °C) up to the testing ages, which were 28 days and 1 year.

2.2. Mercury Intrusion Porosimetry

Mercury intrusion porosimetry is commonly used to characterize the pore network of different materials [44–49]. In the current research, the porosimetry test was performed with a porosimeter model Poremaster-60 GT, of the manufacturer Quantachrome Instruments. The specimens were dried at 50 °C at 48 h prior to the test. In the current research, the total porosity and pore size distributions of the mortars were analyzed. The distributions of the pores by sizes of the samples were represented using the cumulative volume of pores in percentage versus pore size curves. Two tests were carried out for each mortar series at the ages of 28 days and 1 year, on pieces taken from 4 cm × 4 cm × 16 cm prisms.

2.3. Forced Migration Test

The forced chloride migration test was carried out following the NT Build 492 standard [50]. The non-steady-state chloride migration coefficient (D_{NTB}) is obtained from the test. Three different cylinders of a 5 cm height and 10 cm diameter were tested for each mortar series at 28 days and 1 year.

2.4. Mechanical Strengths

Both the compressive and flexural strengths were obtained following the steps established by the European standard EN 1015-11 [51]. Three prismatic specimens were tested for each mortar series at 28 days and 1 year.

3. Results and Discussion

3.1. Mercury Intrusion Porosimetry

The total porosity results are depicted in Figure 2. The study of this parameter is relevant because the pores are the main way for the ingress of aggressive substances into cement-based materials. Then, the determination of the total volume of pores could be useful, especially when those materials could be used in marine works. In addition, the characterization of the total porosity is also interesting because it also provides information about the suitability of the material in relation to other engineering properties, such as thermal and acoustic properties.

The reference series presented smaller values of porosity than mortars containing natural cork and expanded clay. This could be related to the greater porosity of lightweight aggregates [52] compared with sand, which could influence the global solid fraction of the mortar, as a composite material. This parameter was very similar for the reference series at both ages, independently of the cement used, although a scarce reduction in the porosity was observed between 28 days and 1 year. This would be compatible with the results obtained by other authors [53,54], which reported that several active additions, and in particular fly ash, would not produce a noteworthy lessening of the porosity in cementitious materials, despite providing other benefits in their microstructure and properties. The small decrease with time in the total porosity could be related to the development of slag and clinker hydration [25,55] and the pozzolanic reactions of fly ash [12,24], which would lead to the formation of new solids, closing the microstructure, and thus reducing the porosity of the mortars.

With regard to the natural cork series, at 28 days, this parameter was smaller for the 1NCK and 2NCK series in comparison with the 3NCK and 4NCK series. This could be a consequence of the clinker hydration development [55,56], which would have more notable effects in the total porosity at initial ages compared with slag hydration [25,57] and the pozzolanic reactions of fly ash [12,57]. Moreover, the filler effect of the limestone addition [30,31,58] could also have an influence, and particularly in 2NCK mortars. This was more remarkable in the mortars with natural cork, due to their greater porosity, than the reference series. The reduction in the total porosity with time was more noticeable for the 1NCK and 3NCK series, showing at 1 year, lower porosities than those obtained for the 2NCK and 4NCK series. This would have a relation with the effects in the long term of the previously explained slag and clinker hydration [25,55] and fly ash pozzolanic reactions [12,24], which would produce a reduction in the pore volume in the 1NCK, 3NCK and 4NCK samples, as previously explained. In the case of 2NCK mortars, their content of clinker was lower, and limestone is an inert addition [29], which lacks reactivity, so its influence at later ages in the pore network would be less significant. According to several authors [53,54], the addition of fly ash would not produce a high porosity decrease with time, as already indicated, so the results of the 4NCK series would agree with them.

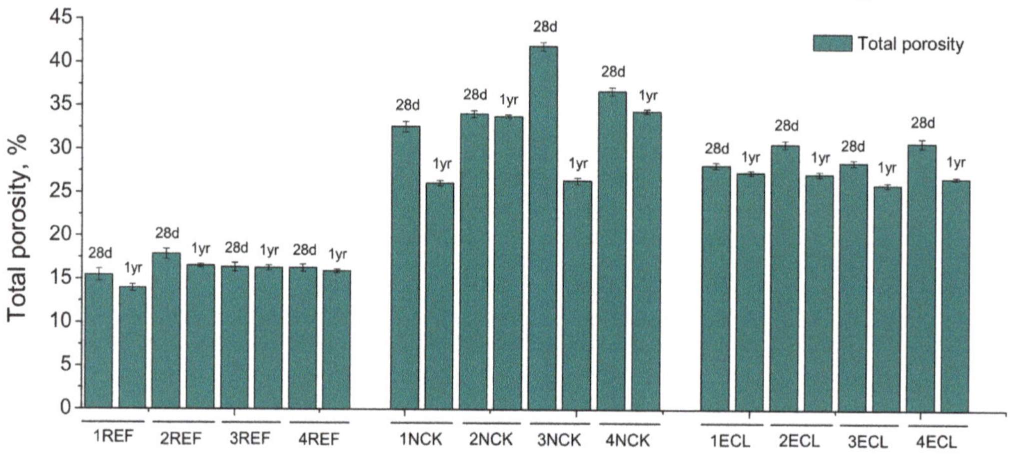

Figure 2. Total porosities obtained for the tested mortars.

Regarding the mortars with expanded clay, at 28 days, the porosity was slightly smaller for the 1ECL and 3ECL series than that observed for the 2ECL and 4ECL series, being higher for this last series with fly ash. This can be explained in relation to the hydration of clinker [25,55,56] and the hydration of slag [25,57], the effects of which were more noteworthy in the short-term total porosity of the expanded clay mortars than those due to the filler effect of the limestone addition [30,31,58] and fly ash pozzolanic activity [12,24,57]. This parameter decreased with time for all series with expanded clay, showing for the 3ECL and 4ECL mortars, the lowest values of the total porosity at one hardening year. Similar trends were observed for the reference and natural cork mortars, which were mainly associated with the influence at later maturation times of the new solid phases formation as products of the hydration reactions of the clinker and slag components [25,55], and the pozzolanic reactions of fly ash [12,24].

Comparing the results obtained for the series with natural cork and expanded clay in the long term, it is noteworthy to underline the similar total porosities noted for those made with ordinary Portland cement (1NCK and 1ECL mortars) and for those that incorporated slag (3NCK and 3ECL series). However, when cements with limestone and fly ash were used, this parameter was lower for mortars with expanded clay (2ECL and 4ECL series) compared with the specimens with natural cork (2NCK and 4NCK series).

The pore size distributions for the reference series can be observed in Figure 3. At 28 days, a higher percentage of pores with sizes under 100 nm was observed for the 3REF mortars, representing approximately 60% of the total pore volume. This result could be related to the slag hydration, already described, which could produce the formation of solids, thus reducing the pore sizes, and it is in agreement with the results of other researchers [57,59,60].

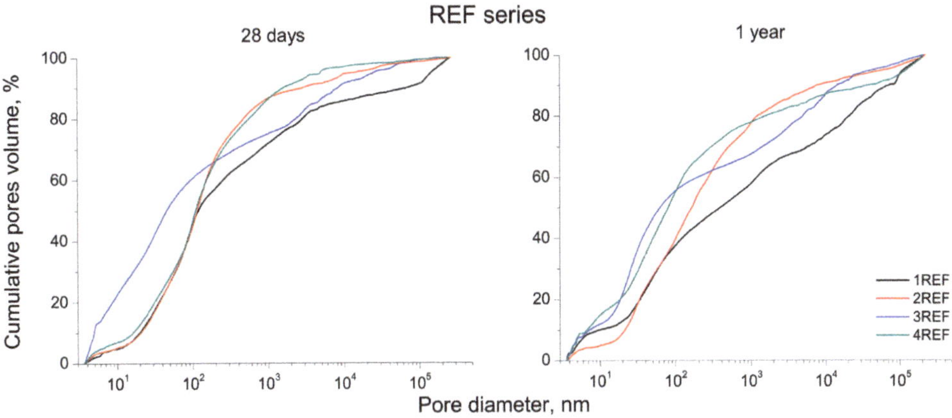

Figure 3. Cumulative volume of pores in percentage versus pore size curves for reference series.

The pore size distributions at that age were very similar for the 2REF and 4REF series, showing lower cumulative percentages of pores with sizes under 100 nm compared with the 3REF mortars, although they had around 90% of pores volume with diameters less than 10 μm. On the one hand, the results of the 2REF series could be justified regarding the filler effect of the limestone addition [30,31,58], which also contributed to the increase in the proportion of finer pores, although its lack of activity [29] would entail the lower pore refinement of these 2REF series in comparison with those with slag. Additionally, the less refined microstructure in the short term for series with fly ash than that observed for 3REF mortars would be a consequence of the delay in the development of the pozzolanic reactions of fly ash in comparison with slag hydration [12,61]. Slag can directly react with setting water [25,55], and fly ash needs enough portlandite produced along the hydration of clinker in order to start its pozzolanic reactions [12]; thus, more time is generally needed for observing the positive influence of fly ash in the pore network than for slag. For the 1REF series, at 28 days, the pores distribution curve until 100 nm was very similar to those noted for the 2REF and 4REF series, although for pore diameters greater than the abovementioned value, the growth of the curve of the 1REF mortars slowed down, which could indicate a lower microstructure refinement, which is in agreement with other works [53,55,62].

At one hardening year, a rise in the proportion of fine pores was observed overall for all the reference series in comparison with the pore size distributions observed at 28 days, which would be a consequence of the abovementioned effects at later ages of hydration [59,60,63] and pozzolanic reactions [12,24]. This increase was specifically notable for specimens with fly ash (4REF series). In the long term, the highest percentage of pores under 100 nm was noted for mortars with slag and fly ash (3REF and 4REF series); however, the 4REF mortars presented greater percentages of pores smaller than 1 μm, which would suggest that they have a more refined microstructure compared with the 3REF specimens. This would indicate the best performance of active additions at later ages [53], and in particular fly ash, whose pozzolanic activity notably improved the pore structure with time [12,24]. The pore size distribution curve hardly changed from 28 days to 1 year for the reference mortars with limestone, showing a less refined microstructure than the 3REF and 4REF series, which is in keeping with other authors [53–55]. In the case of the 1REF mortars, an increase in the percentage of pores lower than 20 nm was observed at 1 year compared with 28 days, which was probably produced by the formation of solids as products of clinker hydration [55,56], although its pore structure was the least refined of the reference series in the long term.

The pore size distributions observed for the series with natural cork are represented in Figure 4. The curve cumulative pore volumes versus pore diameters at 28 days were very similar for all these series, which would suggest that the short-term pore size distributions were mainly influenced by the natural cork aggregate, having the binder less influence. Nevertheless, at that age, the 2NCK mortars showed slightly higher percentages of pores lower than 100 nm, and this could be justified in relation to the filler effect of limestone [30,31,58], which would be more noticeable in the pore structure of natural cork mortars at the initial ages.

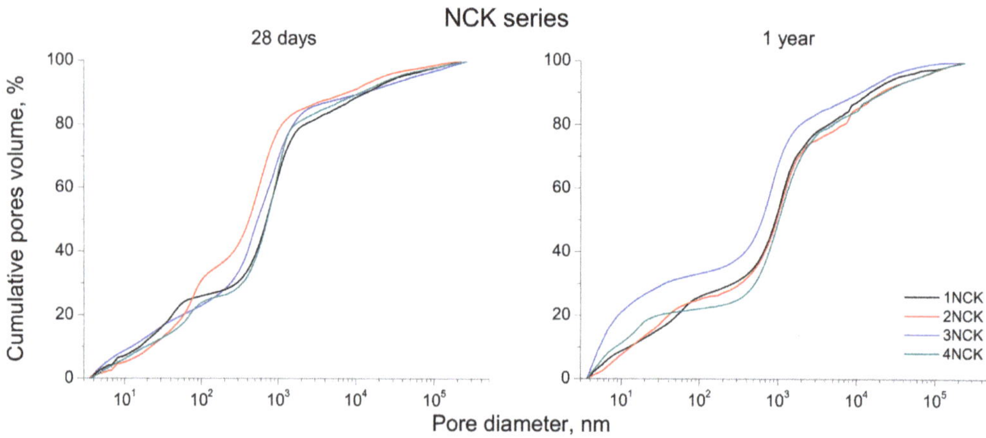

Figure 4. Cumulative volumes of pores in percentage versus pore size curves for mortars with natural cork.

At 1 year, it is noteworthy to underline that the pore structure of the natural cork series with slag (3NCK series) presented a higher refinement in comparison with the other series with this aggregate. This series contained higher proportions of pores under 100 nm and 1 μm, which were probably caused by the abovementioned influence of the hydration of the slag at later ages [25,55,57]. For the 4NCK mortars, between 28 days and 1 year, it has been observed an increase in the percentage of pores lower than 30 nm, which would show the gradual closure of the pore network because of the pozzolanic activity of fly ash [12,24], which was also noticeable when the natural cork aggregate was used. Finally, the long-term pore size distribution was very similar for the 1NCK and 2NCK series. In this result, the lack of activity of the limestone [29] could have had an influence, thus not providing any advantage in the long-term distributions of the pores by size of the natural cork mortars, compared with ordinary Portland cement with no additions, which is contrary to what happened with the active additions of fly ash and slag.

The cumulative volume of pores in percentage versus pore size curves noted for mortars with expanded clay can be observed in Figure 5. The pore network was more refined in the specimens with slag (3ECL series) at 28 days, as suggested by the greater proportion of pores with diameters less than 100 nm compared with the other series with expanded clay. This could be caused by the hydraulic activity of slag [25,55], explained before, which was more remarkable in the short term for the expanded clay series than for those with natural cork. In addition, at that age, smaller differences between the 1ECL, 2ECL and 4ECL mortars in the cumulative volume of pores in percentage versus pore size curves were noted, although the greater proportion of pores under 30 nm observed for the 4ECL mortars was noticeable, which would show the incipient effects of the pozzolanic reactions of fly ash [12,24] in the microstructure.

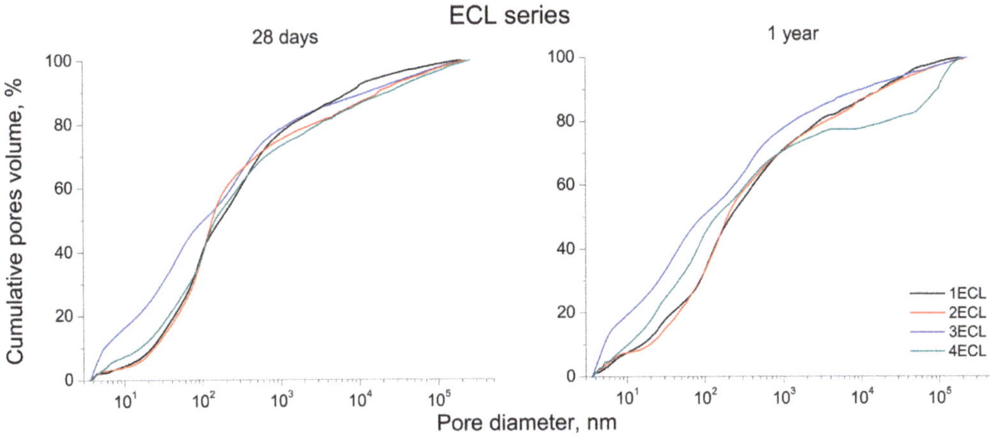

Figure 5. Cumulative volumes of pores in percentage versus pore size curves for series with expanded clay.

At one hardening year, the specimens of the 3ECL series still showed the highest microstructure refinement, with greater percentages of pores smaller than 10 nm, 100 nm, 1 μm and 10 μm compared with the other series with expanded clay. Furthermore, from 28 days to 1 year, a noticeable increase in the percentage of fine pores was observed for these samples with slag. A similar evolution was also noted for the natural cork mortars, as previously explained, so this result obtained for those mortars with expanded clay would confirm the adequate performance in the long-term microstructure development of combining a binder with a high content of slag with the lightweight aggregates studied in this work, being mainly associated with the abovementioned slag hydration [25,55].

This was likewise noted for the fly ash series, which showed a notable growth in the percentages of pores under 100 nm between both testing ages. This tendency again coincided with that observed for the natural cork mortars with this addition, although it was more remarkable when expanded clay was used, and it may be a consequence of the overall reduction in the pore sizes caused by the new solid phases formed in the pozzolanic reactions of fly ash [12,24], already discussed. Regarding the 1ECL and 2ECL series, their cumulative pores volume versus pore diameter curves were relatively similar at 1 year, and their pore structure was overall less refined than that obtained for the mortars with fly ash and slag, with a lower proportion of pores finer than 100 nm. This was also noted for natural cork mortars, in which it was linked to the long-term effects of active additions [12,24,25,55], in comparison with ordinary Portland cement, and particularly with the inert character of limestone addition [29].

Finally, in comparing the different aggregates used, it was observed that the series with natural cork presented a high proportion of coarser pores, and particularly in the range between 400 nm and 2 μm, which would indicate the high rise in their cumulative pore volume versus pore size curves in that range of pores. This was observed independently of the binders and at both testing ages, and it could be related to the effect of the natural cork aggregate. For the reference mortars and those that incorporated expanded clay, this growth of the curve up to a size of 2 μm was more progressive, and therefore, they showed greater percentages of finer pores.

3.2. Non-Steady-State Chloride Migration Coefficient

The results of the non-steady-state chloride migration coefficient are shown in Figure 6. The study of this parameter is important because chloride ions are the most important aggressive agent that can produce the development of corrosion pathologies in reinforced concrete elements. Then, it is important to assess the effects of the natural cork aggregate in combination with eco-friendly cements in the chloride ingress resistance of the analyzed materials.

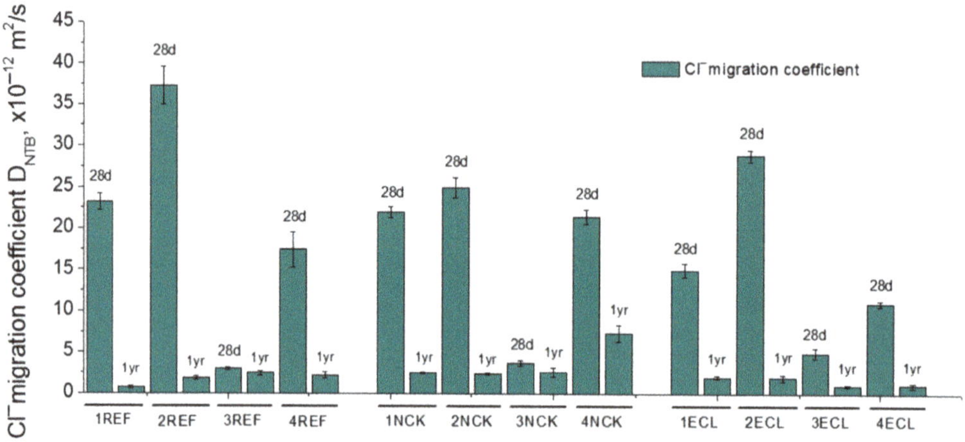

Figure 6. Non-steady-state chloride migration coefficient results for the series tested.

As can be observed in Figure 6, the binder used in the mortars played an important role in the results of this parameter. In general, relatively low migration coefficients were obtained in mortars with lightweight aggregates compared with the reference series. With regard to these reference mortars, the highest migration coefficient at 28 days was observed for the 2REF series, followed by the 1REF series, while it was smaller for the mortars with fly ash and slag, being lower for the 3REF mortars compared with the 4REF mortars. At 1 year, the migration coefficient was very similar for all the reference series, independently of the binder used, showing the 1REF mortars scarce lower values.

For the natural cork mortars, this coefficient was higher for the 1NCK and 2NCK series at 28 days in comparison with the 4NCK and 3NCK series, and the last one with slag presented the lowest values in the short term. After 1 year, the migration coefficient was higher for the 4NCK mortars, whereas small differences were noted for the 1NCK, 2NCK and 3NCK series. It is interesting to highlight that, at both testing ages, the series with natural cork showed similar or slightly higher migration coefficients than the reference mortars for each one of the binders used, with the exception of the 4NCK series, which had greater values of this parameter than the 4REF series.

In relation to the specimens with expanded clay, similar trends were observed in comparison with those described for the natural cork and reference series. The migration coefficient at 28 days was higher for the 2ECL specimens, followed by the 1ECL specimens, while their values were more reduced for the 3ECL and 4ECL series, showing the mortars with slag the lowest coefficient. Scarce differences between the series with expanded clay were noted at 1 year, being slightly lower the migration coefficient for the 3ECL and 4ECL mortars compared with the 1ECL and 2ECL mortars.

Regarding the comparison between the coefficients noted for the natural cork and expanded clay mortars, it is relevant to point out that no substantial differences were detected between them for each binder. In the short term, the 1ECL and 4ECL series presented smaller migration coefficients than the 1NCK and 4NCK series, while this parameter was greater for the 2ECL and 3ECL mortars when compared with the 2NCK and 3ECK mortars. Additionally, in the long term, this coefficient was slightly lower for the series with expanded clay than for those with natural cork, with the greatest difference observed between the 4NCK and 4ECL mortars. These small differences at one hardening year could be related to the less refined microstructure of the natural cork mortars compared with the expanded clay and reference series, especially due to the higher proportion of coarser pores for the natural cork specimens, previously described in the results of mercury intrusion porosimetry. Despite this, the adequate long-term performance of the studied mortars with natural cork in relation to chloride ingress resistance is remarkable, and particularly in comparison with a standardized and commonly used lightweight aggregate, such as expanded clay.

The lower chloride migration coefficients at 28 days obtained for the series with slag and fly ash, independently of the aggregate used, could be a consequence of the greater pore refinement caused in the short term by these active additions [53,64–67], already analyzed in the microstructural results. Moreover, this result could also be influenced by the higher binding capacity of cements with fly ash and slag [68], due to the high content of calcium aluminates provided by these additions, which would hamper the movement of the aggressive chloride ions within the pore network of the mortars, thus decreasing their migration coefficients.

The reduction in this coefficient with time noted for all series tested would agree with the refinement with age of the pore structure observed with mercury intrusion porosimetry, produced by the development of the hydration reactions of slag and clinker hydration and the pozzolanic reactions of fly ash, as previously explained. Lastly, after a very long maturation period, a high degree of the development of these reactions [12,24,25,55] would be expected, which would result in the smaller differences in the migration coefficient observed after 1 year.

3.3. Mechanical Strengths

The results of the compressive and flexural strength are represented in Figures 7 and 8, respectively. As can be observed, both parameters were greatly affected by the aggregate used. The mortars containing natural cork had much lower strengths than the reference series and those prepared with expanded clay, for all the studied binders. This result could be expected, according to the results reported in the literature [43,69] regarding the mechanical performance of cementitious materials with cork aggregates. Additionally, the higher total porosity and greater proportion of pores with diameters greater than 100 nm observed for the natural cork mortars compared with reference ones and those with expanded clay, according to the porosimetry results already discussed, could also have influenced the reduction in the mechanical behavior of the natural cork series. Furthermore, in general, the expanded clay mortars also showed lower mechanical strength than the reference series. This result would be also in agreement with the total porosity results because it was higher for the expanded clay series than for the reference series. The low mechanical strengths observed for the natural cork mortars, combined with their adequate chloride ingress resistance, would recommend their possible use as non-structural material in civil engineering works exposed to maritime environments.

With regard to the reference series, at 28 days, the compressive strength was very similar for the 1REF, 2REF and 4REF series, and it was higher for the 3REF series. The flexural strength at that age was greater for the specimens with active additions (3REF and 4REF series) compared with the 1REF and 2REF mortars. From 28 days to 1 year, an increase in the compressive and flexural strengths was generally observed, showing the 4REF mortars the highest values of both parameters in the long term, followed by the 1REF and 3REF series. The lowest strengths at 1 year were noted for the 2REF series.

Regarding the natural cork mortars, low strengths were observed along the studied time period, regardless of the binder used, below 6 MPa for the compressive strength, and under 2 MPa for the flexural strength. The compressive strength scarcely changed with time for the natural cork series, being relatively similar for all of them, although at 1 year, its greatest values were obtained for the 4NCK series, and the lowest corresponded to the 2NCK series. The rise with time was more noticeable for the flexural strength, and particularly for cork mortars containing additions (2NCK, 3NCK and 4NCK series). In the long term, the highest flexural strength of the natural cork series was noted for the 4NCK mortars, which were closely followed by the 3NCK and 1NCK mortars, whereas the lowest flexural strength was noted for the 2NCK samples.

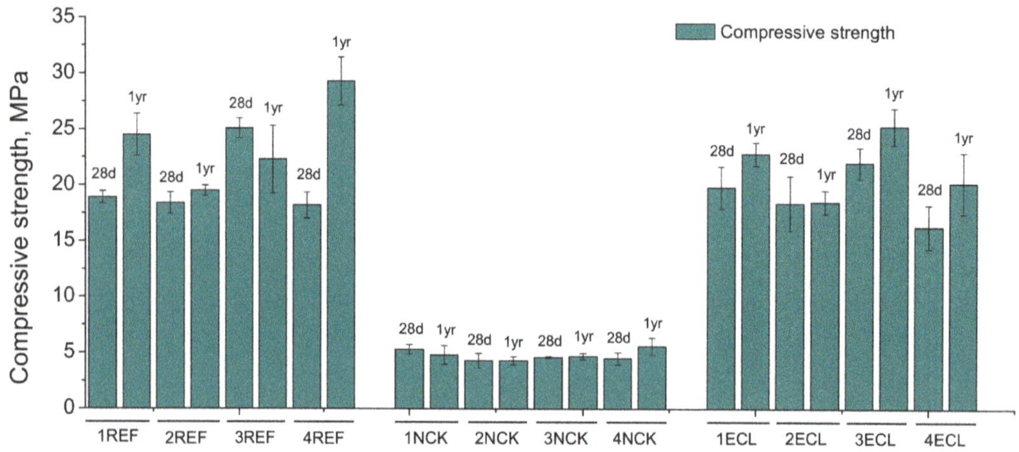

Figure 7. Compressive strength results for the analyzed series.

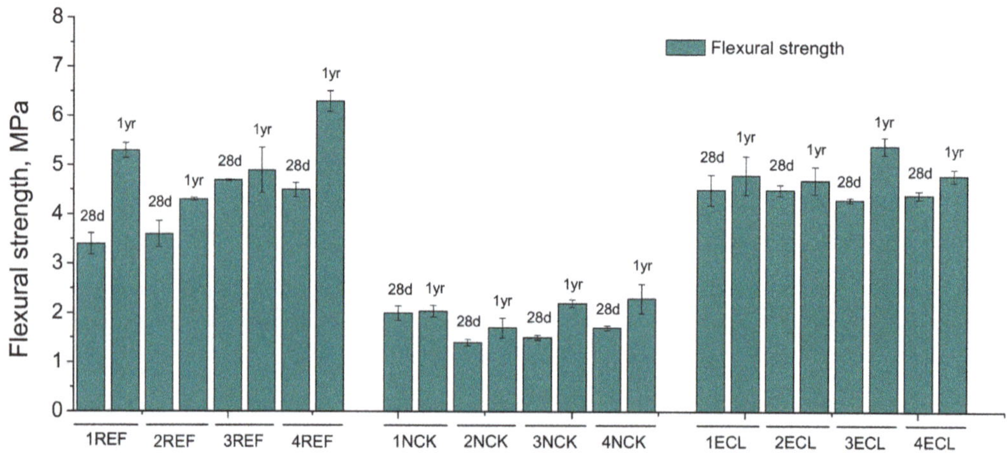

Figure 8. Flexural strength results for the analyzed series.

With respect to the series with expanded clay, the flexural strength at 28 days was very similar for all of them, while the compressive strength was scarcely higher when binder with slag was used (3ECL series), followed by the 1ECL and 2ECL mortars, showing the 4ECL series the lowest values of this strength in the short term. For most of the series with expanded clay, an increase in both strengths with the hardening time was also observed. The greatest compressive strength at 1 year was obtained for the 3ECL series, followed by the 1ECL and 4ECL mortars, whereas the smallest values were noted for the 2ECL series. The long-term results of the flexural strength were similar to those described for the compressive strength, although with lower differences between the different binders.

The improvement with time in the mechanical performance of the majority of the studied series, independently of the binder and aggregate used, could be caused by the solid-phase formation due to the progressive development of slag and clinker hydration [25,55] and fly ash pozzolanic reactions [12,24], already explained. This result would be in keeping with the lessening with time of the total porosity and chloride migration coefficient, as well as with the pore refinement noted for most of the analyzed series, previously discussed. The lower strengths shown overall after 1 year by binders with limestone, for all the aggregates studied, could be caused by the abovementioned lack of the pozzolanic or hydraulic activity of this addition [29], which would result in a smaller improvement in the long-term mechanical performance. The better performance of the binders with slag at 28 days could be due to its hydraulic activity [25,55], which makes their beneficial effects noticeable from the early ages compared with other additions with pozzolanic activity, such as fly ash [70]. Finally, as was mentioned in the microstructure characterization, the effects of fly ash are generally delayed [12,61] because sufficient portlandite is needed for the development of the reactions of this addition [12]. This product is formed during clinker hydration. For this reason, the mortars with fly ash showed flexural and compressive strengths at 1 year that were comparable to those obtained for the mortars with slag for the same aggregate, and they were even higher, in some cases, than the values noted for ordinary Portland cement mortars. The development of a relatively high strength over the very long term in cement-based materials that incorporate fly ash has also been reported by other authors [24,70].

4. Conclusions

The main conclusions obtained from this work are summarized as follows:

1. The lowest total porosity in the long term was generally obtained for reference mortars without lightweight aggregates. The total porosity differences between the series with natural cork and with expanded clay were not high, depending on the binder used.
2. The series with natural cork showed a lower refinement of the microstructure compared with the reference mortars and those with expanded clay. Furthermore, it was observed that the pore size distributions of the natural cork mortars in the short term were mainly affected by the aggregate, with the binder having less influence.
3. A decrease in the total porosity and rise in the refinement of pores with time were observed for most of the mortar series studied, independently of the aggregate used. This evolution of the microstructure would improve the other parameters analyzed, resulting in a lessening with the hardening time of the non-steady-state chloride migration coefficient and a rise in the mechanical strengths.
4. The binder used in the mortars played an important role in the results of the nonsteady-state chloride migration coefficient, especially at 28 days. It is remarkable that the studied mortars with natural cork showed similar or slightly higher migration coefficients than the reference and expanded clay mortars for each one of the binders studied at 1 year.
5. Both the compressive and flexural strengths were greatly influenced by the type of aggregate. The mortars with natural cork had lower mechanical strengths than the reference series and those prepared with expanded clay, for all the studied binders.
6. For each one of the aggregates studied, at the initial ages, the mortars with slag overall showed higher microstructure refinement, a lower chloride migration coefficient and greater compressive strength compared with the other studied binders.
7. In view of the results obtained, it is noteworthy to underline the adequate long-term performance of the studied mortars with natural cork in relation to chloride ingress resistance, and in comparison with a standardized and commonly used lightweight aggregate, such as expanded clay. This adequate chloride ingress resistance, combined with the low mechanical strengths produced by the incorporation of natural cork, would recommend the possible use of this lightweight aggregate in non-structural cement-based materials for civil engineering works exposed to maritime environments. Furthermore, the use of binders with active additions, and especially blast-furnace slag, overall improved the behavior of mortars with natural cork at later ages.

Author Contributions: Conceptualization, J.M.O. and F.G.B.; methodology, J.M.O., F.G.B. and L.P.; investigation, J.M.O., F.G.B., L.P. and L.M.; data curation, J.M.O., F.G.B., L.P. and L.M.; writing—original draft preparation, J.M.O.; writing—review and editing, F.G.B., L.P. and L.M.; supervision, J.M.O. and F.G.B.; funding acquisition, J.M.O. and F.G.B. This research includes part of the work developed along the research stays of J.M.O. in the ISISE-Functional Performance research group in the Department of Civil Engineering of the University of Coimbra in the periods May–July 2019, July–October 2020 and June–September 2021, under the supervision of F.G.B. All authors have read and agreed to the published version of the manuscript.

Funding: This work was partly funded by FCT/MCTES through national funds (PIDDAC) under the R&D Unit Institute for Sustainability and Innovation in Structural Engineering (ISISE), under reference UIDB/04029/2020, and by the FEDER through the COMPETE2020 program, Portugal 2020, within the scope of project CENTRO-01-0247-FEDER-047067 (LIFE4STONE). José Marcos Ortega is indebted to the Conselleria de Innovación, Universidades, Ciencia y Sociedad Digital de la Generalitat Valenciana (Spain), for a fellowship of the BEST/2020 program (reference BEST/2020/079).

Institutional Review Board Statement: Not applicable.

Informed Consent Statement: Not applicable.

Data Availability Statement: The data that support the findings of this study are available from the corresponding author, J.M.O., upon reasonable request.

Conflicts of Interest: The authors declare no conflict of interest.

References

1. Hammond, G.P.; Jones, C.I. Embodied energy and carbon in construction materials. *Proc. Inst. Civ. Eng. Energy* **2008**, *161*, 87–98. [CrossRef]
2. Qualharini, E.L.; Oscar, L.H.C.; Silva, M.R.D. Rehabilitation of buildings as an alternative to sustainability in Brazilian constructions. *Open Eng.* **2019**, *9*, 139–143. [CrossRef]
3. Mália, M.; de Brito, J.; Pinheiro, M.D.; Bravo, M. Construction and demolition waste indicators. *Waste Manag. Res.* **2013**, *31*, 241–255. [CrossRef]
4. de Brito, J.; Kurda, R. The past and future of sustainable concrete: A critical review and new strategies on cement-based materials. *J. Clean. Prod.* **2021**, *281*, 123558. [CrossRef]
5. Almutairi, A.L.; Tayeh, B.A.; Adesina, A.; Isleem, H.F.; Zeyad, A.M. Potential applications of geopolymer concrete in construction: A review. *Case Stud. Constr. Mater.* **2021**, *15*, e00733. [CrossRef]
6. Fraj, A.B.; Idir, R. Concrete based on recycled aggregates—Recycling and environmental analysis: A case study of paris' region. *Constr. Build. Mater.* **2017**, *157*, 952–964. [CrossRef]
7. Puthussery, J.V.; Kumar, R.; Garg, A. Evaluation of recycled concrete aggregates for their suitability in construction activities: An experimental study. *Waste Manag.* **2017**, *60*, 270–276. [CrossRef] [PubMed]
8. Nikbin, I.M.; Aliaghazadeh, M.; Charkhtab, S.; Fathollahpour, A. Environmental impacts and mechanical properties of lightweight concrete containing bauxite residue (red mud). *J. Clean. Prod.* **2016**, *172*, 2683–2694. [CrossRef]
9. Teixeira, E.R.; Camões, A.; Branco, F.G. Synergetic effect of biomass fly ash on improvement of high-volume coal fly ash concrete properties. *Constr. Build. Mater.* **2022**, *314*, 125680. [CrossRef]
10. Fonseca, T.V.; dos Anjos, M.A.S.; Ferreira, R.L.S.; Branco, F.G.; Pereira, L. Evaluation of self-compacting concretes produced with ternary and quaternary blends of different SCM and hydrated-lime. *Constr. Build. Mater.* **2022**, *320*, 126235. [CrossRef]
11. Malhotra, V.M. Fly ash, slag, silica fume, and rice-husk ash in concrete. A review. *Concr. Int.* **1993**, *15*, 23–28.
12. Wang, A.; Zhang, C.; Sun, W. Fly ash effects. *Cem. Concr. Res.* **2004**, *34*, 2057–2060. [CrossRef]
13. Kupwade-Patil, K.; Palkovic, S.D.; Bumajdad, A.; Soriano, C.; Büyüköztürk, O. Use of silica fume and natural volcanic ash as a replacement to Portland cement: Micro and pore structural investigation using NMR, XRD, FTIR and X-ray microtomography. *Constr. Build. Mater.* **2018**, *158*, 574–590. [CrossRef]
14. Elaqra, H.; Rustom, R. Effect of using glass powder as cement replacement on rheological and mechanical properties of cement paste. *Constr. Build. Mater.* **2018**, *179*, 326–335. [CrossRef]
15. Kamali, M.; Ghahremaninezhad, A. Effect of glass powders on the mechanical and durability properties of cementitious materials. *Constr. Build. Mater.* **2015**, *98*, 407–416. [CrossRef]
16. Thomas, M.D.A.; Scott, A.; Bremner, T.; Bilodeau, A.; Day, D. Performance of slag concrete in marine environment. *ACI Mater. J.* **2008**, *105*, 628–634.
17. Faleschini, F.; Zanini, M.A.; Brunelli, K.; Pellegrino, C. Valorization of co-combustion fly ash in concrete production. *Mater. Des.* **2015**, *85*, 687–694. [CrossRef]
18. Elahi, A.; Basheer, P.A.M.; Nanukuttan, S.V.; Khan, Q.U.Z. Mechanical and durability properties of high performance concretes containing supplementary cementitious materials. *Constr. Build. Mater.* **2010**, *24*, 292–299. [CrossRef]
19. Ortega, J.M.; Albaladejo, A.; Pastor, J.L.; Sánchez, I.; Climent, M.A. Influence of using slag cement on the microstructure and durability related properties of cement grouts for micropiles. *Constr. Build. Mater.* **2013**, *38*, 84–93. [CrossRef]
20. Khayat, K.H.; Yahia, A.; Sayed, M. Effect of supplementary cementitious materials on rheological properties, bleeding, and strength of structural grout. *ACI Mater. J.* **2008**, *105*, 585–593.

21. Fernandez, A.; Alonso, M.C.; García-Calvo, J.L.; Lothenbach, B. Influence of the synergy between mineral additions and Portland cement in the physical-mechanical properties of ternary binders. *Mater. Constr.* **2016**, *66*, e097. [CrossRef]
22. Bilek, V.; Fiala, C. Sustainability of concrete elements from concrete with ternary binders. In *Sustainable Construction Materials and Technologies*; Routledge: London, UK, 2013; Volume 2013.
23. Anastasiou, E.; Georgiadis Filikas, K.; Stefanidou, M. Utilization of fine recycled aggregates in concrete with fly ash and steel slag. *Constr. Build. Mater.* **2014**, *50*, 154–161. [CrossRef]
24. Papadakis, V.G. Effect of fly ash on Portland cement systems. *Cem. Concr. Res.* **1999**, *29*, 1727–1736. [CrossRef]
25. Bouikni, A.; Swamy, R.N.; Bali, A. Durability properties of concrete containing 50% and 65% slag. *Constr. Build. Mater.* **2009**, *23*, 2836–2845. [CrossRef]
26. Dadsetan, S.; Bai, J. Mechanical and microstructural properties of self-compacting concrete blended with metakaolin, ground granulated blast-furnace slag and fly ash. *Constr. Build. Mater.* **2017**, *146*, 658–667. [CrossRef]
27. Wedding, P.; Manmohan, D.; Mehta, P. Influence of Pozzolanic, Slag, and Chemical Admixtures on Pore Size Distribution and Permeability of Hardened Cement Pastes. *Cem. Concr. Aggreg.* **1981**, *3*, 63. [CrossRef]
28. Chindaprasirt, P.; Jaturapitakkul, C.; Sinsiri, T. Effect of fly ash fineness on compressive strength and pore size of blended cement paste. *Cem. Concr. Compos.* **2005**, *27*, 425–428. [CrossRef]
29. Dhandapani, Y.; Santhanam, M.; Kaladharan, G.; Ramanathan, S. Towards ternary binders involving limestone additions—A review. *Cem. Concr. Res.* **2021**, *143*, 106396. [CrossRef]
30. Hadji, T.; Guettala, S.; Quéneudec, M. Mix design of high performance concrete with different mineral additions. *World J. Eng.* **2021**, *18*, 767–779. [CrossRef]
31. Marchetti, G.; Irassar, E.F.; Rahhal, V.F. Effects of packing density and water film thickness on fresh and hardened properties of ternary cement pastes. *Adv. Cem. Res.* **2020**, *32*, 444–455. [CrossRef]
32. Branco, F.G.; Godinho, L. On the use of lightweight mortars for the minimization of impact sound transmission. *Constr. Build. Mater.* **2013**, *45*, 184–191. [CrossRef]
33. Belgas, M.d.L.; Branco, F.G. Cork concrete mechanical behavior under high temperatures. *Int. J. Hous. Sci. Its Appl.* **2013**, *37*, 207–215.
34. Souza, N.S.L.D.; Anjos, M.A.S.D.; Sá, M.D.V.V.A.D.; Farias, E.C.D.; Souza, M.M.D.; Branco, F.G.; Pereira, A. Evaluation of sugarcane bagasse ash for lightweight aggregates production. *Constr. Build. Mater.* **2021**, *271*, 121604. [CrossRef]
35. Baroghel-Bouny, V. Water vapour sorption experiments on hardened cementitious materials. *Cem. Concr. Res.* **2007**, *37*, 414–437. [CrossRef]
36. Poon, C.S.; Kou, S.C.; Lam, L. Compressive strength, chloride diffusivity and pore structure of high performance metakaolin and silica fume concrete. *Constr. Build. Mater.* **2006**, *20*, 858–865. [CrossRef]
37. Gurdián, H.; García-Alcocel, E.; Baeza-Brotons, F.; Garcés, P.; Zornoza, E. Corrosion behavior of steel reinforcement in concrete with recycled aggregates, fly ash and spent cracking catalyst. *Materials* **2014**, *7*, 3176–3197. [CrossRef]
38. Kwon, S.-J.; Lee, H.-S.; Karthick, S.; Saraswathy, V.; Yang, H.-M. Long-term corrosion performance of blended cement concrete in the marine environment—A real-time study. *Constr. Build. Mater.* **2017**, *154*, 349–360. [CrossRef]
39. Vázquez, K.; Rodríguez, R.R.; Esteban, M.D. Corrosion Prediction Models in the Reinforcement of Concrete Structures of Offshore Wind Farms. *J. Mar. Sci. Eng.* **2022**, *10*, 185. [CrossRef]
40. Ju, X.; Wu, L.; Liu, M.; Jiang, H.; Zhang, W. Modelling of Chloride Concentration Profiles in Concrete by the Consideration of Concrete Material Factors under Marine Tidal Environment. *J. Mar. Sci. Eng.* **2022**, *10*, 917. [CrossRef]
41. *EN 1992-1-1*; Eurocode 2: Design of Concrete Structures—Part 1-1: General Rules and Rules for Buildings. European Committee for Standardization: Brussels, Belgium, 2004.
42. *EN 197-1:2011*; Cement—Part 1: Composition, specifications and conformity criteria for common cements. European Committee for Standardization: Brussels, Belgium, 2011; p. 30.
43. Branco, F.G.; Tadeu, A.; Belgas, M.d.L. Can cork be used as a concrete aggregate? *Int. J. Hous. Sci. Its Appl.* **2007**, *31*, 1–11.
44. Diamond, S. Mercury porosimetry. *Cem. Concr. Res.* **2000**, *30*, 1517–1525. [CrossRef]
45. Ouellet, S.; Bussière, B.; Aubertin, M.; Benzaazoua, M. Microstructural evolution of cemented paste backfill: Mercury intrusion porosimetry test results. *Cem. Concr. Res.* **2007**, *37*, 1654–1665. [CrossRef]
46. Horpibulsuk, S.; Rachan, R.; Chinkulkijniwat, A.; Raksachon, Y.; Suddeepong, A. Analysis of strength development in cement-stabilized silty clay from microstructural considerations. *Constr. Build. Mater.* **2010**, *24*, 2011–2021. [CrossRef]
47. Ho, L.S.; Nakarai, K.; Duc, M.; Kouby, A.L.; Maachi, A.; Sasaki, T. Analysis of strength development in cement-treated soils under different curing conditions through microstructural and chemical investigations. *Constr. Build. Mater.* **2018**, *166*, 634–646. [CrossRef]
48. Horpibulsuk, S.; Rachan, R.; Raksachon, Y. Role of fly ash on strength and microstructure development in blended cement stabilized silty clay. *Soils Found.* **2009**, *49*, 85–98. [CrossRef]
49. Yu, Z.; Ye, G. The pore structure of cement paste blended with fly ash. *Constr. Build. Mater.* **2013**, *45*, 30–35. [CrossRef]
50. *NT Build 492*; Concrete, Mortar and Cement-Based Repair Materials: Chloride Migration Coefficient from Non-Steady-State Migration Experiments. Nordtest: Oslo, Norway, 1999; p. 8.
51. *EN 1015-11:1999*; Methods of test for mortar for masonry—Part 11: Determination of flexural and compressive strength of hardened mortar. European Committee for Standardization: Brussels, Belgium, 1999; p. 16.

52. Adhikary, S.K.; Rudžionis, Ž.; Tučkutė, S. Characterization of novel lightweight self-compacting cement composites with incorporated expanded glass, aerogel, zeolite and fly ash. *Case Stud. Constr. Mater.* **2022**, *16*, e00879. [CrossRef]
53. Bijen, J. Benefits of slag and fly ash. *Constr. Build. Mater.* **1996**, *10*, 309–314. [CrossRef]
54. Ibáñez-Gosálvez, J.; Real-Herraiz, T.; Ortega, J.M. Microstructure, durability and mechanical properties of mortars prepared using ternary binders with addition of slag, fly ash and limestone. *Appl. Sci.* **2021**, *11*, 6388. [CrossRef]
55. Ortega, J.M.; Sánchez, I.; Climent, M.A. Impedance spectroscopy study of the effect of environmental conditions in the microstructure development of OPC and slag cement mortars. *Arch. Civ. Mech. Eng.* **2015**, *15*, 569–583. [CrossRef]
56. Escalante-García, J.I.; Sharp, J.H. Effect of temperature on the hydration of the main clinker phases in Portland cements: Part I, neat cements. *Cem. Concr. Res.* **1998**, *28*, 1245–1257. [CrossRef]
57. Escalante-García, J.I.; Sharp, J.H. Effect of temperature on the hydration of the main clinker phasesin Portland cements: Part II, blended cements. *Cem. Concr. Res.* **1998**, *28*, 1259–1274. [CrossRef]
58. Meddah, M.S.; Lmbachiya, M.C.; Dhir, R.K. Potential use of binary and composite limestone cements in concrete production. *Constr. Build. Mater.* **2014**, *58*, 193–205. [CrossRef]
59. Duan, P.; Shui, Z.; Chen, W.; Shen, C. Effects of metakaolin, silica fume and slag on pore structure, interfacial transition zone and compressive strength of concrete. *Constr. Build. Mater.* **2013**, *44*, 1–6. [CrossRef]
60. Çakır, Ö.; Aköz, F. Effect of curing conditions on the mortars with and without GGBFS. *Constr. Build. Mater.* **2008**, *22*, 308–314. [CrossRef]
61. Nochaiya, T.; Wongkeo, W.; Chaipanich, A. Utilization of fly ash with silica fume and properties of Portland cement–fly ash–silica fume concrete. *Fuel* **2010**, *89*, 768–774. [CrossRef]
62. Erdem, T.K.; Kirca, O. Use of binary and ternary blends in high strength concrete. *Constr. Build. Mater.* **2008**, *22*, 1477–1483. [CrossRef]
63. Tsakiridis, P.E.; Agatzini-Leonardou, S.; Oustadakis, P. Red mud addition in the raw meal for the production of Portland cement clinker. *J. Hazard. Mater.* **2004**, *116*, 103–110. [CrossRef]
64. Detwiler, R.J.; Fapohunda, C.A.; Natale, J. Use of supplementary cementing materials to increase the resistance to chloride ion penetration of concretes cured at elevated temperatures. *ACI Mater. J.* **1994**, *91*, 63–66.
65. Moon, H.Y.; Kim, H.S.; Choi, D.S. Relationship between average pore diameter and chloride diffusivity in various concretes. *Constr. Build. Mater.* **2006**, *20*, 725–732. [CrossRef]
66. Geiseler, J.; Kollo, H.; Lang, E. Influence of blast furnace cements on durability of concrete structures. *ACI Mater. J.* **1995**, *92*, 252–257.
67. Jain, J.A.; Neithalath, N. Chloride transport in fly ash and glass powder modified concretes—Influence of test methods on microstructure. *Cem. Concr. Compos.* **2010**, *32*, 148–156. [CrossRef]
68. Leng, F.; Feng, N.; Lu, X. An experimental study on the properties of resistance to diffusion of chloride ions of fly ash and blast furnace slag concrete. *Cem. Concr. Res.* **2000**, *30*, 989–992. [CrossRef]
69. Brás, A.; Leal, M.; Faria, P. Cement-cork mortars for thermal bridges correction. Comparison with cement-EPS mortars performance. *Constr. Build. Mater.* **2013**, *49*, 315–327. [CrossRef]
70. Demirboğa, R. Thermal conductivity and compressive strength of concrete incorporation with mineral admixtures. *Build. Environ.* **2007**, *42*, 2467–2471. [CrossRef]

Journal of
Marine Science and Engineering

MDPI

Article

Parametric Design and Numerical Investigation of Hydrodynamic Characteristics of a New Type of Armour Block TB-CUBE Based on SPH Method

Cheng Peng, Hao Wang *, Huaqing Zhang and Hanbao Chen

National Engineering Research Center of Port Hydraulic Construction Technology, Tianjin Research Institute for Water Transport Engineering, Tianjin 300456, China
* Correspondence: wanghaozj@gmail.com

Abstract: Based on the open-source code DualSPHysics, a numerical model was conducted to simulate the regular wave transformation on the slope breakwater with artificial block, and the simulation results were verified according to the measured data from the physical experiment. The deviation between the numerical model and the measured data was less than 6% and 9% in wave run-up and overtopping, respectively, which demonstrated the model can reliably capture the wave evolution on the breakwater with an artificial block. Based on this verified model, the size of the artificial block was adjusted to obtain optimal wave-damping effects. Once obtained, the hydrodynamic characteristics of the optimized new artificial block TB-CUBE were further studied, and the effects of the breakwater slope, water depth in front of the breakwater, incident wave period, and the height on wave run-up were all analyzed. Finally, an empirical formula for wave run-up on this type of article block was suggested through data-fitting, for which the correlation coefficient is 0.981.

Keywords: SPH; DualSPHysics; new type of artificial block; TB-CUBE; wave run-up; overtopping; empirical formula

Citation: Peng, C.; Wang, H.; Zhang, H.; Chen, H. Parametric Design and Numerical Investigation of Hydrodynamic Characteristics of a New Type of Armour Block TB-CUBE Based on SPH Method. *J. Mar. Sci. Eng.* **2022**, *10*, 1116. https://doi.org/10.3390/jmse10081116

Academic Editors: M. Dolores Esteban, José-Santos López-Gutiérrez, Vicente Negro and M. Graça Neves

Received: 6 June 2022
Accepted: 5 August 2022
Published: 13 August 2022

1. Introduction

The implementation of an armor unit or artificial block [1] on rubble mound breakwaters is a typical approach for wave mitigation and ocean disaster prevention. It is of great scientific significance and practical engineering value to comprehend the hydrodynamic characteristics of the armor unit under wave force and harness its wave damping effect. Numerical simulation methods, not limited by time and space, to unravel this armor unit's performance have been widely applied in coastal engineering around the world, and a large number of numerical methods for the interaction between waves and coastal structures have been developed, including the Boundary Element Method [2,3], the Finite Volume Method [4,5], the Arbitrary Lagrangian Eulerian Method [6,7], the Finite Element Method [8,9], and also particle-based methods, such as, Moving Particle Semi-implicit method (MPS) [10] and Smoothed-Particle-Hydrodynamics (SPH) [11]. Open source code including OpenFOAM [12], SWASH [13] and SPHysics [14], have been extensively investigated and adopted during the last several decades.

Among them, the pure particle method that is based on SPH [11], not constrained by the fixed grid, is feasible for solving large deformation free surface flow problems [15] such as wave breaking [16], liquid splashing [17], the interaction between waves and porous media [18], and other fields that the traditional grid simulation methods cannot accurately describe. When SPH was created, the simulation of the real sea state and wave conditions was restricted by limited computer resources. Then, due to algorithm improvements, such as the use of graphics processors, the calculation efficiency has been substantially improved [19]. It began to be widely used in coastal engineering practices [20–23], making it feasible to simulate a wave overtopping scenario with a duration of 1000 irregular wave cycles under real sea conditions [24].

Since these improvements, the SPH has been applied to a variety of uses. Dominguez et al. [25] used DualSPHysics and MoorDyn [26], a coupled model, to simulate the dynamic response of the anchored floating structure under the regular wave forcing, which would been further applied in security assessments of offshore structures such as floating offshore wind turbines, offshore oil platforms, and wave energy converters under real sea conditions. Additionally, SPH has been coupled with the Finite Element Method to investigate the wave-debris-structure interaction for the risk assessment of coastal structures during extreme tsunami events [27]. Crespo et al. [28] used SPH to explore the interaction between waves and floating oscillating water column wave energy converters, but due to the lack of an effective wave-making and wave-absorbing systems, it was limited in practical operation [29]. Meringolo et al. [30] simulated the interaction between waves and permeable caisson breakwaters based on a two-dimensional SPH model with artificial viscous terms. The simulation results showed the reduction of wave pressure at the sea wall holes, and revealed that the empirical formula [31] inaccurately estimated the wave pressure previously due to neglect the existence of sea wall holes. Monaghan et al. [32] and Dalrymple et al. [21] studied the propagation of solitary waves on the beach, and the climbing and overtopping of waves on breakwaters with different roughness, but they did not consider the effect of an armor block. Altomare et al. [33] used the DualSPHysics [19,34] open-source code for the first time to gain insights for the interaction between waves and a 3D breakwater with armor units, and their numerical model results are in good agreement with the measurement data and empirical formula.

The use of armor blocks, including Accropode®, Xbloc®, and Core-Loc™ has been a classical solution for coastal structure construction [1]. There has been a sustaining effect for new type of armor block development [35–37] to achieve better wave damping and low consumption of concrete. Suffice it to say, there is considerable uncertainty in how to effectively protect the coastal structure when facing of longer and larger waves in further deeper coastal waters where more construction has taken place, and this coastal structure is more vulnerable in confronting ocean disasters that are associated with more concern for global warming [38].

High porosity block is promising in coping with this long period wave. The interactions between the armor blocks and the waves on breakwater could be categorized as the interaction of a wave and solid objects, which has been extensively studied numerically [39,40]. The wave run-up and wave overtopping of various blocks under different breakwater structures are quite different, and the empirical formulas that are given by standards could hardly cover all of the structures and blocks, which limits its application. Therefore, it is still necessary to develop new types of amour units and evaluate their function.

Massive physical experiment scenarios would be required for the size parametric selection before the generation of a new type of armor unit, which is largely inefficient. Meanwhile, through the establishment of a numerical model that describes the wave-breakwater-armor-unit interaction could be a solid, low-cost, and efficient tool for abundant parametric selection scenarios under various wave input conditions.

The chief objective of this paper is to optimize the size of a new type of armor block and to gain insight into its wave damping effects. To achieve that, a wave-breakwater-block numerical model that was based on the open-source code DualSPHysics was established to simulate the climbing and overtopping of regular waves on a rubble mound breakwater. First, the numerical results are verified with the results of the physical model test. After that, with this verified model, sensitivity studies are conducted to evaluate the effect of blocks on wave damping with a diversity of design sizes to provide the optimal size. After optimization, the hydrodynamic characteristics of the new artificial block are further explored and a wave run-up empirical formula for this block is provided. Finally, the conclusion and research limitations are given.

2. Materials and Methods

2.1. Block Description

The block TB-CUBE [37], the research object of this paper, was independently developed by the Tianjin Research Institute for Water Transport Engineering (TIWTE). The block is based on a cube, including four 1/4 spherical cavities, eight 1/4 cylindrical cavities, and four 1/2 cylindrical cavities. The shape of the block and the definition of the dimensions are shown in Figure 1. When a wave enters the chamber that is formed in the block during the climbing process on the slope, a vortex is initiated. During the wave's interaction with the concreate surface in the chamber of the hole, energy is dissipated, thus achieving its purpose of wave attenuation. The TB-CUBE block features the advantages of high porosity, flexible replacement adaptability, high stability, and solid structural strength.

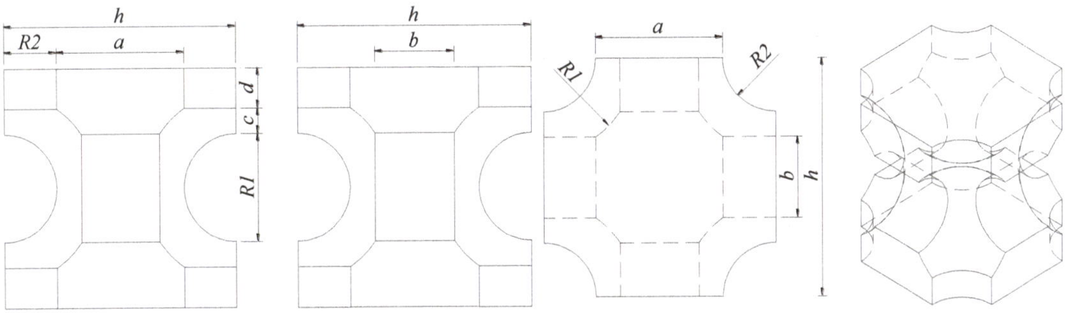

Figure 1. Sketch of the new artificial block TB-CUBE.

2.2. Physical Model Experiment

2D physical model tests for wave run-up of this new artificial block (TB-CUBE) on slope breakwater have been carried out in the wave tank at TIWTE, which is 70 m long, 1 m wide, and 1.5 m height. The geometric scale of the model was 1:40, the test section was a slope structure, and the weight of the TB-CUBE block was 152 g. There was a block stone cushion with a thickness of 25 mm that weighed around 1/10 to 1/20 of the weight of the TB-CUBE block. The layout of the physical model is shown in Figure 2. A total of 20 blocks were installed across the flume width during the experiments. In the test, the climbing and reflection of the TB-CUBE under different combinations of water levels, periods, wave heights, and slopes were experimentally studied. The physical experiment wave parameters are shown in Table 1, which is also the input conditions for the numerical experiment.

$$\mathrm{Kr} = Hr/Hi \tag{1}$$

where Kr is the reflection, and Hr and Hi are the reflect wave and the incident wave, respectively.

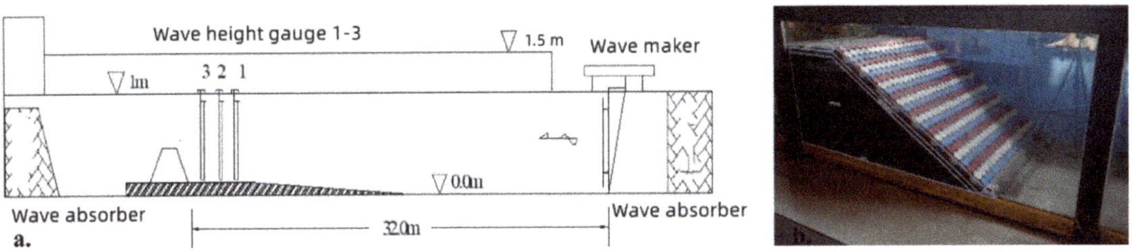

Figure 2. (**a**) Sketch of model layout; (**b**) Model photo with breakwater slope, m = 2.

Table 1. Physical experiment wave parameters, where m is slope, D is water depth, *H* is wave height, and T is period.

D (m)	m		*T* (s)				
			1.2	**1.5**	**1.8**	**2.2**	**2.5**
0.3/0.4	1.5/2.0/2.5	H (m)	0.06 0.08 0.10 0.12	0.06 0.08 0.10 0.12	0.06 0.08 0.10 0.12	0.06 0.08 0.10 0.12	0.06 0.08 0.10 0.12

Armor unit #4 in Table 2 was employed for the physical experiment, while other armor units would be used for the numerical model test in Section 2.3.3.

Table 2. Size of the TB-CUBE blocks.

Block	Size						Porosity ρ_p (%)
	R1	*R2*	*a*	*b*	*c*	*d*	
1#	0.450 h	0.225 h	0.550 h	0.221 h	0.165 h	0.110 h	46
2#	0.450 h	0.315 h	0.370 h	0.357 h	0.006 h	0.179 h	59
3#	0.400 h	0.225 h	0.550 h	0.339 h	0.106 h	0.169 h	40
4#	0.400 h	0.275 h	0.450 h	0.419 h	0.015 h	0.210 h	48
5#	0.475 h	0.225 h	0.550 h	0.163 h	0.193 h	0.082 h	50
6#	0.475 h	0.315 h	0.370 h	0.289 h	0.041 h	0.144 h	60

2.3. Numerical Model

2.3.1. SPH Equations

The main features of the DualSPHysics code are described in other references [19,29,34]. Here, the main features of the solution would be presented. The governing equations for discrete SPH Lagrangian system that simulates weakly compressible fluid, following Monaghan [41], are:

$$\frac{d\rho_i}{dt} = \sum_j m_j \left(V_i - V_j\right) \nabla_i W_{ij} \tag{2}$$

$$\frac{dV_i}{dt} = -\sum_j m_j \left(\frac{P_i + P_j}{\rho_i \rho_j} + E_{ij}\right) \nabla_i W_{ij} + g \tag{3}$$

where *t* is the time, ***V*** is the velocity, *P* is the pressure, ρ is the density, m is the mass, and ***g*** is the gravitational acceleration. W_{ij} is the kernel function that depends on the distance between particles *i* and *j*, as presented in (4). The quintic kernel [42], where the weighting function vanishes for inter-particle distances that are longer than 2 h_{SL} (h_{SL} is the smoothing length), was employed in this study. E_{ij} is the artificial viscosity term provided in DualSPHysics from Monaghan [41].

$$W(r, h_{SL}) = \alpha_D (1 - 0.5q)^4 (2q + 1),\ 0 \leq q \leq 1 \tag{4}$$

where $q = r/h_{SL}$, *r* is the distance between any given two particles *i* and *j*, α_D is equal to $(h_{SL})^3/\pi$. In DualSPHysics, it is based on weakly compressible fluid assumption, allowing the use of state equation to determine the fluid pressure [11].

$$P = B\left[\left(\frac{\rho}{\rho_r}\right)^7 - 1\right] \tag{5}$$

where $B = c_0^2\,\rho_0/7$, ρ_0 is the reference density and c_0 the numerical speed of sound.

2.3.2. Boundary Conditions

The default dynamic boundary condition from Crespo et al. [43] was used except two sidewalls in the following numerical wave flume. The boundary particles meet the same equations as other fluid particles. However, they would not move according to the forces that were imposed on them. Instead, they remain either fixed in position or move according to the assigned function. Once a particle moves close to a boundary and the distance between the boundary particles and the fluid particles is shorter than 2 h_{SL}, the density of the affected boundary particles rises, leading to the pressure enhancement. Consequently, there would be a repulsive force that was imposed on the fluid particle due from the pressure term in the momentum equation.

2.3.3. Numerical Wave Flume Description

The side length of the TB-CUBE block was set at 2 m, and the TB-CUBE block structure was established by 3D AutoCAD, which was imported into DualSPHysics in the form of an STL file to generate discrete particles. To reproduce the elaborated shape of the block and the inter-particle size, dp was set at 0.001 m after tuning. The method of establishing the numerical model of the breakwater was same as the modeling method of the aforementioned block, as displayed in Figure 3.

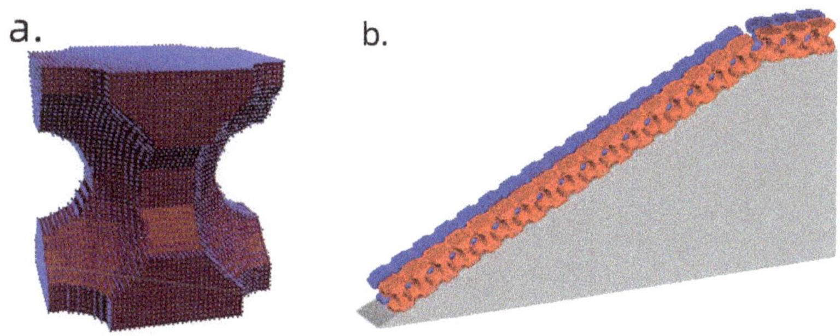

Figure 3. (**a**) Structure of the artificial block, TB-CUBE; (**b**) Artificial block on the breakwater in the numerical model (two blocks in the flume width direction).

In the DualSPHysics model, the active absorbing wave-making program was introduced through C++ programming to provide dynamic boundary conditions and periodic boundary conditions. In the 3D wave tank model, since the two sidewalls of the tank were in the same direction as the wave propagation, the sidewall particles exerted a particularly large repulsive force on the adjacent fluid particles, resulting in incorrect numerical dissipation. This problem was avoided by using periodic boundary conditions, while the rest of the walls were still set as default dynamic boundary conditions.

The three-dimensional numerical wave tank has a total length of 8 m, a width of 0.1 m, and a height of 1 m. The wave-making plate is 0.1 m wide, 0.6 m high, and is located 0.5 m from the left end of the tank.

The constructed numerical wave model was used to verify the TB-CUBE block climbing and wave overtopping to ensure the reliability of the numerical model. The geometric scale of 1:40 was selected for numerical simulation, consistent with the physical experiment. The slope breakwater was located at a distance of 6 m from the wave-making plate, and two rows of TB-CUBE blocks were laid on the simplified impervious slope. The block model had a side length of 0.05 m. A water collecting tank with a length of 0.3 m, a width of 0.1 m, and a height of 0.4 m, was arranged behind the breakwater to collect the overtopped water. In order to take the requirements of accurate simulation of block shape and calculation accuracy into account, the particle spacing in this experiment was set to 0.003 m, and the model layout is shown in Figure 4.

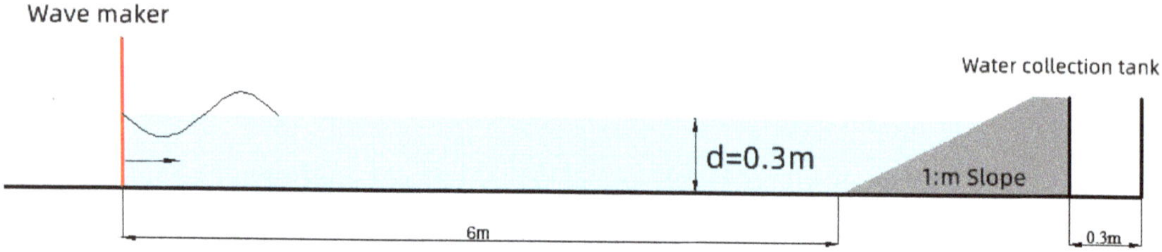

Figure 4. Layout of the numerical wave flume for wave run-up and overtopping, with slope m = 1.5, 2 and 2.5. TB-CUBE armor units were installed on the slope (not shown).

The TB-CUBE block is a specially shaped block with involved porosity that was obtained by digging out spheres and cylinders of a certain size from the cube-based block. Therefore, the porosity can be adjusted by changing the size of the spheres and cylinders. The use of the numerical wave model was to evaluate its run-up and overtopping, to explore its hydrodynamic properties, and to optimize its structural size. To determine the block shape with the optimal wave elimination effect, six representative blocks with different detailed sizes were selected, as shown in Table 2.

3. Results

3.1. Model Validation

Tables 3 and 4 presented the comparison of wave climb-up and wave overtopping from the numerical model and wave flume physical experiment. The deviations of the numerical results of wave run-up and overtopping and results from the physical model have not exceeded 6% and 9%, respectively, and the numerical model also reproduced the trend of the reflection coefficient variation. It can be demonstrated that the numerical model can accurately simulate the wave run-up and overtopping of waves along slope breakwater that was covered with the TB-CUBE.

Table 3. Comparison of wave run-up and reflection coefficient between the numerical model and physical model. Deviation = (numerical model value-physical model value)/physical model value × 100%.

No.	Period (s)	H (m)	Run-Up (m)			Kr	
			Physical Model	Numerical Model	Deviation (%)	Physical Model	Numerical Model
1	1.2	0.06	0.089	0.088	1.2	0.27	0.24
2		0.12	0.210	0.198	5.7	0.25	0.23
3	1.5	0.06	0.102	0.098	3.9	0.28	0.33
4		0.12	0.228	0.233	2.3	0.21	0.23
5	1.8	0.06	0.102	0.098	4.4	0.44	0.51
6		0.12	0.210	0.199	5.2	0.44	0.55

Table 4. Comparison of wave overtopping between the numerical model and physical model.

No.	H (m)	Period (s)	Overtopping L/(m·s)		
			Physical Model	Numerical Model	Deviation (%)
1	0.12	1.2	0.175	0.160	8.6
2		1.5	0.225	0.219	2.7
3		1.8	0.286	0.261	8.7
4		2.2	0.810	0.813	0.4

3.2. Block Structure Optimization Study

Numerical calculations of each type of block to evaluate its run-up and overtopping have been carried out and the case for d = 0.3 m, H = 0.12 m, T = 1.8 s was selected for results discussion.

Figure 5 is a snapshot of the climbing waves (at t = 10.8 s) and the overtopping of the waves (at t = 26.05 s) on the slopes of the six types of armor units, respectively. During the climb and fall of the wave, water particles penetrate into the pores of the block, dissipating partial wave energy. Adjusting the relative radius of the cylinder (*R2*) while keeping the relative radius of the sphere (*R1*) constant allows for a higher or lower amount of water to fall into the pores of the block. For example, in Figure 5, block #2 has a greater *R2* than block #1 (Table 2), which led to more water entering the block and a more rapid dissipation of energy. Due to this difference, the climbing height of block #2 was shorter than that of block #1, with its values at 0.138 m (#2) and 0.178 m (#1), respectively. However, the smaller circular pore size of block #1 reduces the ability for fluid particles to jump out of the block pores, leading to the threshold effect. Consequently, more water is held in the block cavity without overtopping the breakwater, which eventually led to less overtopping from block #1 when compared to block #2. Their values were 0.659 (#1), and 0.758 (#2) L/(m·s), respectively.

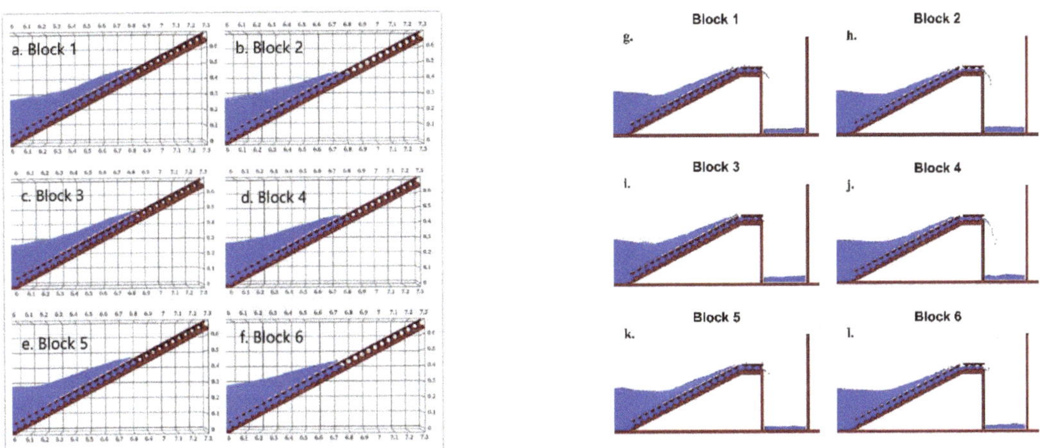

Figure 5. Wave run-up (**a**–**f**) and overtopping (**g**–**l**) snapshots for different type of blocks.

Alternatively, adjusting *R1* while keeping *R2* invariant would change the porosity of the block. Looking at blocks #1, #3, and #5 as examples, their *R1* values were set at 0.45 h, 0.40 h, and 0.475 h (Table 2), respectively, with diversity in porosity (46%, 40%, and 50%). As the porosity increases, the energy dissipation of the wave during the climbing process was greater, and the final performance was such that the climb of block #5 were the smallest values among the three, with block #1 as the second and block #3 the largest. The climb values were 0.178, 0.188, and 0.168 m for #1, #3, and #5, respectively.

The results of the climbing heights, reflection coefficients, and overtopping of different blocks from numerical solutions are plotted in Figure 6. It can be seen from the figure that when the radius of the sphere (*R1*) was constant, the climbing height and reflection coefficient decreased with the increase of the cylinder size (*R2*), while the amount of overtopping wave increased. When the cylinder radius (*R2*) was constant, the climbing height and wave volume decreased with the increase of the sphere size (*R1*). Given the above factors such as block climb, reflection coefficient, overtopping waves, block porosity, and material consumption, block #6 had the best overall performance and block

#6 would be used for TB-CUBE block design parameter and for further hydrodynamic characteristics investigation.

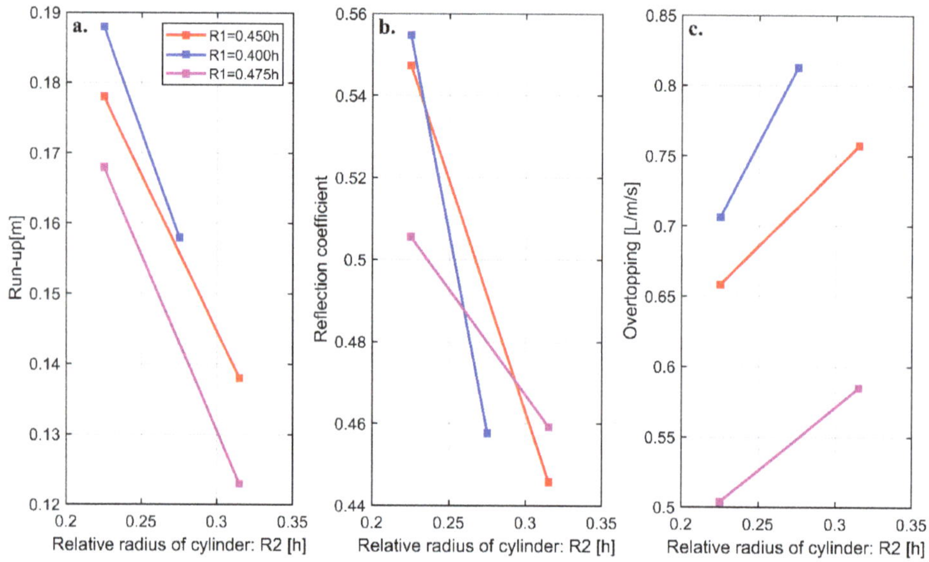

Figure 6. Numerical results of wave performance in the relative wave run-up (**a**), reflection coefficient (**b**), and overtopping (**c**) under various *R1* (relative radius of the sphere) and *R2* (relative radius of the cylinder), as stated in Figure 1 and Table 2.

3.3. Regular Wave Run-Up

Using the numerical wave tank that was constructed, a TB-CUBE hydrodynamic characteristics simulation study was carried out for block #6 with the best performance by the above-mentioned optimization comparison. The distribution of the flow velocity of the wave on the smooth slope versus the TB-CUBE and the climbing and falling on the block slope are shown from Figures 7–11 for t = 5.25 s, 5.45 s, 5.75 s, 6.15 s, and 6.45 s, respectively (a and b for without TB-CUBE; c and d for with TB-CUBE). These figures depict the complete process of wave propagation to the front of the breakwater, the wave climbing on the slope, the exhaustion of the wave energy when climbing to the highest point, and the wave falling back.

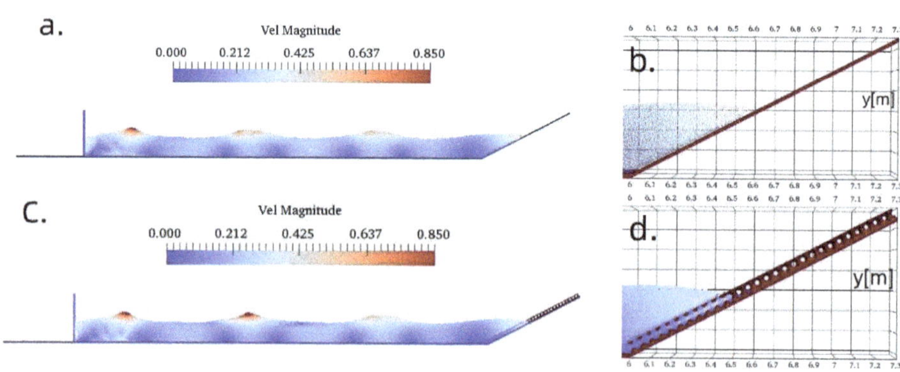

Figure 7. Wave patterns and velocity distribution at t = 5.25 s when thewave propagates to the breakwater: (**a**) Without article block; (**b**) Zoom in; (**c**) With article block; (**d**) Zoom in.

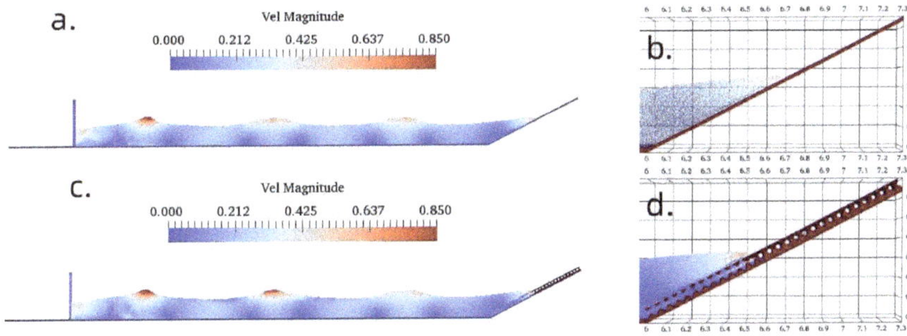

Figure 8. The same as Figure 7, but for wave climbing along the breakwater at t = 5.45 s.

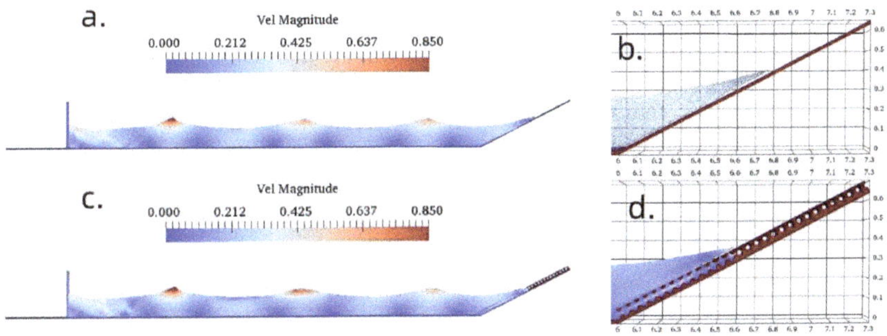

Figure 9. The same as Figure 7, but for wave climbing to the top along the breakwater at t = 5.75 s.

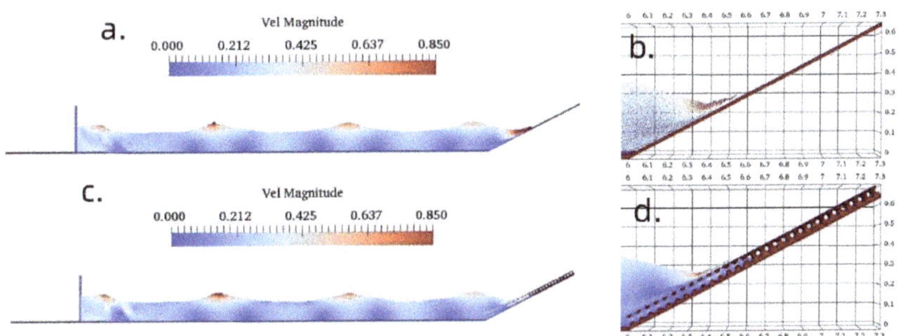

Figure 10. The same as Figure 7, but for wave retreating at t = 6.15 s.

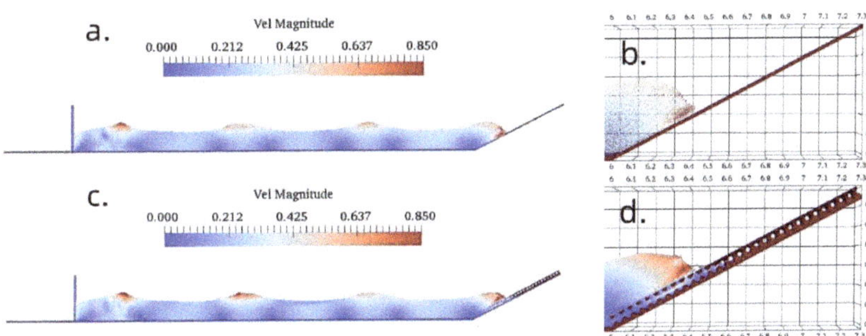

Figure 11. The same as Figure 7, but for wave breaking at t = 6.45 s.

Figure 7 showed that when the wave reached the front of the breakwater, the wave was in a regular shape, with some reflection from the breakwater. The cases with TB-CUBE (Figure 7c,d) showed more reflection than the case without one (Figure 7a,b). The case with TB-CUBE revealed less climbing up energy due to the resistance from the armor unit. Figure 8 shows the moment that the wave touched the slope and began to climb along the breakwater. At this time, the velocity of the water particle on the slope was relatively large. For the smooth slope, the wave is more energetic to achieve high placement (Figure 8b). Figure 9 showed that as the wave gradually climbed on the breakwater, the flow velocity gradually decreased to zero, and reached the maximum height for both the cases with TB-CUBE (Figure 9c,d)and without one (Figure 9a,b). When the wave reached the location of the TB-CUBE block, part of the water also entered the block hole, leading to wave energy and a pronouncedly shorter climb distance on the TB-CUBE breakwater surface (Figure 9c,d) than the smooth slope (Figure 9a,b). The discrepancy in the climbing distance has been clearly displayed in Figure 9b,d. The reflective wave was still stronger in the cases with TB-CUBE block (Figure 9c,d). Figure 10 shows that when the fluid began to fall back along the breakwater due to its own gravity, the flow velocity gradually fortified. At this time, the TB-CUBE had a certain hindering effect on the backflow, so the backflow velocity was lower than that on the smooth slope (Figure 10a,b). Substantial water was trapped in the hole of the block. Figure 11 shows that the falling water met the following wave front, breaking before the breakwater. The flow velocity in the breaking area was large and disorderly, and then the next cycle of climbing began.

3.4. Regular Wave Overtopping

The wave overtopping process on slope breakwater laid with TB-CUBE under regular waves roughly contained five stages: wave propagation to the front of the breakwater, wave breaking, wave climbing along the breakwater, wave overtopping over the breakwater, and fluid falling back. Figure 12 displays the flow velocity at t = 7.70 s, t = 7.80 s, t = 7.90 s, t = 8.05 s, t = 8.30 s, and t = 9.00 s, respectively. Figure 12a describes the wave breaking on the breakwater. Figure 12b shows the wave climbing up the slope and the fluid with higher flow velocity climbing higher. Figure 12c shows that when the waves climbed to the top of the breakwater, the fluid with a high velocity jumped up to the top surface of the breakwater, while the water particles with low velocity in the lower layer flowed into the pores of the block on the top of the breakwater. Figure 12d presents that the waves passed over the top of the breakwater to form overtopping, but it is worth noting that partial water was trapped in the block cavity on the top of the breakwater due to the threshold effect that was generated by the structural properties of the TB-CUBE. Figure 12e,f shows that the water fell back along the slope, and a vortex was formed in front of the breakwater where the flow velocity was disordered and turbulent, and then the next cycle of the wave overtopping process began.

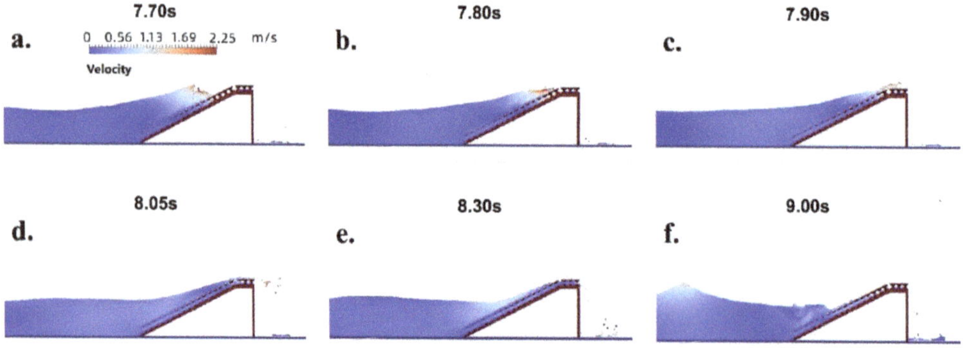

Figure 12. Wave overtopping and velocity distribution at time snapshots 7.70 s (**a**), 7.80 s (**b**), 7.90 s (**c**), 8.05 s (**d**), 8.30 s (**e**), and 9.00 s (**f**).

4. Discussion

4.1. Roughness of Block

For further insights on the wave dissipation characteristics of the TB-CUBE block, the roughness coefficient K_Δ from the numerical experiment results was obtained from the ratio of wave run-up level on slope breakwater with TB-CUBE block and smooth slope breakwater. The mean K_Δ for the TB-CUBE block is 0.462 (Table 5). Comparing the roughness coefficients of various blocks [1], the roughness coefficient of the TB-CUBE is less than 0.5 as well, indicating its preferable performance in wave damping.

$$K_\Delta = R_{block}/R_{smooth},$$

where K_Δ is the roughness coefficient, and R_{block} and R_{smooth} are the wave run-up values on slope breakwater with this TB-CUBE and that on a smooth breakwater, respectively.

Table 5. TB-CUBE roughness-permeation parameter from the numerical model results.

No	Wave Input		Run-Up [m]		Roughness Coefficient K_Δ		
						TB-CUBE	
	T (s)	H (m)	Smooth Interface	with TB-CUBE Block	Smooth Interface	Experiment	Mean
1	1.2	0.06	0.158	0.073	1	0.462	0.462
2		0.12	0.323	0.143	1	0.443	
3	1.8	0.06	0.170	0.093	1	0.547	
4		0.12	0.353	0.158	1	0.448	

4.2. Effect of Slope

Figure 13 shows the relationship between the relative wave run-up (R/H) and the breakwater slope with different incident wave heights from the numerical results. The relationship between R/H and the breakwater slope displays a single-peak curve trend, meaning that the run-up value increases first and then decreases as the slope increases. When the slope is steeper, most of the wave energy has been consumed by the reflection, so the climb is small. As the slope varies from 1.5 to 2, moving toward mildness, the reflection decreases. During that, some amount of the wave energy is converted into potential energy, so the climbing increases substantially. Taking H = 4.8 m and T = 6 s (Figure 13a) as an example, the relative climbing height of the wave increases from 1.23 to 1.67. When the slope changes from 2 to 2.5, the slope becomes milder and the waves break in front of the breakwater, with the dissipation of abundant wave energy. In addition, the climbing net distance of the wave increases on the mild slope, along with the enhanced wave energy dissipation due to increased friction and turbulent energy dissipation. Therefore, the run-up value becomes smaller. For the case of H = 4.8 m and T = 6 s, the relative wave run-up drops from 1.67 to 0.99 (Figure 13a).

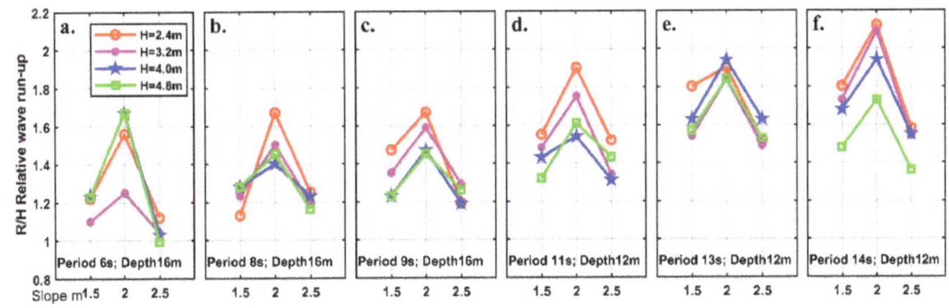

Figure 13. Numerical results for slope versus the wave run-up under various wave heights (**a**) T = 6 s, D = 16 m; (**b**) T = 8 s, D = 16 m; (**c**) T = 9 s, D = 16 m; (**d**) T = 11 s, D = 12 m; (**e**) T = 13 s, D = 12 m; (**f**) T = 14 s, D = 12 m.

4.3. Effect of Inverse Wave Steepness

Figure 14a illustrates the relationship between the reflection coefficient and inverse wave steepness under different slopes from the numerical solutions. The reflection coefficient decreases as the slope becomes gentler. When the slope is constant, and the wave steepness is small, a large amount of wave break occurs. As the wave steepness increases, the wave breaking decreases and is replaced by the reflection wave that increases gradually. When the wave steepness reaches a certain value, the reflection coefficient tends to be stable. Figure 14b,c display the relationship between the height and inverse wave steepness at various periods. The relative wave run-up (R/H) and inverse wave steepness (L/H) are generally positively correlated, meaning that the relative wave run-up elevates with the increase of inverse wave steepness. When inverse wave steepness is small, abundant waves are broken in front of the breakwater, and the wave energy is dissipated. Nevertheless, if it increases further, the amount of breaking wave would also further reduce and meanwhile the wave reflection is enhanced, contributing to a slow rise in the climbing.

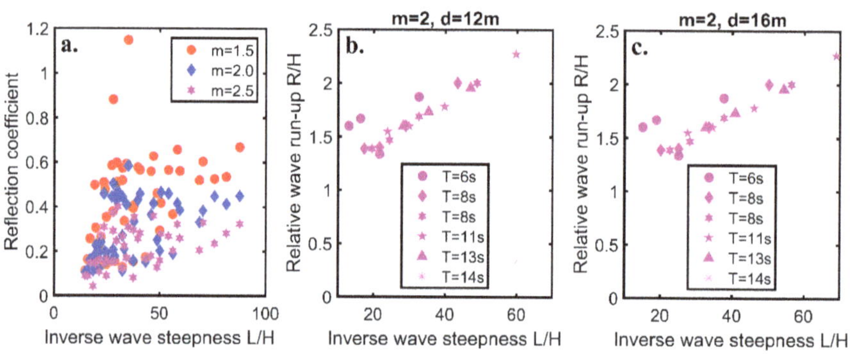

Figure 14. L/H versus reflection coefficient (**a**); L/H versus relative wave run-up (**b**,**c**) from numerical model results.

4.4. Wave Breaking Parameter

The well-accepted wave breaking parameter Iribarren number ζ_0 [44,45] is employed to describe the wave breaking, including three types: spilling breaker ($\zeta_0 < 0.5$), surging breaker ($0.5 < \zeta_0 < 3.3$), or plunging breaker ($\zeta_0 > 3.3$). Figure 15 shows the relationship between the relative wave run-up and the breaking wave parameters from the numerical simulations. When the slope m = 1.5, the minimum breaking wave parameter is 2.33, and the waves are mostly plunging wave. When m = 2, and the wave height is greater than 3.2 m, the waves appear as a surging wave. When the wave height is less than 3.2 m, it is mainly in the form of plunging wave. When m = 2.5 and the breaking wave parameter is in the range of 1.40 to 3.31, the breaking wave becomes a surging breaker.

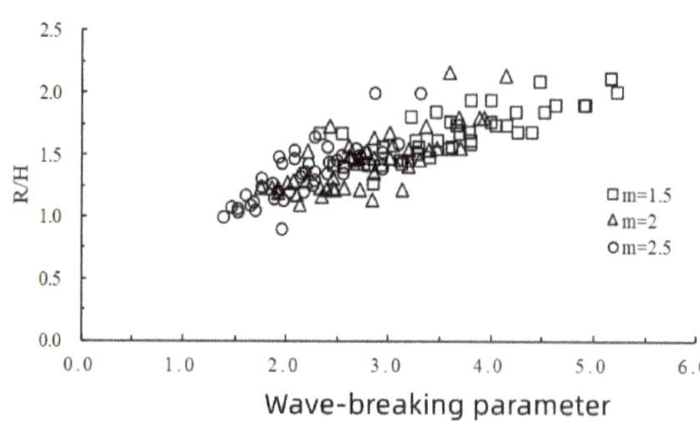

Figure 15. Wave-breaking parameter versus wave run-up from the numerical model results.

In general, the relative climbing height increases with the increase of the breaking wave parameter. When the breaking wave parameter is less than 2.5, the relative climbing height increases rapidly with the increase of the breaking wave parameter and the wave mainly appears as a surging breaking wave. The wave is completely broken, the reflected wave decreases, and the climbing height and the breaking wave parameter show a good linear relationship.

When the breaking wave parameter is greater than 2.5, the relative climbing height increase displays a blunt trend. When the wave parameters become larger, the waves break in the form of plunging breakers in front of the breakwater, and together with the accumulation of the reflected waves, the wave run-up and breaking wave parameters increase nonlinearly.

4.5. Empirical Formula for Wave Run-Up

Using the dimensionless parameter relative to the wave-runup R/H and the inverse wave steepness L/H as parameters, the relationship between R/H and L/H at different slopes is nonlinearly fitted, so as to determine the corresponding slope of the breakwater at different slopes. The fitting results are shown in Figure 16, and the values of A(m), B(m), and C(m) are shown in Table 6.

Table 6. Fitting coefficients.

Function	Slope *m*		
	1.5	2	2.5
A(m)	0.4319	0.2550	0.04805
B(m)	−0.0069	−0.0055	−0.01068
C(m)	0.000282	0.000263	0.000021

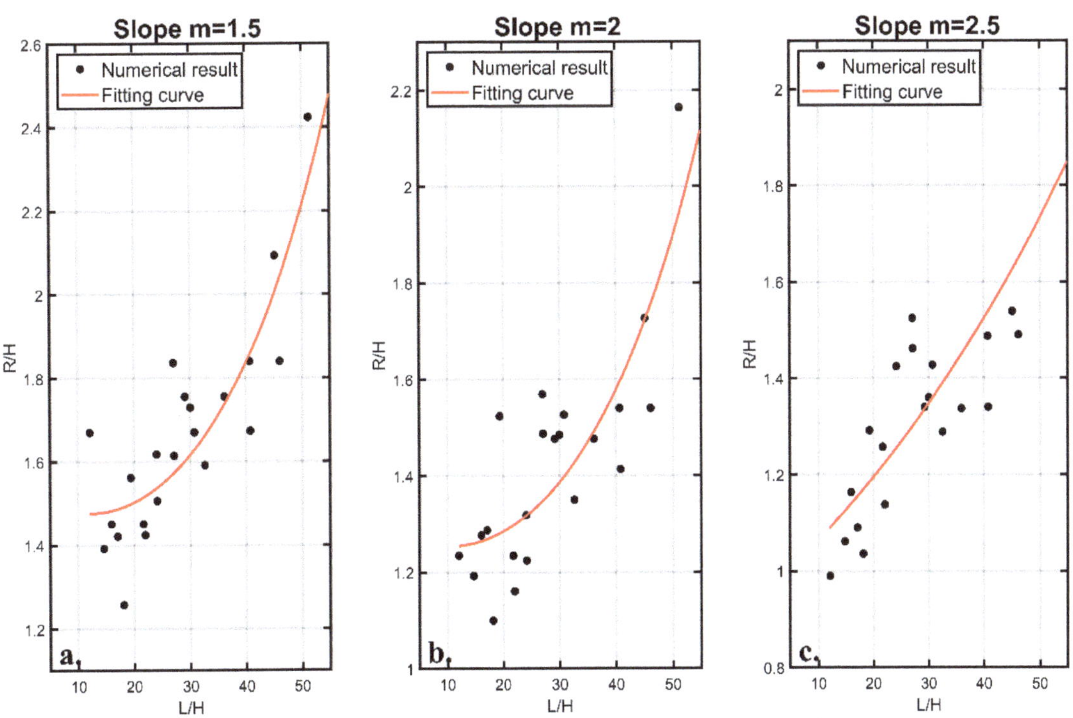

Figure 16. Fitting line for m = 1.5 (**a**); m = 2.0 (**b**); m = 2.5 (**c**) from the numerical model results.

Then, the coefficients $A(\mathrm{m})$, $B(\mathrm{m})$, and $C(\mathrm{m})$ under different slopes in Table 6 were fitted with the slope m value, and the relationship between the slope function and the slope was obtained through analysis. After fitting each formula, the empirical formula of wave climbing on the new armor block TB-CUBE was finally obtained, as shown in the following Formulas (6) to (9):

$$R = H\exp\left[A(\mathrm{m}) + B(\mathrm{m})\frac{L}{H} + C(\mathrm{m})(\frac{L}{H})^2\right] \quad (6)$$

$$A(\mathrm{m}) = 0.677 - 0.048\, exp(\frac{m}{0.92}) \quad (7)$$

$$B(\mathrm{m}) = -0.007 + 7.14 \times 10^{-8}\, exp(\frac{m}{0.201}) \quad (8)$$

$$C(\mathrm{m}) = 2.84 \times 10^{-4} - 7.15 \times 10^{-10} 1.39\, exp(\frac{m}{0.195}) \quad (9)$$

in which R is the wave run-up value, L is the wavelength (m), H is the average wave height in front of the slope breakwater (m), and m is the slope of the breakwater. Numerical experimental data was compared with the calculated value from the formula. The correlation coefficient is 0.981. The formula produced a strong relationship with the data from the numerical experiment, indicating its high accuracy in reflecting the wave run-up phenomenon as displayed in Figure 17.

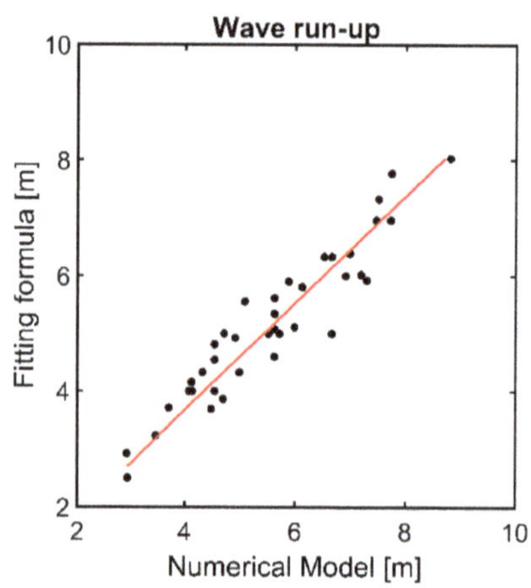

Figure 17. Wave run-up from the numerical model results versus from the fitting formula.

5. Conclusions

Based on the open-source code DualSPHysics, a numerical investigation has been conducted for the parameter design and the understanding of the wave damping effects and hydrodynamic characteristics of one new type of armor block, TB-CUBE. The main conclusions are summarized as follows:

(1) The three-dimensional, SPH-based active absorption wave-making flume has been constructed to reproduce wave evolution on rubble mound breakwaters and to obtain the optimal wave damping effects under various design combinations. The results have proven that when the radius of the sphere is constant, both the wave run-up and reflection coefficient shrink with the increase of the cylinder size, and the overtopping rises. The amount of wave run-up and overtopping decreases with the increase of the size of the sphere under invariant radius of the sphere.

Given the factors of wave run-up, overtopping, wave reflection, porosity, and concrete consumption, the TB-CUBE block with $R1$ and $R2$ values of 0.475 h and 0.315 h, respectively, outperforms others, and thus has been selected for further study.

(2) The mean roughness coefficient that was given by numerical simulations for this optimized TB-CUBE block is 0.462, indicating a promising capability of wave damping. It could also be viewed from the three-dimensional visualization that the structural characteristics of the TB-CUBE block itself have exerted large impacts on the interaction between the wave and the breakwater. It can accommodate the water body and form a turbulent flow effect. The threshold effect makes the quantity of overtopping decrease.

(3) The results from the wave-block-breakwater model demonstrate that the wave run-up value increases along with the increase of wave height, displaying a unimodal distribution trend with the growth of slope, and rises with the increase of inverse wave steepness. The empirical formula for wave run-up on this TB-CUBE has been provided, which would be further employed in engineering practice.

(4) This study has paid attention to the interaction of non-breaking waves on slope breakwater with new type of armor unit, the TB-CUBE, in a 2D wave flume experiment and numerical investigation. Previous studies on unbroken and broken solitary waves highlighted that the interaction of these two types of waves with a coastal deck and the associated hydrodynamic effects were totally different, with the broken waves being more turbulent and introducing more impulsive effects [46]. It is reported that the "aleatory uncertainty" that is involved in the wave breaking and the wave loads on coastal structures, since a minor change in the wave height offshore (e.g., 1%) can lead to a variability of 35% in the pressures and forces that are applied to coastal structures [47]. In fact, this uncertainty is one of the main parameters that is related to the impact of broken waves on structures that is hard to capture via numerical simulations [47]. Further study should consider the effects of wave breaking effects on the stability of the armor unit on slope breakwater. In addition, kinetic energy and velocity or momentum flux of the flow that overtops the breakwater variations are valuable information for this process.

This study used normal incidence of waves. However, the dynamic process of the interaction between breakwater and incoming oblique waves is inherently a three-dimensional problem. Previous research also reported the reduction in overtopping when wave is oblique [48,49]. However, it is still necessary to evaluate the inundation process of the structure and the applied hydrodynamic loads, which includes also yaw and roll moments that further increases the risk for structural damage [9].

Our research is from a small scale (1:40) wave flume experiment which is a common experimental scale as other studies [50–52], and scale effects exist. It is well-accepted that solitary waves normally decay faster in the model than in the prototype due to boundary layer effects and fluid viscosity [53]. Full scale research that is not limited in wave-breakwater-armor-unit interaction has been reported in literature [54–56], and more numerical studies that are based on that would be feasible to be performed in the future.

Author Contributions: Conceptualization, C.P., H.W., H.Z. and H.C.; methodology, C.P., H.W., H.Z. and H.C.; experiment and validation, C.P. and H.W.; formal analysis, C.P. and H.W.; original draft preparation, C.P. and H.W.; review and editing, H.Z. and H.C. funding acquisition, H.Z. and H.C. All authors have read and agreed to the published version of the manuscript.

Funding: This research was funded the National Natural Science Foundation of China (28119002); Fundamental Research Funds for the Central Institutes (TKS20200204; TKS20220602).

Institutional Review Board Statement: Not applicable.

Informed Consent Statement: Not applicable.

Data Availability Statement: Not applicable.

Acknowledgments: We thank three anonymous reviewers for peer reviewing an earlier version of this manuscript, and the editor and reviewer for their constructive feedback that helped us improve the presentation of the manuscript.

Conflicts of Interest: The authors declare no conflict of interest.

References

1. Bruce, T.; van der Meer, J.W.; Franco, L.; Pearson, J.M. Overtopping performance of different armour units for rubble mound breakwaters. *Coast. Eng.* **2009**, *56*, 166–179. [CrossRef]
2. Demirel, V.; Wang, S. An efficient boundary element method for two-dimensional transient wave propagation problems. *Appl. Math. Model.* **1987**, *11*, 411–416. [CrossRef]
3. Belibassakis, K.A. A boundary element method for the hydrodynamic analysis of floating bodies in variable bathymetry regions. *Eng. Anal. Bound. Elem.* **2008**, *32*, 796–810. [CrossRef]
4. Troch, P.; Li, T.Q.; De Rouck, J.; Ingram, D. Wave interaction with a sea dike using a vof finite-volume method. In Proceedings of the 13th International Offshore and Polar Engineering Conference, Honolulu, HI, USA, 25–30 May 2003; pp. 325–332.
5. Jose, J.; Choi, S.J.; Giljarhus, K.E.T.; Gudmestad, O.T. A comparison of numerical simulations of breaking wave forces on a monopile structure using two different numerical models based on finite difference and finite volume methods. *Ocean Eng.* **2017**, *137*, 78–88. [CrossRef]
6. Xiang, T.; Istrati, D. Assessment of extreme wave impact on coastal decks with different geometries via the arbitrary lagrangian-eulerian method. *J. Mar. Sci. Eng.* **2021**, *9*, 1342. [CrossRef]
7. Xiang, T.; Istrati, D.; Yim, S.C.; Buckle, I.G.; Lomonaco, P. Tsunami loads on a representative coastal bridge deck: Experimental study and validation of design equations. *J. Waterw. Port Coast. Ocean Eng.* **2020**, *146*, 04020022. [CrossRef]
8. Istrati, D.; Hasanpour, A.; Buckle, I. Numerical investigation of tsunami-borne debris damming loads on a coastal bridge. In Proceedings of the 17th World Conference on Earthquake Engineering, Sendai, Japan, 13–18 September 2020.
9. Istrati, D.; Buckle, I.G. *Tsunami Loads on Straight and Skewed Bridges—Part 2: Numerical Investigation and Design Recommendations;* FHWA-OR-RD-21-13; Federal Highway Administration Research and Technology: Washington, DC, USA, 2021.
10. Arikawa, T.; Ishikawa, N.; Beppu, M.; Tatesawa, H. Collapse mechanisms of seawall due to the march 2011 japan tsunami using the mps method. *Int. J. Prot. Struct.* **2012**, *3*, 457–476. [CrossRef]
11. Monaghan, J.J. Simulating free surface flows with sph. *J. Comput. Phys.* **1994**, *110*, 399–406. [CrossRef]
12. Higuera, P.; Lara, J.L.; Losada, I.J. Simulating coastal engineering processes with openfoam (r). *Coast. Eng.* **2013**, *71*, 119–134. [CrossRef]
13. Zijlema, M.; Stelling, G.; Smit, P. Swash: An operational public domain code for simulating wave fields and rapidly varied flows in coastal waters. *Coast. Eng.* **2011**, *58*, 992–1012. [CrossRef]
14. Gomez-Gesteira, M.; Rogers, B.D.; Crespo, A.J.C.; Dalrymple, R.A.; Narayanaswamy, M.; Dominguez, J.M. Sphysics—Development of a free-surface fluid solver—Part 1: Theory and formulations. *Comput. Geosci.* **2012**, *48*, 289–299. [CrossRef]
15. Luo, M.; Khayyer, A.; Lin, P. Particle methods in ocean and coastal engineering. *Appl. Ocean Res.* **2021**, *114*, 102734. [CrossRef]
16. Lowe, R.J.; Buckley, M.L.; Altomare, C.; Rijnsdorp, D.P.; Yao, Y.; Suzuki, T.; Bricker, J.D. Numerical simulations of surf zone wave dynamics using smoothed particle hydrodynamics. *Ocean Model.* **2019**, *144*, 101481. [CrossRef]
17. Wei, Z.P.; Dalrymple, R.A.; Rustico, E.; Herault, A.; Bilotta, G. Simulation of nearshore tsunami breaking by smoothed particle hydrodynamics method. *J. Waterw. Port Coast. Ocean Eng.* **2016**, *142*, 05016001. [CrossRef]
18. Ren, B.; Wen, H.; Dong, P.; Wang, Y. Improved sph simulation of wave motions and turbulent flows through porous media. *Coast. Eng.* **2016**, *107*, 14–27. [CrossRef]
19. Dominguez, J.M.; Fourtakas, G.; Altomare, C.; Canelas, R.B.; Tafuni, A.; Garcia-Feal, O.; Martinez-Estevez, I.; Mokos, A.; Vacondio, R.; Crespo, A.J.C.; et al. Dualsphysics: From fluid dynamics to multiphysics problems. *Comput. Part. Mech.* **2021**. [CrossRef]
20. Gomez-Gesteira, M.; Cerqueiroa, D.; Crespoa, C.; Dalrymple, R.A. Green water overtopping analyzed with a sph model. *Ocean Eng.* **2005**, *32*, 223–238. [CrossRef]
21. Dalrymple, R.A.; Rogers, B.D. Numerical modeling of water waves with the sph method. *Coast. Eng.* **2006**, *53*, 141–147. [CrossRef]
22. Crespo, A.J.C.; Gomez-Gesteira, M.; Dalrymple, R.A. Modeling dam break behavior over a wet bed by a sph technique. *J. Waterw. Port Coast. Ocean Eng.* **2008**, *134*, 313–320. [CrossRef]
23. Barreiro, A.; Crespo, A.J.C.; Dominguez, J.M.; Gomez-Gesteira, M. Smoothed particle hydrodynamics for coastal engineering problems. *Comput. Struct.* **2013**, *120*, 96–106. [CrossRef]
24. Altomare, C.; Gironella, X.; Crespo, A.J.C. Simulation of random wave overtopping by a wcsph model. *Appl. Ocean Res.* **2021**, *116*, 102888. [CrossRef]
25. Dominguez, J.M.; Crespo, A.J.C.; Hall, M.; Altomare, C.; Wu, M.H.; Stratigaki, V.; Troch, P.; Cappietti, L.; Gomez-Gesteira, M. Sph simulation of floating structures with moorings. *Coast. Eng.* **2019**, *153*, 103560. [CrossRef]
26. Hall, M.; Goupee, A. Validation of a lumped-mass mooring line model with deepcwind semisubmersible model test data. *Ocean Eng.* **2015**, *104*, 590–603. [CrossRef]
27. Hasanpour, A.; Istrati, D.; Buckle, I. Coupled sph–fem modeling of tsunami-borne large debris flow and impact on coastal structures. *J. Mar. Sci. Eng.* **2021**, *9*, 1068. [CrossRef]
28. Crespo, A.J.C.; Altomare, C.; Dominguez, J.M.; Gonzalez-Cao, J.; Gomez-Gesteira, M. Towards simulating floating offshore oscillating water column converters with smoothed particle hydrodynamics. *Coast. Eng.* **2017**, *126*, 11–26. [CrossRef]
29. Altomare, C.; Dominguez, J.M.; Crespo, A.J.C.; Gonzalez-Cao, J.; Suzuki, T.; Gomez-Gesteira, M.; Troch, P. Long-crested wave generation and absorption for sph-based dualsphysics model. *Coast. Eng.* **2017**, *127*, 37–54. [CrossRef]

30. Meringolo, D.D.; Aristodemo, F.; Veltri, P. Sph numerical modeling of wave-perforated breakwater interaction. *Coast. Eng.* **2015**, *101*, 48–68. [CrossRef]
31. Tabet-Aoul, E.H.; Lambert, E. Tentative new formula for maximum horizontal wave forces acting on perforated caisson. *J. Waterw. Port Coast. Ocean Eng.* **2003**, *129*, 34–40. [CrossRef]
32. Monaghan, J.J.; Kos, A. Solitary waves on a cretan beach. *J. Waterw. Port Coast. Ocean Eng.* **1999**, *125*, 145–154. [CrossRef]
33. Altomare, C.; Crespo, A.J.C.; Rogers, B.D.; Dominguez, J.M.; Gironella, X.; Gomez-Gesteira, M. Numerical modelling of armour block sea breakwater with smoothed particle hydrodynamics. *Comput. Struct.* **2014**, *130*, 34–45. [CrossRef]
34. Crespo, A.J.C.; Dominguez, J.M.; Rogers, B.D.; Gomez-Gesteira, M.; Longshaw, S.; Canelas, R.; Vacondio, R.; Barreiro, A.; Garcia-Feal, O. Dualsphysics: Open-source parallel cfd solver based on smoothed particle hydrodynamics (sph). *Comput. Phys. Commun.* **2015**, *187*, 204–216. [CrossRef]
35. Safari, I.; Mouazé, D.; Ropert, F.; Haquin, S.; Ezersky, A. Hydraulic stability and wave overtopping of starbloc®armored mound breakwaters. *Ocean Eng.* **2018**, *151*, 268–275. [CrossRef]
36. Park, Y.H.; Oh, Y.M.; Ahn, S.M.; Han, T.H.; Kim, Y.T.; Suh, K.D.; Won, D. Development of a new concrete armor unit for high waves. *J. Coast. Res.* **2019**, *35*, 719–728. [CrossRef]
37. Tianjin Research Institute for Water Transport Engineering. A Comprehensive Type of Wave Damping Artificial Block and Embankment. CN 108385609 B, 27 October 2020. (In Chinese)
38. Isobe, M. Impact of global warming on coastal structures in shallow water. *Ocean Eng.* **2013**, *71*, 51–57. [CrossRef]
39. Colling, J.K.; Jafari Kang, S.; Dehdashti, E.; Husain, S.; Masoud, H.; Parker, G.G. Free-decay heave motion of a spherical buoy. *Fluids* **2022**, *7*, 188. [CrossRef]
40. Windt, C.; Davidson, J.; Ringwood, J.V. High-fidelity numerical modelling of ocean wave energy systems: A review of computational fluid dynamics-based numerical wave tanks. *Renew. Sustain. Energy Rev.* **2018**, *93*, 610–630. [CrossRef]
41. Monaghan, J.J. Smoothed particle hydrodynamics. *Annu. Rev. Astron. Astrophys.* **1992**, *30*, 543–574. [CrossRef]
42. Wendland, H. Piecewise polynomial, positive definite and compactly supported radial functions of minimal degree. *Adv. Comput. Math.* **1995**, *4*, 389–396. [CrossRef]
43. Crespo, A.-J.-C.; Gómez-Gesteira, M.; Dalrymple, R.-A. Boundary conditions generated by dynamic particles in sph methods. *Comput. Mater. Contin.* **2007**, *5*, 173–184.
44. Baldock, T.E.; Cox, D.; Maddux, T.; Killian, J.; Fayler, L. Kinematics of breaking tsunami wavefronts: A data set from large scale laboratory experiments. *Coast. Eng.* **2009**, *56*, 506–516. [CrossRef]
45. Hsu, T.W. Geometric characteristics of storm-beach profiles caused by inclined waves. *Ocean Eng.* **1998**, *25*, 69–84. [CrossRef]
46. Istrati, D.; Buckle, I.; Lomonaco, P.; Yim, S. Deciphering the tsunami wave impact and associated connection forces in open-girder coastal bridges. *J. Mar. Sci. Eng.* **2018**, *6*, 148. [CrossRef]
47. Istrati, D. Large-Scale Experiments of Tsunami Inundation of Bridges Including Fluid-Structure-Interaction. Ph.D. Thesis, University of Nevada, Reno, NV, USA, 2017.
48. Van Gent, M.R.A. Influence of oblique wave attack on wave overtopping at caisson breakwaters with sea and swell conditions. *Coast. Eng.* **2021**, *164*, 103834. [CrossRef]
49. Van Gent, M.R.A.; van der Werf, I.M. Influence of oblique wave attack on wave overtopping and forces on rubble mound breakwater crest walls. *Coast. Eng.* **2019**, *151*, 78–96. [CrossRef]
50. Koosheh, A.; Etemad-Shahidi, A.; Cartwright, N.; Tomlinson, R.; van Gent, M.R.A. Experimental study of wave overtopping at rubble mound seawalls. *Coast. Eng.* **2022**, *172*, 104062. [CrossRef]
51. Zheng, J.; Yao, Y.; Chen, S.; Chen, S.; Zhang, Q. Laboratory study on wave-induced setup and wave-driven current in a 2dh reef-lagoon-channel system. *Coast. Eng.* **2020**, *162*, 103772. [CrossRef]
52. Vieira, F.; Taveira-Pinto, F.; Rosa-Santos, P. New developments in assessment of wave overtopping on single-layer cube armoured breakwaters based on laboratory experiments. *Coast. Eng.* **2021**, *166*, 103883. [CrossRef]
53. Heller, V. Scale effects in physical hydraulic engineering models. *J. Hydraul. Res.* **2011**, *49*, 293–306. [CrossRef]
54. Chen, S.-g.; Gong, E.-y.; Zhao, X.; Arikawa, T.; Chen, X. Large-scale experimental study on scour around offshore wind monopiles under irregular waves. *Water Sci. Eng.* **2022**, *15*, 40–46. [CrossRef]
55. Chen, B.; Li, S. Experimental study of local scour around a vertical cylinder under wave-only and combined wave-current conditions in a large-scale flume. *J. Hydraul. Eng.* **2018**, *144*, 04018058. [CrossRef]
56. Schoonees, T.; Kerpen, N.B.; Schlurmann, T. Full-scale experimental study on wave overtopping at stepped revetments. *Coast. Eng.* **2021**, *167*, 103887. [CrossRef]

Journal of
Marine Science and Engineering

Article

Modelling of Chloride Concentration Profiles in Concrete by the Consideration of Concrete Material Factors under Marine Tidal Environment

Xueli Ju, Linjian Wu *, Mingwei Liu, Han Jiang and Wenxiao Zhang

National Engineering Research Center for Inland Waterway Regulation, School of River and Ocean Engineering, Chongqing Jiaotong University, 66 Xuefu Road, Nan'an District, Chongqing 400074, China; juxueli97@126.com (X.J.); mingwei_liu@126.com (M.L.); 15213118329@163.com (H.J.); zwx126737@126.com (W.Z.)
* Correspondence: wljabgf@126.com

Abstract: The corrosion of reinforcement induced by chloride ions is one of the most significant causes of durability deterioration for reinforced concrete (RC) buildings. The concrete material factors, including the water-to-cement ratio (w/c) of concrete, as well as the content, shape, particle grading, and random distribution of coarse aggregate embedded in mortar, have a marked effect on chloride transport performance within concrete. However, comprehensive consideration for the effects of both w/c and coarse aggregate performances on chloride diffusion characteristics in concrete is scarce, especially regarding the chloride diffusion model of concrete. In this paper, an indoor exposure experiment exploring chloride ions intruding into mortar and concrete specimens with w/c = 0.4, 0.5 and 0.6 was carried out, in order to acquire the chloride diffusion parameters for concrete three-phases composites. Based on the numerical algorithm of random generation and placement of two-dimensional random convex polygon coarse aggregate, mesoscopic numerical models for concrete, considering various coarse aggregate contents as well as grading, were established. Using the numerical simulation method of finite element analysis for chloride transport in cement-based materials, which can replace some of the exposure tests, the influences of w/c, coarse aggregate content and grading on chloride diffusion performance in concrete mesoscopic models were systematically probed. According to the Fick's second law, a chloride diffusion model by the consideration of w/c, volume fraction of coarse aggregate (VFCA), and maximum size of coarse aggregate (MSCA) was developed to assess the chloride concentration profiles in concrete under arbitrary w/c, coarse aggregate content, and coarse aggregate grading conditions. Certainly, the precision accuracy for this proposed chloride diffusion model was validated. The research results can provide theoretical support for chloride erosion behavior and structural durability assessment of concrete with different mix proportions.

Keywords: concrete composites; impact factors of concrete material; chloride diffusion characteristics; prediction model

Citation: Ju, X.; Wu, L.; Liu, M.; Jiang, H.; Zhang, W. Modelling of Chloride Concentration Profiles in Concrete by the Consideration of Concrete Material Factors under Marine Tidal Environment. *J. Mar. Sci. Eng.* **2022**, *10*, 917. https://doi.org/10.3390/jmse10070917

Academic Editors: M. Dolores Esteban, José-Santos López-Gutiérrez, Vicente Negro and M. Graça Neves

Received: 30 May 2022
Accepted: 30 June 2022
Published: 2 July 2022

1. Introduction

Reinforced concrete is extensively applied in water conservancy infrastructure buildings because of its low cost and excellent mechanical properties [1–3]. The durability of RC is closely related to various impact factors, such as chloride-induced steel corrosion, sulfate corrosion, carbonation, freezing-thawing cycles for concrete deterioration, and the influences of damage cracks, cover thickness, mix proportion, environmental temperature, and humidity [4–8]. In particular, chloride intruding into concrete, which induces steel corrosion, is considered to be an important factor to the declining durability of marine RC buildings [8–12]. At the mesoscopic level, concrete is a three-phase composite composed of mortar, coarse aggregate, and an interfacial transition zone (ITZ) [13,14]. In particular, the material properties of each phase in concrete include hardness, compactness, porosity, permeability, etc. [2,15]. The mortar and ITZ phases can be regarded as diffusion media for chloride ion transport, while the coarse aggregate phase can be regarded as a non-diffusion medium for chloride ion transport [16,17]. For concrete composites, the concrete penetrability principally relies on its composition materials and corresponding geometric arrangement, which include the water-to-cement ratio (w/c), coarse aggregate content (named VFCA, V_{ca}), shape, grading (characterized by MSCA, S_{max}), and random distributions of aggregates in mortar [18]. These typical material factors will change the meso- and microstructure of concrete to varying degrees and affect the resistance of concrete to chloride ion erosion. The water-to-cement ratio w/c is a crucial parameter in concrete structure service life. A larger w/c leads to more capillary pores and microcrack channels in mortar [19], which is more conducive to chloride transport in concrete. As the representative medium for chloride transport in concrete, the coarse aggregate content, particle grading, shape and random distribution in mortar will change the microstructures and capillary porosity of concrete to varying degrees and exert an impact on chloride transport behavior in heterogeneous materials [20]. In summary, the contents and proportions of water and cement in mortar and the characteristics of aggregate embedded in cement paste will influence the transport behaviors of chloride ions in concrete. Consequently, the characteristics of the w/c as well as the coarse aggregate are regarded as vital material factors affecting chloride transport behaviors in concrete. Previous studies lack relevant results that comprehensively consider the influences of the above material factors on chloride transport behavior; therefore, further investigation needs to be conducted.

There have been some attempts to study the transport behaviors of chloride ions by considering various concrete material factors. Golewski and Szostak [21–23] indicated that the nano-admixtures in the form of the active C-S-H seeds is an innovative form for accelerating the curing of concrete with fly ash (FA) at a very early stage, which can lead to the formation of a more homogeneous and compact cement matrix structure. By this token, the chloride penetration resistance for the aforementioned concrete materials can be greatly increased. Considering the effect of w/c on chloride transport behaviors within concrete materials, the physical experiments carried out by Mangat et al. [24] and Costa et al. [25] showed that the smaller the water-binder ratio (w/b) of ordinary Portland cement concrete is, the smaller the apparent chloride diffusion coefficient obtained by regression. Chen et al. [26] measured the resistivity of cement-based materials through the slice-resistivity method and evaluated the chloride permeation of concrete with diverse w/c at exposure times without regard to polarization. The results indicated that the chloride diffusivity decreased as the w/c decreased. Chen et al. [27] conducted a rapid chloride migration (RCM) test to measure the chloride diffusivity of cement materials. Moreover, a chloride diffusion coefficient model accounting for the effects of w/c and curing age was proposed. Gao et al. [28] employed an indoor exposure experiment to reveal that a lower w/c can observably decrease the chloride concentration peak value of concrete and its chloride diffusion coefficient. Khanzadeh et al. [29] conducted a field exposure experiment under the marine tidal zone to explore the influence of the w/b on the maximum phenomenon of chloride concentrations in concrete. The results showed that the maximum depth of chloride diffusion increased as the w/c increased. Higashiyama et al. [30] reported that a lower w/c is more effective in resisting chloride ingress (having a lower chloride diffusion coefficient). Similarly, Ribeiro et al. [31] described that a low w/b can effectively decrease the porosity and chloride diffusion coefficient in cement-based composites. In addition, to reach an extremely high resistance to chloride penetration, the w/b must be less than 0.40. Du et al. [32] demonstrated that the w/c can be considered a microscale calculating parameter because it determines the pore structure and porosity of mortar at the microscale. Furthermore, increasing the w/c can cause the chloride diffusivity of concrete to increase sharply; thus, reducing the w/c is the most effective way to decrease chloride ion intrusion into concrete.

In most actual engineering projects, the properties of concrete materials are subject to deterioration and variation, particularly in marine environments. To satisfy economic, mechanical, and reliability requirements, VFCAs are usually approximately 40% within concrete composites [33]. The properties of aggregates are considered an important factor that affects the corrosion resistance and durability of concrete structures. Wang et al. [33] probed the influence of VFCA on chloride ion distribution and diffusion behavior in concrete by means of an indoor drying-wetting cycle experiment. The chloride concentration and chloride diffusivity decreased with increasing VFCA, while the surface chloride concentration was not disturbed by the coarse aggregate. Yang et al. [34] used the accelerated chloride migration test (ACMT) to reveal that the chloride diffusivity of cement-based materials decreased with increasing aggregate content. Li et al. [35] explored the influence of aggregate content on chloride diffusion in concrete by simulating concrete with different VFCAs. The results showed that the larger the VFCA was, the smaller the chloride diffusivity was in concrete. Wu et al. [37] carried out an indoor real-time tidal cycle experiment, quantified the effect of VFCA on chloride diffusion parameters, and established a two-dimensional chloride diffusion model taking into account the effect of VFCA. Wu et al. [38] obtained the chloride diffusivity in concrete with different VFCAs by RCM and further revealed that the chloride diffusivity for concrete decreases with increasing coarse aggregate volume fraction. Du et al. [32] employed the finite element method to quantify the influence of VFCA on chloride diffusivity for concrete models, and the results indicated that an increasing VFCA can cause an obvious decrease in the chloride diffusivity of concrete. Comprehensively, the chloride diffusivity of concrete determined by the experimental and numerical methods declined with increasing VFCA.

The sizes of concrete aggregates also affect concrete properties [39,40]. Previous studies have also considered the effects of aggregate particle size or grading on chloride diffusion characteristics in concrete. Basheer et al. [41] conducted air penetrability experiments, accelerated carbonation tests, and freeze–thaw/salt scaling resistance experiments to explore the effects of aggregate size and gradation on the microstructural characteristics and durability of concrete. The results demonstrated that as the proportion of larger aggregates increased in the mixture, the local porosity at the ITZ increased, and the overall durability of concrete decreased. Wang et al. [33] elaborated the influence of the MSCA on the chloride diffusion behavior of concrete via an indoor drying-wetting cycle experiment. It was concluded that the chloride diffusion concentration increased with decreasing MSCA, and the surface chloride concentration had nothing to do with the coarse aggregate factor. Wu et al. [38] indicated that a larger MSCA led to a larger chloride diffusion coefficient in the interface transition zone (ITZ), which represents the chloride diffusion capacity of the ITZ. In addition, an empirical model of the ITZ chloride diffusion coefficient accounting for the MSCA was proposed. Altogether, the coarse aggregate size/grading can be regarded as one of the vital impact factors on chloride transport performance for concrete composites.

In summary, previous studies determined that the w/c, VFCA, and MSCA significantly affect the chloride diffusion behavior in cement matrix composites by changing the microstructure and capillary porosity of concrete [42]. However, most of the previous achievements have used a single concrete material factor alone in considering the effects of the w/c or VFCA and MSCA on chloride diffusion characteristics. Few studies have considered the influences of two or more of the above factors on chloride diffusion behavior and concrete structure durability. Moreover, the chloride concentration distribution, diffusion characteristics and corresponding prediction models in concrete have not been reported when comprehensively considering the combined effects of the w/c and coarse aggregate characteristics (content, shape, particle grading, and random distribution).

Therefore, the research significance of this paper is to further investigate and quantify the influence of comprehensive concrete material factors on chloride transport characteristics. Specifically, an exposure experiment for chloride intruding into mortar and concrete materials with various w/c is carried out to determine the important diffusion parameters of chloride ions for the mortar, coarse aggregate, and ITZ phases in concrete composites. The chloride diffusion characteristics are deeply investigated by the numerical simulation method for chloride ion diffusion in mesoscopic models of concrete, including the w/c, coarse aggregate content and particle grading. Finally, the chloride diffusion prediction model for concrete considering the concrete material factors of the w/c, VFCA, and MSCA is established. The above research results can provide necessary theoretical support for the evaluation of chloride erosion behavior and structural durability assessment of concrete with different mix proportions.

2. Experimental Procedure

2.1. Materials and Mix Proportions

During the experimentation described in this paper, the ordinary Portland cement (OPC) (P.O. 42.5) with a density of 3100 kg/m^3 was used as the cementitious material. Crushed limestone (apparent density of 2690 kg/m^3) with continuous grading and a size range of 5–40 mm was considered as the coarse aggregate. River sand without chloride (fineness modulus of 2.4 and apparent density of 2610 kg/m^3) was adopted as the fine aggregate. To avert the influences of external chloride ions on measurements, distilled water was used for preparation and testing during this paper's whole experimental progress. In this paper, two types of mortar and concrete mixture proportions were prepared on the basis of $w/c = 0.4$, 0.5, and 0.6; and the mix proportion details are exhibited in Tables 1 and 2. For the concrete, the VFCA is approximately 0.4–0.45.

Table 1. Mortar mixture details.

w/c	Water (kg/m^3)	Cement (kg/m^3)	Sand (kg/m^3)
0.4	325	813	1002
0.5	342	684	1114
0.6	354	590	1195

Table 2. Concrete mixture details.

w/c	Water (kg/m^3)	Cement (kg/m^3)	Sand (kg/m^3)	Coarse Aggregate (kg/m^3)	V_{ca}
0.4	195	488	601	1116	0.415
0.5	195	390	635	1180	0.439
0.6	195	325	658	1222	0.454

2.2. Specimen Preparation and Exposure Conditions

The specimen preparation of concrete was done according to SL/T352-2020 (Chinese standard). Cubic mortar and concrete specimens were cast in $100 \times 100 \times 100$ mm^3 PVC molds and subsequently moved to a vibration table for compaction. After 24 h initial curing and 28 days final curing for all the specimens, the epoxy polyurethane coating was used to seal five surfaces of all the experimental specimens except for one side to ensure that the chloride ions can only intrude into the mortar and concrete materials via one-dimensional diffusion. For an arbitrary exposure time, three mortar and concrete specimens were prepared to acquire the tested results as accurately as possible by averaging the aforementioned three samples.

After the preparation, curing, covering and other steps for the experimental specimens, they were moved and exposed to a man-made 3.5% NaCl solution under the condition of a drying-wetting cycle. During the drying-wetting alternation of chlorine salt solutions, each drying-wetting period and its corresponding dry-wet ratio were set as 24 h and 1:1, i.e., the experimental specimens were exposed in a drying environment along sustaining 12 h, in which the environmental temperature and relative humidity are 25 °C and 70%, respectively; and during the rest of 12 h, the specimens were immersed in 3.5% NaCl solution. The 3.5% NaCl solution in this simulated marine tidal environment was replaced once every 10 drying-wetting cycles, and the exposure times during this experiment were 60, 120, 240, 360, 480, and 600 consecutive days.

2.3. Powder Sampling and Chloride Ion Testing

Sampling was conducted at different exposure periods. For each exposure time, six types of experimental specimens, including three mortars and three concrete specimens, were used to test the chloride concentrations. Many powders were acquired by grinding the experimental mortar and concrete specimens with a precision of 2 mm along the depth direction. Subsequently, the free chloride concentrations within the powder samples were tested by using the method of rapid chloride ion content determination based on the standard [43].

2.4. Results and Discussion

2.4.1. Chloride Concentration Profiles

For different exposure times, the chloride concentration profiles of mortar and concrete specimens with different w/c are exhibited in Figures 1 and 2. The tested chloride concentrations are all expressed as a percentage of the concrete mass (%) during this paper's descriptions.

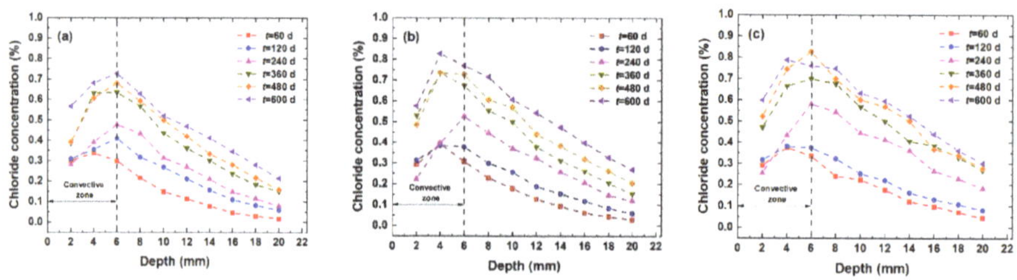

Figure 1. Chloride concentration profiles in mortar specimen at different exposure times: (**a**) $w/c = 0.4$; (**b**) $w/c = 0.5$; (**c**) $w/c = 0.6$.

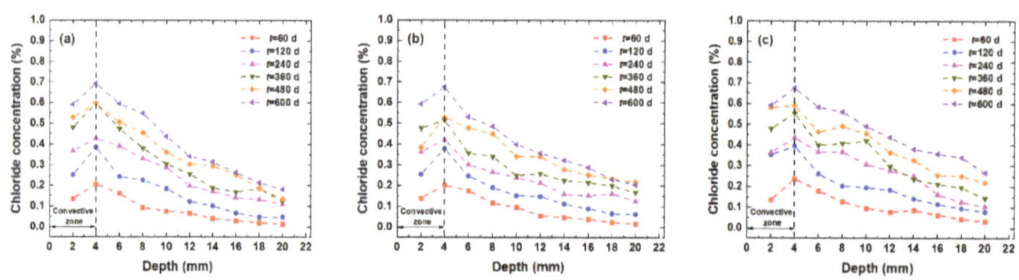

Figure 2. Chloride concentration profiles in concrete specimen versus depths at different exposure times: (**a**) $w/c = 0.4$; (**b**) $w/c = 0.5$; (**c**) $w/c = 0.6$.

As shown in Figures 1 and 2, the tested chloride concentrations increase with increasing exposure time and w/c. The chloride concentrations first increase and then decrease with increasing depth in the experimental specimens. The aforementioned variation laws of chloride diffusion concentrations are consistent with the conclusions reported by literatures [28,33]. In addition, the convection zone affected by drying-wetting cycles can be clearly observed in Figures 1 and 2. The convection zone widths for the experimental mortar and concrete specimens are 6 mm and 4 mm, respectively. This is because the coarse aggregate hardness and compactness exceed those of the mortar [38]. If the coarse aggregate proportion in the convection zone is larger, then the available chloride transport medium in the concrete and the overall porosity within the region will also be reduced accordingly [33,41], and convection will be inhibited. Therefore, the convection zone of the concrete specimen is shorter than that of the mortar.

2.4.2. Chloride Diffusion Parameters

During chloride transport in mortar or concrete materials, only the diffusion mechanism is considered to elaborate the chloride movement behaviors. Therefore, the convection mechanism of chloride ion transport into concrete can be temporarily neglected. Utilizing the closed-form solution of Fick's second law, the chloride profiles within the stable diffusion zone of cement-based materials, including the mortar and concrete, can be quantified as follows:

$$C(x,t) = C_s(t)\cdot\left[1-\mathrm{erf}\left(\frac{x}{2\sqrt{D_a(t)\cdot t}}\right)\right], \tag{1}$$

where $C(x, t)$ is the chloride concentration (%) of the cement-based materials (mortar and concrete) at arbitrary depth x and exposure time t. $C_s(t)$ is the time-dependent surface chloride concentration (%). $D_a(t)$ is the time-dependent apparent chloride diffusion coefficient (m^2/s). $\mathrm{erf}(\cdot)$ is the error function.

$C_s(t)$ and $D_a(t)$, which are both influenced by multiple factors, such as exposure times, service environments, and properties of the concrete composites, are regarded as the two most significant parameters affecting the chloride diffusion behavior in concrete. The existing research results consider that $C_s(t)$ is usually quantified in the form of a logarithmic function [44], and the time-dependent $D_a(t)$ is quantified via a power function proposed by Thomas et al. and is shown in the literature [33]. The two expressions are represented as follows:

$$C_s(t) = A\cdot\ln(t) + B, \tag{2}$$

$$D_a(t) = \frac{D_{28}}{1-m}\cdot\left(\frac{t_{28}}{t}\right)^m, \tag{3}$$

where A and B are the regression parameters; D_{28} is the apparent chloride diffusion coefficient at t = 28 days, i.e., named as the reference chloride diffusion coefficient (m^2/s); and m denotes the aging factor; t_{28} means the reference diffusion time, i.e., t_{28} = 28 days.

Combined with the chloride profiles for the mortar specimens exhibited in Figure 1, the values of C_s, D_{28} and m for mortars at different exposure times with different w/c can be confirmed according to fitting Equations (1) and (3) by means of the nonlinear least squares method, as elaborated in Figure 3.

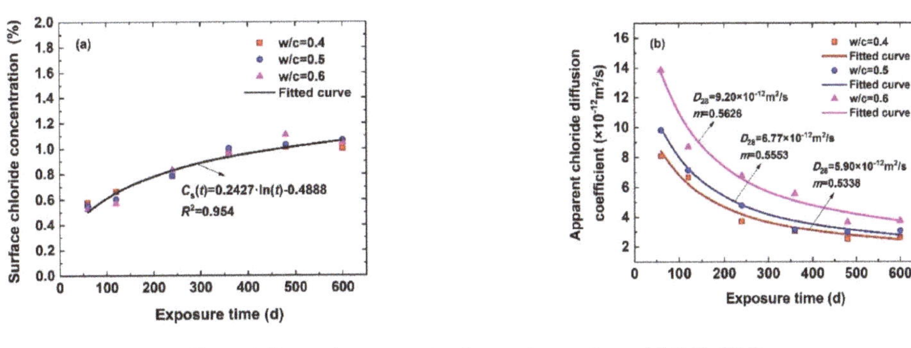

Figure 3. Regression parameters for mortar specimen: (**a**) $C_s(t)$; (**b**) D_a.

The values of C_s are almost constant with increasing w/c, so it can be considered that the C_s values in the mortar have nothing to do with the w/c. Therefore, in this paper, the average values of C_s with different w/c are considered representative values. The empirical expression of C_s with respect to t can be determined as follows: $C_s(t) = 0.2427\cdot\ln(t) - 0.4888$ (for mortar specimens). The time-dependent chloride diffusion coefficients for the mortar, i.e., chloride diffusivities for the homogeneous mortar phase within concrete composites, are set as $D_m(t) = D_{28m}\cdot(t_{28}/t)^m$, and the detailed parameters are shown in Table 3.

Table 3. Chloride diffusion parameters for mortars with different w/c.

w/c	$C_s(t)$	D_{28m} ($\times 10^{-12}$ m^2/s)	t_{28} (Day)	m
0.4		5.90		0.5338
0.5	0.2427·ln(t) − 0.4888	6.77	28	0.5553
0.6		9.20		0.5626

3. Numerical Simulation Method for Chloride Diffusion in a Mesoscopic Model of Concrete

In practical engineering, it can be considered that chloride intrusion into concrete is a natural and relatively long-term behavior. Physical experimental studies on long-term chloride diffusion will not only require substantial time but also consume more manpower, material, and financial resources [33,45]. Consequently, a mesoscopic simulation method for numerical models of concrete is suggested to replace some physical experiments to investigate the long-term diffusion progress of chloride ions in concrete [45]. Most of the existing concrete numerical models based on the finite element method simplify heterogeneous concrete composite materials into isotropous uniform diffusion media. This assumption is quite different from the actual situation.

Coarse aggregate is generally one of the main components for concrete composites, and its shape, volume fraction, particle grading, and random distribution within mortar should be comprehensively considered in the mesoscopic numerical model for concrete. In addition, the ITZ is regarded as the loosest zone in concrete, which easily cracks under a series of external loads compared with the mortar phase [38,46].

For this paper's studies, considering the mesoscopic structure of multiphase composites, mesoscopic models of concrete with convex polygonal coarse aggregates are established based on the Monte Carlo stochastic simulation algorithm. In particular, the ITZs are thought to have a uniform distribution on the surface of the coarse aggregate, the ITZ thicknesses remain consistent, and the two phases of the coarse aggregate and ITZ are completely wrapped inside the mortar phase in the model, as shown in Figure 4.

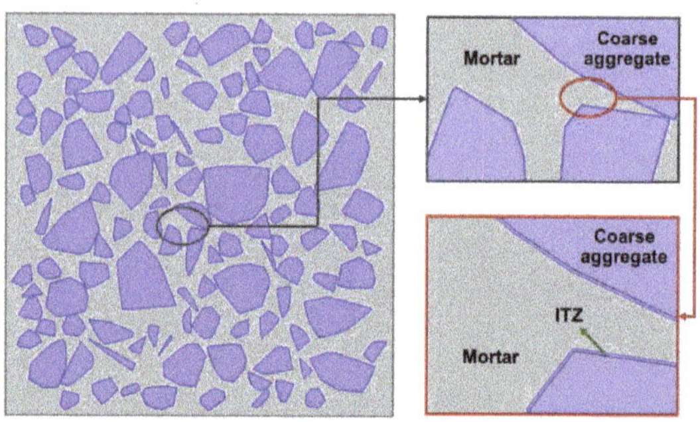

Figure 4. Three-phase concrete composites composed by mortar, coarse aggregate, and ITZ.

3.1. Mesoscopic Numerical Model for Concrete

The mesoscopic numerical model for concrete with different w/c is built in the form of a square of 100×100 mm^2, as shown in Figure 5. The w/c values and VFCAs within the mesoscopic numerical models of concrete in Section 3 are all consistent with the concrete specimens of the indoor exposure experiment elaborated in Table 2 of Section 2. Noticeably, 10 parallel samples are taken for any mesoscopic numerical model of concrete; namely, each of the simulation analysis values of chloride concentration represent an average chloride concentration of 10 parallel samples.

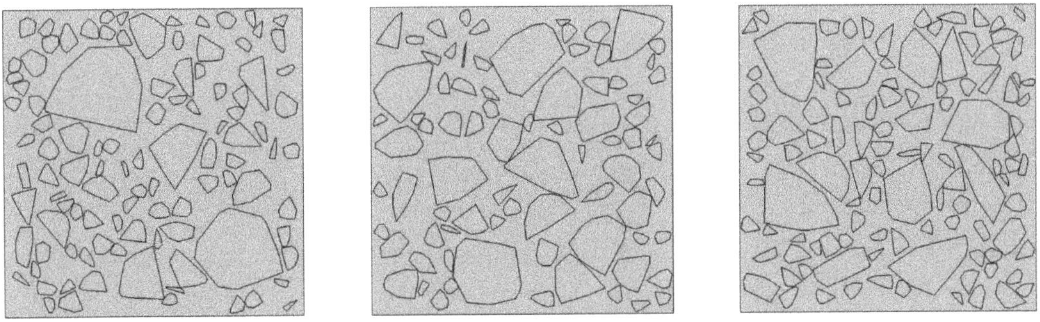

(**a**) (**b**) (**c**)

Figure 5. Mesoscopic numerical model of concrete: (**a**) w/c = 0.4, V_{ca} = 0.42; (**b**) w/c = 0.5, V_{ca} = 0.44; (**c**) w/c = 0.6, V_{ca} = 0.45.

3.2. Initial Conditions, Boundary Conditions, and Chloride Diffusion Coefficients in Various Phases for the Mesoscopic Numerical Model of Concrete

3.2.1. Initial and Boundary Conditions for the Numerical Concrete Model

During the numerical simulation process, it is assumed that there is no initial chloride content in the mesoscopic numerical model of concrete. Therefore, the initial chloride concentration at any concrete model is set as $C(x > 0, t = 0) = C_0 = 0$. In addition, the left border position of the numerical model is set as the chloride diffusion surface with C_s, and the other three borders are set according to the flux-free property to ensure one-dimensional chloride diffusion behavior in the mesoscopic numerical model for concrete. Based on the physical experiment in Section 2, the time-dependency of C_s is $C_s(t) = 0.2427 \cdot \ln(t) - 0.4888$, as shown in Table 3.

3.2.2. Chloride Diffusion Coefficients in Various Phases for the Numerical Model of Concrete

The coarse aggregate phase in concrete composites is generally regarded as impermeable; thus, its apparent chloride diffusion coefficient is set as $D_{ca} = 0\ \mathrm{m^2/s}$. For the ITZ phase, the ITZ thickness, i.e., t_{itz} = 0.05 mm, is selected as representative during the numerical simulation process [38]. The ITZ chloride diffusion coefficient is defined as $D_{\mathrm{itz}}(t) = N_{\mathrm{itz}} \cdot D_{\mathrm{m}}(t)$, where $D_{\mathrm{itz}}(t)$ is the time-dependent chloride diffusion coefficient of the ITZ and N_{itz} denotes the normalized chloride diffusion coefficient of the ITZ. According to the normalized ITZ chloride diffusion coefficient model proposed by the literature [38], when the ITZ thickness t_{itz} = 0.05 mm and the MSCA S_{max} = 40 mm, the normalized ITZ chloride diffusivities with different w/c values are $N_{\mathrm{itz}}(w/c = 0.4, V_{ca} = 0.415) = 45.483$; $N_{\mathrm{itz}}(w/c = 0.5, V_{ca} = 0.439) = 44.567$; and $N_{\mathrm{itz}}(w/c = 0.6, V_{ca} = 0.454) = 43.913$. Ultimately, the chloride diffusion coefficient for the mortar phase in concrete is $D_{\mathrm{m}}(t) = [D_{28\mathrm{m}}/(1 - m)] \cdot (t_{28}/t)^m$, where the $D_{28\mathrm{m}}$ and m values are shown in Table 3.

3.3. Validation of the Mesoscopic Numerical Simulation Method for Chloride Diffusion in Concrete Models

The simulated calculating time is set from t_i = 0 day to t_e = 700 days, and the time step is 1 d. After the computation process is finished, the simulated chloride concentrations at various depths in the numerical models of concrete are determined. The diffusion depths are the same as those in the indoor exposure experiment in Section 2.3. Moreover, the chloride profiles determined by numerical simulation (NSVs) and the experiment (EVs) are shown in Figure 6. In addition, the chloride concentration results confirmed by the mesoscopic models of concrete versus the tested values via indoor experiments are plotted in Figure 7.

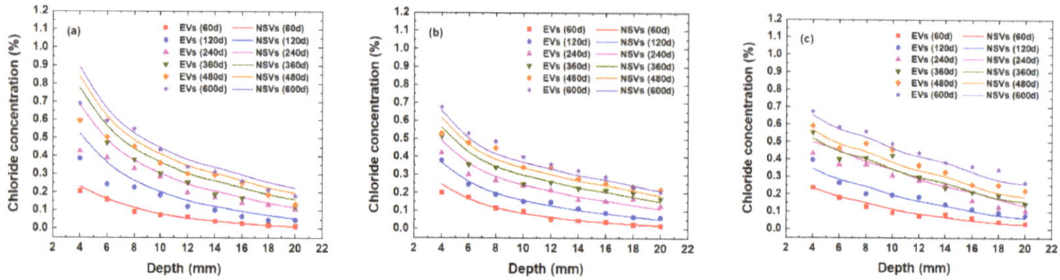

Figure 6. Comparison of the chloride concentration profiles in concrete based on the physical experiment and the mesoscopic numerical simulation method: (**a**) $w/c = 0.4$; (**b**) $w/c = 0.5$; (**c**) $w/c = 0.6$.

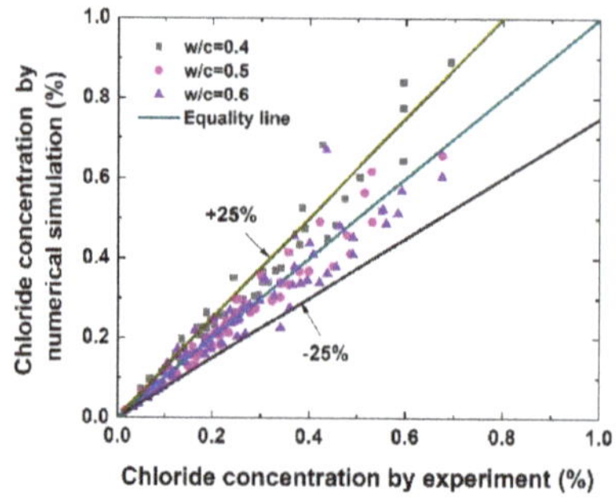

Figure 7. Comparison of the chloride concentrations determined by mesoscopic numerical simulation method versus those of the experiment.

From Figures 6 and 7, we can see that the numerical simulation values of chloride concentration exhibit good agreement with the experimental measurements, and the relative errors between the two methods are all within a ±25% range, which validates the accuracy of the mesoscopic numerical simulation method.

Therefore, in practical engineering, the mesoscopic numerical simulation method can be reasonably adopted to assess the chloride distribution in RC structures. In addition, if the significant chloride diffusion parameters of the uniform mortar phase in concrete multiphase composites are obtained, the chloride diffusion behavior in concrete may be accurately estimated. Furthermore, the mesoscopic numerical simulation method is suggested to replace some physical experiments for exploring the chloride distribution within concrete while considering the influences of material factors (including w/c, coarse aggregate shape, coarse aggregate volume fraction, coarse aggregate particle grading, and random distribution) to improve the efficiency of scientific research.

4. Numerical Simulation Experiment of Chloride Transport in a Mesoscopic Concrete Model Considering Material Factors

On the basis of the mesoscopic numerical simulation method for concrete, this paper intends to carry out mesoscale numerical simulation experiments for concrete considering concrete material factors and to further explore the influences of a series of concrete multiphase composite material factors on the chloride transport behavior and diffusion characteristics.

4.1. Impact Factors for the Numerical Simulation Experiment

Considering the concrete material factors, mesoscopic numerical models of concrete with different water-to-cement ratios $w/c = 0.4$, 0.5, and 0.6, different VFCAs $V_{ca} = 0$, 0.1, 0.2, 0.3, and 0.4, and MSCAs $S_{max} = 10$, 16, 20, 25, 31.5, and 40 mm were established, and mesoscopic numerical simulation experiments of chloride diffusion in concrete models were carried out. The specific simulation conditions are as follows in Table 4.

Table 4. Conditions of the mesoscopic numerical simulation experiment.

w/c	V_{ca}	S_{max} (mm)					
0.4	0	-	-	-	-	-	-
	0.1	5–10	5–16	5–20	5–25	5–31.5	5–40
	0.2	5–10	5–16	5–20	5–25	5–31.5	5–40
	0.3	5–10	5–16	5–20	5–25	5–31.5	5–40
	0.4	5–10	5–16	5–20	5–25	5–31.5	5–40
0.5	0	-	-	-	-	-	-
	0.1	5–10	5–16	5–20	5–25	5–31.5	5–40
	0.2	5–10	5–16	5–20	5–25	5–31.5	5–40
	0.3	5–10	5–16	5–20	5–25	5–31.5	5–40
	0.4	5–10	5–16	5–20	5–25	5–31.5	5–40
0.6	0	-	-	-	-	-	-
	0.1	5–10	5–16	5–20	5–25	5–31.5	5–40
	0.2	5–10	5–16	5–20	5–25	5–31.5	5–40
	0.3	5–10	5–16	5–20	5–25	5–31.5	5–40
	0.4	5–10	5–16	5–20	5–25	5–31.5	5–40

Figure 8 exhibits the mesoscopic numerical model examples of concrete for V_{ca} = 0, 0.3, 0.4 and S_{max} = 20, 25 mm. To ensure the accuracy of the numerical simulation results, when $V_{ca} \neq 0$, the average chloride concentrations for 10 model parallel samples are considered as the final simulation results.

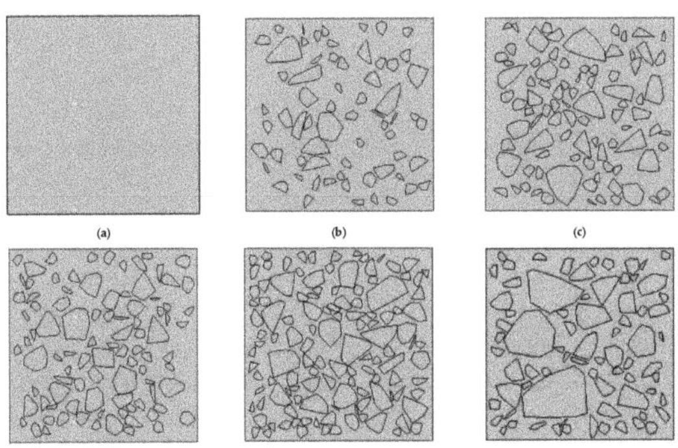

Figure 8. Mesoscopic numerical models of concrete: (**a**) V_{ca} = 0; (**b**) V_{ca} = 0.2, S_{max} = 25 mm; (**c**) V_{ca} = 0.3, S_{max} = 20 mm; (**d**) V_{ca} = 0.3, S_{max} = 31.5 mm; (**e**) V_{ca} = 0.4, S_{max} = 25 mm; (**f**) V_{ca} = 0.4, S_{max} = 40 mm.

4.2. Chloride Diffusion Coefficients for the Mesoscopic Numerical Models of Concrete

In the numerical simulation experiment for chloride diffusion in mesoscopic models of concrete, the initial conditions, boundary conditions, and the chloride diffusion coefficients of the mortar phase are still consistent with Section 3.2; in particular, the normalized ITZ chloride diffusivities are as shown in Table 5, which were determined by Equation (8) in the literature [38].

Table 5. Normalized ITZ chloride diffusivities, D_{nitz}.

V_{ca}	S_{max} (mm)					
	10	**16**	**20**	**25**	**31.5**	**40**
0.1	9.05	12.77	16.06	21.39	31.06	50.56
0.2	8.18	11.75	14.96	20.25	29.99	50.13
0.3	7.23	10.59	13.66	18.77	28.37	48.70
0.4	6.23	9.29	12.13	16.92	26.10	46.01

5. Results and Analysis

5.1. Influences of Concrete Material Factors on Chloride Profiles

5.1.1. Influence of Water-to-Cement Ratio, w/c

Figure 9 illustrates the chloride profiles in mesoscopic numerical models of concrete with different w/c values for exposure times t = 360 and 600 days, MSCAs S_{max} = 20 and 25 mm, and VFCAs V_{ca} = 0, 0.1, 0.2, 0.3, 0.4.

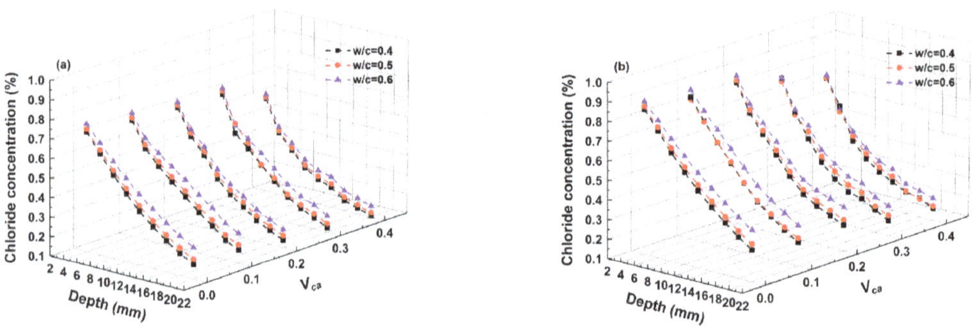

Figure 9. Chloride profiles in mesoscopic numerical models of concrete: (**a**) S_{max} = 25 mm, t = 360 d; (**b**) S_{max} = 20 mm, t = 600 d.

It can be determined from the figures that under the same exposure time t, the chloride concentration in the mesoscopic numerical model for concrete increases gradually with increasing w/c, which fully demonstrates that the w/c factor will significantly affect the chloride concentration distribution. Quantitatively, when V_{ca} = 0.4, S_{max} = 20 mm, the chloride concentration $C(x = 4$ mm, $t = 600$ days) increases almost 2.3% with the increasing of w/c from 0.4 to 0.6. When the w/c is larger, the capillary pore structure and microfracture channels in mortar are greater, which is more conducive to the transmission and chloride diffusion in concrete.

5.1.2. Influence of VFCA, V_{ca}

Figure 10 shows that the chloride concentrations gradually decrease with increasing V_{ca}, and the greater V_{ca} is, the greater the reduction in the chloride concentration in concrete. Especially, when w/c = 0.5, S_{max} = 20 mm, the chloride concentration $C(x = 4$ mm, $t = 600$ days) decreases almost 16.4% with the increasing of V_{ca} from 0 to 0.4.

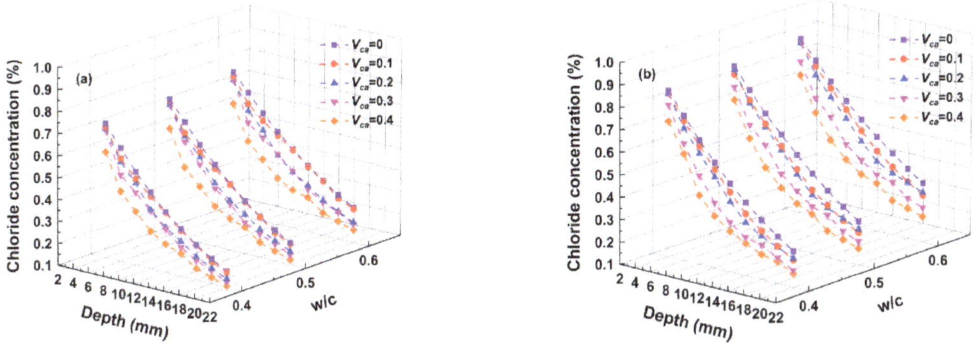

Figure 10. Chloride profiles in mesoscopic numerical models of concrete: (**a**) S_{max} = 25 mm, t = 360 d; (**b**) S_{max} = 20 mm, t = 600 d.

5.1.3. Influence of MSCA, S_{max}

Figure 11 shows the variation trend of the chloride concentrations in the mesoscopic numerical models for concrete with different S_{max} and w/c values, taking the calculated exposure times of t = 360 and 600 days, with V_{ca} = 0.3 and 0.4. Obviously, under conditions of the same V_{ca}, the chloride concentrations with different S_{max} values in concrete are not significantly different.

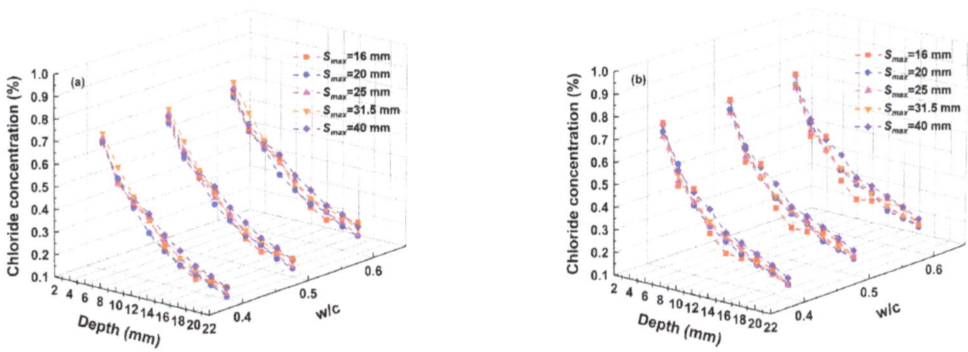

Figure 11. Chloride profiles in mesoscopic numerical model of concrete: (**a**) V_{ca} = 0.3, t = 360 d; (**b**) V_{ca} = 0.4, t = 600 d.

5.2. *Influences of Concrete Material Factors on Surface Chloride Concentrations*

The surface chloride concentration C_s values of the mesoscopic numerical model for concrete with different exposure times t, V_{ca}, and S_{max} are all determined in accordance with fitting the closed-form solution of Fick's second law, as expressed by Equation (1), to the tested chloride profiles shown in Figure 11 by using nonlinear regression analysis.

5.2.1. Influence of Water-to-Cement Ratio, w/c

The variations in C_s with increasing w/c are plotted in Figure 12. The figures include the computational exposure times t = 120, 360, and 600 days, V_{ca} = 0.3, 0.4, and S_{max}= 20, 25 mm. Obviously, C_s is almost unchanged with the increase in w/c, which is consistent with the conclusion in Section 2.4 of this paper. Therefore, it is considered that the surface chloride concentration has no specific relation with w/c.

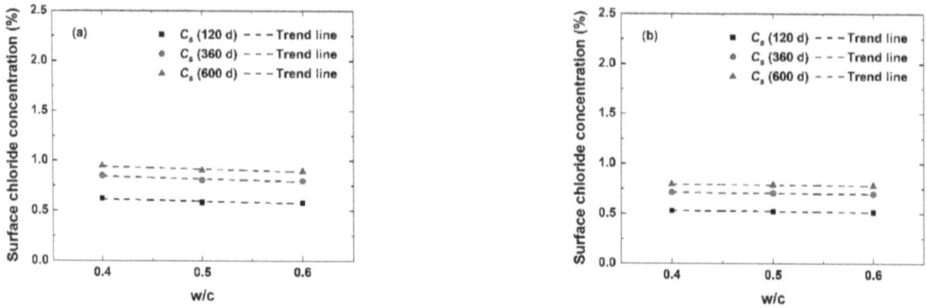

Figure 12. Surface chloride concentrations of mesoscopic numerical model for concrete: (**a**) V_{ca} = 0.3, S_{max} = 20 mm; (**b**) V_{ca} = 0.4, S_{max} = 25 mm.

5.2.2. Influence of VFCA, V_{ca}

Figure 13 illustrates the variation in C_s with V_{ca}. In the figure, the computational exposure times are t = 120, 360, and 600 days, w/c = 0.4, 0.5, and 0.6, and S_{max} = 16 mm, 20 mm, and 25 mm. Figure 13 shows that the C_s of the mesoscopic numerical model for concrete decreases linearly with increasing V_{ca}. This trend shows that the V_{ca} factor will affect the C_s values. Especially, when w/c = 0.5, S_{max} = 20 mm, the $C_s(t = 600 \text{ days})$ decreases almost 18.5% with the increasing of V_{ca} from 0 to 0.4.

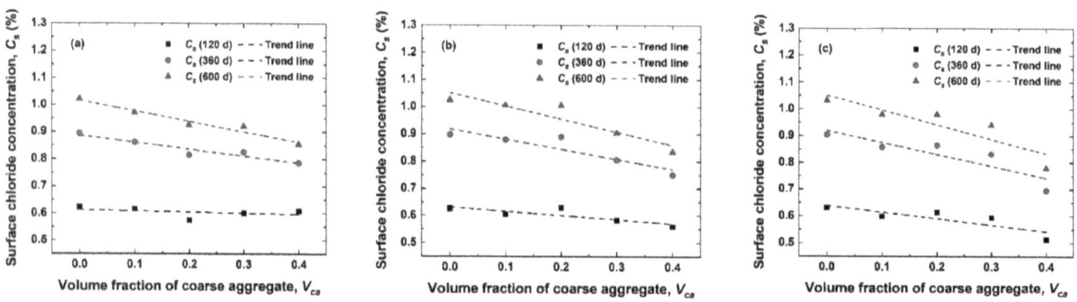

Figure 13. Surface chloride concentration of mesoscopic numerical model of concrete: (**a**) w/c = 0.4, S_{max} = 16 mm; (**b**) w/c = 0.5, S_{max} = 20 mm; (**c**) w/c = 0.6, S_{max} = 25 mm.

5.2.3. Influence of MSCA, S_{max}

Similarly, the variation in C_s with S_{max} in the mesoscopic numerical model for concrete is plotted in Figure 14. The computational exposure times are t = 120, 360, and 600 days, w/c = 0.4, 0.5, and 0.6, and V_{ca} = 0.2, 0.3, and 0.4. Clearly, C_s increases nearly linearly with increasing S_{max}. This shows that S_{max} can affect the C_s values. Quantitatively, when w/c = 0.5, V_{ca} = 0.4, the $C_s(t = 600 \text{ days})$ increases almost 23.4% with the increasing of S_{max} from 10 mm to 40 mm. In addition, C_s increases with increasing exposure time t at different mesoscopic numerical models for concrete.

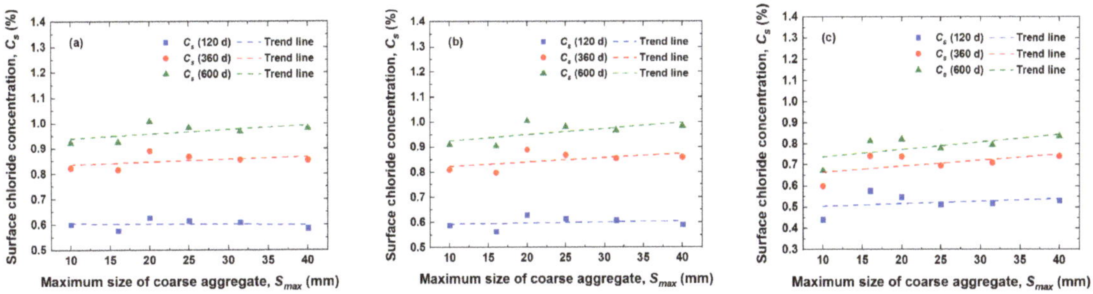

Figure 14. Surface chloride concentration of mesoscopic numerical model of concrete: (**a**) $w/c = 0.4$, $V_{ca} = 0.2$; (**b**) $w/c = 0.5$, $V_{ca} = 0.3$; (**c**) $w/c = 0.6$, $V_{ca} = 0.4$.

5.2.4. Prediction Empirical Model of C_s Considering Concrete Material Factors

According to the variation law of the C_s values with the concrete material factors mentioned above, the C_s values are independent of w/c but have a certain quantitative relation with V_{ca} and S_{max}. Therefore, to further quantify the influences of V_{ca} and S_{max} on C_s, the average value of C_s with different w/c is considered to be the representative C_s value for the concrete mesoscopic numerical model, and the trend of C_s with exposure time t at different V_{ca} and S_{max} is fitted by using the logarithmic function form (Equation (2)).

The variation in the correction parameters A and B in Equation (2) at different values of V_{ca} and S_{max} can be obtained, as shown in Figure 15. A and B show linear trends with increasing V_{ca}, which are fitted by the linear functions shown in Equations (4) and (5).

$$A(V_{ca}) = C \cdot V_{ca} + D, \tag{4}$$

$$B(V_{ca}) = E \cdot V_{ca} + F, \tag{5}$$

where $C(x, t)$ is the chloride concentration (%) of the cement-based materials (mortar and to simplify the calculation, the normalized maximum size of coarse aggregate is defined as $N_{Smax} = S_{max}/(S_{max} = 10 \text{ mm})$, and the variations in parameters C, D, E, and F within Equations (4) and (5) versus the N_{Smax} values are plotted in Figures 16 and 17.

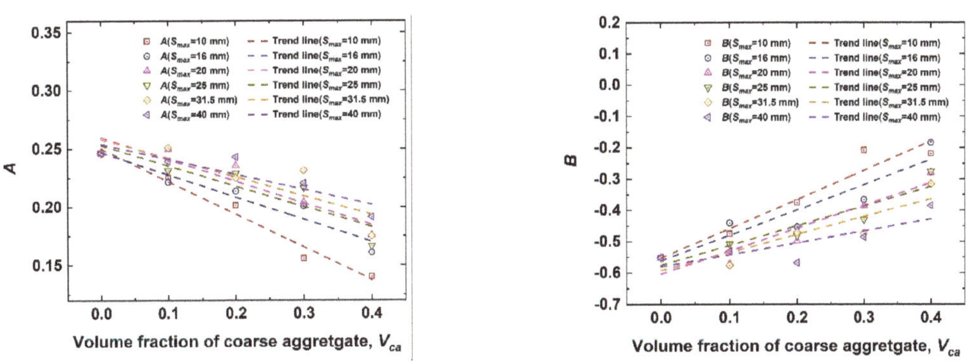

(**a**) (**b**)

Figure 15. Regression parameters of A and B with different S_{max} values versus V_{ca}: (**a**) Variation trend of parameter A; (**b**) Variation trend of parameter B.

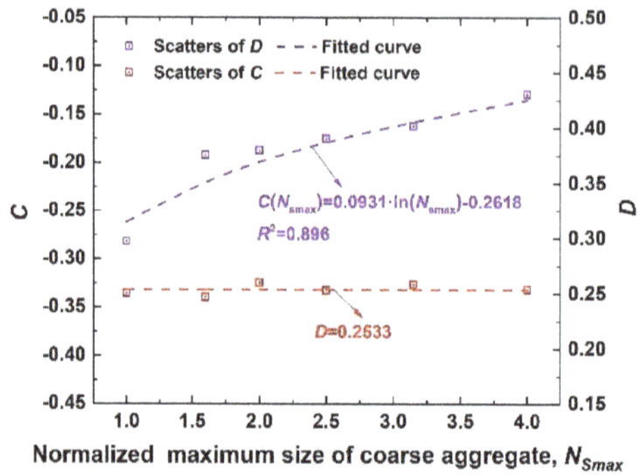

Figure 16. Regression parameters C and D versus N_{Smax}.

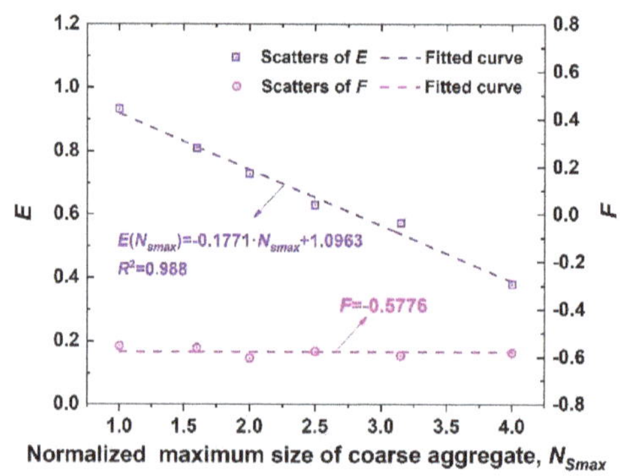

Figure 17. Regression parameters E and F versus N_{Smax}.

As observed in Figure 16, C increases logarithmically with increasing N_{Smax}. The variation in C with N_{Smax} is fitted by a logarithmic function, as shown in Equation (6). Obviously, D exhibits no difference with the increase of the N_{Smax}, and the D scatters present slightly random fluctuation with varying N_{Smax}; in addition, the relative errors between the scatters of D and its corresponding average values are all in a ±5% margin. As a consequence, it is reasonable to take the average value of D scatters as the representative value, namely, $D = 0.2533$.

$$C(N_{Smax}) = 0.0931 \cdot \ln(N_{Smax}) - 0.2618. \quad (6)$$

Figure 17 shows that the correction parameter E decreases linearly with increasing N_{Smax}. Thus, the linear function form of the correction parameter E with N_{Smax} can be obtained through Equation (7). Parameter F shows no obvious difference with increasing N_{Smax}. The relative errors between the scatters of F and their corresponding mean are included in ±5%, so the average value of $F = -0.5776$ is reasonable as the representative value under different w/c and N_{Smax}.

$$E(N_{Smax}) = -0.1771 \cdot N_{Smax} + 1.0963. \quad (7)$$

In summary, by substituting Equations (4)–(7) into Equation (2), the empirical model of C_s for concrete by considering the material factors (V_{ca} and S_{max}) can be established, as shown in Equation (8). The time dependency of C_s is also considered in this model, as follows:

$$\begin{cases} C_s(t, V_{ca}, S_{max}) = A(V_{ca}, N_{Smax})\cdot \ln(t) + B(V_{ca}, N_{Smax}) \\ A(V_{ca}, N_{Smax}) = C(N_{Smax})\cdot V_{ca} + 0.2533 \\ B(V_{ca}, N_{Smax}) = E(N_{Smax})\cdot V_{ca} - 0.5776 \\ C(N_{Smax}) = 0.0931\cdot \ln(N_{Smax}) - 0.2618 \\ E(N_{Smax}) = -0.1771\cdot N_{Smax} + 1.0693 \end{cases} \quad . \tag{8}$$

5.3. *Influences of Concrete Material Factors on the Chloride Diffusion Coefficients*

In this paper, the apparent chloride diffusion coefficients D_a can also be determined by fitting Equation (1) using regression analysis. D_a, which reflects the chloride ion diffusion ability in concrete, can be obtained. Combined with the power function of Equation (3), the reference chloride diffusion coefficient D_{28} for the concrete mesoscopic numerical model with different values of V_{ca}, w/c and S_{max} can be regressed.

5.3.1. Influence of Water-to-Cement Ratio, w/c

The variation in the reference chloride diffusion coefficient D_{28} with w/c in the mesoscale numerical model for concrete is shown in Figure 18. When V_{ca} and S_{max} are constants, the D_{28} values both increase with increasing w/c. Quantitatively, when V_{ca} = 0.4, S_{max} = 20 mm, the D_{28} increases almost 60.1% with the increasing of w/c from 0.4 to 0.6.

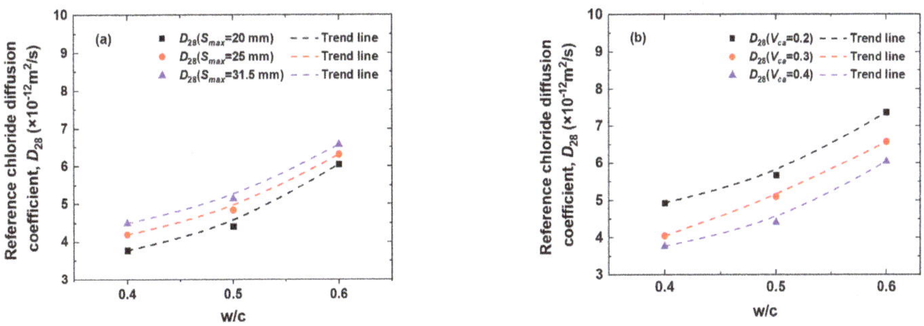

Figure 18. Reference chloride diffusion coefficient D_{28} versus w/c: (**a**) V_{ca} = 0.4; (**b**) S_{max} = 20 mm.

5.3.2. Influence of VFCA, V_{ca}

Figure 19 shows the variation in D_{28} in the mesoscopic numerical model for concrete with respect to V_{ca}. At arbitrary S_{max} and w/c, the D_{28} values decrease with increasing V_{ca}. This phenomenon illustrates that the concrete material factor V_{ca} will reduce the chloride diffusion coefficient D_a, and the greater V_{ca} is, the more obvious the reduction of D_a. Especially, when w/c = 0.5, S_{max} = 20 mm, the D_{28} decreases almost 35.8% with the increasing of V_{ca} from 0 to 0.4.

5.3.3. Influence of MSVA, S_{max}

Figure 20 depicts the D_{28} values in the mesoscopic numerical model for concrete with S_{max}. Obviously, at an arbitrary V_{ca} (Figure 20a) and w/c (Figure 20b), D_{28} shows a quadratic function trend of first decreasing and then increasing with S_{max}. The minimum D_{28} value appears in the range of S_{max} = 16–20 mm. Respectively, when w/c = 0.5, V_{ca} = 0.4, D_{28}(S_{max} = 40 mm) increases approximately 43.4% and 30.2% than those of S_{max} = 16 mm and 20 mm. It is noteworthy that D_{28} has the largest discreteness compared with the overall trend line when S_{max} = 16 mm.

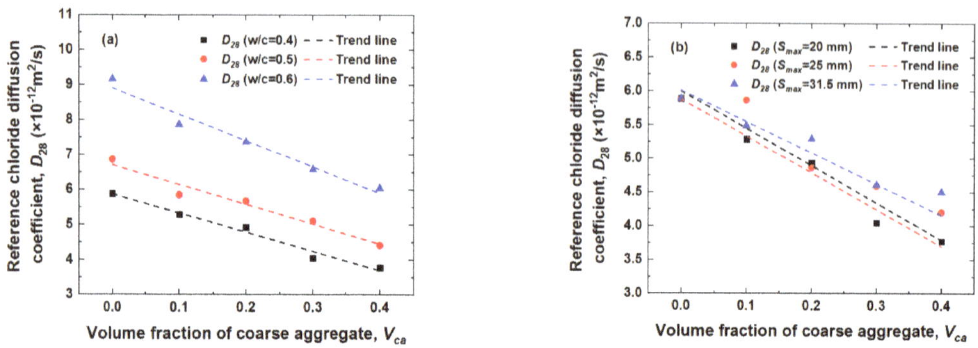

Figure 19. Reference chloride diffusion coefficient D_{28} versus V_{ca}: (**a**) S_{max} = 20 mm; (**b**) w/c = 0.4.

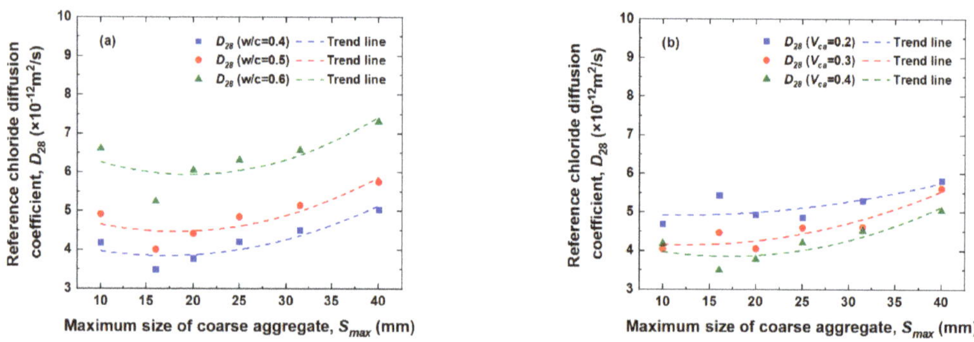

Figure 20. Reference chloride diffusion coefficient D_{28} versus S_{max}: (**a**) V_{ca} = 0.4; (**b**) w/c = 0.4.

5.3.4. Prediction Empirical Model of D_{28} Considering Concrete Material Factors

To quantitatively evaluate the influences of concrete material factors on the chloride diffusion coefficient, the w/c-corrected parameter a and the V_{ca}-corrected parameter b are defined. These two expressions are both exhibited as follows:

$$a = \frac{D_{28}(w/c, V_{ca} = 0)}{D_{28}(w/c = 0.4, V_{ca} = 0)}, \tag{9}$$

$$b = \frac{D_{28}(w/c, V_{ca}, S_{max})}{D_{28}(w/c, V_{ca} = 0)}, \tag{10}$$

where $D_{28}(w/c, V_{ca} = 0)$ represents the reference chloride diffusion coefficient (m^2/s) of the mesoscopic numerical model for concrete when $V_{ca} = 0$; $D_{28}(V_{ca} = 0, w/c = 0.4)$ represents the reference chloride diffusion coefficient (m^2/s) when $V_{ca} = 0$, $w/c = 0.4$; and $D_{28}(V_{ca} = 0, w/c = 0.4) = 5.90 \times 10^{-12}$ (m^2/s), as shown in Table 3.

Based on Equation (9), the w/c-corrected parameter a can be quantitatively calculated, and the results are shown in Figure 21.

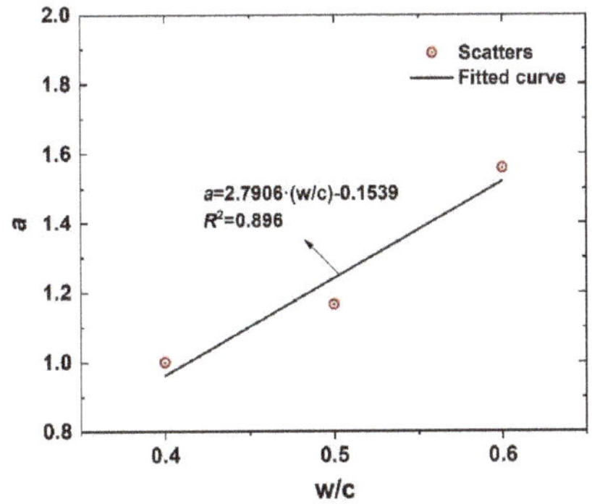

Figure 21. w/c impact parameter a in mesoscopic numerical model for concrete.

According to Figure 21, the linear formula is used to regress and determine the scatters of the w/c-corrected parameter a, and the linear function expression of a is obtained by regression analysis, as shown in Figure 21 and expressed in Equation (11):

$$a(w/c) = 2.7906\cdot(w/c) - 0.1539. \tag{11}$$

The V_{ca}-corrected parameter b (Equation (10)) of the chloride diffusion coefficient in the mesoscopic numerical model for concrete with increasing V_{ca} is shown in Figure 22. It can be observed that b decreases linearly as V_{ca} increases at an arbitrary w/c. The corrected parameters c and d in Equation (12) fitted by a linear function form with the values of N_{Smax} are depicted in Figure 23.

$$b(V_{ca}) = c\cdot V_{ca} + d. \tag{12}$$

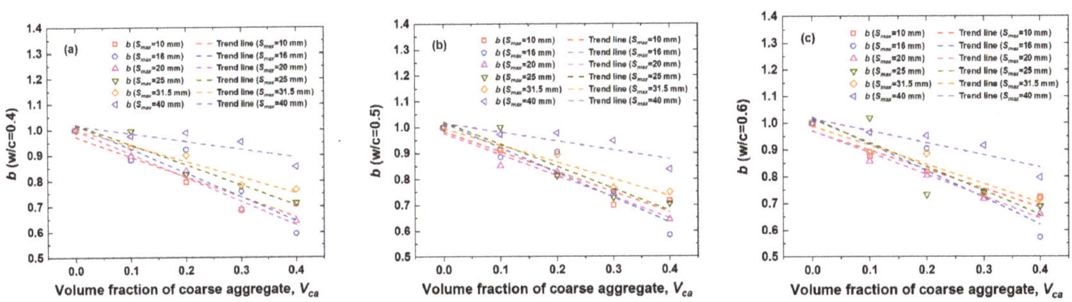

Figure 22. Corrected parameter b versus V_{ca}: (**a**) w/c = 0.4; (b) w/c = 0.5; (**c**) w/c = 0.6.

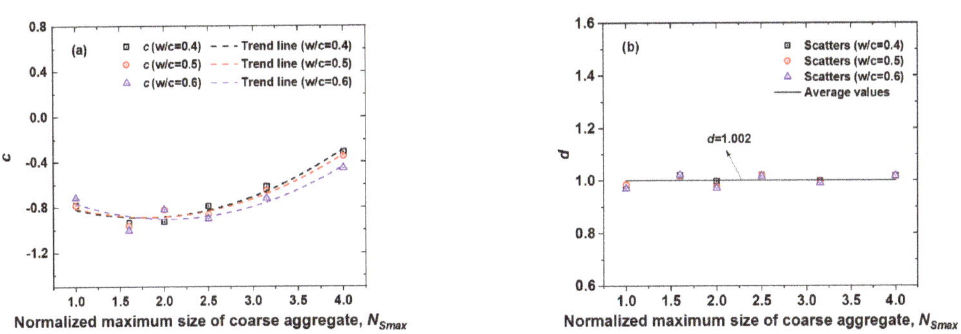

Figure 23. Corrected parameters c, d versus N_{Smax}: (**a**) parameter c; (**b**) parameter d.

From Figure 23, parameter c shows a quadratic function trend with increasing N_{Smax}. The function form shown in Equation (13) is used to fit the curve of c with N_{Smax}, where e, f and g are the coefficients related to w/c. Evidently, parameter d exhibits no significant trend with increasing N_{Smax}. The quantitative calculation shows that the relative error between the d scatters and the average value is within the range of $\pm 5\%$, so it is reasonable to take the average value of d with different values of N_{Smax} as the representative value, namely, $d = 1.002$.

$$c(N_{Smax}) = e \cdot N_{Smax}^2 + f \cdot N_{Smax} + g. \quad (13)$$

Figure 24 depicts the variation of correction parameters e, f, and g versus w/c, and the three exhibit linear function relationships with respect to w/c. The linear function is used to fit e, f, and g, and the expressions are determined, as shown in Figure 24 and Equation (14):

$$\begin{cases} e(w/c) = 0.095 \cdot (w/c) + 0.0825 \\ f(w/c) = -0.785 \cdot (w/c) - 0.1058 \\ g(w/c) = 0.9215 \cdot (w/c) - 0.8947 \end{cases}. \quad (14)$$

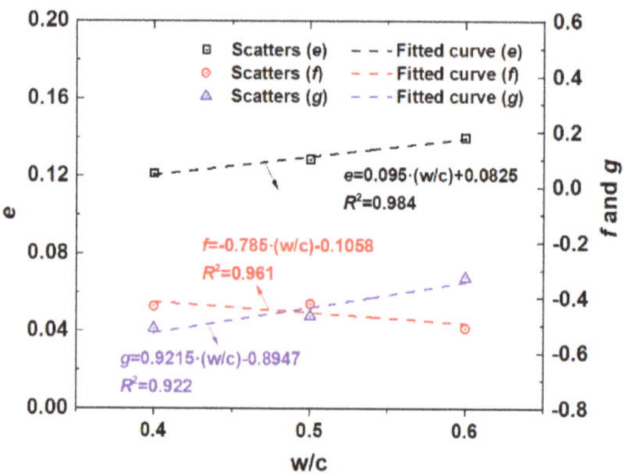

Figure 24. Corrected parameters e, f, g versus w/c.

In summary, by substituting Equations (11)–(14) into Equation (3), the model of D_a in the mesoscopic numerical model for concrete considering the concrete material factors (V_{ca}, S_{max}, w/c) is established, as shown in Equation (15).

$$\begin{cases} D_a(V_{ca}, S_{max}, w/c) = \frac{D_{28}(V_{ca}, S_{max}, w/c)}{1-m} \cdot \left(\frac{t_{28}}{t}\right)^m \\ D_{28}(V_{ca}, S_{max}, w/c) = a(V_{ca}, N_{Smax}, w/c) \cdot D_{28}(V_{ca} = 0, w/c) \\ D_{28}(V_{ca} = 0, w/c) = b(w/c) \cdot D_{28}(V_{ca} = 0, w/c = 0.4) \\ D_{28}(V_{ca} = 0, w/c = 0.4) = 5.90 \times 10^{-12} \mathrm{m}^2/s \\ a(w/c) = 2.7906 \cdot (w/c) - 0.1539 \\ b(V_{ca}, N_{Smax}, w/c) = c(N_{Smax}, w/c) \cdot V_{ca} + 1.002 \\ c(N_{Smax}, w/c) = e(w/c) \cdot {N_{Smax}}^2 + f(w/c) \cdot N_{Smax} + g(w/c) \\ e(w/c) = 0.095 \cdot (w/c) + 0.0825 \\ f(w/c) = -0.785 \cdot (w/c) - 0.1058 \\ g(w/c) = 0.9215 \cdot (w/c) - 0.8947 \end{cases}. \quad (15)$$

5.4. Influences of Concrete Material Factors on Aging Factor

In Equation (3), the aging factor m is an important parameter for quantifying chloride transportation in concrete composites [33]. The aging factor was found not to be constant with time in concrete specimens with mineral admixtures. It was also found that the exposure conditions and concrete material factors also affect the aging factor.

5.4.1. Influence of Water-to-Cement Ratio, w/c

The variation trend of D_a versus t is fitted by using a power function of Equation (3), and the aging factor m with different values of V_{ca}, w/c, and S_{max} can be obtained by regression analysis. The variation is shown in Figure 25, and the m values increase linearly with increasing w/c. Particularly, when $V_{ca} = 0.4$, $S_{max} = 20$ mm, the m increases almost 10.5% with the increasing of w/c from 0.4 to 0.6.

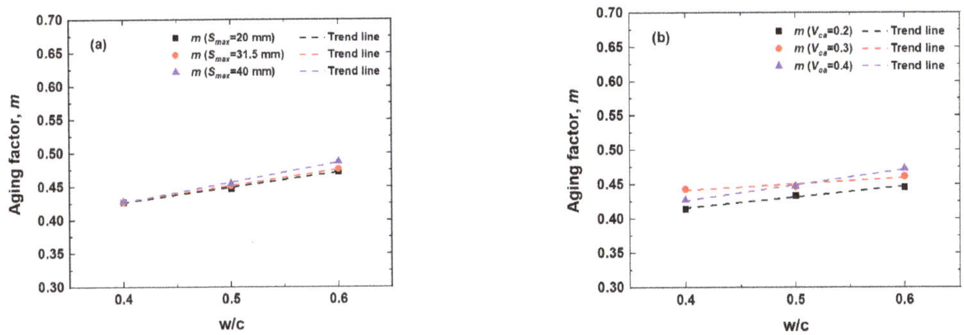

Figure 25. Aging factor m versus w/c: (**a**) $V_{ca} = 0.4$; (**b**) $S_{max} = 20$ mm.

5.4.2. Influence of VFCA, V_{ca} and MSCA, S_{max}

Figures 26 and 27 depict the variation in the aging factor m with V_{ca} and S_{max} in the mesoscopic numerical model for concrete, respectively. It can be observed from the figures that m has no obvious change in trend with increasing V_{ca} and S_{max}. It can be considered that m is unrelated to V_{ca} and S_{max}.

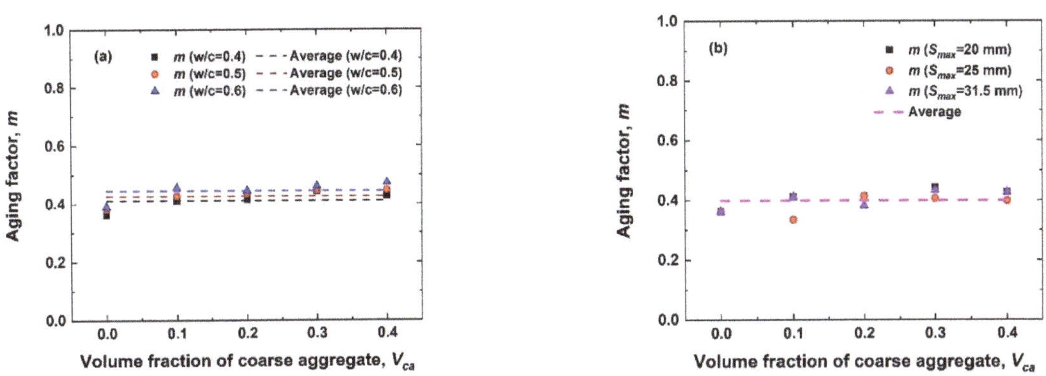

Figure 26. Aging factor m versus V_{ca}: (**a**) $S_{max} = 20$ mm; (**b**) $w/c = 0.4$.

Therefore, the average value of m corresponding to the different V_{ca} and S_{max} can be used as the representative value of the aging factor for the concrete mesoscopic numerical model with different w/c. The quantitative calculation testifies that the relative error between the m scatters and their mean is in a range of ±5% for any w/c. Thus, it is thought that the aging factor is independent of V_{ca} and S_{max}, and it is reasonable to take the average value as the representative value of the aging factor corresponding to different w/c.

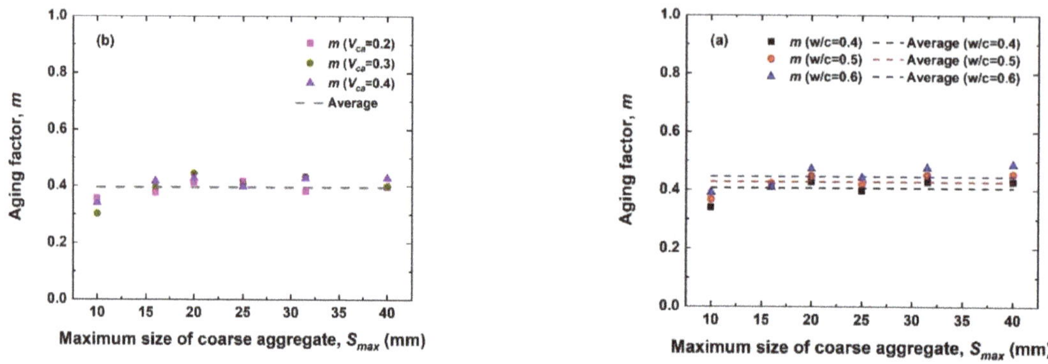

Figure 27. Aging factor m versus S_{max}: (**a**) V_{ca} = 0.4; (**b**) w/c = 0.4.

Based on the aforementioned analysis in this paper, the variation in aging factor m with w/c is fitted by using a linear function, and the empirical expression of the aging factor m of the mesoscopic numerical model for concrete is shown in Figure 28 and Equation (16).

$$m(w/c) = 0.1726 \cdot (w/c) + 0.3184. \tag{16}$$

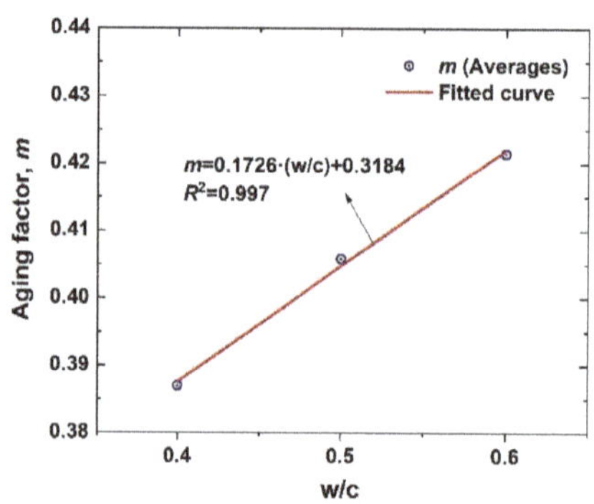

Figure 28. Average aging factor m versus w/c.

6. Modelling of Chloride Concentration Considering Concrete Material Factors

6.1. Model Establishment

According to the aforementioned $C_s(t, V_{ca}, S_{max})$, $D_a(t, V_{ca}, S_{max}, w/c)$, and $m(w/c)$, a chloride diffusion model of concrete that accounts for the concrete material factors can be established based on Fick's second law of diffusion. For this model, the time-dependency of C_s and D_a are both considered as well. The expression is shown in Equation (17):

$$\left\{\begin{array}{l}
C(x,t,V_{ca},S_{max},w/c) = C_s(t,V_{ca},S_{max})\cdot\left[1-\mathrm{erf}\left(\frac{x}{2\sqrt{D_a(t,V_{ca},S_{max},w/c)\cdot t}}\right)\right] \\
D_a(t,V_{ca},S_{max},w/c) = \frac{D_{28}(V_{ca},S_{max},w/c)}{1-m}\cdot\left(\frac{t_{28}}{t}\right)^{m(w/c)} \\
C_s(t,V_{ca},S_{max}) = A(V_{ca},N_{Smax})\cdot\ln(t) + B(V_{ca},N_{Smax}) \\
A(V_{ca},N_{Smax}) = C(N_{Smax})\cdot V_{ca} + 0.2536 \\
B(V_{ca},N_{Smax}) = E(N_{Smax})\cdot V_{ca} - 0.5776 \\
C(N_{Smax}) = 0.0931\cdot\ln(N_{Smax}) - 0.2613 \\
E(N_{Smax}) = -0.1717\cdot N_{Smax} + 1.0693 \\
D_{28}(V_{ca},S_{max},w/c) = a(V_{ca},N_{Smax})\cdot D_{28}(V_{ca}=0) \\
D_{28}(V_{ca}=0) = b(w/c)\cdot D_{28}(V_{ca}=0,w/c=0.4) \\
D_{28}(V_{ca}=0,w/c=0.4) = 5.90\times10^{-12}\mathrm{m}^2/s \\
a(w/c) = 2.7906\cdot(w/c) - 0.1539 \\
b(V_{ca},N_{Smax}) = c(N_{Smax})\cdot V_{ca} + 1.002 \\
c(N_{Smax},w/c) = e(w/c)\cdot {N_{Smax}}^2 + f(w/c)\cdot N_{Smax} + g(w/c) \\
e(w/c) = 0.095\cdot(w/c) + 0.0825 \\
f(w/c) = -0.785\cdot(w/c) - 0.1058 \\
g(w/c) = 0.9215\cdot(w/c) - 0.8947 \\
m(w/c) = 0.1726\cdot(w/c) + 0.3184
\end{array}\right. . \quad (17)$$

The model (Equation (17)) can be used to efficiently estimate and predict the spatial and temporal distribution of chloride ion diffusion concentration in concrete under the conditions of arbitrary w/c, V_{ca} and S_{max}.

6.2. Model Validation

To verify the reasonability of the estimated model in line with Equation (17), the mesoscopic numerical simulation values (MNSVs) of the chloride concentrations determined by the numerical simulation experiment (Figures 9–11) are compared with the predicted chloride results by the prediction model of Equation (17), and the error analysis was carried out. The results are shown in Figure 29.

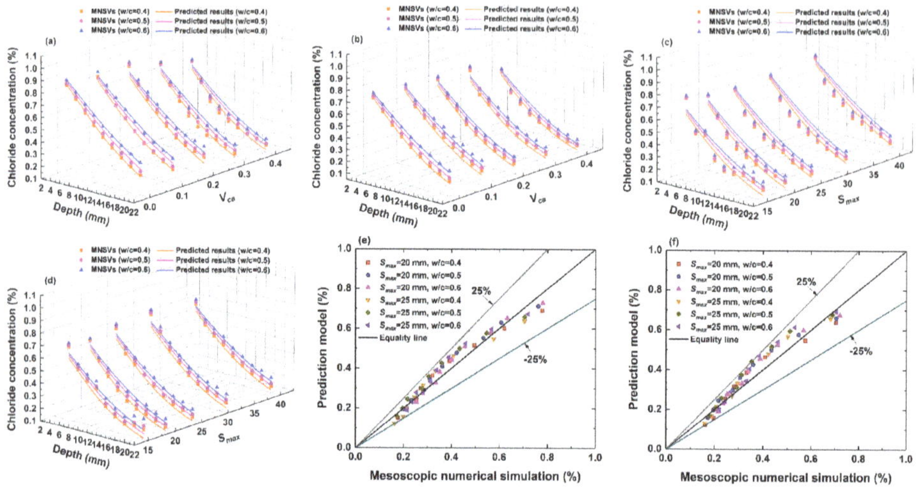

Figure 29. Comparison of the chloride concentration results by mesoscopic numerical simulation and those via prediction model Equation (17): (**a**) S_{max} = 20 mm, t = 600 d; (**b**) S_{max} = 25 mm, t = 360 d; (**c**) V_{ca} = 0.4, t = 600 d; (**d**) V_{ca} = 0.3, t = 360 d; (**e**) predicted results vs. MNSVs: S_{max} = 20 mm, t = 600 d; (**f**) predicted results vs. MNSVs: V_{ca} = 0.4, t = 600.

As shown in Figure 29, the estimated results of the chloride diffusion model in Equation (17) are consistent with the chloride concentrations via mesoscopic numerical simulation, and the relative errors between the two results are within the range of ±25%, thereby verifying the accuracy of the model (Equation (17)) considering the concrete material factor in Equation (17), as presented in this paper's research work.

7. Conclusions

In this paper, mesoscopic numerical models for concrete with different values of V_{ca} and S_{max} are constructed, and a large number of chloride ion diffusion simulation experiments are conducted based on these models. The influences of concrete material factors (including w/c, V_{ca}, and S_{max}) on chloride diffusion characteristics within concrete are quantified. On this basis, a chloride diffusion model considering the concrete material factors is established. The specific conclusions are as follows:

(1) It is emphasized and proven that the mesoscopic numerical simulation method for concrete can be used to partially replace physical experimentation to investigate the influences of concrete material factors on chloride diffusion behaviors;

(2) For the impact factor of the w/c, the chloride concentration, D_{28} and m of concrete increase with increasing w/c, but C_s is unrelated to the w/c. Quantitatively, when V_{ca} = 0.4, S_{max} = 20 mm, the chloride concentration $C(x$ = 4 mm, t = 600 days), D_{28}, m increase almost 2.3%, 60.1%, and 10.5% with the increasing of w/c from 0.4 to 0.6;

(3) For the impact factor of V_{ca}, the chloride concentration, C_s and D_{28} of concrete show decreasing trends with increasing V_{ca}. Especially, when w/c = 0.5, S_{max} = 20 mm, the chloride concentration $C(x$ = 4 mm, t = 600 days), $C_s(t$ = 600 days), D_{28}, decrease almost 16.4%, 18.5%, and 35.8% with the increasing of V_{ca} from 0 to 0.4. m is unrelated to the change in V_{ca} but is a linear function of the w/c;

(4) For the impact factor of S_{max}, there is no significant effect of S_{max} on chloride concentration and m. C_s is positively correlated with S_{max}, when w/c = 0.5, V_{ca} = 0.4, the $C_s(t$ = 600 days) increases almost 23.4% with the increasing of S_{max} from 10 mm to 40 mm. D_{28} shows a quadratic curve trend of first decreasing and then increasing when S_{max} is in the range of 16–20 mm. Respectively, when w/c = 0.5,

V_{ca} = 0.4, D_{28}(S_{max} = 40 mm) increases approximately 43.4% and 30.2% than those of S_{max} = 16 mm and 20 mm;

(5) According to the quantitative relationship between the chloride diffusion parameters and w/c, V_{ca}, and S_{max}, a chloride diffusion model that considers the influences of important concrete material factors is established. The accuracy of the model proposed in this paper is verified by comparing the predicted results with mesoscopic numerical simulation values. This model is used to predict the spatial and temporal distribution of the chloride diffusion concentration in concrete for any w/c, V_{ca} and S_{max}.

Author Contributions: Investigation, data curation, formal analysis, writing—original draft preparation, X.J.; conceptualization, methodology, L.W.; supervision, resources, M.L.; validation, H.J.; data curation, visualization, W.Z. All authors have read and agreed to the published version of the manuscript.

Funding: This research was funded by supported by the Open-end Fund of Key Laboratory of Hydraulic and Waterway Engineering of the Ministry of Education in Chongqing Jiaotong University (SLK2021B13), the Science and Technology Innovation Project of Chongqing Education Commission (KJCX2020030), and the Talents Plan Project in Chongqing (cstc2021ycjh-bgzxm0053).

Institutional Review Board Statement: Not applicable.

Informed Consent Statement: Not applicable.

Data Availability Statement: Not applicable.

Conflicts of Interest: The authors declare no conflict of interest.

References

1. Nguyen, Q.D.; Afroz, S.; Castel, A. Influence of Calcined Clay Reactivity on the Mechanical Properties and Chloride Diffusion Resistance of Limestone Calcined Clay Cement (LC3) Concrete. *J. Mar. Sci. Eng.* **2020**, *8*, 301. [CrossRef]
2. Zhang, J.Z.; Guo, J.; Li, D.H.; Zhang, Y.R.; Bian, F.; Fang, Z.F. The influence of admixture on chloride time-varying diffusivity and microstructure of concrete by low-field NMR. *Ocean Eng.* **2017**, *142*, 94–101. [CrossRef]
3. Chen, X.; Shen, J. Experimental Investigation on Deterioration Mechanisms of Concrete under Tensile Stress-Chloride Ion-Carbon Dioxide Multiple Corrosion Environment. *J. Mar. Sci. Eng.* **2022**, *10*, 80. [CrossRef]
4. Oktavianus, Y.; Sofi, M.; Lumantarna, E.; Kusuma, G.; Duffield, C. Long-Term Performance of Trestle Bridges: Case Study of an Indonesian Marine Port Structure. *J. Mar. Sci. Eng.* **2020**, *8*, 358. [CrossRef]
5. Balestra, C.E.T.; Reichert, T.A.; Pansera, W.A.; Savaris, G. Evaluation of chloride ion penetration through concrete surface electrical resistivity of field naturally degraded structures present in marine environment. *Constr. Build. Mater.* **2020**, *230*, 116979. [CrossRef]
6. Balestra, C.E.T.; Reichert, T.A.; Savaris, G. Contribution for durability studies based on chloride profiles analysis of real marine structures in different marine aggressive zones. *Constr. Build. Mater.* **2019**, *206*, 140–150. [CrossRef]
7. Valipour, M.; Shekarchi, M.; Arezoumandi, M. Chlorine diffusion resistivity of sustainable green concrete in harsh marine environments. *J. Clean. Prod.* **2017**, *142*, 4092–4100. [CrossRef]
8. Balestra, C.E.T.; Reichert, T.A.; Pansera, W.A.; Savaris, G. Chloride profile modeling contemplating the convection zone based on concrete structures present for more than 40 years in different marine aggressive zones. *Constr. Build. Mater.* **2019**, *198*, 345–358. [CrossRef]
9. Zhang, Z.M.; Chen, R.; Hu, J.; Wang, Y.Y.; Huang, H.L.; Ma, Y.W.; Zhang, Z.H.; Wang, H.; Yin, S.H.; Wei, J.X.; et al. Corrosion behavior of the reinforcement in chloride-contaminated alkali-activated fly ash pore solution. *Compos. Part B Eng.* **2021**, *224*, 109215. [CrossRef]
10. Shao, W.; Nie, Y.H.; Shi, D.D.; Xu, Y.Z. Probabilistic analysis of the behaviour of laterally loaded piles in chloride environments. *Ocean Eng.* **2020**, *217*, 107872. [CrossRef]
11. Vázquez, K.; Rodríguez, R.R.; Esteban, M.D. Corrosion Prediction Models in the Reinforcement of Concrete Structures of Offshore Wind Farms. *J. Mar. Sci. Eng.* **2022**, *10*, 185. [CrossRef]
12. Zhang, Z.M.; Hu, J.; Ma, Y.W.; Wang, Y.Y.; Huang, H.L.; Zhang, Z.H.; Wei, J.X.; Yin, S.H.; Yu, Q.J. A state-of-the-art review on Ag/AgCl ion-selective electrode used for non-destructive chloride detection in concrete. *Compos. Part B Eng.* **2020**, *200*, 108289. [CrossRef]
13. Shen, X.H.; Liu, Q.F.; Hu, Z.; Jiang, W.Q.; Lin, X.S.; Hou, D.S.; Hao, P. Combine ingress of chloride and carbonation in marine-exposed concrete under unsaturated environment: A numerical study. *Ocean Eng.* **2019**, *189*, 106350. [CrossRef]
14. De Medeiros-Junior, R.A.; de Lima, M.G.; de Brito, P.C.; de Medeiros MH, F. Chloride penetration into concrete in an offshore platform-analysis of exposure conditions. *Ocean Eng.* **2015**, *103*, 78–87. [CrossRef]

15. Afroughsabet, V.; Biolzi, L.; Monteiro, P.J.M. The effect of steel and polypropylene fibers on the chloride diffusivity and drying shrinkage of high-strength concrete. *Compos. Part B Eng.* **2018**, *139*, 84–96. [CrossRef]
16. Li, K.; Stroeven, P.; Stroeven, M.; Sluys, L.J. A numerical investigation into the influence of the interfacial transition zone on the permeability of partially saturated cement paste between aggregate surfaces. *Cem. Concr. Res.* **2017**, *102*, 99–108. [CrossRef]
17. Jiang, W.Q.; Shen, X.H.; Hong, S.X.; Wu, Z.Y.; Liu, Q.F. Binding capacity and diffusivity of concrete subjected to freeze-thaw and chloride attack: A numerical study. *Ocean Eng.* **2019**, *186*, 106093. [CrossRef]
18. Wang, Y.Z.; Liu, C.X.; Li, Q.M.; Wu, L.J. Chloride ion concentration distribution characteristics within concrete covering-layer considering the reinforcement bar presence. *Ocean Eng.* **2019**, *173*, 608–616. [CrossRef]
19. Valipour, M.; Pargar, F.; Shekarchi, M.; Khani, S.; Moradian, M. In situ study of chloride ingress in concretes containing natural zeolite, metakaolin and silica fume exposed to various exposure conditions in a harsh marine environment. *Constr. Build. Mater.* **2013**, *46*, 63–70. [CrossRef]
20. Jamkar, S.S.; Rao, C.B.K. Index of Aggregate Particle Shape and Texture of coarse aggregate as a parameter for concrete mix proportioning. *Cem. Concr. Res.* **2004**, *34*, 2021–2027. [CrossRef]
21. Golewski, G.L.; Szostak, B. Application of the C-S-H Phase Nucleating Agents to Improve the Performance of Sustainable Concrete Composites Containing Fly Ash for Use in the Precast Concrete Industry. *Materials* **2021**, *14*, 6514. [CrossRef] [PubMed]
22. Golewski, G.L.; Szostak, B. Strengthening the very early-age structure of cementitious composites with coal fly ash via incorporating a novel nanoadmixture based on C-S-H phase activators. *Constr. Build. Mater.* **2021**, *312*, 125426. [CrossRef]
23. Golewski, G.L. Evaluation of morphology and size of cracks of the Interfacial Transition Zone (ITZ) in concrete containing fly ash (FA). *J. Hazard. Mater.* **2018**, *357*, 298–304. [CrossRef]
24. Mangat, P.S.; Molloy, B.T. Prediction of long term chloride concentration in concrete. *Mater. Struct.* **1994**, *27*, 338–346. [CrossRef]
25. Costa, A.; Appleton, J. Chloride penetration into concrete in marine environment-Part I: Main parameters affecting chloride penetration. *Mater. Struct.* **1999**, *32*, 252–259. [CrossRef]
26. Tian, C.; Chen, J.W.; Wei, X.S. The Slice-Resistivity method to measure the chloride ion diffusion coefficient of cementitious materials. *Constr. Build. Mater.* **2020**, *243*, 118155. [CrossRef]
27. Chen, J.W.; Tian, C.; Wei, X.S. Experimental and simulation study on chloride permeability in cement paste. *Constr. Build. Mater.* **2020**, *262*, 120600. [CrossRef]
28. Gao, Y.H.; Zhang, J.Z.; Zhang, S.; Zhang, Y.R. Probability distribution of convection zone depth of chloride in concrete in a marine tidal environment. *Constr. Build. Mater.* **2017**, *140*, 485–495. [CrossRef]
29. Mehdi, K.M.; Seyedhamed, S.; Mohammad, S. Quantifying maximum phenomenon in chloride ion profiles and its influence on service-life prediction of concrete structures exposed to seawater tidal zone—A field oriented study. *Constr. Build. Mater.* **2018**, *180*, 109–116.
30. Higashiyama, H.; Yamauchi, K.; Sappakittipakorn, M.; Sano, M.; Takahashi, O. A visual investigation on chloride ingress into ceramic waste aggregate mortars having different water to cement ratios. *Constr. Build. Mater.* **2013**, *40*, 1021–1028. [CrossRef]
31. Ribeiro, D.V.; Pinto, S.A.; Amorim Júnior, N.S.; Andrade Neto, J.S.; Santos, I.H.L.; Marques, S.L.; França, M.J. Effects of binders characteristics and concrete dosing parameters on the chloride diffusion coefficient. *Cem. Concr. Compos.* **2021**, *122*, 104114. [CrossRef]
32. Du, X.L.; Jin, L.; Ma, G.W. A meso-scale numerical method for the simulation of chloride diffusivity in concrete. *Finite Elem. Anal. Des.* **2014**, *85*, 87–100. [CrossRef]
33. Wang, Y.Z.; Wu, L.J.; Wang, Y.C.; Liu, C.X.; Li, Q.M. Effects of coarse aggregates on chloride diffusion coefficients of concrete and interfacial transition zone under experimental drying-wetting cycles. *Constr. Build. Mater.* **2018**, *185*, 230–245. [CrossRef]
34. Yang, C.C. Effect of the percolated interfacial transition zone on the chloride migration coefficient of cement-based materials. *Mater. Chem. Phys.* **2005**, *91*, 538–544. [CrossRef]
35. Li, L.Y.; Xia, J.; Lin, S.-S. A multi-phase model for predicting the effective diffusion coefficient of chlorides in concrete. *Constr. Build. Mater.* **2012**, *26*, 295–301. [CrossRef]
36. Wu, L.J.; Wang, Y.Z.; Wang, Y.C.; Ju, X.L.; Li, Q.M. Modelling of two-dimensional chloride diffusion concentrations considering the heterogeneity of concrete materials. *Constr. Build. Mater.* **2020**, *243*, 118213. [CrossRef]
37. Wu, L.J.; Wang, Y.Z.; Wang, Y.C.; Ju, X.L.; Li, Q.M. Modelling of two-dimensional chloride diffusion concentrations considering the heterogeneity of concrete materials. *Constr. Build. Mater.* **2020**, *243*, 118213. [CrossRef]
38. Wu, L.J.; Ju, X.L.; Liu, M.W.; Guan, L.; Ma, Y.F.; Li, M.L. Influences of multiple factors on the chloride diffusivity of the interfacial transition zone in concrete composites. *Compos. Part B Eng.* **2020**, *199*, 108236. [CrossRef]
39. Grabiec, A.M.; Zawal, D.; Szulc, J. Influence of type and maximum aggregate size on some properties of high-strength concrete made of pozzolana cement in respect of binder and carbon dioxide intensity indexes. *Constr. Build. Mater.* **2015**, *98*, 17–24. [CrossRef]
40. Gao, P.; Chen, Y.; Huang, H.; Qian, Z.; Schlangen, E.; Wei, J.; Yu, Q. Effect of coarse aggregate size on non-uniform stress/strain and drying-induced microcracking in concrete. *Compos. Part B Eng.* **2021**, *216*, 108880. [CrossRef]
41. Basheer, L.; Basheer, P.A.M.; Long, A.E. Influence of coarse aggregate on the permeation, durability and the microstructure characteristics of ordinary Portland cement concrete. *Constr. Build. Mater.* **2005**, *19*, 682–690. [CrossRef]
42. Pang, L.; Li, Q. Service life prediction of RC structures in marine environment using long term chloride ingress data: Comparison between exposure trials and real structure surveys. *Constr. Build. Mater.* **2016**, *113*, 979–987. [CrossRef]

43. *ASTM C1218*; Standard Test Method for Water-Soluble Chloride in Mortar and Concrete. ASTM International (ASTM): West Conshohocken, PA, USA, 2020.
44. Pack, S.W.; Jung, M.S.; Song, H.W.; Kim, S.H.; Ann, K.Y. Prediction of time dependent chloride transport in concrete structures exposed to a marine environment. *Cem. Concr. Res.* **2010**, *40*, 302–312. [CrossRef]
45. Wang, Y.Z.; Wu, L.J.; Wang, Y.C.; Li, Q.M.; Xiao, Z. Prediction model of long-term chloride diffusion into plain concrete considering the effect of the heterogeneity of materials exposed to marine tidal zone. *Constr. Build. Mater.* **2018**, *159*, 297–315. [CrossRef]
46. Gao, Y.; de Schutter, G.; Ye, G.; Tan, Z.; Wu, K. The ITZ microstructure, thickness and porosity in blended cementitious composite: Effects of curing age, water to binder ratio and aggregate content. *Compos. Part B Eng.* **2014**, *60*, 1–13. [CrossRef]

Journal of
Marine Science and Engineering

MDPI

Article

A Flexible Scheduling for Twin Yard Cranes at Container Terminals Considering Dynamic Cut-Off Time

Junjun Li [1,*], Jingyu Yang [2], Bowei Xu [2], Wenjun Yin [2], Yongsheng Yang [2], Junfeng Wu [3], Ye Zhou [4] and Yi Shen [4]

[1] Merchant Marine College, Shanghai Maritime University, Shanghai 201306, China
[2] Institute of Logistics Science and Engineering, Shanghai Maritime University, Shanghai 201306, China; 201930510021@stu.shmtu.edu.cn (J.Y.); bwxu@shmtu.edu.cn (B.X.); yinwenjuntao@163.com (W.Y.); yangys@shmtu.edu.cn (Y.Y.)
[3] Office of General Manager, Shanghai WinJoin Information Technology Co., Ltd., Shanghai 200126, China; wujf@winjoinit.com
[4] Project Department, Shanghai WinJoin Information Technology Co., Ltd., Shanghai 200126, China; zhouye@winjoinit.com (Y.Z.); shenyi@winjoinit.com (Y.S.)
* Correspondence: lijj@shmtu.edu.cn

Abstract: Yard handling is an important part of port logistics and affects the overall efficiency of the container port. The yard crane scheduling is affected by various external factors. For example, the dynamic cut-off time makes the release time of yard cranes variable, and the yard crane task arrangement will change frequently, resulting in a lot of computational time. To increase the flexibility of container yard handling, a twin yard cranes scheduling model is established considering the no-crossing constraints and the dynamic cut-off time. A joint scheduling of PSO and local re-scheduling strategy (LRPSO) is put forward to deal with the problem faster and more effectively. Small-scale and large-scale experiments are simulated to verify the performance of the proposed method. Results show that the scheduling method is more efficient.

Keywords: container ports; twin yard cranes scheduling; dynamic cut-off time; no-crossing constraints; flexible

Citation: Li, J.; Yang, J.; Xu, B.; Yin, W.; Yang, Y.; Wu, J.; Zhou, Y.; Shen, Y. A Flexible Scheduling for Twin Yard Cranes at Container Terminals Considering Dynamic Cut-Off Time. *J. Mar. Sci. Eng.* **2022**, *10*, 675. https://doi.org/10.3390/jmse10050675

Academic Editors: M. Dolores Esteban, José-Santos López-Gutiérrez, Vicente Negro and M. Graça Neves

Received: 29 March 2022
Accepted: 11 May 2022
Published: 16 May 2022

1. Introduction

In recent years, the growing demand of shipping has had higher requirements on port handlings, and the ability of terminals to respond to emergencies needs to be improved. The container terminal yard undertakes both landing and shipping transportation. The yard crane is one of the important types of transportation equipment in the terminal yard, and its work efficiency directly affects the overall carrying efficiency of the terminal. In each container block, twin yard cranes are generally used for common rail or multi-track cooperation. Usually, the twin yard cranes are identical and move on a common rail, and collisions between them need to be avoided. This indicates that a certain safe distance should be maintained between yard cranes. In addition to internal constraints, yard cranes are often affected by other aspects, such as port congestion (Shanghai Port congestion in 2018, etc.) or sudden health conditions (2019 New Crown Epidemic, etc.) and delays in arrival time of ships. Heuermann, A. et al. [1] proposed a concept about dynamic cut-off time (cut-off time is the latest time a container may be delivered to a terminal for loading to the scheduled container vessel) and also pointed out that it is an influencing factor of yard crane handling that cannot be ignored. To improve the turnover rate of ships, the terminal yard will arrange scheduling in advance, resulting in frequent changes in the yard crane task schedule when the dynamic cut-off time occurs. If the dynamic changes increase, it is often not cost-effective to continue global scheduling. Recalculation often means a lot of time cost in a large-scale scheduling solution. In this paper, the twin yard cranes scheduling considering the dynamic cut-off time and the non-crossing constraints of yard cranes is studied. A joint scheduling of PSO and local re-scheduling strategy (LRPSO), which includes a particle swarm optimization (PSO) algorithm for initial scheduling and a local optimization strategy for re-scheduling, is proposed to shorten the computing time and improve the ability of port to respond to emergencies. Experiment results validate that the proposed method can deal with more dynamic changes at the same time and improve the flexibility of the container yard.

2. Literature Review

Improving the efficiency of container yard in port can effectively shorten the time of ships in port and improve the efficiency of port. Kemme, N. [2] pointed out that, in fact, shipping lines tend to make fewer but bigger calls with their larger vessels, i.e., increasing container volumes have to be transported in short periods of time, thus inducing increasing peak requirements in storage and handling capacity. As an important transit station for container transport at the port, the work arrangement of the container yard at the terminal is generally determined by various factors, such as the cut-off time, port resources arrangement, policy, and other factors. As an important part of the container yard, yard crane scheduling will also be affected by multiple factors.

At present, many researchers have studied the yard crane scheduling. Some researchers are concerned with the internal cross-constraint problem in multiple cranes' cooperation. Nils, B. et al. [3] provided a classification scheme for crane scheduling problems with crane interference, and in their classification scheme, any crane scheduling problem is described by three basic elements: the terminal layout (including the available cranes), the characteristics of container moving, and an objective to be followed. Ehleiter, A. et al. [4] introduced that by considering the number of cranes, the crane scheduling can be classified as one crane, twin cranes, and triple cranes scheduling problems. Additionally, the crane scheduling problems can also be classified by considering whether the cranes can crossover each other or not. Based on the above classification ideas, this paper reviews some literature about different types of yard crane scheduling.

In the research about yard crane scheduling algorithms, some researchers proposed scheduling solution methods for different yard crane specifications. Amelie, E. et al. [5] proposed a polynomial-time algorithm to solve the twin crane scheduling problem with no-crossing constraints. Zey, L. et al. [6] solved the yard crane scheduling problem by using branch and bound algorithm for triple-crossover-cranes. Zheng, F. et al. [7] considered that the task processing time would change dynamically due to the arrangement of task retrieval and storage in the twin yard cranes scheduling problem with an inter-crane interference constraint. They used heuristics, genetic algorithms, and other methods to solve the problem. Chen, T.C. et al. [8] were concerned with the location problem of the signal transmitting station and the allocation and scheduling. They used modern network communication technology, and proposed a hybrid evolutionary algorithm IAPSO combining immune algorithm and particle swarm algorithm, which could simultaneously determine the number of signal transmission stations and the category of each station. Guo, P. et al. [9] considered the twin cranes scheduling in an automated railway container yard with a handover area, and proposed four FCFS dispatching rules to shorten the overall transportation distance of the yard crane, thereby saving the energy consumption. Lei, D. et al. [10] studied the container dispatch problem in railway container yards, proposed the concept of "dig box coefficient", and used a multi-stage genetic algorithm considering yard crane storing containers, container sequence of the target position and the working process. Yu, M. et al. [11] proposed a variety of types of iterative-related learning particle swarm optimization (IVLPSO) to solve the quay crane scheduling and avoid the local excessively fast convergence problem of PSO.

Some researchers have also researched different perspectives. Briskorn, D [12] provided a polynomial programming framework for twin cranes scheduling to facilitate subsequent research, and proved that the problem is NP-hard. Zey, L. et al. [13] considered the priority constraints caused by stacking containers in the handover area and provided insights about the placement of the handover area. Ehleiter, A. et al. [4] aimed at improving the efficiency of yard crane scheduling at seaside peak times (when a container ship is berthing), and proposed a heuristic algorithm for two crossover cranes scheduling. Without considering the release time and the deadline of the crane task, the algorithm minimized the completion time of a set of containers storage and retrieval request at seaside peak times. Guvenc, D. et al. [14] proposed a scheduling method for a container yard crane using tabu search algorithm. Gharehgozli, A. et al. [15] studied the crane scheduling problem under a new entry and exit mode of multi-containers, and proposed a three-stage solution method and a heuristic algorithm to solve the problem. Kizilay, D. et al. [16] comprehensively integrated the port container terminal problem, considering the joint optimization of quay crane allocation and scheduling, yard crane allocation and scheduling, yard location allocation, and yard truck allocation and scheduling. Zhou, C.H. et al. [17] considered the cooperation problem between the yard crane and other terminal carrying equipment, and established a mixed scheduling problem model.

In a word, it is found that PSO and other heuristic algorithms have been widely used to solve the yard crane scheduling problem, but these methods have low flexibility. When the release times of crane tasks are randomly changed by liner cut-off time or other factors, they cannot meet the flexibility requirements of today's port. Some researchers may have discussed this issue, such as Zheng, F. et al. [7], who considered the change in the processing time of the yard crane task, but the release time of the task was regarded as static. It is necessary to take the task release time as a dynamic variable to improve the flexibility of twin yard cranes scheduling with no-crossing constraints. This paper takes the dynamic release time into consideration and builds a mathematical model for it. Then, a joint scheduling method consisting of PSO and local re-scheduling algorithm is designed, and its effectiveness and efficiency are validated by experiments. A global strategy algorithm and PSO are used for comparison experiments. Conclusions are made in the end.

3. Twin Yard Cranes Scheduling Model with Dynamic Cut-Off Time

3.1. Problem Statement

Usually, there are two identical yard cranes per block, called twin cranes. For these twin cranes, no-crossing constraints should be taken into consideration. The cranes can never occupy the same bay and therefore cannot cross each other. In addition, a safety distance constraint should be satisfied. The two cranes are called crane 0 and crane 1.

There are some assumptions: (1) Yard cranes move along the bay dimension with a constant speed. (2) Containers enter or leave the container block through the handover points (HP) of the block, and there is one HP at each end of the yard block. (3) Crane 0 moves along bay $0 \sim bay_{max} - 1$ and crane 1 moves along bay $1 \sim bay_{max}$ when there is no-crossing constraint, where bay_{max} denotes the bay with the maximum number. (4) Set bay 0 and bay_{max} are the initial locations of cranes 0 and 1.

The crane spreader which moves along the row dimension loads and unloads containers along the tiers dimension. No interference is deemed to occur between the two yard cranes spreaders.

The crane task is divided into three stages: sliding, waiting, and handling, as shown in Figure 1. During sliding, the crane moves from the current position to the target position; during waiting, the crane waits and starts to handle until the task is released; during handling, the crane loads or unloads the container. The crane finishes one task after the above three stages. Each job of a yard crane consists of two tasks, one task for loading and the other for unloading.

Each yard crane has an initial task list, and an example is shown in Table 1. Referring to Amelie, E. et al. [5], a dummy start and a dummy end task are set at the beginning and end of the crane task list, respectively. The dummy tasks' target locations are set as the initial locations of the yard crane. The "Release Time" represents the task release time of the yard crane; the current task can only be executed after the task release time. In Table 1,

"Task" represents the crane task number; "Bay" and "Row" represent the target location of the crane and its spreader, respectively; "Time" represents the handling time consumed by crane, and "Release time" represents the release time of the task.

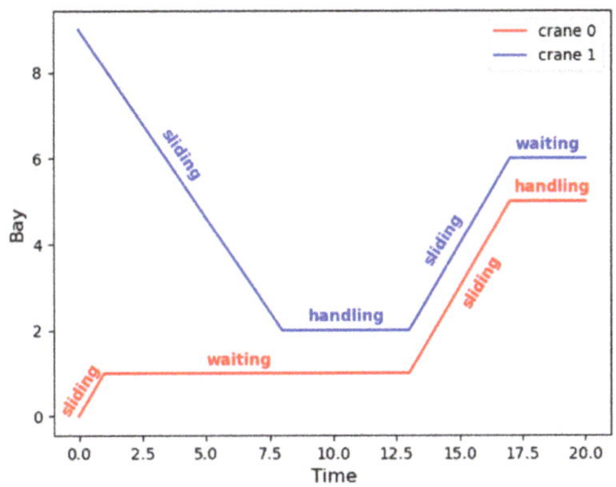

Figure 1. Spatial-temporal graphic of twin yard cranes operational plan.

Table 1. An example of initial task.

Task	Handling			Release Time
	Bay	Row	Time	
0	0	0	0	0
1	3	0	3	4
2	1	4	4	0
3	2	3	2	6
4	0	3	2	0
5	0	0	0	0

Crane movements (e.g., detours) that are not part of direct handling are listed in the information row of the task list. Next, the information row will be sorted by the task starting time. This is different from other research on yard crane scheduling, whose results generally consist of the initial crane task list and the detour scheduling of the crane [5–7]. The advantages of the information row in this paper are not only that the final result can be fully reflected in the task list after continuous updating, but also that the presentation of the task list is more convenient for future optimization and algorithm expansion. The presentation of the results in this paper will be described in detail in Section 5.

3.2. Mathematical Formulation

3.2.1. Basic Notations

Decision variables and input Parameters are described in Tables 2 and 3, respectively.

Table 2. Descriptions of decision variables.

Decision Variables	Description
F	Completion time
i	Task sequence number, $i \in [1, N]$
τ_i^k	Time consumed by crane k to process task i
$\tau_i^k : p$	Sliding time of crane k when processing task i
b_t^k	Bay location of crane k at time t
j_i^k	Task i handled by crane k

Table 2. *Cont.*

Decision Variables	Description
j_e^k	Dummy end task of crane k
$j_i^k : ST$	Start time of crane k to process task i
$j_i^k : Wait$	Waiting time of crane k when processing task i

Table 3. Descriptions of input parameters.

Input Parameters	Description
J	Total number of crane tasks
N	Total number of tasks, $N \geq 2J$
k	Crane number, $k \in \{0,1\}$
$j_i^k : Bay$	Target bay location of crane k to handle task i
$j_i^k : Row$	Target row location of crane k to handle task i
$j_i^k : Rd$	Release time of crane k to process task i
$j_i^k : T$	Working time of crane k to process task i

3.2.2. Mathematical Model

The objective function is to minimize the maximum completion time of the yard crane tasks, as shown in Equations (1) and (2).

$$\min F \tag{1}$$

$$F = \max\left(\sum_{1}^{N} \tau_i^0, \sum_{1}^{N} \tau_i^1\right) \tag{2}$$

The constraint module includes three parts: the task completion time constraint, the task constraint, and the no-crossing and safety distance constraint. They are detailed as follows:

$$\tau_i^k : p = \max\left(\begin{array}{c} (|j_i : Bay - j_{i-1} : Bay|), \\ (\rho |j_i : Row - j_{i-1} : Row|) \end{array} \right) \tag{3}$$

$$j_i^k : wait = \begin{cases} j_{i+1}^k : Rd - j_i^k : ST - \tau_i^k : p - j_i^k : T, j_{i+1}^k : Rd \\ \geq j_i^k : ST + \tau_i^k : p + j_i^k : T \\ 0, \text{others} \end{cases} \tag{4}$$

$$\tau_i^k = \tau_i^k : p + j_i^k : wait + j_i^k : T \tag{5}$$

$$j_i^k : ST = j_{i-1}^k : ST + \tau_{i-1}^k \tag{6}$$

$$F = \max\left(j_e^0 : ST, j_e^1 : ST\right) \tag{7}$$

$$j_i^k : ST \geq j_i^k : Rd \tag{8}$$

$$b_t^k = \begin{cases} j_{i-1}^k : Bay + sgn(j_i : Bay - j_{i-1} : Bay) \times \\ |t - j_{i-1} : ST|, t \in \left(j_i^k : ST, j_i^k : ST + \tau_i^k : p\right] \\ j_i^k : Bay, t \in \left[j_i^k : ST + \tau_i^k : p, j_{i+1}^k : ST\right) \end{cases} \tag{9}$$

$$b_t^0 \leq b_t^1 - 1 \tag{10}$$

$$0 \leq b_t^0 \leq bay_{max} - 1 \tag{11}$$

$$1 \leq b_t^1 \leq bay_{max} \tag{12}$$

Equations (3) and (4) give the calculation method of sliding time and waiting time of crane k when processing task i, respectively. It is assumed that the speed of the crane in bay is ρ times that of the crane spreader in row. In Equation (5), the task timeline of the yard crane is composed of three parts: the sliding time, the waiting time, and the handling time of the task. It is easy to deduce Equation (6) from Equations (2)–(4), and Equation (6) shows that the current task completion time is the start time of the next task. Thus, the makespan of the whole crane tasks is equal to the start time of the crane dummy end task, as shown in Equation (7).

Equation (8) indicates that the start time of the task is later than the release time of the task. Equation (9) indicates that the completion time of the task will be later than the deadline, where $sgn(x)$ is the sign function. Equation (10) defines the safety distance between the yard cranes to be at least one bay; Equations (11) and (12) define the moving range of crane 0 and 1 to be bay 0 to $bay_{max} - 1$ and bay 1 to bay_{max}, respectively.

4. Joint Scheduling of PSO and Local Re-Scheduling

4.1. Joint Scheduling Idea

From Section 2, heuristic algorithms such as PSO have been applied many times to yard crane scheduling problems; however, these studies have not considered the dynamic variables. When the cut-off time changes, the yard crane tasks' release times will change accordingly. The initial scheduling no longer meets the demand. If the global algorithm is used to re-schedule the current yard crane tasks, it will cost a lot of time. When the yard crane tasks changes frequently, especially, the time cost will increase greatly. Therefore, it is not efficient to use a global scheduling method to solve the yard crane scheduling problem considering dynamic cut-off time. A local re-scheduling method joint with PSO initial scheduling called LRPSO in the paper is used to solve the problem. The PSO initial scheduling ensures that the initial solution is optimal or close to optimal. Then, the local re-scheduling strategy is used to deal with the frequently changing yard crane tasks, so as to reduce the overall solution time on the basis of ensuring that the solution is optimal or close to optimal.

4.2. PSO for Initial Scheduling

Some researchers use the PSO characteristics of rapid convergence to solve the scheduling problem, and have confirmed the advantages of PSO method for many scheduling problems [18–20]. PSO is used in this paper to solve the initial scheduling as a whole, so as to ensure that the final results are optimal or nearly optimal. The velocity and position are normally updated by the following:

$$v_{i+1} = wv_i + c_1r_1(pbest_i - p_i) + c_2r_2(gbest - p_i) \tag{13}$$

$$p_{i+1} = p_i + v_{i+1} \tag{14}$$

where

v_i: Velocity of particle at i-th iteration,
w: Inertia weight,
c_1: Cognitive coefficient,
c_2: Social coefficient,
r_1, r_2: Random numbers in (0, 1), regenerated in each iteration
$pbest_i$: Local best position of particle
$gbest$: Global best position of swarm
p_i: Position of particle

In PSO solving, there are no changes in the release time of yard crane tasks. The algorithm steps are described in Algorithm 1.

Algorithm 1: The Algorithm Steps

Step 1. Initialize the task parameters.
Step 2. Assign tasks to the yard cranes according to the task information, randomly generate task scheduling sequences after allocation, then generate initial scheduling particles based on continuous integer task sequence.
Step 3. According to the set initial number of particle swarms, generate the corresponding number of scheduled particles and initial swarm parameters: particle position, local optimal position, global optimal position, global optimal value, and particle velocity.
Step 4. Perform particle swarm iteration.
Step 5. Stop the iteration when the set number of iterations is met, otherwise turn to Step 4.
Step 6. End.

4.3. Local Re-Scheduling for Dynamic Cut-Off Time

Although an optimal or nearly optimal result has already been scheduled for the initial static crane task list, re-scheduling is still needed according to the constraints of the crane scheduling model when some task release times change. It will be time-consuming if the global solution method is directly used, and it is obviously not cost-effective. Therefore, based on the static initial optimal scheduling, a sort of local re-scheduling is proposed to solve this problem, so that the solving range is local.

The task types and its corresponding task number are set in Table 4. Local re-scheduling of crane 0 is used to ensure that no new cross problem will appear in local adjustments. Local re-scheduling of crane 1 represents a detour task to avoid new interferences between yard cranes.

Table 4. Task type and its number.

Task Type	Task Number
The actual task	j
Originally planned task	j^*
Detour task of crane 0	−2
Detour task of crane 1	−1
Local re-scheduling of crane 0	−3
Local re-scheduling of crane 1	−4

Assume that the task release time of task I of yard crane 0 changes, as shown in Figure 2, which is a time-domain diagram of local re-scheduling strategy. In Figure 2, "W_0" and "W_1" represent the initial scheduling timelines of cranes 0 and 1, respectively. Labels 1 to 5 are the specific time periods on the timeline "W_0" and "W_1" that need to be partially rescheduled. The gray part on the timeline represents the time occupancy of the corresponding task.

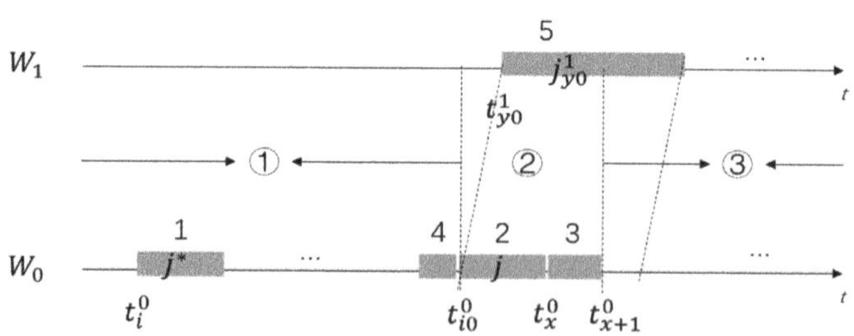

Figure 2. A time-domain diagram of local re-scheduling strategy.

Label 1 corresponds to the original handling period of task i, and t_i^0 represents the original $j_i^0 : ST$ on the timeline. Label 2 corresponds to the new handling period of task i after rescheduling, which meets the constraint of release time. Label t_{i0}^0 indicates the time when task i needs to be re-inserted. After the insertion, the new number for task i is i_0. Then, the time period 1 would become a no-handling period of crane 0 to avoid increasing interference (crane 0 waits after reaching the destination position while the spreader does not move).

The time period of label 3 ensures that yard crane 0 returns to the target location of task j^0_{i0-1} when the newly inserted task is handled. During t^0_x ~ t^0_{x+1}, yard crane 0 will move to the target location of task j^0_{i0-1}.

After the local re-scheduling of labels 1 to 3 is completed, add detour task to yard crane 0 if there is interference at label 3. Additionally, there will be a local re-scheduling task at label 4, in which the yard crane 0 gives way to yard crane 1 until yard crane 1 completes the current task. In this way, there are no other changes for yard crane 0 except that the locations of labels 2, 3, and 4 are changed. It makes the non-crossing constraints between the yard cranes satisfied in the time period at the time period marked ① in Figure 2.

Label 5 corresponds to the detour task of yard crane 1. Finally, yard crane 1 will return to the initial location, that is, the starting location of task j^1_{y0}. The detour task time is determined by the extension time of yard crane 0. It keeps yard crane free of interference at the time periods of labels ② and ③. By means of above steps, a new scheduling is obtained based on the original optimal scheduling.

As mentioned above, the algorithm starts from a static initial optimal scheduling. It is assumed that the release time of crane 0 task i is delayed, $j^0_i : Rd \rightarrow j^0_i : Rd^*$. The steps of the algorithm are described in Algorithm 2:

Algorithm 2: The Algorithm Steps

Step 1. Find the insertion point i_0. Take the minimum value of i_0,
$i_0 \in \{i_0 = e-1\} \cup min\{j^0_{i0} : Rd > 0 \cap j^0_i : Rd^* \geq j^0_{i0} : ST\}$. If $i_0 \leq i$, turn to step 5; otherwise, turn to the next step.
Step 2. Insert task i ($i \in \{j^0_i : Job = j\}$) into position of task i_0 in the task list. Revalue $j^0_{e-1} : Rd = j^0_i : Rd^*, j^0_i : Job = -3$. If $i_0 = e-1$, turn to step 4; otherwise, turn to the next step.
Step 3. As label "j^0_{x-1}" and label "j^0_{i0-1}" are shown in Figure 2, if $j^0_{x-1} : Bay \neq j^0_{i0-1} : Bay$, insert a task j^0_x into task x in the task list, set $j^0_x : Bay = j^0_{i0-1} : Bay, j^0_x : Row = j^0_{i0-1} : Row, j^0_x : Job = -4$, and set the remaining parameters to zero. If $j^0_{x-1} : Bay = j^0_{i0-1} : Bay$, turn to the next step.
Step 4. As label "y_0" shown is in Figure 2, get the value of y_0,
$y_0 \in \{y_0 = e-1\} \cup min\left\{j^1_{y0} : ST \geq j^0_{i0}\right\}$. When there is interference between crane 0′s whole tasks and crane 1′s tasks 0 to j^1_{y0-1}, crane 0 makes way. Crane 1 gives way to the bay 9 of the yard, and returns to the initial position $j^1_{y0-1} : Bay$ at the end of the detour. The time it takes for crane 1 to give way is equal to the delay of crane 0. Then, insert the detour task for crane 1 into task y_0 in the task list of crane 1.
Step 5. Update the release time of task i, $j^0_i : Rd = j^0_i : Rd^*$.
Step 6. End.

5. Computational Experiments

5.1. Parameter Setting

The size of each yard is set to bay ∈ [0, 9], row ∈ [0, 5] (the yard data refer to the parameters of Qingdao Xinqianwan automated terminal). Yard crane ∈ {0,1}. The particle swarm size is set to 10, the weight is 0.729, the cognitive coefficient and the social coefficient are 1.494, and the number of iterations is 50.

It adopts MacBook Pro 13 (from Apple, Cupertino, CA, USA) with 2.3 GHz quad-core Intel Core i5 processor, 8 GB 2133 MHz LPDDR3 memory, and the programming language is Python 3.8 (from Python Software Foundation, Wilmington, DE, USA).

5.2. The Effectiveness Experiments

Randomly generate a task group with 20 tasks, as shown in Table 5. In the table, the representations of "Task", "Bay", "Row", "Time", and "Release Time" are the same as those in Section 3.1. The particle swarm algorithm is used to find the initial global solution.

Firstly, the algorithm generates scheduling particles according to the task information provided in Table 5. The scheduling particles include the task allocation of yard cranes 0 and 1, the scheduling sequence, and the priority of each yard crane task. Then, the particle swarm algorithm is used to obtain an optimal or near-optimal solution. The best result obtained by PSO is shown in Figure 3. Figure 3 gives a spatiotemporal diagram of yard cranes 0 (red line) and 1 (blue line). The horizontal axis represents the time, and the vertical axis represents the bay location of the yard crane. All tasks' numbers are displayed in the figure according to their handling time and bay locations. The red/blue number indicates that the task was handled by crane 0/1. It can be seen that there is no collision between yard cranes 0 and 1; that is, the algorithm can obtain a solution that meets the constraints.

Table 5. Initial tasks.

Task	Bay	Row	Time	Release Time
1	1	2	4	0
2	5	3	4	0
3	6	1	4	0
4	1	3	4	0
5	4	4	2	0
6	5	2	4	0
7	2	1	4	0
8	7	2	1	0
9	8	3	4	0
10	7	4	3	0
11	8	1	6	0
12	1	2	4	0
13	6	3	4	0
14	2	3	3	0
15	6	4	2	0
16	4	3	4	0
17	5	2	4	0
18	7	1	4	0
19	2	4	1	0
20	8	4	2	0

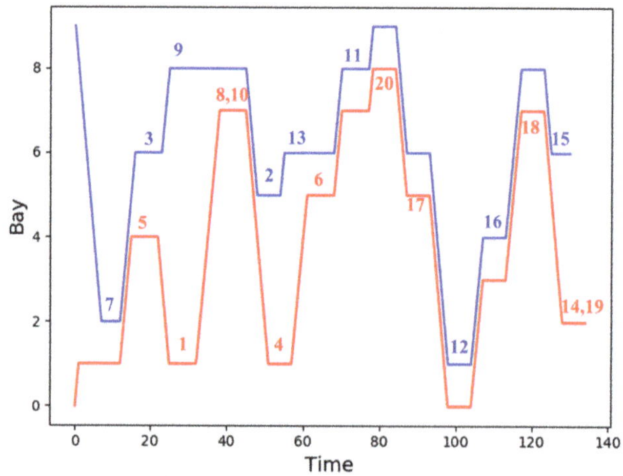

Figure 3. Spatiotemporal diagrams of twin yard cranes operation.

Table 6 shows the changes in the "Release Time" of Table 5. Figure 4 shows the results obtained by LRPSO in Table 6. Same as Figure 3, the task numbers are displayed in the figure. From Figure 4, it can be seen that all tasks are handled after their release time, and the overall completion time becomes longer after the task release time changes. The yard cranes complete all tasks without collisions, so LRPSO meets the constraints.

Table 6. Tasks' release time renewal.

Number	Bay	Row	Time	Release Time
1	1	2	4	27
2	5	3	4	17
3	6	1	4	35
4	1	3	4	11
5	4	4	2	48
6	5	2	4	20
7	2	1	4	55
8	7	2	1	14
9	8	3	4	28
10	7	4	3	14
11	8	1	6	30
12	1	2	4	43
13	6	3	4	10
14	2	3	3	23
15	6	4	2	19
16	4	3	4	61
17	5	2	4	48
18	7	1	4	19
19	2	4	1	13
20	8	4	2	32

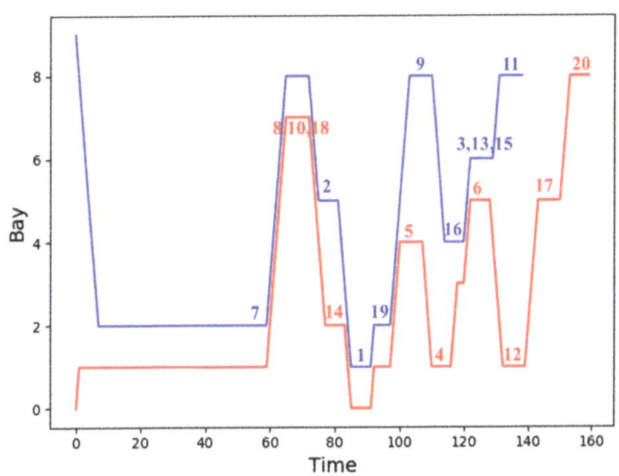

Figure 4. Twin yard cranes operation after Tasks' release time renewal.

It can be seen that the joint scheduling method including PSO and local re-scheduling is effective.

5.3. Simulation Comparison

Simulation experiments are carried out for different task scales and different task release time ranges (variation ranges). The compared algorithms are global strategy algorithm and PSO algorithm. After the release time of tasks changed, the global strategy algorithm always re-solves the fitness value (completion time) according to the original optimal plan. The PSO algorithm re-solves iteratively according to the new tasks.

In each set of simulation experiments, the PSO algorithm is first used to generate an initial optimal planning solution. Then, the release time of some tasks in the original task group are randomly changed according to the variation range, and three algorithms are used to solve the dynamic task group. Finally, the performance of the algorithms is compared in terms of calculation time and fitness value results.

Comparison results are given in Table 7, where "N" represents the total number of tasks; "variation range" refers to the variation range of the overall tasks' release time, "0" means that there is no change, and "100%" means that all the tasks' release time are changed. "CPU Time" represents the calculation time of algorithms in seconds, and the recorded results are kept to one decimal place.

Table 7. Simulation comparison data.

N	Variation Range	CPU Time (s)			Completion Time		
		LRPSO	Global Strategy	PSO	LRPSO	Global Strategy	PSO
500	5%	0.5	2.1	626.3	5058	5983	5021
	15%	0.2	2.5	646.2	5682	6389	5613
	25%	0.8	2.3	651.3	5467	6526	5481
	50%	1.1	2.6	677.5	7907	6596	6234
	75%	1.4	2.3	689.3	6533	6549	6399
1000	5%	1.1	4.4	1292.4	8795	13,318	11,731
	15%	1.5	4.7	1260.3	9895	13,052	12,249
	25%	1.6	4.9	830.9	10,770	12,972	12,373
	50%	2.1	4.3	840.9	13,843	13,299	12,602
	75%	2.9	4.2	858.3	16,050	13,170	12,869
1500	5%	1.6	6.4	1366.4	8651	13,120	11,413
	15%	1.8	6.5	1163.5	9288	11,482	11,230
	25%	2.1	6.5	1345.7	10,105	11,515	11,586
	50%	2.8	6.6	1325.5	12,133	12,113	11,820
	75%	3.7	6.5	1344.1	14,194	12,149	11,884
2000	5%	2.2	8.7	1833.5	11,339	14,962	14,164
	15%	2.8	9.1	1766.6	12,167	15,253	14,954
	25%	3.4	8.8	1759.5	15,841	22,105	21,564
	50%	5.4	9.0	1736.3	18,800	18,217	18,243
	75%	8	9.0	1985.5	23,309	18,537	18,303

Figure 5 summarizes the CPU time and the solution results of LRPSO, global strategy algorithm, and PSO algorithm under different task scales based on the data in Table 7. The histogram shows the CPU time, and the line graph shows the fitness values. Since the CPU time of PSO far exceeds that of the other two algorithms, it is not shown. On the whole, as the task scale increases, the solution time of each algorithm increases. The CPU time of the LRPSO is much shorter than other two algorithms. When the release time variation range is less than 50%, the solutions of the LRPSO are better than other algorithms. Under the same task scale, the solution time of the LRPSO increases with the increase of the task variation range. As the task scale increases, the trend of time spent and solution results shows obvious regularity: when the release time variation range exceeds 50%, the increase rate of the solution time and fitness value solution results is accelerated, and the slope becomes steeper. It can be seen that the performance of the LRPSO is better than other algorithms within the release time variation range of 50%. When the release time variation range exceeds 50%, the performance of the LRPSO is greatly reduced compared with other algorithms and is no longer applicable.

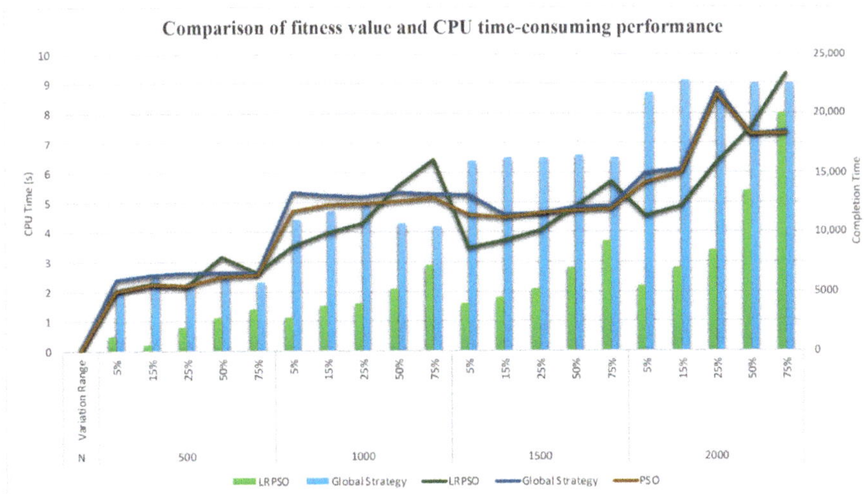

Figure 5. Comparison of fitness value and CPU time.

Figure 6 shows the average CPU time and average solutions of each algorithm under different release time variation ranges. In the figure, the left half shows the CPU time comparison, and the right half shows the completion time comparison. The same is shown with Figure 5: since the average CPU time of PSO far exceeds that of the other two algorithms, it is not shown. It can be seen that when the release time variation range does not exceed 50%, the LRPSO has a better performance.

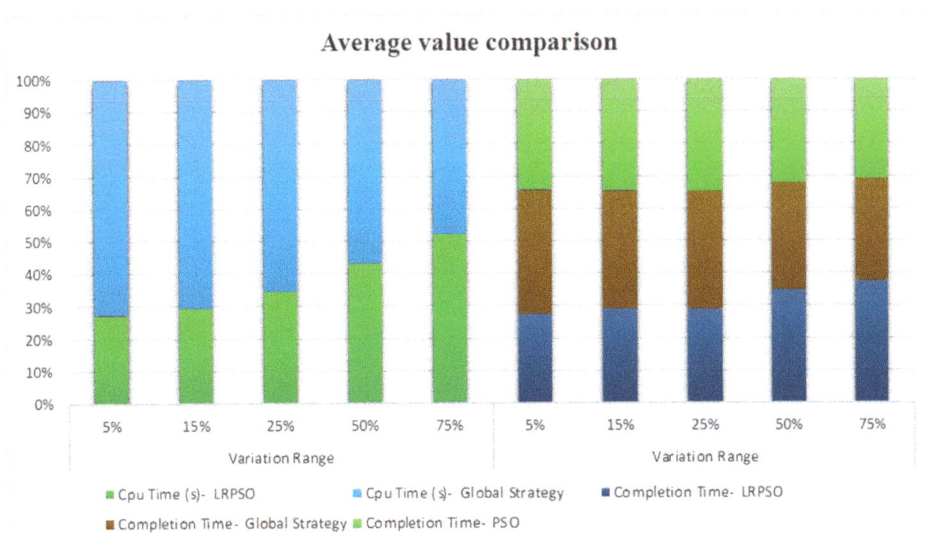

Figure 6. Comparison chart of average result values.

From the comparison results, it can be seen that the LRPSO can be applied to large-scale dynamic yard crane scheduling with a release time variation range less than 50%.

6. Conclusions

Aiming at improving the flexibility of the terminal yard, this paper studies the twin yard cranes scheduling problem, considering the dynamic cut-off time and no-crossing constraint. The dynamic cut-off time makes the release time of the yard crane variable, and the yard crane task arrangement will change frequently, resulting in a lot of computing time. To decrease the computing time and improve the terminal yard efficiency, a joint scheduling strategy of PSO and local re-scheduling is proposed. This solution method is proved to be effective and practical, and it is easier to be optimized and extended. The contributions of this paper are as follows: (1) a twin yard cranes scheduling model is built considering environment variables, which is closer to reality; (2) a local re-scheduling strategy joint with PSO is proposed, which can improve the adaptability of yard crane scheduling to environmental variables and other influencing factors. The experiment results show that LRPSO has better performance compared with the global strategy and PSO in the large-scale scheduling with the release time variable variation range less than 50%.

We will use more scheduling methods for comparison in future research, such as polynomial-time, heuristics, genetic algorithms, and so on. Another further study is to set both task release time and task processing time as dynamic variables. Furthermore, we will also concentrate on optimizing the schedule of the initial retrieval and storage of yard cranes, taking container storage scheduling of the yard into consideration, and optimizing the cooperation of automated guided vehicles and yard cranes.

Author Contributions: J.L. and J.Y. contributed conceptualization and methodology; J.L., J.Y. and B.X. wrote the paper; W.Y. and Y.Y. contributed English editing and correction; B.X., J.W., Y.Z. and Y.S. contributed validation; J.L. and B.X. contributed project administration; B.X. contributed funding acquisition. All authors have read and agreed to the published version of the manuscript.

Funding: This research was funded by the National Natural Science Foundation of China, grant number 52102466, and the Natural Science Foundation of Shanghai, grant number 21ZR1426900.

Data Availability Statement: The data used to support the findings of this study are included within the article.

Conflicts of Interest: The authors declare no conflict of interest.

References

1. Heuermann, A.; Duin, H.; Gorldt, C.; Thoben, K.D. A concept for predictability and adaptability in maritime container supply chains. In *International Conference on Dynamics in Logistics*; Freitag, M., Kotzab, H., Pannek, J., Eds.; Springer: Cham, Switzerland, 2018; pp. 243–249.
2. Kemme, N. State-of-the-Art Yard Crane Scheduling and Stacking. In *Operations Research/Computer Science Interfaces Series, Handbook of Terminal Planning*; Böse, J.W., Ed.; Springer: Cham, Switzerland, 2020; pp. 383–413.
3. Nils, B.; Dirk, B.; Frank, M. A generalized classification scheme for crane scheduling with interference. *Eur. J. Oper. Res.* **2017**, *258*, 343–357.
4. Ehleiter, A.; Jaehn, F. Scheduling crossover cranes at container terminals during seaside peak times. *J. Heuristics* **2018**, *24*, 899–932. [CrossRef]
5. Amelie, E. A decomposition-based approach to the scheduling of identical automated yard cranes at container terminals. *J. Sched.* **2019**, *22*, 517–541.
6. Zey, L.; Briskorn, D. Interference aware scheduling of triple-crossover-cranes. *J. Sched.* **2019**, *23*, 465–485.
7. Zheng, F.; Man, X.; Chu, F.; Liu, M.; Chu, C. Two Yard Crane Scheduling with Dynamic Processing Time and Interference. *IEEE Trans. Intell. Transp. Syst.* **2018**, *19*, 3775–3784. [CrossRef]
8. Chen, T.C.; Wang, S.C.; Tseng, W.C. Using a Hybrid Evolutionary Algorithm for Solving Signal Transmission Station Location and Allocation Problem with Different Regional Communication Quality Restriction. *Int. J. Eng. Technol. Innov.* **2020**, *10*, 165–178. [CrossRef]
9. Guo, P.; Wang, L.; Xue, C.; Wang, Y. Dispatching Rules for Scheduling Twin Automated Gantry Cranes in an Automated Railroad Container Terminal. *Arab. J. Sci. Eng.* **2020**, *45*, 2205–2217. [CrossRef]
10. Lei, D.; Zhang, P.; Zhang, Y.; Xia, Y.; Zhao, S. Research on optimization of multi stage yard crane scheduling based on genetic algorithm. *J. Ambient. Intell. Humaniz.* **2020**, *11*, 483–494. [CrossRef]

11. Yu, M.; Cong, X.; Niu, B.; Qu, R. Iteration-Related Various Learning Particle Swarm Optimization for Quay Crane Scheduling Problem. In *Bio-Inspired Computing: Theories and Applications, Communications in Computer and Information Science, Proceedings of the 13th International Conference BIC-TA 2018, Beijing, China, 2–4 November 2018*; Qiao, J., Zhao, X., Pan, L., Zuo, X., Zhang, Q., Huang, S., Eds.; Springer: New York, NY, USA, 2018; Volume 952, pp. 201–202.
12. Briskorn, D. Routing two stacking cranes with predetermined container sequences. *J. Sched.* **2021**, *24*, 367–380. [CrossRef]
13. Zey, L.; Briskorn, D.; Boysen, N. Twin-crane scheduling during seaside workload peaks with a dedicated handshake area. *J. Sched.* **2022**, *25*, 3–34. [CrossRef]
14. Guvenc, D.; Erhan, K. A flexible crane scheduling methodology for container terminals. *Flex. Serv. Manuf. J.* **2017**, *29*, 64–96.
15. Gharehgozli, A.; Yu, Y.; de Koster, R.; Du, S. Sequencing storage and retrieval requests in a container block with multiple open locations. *Transp. Res. Part E Logist. Transp. Rev.* **2019**, *125*, 261–284. [CrossRef]
16. Kizilay, D.; Van Hentenryck, P.; & Eliiyi, D.T. Constraint programming models for integrated container terminal operations. *Eur. J. Oper. Res.* **2020**, *286*, 945–962. [CrossRef]
17. Zhou, C.H.; Lee, B.K.; Li, H.B. Integrated optimization on yard crane scheduling and vehicle positioning at container yards. *Transp. Res. Part E Logist. Transp. Rev.* **2020**, *138*, 101966. [CrossRef]
18. He, J.; Huang, Y.; Yan, W.; Wang, S. Integrated internal truck, yard crane and quay crane scheduling in a container terminal considering energy consumption. *Expert Syst. Appl.* **2015**, *42*, 2464–2487. [CrossRef]
19. Zarrouk, R.; Bennour, I.E.; Jemai, A. A two-level particle swarm optimization algorithm for the flexible job shop scheduling problem. *Swarm Intell.-US.* **2019**, *13*, 145–168. [CrossRef]
20. Vairam, T.; Sarathambekai, S.; Umamaheswari, K. Multiprocessor task scheduling problem using hybrid discrete particle swarm optimization. *Sādhanā* **2018**, *43*, 206. [CrossRef]

Journal of
Marine Science and Engineering

MDPI

Review

Review of the Influence of Oceanographic and Geometric Parameters on Oscillating Water Columns

Nerea Portillo Juan *, Vicente Negro Valdecantos, M. Dolores Esteban * and José Santos López Gutiérrez

Environment, Coast and Ocean Research Laboratory, Universidad Politécnica de Madrid, 28040 Madrid, Spain; vicente.negro@upm.es (V.N.V.); josesantos.lopez@upm.es (J.S.L.G.)
* Correspondence: nf.portillo@alumnos.upm.es (N.P.J.); mariadolores.esteban@upm.es (M.D.E.); Tel.: +34-910-674-352 (N.P.J. & M.D.E.)

Abstract: Wave energy is one of the most powerful sources of energy on our planet, but its exploitation is difficult. Much current research on renewable energy is focused on how to harness ocean energy. However, wave energy converter (WEC) technology is still immature and how to reach high levels of efficiency is still unknown. In coming years, this field is likely to reach a high level of development, so it is important to continue research on the improvement of the performance of these devices. One of the most important wave energy converters is the oscillating water column (OWC). The main difficulty of OWCs is that they have to provide good rates of hydrodynamic efficiency for many different types of sea states (different periods, heights, wavelengths, etc.). The other big concern is the optimization of the geometric parameters of the device. This research paper is focused on these two big concerns: how oceanographic parameters affect the hydrodynamic behavior of an OWC and its geometric optimization. Different studies about how wave and geometric characteristics affect the performance of an OWC are reviewed and relationships between these and the hydrodynamic performance of an OWC are finally outlined and summed up.

Keywords: oscillating water column; hydrodynamic efficiency; geometric optimization; OWC turbines; multipurpose platforms; breakwater-integrated OWCs; dual chambers OWCs; wave climate

Citation: Portillo Juan, N.; Negro Valdecantos, V.; Esteban, M.D.; López Gutiérrez, J.S. Review of the Influence of Oceanographic and Geometric Parameters on Oscillating Water Columns. *J. Mar. Sci. Eng.* **2022**, *10*, 226. https://doi.org/10.3390/jmse10020226

Academic Editor: Luca Cavallaro

Received: 31 December 2021
Accepted: 3 February 2022
Published: 8 February 2022

1. Introduction

Since 1880, atmospheric levels of carbon dioxide (CO_2), methane (CH_4), nitrogen dioxide (NO_2) and other greenhouse gases have been increasing and this is due to anthropogenic causes. It would have taken thousands of years for nature to produce current levels of greenhouse gases, but humankind has produced them in only a few decades. Climate change is, definitely, one of the biggest threats of our century [1].

The development of new renewable energies is a key to reducing the consumption of fossil fuels and to fighting against climate change. One of the main sources of renewable power is ocean waves. While wave energy technology is still immature and is not cost efficient yet, it is very likely that in the coming decades this new source of energy will reach a high level of development [1].

Ocean wave energy is based on the power of surface gravity. Waves are formed due to the imbalance between gravitational force and the shear force produced by wind [2].

Wave energy has many advantages: it has one of the highest rates of energy density and its environmental impact is minimal. Moreover, waves will always exist, and they are a reliable and predictable source of energy that can travel huge distances without losing energy [2,3].

Despite these big advantages, there are still some issues that must be solved in order to make wave energy profitable: waves are not regular, they have different heights, periods, directions, phases, etc. This broad range of possibilities makes it very difficult to operate energy devices efficiently most of the time. In many cases, only low energy conversion efficiency levels can be reached. Moreover, the loading on the devices can be 100 times higher than the average loading in extreme weather conditions, which significantly increases the difficulty of designing the technology. A balance between money invested and structural safety is needed. The devices used to extract wave energy need a complex and expensive process of evaluation [2,3].

Furthermore, using wave energy is not affordable for all countries. Only in areas where wave activity is intense are wave energy devices cost efficient. These areas are between the latitudes of 30° and 60° in both hemispheres [3], as can be seen in Figure 1.

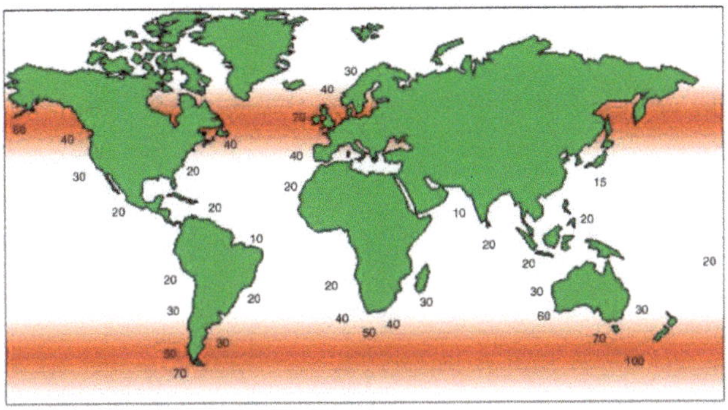

Figure 1. Global wave power distribution by kW/m of crest length (Centre for Renewable Energy Sources (CRES) [4]).

The devices used to extract wave energy are called wave energy converters (WECs). The main problem of WECs is that they have to operate efficiently with a great variability of wave parameters. Each wave is unique: there are no two waves in the ocean with the same period, length, and height and WECs have to be efficient for all of them. There are different types of WECs, but one of the most promising is the oscillating water column (OWC). The purpose of this article is to define the relationships between wave characteristics and the geometric parameters of the OWC and its performance and hydrodynamic efficiency. In order to extrapolate these relationships, some of the most relevant studies about OWCs are reviewed and analyzed.

The structure of the article is as follows: Section 2 describes the methodology; Section 3 presents the basic concepts about OWCs; Section 4 presents the relationships between wave characteristics and the performance of the OWC; Section 5 looks at the relationship between the geometric characteristics of the OWC and its hydrodynamic performance; Section 6 is on the influence of air and turbine properties; and, finally, Section 7, presents the conclusions.

2. Research Methodology

In this research paper a structured literature review based on Adebisi et al. (2019) [5] was conducted. The structured literature review had three stages: planning, review, and information synthesis. In the first stage, planning, the searching criteria and databases were defined: the publications were obtained from reputable databases, mainly Web of Science (WOS), Scopus, Google Scholar, and the MDPI search engine. The searching criteria selected was the topic of the articles. The search keywords were: *OWC classification, wave influence OWC, hydrodynamic optimization OWC, geometric optimization OWC, OWC turbines, air compressibility OWC, air humidity OWC, biradial turbines, dual-chamber OWC, U-shaped OWC, L-shaped OWC, V-shaped OWC, breakwater-integrated OWC*, and *multipurpose platforms*.

In the second stage, review, the articles were searched and filtered, excluding the ones that did not fit with the purpose of the research. A total of 133 papers were reviewed.

Finally, in the third stage, all the information of these 133 articles was analyzed and synthesized.

Once the state of the art was studied, different relationships between wave characteristics and geometric parameters and the hydrodynamic performance of the OWC were defined and summed up in tables (Tables 1–3).

Table 1. Influence of wave characteristics on the performance of OWCs.

	Forces	Efficiency	Comments
Wave period *T*	Less sensitive	-	
Wave Height *H*	Positive relationship	Positive relationship	More important near the breaking point
Wave Length *L*	Negative relationship	Positive relationship	Better to not be affected by bottom friction
Wave Steepness *H/L*	Positive relationship	Negative relationship	Increases nonlinearities
Relative depth *d/L*	Increases until 0.16 and decreases	Maximum around 0.131	
Comments	Peak horizontal force is 2.5-3 times the peak vertical force	Only vertical velocity is useful, the horizontal component is wasted	

Table 2. Influence of the geometry of OWCs on their performance.

	Efficiency	Frequency Bandwidth	Resonant Frequency	Comments
Relative Chamber Width *(Divided by water depth)*	Positive for low frequencies and long waves Negative for high frequencies and short waves	Negative Relationship	Positive relationship for long waves Negative relationship for short waves	Optimum value 0.8–1
Relative Chamber Length *(Divided by wavelength)*	Follows an U-inverted shape	Follows a concave parabolic shape	Negative relationship	More important for great values of periods and heights
Relative submergence *(Divided by water depth)*	Better behavior at resonant frequency for low values	Negative relationship	Negative relationship	Optimum value 0.3–80.44 depends on the height and period of the wave
Opening ratio	Follows an U-inverted shape	-	-	Significantly influenced by wave height
Thickness of the wall	Indifferent	Negative relationship	-	-

Table 3. Influence of geometry of dual chamber on the performance of OWCs.

	Wave Forces	Efficiency	Frequency Bandwidth	Resonant Frequency	Comments
Inner chamber draft	Positive relationship	Negative relationship	Positive relationship	Negative relationship	Must be bigger than the outer draft
Outer chamber draft	Positive relationship	Negative relationship	Negative relationship	Negative relationship	Must be smaller than the inner draft
Chamber breadth	-		Positive relationship	-	-
Thickness of the wall	-	Negative relationship for high frequencies	-	-	-
Volume of the chamber	Negative relationship	Positive relationship	Negative relationship	-	-

In these tables, a *positive relationship* means that if one variable increases, the other also does; and a *negative relationship* means that if one variable increases, the other decreases.

3. OWCs: Concept, Elements, and Classification

OWCs are a type of WEC that extracts the energy from a bidirectional air flow generated by an oscillating sea-level chamber [3,6]. They are one of the WECs that are being studied more deeply and that are likely to be highly developed in the next few years. The OWC's principle of action is presented in Figure 2.

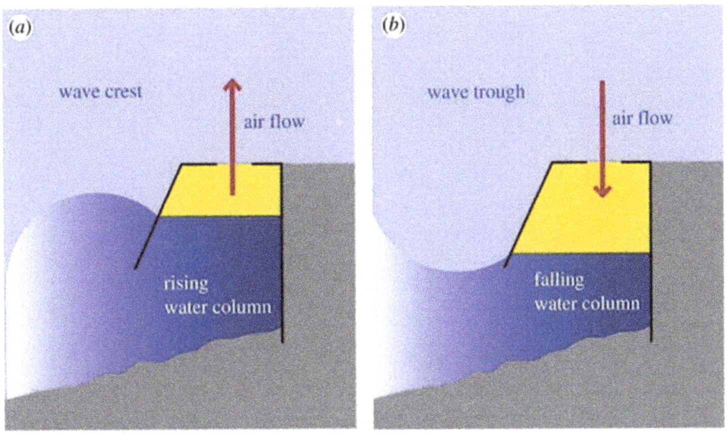

Figure 2. Principle of OWC action (Heath, T.V., 2012 [6]). (**a**) shows how the water rises pressurizing the collector and (**b**) how the water falls rarefying the collector.

OWCs are composed of two main parts that are the air chamber and the PTO, which is usually a self-rectifying axial flow air turbine. The most common ones are the Wells turbines [3,6].

OWCs have many advantages: there are not many moving parts and they have no moving parts in the water, which improves their durability and makes their maintenance cheaper and easier, and they do not have gearboxes, which also facilitates their maintenance. In addition, the concept of the OWC is versatile and can be used on the coastline, in the nearshore region or floating offshore. They are solid and secure and they use sea space efficiently [6].

3.1. OWC Elements

OWCs have two main parts: the chamber and the PTO system. As Section 5, on the influence of OWC geometry, refers to different elements of the OWC, this subsection shows a review of these elements with the aim of facilitating the understanding of the results presented.

3.1.1. OWC Chamber

The OWC chamber is formed by walls of different materials and its main geometric characteristics are the following:

The chamber length (a in Figure 3) and the chamber width, which is the dimension of the chamber in the direction perpendicular to the paper.

The front wall width or thickness (t in Figure 3)

The front lip submersion depth, also named as front wall submergence depth [7], or immersion of the front lip [8] (d in Figure 3).

The opening ratio, which is defined as the ratio of the orifice area and the cross-sectional area of the chamber [9]. The area of the orifice is determined by the parameter r in Figure 3.

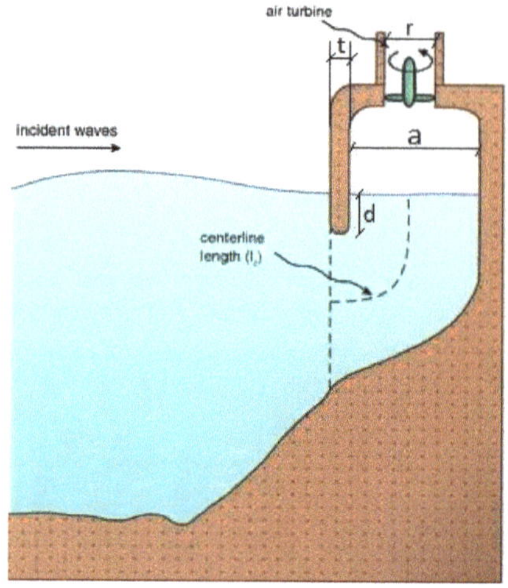

Figure 3. Sketch of the geometric parameters of an OWC. Image source: [10].

3.1.2. PTO Systems

The most common PTO systems used in OWCs are air turbines. Of the three types of air turbines (Wells turbines, impulse turbines and Denniss-Auld turbines), Wells turbines and impulse turbines are the most used turbines in OWCs [11].

Wells Turbines

These were the most used turbines in OWCs until the development of biradial turbines. The most remarkable characteristic of these turbines is that they are self-rectifying turbines: they rotate in a single direction in a bidirectional flow. Although they reach only a moderately high peak efficiency as compared with conventional turbines, they can operate in reciprocating flow without the need of a rectifying valve system, which facilitates their maintenance and operation. Wells turbines have a linear pressure drop-flow rate which is well suited to energy conversion in OWCs. The type of Wells turbine most used is the multiplane turbine. Its power output operates at low flow rates and it drops sharply for flow rates above a critical value due to aerodynamic losses. Therefore, these turbines are not appropriate for very energetic seas, and if they are used, a bypass pressure-relief valve is needed [9,12]. Their main manufacturer is Voith Hydro Wavegen [13] (Figure 4).

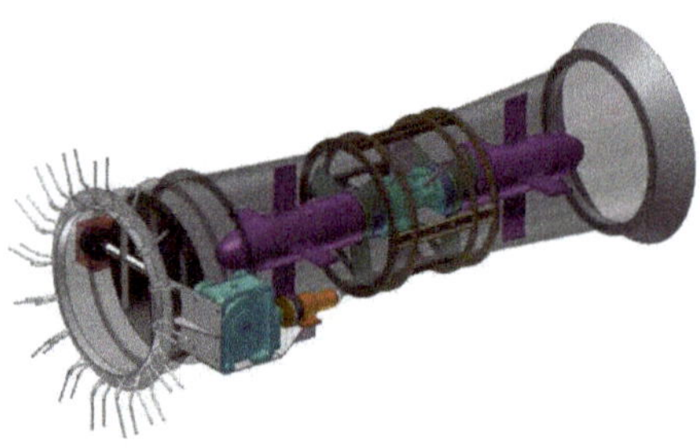

Figure 4. Wells turbine tested at LIMPET (Torre-Enciso et al. 2009 [14]).

Impulse Turbines

Impulse turbines are self-pitched, controlled turbines with guided vanes [15]. The most common types used in OWCs are the axial-flow impulse turbine and the recently patented biradial turbine [16].

- Axial-flow impulse turbines

This type of turbine is the one used in the Irish OE buoy and the Jeju island power plant [16]. The main problem of this type of turbine is the large aerodynamic losses due to the excessive incidence flow angle at the entry to the second row of guide vanes. To reduce these losses, guided vanes of variable geometry can be used. Another option is to increase the distance between the guided vane rows and the rotor blades. Although the operational flow range of the impulse turbine is wider compared with that of the Wells turbine, its peak efficiency hardly exceeds 50% [11,15].

- Biradial impulse turbines (Figure 5)

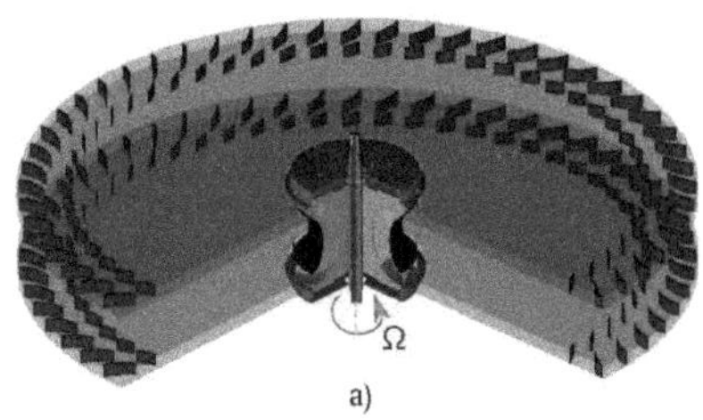

Figure 5. Biradial turbine (Carrelhas et al. 2019 [16]).

There are two types of biradial impulse turbine: those with axially-sliding guide vanes and those with fixed guide vanes. The first type was model tested and found to be the most efficient self-rectifying turbine known so far, with a peak efficiency of about 78%, but it has the disadvantage of being more complex [16].

Biradial impulse turbines are the most recent turbines developed for OWCs and they have proved to be more efficient than Wells turbines and overcome some of their main limitations [17]. Falcão et al. (2014) [18] showed that they were the best option for spar buoys and they were the turbines chosen for the MARMOK-A-5 OWC spar buoy [16].

Apart from these three most common turbines used in OWCs as PTOs, there is current research on dielectric elastomeric generators that can be implemented in OWCs [19]. The dielectric elastomeric generator (DEG) is a compliant polymeric generator that makes it possible to convert mechanical energy into electrical energy by exploiting the large deformations of elastomeric membranes, and has several advantages compared to more conventional PTOs [20].

3.2. OWC Classification

- Fixed structures
  - Separate or isolated

There are a few examples of this kind of OWC in Europe. In 1991, the European Commission decided to include wave energy in the program of renewable energy. This led to new projects and studies in this field. Two full-sized isolated fixed-structure OWC plants were built [2,11]:

- The Pico plant in Azores, Portugal (400 kW) (Figure 6) has been operating since 1999 with Wells turbines built on the sea bottom next to a vertical cliff [11,21].
- The Islay LIMPET plant in Scotland, UK (500 kW) (Figure 7). It was completed in 2000, also with Wells turbines, and is located in a recess carved into a rocky cliff [11,23].

Figure 6. Pico OWC plant (TETHYS [22]).

Figure 7. LIMPET OWC plant (Sun et al. 2018 [24]).

In Guangdong Province (China) another shoreline OWC plant (100 kW) was built in 2001 [11].

In 2015, a bottom-standing OWC (500 kW) was completed at Yongsoo, Jeju Island, South Korea [11] (Figure 8).

Figure 8. Yongsoo OWC (Curto et al. 2021 [25]).

In 2019, a similar concept was developed on King Island, Australia (200 kW) [25].

There have been other OWC projects that failed, such as the OSPREY plant (1 MW) that was located on the Scottish coast or the greenWAVE (1 MW) plant built by Oceanlinx in Port Adelaide [11].

○ Integrated in a breakwater [11,26]

When building an OWC, one option is to integrate it into a breakwater. This option is very convenient for several reasons: construction costs are shared, operation and maintenance are easier, and its efficiency is usually higher.

Some examples are the OWCs installed at the Sakata port (Japan) and in Mutriku (Spain) (Figure 9).

Figure 9. Mutriku OWC plant.

The REWEC3 is another example. It is a U-shaped OWC installed at the Port of Civitavecchia, Italy. It is a fixed oscillating OWC incorporated in upright breakwaters. The main feature of the REWEC3 is the possibility of tuning the natural period of the water column in order to match a desired wave period through the size of the U-duct [16,27].

- Floating structures

This type of structure was proposed for the first time by Masuda in 1986 [28], who developed a floating device known as a backward bent duct buoy (BBDB) that consists of an L-shaped OWC, a buoyancy caisson-type module, an air chamber, and an air turbine driving an electrical generator [11,29]. Since then, several countries in Europe, China, and the USA have studied this new concept [11,29].

Some years later, other similar systems were developed. Among these, the best known are the sloped buoy, the spar buoy, and the Mighty Whale [25].

The Mighty Whale (110 kW) was developed in Japan in 1998 [11] (Figure 10).

Figure 10. The Mighty Whale (JAMSTEC [30]).

The sloped buoy is composed of three parallel pipes installed on a floating buoy with a tilt angle of 45° [25].

The spar buoy was initially developed by McCormick (Figure 11) and it consists in an axisymmetric floating OWC open at both ends and attached to a floater [11,31].

Figure 11. Spar buoy.

Between 2008 and 2011, the OE buoy (a 1:4 scale model of a BBDB) was tested in Spiddal, in Galway Bay [11,32].

Oceanlinx also developed a floating OWC model in 2010 in Port Kembla, Australia (Figure 12). It was called the MK3 and it consisted in a floating platform with different chambers with an air turbine each [11].

Figure 12. Oceanlinx MK3 (TETHYS [33]).

The Instituto Superior Técnico (IST) is one of the most active research institutions in wave energy and has worked in the development of spar buoys and biradial turbines [34].

In 2012, it developed a prototype of a spar buoy in Narec and tested it at a scale of 1:16 [35].

In 2014, it also conducted different experiments with arrays and extreme wave conditions in Plymouth [35].

In 2016, it developed with IDOM the MARMOK-A-5 OWC, a spar buoy that was deployed at the Biscay Marine Energy Platform (BiMEP) test site [16]. It was the first WEC connected to the electricity grid in Spain and one of the first in the world [36].

- Special OWCs
  - Multi-OWCs

The Seabreath is one example of this type of device (Figure 13). It consists in a floating attenuator with a set of rectangular chambers and an open bottom [11,37].

The LEANCON is another multi-OWC device with two rows under two beams connected to each other in a V shape [11].

- Multipurpose platforms

In this group, multifunctional platforms can be included. The suitable combination of different renewable energy conversion systems on the same platform offers significant opportunities to reduce the cost of energy and can notably contribute to the sustainable exploitation of the marine natural resource [38]. There are many research projects in this field. Research on hybrid wind–wave systems has been driven primarily by a number of European research projects aimed at developing the concept of hybrid and multiplatform systems (Marina Platform, ORECCA, TROPOS, H2OCEAN, and MERMAID) [39]. Some examples of the new concepts proposed are:

○ The WindWaveFloat that intends to equip the WindFloat, a semisubmersible three-column floating platform, with different types of WECS, such as OWCs and point absorbers [38,40].
○ The BlueGrowthFarm, a project founded by the European Union that proposes a multifunctional platform with 10 MW wind turbines and several OWCs [41].
○ The hybrid wind–wave energy converter developed by Pérez-Collado et al. in 2018 [39,42].

Figure 13. Seabreath OWC (Martinelli et al. 2013 [37]).

A scheme that sums up the classification of the OWCs is presented below (Figure 14).

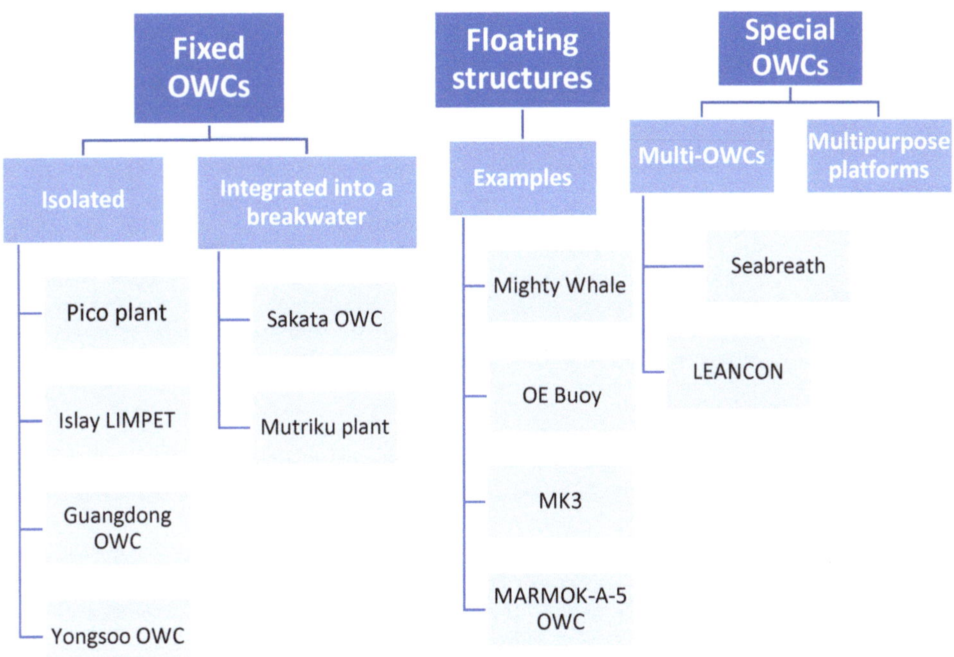

Figure 14. Scheme classification of OWCs.

4. Influence of Wave Characteristics

The performance of OWCs is mainly determined by wave characteristics and the geometric characteristics of the device [43].

The basic geometric characteristics of waves are height, length, and period.

Ning et al. (2016) showed that horizontal wave forces in OWCs increase with the wave height and decrease with the wave length [44]. Rezanejad et al. (2017) observed that the influence of the wave height on the performance of the OWC device is of lesser importance than the influence of the wave period [45], and Huang et al. (2019) reported that horizontal and vertical wave forces are less sensitive to the wave period, but increase with the wave height [46].

Apart from studying the basic characteristics of the waves, other research has been carried out to study the relationship between different monomials that characterize waves and the performance of an OWC.

4.1. Relative Water Depth d/L

Sundar et al. (2014) studied the wave forces that acted on an OWC device and concluded that an OWC reaches the optimum value of its hydrodynamic efficiency when the relative water depth is around 0.131 [47].

One year later, they carried out an experimental study on a circular curve bottom profile OWC. They observed that vertical forces increased with the relative water depth up to $d/L = 0.16$ for all the different ranges of steepness, but that for values of d/L higher than 0.16, vertical forces decreased. Moreover, it was found that this reduction increases as the steepness of the wave decreases [48].

In addition, for $d/L = 0.131$, where the maximum absorption of the incident wave energy occurs, vertical forces are lower and for d/L greater than 0.15, nonlinearities dominate [48].

Referring to horizontal forces, the trend is similar to that in the behavior of vertical forces, but nonlinearities are much less important [48]. They also discovered that the peak horizontal force is more than 2.5–3 times the peak vertical force [48].

4.2. Water Steepness H/L

Thiruvenkatasamy and Neelamani (1997) found that an increase in wave steepness causes a decrease in the efficiency of the OWC [49]. Ashlin et al. (2015) discovered that wave forces acting on the OWC increase with H/L [48]. Jasron et al. (2019) showed that an increase in wave steepness leads to an increase in the maximum air pressure inside the chamber, which gives a lower hydrodynamic efficiency [43], and Zhou et al. (2020) reached the same conclusion after investigating the hydrodynamic performance of an OWC integrated into a monopile offshore wind turbine (OWT). The efficiency of the device decreases as wave steepness increases, especially next to the resonant frequency. As the wave steepness increases, nonlinearities also do, which causes a loss of energy [50].

4.3. Other Considerations

D'Aquino et al. (2018) [51] studied the correlation for two depths (4 and 11 m) between wavelength (L), wave period (T), wave height (H), and the velocity of the wave (V). Their most important findings were that:

- The influence of H on the performance of the OWC becomes much more important near the breakpoint of the wave (at depth of 4 m in their study).
- For a better performance of the device, it is recommended to locate the OWC where it is not affected by bottom friction.
- The highest value of wave power does not necessarily correspond to the highest value of mechanical power extracted by the device. Only vertical velocity of the waves is useful and the wave power dissipated by the horizontal component of the velocity is wasted.

Considering all these studies, the relationship between the wave characteristics and the performance of an OWC are summed up in Table 1.

5. Influence of OWC Geometry

The geometry of the OWC is probably the most popular topic about OWCs due to its strong influence on the behavior of the WEC. The concept of geometry is very broad, and includes many aspects. The most significant are the dimensions and the shape of an OWC; the number of chambers; the configuration of the OWC; multi-OWCs and arrays. In this section all of them are studied.

5.1. Geometric Parameters

Optimizing the geometric parameters of an OWC is usually the main purpose of its designer. Therefore, many investigations have been carried out to obtain the best hydrodynamic performance of these devices. Malmo and Reitan (1985) proved that the natural frequency of an OWC is primarily determined by the immersion of the front lip [8]. Zheng et al. (1989) showed that flared harbor walls have better efficiency than rectangular walls [52]. Evans and Porter (1995) proved that for low values of the chamber length divided by the water depth, the water can be treated as a rigid body and that for high values of the submergence of the wall divided by the water depth, the capture width ratio (CWR) becomes narrower [53]. Thomas et al. (2007) showed that the thickness of the front wall does not have a remarkable influence on the hydrodynamic performance of the device energy conversion capacity [54]. Bouali et al. (2013) discovered that the front wall submergence depth is a key in the design of an OWC and its optimal value is between 0.38 and 0.44 times the water depth. They also proved that the width of the OWC is another important parameter and its optimum value is between 0.8 and 1 times the water depth [7]. Ashlin et al. (2015) and Sundar et al. (2014) proved that the circular bottom profile of the OWC is more efficient [47,48]. Ning et al. (2016) [44] studied the effect of the chamber width, the front wall draught, the orifice scale, and the bottom slope on the hydrodynamic efficiency of the OWC. They concluded that the water motion highly depends on the wavelength divided by the chamber width and, finally, Simonetti et al. (2017) [55] studied the geometric optimization of the OWC and the turbine damping and pointed out that the CWR of the OWC follows a concave parabolic shape with the chamber length.

Some of the most recent studies are explained in more detail below.

Tsai et al. (2018) studied the performance of a modified breakwater-integrated OWC that was supposed to be installed at Taichung Harbor, Taiwan. They obtained [56]:

- Effect of chamber width: The hydrodynamic efficiency increases as the chamber width increases for low frequencies, but it follows an opposite trend for high frequencies. In addition, the resonant frequency decreases as the chamber width increases, in line with Ning et al.'s results [44]).
- Effect of opening ratio: The hydrodynamic efficiency of the OWC increases with the opening ratio until reaching a maximum, and then decreases.

Hsien et al. (2020) analyzed different parameters of an OWC divided into three chambers by two inner walls and they obtained the following results [26]:

- Effect of relative chamber length: It is defined as the length of the chamber divided by the wavelength. As this ratio increases, the efficiency also does, until it reaches a maximum, and then decreases. It becomes more important with the period and height of the wave.
- Effect of submergence depth ratio: It is defined as the submersion depth of the OWC divided by the total depth. Its effect depends on the height and period of the wave.
- Effect of opening ratio: The results obtained were similar to the results obtained by Tsai et al. in 2018.

Medina et al. (2020) focused on the effects of the front wall thickness and the bottom profile of the OWC on its efficiency. They concluded that [57]:

- Effect of nonlinearities: Nonlinear effects can significantly change the power absorption of an OWC.
- Effect of wall thickness: The efficient bandwidth of the OWC reduces as the thickness of the front wall increases.
- Effect of the bottom of the chamber: The efficiency band slightly shifts to longer periods as the bottom of the chamber becomes steeper. Therefore, for small periods it is better to use a flat bottom. The period in which resonance occurs is almost independent of the bottom geometrical configuration and it is mostly determined by the natural frequency of the water column.

Koley et al. (2020) [58] also studied the geometric optimization of OWCs and stated that for high values of the wave length, wider chambers should be used and that for short waves, narrower chambers are more efficient, which was in line with previous studies (Evans and Porter, 1995 [53] and Rezanejad et al., 2013 [59])

Chen et al. (2021) [60] analyzed the influence of the chamber width and the submergence depth of the OWC. They showed that the efficiency of the chamber follows a concave parabolic shape with the chamber width divided by the wave length, and pointed out that the front wall draught corresponds to a unique chamber width, so that the hydrodynamic efficiency reaches the maximum value under certain wave conditions.

Guo et al. (2021) [61] developed an analytical model to study the CWR of an OWC integrated in a breakwater and concluded that bandwidth becomes narrower as the width increases.

A table that sums up all the geometric properties and how they affect the performance of an OWC is presented below (Table 2), but it is important to say that these relationships do not have to be interpreted as fixed rules. To optimize the behavior of an OWC, each specific case should be studied in detail.

5.2. U-Shaped, L-Shaped, and V-Shaped OWCs

To improve the performance of OWCs, several design concepts diverging from the conventional OWC shape have been investigated. The most common designs proposed are the U-shaped and the L-shaped OWC [62].

Bocotti (2006) [63] was the first researcher that showed that U-shaped OWCs give better performance for both swells with large waves and small waves. This is because the natural period and the amplitude of the pressure fluctuations on the opening of a U-shaped OWC are greater than those in a conventional OWCs. Seven years later, Malara and Arena proved the same as Bocotti [64].

Vyzikas et al. (2017) [65] developed a physical model to study the performance of four three-chamber OWC models. They investigated the common U-shaped OWC, one version that resembled the U-shaped Bocotti OWC, the conventional OWC, and an adapted version of the conventional OWC. In line with previous studies, they saw that U-shaped models performed better than conventional ones, especially when they were close to the peak of their performance.

Fox et al. (2021) [62] developed a numerical model in which they studied the performance of a conventional, a U-shaped, and an L-shaped OWC and also concluded that the U-shaped gave better hydrodynamic performance.

López et al. (2021) [66] also studied the performance of different OWC configurations (conventional, stepped-bottom, U-shaped, and L-shaped) with a numerical model using the case study of Vigo port (Spain), but they found, in contrast to Fox et al., that the OWC that performed better was the L-shaped one.

5.2.1. U-Shaped OWCs

The U-shaped OWC is an OWC with an additional vertical duct at the wave-beaten side [63]. The shape modification of the U-duct induces an increase of the OWC's natural period when compared with the conventional shape, making the energy conversion more efficient for the most common wave periods [62] (Figure 15).

Ning et al. (2020) [67] developed a numerical model to study the geometrical optimization of a U-shaped OWC. They found that:

- The air pressure inside the chamber and, consequently, the hydrodynamic efficiency of the OWC increases with vertical duct height (d_1 in Figure 15). In addition, the effect of the vertical duct height on the maximum pressure increases with *kh* increasing in relatively low frequency region.
- The duct width (*b* in Figure 15) has also a great influence on the air pressure inside the chamber. A larger vertical duct width gives a larger hydrodynamic efficiency in a relatively low frequency domain. Moreover, the total hydrodynamic force exerted on the U-shaped water column almost increases linearly with the vertical duct width.
- The wall width influences the maximum air pressure and the hydrodynamic efficiency too. The higher the wall width, the higher the maximum air pressure and efficiency.

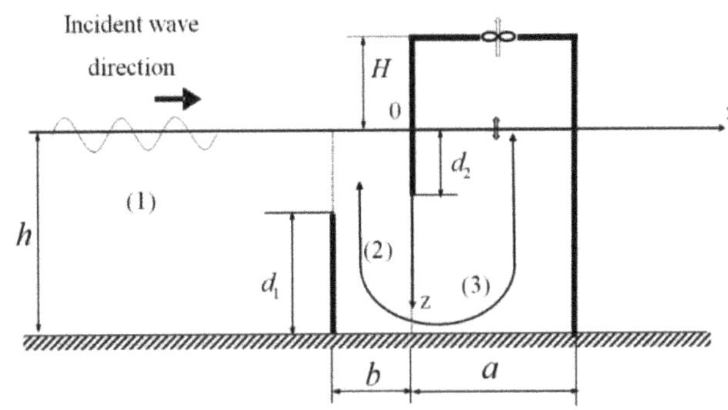

Figure 15. U-shape OWC sketch (George et al. 2021 [68]).

More recently, George et al. (2021) [68] used artificial neural networks to optimize the design of a U-shaped OWC. They developed a method using machine learning to obtain the optimal design.

5.2.2. L-Shaped OWCs

Rezanejad et al. (2019) [69] conducted an experiment in which they studied the hydrodynamic performance of an L-shaped OWC. They concluded that the amplification factor and the performance of the device are enhanced by reducing the immersion depth of the device.

López et al. (2021) [66] pointed out that:

- A wider vertical section of the chamber improves the performance of the L-shaped OWC, especially for periods different from the natural period of oscillation of the chamber.
- The greater the height of the horizontal duct of the chamber, the higher the captured energy.
- A shallower entrance enhances the efficiency of the OWC.

Samak et al. (2021) [70] studied how to improve the performance of an OWC with an L-shaped wall and concluded that the L-shaped front wall reduced the air pressure, but it enhanced the hydrodynamic efficiency of the device, especially for long waves.

5.2.3. V-shaped OWCs

The V-shape minimizes the vertical forces that act on the OWC. These devices have the highest efficiencies for the lowest wave states [71].

Kelly et al. (2013) [72] proposed a V-shape platform with 32 OWCs and highlighted the advantages of this new concept.

5.3. Use of Dual Chambers

The use of dual chambers in an OWC is very common in the design of OWCs. Rezanejad et al. (2015) [73], and later Elhanafi et al. (2018) [74] and Ning et al. (2019) [75,76], found that the capture width ratio of the OWC device was highly improved by introducing a dual chamber. Some of these studies are presented below.

Ning et al. have conducted several experiments to study the performance of dual chambers. In 2018 they developed different computational models [77] and in 2019 they studied the performance of these devices experimentally [75]. They obtained the following results:

- Efficiency: The results obtained in the models showed that the effective bandwidth of the dual chamber was about three times larger than in the single chamber. In addition, the peak efficiency was around 8% greater in the models (2018) and about 4% greater in the experiment setup (2019).
- Effect of submergence: A decrease in the outer chamber submergence causes an increase in both the maximum efficiency and the effective frequency bandwidth.
- Effect of chamber width: It increases the hydrodynamic performance of the device.
- Effect of the thickness of the shell: It was not very significant in the low-frequency domain, but in the high-frequency domain, thinner shells were better.
- Effect of the volume of the chamber: As it increases, the effective frequency bandwidth gets smaller, but the efficiency increases. An optimum chamber volume exists and it should be calculated for each case.

Wang et al. (2020) studied the wave loads on a land-based dual chamber OWC using a nonlinear numerical model and they reached the following conclusions [78]:

- Wave forces: They only studied the effect of horizontal forces because they are much bigger than vertical forces. The outer chamber is the chamber that suffers the highest wave loads, and these forces increase with frequency. However, forces in inner chamber increase with kh until they reach a peak, and then decrease.
- Effect of submergence: Horizontal forces in the outer chamber increase with its submergence. Increasing the curtain wall draft does not only reduce the power capture efficiency, as pointed out by Ning et al. [75,76], but also enlarges the forces on the OWC. Therefore, small curtain drafts are highly recommended.
- Effect of sub-chamber width ratio: Reducing the sub-chamber width ratio can help to reduce forces on the structure. Therefore, small sub-chamber width ratios are suggested.

5.4. Use of Arrays of OWCs

Another field of interest is how the use of multi-column and multi-OWC plants affects hydrodynamic performance, as well as what is their optimum configuration.

Hsieh et al. (2010) analyzed the use of multi-column devices and showed that they have better energy conversion effectiveness than isolated OWCs [79]. Tseng et al. (2000) reported that multi-resonant OWCs yield only 28.5% efficiency due to high energy losses [80]. Atan et al. (2019) studied the effect of arrays of OWCs on the nearshore wave climate and concluded that they do not significantly alter the nearshore climate [81], and finally, Jasron et al. (2020) studied three different arrangements of OWC devices to optimize their performance [82]: a single water column, a double water column in series arrangement perpendicular to the direction of wave propagation, and a double water column parallel to the direction of wave propagation. The results showed that the double OWC configuration operates more efficiently at deeper water depths than the single device and that the efficient frequency bandwidth is broader, especially for an arrangement in which the OWCs are in parallel, which showed a more stable behavior.

Another important fact is that when the relative depth is greater than 0.9 the effectiveness of all the devices significantly decreases due to the presence of nonlinearities.

5.5. Influence of the Integration of a Breakwater in the OWC System

One of the aspects of most interest when studying the design and construction of OWCs, and in general of WECs, is if integrating them into an existing breakwater is beneficial or not.

In the case of WECs, Zhao et al. (2019) studied the performance of a breakwater-integrated WEC system with oscillating buoys and compared it to that of an isolated WEC by studying the heave-response-amplitude operator values and wave forces on the device. They concluded that the integrated WEC system had larger HRAO (heave-response-amplitude operator) values and wave forces in the heave mode. This improves the hydrodynamic efficiency of the oscillating buoy, but should also be considered when designing the device. The presence of the breakwater amplifies the energy conversion performance of the WEC and the forces that act on it [83].

Several studies and experiments have been carried out to study the effect of breakwaters specifically on OWCs, with both single-chamber OWCs and dual chambers (Wan et al. (2020) [84]).

Falcão and Henriques (2016) pointed out that the integration of an OWC into a breakwater has many advantages, such as better access for construction and easier operation and maintenance [11].

Viviano et al. (2016) showed that integrating an OWC into a breakwater helps to reduce wave reflection and, consequently, to improve the hydrodynamic performance of the device [85].

In 2019, Reabroy et al. did a numerical model to investigate the hydrodynamic performance of an asymmetric floating device WEC. They studied two magnitudes: the RAO (response amplitude operator) and the CWR (capture width ratio). Both magnitudes were greater for the breakwater-integrated WEC system for all the wave periods studied, which means higher energy efficiency, but also higher wave loads [86]. These results are in line with Howe et al.'s (2017) [87] results.

Howe et al. (2020) [88] developed a proof-of-concept for a floating breakwater integrated with an OWC. They carried out an experiment in which they tested the performance of a bent-duct type OWC in four main aspects: device configuration, breakwater width, pneumatic damping, and structure motion influence. They concluded that device configuration, especially device spacing, really influences the performance of the OWC. They also pointed out that device integration is very beneficial and that pneumatic damping and motion influence are key aspects in the design of the OWC, while breakwater width does not play a key role in OWC performance.

Konispolaties et al. (2020) studied the configuration of different OWCs in front of a vertical breakwater. They studied three configurations: parallel to the wall, perpendicular, and rectangular arrangement. The results obtained showed that the installation of an OWC array in front of a vertical breakwater can be an effective way to improve its power absorption efficiency [89].

In 2021 they carried out another investigation studying the hydrodynamic efficiency of a WEC in front of an orthogonal breakwater, and as in the previous year, they concluded that the presence of a breakwater enhances the efficiency of WECs. However, this enhancement strongly depends on the distance between the device and the walls, the draught, the wave number, and the heading angle [90]. These findings are in line with Howe et al.'s (2020) results [88].

Wang and Zhang (2021) [91] developed a numerical in model in which they studied the performance of an OWC mounted over an immersed horizontal plate and compared it with that of an OWC breakwater integrated system. They concluded that the integration of a horizontal plate can significantly enhance the performance of an OWC, especially if the immersion depth of the plate is small, and that this system is more efficient than the common breakwater-integrated system. One year later, in 2022 [92], they proposed another novel system composed by an offshore heaving OWC device and a floating stationary breakwater with a certain length of gap between each other. They proved that the inclusion of a floating stationary breakwater is beneficial for enhancing the peak efficiency to a much higher extent, especially with a relatively larger immersed depth of the breakwater, and that the gap distance between the heaving OWC and breakwater has a dominant effect on the overall extraction efficiency.

5.6. Use of Multifunctional Platforms

Another field of interest in renewable energy resources is the design of multifunctional platforms, because their use could significantly reduce the cost and enhance the behavior of different structures. Sharing the infrastructure for different energy converters can reduce the investment and optimize their hydrodynamic efficiency.

Zhou et al. (2020) [50] studied the use of multifunctional platforms. They concluded that if an OWC is constructed next to an OWT, the forces and overturning moment acting on the monopile are lower, especially for high-frequency waves. This is due to the redistribution of the wave potential around the OWT caused by the OWC [50].

6. Influence of Air and Turbine Properties

As well as the main geometric parameters and the moisture of the air, air compressibility and turbine damping also have a great influence on the performance of an OWC.

6.1. Air Properties

When modeling the air inside the OWC, there are two main air properties that affect its performance: humidity and compressibility. Not considering these properties properly in the numerical models can lead to an overestimation of the performance of the OWC.

Referring to the humidity of the air, Medina-López et al. (2019) [93] studied it numerically and analytically and showed that the humid air model had an efficiency of 50–70% compared to the dry air model. One year later, Moñino et al. (2020) [94] showed that an increase in moisture of 45–70% in the air-water vapor mixture of the OWC induces reductions in the power input to the turbine by around 30%.

With respect to air compressibility, many numerical models assume the incompressibility of the air. However, this assumption is not real and, as happens with air moisture, it can lead to an overestimation of the performance of the OWC. Thakker et al. (2003) [95] were one of the first to study this effect in OWCs with impulse turbines and proved that not considering air compressibility can result in overestimation of the capture width ratio (CWR) of the device. Teixeira et al. (2013) [96] studied it with Wells turbines and reached the same conclusion. More recently, Gonçalves et al. (2020) [97] studied this compressibility effect in both Wells and impulse turbines, and concluded that not considering the compressibility of the air can result in overestimation of hydrodynamic efficiency of the OWC by 16–20% in both cases. In addition, this overestimation increases with the flow rate of the turbine.

6.2. Turbine Properties

One of the most important characteristics to consider when designing an OWC is the coupling between the OWC and the PTO [98]. To obtain good rates of efficiency it is necessary to use a turbine damping that fits the damping of the chamber. López et al. (2014) [99] proved that a correct selection of the damping supposes an increment of the capture factor of more than 10 points, independently of the incident wave conditions. Therefore, although the damping of the chamber varies with wave properties, studying the damping of the chamber is essential for an efficient performance of the OWC.

Among others, Pereiras et al. (2015) [98] developed a methodology to obtain the best impulse turbine for a given OWC. It consisted in obtaining the optimum damping of the chamber and once it was obtained, selecting the best turbine for this damping.

López et al. (2016) [100] also developed a methodology to obtain the optimum turbine for a given OWC and sea state. They also proved that to get the maximum pneumatic energy, turbine-induced damping should be close to the optimum chamber damping.

Simonetti et al. (2017) [55] studied the damping of the OWC and its relationship with the geometry of the chamber. They concluded the following remarkable points:

- The effect of the damping of the OWC CWR mainly depends on the chamber length and its relative water depth.
- Near the resonant condition the effect of the optimal damping of the OWC is insensitive to the chamber geometry.
- For other frequencies, higher values of optimal damping are found for decreasing chamber length.
- The optimal damping is minimal for the resonant condition and increases for the remaining values of relative water depth.
- Underdamping (damping coefficient lower than the optimal) reduces the device performance at a higher rate than overdamping.

Bouali et al. (2017) [101] also studied this problem and stated that for each size of the OWC chamber there is an optimal value of the PTO damping, and there is only one pair, chamber size and corresponding damping, that obtains the maximum energy conversion for given wave conditions.

7. Conclusions

After analyzing research on the performance of OWCs, some general patterns can be remarked:

- Wave characteristics
    - Horizontal wave forces increase with the wave height and decrease with the wave length.
    - The peak horizontal force is more than 2.5–3 times the peak vertical force.
    - The maximum absorption of wave energy occurs when the relative water depth is around 0.131.
    - An increase in wave steepness means an increase in the nonlinearities of the waves, which gives lower efficiencies of the OWC and larger forces.
    - OWCs should be located where the bottom friction does not affect their performance.
    - Neither the wave height nor the wave period may totally represent the nature of a wave; a design based on wave energy is recommended for OWCs.
    - A large amount of wave energy does not always imply a good conversion efficiency.
- Geometric parameters
    - Natural frequency of an OWC mainly depends on its chamber width and it is influenced by the front lip depth.
    - Wider chambers should be used for longer waves and narrower chambers for shorter waves.
    - For high values of the relative submergence of the lip wall, the CWR becomes narrower.
    - Hydrodynamic efficiency increases as the breadth increases for low frequencies, but it follows an opposite trend for high frequencies.
    - The CWR mainly depends on the bottom slope and it shifts to longer periods, as the bottom of the chamber becomes steeper. The CWR of the OWC reduces as the thickness of the front barrier wall is increased.
    - As the chamber length–water depth ratio decreases, the period of maximum hydrodynamic efficiency becomes shorter.
    - Circular bottom profile of the OWC is more efficient in terms of hydrodynamic performance.
- U-shaped, L-shaped, and V-shaped OWCs
    - They usually perform better than conventional OWCs.

    - Most studies point out the U-shaped OWC is the most efficient model. Its efficiency increases with the duct height and width and the wall width.
- Dual chambers
    - They usually give higher peak efficiencies and a much larger CWR than single chambers.
    - The outer shell draft should be smaller than the inner one.
    - Thinner shells are recommended because they have a better performance in the high-frequency region. However, the design of a dual chamber needs a balance between strength and hydrodynamic performance.
    - The seaside wall is the wall that suffers the highest forces. Therefore, it should be built with the best materials.
    - A smaller sub-chamber width ratio is suggested.
- Breakwater-integrated OWCs
    - These have better access for construction, and easier operation and maintenance.
    - It helps to reduce wave reflection.
    - The presence of the breakwater enhances the energy conversion performance of the OWC, but also amplifies the forces, which should be considered in the design process.
    - It increases the CWR of the OWC.
    - There are current research lines on new innovative systems.
- Multifunctional platforms
    - These can reduce the investment, forces that act on the OWC and optimize its hydrodynamic efficiency.
    - They are very convenient solutions.
    - Research is currently being carried out in this field.
- Air and turbine properties
    - Not considering air compressibility and humidity can lead to overestimating the hydrodynamic performance of the OWC.
    - There is only one pair, chamber size, and corresponding damping that allows obtaining the maximum energy conversion for given wave conditions.
    - The effect of the damping of the OWC CWR mainly depends on the chamber length and its relative water depth. At the resonant point it is insensitive to geometric characteristics and the value of the optimum damping is minimal.
    - Overdamping is better than underdamping.

Although these general patterns give an idea of the performance of an OWC, it is important to say that a universal formula that could be applied in every kind of OWC would be very difficult, if not impossible, to obtain. Similar types of OWC wave energy conversion systems may have some design guidelines in common, but there are some uncertainties, such as sea bottom conditions, slopes in the way of incident waves, and other environmental variables that can make them work completely differently.

Therefore, future research should still focus on the process of optimization of OWC devices, on the question of how all these parameters affect the hydrodynamic performance of OWCs, and on how to make OWCs efficient for a broad range of parameters. Special efforts must be made to improve the most promising and innovative systems, which are multipurpose platforms, breakwater-integrated OWCs, U- and L-shaped OWCs, and biradial turbines. Although in recent years many advances have been made, OWC technology is still immature and great developments are needed to make it profitable and cost efficient.

Author Contributions: All the authors contributed to choosing data, discussion, and methodology, figures, and references to provide an accurate paper. Conceptualization, N.P.J., V.N.V., M.D.E. and J.S.L.G.; funding acquisition, N.P.J.; investigation, N.P.J., V.N.V., M.D.E. and J.S.L.G.; methodology, N.P.J., V.N.V., M.D.E. and J.S.L.G.; writing—original draft, N.P.J.; writing—review and editing, N.P.J., V.N.V., M.D.E. and J.S.L.G. All authors have read and agreed to the published version of the manuscript.

Funding: This research received no external funding.

Conflicts of Interest: The authors declare no conflict of interest.

References

1. IPCC. *Las Evaluaciones del IPCC de 1990 y 1992*; Cambridge University Press: Cambridge, UK, 1992; p. 196.
2. Falnes, J. A review of wave-energy extraction. *Mar. Struct.* **2007**, *20*, 185–201. [CrossRef]
3. Czech, B.; Bauer, P. Wave Energy Converter Concepts, Design Challenge and Classification. *IEEE Ind. Electron. Mag.* **2012**, *6*, 4–16. [CrossRef]
4. Centre for Renewable Energy Sources (CRES). *Wave Energy Utilization in Europe*; Centre for Renewable Energy Sources Publication Service: Athens, Greece, 2002; p. 32, ISBN 9608690714.
5. Adebisi, N.; Balogun, A.-L.; Min, T.H.; Tella, A. Advances in estimating Sea Level Rise: A review of tide gauge, satellite altimetry and spatial data science approaches. *Ocean. Coast. Manag.* **2021**, *208*. [CrossRef]
6. Heath, T.V. A review of oscillating water columns. *Philos. Trans. R. Soc.* **2012**, *370*, 235–245. [CrossRef] [PubMed]
7. Bouali, B.; Larbi, S. Contribution to the geometry optimization of an oscillating water column wave energy converter. *Energy Procedia* **2013**, *36*, 565–573. [CrossRef]
8. Malmo, O.; Reitan, A. Wave power absorption by an oscillating water column in a channel. *J. Fluid Mech.* **1985**, *158*, 153–175. [CrossRef]
9. Raghunathan, S. The wells air turbine for wave energy conversion. *Prog. Aerosp. Sci.* **1995**, *31*, 335–386. [CrossRef]
10. Todalshaug, J.H. Hydrodynamics of WECs. In *Handbook of Ocean Wave Energy*; Pecher, A., Kofoed, J.P., Eds.; Springer International Publishing: Berlin, Germany, 2017; pp. 139–158.
11. Falcão, A.; Henriques, J. Oscillating-water-column wave energy converters and air turbines: A review. *Renew. Energy* **2016**, *85*, 1391–1424. [CrossRef]
12. Brito-Melo, A.; Gato, L.M.C.; Sarmento, A. Analysis of Wells turbine design parameters by numerical simulation of the OWC performance. *Ocean Eng.* **2002**, *29*, 1463–1477. [CrossRef]
13. TETHYS. Voith Hydro. Available online: https://tethys.pnnl.gov/organization/voith-hydro (accessed on 26 December 2021).
14. Torre-Enciso, Y.; Ortubia, I.; Aguileta, L.I.L.D.; Marqués, J. Mutriku wave power plant: From the thinking out to the reality. In Proceedings of the 8th European Wave and Tidal Energy Conference (EWTEC 2009), Uppsala, Sweden, 10 September 2009.
15. Têtu, A. Power Take-off systems for WECs. In *Handbook of Ocean Wave Energy*; Pecher, A., Kofoed, J.P., Eds.; Springer International Publishing: Berlin, Germany, 2017; pp. 203–220.
16. Carrelhas, A.A.D.; Gato, L.M.C.; Henriques, J.C.C.; Falcão, A.F.O.; Varandas, J. Test results of a 30 kW self-rectifying biradial air turbine-generator prototype. *Renew. Sustain. Energy Rev.* **2019**, *109*, 187–198. [CrossRef]
17. Henriques, J.C.C.; Portillo, J.C.C.; Gato, L.M.C.; Gomes, R.P.F.; Ferreira, D.N.; Falcão, A.F.O. Design of oscillating-water-column wave energy converters with an application to self-powered sensor buoys. *Energy* **2016**, *112*, 852–867. [CrossRef]
18. Falcão, A.F.O.; Henriques, J.C.C.; Gato, L.M.C.; Gomes, R.P.F. Air turbine choice and optimization for floating oscillating-water-column wave energy converter. *Ocean Eng.* **2014**, *75*, 148–156. [CrossRef]
19. Romolo, A.; Henriques, J.; Gato, L.M.C.; Malara, G.; Laface, V.; Gomes, R.; Portillo, J.C.C.; Falcão, A.; Arena, F. Power Take-Off Selection for a U-Shaped OWC Wave Energy Converter. *Energy* **2020**, *215*. [CrossRef]
20. Moretti, G.; Malara, G.; Scialò, A.; Daniele, L.; Romolo, A.; Vertechy, R.; Fontana, M.; Arena, F. Modelling and field testing of a breakwater-integrated U-OWC wave energy converter with dielectric elastomer generator. *Renew. Energy* **2020**, *146*, 628–642. [CrossRef]
21. Falcao, A. The shoreline OWC wave power plant at the Azores. In Proceedings of the 4th European Wave Energy Conference, Aalborg, Denmark, 4–6 December 2000; pp. 42–47.
22. TETHYS. Pico Power Plant. Available online: https://tethys.pnnl.gov/project-sites/pico-power-plant (accessed on 24 December 2021).
23. Heath, T.V.; Whittaker, T.J.T.; Boake, C.B. The design, construction and operation of the LIMPET wave energy converter (Islay, Scotland). *Renew. Energy* **2016**, *85*, 49–50. [CrossRef]
24. Sun, C.; Shang, J.; Luo, Z.; Lu, Z.; Wang, R. A Review of Wave Energy Extraction Technology. *IOP Conf. Ser. Mater. Sci. Eng.* **2018**, *394*, 042038. [CrossRef]
25. Curto, D.; Franzitta, V.; Guercio, A. Sea Wave Energy. A Review of the Current Technologies and Perspectives. *Energies* **2021**, *14*, 6604. [CrossRef]
26. Hsien, H.L.; Cheng-Han, C. Parametric Study for an Oscillating Water Column Wave Energy Conversion System Installed on a Breakwater. *Energies* **2020**, *13*, 1926. [CrossRef]

27. Arena, F.; Romolo, A.; Malara, G.; Fiamma, V.; Laface, V. The first full operative U-OWC plants in the port of Civitavecchia. In Proceedings of the 36th International Conference on Ocean, Offshore and Arctic Engineering, Trondheim, Norway, 25–30 June 2017; p. V010T009A022.
28. Bijun, W.; Meng, L.; Rukang, W.; Tianxiang, C.; Yunqiu, Z.; Yin, Y. BBDB wave energy conversion technology and perspective in China. *Ocean Eng.* **2018**, *69*, 281–291. [CrossRef]
29. Falcão, A. Wave energy utilization: A review of the technologies. *Renew. Sustain. Energy Rev.* **2010**, *14*, 899–918. [CrossRef]
30. JAMSTEC. JAMSTEC Gallery. Mighty Whale. Available online: http://www.jamstec.go.jp/gallery/j/research/system/images/system_002_l.jpg (accessed on 24 December 2021).
31. McCormick, M. E Analysis of a Wave Energy Conversion Buoy. *J. Hydronautics* **2012**, *8*, 77–82. [CrossRef]
32. Alcorn, R.; Bravette, A.; Healy, M.; Lewis, A. FP7 EU funded CORES wave energy project: A coordinators' perspective on the Galway Bay sea trials. *Underw. Technol.* **2014**, *32*, 51–59. [CrossRef]
33. TETHYS. Oceanlinx MK3. Available online: https://tethys.pnnl.gov/project-sites/oceanlinx-mk3 (accessed on 24 December 2021).
34. Zabala, I.; Henriques, J.; Gomez, A.; Falcão, A.; Amezaga, A.; Gomes, R.; Gato, L.M.C. Assessment of a spar buoy oscillating-water-column wave energy converter including a fully dynamic model. In Proceedings of the 12th European Wave and Tidal Energy Conference, Cork, Ireland, 27 August–1 September 2017.
35. Henriques, J. The (Integrated) Development of the Spar-Buoy OWC Wave Energy Converter at IST. Master in Renewable Energy in the Marine Environment. 2021. Available online: https://renovables-eurorregion.com/wp-content/uploads/MORE__The_integrated_development_spar_buoy_OWC_20210622.pdf (accessed on 30 December 2021).
36. IDOM. Prototipo WEC MARMOK A-5. Available online: https://www.idom.com/en/project/wave-energy-converter-prototype-marmok-a-5-2/ (accessed on 30 December 2021).
37. Martinelli, L.; Pezzutto, P.; Ruol, P. Experimentally Based Model to Size the Geometry of a New OWC Device, with Reference to the Mediterranean Sea Wave Environment. *Energies* **2013**, *6*, 4696–4720. [CrossRef]
38. Sarmiento, J.; Iturrioz, A.; Ayllón, V.; Guanche, R.; Losada, I.J. Experimental modelling of a multi-use floating platform for wave and wind energy harvesting. *Ocean Eng.* **2019**, *173*, 761–773. [CrossRef]
39. Perez-Collazo, C.; Pemberton, R.; Greaves, D.; Iglesias, G. Monopile-mounted wave energy converter for a hybrid wind-wave system. *Energy Convers. Manag.* **2019**, *199*, 111971. [CrossRef]
40. Principle Power. WindFloat. Available online: https://www.principlepower.com/windfloat (accessed on 30 December 2021).
41. Ruzzo, C.; Muggiasca, S.; Malara, G.; Taruffi, F.; Belloli, M.; Collu, M.; Li, L.; Brizzi, G.; Arena, F. Scaling strategies for multi-purpose floating structures physical modeling: State of art and new perspectives. *Appl. Ocean. Res.* **2021**, *108*, 102487. [CrossRef]
42. Pérez-Collazo, C.; Greaves, D.; Iglesias, G. Proof of concept of a novel hybrid wind-wave energy converter. In Proceedings of the 37th International Conference on Ocean, Offshore and Arctic Engineering, Madrid, Spain, 17–22 June 2018; p. V010T009A019.
43. Jasron, J.; Soeparmani, S.; Yuliati, L.; Darmadi, D. Experimental Study on the Effect of Ocean Wave Characteristics on Air Pressure in Oscillating Water Column Device. *IOP Conf. Ser. Mater. Sci. Eng.* **2019**, *494*. [CrossRef]
44. Ning, D.Z.; Wang, R.Q.; Zou, Q.P.; Teng, B. An experimental investigation of hydrodynamics of a fixed OWC wave energy converter. *Appl. Energy* **2016**, *168*, 636–648. [CrossRef]
45. Rezanejad, K.; Guedes Soares, C.; Lopez, I.; Carballo, R. Experimental and numerical investigation of the hydrodynamic performance of an oscillating water column wave energy converter. *Renew. Energy* **2017**, *106*, 1–16. [CrossRef]
46. Huang, Z.; Xu, C.; Huang, S. A CFD simulation of wave loads on a pile-type oscillating-water-column device. *J. Hydrodyn.* **2019**, *31*, 41–49. [CrossRef]
47. Sundar, V.; Sannasiraj, S.A.; Ashlin, J.; Jegatheeswaran, B. An experimental study of an oscillating water column with different bottom configuration. In Proceedings of the 3rd IAHR Europe Congress, Porto, Portugal, 14–16 April 2014.
48. Ashlin, J.; Sannasiraj, S.; Sundar, V. Wave forces on an Oscillating Water Column Device. *Procedia Eng.* **2015**, *116*, 1019–1026. [CrossRef]
49. Thiruvenkatasamy, K.; Neelamani, S. On the efficiency of wave energy caisson in array. *Appl. Ocean. Res.* **1997**, *19*, 61–72. [CrossRef]
50. Zhou, Y.; Ning, D.; Shi, W.; Johanning, L.; Liang, D. Hydrodynamic investigation on an OWC wave energy converter integrated into an offshore wind turbine monopile. *Coast. Eng.* **2020**, *162*, 103731. [CrossRef]
51. D'Aquino, C.; Cataldo, C.; Casagrande, L. Evaluation of the energy extraction of a small-scale wave energy converter. *Braz. J. Water Resour.* **2019**, *24*, e13. [CrossRef]
52. Zheng, W. *Experimental Research and Parameters Optimization of a Prototype OWC Wave Power Device*; American Society of Civil Engineers: Reston, VA, USA, 1989; pp. 43–50.
53. Evans, D.V.; Porter, R. Hydrodynamic characteristics of an oscillating water column device. *Appl. Ocean. Res.* **1995**, *17*, 155–164. [CrossRef]
54. Morris-Thomas, M.T.; Irvin, R.J.; Thiagarajan, K.P. An Investigation into the Hydrodynamic Efficiency of an Oscillating Water Column. *J. Offshore Mech. Arct. Eng.* **2007**, *129*, 273–278. [CrossRef]
55. Simonetti, I.; Cappietti, L.; Elsafti, H.; Oumeraci, H. Optimization of the geometry and the turbine induced damping for fixed detached and asymmetric OWC devices: A numerical study. *Energy* **2017**, *139*, 1197–1209. [CrossRef]

56. Tsai, C.; Chun-Han, K.; Ying-Chi, C. Investigation on Performance of a Modified Breakwater-Integrated OWC Wave Energy Converter. *Sustainability* **2018**, *10*, 643. [CrossRef]
57. Medina, A.; Blanco, J.M.; Silva, R.; Izquierdo, U. The Influence of the Chamber Configuration on the Hydrodynamic Efficiency of Oscillating Water Column Devices. *J. Mar. Sci. Eng.* **2020**, *8*, 751. [CrossRef]
58. Koley, S.; Trivedi, K. Mathematical modeling of oscillating water column wave energy converter devices over the undulated sea bed. *Eng. Anal. Bound. Elem.* **2020**, *117*, 26–40. [CrossRef]
59. Rezanejad, K.; Bhattacharjee, J.; Guedes Soares, C. Stepped sea bottom effects on the efficiency of nearshore oscillating water column device. *Ocean Eng.* **2013**, *70*, 25–38. [CrossRef]
60. Chen, J.; Wen, H.; Wang, Y.; Wang, G. A correlation study of optimal chamber width with the relative front wall draught of onshore OWC device. *Energy* **2021**, *225*, 120307. [CrossRef]
61. Guo, B.; Ning, D.; Wang, R.; Ding, B. Hydrodynamics of an oscillating water column WEC—Breakwater integrated system with a pitching front-wall. *Renew. Energy* **2021**, *176*, 67–80. [CrossRef]
62. Fox, B.N.; Gomes, R.P.F.; Gato, L.M.C. Analysis of oscillating-water-column wave energy converter configurations for integration into caisson breakwaters. *Appl. Energy* **2021**, *295*. [CrossRef]
63. Boccotti, P. Comparison between a U-OWC and a conventional OWC. *Ocean Eng.* **2007**, *34*, 799–805. [CrossRef]
64. Malara, G.; Arena, F. Analytical modelling of an U-Oscillating Water Column and performance in random waves. *Renew. Energy* **2013**, *60*, 116–126. [CrossRef]
65. Vyzikas, T.; Deshoulieres, S.; Barton, M.; Giroux, O.; Greaves, D.; Simmonds, D. Experimental investigation of different geometries of fixed oscillating water column devices. *Renew. Energy* **2017**, *104*, 248–258. [CrossRef]
66. Lopez, I.; Carballo, R.; Fouz, D.M.; Iglesias, G. Design Selection and Geometry in OWC Wave Energy Converters for Performance. *Energies* **2021**, *14*, 1707. [CrossRef]
67. Ning, D.; Guo, B.; Wang, R.; Vyzikas, T.; Greaves, D. Geometrical investigation of a U-shaped oscillating water column wave energy device. *Appl. Ocean. Res.* **2020**, *97*. [CrossRef]
68. George, A.; Cho, I.-H.; Kim, M.-H. Optimal Design of a U-Shaped Oscillating Water Column Device Using an Artificial Neural Network Model. *Processes* **2021**, *9*, 1338. [CrossRef]
69. Rezanejad, K.; Souto-Iglesias, A.; Guedes Soares, C. Experimental investigation on the hydrodynamic performance of an L-shaped duct oscillating water column wave energy converter. *Ocean Eng.* **2019**, *173*, 388–398. [CrossRef]
70. Samak, M.M.; Elgamal, H.; Nagib Elmekawy, A.M. The contribution of L-shaped front wall in the improvement of the oscillating water column wave energy converter performance. *Energy* **2021**, *226*, 120421. [CrossRef]
71. Kofoed, J.; Frigaard, P. *Hydraulic Evaluation of the LEANCON Wave Energy Converter*; DCE Technical Reports No. 45; Department of Civil Engineering, Aalborg University: Aalborg, Denmark, 2008.
72. Kelly, T.; Dooley, T.; Campbell, J.; Ringwood, J.V. Comparison of the Experimental and Numerical Results of Modelling a 32-Oscillating Water Column (OWC), V-Shaped Floating Wave Energy Converter. *Energies* **2013**, *6*, 4045–4077. [CrossRef]
73. Rezanejad, K.; Bhattacharjee, J.; Soares, C.G. Analytical and numerical study of dual chamber oscillating water columns on stepped bottom. *Renew. Energy* **2015**, *75*, 272–282. [CrossRef]
74. Elhanafi, A.; Macfarlane, G.; Ning, D. Hydrodynamic performance of single-chamber and dual-chamber offshore-stationary Oscillating Water Column devices using CFD. *Appl. Energy* **2018**, *228*, 82–96. [CrossRef]
75. Ning, D.; Zhou, Y.; Mayon, R.; Johanning, L. Experimental investigation on the hydrodynamic performance of a cylindrical dual-chamber Oscillating Water Column device. *Appl. Energy* **2020**, *260*. [CrossRef]
76. Zhao, X.L.; Ning, D.Z.; Zou, Q.P.; Qiao, D.S.; Cai, S.Q. Hybrid floating breakwater-WEC system: A review. *Ocean Eng.* **2019**, *186*, 106–126. [CrossRef]
77. Ning, D.; Zhou, Y.; Zhang, C. Hydrodynamic modeling of a novel dual-chamber OWC wave energy converter. *Ocean. Res.* **2018**, *78*, 180–191. [CrossRef]
78. Wang, R.; Ning, D.; Zou, Q. Wave loads on a land-based dual-chamber Oscillating Water Column wave energy device. *Coast. Eng.* **2020**, *160*. [CrossRef]
79. Hsieh, M.F.; Dorrel, D.G.; Lin, C.C. A Multichamber Oscillating Water Column using Cascaded Savonius Turbines. *IEEE Trans. Ind. Appl.* **2010**, *46*, 2372–2380. [CrossRef]
80. Tseng, R.S.; Wu, R.H.; Huang, C.C. Model study of a shoreline wave-power system. *Ocean Eng.* **2000**, *27*, 801–821. [CrossRef]
81. Atan, R.; Finnegan, W.; Nash, S.; Goggins, J. The effect of arrays of wave energy converters on the nearshore wave climate. *Ocean Eng.* **2019**, *172*, 373–384. [CrossRef]
82. Jasron, J.; Soeparmani, S.; Yuliati, L.; Darmadi, D. Comparison of the performance of oscillating water column devices based on arrangements of water columns. *J. Mech. Eng. Sci.* **2020**, *14*, 7082–7093. [CrossRef]
83. Zhao, X.L.; Ning, D.Z.; Liang, D.F. Experimental investigation on hydrodynamic performance of a breakwater integrated WEC system. *Ocean Eng.* **2019**, *171*, 25–31. [CrossRef]
84. Wan, C.; Yang, C.; Fang, Q.; You, Z.; Geng, J.; Wang, Y. Hydrodynamic Investigation of a Dual-Cylindrical OWC Wave Energy Converter Integrated into a Fixed Caisson Breakwater. *Energies* **2020**, *13*, 896. [CrossRef]
85. Viviano, A.; Naty, S.; Foti, E.; Bruce, T.; Allsop, W.; Vicinanza, D. Large-scale experiments on the behaviour of a generalised Oscillating Water Column under random waves. *Renew. Energy* **2016**, *29*, 875–887. [CrossRef]

86. Reabroy, R.; Zheng, X.; Zhang, L.; Zang, J.; Yuan, Z.; Liu, M.; Sun, K.; Tiaple, Y. Hydrodynamic response and power efficiency analysis of heaving wave energy converter integrated with breakwater. *Energy Convers. Manag.* **2019**, *195*, 1174–1186. [CrossRef]
87. Howe, D.; Nader, J.-R. OWC WEC integrated within a breakwater versus isolated: Experimental and numerical theoretical study. *Int. J. Mar. Energy* **2017**, *20*, 165–182. [CrossRef]
88. Howe, D.; Nader, J.-R.; Macfarlane, G. Experimental investigation of multiple Oscillating Water Column Wave Energy Converters integrated in a floating breakwater: Energy extraction performance. *Appl. Ocean. Res.* **2020**, *97*, 102086. [CrossRef]
89. Konispoliatis, D. Performance of an Array of Oscillating Water Column Devices in Front of a Fixed Vertical Breakwater. *J. Mar. Sci. Eng.* **2020**, *8*, 912. [CrossRef]
90. Konispoliatis, D.; Mavrakos, S. Hydrodynamic Efficiency of a Wave Energy Converter in Front of an Orthogonal Breakwater. *J. Mar. Sci. Eng.* **2021**, *9*, 94. [CrossRef]
91. Wang, C.; Zhang, Y. Hydrodynamic performance of an offshore Oscillating Water Column device mounted over an immersed horizontal plate: A numerical study. *Energy* **2021**, *222*, 119964. [CrossRef]
92. Wang, C.; Ma, T.; Zhang, Y. Semi-analytical study on an integrated-system with separated heaving OWC and breakwater: Structure size optimization and gap resonance utilization. *Ocean Eng.* **2022**, *245*, 110319. [CrossRef]
93. Medina-Lopez, E.; Borthwick, A.G.L.; Moñino, A. Analytical and numerical simulations of an oscillating water column with humidity in the air chamber. *J. Clean. Prod.* **2019**, *238*, 117898. [CrossRef]
94. Moñino, A.; Quirós, C.; Mengíbar, F.; Medina-Lopez, E.; Clavero, M. Thermodynamics of the OWC chamber: Experimental turbine performance under stationary flow. *Renew. Energy* **2020**, *155*, 317–329. [CrossRef]
95. Thakker, A.; Dhanasekaran, T.S.; Takao, M.; Setoguchi, T. Effects of compressibility on the performance of a wave-energy conversion device with an impulse turbine using a numerical simulation technique. *Int. J. Rotating Mach.* **2003**, *9*, 443–450. [CrossRef]
96. Teixeira, P.R.F.; Davyt, D.P.; Didier, E.; Ramalhais, R. Numerical simulation of an oscillating water column device using a code based on Navier-Stokes equations. *Energy* **2013**, *61*, 513–530. [CrossRef]
97. Goncalves, R.A.A.C.; Teixeira, P.R.F.; Didier, E.; Torres, F.R. Numerical analysis of the influence of air compressibility effects on an oscillating water column wave energy converter chamber. *Renew. Energy* **2020**, *153*, 1183–1193. [CrossRef]
98. Pereiras, B.; Lopez, I.; Castro, F.; Iglesias, G. Non-dimensional analysis for matching an impulse turbine to an OWC (oscillating water column) with an optimum energy transfer. *Energy* **2015**, *87*, 481–489. [CrossRef]
99. López, I.; Pereiras, B.; Castro, F.; Iglesias, G. Optimisation of turbine-induced damping for an OWC wave energy converter using a RANS–VOF numerical model. *Appl. Energy* **2014**, *127*, 105–114. [CrossRef]
100. López, I.; Pereiras, B.; Castro, F.; Iglesias, G. Holistic performance analysis and turbine-induced damping for an OWC wave energy converter. *Renew. Energy* **2016**, *85*, 1155–1163. [CrossRef]
101. Bouali, B.; Larbi, S. Sequential optimization and performance prediction of an oscillating water column wave energy converter. *Ocean Eng.* **2017**, *131*, 162–173. [CrossRef]

Journal of
Marine Science and Engineering

Article

Corrosion Prediction Models in the Reinforcement of Concrete Structures of Offshore Wind Farms

Kerman Vázquez *, Raúl Rubén Rodríguez * and M. Dolores Esteban *

Department of Civil Engineering, Universidad Europea de Madrid, 28670 Madrid, Spain
* Correspondence: kermanvf1994@gmail.com (K.V.); raulruben.rodriguez@universidadeuropea.es (R.R.R.); mariadolores.esteban@universidadeuropea.es (M.D.E.)

Abstract: The growth of offshore wind farms (OWF's) is expected to be significant. Reducing operation and maintenance (O&M) costs will be important to ensure its development. The foundation is the most important structural element, with concrete as its main constituent. With concrete structures, particular attention must be paid to corrosion of embedded steel especially in marine environments, as poor maintenance management can have significant economic and structural safety consequences. This article presents a systematic analysis of prevalent corrosion prediction models and the subsequent development of a tool for estimating the diameter loss in the reinforcement of concrete structures in OWFs. For validation, the tool methodology is applied to 32 real cases to evaluate the difference between the calculated and the real diameter loss. The results indicate that the combination between the chloride diffusion model of the Spanish code on structural concrete (EHE-08) and the corrosion rate model of Li (2004) guarantees favourable diameter loss prediction results. The ability to rapidly calculate the diameter loss of reinforcement in concrete structural elements as a function of time, provides OWF operators with a valuable tool for the planning of maintenance strategies and cost optimisation.

Keywords: offshore wind; maintenance; foundation; concrete; corrosion; models

Citation: Vázquez, K.; Rodríguez, R.R.; Esteban, M.D. Corrosion Prediction Models in the Reinforcement of Concrete Structures of Offshore Wind Farms. *J. Mar. Sci. Eng.* **2022**, *10*, 185. https://doi.org/10.3390/jmse10020185

Academic Editor: Carlos Guedes Soares

Received: 22 December 2021
Accepted: 26 January 2022
Published: 29 January 2022

1. Introduction

Currently, offshore wind installations worldwide have a total installed capacity of 35.3 GW. Europe (UK+27) continues to lead the ranking with 26 GW of installed capacity, which accounts for 70% of global offshore wind installations [1]. China, with 3 GW of installed capacity, has lead offshore installations during 2020, despite the COVID-19 situation [2]. Figure 1 shows the distribution of offshore installations by country in 2020 according to statistics published in [3]. Given the low individual development in some countries, a group called rest of the world (ROW) has been created, which accumulates about 1% of the world's installed offshore wind capacity. This group includes countries such as Spain and Portugal, which only have a few megawatts of installed capacity, with 5 and 25 MW respectively.

To meet the requirements of a climate-neutral energy sector by 2050 and achieve the partial target recently established by the European Union (EU) [4] of reducing greenhouse gas emissions by 55% by 2030, the rate of installations in the wind industry will have to increase significantly in the coming decades. The long-term strategy presented by the European Union in 2018 highlighted the need for 88 GW of installed offshore wind capacity by 2030, and 450 GW (under the maximum scenario) by 2050 in order to reach these targets [5,6]. With the new objectives mentioned above, it necessary to review the 2018 targets and explore options for accelerating the rate of expansion of offshore wind energy. Different references highlight the importance of offshore wind as the alternative technology to achieve decarbonization, as well as the need to industrialize and develop floating offshore wind to drive the energy transition [7]. Exponential growth is forecast for both technologies in the coming decades, with offshore wind expected to grow to 270 GW worldwide by 2030 and around 16.5 GW to be built globally in the next 10 years.

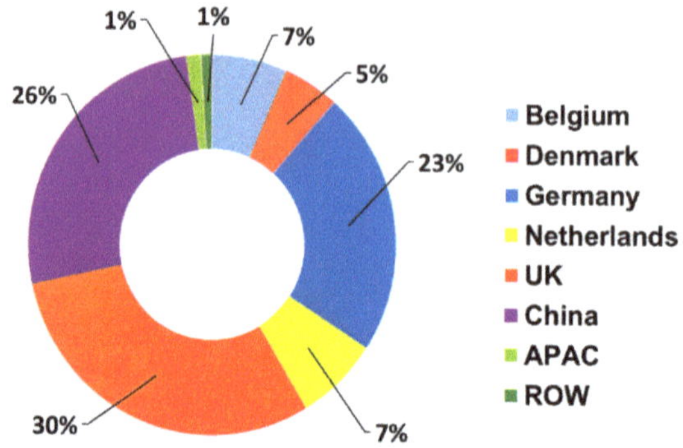

Figure 1. Distribution of offshore installations by country in 2020 (Where ROW means rest of the world and it is formed by the following countries: Finland, France, Ireland, Norway, Sweden, Portugal, and Spain).

Achieving this growth entails a number of challenges. In the case of offshore wind, the main obstacle is its high generation cost [8]. One of the elements with the greatest impact on the levelized cost of energy (LCOE) is the wind turbine foundation. The cost of production and installation of the foundation accounts for 25% of the capital expenditure (CAPEX), being together with the turbine one of the most expensive elements among the non-recurring costs of an offshore wind farm (OWF) [9]. Although data on the operation and maintenance (O&M) cost of offshore wind are limited, most references agree that they range between 20 to 30% of the LCOE [10]. No solid references have been found on the maintenance cost of offshore foundations, although some studies estimate that 38% of the operational expenditure (OPEX) is spent on various maintenance tasks, including visual inspections and other activities [11,12].

Currently, in the offshore wind industry there are various types of foundations [13–15]. However, the materials used in these support structures are mainly concrete and steel, sometimes even a combination of both. Although most offshore wind structures and components are made of steel, concrete also has an important role in the offshore wind industry [16]. In Europe, several wind farms have concrete foundations, mostly formed with gravity-based structures [17–19]. Recently, Royal Boskalis Westminster N.V. (Boskalis) in consortium with Bouygues Travaux Publics (Bouygues) and Saipem has been awarded the design, construction, and installation scope for 71 concrete gravity-based structures (GBS) as foundation for the Fécamp OWF in Normandy, France. The development of offshore wind energy in the coming years will require sites located at different distances from shore and water depths to be able to host all offshore wind projects. For this reason, it is expected that opportunities for foundation types such as GBS and others will play a greater role [20]. According to the interactive data tool developed by [21], it is estimated that GBS will remain the second most widely used foundation type between 2026–2030 after steel monopiles, and the importance of concrete in the future of offshore wind support structures is highlighted in [22].

Concrete is also used as the constitutive material of other elements of an offshore foundation, such as the working platform of the transition piece (TP). This type of TP has been used in the foundations of Fryslan wind farm (WF), currently under construction, or for the working platform of the TP in West of Duddon Sands WF currently in operation. Going one step further, it is a possibility that, during future development, concrete could be also used in offshore wind turbine towers [23].

The future trend will be towards larger wind turbines and deeper waters, where floating foundations are the best alternative. In the field of floating wind turbines, there are numerous solutions and proposals under development whose main material is concrete, for example the ones mentioned in [24]. Furthermore, now is the time for developed and commercialized innovative floating foundation solutions and other technologies in order to achieve the targets globally set for offshore wind industry [25].

From the above, it can be concluded that concrete is a material that is widely used in the different elements of an OWF, and it seems that it will continue to be useful in the future of the offshore wind industry.

Like any other structure, concrete foundations and elements require periodic maintenance activities to ensure the durability of their materials during their useful life [26]. Particular attention must be paid to this aspect in the case of concrete structures in marine environments, as they face one of their biggest durability problems: corrosion of embedded steel. Corrosion of steel embedded in concrete is an electrochemical process where a flow of electrons from the anode to the cathode occurs through a conductive element (the steel reinforcement) caused by a potential differential [27]. For cathodic reduction to take place, the presence of oxygen and suitable humidity conditions are essential. Fortunately, concrete has a high alkalinity as a result of the products generated during the hydration of the cement [28]. Under these conditions, a passive layer is created around the reinforcement; a microscopic film (free of porosity) of iron oxide and hydroxide residues that protects the steel from corrosion attack [29].

The most common cause of corrosion in steel embedded in concrete located in marine environments is the presence of chloride ions. Chloride ions penetrate the concrete, causing localized breaks in the protective layer. The formation of small anodic zones creates a potential differential, which in the presence of oxygen and humidity conditions initiates the corrosion process [30]. Ultimately, the corrosion process results in a loss of material of the steel bar progressive over time, which causes a loss of strength capacity in the structure. It is therefore crucial to know the evolution of the corrosion of steel embedded in concrete in offshore structures, as this will make it possible to provide the owner of an OWF with a useful data tool for decision-making, managing available resources, and being able to develop different maintenance strategies in order to reduce O&M costs.

In situ tests are now available to determine the corrosion rate in reinforced concrete structures [31], as well as a number of advanced techniques that provide insight into what is happening in embedded reinforcement [32]. These tests and techniques usually require expensive technological equipment or even the need to extract samples of the reinforcement in order to determine parameters such as the corrosion rate in this case, from which the evolution of the diameter loss in the reinforcement can be calculated. The complexity of these activities in the marine environment, the distance from the coast, and the rental cost of equipment for carrying out any type of repair activity on a corrosion-affected element all significantly increase the overall cost of maintenance if they are not properly managed [33,34].

For this purpose, the alternative of corrosion rate calculation models is presented [35], which consists of equations that allow the calculation of the value of the corrosion rate through some parameters of the environmental conditions of the structure and its materials. The characteristics and equations of these models will be developed in more detail in Section 2 of this article. Unlike in situ tests and previous techniques, these models do not require advanced technological equipment, and the work to be carried out to obtain some parameters is simpler and less costly.

The use of mathematical models for the calculation of corrosion rates is not widespread and often generates a feeling of mistrust among technicians specializing in the durability of concrete structures, which often leads to destructive testing and more complex and costly techniques on the structures, and even sometimes not to act on the problem until it reaches a magnitude that puts the safety of the structure at real risk.

This article presents the result of research work where the objective is to develop a tool for calculating the loss of diameter of reinforcement embedded in concrete in foundations or structures of OWF. To this end, existing prediction models for characterizing the useful life (initiation and propagation period) of structures affected by corrosion, which are simple to apply and do not require major actions or extensive testing campaigns on the structure under study, will be applied to a total of 32 real case studies. The results and conclusions will determine the method that best fits reality, whether the use of predictive models is valid or not and reflects reality. It will also allow to know the loss of diameter in a reinforced concrete structure throughout its service life, which is a key aspect for optimizing and programming complex O&M strategies, as well as making better use of the financial and material resources available at this stage in offshore wind installations.

2. Materials and Methods

2.1. General Methodology

This section describes the methodology used to achieve the objectives of this work (Figure 2). As mentioned above, the development and validation of a tool for calculating the loss of reinforcement diameter in concrete structures, is based on the results and conclusions obtained after applying this work methodology to 32 real case studies, thus the first step was to select the case studies, analyse them, and collect the following data for each of them: general data of the structure, real loss of diameter, parameters that depend on the type of concrete, and parameters that depend on the exposure environment of the structure. All these data will be needed in the following steps.

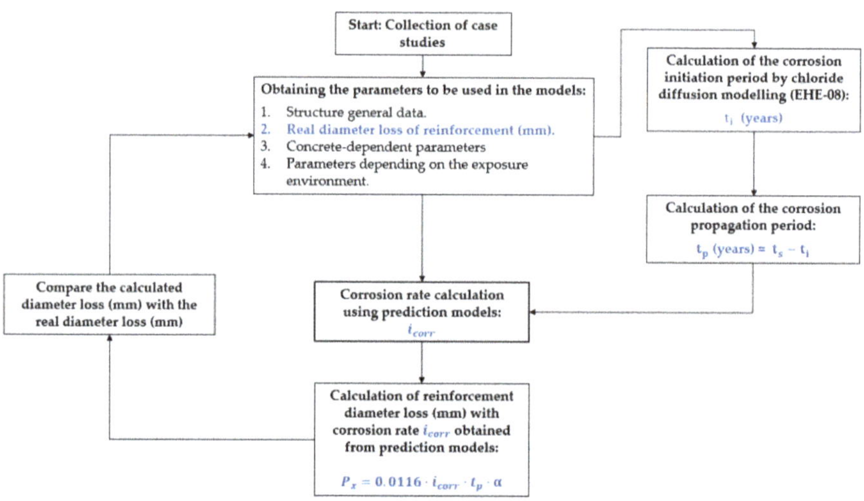

Figure 2. General methodology flowchart to calculate the diameter loss with the proposed corrosion prediction models and validation using real study cases.

The next step is to characterize the corrosion phases. Corrosion of steel reinforcement embedded in concrete consists of two phases: initiation and propagation (Figure 3) [36]. The initiation period is the necessary time for the chlorides, or other aggressive substances, to reach the critical concentration at the reinforcement position and for corrosion to start. Once this threshold is reached, corrosion will start, and the period of the propagation phase will directly depend on the corrosion rate. The sum of both periods estimates the service life of the structure.

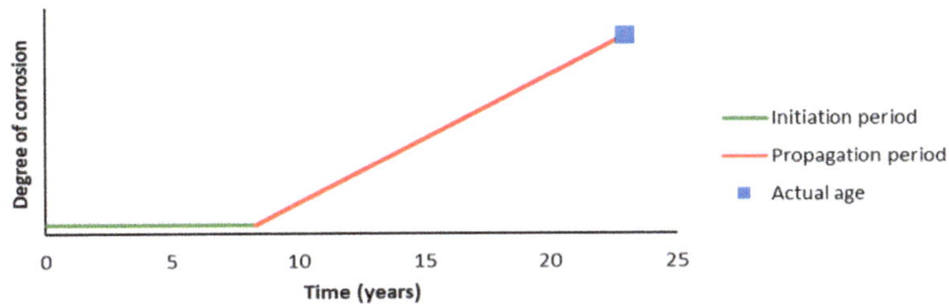

Figure 3. Qualitative representation of the service life of a structure affected by corrosion (degree of corrosion usually measured in loss of reinforcement section).

Based on the above figure, the first step is to calculate the initiation period (t_i). The initiation period is calculated in years, and the simplest and most widely used models are the Fickian diffusion models. Although there are currently several chloride diffusions models as listed in the reference [37], in this case the one proposed by the Spanish code on structural concrete (EHE-08) [38] has been selected. The parameters and equation of this diffusion model are very similar to the one proposed by the International Federation of Structural Concrete (FIB), which is more extended at an international level [39]. The calculation of the initiation period (t_i) in the real cases of study will be calculated using Equation (3), for which it will be necessary to know the concrete cover and the chloride concentrations at different depths in each structure.

Once the initiation period has been calculated, the propagation period (t_p) is calculated. From the graph in Figure 3, the propagation period can be obtained as the difference between the actual age of the structure (t_s) and the corrosion initiation period previously calculated (see Equation (1)). The current years of the structure in service correspond to the moment when a durability study is carried out from which the parameters indicated in the previous point of the flow diagram are extracted. The current age of the structure is obtained from the documentation of each case studied.

$$t_p = t_s - t_i \quad (1)$$

The next step is to calculate the corrosion rate i_{corr}, which is defined as the rate at which electrons move away from iron in anodic reactions ($\mu A/cm^2$) [40] and will therefore determine the rate at which reinforcement will lose diameter over time. The corrosion rate will not remain constant during the propagation period. This is because there are many factors that vary with time and the characteristics of the structure that cause the corrosion rate to be non-constant, especially environmental factors. The main factors are: chloride concentration, humidity, and temperature [41]. It is usually not a simple task to obtain the corrosion rate as on-site measurements are required. This type of work is even more complex on foundations or offshore concrete structures due to the constraints of the marine environment. Fortunately, empirical models have been developed which, using a series of parameters and equations, allow corrosion rates to be calculated more simply. In this work, six existing models (developed in detail in Section 2.2.2) have been analyzed and applied to calculate the corrosion rate using the parameters obtained from each case study (first step in the methodology) and the previously calculated propagation time.

Once the propagation period and the corrosion rate have been calculated for each of the study cases, the diameter loss is calculated using Equation (2) [28].

$$P_x = 0.0116 \cdot i_{corr} \cdot t_p \cdot \alpha \quad (2)$$

where P_x is the diameter loss (mm), 0.0116 a conversion factor for steel from $\mu A/cm^2$ to mm/year, i_{corr} is the corrosion rate ($\mu A/cm^2$), t_p is the propagation period (years) and α is the factor normally adopted for corrosion caused by chloride attack, in this case a value of 8 is taken as it is the commonly used value [42]. This pitting factor is used to convert a localized section loss due to chloride attack to a generalized section loss of the reinforcement diameter.

Finally, the results of diameter loss in the reinforcement calculated with the corrosion prediction models are compared with the actual diameter loss in the 32 cases studied. The diameter loss data is real and has been obtained in most cases through on-site measurements, which are recorded in the documentation of each of the case studies.

2.2. Systematic Analysis of Corrosion Prediction Models

This section presents the systematic analysis that has been carried out on the existing prediction models for estimating the corrosion initiation period and the corrosion rate in reinforced concrete structures.

2.2.1. Chloride Diffusion Model

This section presents the equations used for the calculation of the corrosion initiation period using the chloride diffusion model.

In this case, the model proposed by the Spanish code on structural concrete (EHE-08) has been used for the calculation of the corrosion initiation period. The equation of the EHE-08 model is as follows:

$$t_i = \left(\frac{d}{K_{cl}}\right)^2 \tag{3}$$

where t_i is the initiation period in years, d is the concrete cover (mm), and K_{cl} is the chloride penetration coefficient, which is calculated by Equation (4):

$$K_{cl} = \alpha \cdot \sqrt{(12 \cdot D(t)} \cdot \left(1 - \left(\sqrt{\frac{C_{th} - C_b}{C_s - C_b}}\right)\right) \tag{4}$$

where α is a unit conversation factor taken as 56157, C_{th} is the critical chloride concentration (% weight of cement), C_{th} critical chloride concentration (% weight of cement), C_s surface chloride concentration (% weight of cement), C_b chloride content contributed by the raw materials (aggregates, cement, water, etc.) during manufacture (% weight of cement), and finally *D(t)* is the effective chloride diffusion coefficient for age *t* calculated by Equation (5) and expressed in (cm^2/s):

$$D_{(t)} = D(t_0) \cdot \left(\frac{t_0}{t}\right)^n \tag{5}$$

where $D(t_0)$ chloride diffusion coefficient at age t whose values have been obtained from the recommendation of the structural code for a reference age t_0 of 28 days ($t_0 = 0.0767$ years, and n is the age factor whose value in the absence of tests is assumed to be 0.5.

2.2.2. Corrosion Rate Calculation Models

This section presents the equations used for the calculation of the corrosion rate. Existing corrosion rate calculation models have been used to calculate this parameter. After an exhaustive study on the state of the art, this research considers six different prediction models for the calculation of the corrosion rate, listed and developed by reference [35]: (1) Liu and Weyers (1998); (2) Vu and Stewart (2000); (3) Li (2004 a); (4) Li (2004 b); (5) Kong et al. (2006); (6) New Empirical Model (2019).

Liu and Weyers (1998)

This model was carried out in 1998 by creating a database (2927 results) obtained from seven-series specimens exposed to high chloride contamination over a time period of 5 years. The model proposes a non-linear regression which varies with the following parameters: chloride content, ambient temperature, ohmic strength of concrete, and the duration of the active phase of corrosion (propagation period).

$$i_{corr(t)} = 0.926 \cdot exp \cdot [7.89 + 0.7771 \cdot ln \cdot (1.69 \cdot C_t) - \frac{3006}{T} - 0.000116 \cdot R_c \quad (6)$$

where $i_{corr(t)}$ is the corrosion rate ($\mu A/cm^2$) at time *t*; C_t is the chloride content at the reinforcement position (kg/m^3). T is the temperature in degrees kelvin at the position of the reinforcement; R_c is the ohmic resistance of concrete (ohms); t_p is the corrosion propagation time (years).

Regarding the ohmic resistance of concrete, Liu (1996) [43] established a relationship between the ohmic strength of concrete and its chloride content, which has been used to calculate the ohmic resistance of concrete. The equation is as follows:

$$R_c = exp \cdot [8.03 - 0.54 \cdot ln \cdot (1 + 1.69 \cdot C_t)] \quad (7)$$

where C_t is the chloride concentration at the reinforcement position (kg/m^3) as mentioned above.

Vu and Stewart (2000)

Vu and Stewart (2000) model is focused on structures located in environments with a relative humidity of 75% and an average annual temperature of approximately 20 °C. This model takes into account the water-to-cement ratio (w/c) and, unlike the previous model, includes among its variables the concrete cover of the reinforcement. The model is defined by the following equation:

$$i_{corr(t)} = 0.85 \cdot t_p^{-0.29} \cdot i_{corr_0} \quad (8)$$

where $i_{corr(t)}$ is the corrosion rate ($\mu A/cm^2$) at time *t*; t_p is the propagation period (years) and i_{corr_0} is the corrosion rate at the beginning of the corrosion propagation period ($\mu A/cm^2$). It is calculated as indicated in Equation (9), where w/c is the water-to-cement ratio and d_c is the concrete cover depth (mm).

$$i_{corr_0} = \frac{37.8 \cdot \left(1 - \frac{w}{c}\right)^{-1.64}}{d_c} \quad (9)$$

Bolomey formula shown in Equation (10) has been used to obtain the approximate w/c ratio, where f_{ck} is the concrete compressive strength (MPa).

$$\frac{w}{c} = \frac{27}{f_{ck} + 13.5} \quad (10)$$

It is a simple model for application in real cases. However, it is important to note that it does not take into account aspects related to the environment in which the structure is located, neither the relative humidity nor the ambient temperature.

Li (2004 a)

This is the first of the models proposed by Li (2004 a). It is a simple corrosion rate prediction model, easy to apply in real cases, in which the corrosion rate is a function of the duration time of the active phase of corrosion; this is the propagation time (t_p).

$$i_{corr} = 0.3683 \cdot \ln(t_p) + 1.1305 \quad (11)$$

As can be seen, this model does not take into account the parameters related to the exposure environment of the structure, nor the physico-chemical characteristics of the concrete, at least not explicitly. However, it does implicitly in this work, since the corrosion propagation period is obtained using the models proposed by the EHE-08, in which the characteristics of the concrete, such as concrete cover, the age of the structure, w/c ratio, chloride concentration, and environmental conditions among others are taken into account. This is mainly reflected in Equations (4) and (5).

Li (2004 b)

Following his first model, Li (2004 b) develops another specific model for structures exposed to environments with high chloride concentrations, considering the following parameters: ambient temperature, relative humidity, w/c ratio, concrete cover, and chloride concentration on the position of the reinforcement. Li (2004 b) proposes the following equation to calculate the corrosion rate:

$$i_{corr\ (t)} = 2.486 \cdot \left(\frac{RH}{45}\right)^{1.6072} \cdot \left(\frac{T}{10}\right)^{0.3879} \cdot \left(\frac{\frac{w}{c}}{0.35}\right)^{0.4447} \cdot \left(\frac{d_c}{10}\right)^{-0.2761} \cdot k_{cl}^{1.7376} \tag{12}$$

where $i_{corr\ (t)}$ is the corrosion rate ($\mu A/cm^2$) at time t, K_{cl} is the chloride concentration on the position of the reinforcement (% weight of the concrete, and limited to 0.14–0.43%), T is the temperature in degrees kelvin at the position of the reinforcement, RH is the relative humidity, d_c is the concrete cover (mm), and w/c is the water-to-cement ratio.

Kong et al. (2006)

This model is based on the study carried out by Liu and Weyers 1998. In this case, the model depends on the same parameters: chloride content, ambient temperature, and concrete resistivity.

$$ln\ i_{corr\ (t)} = 8.617 + 0.618 \cdot ln \cdot C_t - \frac{3034}{T} - 5 \cdot 10^{-3} \cdot \rho \tag{13}$$

where $i_{corr\ (t)}$ is the corrosion rate ($\mu A/cm^2$) at time t; C_t is the chloride concentration at the reinforcement position (kg/m^3). T is the Temperature (K) at the position of the reinforcement; ρ is the ohmic resistance of concrete (ohms), which is given by Equation (14) if the chloride concentration is less than 3.6 kg/m^3, and $\rho = 10$ if the chloride concentration is higher than 3.6 kg/m^3.

$$\rho = \left[27.5 \cdot \left(0.35 - \frac{w}{c}\right) + 11.1\right] \cdot (1.8 - C_t) + \left(1 - \frac{RH}{100}\right)^2 + 40 \tag{14}$$

New Empirical Model (Lu et al., 2019)

This is a recently developed corrosion rate prediction model (2019) and takes into account all factors and parameters involved in the corrosion processes of reinforcement embedded in concrete structural members. In this model, the corrosion rate is obtained by the following equation:

$$i_{corr} = \exp\left[A + 8.617 + 0.6181 \cdot \ln(C_t) - \frac{3034}{T \cdot f(RH)} - 5 \cdot 10^{-3} \cdot \rho\right] \cdot f(t) \tag{15}$$

where $A = -7.387$ is a constant adjustment factor, C_t chloride concentration at the reinforcement position (kg/m^3), T is the temperature (Kelvin) at the position of the reinforcement, $f\ (RH) = 2.5 + RH$ is a factor that is a function of humidity, RH is the relative humidity at the location of the structure, and $f\ (t)$ is calculates as a function of the corrosion propagation time t_p, defined by Equation (15a).

$$f(t) = \frac{1}{\sqrt[3]{1 + t_p}} \tag{15a}$$

The ohmic resistance of concrete (kΩ-cm) is calculated by the method proposed in [44], whose equation is as follows:

$$\rho = \left(2.22{\cdot}10^{-4}{\cdot}c^2 - 0.18848{\cdot}c + 48.3516\right){\cdot}K_{\frac{w}{b}}{\cdot}K_{cl}{\cdot}K_t{\cdot}K_{ma}{\cdot}K_{pe} \quad (16)$$

where c represents the gelled material in concrete (kg/m^3) and K_{wb}, K_{cl}, K_t, K_{ma}, K_{pe} are the correction factors of the water-to-cement ratio (w/b), chloride content, ambient temperature, mineral admixture, and external environment on the concrete resistivity, respectively, which are given by the following equations:

$$K_{w/b} = 20.895{\cdot}\left(\frac{w}{b}\right)^2 - 19.551{\cdot}\left(\frac{w}{b}\right) + 5.4774 \quad (16a)$$

$$K_{cl} = 607480{\cdot}C_t^2 - 2242.3{\cdot}C_t + 2.9926 \quad (16b)$$

$$K_T = -0.0133{\cdot}T + 4.8945 \quad (16c)$$

$$K_{ma} = 1 \quad (16d)$$

where w/b is the water-to-cement ratio, C_t is the chloride content at the reinforcement position (kg/m^3), and T is the temperature at the position of the reinforcement (Kelvin). For Portland cement concretes, the value used for K_{ma} shall be equal to 1. Finally, K_{pe} value shall be adopted depending on how the structure is affected by sea level. K_{pe} shall have a value equal to 1 for the atmospheric zones, a value of 0.88 for the splash zone, and 0.67 for the tidal zone [35].

The correction coefficients introduced by this new model: $K_{w/b}$, K_{cl}, K_t, K_{ma}, K_{pe} are quite easy to obtain, with the exception of the K_{ma} coefficient. Its value depends on the chloride concentration, the type of concrete and its additives, and the mineral mix of gelled material (kg/m^3), the last two being very complex to calculate in practice. Based on the above, it has been decided to use Equation (14) to obtain the ohmic resistance of concrete, as it is less complicated.

3. Case Study Description

The following paragraphs provide an overview of the selected case studies in which the methodology and the corrosion prediction models stated in the previous section will be applied. The parameters to be used in the calculation of the corrosion rate and the initiation period, and how they have been obtained from the case studies, are described.

3.1. Selection of Case Studies

A compilation of durability study reports, technical notes and structural repair projects has been carried out. All the documents analysed were provided by the Technical Institute of Materials and Constructions (INTEMAC). A total of 22 structures were studied from which 32 case studies were obtained. Some real images taken from these reports are shown in Figure 4. In most cases, the documentation from which the data were extracted consisted of technical notes, durability reports, and repair projects, among others. The fact that the number of case studies (32) is greater than the number of structures analysed (22) is because several case studies have been taken from a single structure, for example in the case of large structures such as viaducts or piers, several study areas with corrosion problems have been deemed appropriate to be part of the case studies.

It should be noted that the majority of the documentation corresponds to durability studies of reinforced concrete structures, as well as a short number of repair projects of prestressed concrete structures, in particular five case studies. The main descriptive data of the selected case studies are shown in Table 1. All the structures studied are privately owned and managed, thus for reasons of confidentiality no further details can be provided. A total of five different types of structure have been studied: one (1) concrete lighthouse, four (4) viaducts, one (1) bridge, two (2) piers, and one (1) quay.

Table 1. General data on the structures selected for the case study.

	Structure Typology	Location	Material	Construction Year [1]	Study Year [2]	Structural Element [3]
1	Concrete lighthouse	Canary Islands, Spain	Reinforced concrete	1976	2014	Main structure
2	Viaduct 1	Castilla-León, Spain	Prestressed concrete	1972	2007	Beam
3			Reinforced concrete			Pile 2
4			Reinforced concrete			Pile 3
5			Reinforced concrete			Pile 6
6	Bridge 1	Catalonia, Spain	Reinforced concrete	1974	2019	Beam
7	Viaduct 2	Castilla-León, Spain	Post-tensioned concrete	1972	2007	Beam
8			Reinforced concrete			Beam
9			Reinforced concrete			Pile 3
10			Reinforced concrete			Pile 6
11			Post-tensioned concrete			Beam
12			Reinforced concrete			Pile 5
13	Pier 1	Vizcaya, Spain	Reinforced concrete	1949	2012	Beam 15
14			Reinforced concrete			Pile 22
15			Reinforced concrete			Beam 24
16			Reinforced concrete			Pile 26
17			Reinforced concrete			Beam 28
18	Quay 1	Tarragona, Spain	Reinforced concrete	1998	2016	Beam C-1
19			Reinforced concrete			Beam C-2
20			Reinforced concrete			Beam C-3
21	Pier 2	Canary Islands, Spain	Reinforced concrete	1977	2014	Beam 1
22			Reinforced concrete			Beam 2
23			Reinforced concrete			Beam 3
24			Prestressed concrete			Beam 4
25	Viaduct 3	· Murcia, Spain	Reinforced concrete	1991	2014	Pile 9
26			Reinforced concrete			Pile 20
27			Reinforced concrete			Beam 6
28			Reinforced concrete			Beam 12

Table 1. *Cont.*

	Structure Typology	Location	Material	Construction Year [1]	Study Year [2]	Structural Element [3]
29	Viaduct 4	Ciudad Real, Spain	Reinforced concrete	1980	2016	Pile 8
30			Reinforced concrete			Pile 11
31			Reinforced concrete			Beam 7
32			Prestressed concrete			Beam 7

(1) Year in which the structure was first commissioned. (2) Year in which the durability study of the structure was carried out, and whose information has been collected and used for this study. (3) This is the structural element analyzed in the corresponding structure.

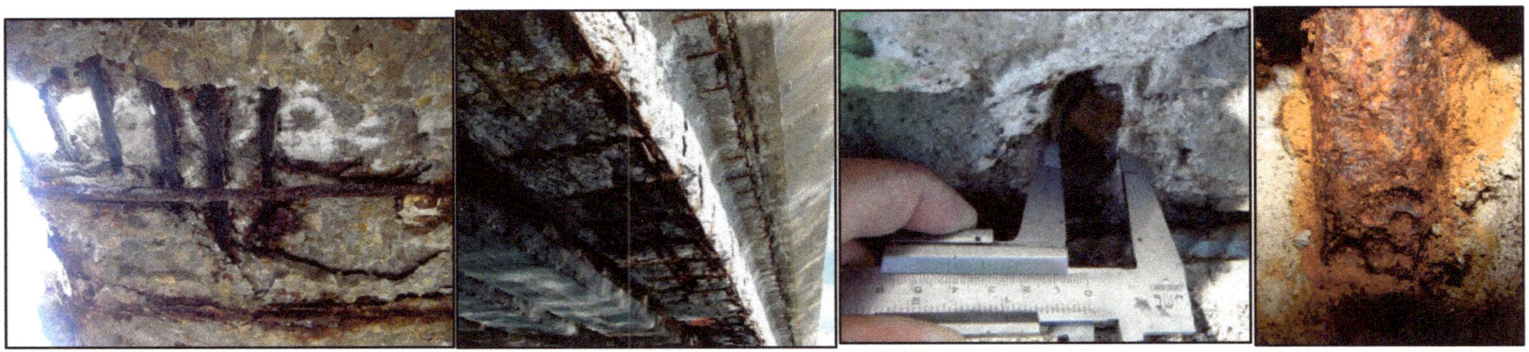

(a) (b) (c) (d)

Figure 4. Structures with severe reinforcement corrosion problems. Photos obtained from the available documentation of some of the case studies (Source: Technical Institute of Materials and Constructions (INTEMAC); different dates). (**a**) Viaduct 4–Beam 7, (**b**) Pier 2–Beam 1, (**c**) Viaduct 3–Pile 20, and (**d**) Viaduct 2.

As explained above, during the application of the prediction models to the case studies, some parameters are necessary to allow the application of the different formulations involved. In order to make it easier, to identify the parameters in the documentation of each case study, the parameters are classified into two main groups

- **Concrete-dependent:** chloride concentration at reinforcement position, w/c ratio, concrete cover, and compressive strength of concrete.
- **Exposure environment of the structure:** temperature and relative humidity.

In the following sections, a brief description of the parameters is given, explaining how they have been obtained in each case study.

3.2. Model Parameters

3.2.1. Concrete-Dependent Parameters

The chloride concentration in the position of the reinforcement is one of the key parameters and necessary in all proposed corrosion rate models. The most commonly used tool to obtain this data is to carry out chloride concentration profiles. For this purpose, test data indicating the chloride concentration at certain depths of concrete samples have been extracted from the documentation (see Table 2 of each case study. In all cases, the chloride profiles have been graphically represented (see Figure 5) in order to clearly define the chloride concentration at different depths. The chloride concentration at the position of the reinforcement, and the chloride concentration on the outer surface of the structure are required, as they are needed to calculate the corrosion rate and the initiation period through the proposed models.

Table 2. Compilation based on chloride concentration data in the position of pile reinforcement of Viaduct 3 (% weight of cement). (Data source: Technical Institute of Materials and Constructions (INTEMAC); May 2014).

Structural Elements	Sample Designation and Location	Chloride Content (%) at Each Depth (mm)		
		0–15	20–35	40–55
VIADUCT PILES	T-32 Pile 2	0.08	0.08	0.08
	T-33 Pile 2	0.49	0.20	0.14
	T-34 Pile 3	0.57	0.39	0.26
	T-35 Pile 3	2.20	1.93	0.38
	T-41 Pile 3	1.65	0.46	0.41
	T-23 Pile 2	0.62	0.42	0.23
	T-38 Pile 2	3.05	2.82	0.70
	T-39 Pile 1	1.99	1.70	0.57

Another important parameter is the w/c ratio. It is a very difficult parameter to obtain in existing structures if there is no available data and requires specific laboratory tests that are not usually carried out as part of the scope in durability studies. When concrete is exposed to chloride attack, it is necessary to guarantee a low permeability of the concrete in order to prevent the advance of chlorides [45]. A low *w/c* ratio guarantees higher permeability of the concrete, although it makes it less workable on site. Based on the above, we can say that porosity is directly related to the w/c ration of the concrete, as a low-porosity concrete will ensure a certain degree of low permeability. To obtain the *w/c* ratio of the case studies from available data, the following considerations have been made:

- Porosities between 12–14%, a *w/c* ratio of 0.45 to 0.50 is established.
- Porosities between 14–16%, a *w/c* ratio of 0.55 to 0.60 is established, although the latter value (0.60) can lead to even somewhat higher porosity values.

In addition to the above, the minimum w/c ratio requirements demanded by the regulations in force closest to the construction of the structures studied have been consulted in references [46,47].

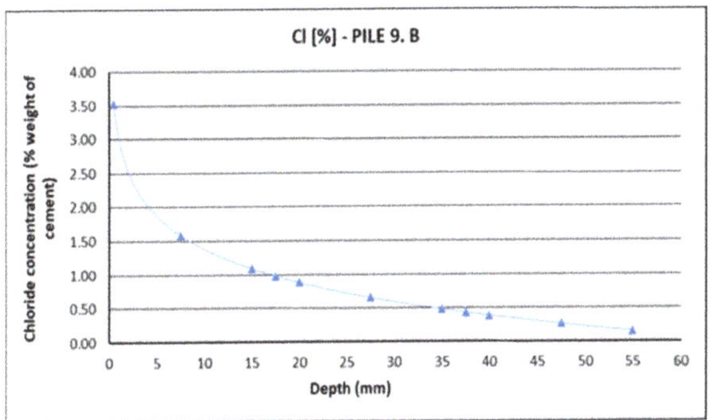

Figure 5. Chloride profile showing the chloride concentration as a function of depth in the test sample.

Another key parameter is the concrete cover, measured in millimetres (mm). The thickness of the concrete cover is the protection of the reinforcement against the penetration of aggressive agents such as chlorides and others and is a decisive parameter which directly conditions the corrosion initiation period. Typically, concrete cover depth is obtained by probing the study area. In this case, the concrete cover has been obtained from the documentation of each case study.

Finally, for the model proposed by Vu and Stewart (2000), the concrete compressive strength (fck) data is required. This parameter is obtained from the results of simple compression tests on concrete specimens. It is very common to carry out this type of tests as part of the materials characterisation campaign in durability studies of structures, thus it was easy to obtain this parameter from the documentation of each case study.

3.2.2. Parameters Depending on the Exposure Environment

The two parameters that depend on the exposure environment in the location of structures are temperature and the relative humidity

Firstly, two important aspects should be highlighted with regard to the parameter of temperature. On the one hand, it is a complex task to measure the temperature at the exact position of the reinforcement as required by some corrosion rate models, thus the value of the average annual temperature where the structure is located has been taken, since the thermal variations in the depth of the concrete up to the position of the reinforcement are minimal according to [42]. As the structures of the 32 cases are located on Spanish territory, data from the State Meteorological Agency [48] have been used to obtain the mean temperature data. As in the corrosion rate models, the unit of temperature is in degrees kelvin (K), it has been necessary to convert the unit from degrees Celsius (°C) to degrees kelvin (K), as the data from the meteorological agency are in degrees Celsius.

The relative humidity parameter is required in several of the proposed models. Average annual relative humidity values have been used for the area around the structure under study. In the same way, these data have been provided by the State Meteorological Agency depending on the location of the structure.

4. Results

Formula application and validity are here presented together with the results obtained for the different case studies, which are shown below according to the methodology and different equations previously exposed.

The section is divided into two parts, where the results obtained from the propagation time in each case study are shown first. The second part of the results shows the calculated diameter loss in each case study with the corrosion rate of each model.

4.1. Propagation Period

The propagation period is an essential parameter to calculate the loss of reinforcement diameter (see Equation (2). In addition, it is a parameter required by the models of Vu and Stewart (2000) and Li (2004) to calculate the corrosion rate. This section explains how to calculate the propagation period with the help of an example.

To obtain the propagation period, the corrosion initiation period is first calculated using the chloride diffusion model proposed by the Spanish code on structural concrete (EHE-08) [38], as explained previously in the methodology. Table 3 shows the calculation of the initiation period for the case study of pier 9 of viaduct 3 using the indications of the structural code and Equations (3)–(5) and all the parameters necessary for its calculation.

Table 3. Calculation of the corrosion initiation period. Example: Pier 9, Viaduct 3.

Calculation of the Initiation Period (Years)		
Parameter	**Source**	**Data**
Concrete Cover	Structure documentation	40 mm
Service life of the structure	Structure documentation	23 years
Cement type	Structure documentation	CEM I
Porosity	Structure documentation	13.4%
w/c ratio	Based on porosity	0.5
$D(t_0)$	Table A.9.4 (EHE-08)	1.58×10^{-11} m^2/s
t_0	Section 1.2.2.2 (EHE-08)	0.0767 years
n	Section 1.2.2.2 (EHE-08)	0.5
C_s (% weight cement)	Chloride profiles	3.53
C_{th} (% weight cement)	EHE-08 recommendation	0.6
C_b (% weight cement)	EHE-08 recommendation	0.4
Reinforcement diameter	Structure documentation	25 mm
K_{cl}	(Equation (4))	13.88
Initiation period	(Equation (3))	8.30 years

The difference between the age of the structure at the time the durability study was carried out and the initiation period is the propagation period. In this case, the initiation period is 8.30 years, and the age of the structure when the study was carried out is 23 years, therefore, the propagation period is 14.7 years. Table 4 shows the propagation time results obtained for each case study.

4.2. Calculation of Diameter Loss

4.2.1. Liu & Weyers (1998)

The 32 case studies (5 prestressed concrete and 27 reinforced concrete) have been considered, comparing the diameter loss values in the reinforcements obtained with the corrosion rate model and the real diameter loss obtained from the documentation of each case study. Table 5 shows the results obtained.

When analyzing the results, only 4 cases (21, 22, 23, and 24 of those described in Table 2) have diameter loss values similar to the real ones. In all these cases, the section loss obtained with the model (Px) has always been slightly higher than the real diameter loss (Pr), guaranteeing an average safety factor of 1.5. In the remaining 28 cases, the values obtained with the model are far from reality, not guaranteeing the necessary safety margin.

Furthermore, from the results, it has been observed that in the case of prestressed and post-tensioned concrete structures (2, 7, 11, 24, and 32) the values obtained with the corrosion rate prediction model are even further away from the reality than in the case of reinforced concrete structures.

Table 4. Propagation period results for each of the case studies. Calculated from the age of the structure at the time of the durability study (*ts*) and the initiation period (*ti*).

Nº	ts (Years)	Initiation Period ti (Years)	Propagation Period tp (Years) = ts–ti
1	38	11.39	26.61
2	35	18.60	16.40
3	35	3.95	31.05
4	35	3.49	31.51
5	35	4.40	30.60
6	45	11.81	33.52
7	35	7.14	27.86
8	35	2.79	32.21
9	35	3.11	31.89
10	35	3.11	31.89
11	35	8.89	26.11
12	35	3.55	31.45
13	63	2.84	60.16
14	63	7.36	55.64
15	63	8.09	54.91
16	63	6.87	56.13
17	63	12.09	50.91
18	18	9.54	8.46
19	18	10.15	7.85
20	18	9.82	8.18
21	37	10.90	26.10
22	37	15.87	21.13
23	37	9.11	27.89
24	37	17.86	19.14
25	23	16.63	6.37
26	23	8.30	14.70
27	23	5.23	17.77
28	23	5.71	17.29
29	36	24.02	11.98
30	36	10.18	25.82
31	36	10.42	25.58
32	36	9.17	26.83

Table 5. Case studies with favourable results using the Liu & Weyers (1998) model.

Nº	Real Diameter Loss Pr (mm)	Diameter Loss Liu & Weyers Px (mm)-Equation (2)	Safety Factor Px/Pr
21	1.20	1.493	1.24
22	0.60	0.856	1.43
23	0.60	1.788	2.98
24	0.80	1.281	1.60

4.2.2. Vu and Stewart (2000)

As far as the data analysis is concerned, of the 32 cases studied, in 18 of them (56.25% of the total sample), the diameter loss value obtained using the model is greater than real diameter loss. The safety factor provided by the model for these 18 cases is approximately 2.29. Table 6 shows the results obtained.

In the 5 cases of prestressed and post-tensioned concrete structures analyzed (2, 7, 11, 24, and 32), in 4 of them the model values are far away from the real values (safety margin 0.40), while in the only case that meets the safety criteria, the result obtained from the model is four times higher than the real value.

Table 6. Case studies with favourable results using the Vu and Stewart (1998) model.

N°	Real Diameter Loss Pr (mm)	Diameter Loss Liu & Weyers Px (mm)-Equation (2)	Safety Factor Px/Pr
6	5.10	9.408	1.84
3	4.00	4.988	1.25
4	4.00	4.667	1.17
5	4.00	5.511	1.38
8	3.60	3.884	1.08
13	6.25	11.575	1.85
14	5.00	7.157	1.43
16	9.00	11.534	1.28
17	5.00	9.469	1.89
18	1.00	1.486	1.49
19	1.50	2.105	1.40
20	1.00	3.200	3.20
21	1.20	5.373	4.48
22	0.60	6.462	10.77
23	0.60	6.027	10.05
24	0.80	3.396	4.25
26	0.50	4.083	8.17
29	2.10	2.155	1.03

4.2.3. Li (2004 a)

Undoubtedly, this model, besides being the simplest for its application to real cases, is also the one that has given the best results. Table 7 shows the results obtained, in which ofthe 32 cases studied, 26 of them are favorable, which means that the diameter loss obtained with the model is greater than the real diameter loss. Only six cases are unfavorable, which means that the diameter loss obtained with the model is smaller than the actual diameter loss, although with values very close to the real ones.

The safety factor obtained is 1.98 on average, which means that, in the favorable cases, the section loss value obtained by the model is twice the real value. The distribution of this average is quite homogeneous, varying between 1.73 and 2.07 in most cases. In the six unfavourable cases, the values obtained are very close to real values of diameter loss.

As for the prestressed and post-tensioned concrete structures analysed with this model, the trend changes with respect to the previous ones. Of the five prestressed cases studied, the results were favourable in three of them, while in the remaining two the results were unfavourable.

4.2.4. Li (2004 b)

The section loss results obtained were not satisfactory, since of the 32 cases studied, 22 of them have been unfavourable, with diameter loss results much lower than the real ones, with a safety coefficient of 0.45. Of the 10 cases where the diameter loss value is higher than the real one, the results are not particularly encouraging, as the safety factor is very tight in some cases—1.51 on average. Table 8 shows the results obtained with this model.

Of the five prestressed and post-tensioned concrete structures cases analysed, in one of them (case number 24), favourable results have been obtained through the model, with a safety factor of 1.37. In the remaining four cases, the results obtained are one below the real values of diameter loss.

4.2.5. Kong et al. (2006)

The results of this model have not been favourable. Only in case number 23, the diameter loss value calculated with the model was higher than the actual diameter loss. In this case, the diameter loss calculated with the corrosion rate of the model was 0.605 mm, while the real diameter loss in the reinforcement is 0.60 mm. These results barely guarantee a safety factor of 1.01.

Table 7. Case studies with favourable results using the Li (2004 a) model.

N°	Real Diameter Loss Pr (mm)	Diameter Loss Li (2004) Px (mm)-Equation (2)	Safety Factor Px/Pr
6	5.10	7.540	1.48
3	4.00	6.903	1.73
4	4.00	7.022	1.76
5	4.00	6.788	1.70
8	3.60	7.202	2.00
9	4.00	7.119	1.78
10	4.00	7.119	1.78
11	5.00	5.650	1.13
13	6.25	14.736	2.36
14	5.00	12.179	2.44
15	6.25	13.278	2.12
16	9.00	13.615	1.51
17	5.00	13.480	2.70
18	1.00	1.505	1.50
20	1.00	1.446	1.45
21	1.20	5.648	4.71
22	0.60	4.420	7.37
23	0.60	6.099	10.16
24	0.80	3.939	4.92
25	2.00	2.893	1.45
26	0.50	1.071	2.14
27	3.00	3.612	1.20
29	2.10	2.274	1.08
30	2.70	5.578	2.07
31	5.00	5.518	1.10
32	4.00	5.831	1.46

Table 8. Case studies with positive results using the Li (2004 b) model.

N°	Real Diameter Loss Pr (mm)	Diameter Loss Liu and Weyers Px (mm)–Equation (2)	Safety Factor Px/Pr
12	10.00	11.270	1.13
13	6.25	12.885	2.06
15	6.25	21.462	3.43
16	9.00	7.287	1.19
17	5.00	10.697	1.46
18	1.00	1.311	1.31
21	1.20	1.637	2.21
22	0.60	1.325	2.91
23	0.60	1.749	1.75
24	0.80	1.094	1.37

4.2.6. New Empirical Model (Lu et al., 2019)

In the results obtained with this model, 25 of the 32 cases analysed were unfavourable. The seven cases shown in Table 9 have positive results, where the diameter loss calculated through the corrosion rate of this model has been higher than the real diameter loss. The average safety factor obtained is 1.5.

In the case of prestressed and post-tensioned concrete structures, the trend is exactly the same as in the rest of the models. All section loss values calculated with this model for this type of structure are much lower than the actual diameter loss.

Table 9. Case studies with favourable results using the New Empirical model (2019).

N°	Real Diameter Loss Pr (mm)	Diameter Loss New Model Px (mm)	Safety Factor Px/Pr
18	1.00	1.118	1.12
20	1.00	1.031	1.03
21	1.20	2.199	1.83
22	0.60	1.389	2.31
23	0.60	2.558	4.26
24	0.80	1.901	2.38
26	0.50	0.571	1.14

5. Discussion

5.1. Comparison of Model Results

To compare the results presented previously, this study analysed which of the corrosion rate calculation models has the highest number of cases where the diameter loss calculated with the proposed methodology is higher than the actual diameter loss.

The graph in Figure 6 shows the distribution between the calculated diameter loss with each of the models and the actual observed diameter loss. When the difference between the two is greater than zero, it means that the value calculated through the model is greater than the actual observed diameter loss.

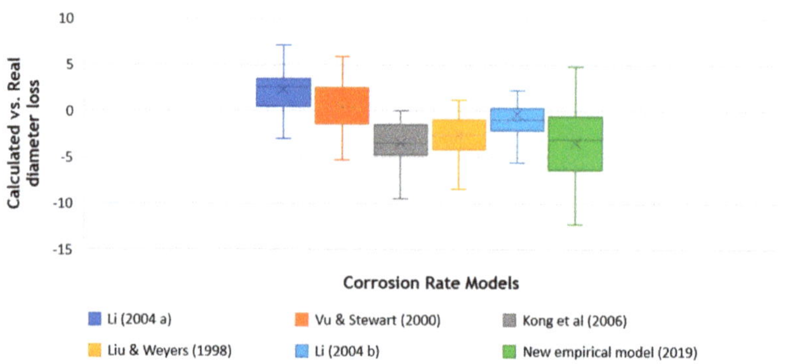

Figure 6. Distribution of estimated vs. real diameter loss in the study cases with different corrosion rate models.

It can be seen that only the Li (2004 a) model has the best performance, as it is the only one with a positive mean of the deviations, and this deviation is positive for more than 75% of the cases studied. In the Vu and Stewart (2000) model, the mean is practically zero and the model has the same probability of hitting and missing, as the median is also at zero.

In the rest of the models, the mean is clearly negative, and as can be seen in more than 75% of the cases, the diameter loss calculations with the model would be wrong.

Therefore, the results of the combination of the chloride diffusion model (EHE-08) for the calculation of the initiation period and the corrosion rate calculation model Li (2004 a) have given the most satisfactory results. In addition to the importance of efficiency in as many cases as possible, ensuring optimal safety margins between the calculated results and reality is also very important.

Figure 7 shows the safety factors obtained for each case study. In 26 of the 32 cases studied, the safety factors are at least greater than one (orange line), and practically all the values are homogeneously between one and two (green line). There are six cases where the diameter loss calculated with the model is less than the actual diameter loss of the reinforcement (red color bar). Only in four of the cases are the safety factors significantly higher than two. This means that in addition to the model having a high probability of success, the safety factors obtained are very reasonable, and similar to the safety coefficients factors used in the design of concrete structures.

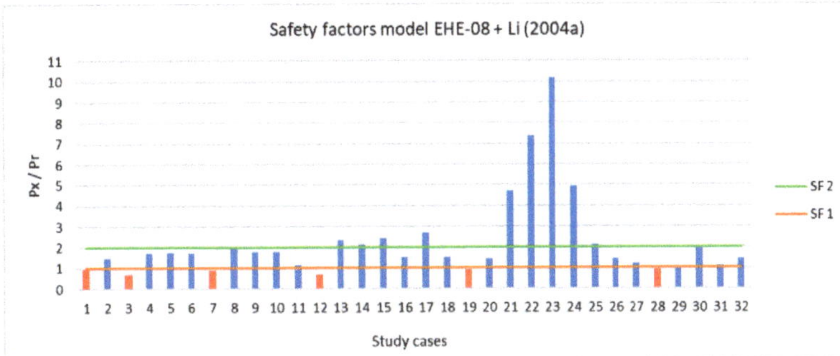

Figure 7. Safety factors obtained for each case study by applying the EHE-08 + Li (2004) model.

In reference to prestressed or post-tensioned cases, none of the models have been very successful. Table 10 shows the corrosion rate models and the number of cases where the result was positive. As can be seen, there are a limited number of cases of prestressed or post-tensioned concrete, where the value of the diameter loss calculated with the models is higher than the actual diameter loss. The best performer is again the model of Li (2004 a), with three cases with favourable results. In the case of the model of Kong et al. (2006), none of the five cases studied is favourable. These poor model results in prestressed or post-tensioned concrete structures may be due to stress corrosion cracking phenomena, which are different from conventional corrosion processes [49,50].

Table 10. Corrosion rate models and number of cases of prestressed and post-tensioned concrete structures with favourable results.

Corrosion Rate Model	Number of Positive Cases	Study Case
Liu & Weyers (1998)	1	24
Kong et al. (2006)	0	0
Vu & Stewart (2000)	1	24
Li (2004 a)	3	11, 24, 32
Li (2004 b)	1	24
New Empirical Model (Lu et al., 2019)	1	24

After analysing the results presented in the previous section and comparing the different corrosion rate models proposed, it can be seen that the model with the best estimates of diameter loss in the 32 cases studied was the corrosion rate model proposed by Li (2004 a). These good results are also related to a good combination between the model of Li (2004 a), which represents the corrosion propagation phase, and the chloride diffusion model of the (EHE-08), which has allowed the characterisation of the corrosion initiation time. In combination, these two proposed models have made it possible to obtain, with simple calculations, an estimated value of diameter loss in the reinforcement that is very close to reality in a representative sample of real cases of reinforced and prestressed concrete structures affected by corrosion.

5.2. Potential Applications

Using the proposed methodology and the validated corrosion prediction models, it is possible to know with a high level of confidence the loss of diameter suffered by a reinforcement affected by corrosion as a function of propagation time and corrosion rate and therefore create a tool to know how the resistance capacity of any structural element affected by corrosion problems varies during lifetime. This has great potential in the field of maintenance of offshore structures such as concrete foundations and is a fundamental tool for any operator to plan appropriate maintenance strategies and to optimise human and material resources while guaranteeing structural safety during lifetime.

For this purpose, and based on the methodology proposed in the previous chapter, spreadsheets have been created as a tool to obtain the coefficient of resistant variation K, which indicates a measure of loss of structural safety, as indicated in the Equation (17).

$$K = \frac{R_a}{R_n} \tag{17}$$

where R_a is the altered resistant capacity (affected by the diameter loss as a result of corrosion) and R_n is the nominal resistance capacity.

When this coefficient is close to 1, the altered resistant capacity of the element is close to the nominal resistant capacity. As the value of the coefficient moves away below the unit value, the altered resistant capacity of the element is less than the nominal resistant capacity, which can be a risk to safety and security. In this case, a limit shall be established that, for those values close to K = 0.85, structural strengthening or other actions shall be considered to ensure the proper functioning of the element and the minimum necessary safety conditions.

Figure 8 shows an example for a reinforced concrete element working in simple bending and affected by corrosion problems in its principal reinforcement. This tool is equally applicable to other structural elements in compression or shear. On the x-axis is the age of the structure, and on the y-axis is the safety loss coefficient (K) expressed as a percentage. As can be seen, at approximately 27 years of age, the main bending reinforcement of the structure would have lost sufficient section to cause a loss of resistance capacity of 15% of the nominal resistance capacity.

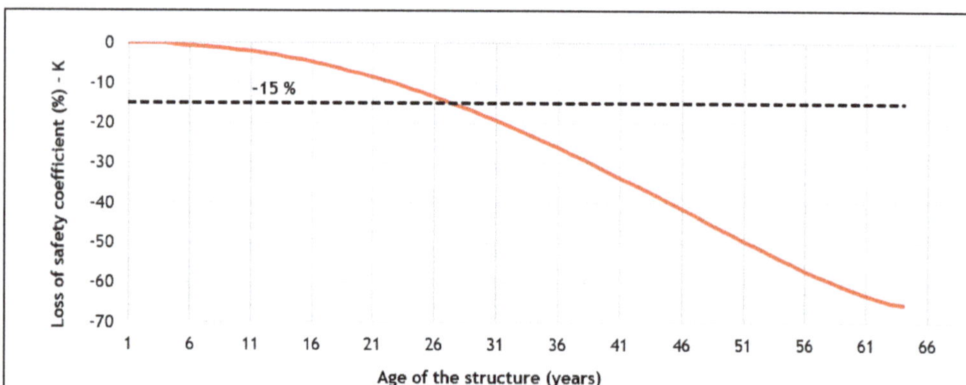

Figure 8. Estimation of the loss of strength capacity over its service life of a structural element working on single flexure with corrosion problems in its main reinforcement.

This means that, under the criterion of considering actions once there has been a 15% loss of strength of the element, at 27 years of age of the structure, it would be necessary to carry out a repair or reinforcement action of the structural element. By the time the structure is 61 years old, it will have lost approximately 60% of its load-bearing capacity, indicating an imminent risk of collapse.

With this analytical data, it is easier to plan and optimise the maintenance of the structures, especially when it comes to offshore wind turbine foundations that are located in marine environments that are very difficult to access, and where any maintenance activity is very costly for the project.

6. Conclusions

Finally, the main conclusions of this paper are set out in the following points:

- In more than 75% of the 32 case studies, the application of the chloride diffusion model of the EHE-08 and the corrosion rate calculation model of Li (2004 a) has resulted in similar diameter loss values when compared to real reinforced and prestressed concrete structures that have been affected by active corrosion processes. The calculated diameter loss results promise reasonable safety coefficients, with a minimum value of 1.73, and the average safety factor of 1.98. This means that the diameter loss value calculated through the models is twice as high as the actual diameter loss.
- The combination of the EHE-08 diffusion model and the Vu and Stewart (2000) corrosion rate calculation model is the next best performer in diameter loss. In this case, the median is zero, which means that the combination of models has approximately the same probability of success as failure in a case study.
- The combination of the EHE-08 chloride diffusion model and the other corrosion rate calculation models: Liu and Weyers (1998), Li (2004 b), Kong et al. (2006), and New Empirical Model (Lu et al., 2019) has not obtained satisfactory results. The calculated reinforcement diameter losses were generally much lower in practically all cases, with the combination with the model of Kong et al. (2006) giving the worst results.
- The application of the EHE-08 diffusion model with none of the corrosion rate calculation models proposed, has proved satisfactory results in the case studies of prestressed and post-tensioned concrete structures analyzed as part of the whole sample. Again, the model of Li (2004 a), together with the chloride diffusion model of the EHE-08, gave the best results in this aspect.
- A tool for rapidly estimating the section loss of reinforcement in offshore concrete structural elements as a function of time provides offshore wind farm operators with a cost-effective approach for planning their maintenance strategies and the optimisation of material costs and human resources. This is essential, considering the exponential expansion of OWF's, which will only be compatible if this type of proposal contributes to the reduction of O&M costs, and where reinforced concrete will continue to be represented.

Author Contributions: Conceptualization, M.D.E., R.R.R. and K.V.; methodology, M.D.E., R.R.R. and K.V.; validation, M.D.E., R.R.R. and K.V.; formal analysis, M.D.E., R.R.R. and K.V.; investigation, K.V.; writing—original draft preparation, M.D.E., R.R.R. and K.V.; writing—review and editing, M.D.E., R.R.R. and K.V.; supervision, R.R.R. and M.D.E. All authors have read and agreed to the published version of the manuscript.

Funding: Not applicable.

Institutional Review Board Statement: Not applicable.

Informed Consent Statement: Not applicable.

Data Availability Statement: Data available on request due to privacy restrictions. The data presented in this study are available on request from the corresponding author. The data are not publicly available due to confidential reasons.

Acknowledgments: K.V.; R.R.R. and M.D.E. are very grateful to the Technical Institute of Materials and Constructions (INTEMAC) for providing us with the case studies and all the documentation necessary to carry out this research work.

Conflicts of Interest: The authors declare no conflict of interest.

References

1. Ramírez, L.; Brindley, G. Offshore Wind Energy: 2021 Mid-Year Statistics. 2021. Available online: Windeurope.org (accessed on 10 November 2021).
2. Lee, J.; Zhao, F. Global Offshore Wind Report 2021. 2021. Available online: http://www.gwec.net/global-figures/wind-energy-global-status/ (accessed on 10 November 2021).
3. International Renewable Energy Agency. Renewable Energy Statistics 2021. 2021. Available online: https://www.irena.org/publications/2021/March/Renewable-Capacity-Statistics-2021 (accessed on 10 November 2021).
4. European Union. Regulation (EU) 2021/1119 of the European Parliament and of the Council of 30 June 2021 Establishing the Framework for Achieving Climate Neutrality and Amending Regulations (EC) No 401/2009 and (EU) 2018/1999 ('European Climate Law'). 2021. Available online: https://eur-lex.europa.eu/legal-content/EN/TXT/PDF/?uri=CELEX:52019DC0640&from=EN (accessed on 10 November 2021).
5. European Commission. In-Depth Analysis in Support of the Commission Communication Com (2018) 773: A Clean Planet for all A European Long-Term Strategic Vision for a Prosperous, Modern, Competitive and Table of Contents. 2018. Available online: https://knowledge4policy.ec.europa.eu/publication/depth-analysis-support-com2018-773-clean-planet-all-european-strategic-long-term-vision_en (accessed on 10 November 2021).
6. European Commission. Guidance Document on Wind Energy Developments and EU Nature Legislation. 2020. Available online: https://ec.europa.eu/environment/nature/natura2000/management/docs/wind_farms_en.pdf (accessed on 15 October 2021).
7. Fraile, I.K.D.; Vandenberghe, A.; Klonari, V.; Ramirez, L.; Pineda, I.; Tardieu, P.; Malvault, B. Getting Fit for 55 and Set for 2050. 2021. Available online: https://windeurope.org/intelligence-platform/product/getting-fit-for-55-and-set-for-2050/# (accessed on 10 November 2021).
8. Johnston, B.; Foley, A.; Doran, J.; Littler, T. Levelised cost of energy, A challenge for offshore wind. *Renew. Energy* **2020**, *160*, 876–885. [CrossRef]
9. Mone, C.; Stehly, T.; Maples, B.; Settle, E. 2017 Cost of Wind Energy Review. 2017. Available online: https://www.nrel.gov/docs/fy18osti/72167.pdf (accessed on 15 March 2021).
10. Crabtree, C.J.; Zappalá, D.; Hogg, S.I. Wind energy: UK experiences and offshore operational challenges. *Proc. Inst. Mech. Eng. Part A J. Power Energy* **2015**, *229*, 727–746. [CrossRef]
11. BVG. Associates. Value Breakdown for The Offshore Wind Sector. 2010. Available online: https://www.gov.uk/government/uploads/system/uploads/attachment_data/file/48171/2806-value-breakdown-offshore-wind-sector.pdf (accessed on 12 April 2021).
12. Miedema, R. Offshore Wind Energy Operations & Maintenance Analysis. Research Thesis, Hogeschool Van Amsterdam, Amsterdam, The Netherlands. 2012. Available online: http://sciencecentre.amccentre.nl/studies/Thesis_Robert_Miedema.pdf (accessed on 15 March 2021).
13. Wu, X.; Hu, Y.; Li, Y.; Yang, J.; Duan, L.; Wang, T.; Adcock, T.; Jiang, Z.; Gao, Z.; Lin, Z.; et al. Foundations of offshore wind turbines: A review. *Renew. Sustain. Energy Rev.* **2018**, *104*, 379–393. [CrossRef]
14. Sánchez, S.; López-Gutiérrez, J.S.; Negro, V.; Esteban, M.D. Foundations in offshore wind farms: Evolution, characteristics and range of use. Analysis of main dimensional parameters in monopile foundations. *J. Mar. Sci. Eng.* **2019**, *7*, 441. [CrossRef]
15. Esteban, M.D.; Matutano, C. Offshore Wind Foundation Design: Some Key Issues. *J. Energy Resour. Technol.* **2015**, *137*, 1–6. [CrossRef]
16. Fernández, R.P.; Pardob, M.L. Offshore concrete structures. *Ocean Eng.* **2013**, *58*, 304–316. [CrossRef]
17. Esteban, M.D.; López-Gutierrez, J.-S.; Negro, V. Gravity-Based Foundations in the Offshore Wind Sector. *J. Mar. Sci. Eng.* **2019**, *7*, 64. [CrossRef]
18. Trust, T.C. Offshore Wind Industry Review of GBS. 2015. Available online: https://prod-drupal-files.storage.googleapis.com/documents/resource/public/Offshore_wind_industry_review_of_Gravity_Based_Structures_REPORT.pdf (accessed on 20 December 2021).
19. Esteban, M.D.; Couñago, B.; López-gutiérrez, J.S.; Negro, V.; Vellisco, F. Gravity based support structures for offshore wind turbine generators: Review of the installation process. *Ocean Eng.* **2015**, *110*, 281–291. [CrossRef]
20. Rodrigues, S.; Restrepo, C.; Kontos, E.; Pinto, R.T.; Bauer, P. Trends of offshore wind projects. *Renew. Sustain. Energy Rev.* **2015**, *49*, 1114–1135. [CrossRef]
21. WindEurope—The Voice of the Wind Energy Industry. Available online: https://windeurope.org/ (accessed on 16 December 2021).
22. Mathern, A.; von der Haar, C.; Marx, S. Concrete support structures for offshore wind turbines: Current status, challenges, and future trends. *Energies* **2021**, *14*, 1995. [CrossRef]
23. ELISA—Elican Project. Esteyco. Available online: https://www.esteyco.com/proyectos/elisa-elican-project/ (accessed on 16 December 2021).
24. James, R.; Ros, M.C. Floating Offshore Wind: Market and Technology Review. *Carbon Trust* **2015**, *439*. [CrossRef]
25. International Renewable Energy Agency (IRENA). Innovation Outlook: Offshore Wind. Available online: https://www.irena.org/publications/2016/oct/innovation-outlook-offshore-wind (accessed on 12 December 2021).
26. DNV GL. DNVGL-ST-0126: Support Structures for Wind Turbines. 2018. Available online: https://rules.dnvgl.com/docs/pdf/DNVGL/ST/2018-07/DNVGL-ST-0126.pdf (accessed on 15 May 2021).
27. ACI Committee 222. Protection of Metals in Concrete Against Corrosion. In *Aci 222R-01*; ACI: Farmington Hills, MI, USA, 2001; pp. 1–41.

28. EC Innovation Programme; Andrade, C.; Izquierdo, D. A Validated Users Manual for Assessing the Residual Service Life of Concrete Structures. Madrid. 2001. Available online: https://www.ietcc.csic.es/wp-content/uploads/1989/02/manual_contecvet_ingles.pdf (accessed on 20 October 2021).
29. Feliu, S.; Andrade, C. Manual: Inspección de Obras Dañadas por Corrosión de Armaduras. In *Acor*; IETCC: Madrid, Spain, 1989; pp. 1–122. [CrossRef]
30. Odriozola, M.Á.B. Corrosión de las Armaduras del Hormigón Armado en Ambiente Marino: Zona de Carrera de Mareas y Zona Sumergida. Ph.D. Thesis, Universidad Politécnica de Madrid, Madrid, Spain, 2007.
31. Pruckner, F. Diagnosis and Protection of Corroding Steel in Concrete. Ph.D. Thesis, Norwegian University of Science and Technology Faculty of Engineering Science and Technology Department of Structural Engineering, Trondheim, Norway, November 2002.
32. Ciolko, A. *Nondestructive Methods for Condition Evaluation of Prestressing Steel Strands Nondestructive Methods for Condition Evaluation of Prestressing Steel Strands in Concrete Bridges Final Report Phase I: Technology Review*; Prepared for National Cooperative Highway Research Program; no. July 2015; NCHRP: Washington, DC, USA, 1999.
33. Liu, L.; Fu, Y.; Ma, S.; Huang, L.; Wei, S.; Pang, L. Optimal scheduling strategy of O&M task for OWF. *IET Renew. Power Gener.* **2019**, *13*, 2580–2586. [CrossRef]
34. Shafiee, M. Maintenance logistics organization for offshore wind energy: Current progress and future perspectives. *Renew. Energy* **2015**, *77*, 182–193. [CrossRef]
35. Lu, Z.H.; Lun, P.Y.; Li, W.; Luo, Z.; Li, Y.; Liu, P. Empirical model of corrosion rate for steel reinforced concrete structures in chloride-laden environments. *Adv. Struct. Eng.* **2019**, *22*, 223–239. [CrossRef]
36. Tutti, K. Corrosion of Steel in Concrete. Swedish Cement and Concrete Research Institute. 1982. Available online: http://www.cbi.se/viewNavMenu.do?menuID=317&oid=857 (accessed on 15 May 2021).
37. García, S.F. Corrosión De Armaduras En El Hormigón Armado En Ambiente Marino Aéreo. Ph.D. Thesis, Universidad Poltécnica de Madrid, Madrid, Spain, 2016. Available online: https://oa.upm.es/39374/1/Susana_Fernandez_Garcia.pdf (accessed on 16 April 2021).
38. Comisión Permanente del Hormigón. Instrucción de Hormigón Estructural (EHE-2008). 2011. Available online: http://www.ponderosa.es/docs/Norma-EHE-08.pdf (accessed on 20 December 2021).
39. Helland, S.; Aarstein, R.; Maage, M. In-field performance of North Sea offshore platforms with regard to chloride resistance. *Struct. Concr.* **2010**, *11*, 15–24. [CrossRef]
40. Markeset, G.; Mytdal, R. Modelling of Reinforcement Corrosion in Concrete-State of the Art. 2008. Available online: http://smartpipe.com/upload/Byggforsk/Publikasjoner/coin-no7.pdf (accessed on 22 February 2021).
41. Bertolini, L.; Elsener, B.; Pedeferri, P.; Redalli, E.; Polder, R. *Corrosion of Steel in Concrete. Prevention, Diagnosis, Repair*; Wiley: Hoboken, NJ, USA, 2013.
42. Zhang, D.; Zeng, Y.; Fang, M.; Jin, W. Service life prediction of precast concrete structures exposed to chloride environment. *Adv. Civ. Eng.* **2019**, *2019*, 3216328. [CrossRef]
43. Liu, Y. Modeling time-to-corrosion cracking in chloride contaminated reinforced concrete structures. *ACI Mater. J.* **1999**, *96*, 611–613. Available online: https://trid.trb.org/view/514367 (accessed on 15 April 2021).
44. Yu, B.; Yang, L.; Wu, M.; Li, B. Practical model for predicting corrosion rate of steel reinforcement in concrete structures. *Constr. Build. Mater.* **2014**, *54*, 385–401. [CrossRef]
45. Gjørv, O.E. *Durability Design of Concrete Structures in Severe Environments*; Routledge: Abingdon-on-Thames, UK, 2014.
46. Ministerio de Obras Públicas. *Instrucción para el Proyecto y la Ejecución de Obras de Hormigón en Masa o Armado (EH-88)*; Ministerio de Obras Públicas: Madrid, Spain, 1988.
47. Ministerio de Obras Públicas. *Instrucción para el Proyecto y Ejecución de Obras de Hormigón*, 1st ed.; Ministerio de Obras Públicas: Santander, Spain, 1939; Volume 1.
48. Agencia Estatal de Meteorología—AEMET. Gobierno de España. Available online: http://www.aemet.es/es/portada (accessed on 18 December 2021).
49. Hurtado, M.A. Corrosion Bajo Tension De Alambre de Acero Pretensado en Medios Neutros con HCO_3 y Alcalinos con SO_4.. Ph.D. Thesis, Universidad Complutense Madrid, Madrid, Spain, 1993.
50. Khalifeh, A. Stress Corrosion Cracking Damages. In *Failure Analysis*; InTech Open: London, UK, 2019. [CrossRef]

Journal of
Marine Science and Engineering

MDPI

Article

Influence of Asphaltenes on the Low-Sulphur Residual Marine Fuels' Stability

Ksenia I. Smyshlyaeva [1], Viacheslav A. Rudko [1,*], Vladimir G. Povarov [1], Alina A. Shaidulina [1], Ignaty Efimov [1], Renat R. Gabdulkhakov [1], Igor N. Pyagay [1] and James G. Speight [2]

1 Saint Petersburg Mining University, 199106 St. Petersburg, Russia; ks.smyshlyaeva@mail.ru (K.I.S.); povarovvg@rambler.ru (V.G.P.); artur2345672@mail.ru (A.A.S.); efimov.ignaty@gmail.com (I.E.); renat18061995@gmail.com (R.R.G.); igor-pya@yandex.ru (I.N.P.)
2 CD and W Incorporated, Laramie, WY 82070, USA; jamessp8@aol.com
* Correspondence: rva1993@mail.ru

Abstract: The effects of asphaltenes from two heavy oil residues on the sedimentation stability of residual marine fuels were assessed and compared. As base components of residual marine fuels, the vacuum residue (VacRes) and visbreaking residue (VisRes) were taken. The heptane-insoluble fractions (HI-fractions), including asphaltenes, isolated from vacuum residue and visbreaking residue, were analyzed to determine the elemental composition (XRF) and cluster parameters (XRD). The results of the analysis of the parameters of the asphaltene cluster (HI-fraction) for vacuum residue and visbreaking residue showed that d_γ – 6.1 and 5.9 Å, L_c – 26.72 and 20.78 Å, and L_a – 7.68 and 7.20 Å. The sedimentation stability of residual marine fuel was determined according to the ISO 10307-1-2009 (TSA) method and described using ternary phase diagrams. The ratio of stable compositions to the total number of possible compositions (with a step of 10 wt%) was 65/66 or 98.5% for residual marine fuel comprising a mixture VacRes/ULSD/LCGO (vacuum residue/ultra-low sulphur diesel/light catalytic gas oil). Meanwhile, the ratio of stable compositions to the total number of possible compositions was 38/66 or 57.6% for residual marine fuel comprising a mixture VisRes/ULSD/LCGO (visbreaking residue/ultra-low sulphur diesel/light catalytic gas oil).

Keywords: heptane-insoluble fraction; SARA; ternary phase diagrams; petroleum residues; TSA

Citation: Smyshlyaeva, K.I.; Rudko, V.A.; Povarov, V.G.; Shaidulina, A.A.; Efimov, I.; Gabdulkhakov, R.R.; Pyagay, I.N.; Speight, J.G. Influence of Asphaltenes on the Low-Sulphur Residual Marine Fuels' Stability. *J. Mar. Sci. Eng.* **2021**, *9*, 1235. https://doi.org/10.3390/jmse9111235

Academic Editors: M. Dolores Esteban, José-Santos López-Gutiérrez, Vicente Negro and M. Graça Neves

Received: 2 October 2021
Accepted: 4 November 2021
Published: 8 November 2021

1. Introduction

From 1 January 2020, the International Maritime Organization (IMO) has tightened the requirements for the sulphur content in residual marine fuel (from 3.50 down to 0.50 weight per cent), which is related to environmental protection and human health concerns and is important in the implementation of the UNESCO's Sustainable Development Goals [1,2]. Petroleum-based fuels have historically been used as marine fuels. However, aiming to decrease the SO_x emissions from vessels and reduce the greenhouse effect, alternative fuels have also been proposed, e.g., LNG as a low-sulphur marine fuel [3,4]. Currently, this has been implemented only on LNG-tankers [5]. The limited distribution of LNG in ports and refuelling bases remains the main issue [6]. The analysis of the results of studies [7–9] shows that the substitution of most of the hydrocarbon resources with the alternative cannot be achieved in the nearest future, thus the processing of oil stock and obtaining residual marine fuels will stay relevant.

The most difficult technical challenge is the production of low-sulphur fuel oils (LSFOs) from petroleum feedstock [10]. The existing methods for LSFO production can be divided into two main classes: direct and indirect hydrogenation [11]. Indirect hydrogenation is the cheapest method that does not require large capital investments (e.g., ExxonMobil [12], Shell [13], Sunoco Partners Marketing & Terminals L.P. [14], Mawetal [15]). In this case, the LSFO is obtained by selective compounding of sulphur residual oil fractions of primary or secondary oil refining processes with low-sulphur middle distillate fractions of hydrogenation processes. As the latter, various types of diesel and gas oil fractions with a high content of paraffinic/naphthenic hydrocarbons (from 50–85 to 100%) in their composition (ExxonMobil [12], Shell [13], Sunoco Partners Marketing & Terminals L.P. [14]) are usually used.

Besides the saturated aromatic compounds and resins, the residual petroleum fractions contain asphaltenes—high molecular weight carbon compounds that are generally identified as substances insoluble in normal alkane solvents, e.g., n-heptane (heptane-insoluble or HI); however, they are soluble in aromatic solvents [16–18].

Thus, obtaining LSFOs by selective compounding requires solving the problem of incompatibility arising when mixing the fuel components of paraffinic/naphthenic and asphaltene base [19,20]. As a result, marine fuel, as a multicomponent petroleum product, loses its sedimentation stability during storage and transportation and asphalt-resinous-paraffin deposits (ARPDs) stratify, coagulate, and precipitate on pipelines and in storage tanks [21–24].

Asphaltenes are high molecular weight components of crude oils and petroleum products that include polyaromatic rings with long or short aliphatic chains, various functional groups, and heteroatoms. The structure and properties of asphaltenes strongly depend on their origin. For example, compared with natural asphaltenes, thermal asphaltenes originating from thermodestructive processes have a lower molecular weight and are predominantly concentrated in the plane [16,18,20,25–29].

With an increase in the concentration of asphaltenes in marine fuel, their molecules form nano-aggregates and, with a further increase in the concentration, the nano-aggregates form clusters and precipitate. Because of the asphaltenes' tendency towards aggregation, precipitation, and deposition on various surfaces, the study of the structure of these materials and of the mechanism of their formation has always remained relevant. Figure 1 shows the modified Yen model, which has the greatest agreement with experimental data among the existing models [16,25].

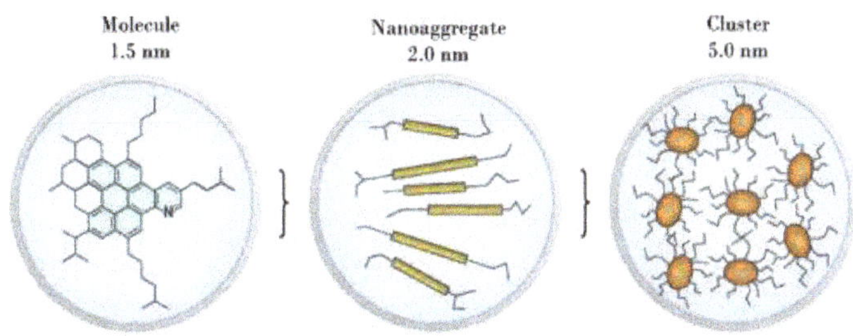

Figure 1. The mechanism of formation of aggregates and clusters of asphaltenes, based on the modified Yen model [16].

In the review by Zhang Y. et al., the mechanism of asphaltenes' aggregation is represented as pancake bonding, occurring through the heteroassociation between radical particles and polycyclic aromatic hydrocarbons (PAHs) and/or PAHs with heteroatoms [16].

Various analytical methods are used to determine the composition and structure of petroleum asphaltenes, e.g., as nuclear magnetic resonance (NMR) measurements, determination of the molecular structure using X-ray diffraction (XRD), and simulated distillation for composition analysis [30,31].

According to the Richardson's solvent method, the group components can be classified into maltenes (γ-fraction, heptane-soluble—HS), asphaltenes (β-fraction, heptane-insoluble and toluene-soluble—HI-TS), carbenes (α_2-fraction, toluene-insoluble and quinoline-soluble—TI-QS), and carbides (α_1-fraction, toluene-insoluble and quinoline-insoluble—TI-QI).

X-ray diffraction analysis (XRD) can be used to quantify the fine structure of petroleum asphaltenes. This method can be used to obtain information on the parameters of the asphaltene cluster. These parameters are the interplanar spacings between the aliphatic chains (d_γ), interplanar spacings between the aromatic shields (d_{002} or d_m), cluster diameter (L_a), cluster height (L_c), number of aromatic sheets in a cluster (NO_{ar}), and aromaticity (f_a) [32].

X-ray diffraction patterns of petroleum asphaltenes consist of four main peaks or bands (reflections), namely, a γ-band, a graphene band or (002), a (100) band, and a (110) band. Figure 2 shows a general image of a typical X-ray diffraction pattern and a cross-sectional view of a petroleum asphaltene model with marked interplanar spacings and cluster sizes.

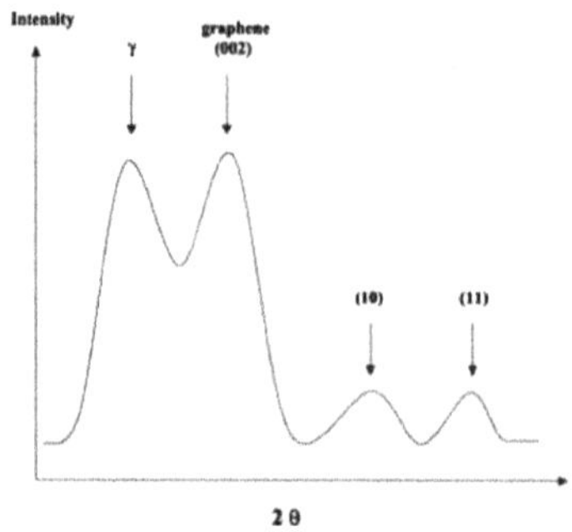

(a)

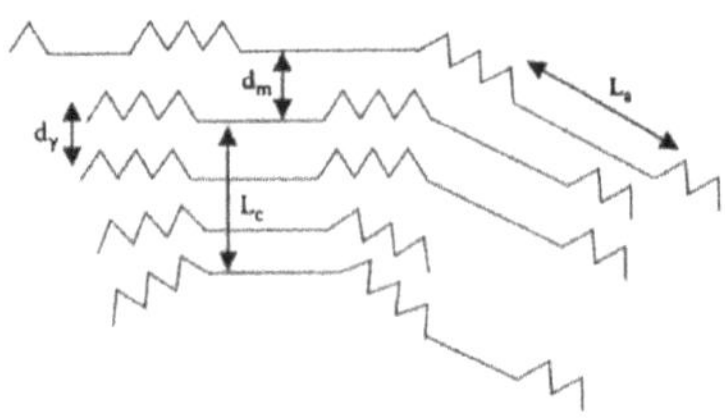

(b)

Figure 2. X-ray diffraction pattern of a petroleum asphaltene sample (**a**) [33] and a cross-sectional view of the asphaltene model (**b**) [34]. Reproduced or adapted from [33,34], with permission from Elsevier, 2002 and CRC Press Taylor & Francis Group, 2009.

The γ-band characterizes the spacing between saturated structures, which arises from X-rays scattered by condensed naphthenic rings or aliphatic chains. The graphene or (002) band, on the other hand, results from X-ray diffraction from a cluster of aromatic sheets located one below the other. The reflections of the (100) and (110) bands come from the in-plan structure of the aromatic sheets, which indicates the average cluster diameter. The size of the aromatic sheet is often calculated from the (100) band because the (110) band is often not visible owing to its very low intensity.

The positions of the bands or reflections in the X-ray diffraction pattern are determined by the double Bragg diffraction angle (2θ) and are about 20° for the γ-band, about 26° for the (002) band, about 43° for the (100) band, and about 77° for the (110) band [32–36].

An important conclusion about the regularities of changes in the structure of asphaltenes under thermal effects is that, with an increase in the degree of conversion, the cluster diameter (L_a) decreases and aromaticity (f_a) increases, which is mainly associated with the loss of aliphatic carbon in the side chains owing to dealkylation reactions and with a decrease in the number of aromatic sheets (NO_{ar}) in the cluster [32].

In residual marine fuel oils, the amount and structure of asphaltenes affect the characteristics relevant to the storage and use of the fuel. For example, in the heavy fuel oil (HFO), the ignition delay time (IDT) in an internal combustion engine decreases with the decrease in the asphaltene amount [37].

The aim of the study is to compare the effects of VacRes and VisRes asphaltenes' structures on the sedimentation stability of residual marine fuels.

2. Materials and Methods

2.1. Materials

Components of residual marine fuels were selected from a Siberian industrial plant processing a mixture of West Siberian oils. Visbreaking residue (VisRes) and vacuum residue (VacRes) were selected as the test asphaltenes carriers. The raw material of the visbreaking unit is the vacuum residue. Ultra-low sulphur diesel (ULSD) from a diesel hydrotreatment unit and light gas oil catalytic cracking (light cycle gasoil) (LCGO)/light gas oil catalytic cracking from a catalytic cracking unit (fluid catalytic cracking) (FCC) for the preparation of the sample models of residual marine fuels were also selected and studied. Table 1 shows the physical and chemical quality indicators and SARA (saturates, aromatics, resins, asphaltenes) analysis of the selected oil products.

Table 1. Physical and chemical characteristics of marine fuel components.

Parameter	VacRes	VisRes	ULSD	LCGO
Density at 15 °C, kg/m^3	990.2	961.2	855.0	956.0
Viscosity, mm^2/s at 40 °C	-	-	2.790	2.400
Viscosity, mm^2/s at 50 °C	19,516.7	1743.9	-	-
Sulphur, %	1.6020	0.9458	0.0007	0.0766
SARA analysis, wt%:				
saturates	14.90	20.35	96.08	24.77
aromatic	48.10	42.65	3.92	75.23
resins	21.00	10.40	-	-
asphaltenes, including	16.00	26.60	-	-
- carboids (TI-QI)	1.1	1.6	-	-
- carbenes (TI-QS)	0.5	0.8	-	-

Two types of heavy oil residues were taken as basic high-viscosity components to identify the effect of petroleum asphaltenes (HI-fraction) on the sedimentation stability of residual marine fuel: (1) straight-run—the residual fraction of the vacuum distillation process of oil stock (VacRes), (2) thermodestructive—the residual fraction of tar visbreaking (VisRes). The HI-fraction also contains carbenes and carbides (as in the product of a thermodestructive process) [38].

2.2. *SARA Analysis of Vacuum and Visbreaking Residues*

SARA analysis is one of the simple and widely used laboratory methods for breaking heavy oil residues into components. In this method, the compounds are separated into four analytic groups by solubility: saturated hydrocarbons, aromatic compounds, resins, and asphaltenes (respectively, saturates, aromatics, resins, asphaltenes—SARA). Figure 3 shows a general flowchart of analysis of group hydrocarbon composition of residual oil fractions.

At the first stage, oil residues (vacuum residue, visbreaking residue) were mixed at an n-heptane/residue ratio = 40:1, then the mixture was stored for 24 h in a dark place at room temperature, after which the β-fraction (asphaltenes) and α-fraction (carbenes and carbides) were filtered from maltenes using a filter. After deasphalting, maltenes were analyzed by liquid adsorption chromatography.

2.3. *XRF Analysis of Petroleum Asphaltene Composition*

The quantitative content of the microelement composition of petroleum asphaltenes was carried out on an XRF-1800 Shimadzu wavelength X-ray fluorescence spectrometer without preliminary ashing of the samples and using the classical method of additions ($CaCl_2$). The cathode current was 90 mA and the tube voltage was 40 kV. For the calculations, a method of fundamental parameters with a standard algorithm taking into account the effect of the carbon matrix of the sample on the absorption of X-ray radiation was used [39]. A sample weighed around 0.1 g. We used a method for calibrating the detector by one element, the essence of which is as follows:

(1) First, 1 ml of a $CaCl_2$ solution in isopropanol (concentration of 2 mg Ca per 1 ml of isoprapanol) was added to a weighed portion of asphaltene and dried at 120 °C. Each sample was shot twice. The first time without additive and the second time with additive. Moreover, in both cases, the carbon content was postulated to be the same and was taken as equal to 98% of the mass. The remaining 2% were distributed according to the standard algorithm by the method of semi-quantitative analysis between the detected elements.

(2) Then, two ratios were calculated: $A = \% \, Ca/\% \, S$ in the original sample and $B = \% \, Ca/\% \, S$ in the sample with the addition of calcium. Naturally, B is always greater than A, for no sulphur was added. An equation of the form $(D + A \cdot X \cdot M)/(X \cdot M) = B$. Here, D is the mass of calcium addition (2 mg), M is the mass of the sample (100 mg), and X is the true mass fraction of sulphur in the sample (it is unknown).

(3) Solving the equation for X, we get $X = D/(M \cdot (B - A))$—the true mass fraction of sulphur. The quantitative content of the remaining elements is calculated by the proportion from the data for the sample without additive. That is, we believe that, if the content of calcium in the results of the analysis of a sample without additive is 10 times less than that of sulphur, then its true content is also 10 times less than that of X.

(4) If the sum of all impurities differs from 2% by more than 10% (that is, more than 2.2 or less than 0.18%), then a correction was made to the carbon content in the sample. Well, that is, if the sum of all elements turns out to be 3%, then the calculations are repeated from the very beginning with $\% \, C = 97$. The procedure converges. Usually, 3–4 iterations are needed.

(5) This method was verified by analyzing artificial carbon mixtures and compared with the classical calibration curve method. It has always turned out to be much better, as the concentration ratios are subject to less variation than the concentrations themselves. A similar technique is widely used in quantitative gas chromatographic analysis.

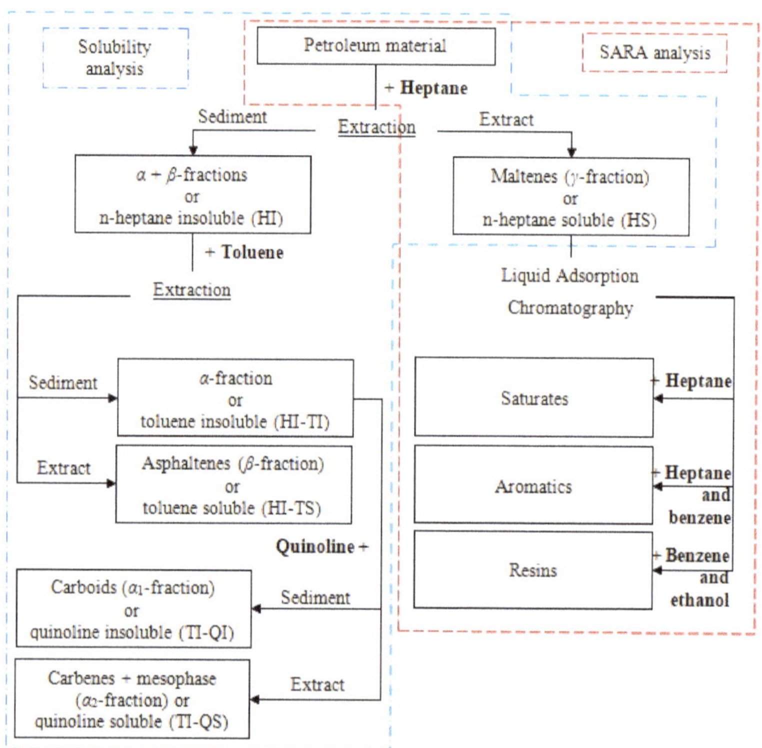

Figure 3. Flowchart of the complex analysis of the group chemical composition by solubility according to the SARA method.

2.4. XRD Analysis of the Structure of Petroleum Asphaltenes

X-ray structural analysis of samples of petroleum asphaltenes was carried out using an XRD-7000 Shimadzu X-ray powder diffractometer (Cu K_α-radiation, 2.7 kW) at room temperature. X-ray diffraction patterns were taken with long integration times (2 s) and a scan step size of 0.02. The obtained peaks (reflections) of petroleum asphaltenes with a maximum at double Bragg diffraction angles 2θ characterize certain structural components.

To identify the sizes of clusters of petroleum asphaltenes, the obtained X-ray diffraction patterns were deconvoluted into individual peaks. To do this, the spectrum was decomposed into individual Gaussians, as given by Equation (1):

$$f_i(x) = A_i \cdot e^{-\frac{(x-\mu_i)^2}{B_i}}, \quad (1)$$

where A_i and B_i are the parameters responsible for the height and width of the peak and μ_i is the coordinate of the peak maximum.

The resulting spectrum was determined as the sum of four Gaussians with maxima around 20° (γ-band), 26° ((002) band), 40° ((100) band), and 21° (paraffin impurity band), as given by Equation (2):

$$g(x) = \sum_{i=1}^{4} f_i(x) + C, \quad (2)$$

where C is the background level.

The first three peaks represent the peaks of the investigated asphaltenes. The fourth peak results from the presence of paraffin in the sample; taking this peak into account improves the quality of spectrum deconvolution. The background signal level (C) was calculated by averaging the signal at the end of the spectrum (85–90°). The parameters A_i, B_i, and μ_i for each Gaussian of the function $g(x)$ were calculated by the confidence interval method from the SciPy package (method='trf') [40]. A restriction $\mu_i \in [2\theta_i - 1; 2\theta_i + 1]$, where $2\theta_i$ corresponds to the maximum of the respective peak, was imposed on the parameter μ_i. The parameters A_i and B_i were calculated without imposing restrictions.

The Wulff–Bragg's equation [41,42] (3) was used to determine the interplanar spacings d_{002} (or d_m), d_{100}, or d_{110} in Å of the obtained samples of petroleum asphaltenes:

$$d = \lambda / 2 \sin\theta, \quad (3)$$

where λ is the X-ray wavelength, e.g., equal to 1.5406 Å for CuKα; θ is the Bragg's diffraction angle, rad.

The spacing between the saturated structures of asphaltenes was measured according to the modified Wulff–Bragg's equation proposed by Yen et al. [33,35,43], as given by (4):

$$d_\gamma = 5\lambda / 8 \sin\theta_\gamma \quad (4)$$

The average linear cluster size L_c (average cluster height) and L_a (average cluster diameter) were determined in Å according to the Scherer (5) and Warren (6) equations, respectively [44,45]:

$$L_c = 0,89\lambda / \beta_{002} \cos\theta_{002}, \quad (5)$$

$$L_a = 1,84\lambda / \beta_{100} \cos\theta_{100} \quad L_a = 1,84\lambda / \beta_{110} \cos\theta_{110}, \quad (6)$$

where 0.89 is the Scherer's constant, which is conventionally set to be the same for the sake of consistency in the published results [46]; 11.84 is the coefficient derived by Warren for two-dimensional particle size; and β is the full width at half maximum (FWHM) of the diffraction band height (in rad) minus the instrumental peak width b of 0.14°.

The number of aromatic sheets per stacked cluster was determined by Equation (7):

$$M = \frac{L_c}{d_m} + 1 \quad (7)$$

The average number of aromatic rings per aromatic sheet (layer) was determined from Equation (8):

$$NO_{ar} = \frac{L_a}{2667} \quad (8)$$

The aromaticity of asphaltene molecules (f_a) was calculated from the area (A) of the peaks for γ-bands and graphene bands (002) using Equation (9):

$$f_a = \frac{A_{002}}{A_{002} + A_\gamma} = \frac{C_A}{C} = \frac{C_A}{C_A + C_S} \tag{9}$$

where C_S, C_A, and C are the number of saturated, aromatic, and total carbon atoms per structural unit, respectively. It should be noted that the calculation of the aromaticity f_a is based only on the aromatic carbon cluster, which contributes to the formation of the graphene peak (002), and not on the total aromatic carbon in asphaltene, thus it does not represent the true aromaticity of the asphaltene molecule [32].

2.5. The Analysis of Stability of Residual Marine Fuels Using Ternary Phase Diagrams

One of the best graphic ways to describe the composition dependence on the relative quantity of phases is the phase diagrams, which were proposed by J. W. Gibbs and H. W. Roozeboom [47]. The method of building and analyzing the composition–property diagrams was formulated by N.S. Kurnakov [48]. The composition of three-component condensed systems on a plane is represented using an equilateral triangle—a triangle of concentrations (Figure 4). Pure components (A, B, and C) are located in the apices of the triangle of concentrations, two-component systems (AB, BC, and CA) correspond to points on the sides of the triangle, and three-component systems correspond to points (e.g., points 1 and 2) within the triangle of concentrations.

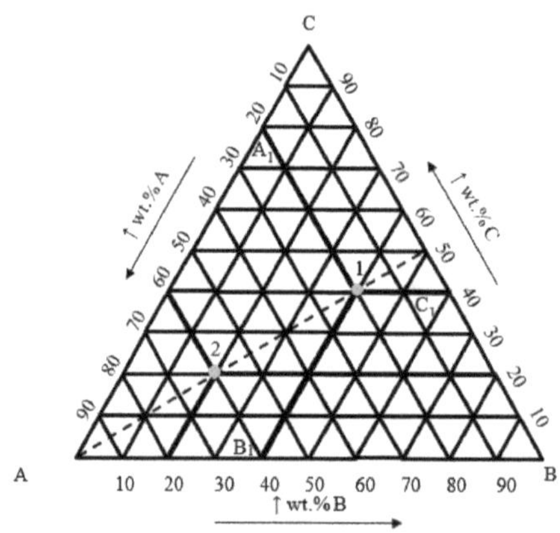

Figure 4. Ternary phase diagram A–B–C.

The amount of all three components of a concentration triangle can be determined by drawing lines parallel to the sides of the triangle through a point selected in the triangle, and using the scale of the triangle sides. Let us, for example, determine the component composition of Sample 1. Let us draw a line parallel to the BC side and passing through the point A—we get an $A_1 1$ line. Point A_1 on the CA side demonstrates the relative amount of the component A. Similarly, we can obtain the amounts of the B and C components.

In this work, as external conditions, the parameters of the experiment in accordance with ISO 10307-1-2009 (analysis of "total sediment after chemical aging"—TSA) are used. The molecules of asphaltene hydrocarbons are either in a colloid dispersed form (liquid phase) or precipitated (solid phase). With a step size of 10 weight percent at each point of the ternary phase diagram, the indicator "total sediment after ageing" characterizing the stability of the fuel was determined. Stable fuel compositions are combined into a "stability area".

2.6. The Effect of Adding VacRes and VisRes Asphaltenes to the Commercial Residual Marine Fuel on the TSA Index

The HI-fraction (asphaltenes) isolated from the vacuum and visbreaking residues was filtered, dried, and crushed. Residual marine fuel was warmed up to 70 °C. Then, asphaltenes were introduced into residual marine fuel in parts during 1 h with heating and stirring. Then, the resulting mixture was stirred for 5 h until the mixture was homogeneous at a temperature of 70 °C.

3. Results and Discussion

3.1. Composition and Morphology of Petroleum Asphaltenes

The HI-fraction content is 16.0% for VacRes and 26.6% for VisRes (Table 1). The results of XRF analysis were used to determine the microelement composition of asphaltene impurities (Table 2). As expected, the main impurity in asphaltenes isolated from both VacRes and VisRes is sulphur, with contents of 4.74 and 3.58 wt%, respectively. The amount of potentially valuable elements of oil stock in the isolated asphaltenes—V and Ni—is from 0.099 to 0.108 wt% and from 0.028 to 0.036 wt%, respectively. The sulphur content of VisRes asphaltenes is 24.4% less than that of VacRes asphaltenes. This is due to the thermal cracking of VisRes asphaltenes, which results in the cracking of aliphatic chains containing sulphur. As a result of cracking, the bonds are broken, which includes sulphur, which is removed in the form of volatile compounds. This amount is comparable in content to petroleum cokes. The content of Ni and V in VacRes and VisRes is insignificant and is within the measurement error of the device. The amount of Ca, Si, and Al is from 0.00812 to 0.01131 wt%, from 0.00903 to 0.01508 wt%, and from 0.00271to 0.00471 wt%, respectively. Ca, Si, and Al are not destroyed during cracking, but remain part of the asphaltene. However, as the mass of the hydrocarbon part decreases, while the masses of Ca, Si, and Al remain constant, the content of these heteroatoms increases.

Table 2. The microelement composition of asphaltenes from VacRes and VisRes (% wt).

Asphaltenes	S	V	Fe	Ni	Ca	Si	Al	Zn	Cl	K
VacRes	4.74	0.10833	0.04514	0.03611	0.00812	0.00903	0.00271	0.00181	0.01625	-
VisRes	3.58	0.09993	0.02734	0.02828	0.01131	0.01508	0.00471	0.00189	-	0.00283

3.2. Petroleum Asphaltenes Structure

Figure 5 shows the diffractograms of VacRes (Figure 5a) and VisRes (Figure 5b) asphaltenes for real spectra and curve deconvolution. After the deconvolution of the XRD patterns of VacRes and VisRes asphaltenes, four peaks (bands) were obtained.

One of the peaks, located at a 2θ angle of about 21°, is not characteristic of asphaltenes and is caused by the presence of long-chain crystallized n-paraffinic hydrocarbons, which, in some cases, co-precipitate with the asphaltenes [49]. The other three peaks are characteristic of asphaltenes and are located at about 20° (γ-band), 26° (band (002) or graphene band), and 40° (band (100)). The results of the values measured during XRD analysis and the calculated parameters of the fine structure of asphaltenes are given in Table 3.

The spacing between the saturated structures (d_γ) is larger for VacRes asphaltenes than for VisRes asphaltenes, and is about 6.1 and 5.9 Å, respectively. This is due to a decrease in the length of aliphatic chains due to cracking reactions.

The interplanar spacing between aromatic sheets d_{002}, on the contrary, is higher for asphaltenes from VisRes (3.88 Å) than for asphaltenes from VacRes (3.57 Å). The increase in the spacing between the neighboring aromatic sheets after thermal cracking is due to the removal of side-chains and, as a result, the cluster compactness loosening.

As a result of cracking, for the cluster size, L_c and L_a are reduced (from 26.72 to 20.78 Å, respectively, and from 7.68 to 7.20 Å for asphaltenes from VacRes and VisRes, respectively) owing to the rupture of aliphatic chains between aromatic sheets in the cluster. Because of the division of the cluster into parts, there is a decrease in the number of aromatic sheets per stacked cluster M cluster (from 8 to 7) and a decrease in the average number of aromatic rings in the NO_{ar} sheets (from 2.9 to 2.7).

As a result of the thermolysis reaction, cracking of aliphatic chains occurs, their number decreases, the proportion of the densification reaction is insignificant, and the number of aromatic rings practically does not change, which is confirmed by an increase in the aromaticity of asphaltene molecules (f_a from 0.04 to 0.10).

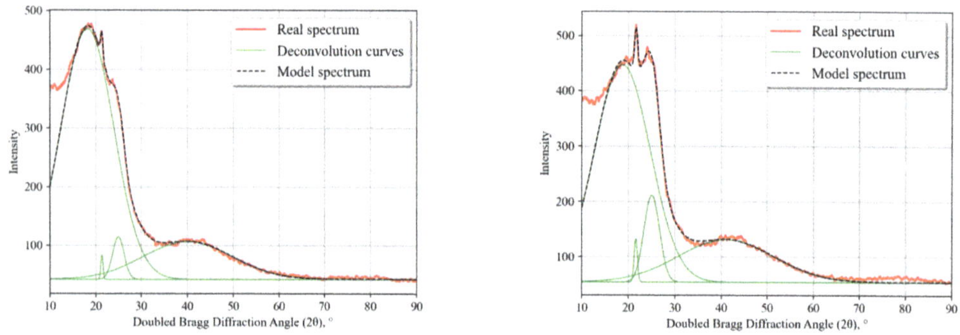

(a) (b)

Figure 5. Comparison of diffractograms of asphaltenes from VacRes and VisRes: (**a**) asphaltenes obtained from VacRes and (**b**) asphaltenes obtained from VisRes.

Table 3. Crystalline parameters derived from XRD data for petroleum asphaltenes.

	γ-band			Graphene or (002) Band			(100) Band	
Estimate	VacRes	VisRes	Estimate	VacRes	VisRes	Estimate	VacRes	VisRes
2θ, °	18.1802	18.7465	2θ, °	24.9476	25.0246	2θ, °	40.0377	41.0000
FWHM, °	13.4967	13.8509	FWHM, °	3.0142	3.8752	FWHM, °	22.5059	24.0985
d_γ, Å	6.0946	5.9121	d_{002}, Å	3.5663	3.8752	d_{100}, Å	2.2502	2.1996
fa	0.04	0.10	L_c, Å	26.7221	20.7793	L_a, Å	7.6808	7.1955
			M	8	7	NO_{ar}	2.9	2.7

3.3. Stability of the Residual Marine Fuel Oil

Figure 6 shows the ternary phase diagrams of stability of the residual marine fuel oil (VacRes–ULSD–LCGO and VisRes–ULSD–LCGO).

It can be seen from the phase diagrams of sedimentation stability (ISO 10307-1-2009) that the obtained results differ dramatically for marine fuels based on primary processing residues (VacRes) and thermodestructive process residues (VisRes). Vacuum and visbreaking residues contain about 37.0 wt% of resinous-asphaltene substances, where vacuum residue contains 16.0 wt% of asphaltenes, and VisRes contains 26.6 wt% of HI-fractions, which, in addition to asphaltenes, also contains carbenes and carbides. Carbenes and caborides are more condensed structures than asphaltenes. In addition, their solubility is lower than the solubility of asphaltenes in hydrocarbons. Thus, carbenes and carbides are more prone to sediment formation. The ratio of the 65 stable compositions (number of green dots in Figure 6) to the total number of possible compositions of 66 was 98.5% for residual marine fuel comprising a mixture VacRes–ULSD–LCGO. Meanwhile, the ratio of the 38 stable compositions (number of green dots in Figure 6) to the total number of possible compositions of 66 was 57.6% for residual marine fuel comprising a mixture VisRes–ULSD–LCGO.

Additional studies have been conducted to support the claim that differences in asphaltene structure have a significant impact on the stability of residual marine fuel. The isolated asphaltenes from VacRes and VisRes were added in amounts of 2.5 and 5.0% wt., respectively, into commercial residual marine fuel of RMG 380. Quality indicators of commercial residual marine fuel RMG 380 are presented in Table 4. The results of determining TSA are presented in Table 5.

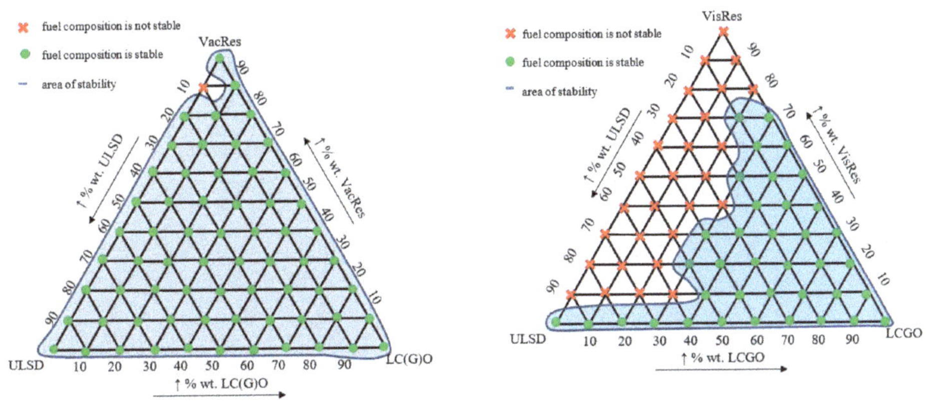

(a) (b)

Figure 6. The ternary phase diagrams of stability of the residual marine fuel oil: (**a**) VacRes–ULSD–LCGO and (**b**) VisRes–ULSD–LCGO.

Table 4. The quality indicators of commercial residual marine fuel of RMG 380.

Parameter	RMG 380
Density at 15 °C, kg/m^3	956.0
Viscosity, mm^2/s at 50 °C	321.5
Flash point, °C	98
Pour point, °C	19
TSA, %wt.	0.0268

Table 5. TSA results for RMG 380 with VacRes and VisRes asphaltenes.

Asphaltenes Content added to RMG 380, %wt.	TSA RMG 380 + Asphaltenes, %wt.	
	VacRes Asphaltenes	VisRes Asphaltenes
0.0	0.0268	
2.5	0.0320	0.0984
5.0	0.1278	0.2653

TSA of residual marine fuel with the addition of 2.5% asphaltenes from VacRes increases 1.2 times, while adding asphaltenes from VisRes increases it 3.7 times. TSA of residual marine fuel with the addition of 5.0% asphaltenes from VacRes increases 4.8 times and, with the addition of asphaltenes from VisRes, it increases 9.9 times compared with the initial TSA value. Thus, the additional study confirms that the structure of asphaltenes is an important factor affecting the change in the stability region of residual marine fuel. The group hydrocarbon composition (SARA) of the fuel has a significant effect on the stability of residual marine fuel, but the degree of condensation of asphaltenes, cluster parameters, microstructure, and morphology also have a huge effect on the stability of the fuel system.

This is because it is the decrease in the solubility of asphaltenes of heavy oil residues that is the main reason for the loss of fuel stability, which in turn depends on these parameters.

4. Conclusions

The stability of residual marine fuel is influenced not only by the component and hydrocarbon composition, but also by the structure of asphaltenes.

The main reason for the instability of residual marine fuel is the component and hydrocarbon composition, which is confirmed using three-component phase diagrams. Vacuum and visbreaking residues contain about 37.0 wt% of resinous-asphaltene substances, where vacuum residue contains 16.0 wt% of asphaltenes and VisRes contains 26.6 wt% of HI-fractions. Paraffin-naphthenic hydrocarbons (96%) contained in ULSD have a significant effect on the stability of the fuel system, especially n-paraffins, the content of which reaches 43%. With an increase in their concentration in the mixture, VacRes/ULSD, and reaching the critical concentration, the hydrocarbon composition of the mixture changes so much that the fuel mixture becomes unstable. LCGO has a high content of aromatic hydrocarbons, which, when added even in an amount of 20%, increase the solubility of asphaltenes and make the entire fuel mixture stable.

In addition, asphaltenes of different nature, which are the main cause of the instability of residual marine fuel, have a different structure.

As a result of cracking, for the cluster size, L_c and L_a are reduced (from 26.72 to 20.78 Å, respectively, and from 7.68 to 7.20 Å for asphaltenes from VacRes and VisRes, respectively) owing to the rupture of aliphatic chains between aromatic sheets in the cluster. Because of the division of the cluster into parts, there is a decrease in the number of aromatic sheets per stacked cluster M cluster (from 8 to 7) and a decrease in the average number of aromatic rings in the NOar sheets (from 2.9 to 2.7).

Author Contributions: Formal analysis, Investigation, Writing—original draft, K.I.S.; Formal analysis, Methodology, Writing—review & editing, V.A.R.; Methodology, Supervision, V.G.P.; Investigation, A.A.S.; Software, I.E.; Investigation, R.R.G.; Formal analysis, Project administration, Supervision, I.N.P.; Methodology, Project administration, J.G.S. All authors have read and agreed to the published version of the manuscript.

Funding: This research received no external funding.

Institutional Review Board Statement: Not applicable.

Informed Consent Statement: Not applicable.

Data Availability Statement: Not applicable.

Acknowledgments: The study was carried out in the Scientific Center "Issues of Mineral and Technogenic Resources Processing" with the involvement of the laboratory base of Center for collective use of Saint Petersburg Mining University.

Conflicts of Interest: The authors declare no conflict of interest.

References

1. Spezzano, P. Mapping the susceptibility of UNESCO World Cultural Heritage sites in Europe to ambient (outdoor) air pollution. *Sci. Total Environ.* **2021**, *754*, 142345. [CrossRef]
2. Buslaev, G.; Morenov, V.; Konyaev, Y.; Kraslawski, A. Reduction of carbon footprint of the production and field transport of high-viscosity oils in the Arctic region. *Chem. Eng. Process.-Process Intensif.* **2020**, *159*, 108189. [CrossRef]
3. El-Houjeiri, H.; Monfort, J.C.; Bouchard, J.; Przesmitzki, S. Life Cycle Assessment of Greenhouse Gas Emissions from Marine Fuels: A Case Study of Saudi Crude Oil versus Natural Gas in Different Global Regions. *J. Ind. Ecol.* **2019**, *23*, 374–388. [CrossRef]
4. Thomson, H.; Corbett, J.J.; Winebrake, J.J. Natural gas as a marine fuel. *Energy Policy* **2015**, *87*, 153–167. [CrossRef]
5. Raitt, D. LNG as marine fuel—Where technology meets logistics. *APPEA J.* **2018**, *58*, 593. [CrossRef]
6. Herdzik, J. Consequences of using LNG as a marine fuel. *J. KONES* **2013**, *20*, 159–166.
7. Litvinenko, V. The Role of Hydrocarbons in the Global Energy Agenda: The Focus on Liquefied Natural Gas. *Resources* **2020**, *9*, 59. [CrossRef]
8. Litvinenko, V.S.; Dvoynikov, M.V.; Trushko, V.L. Elaboration of a conceptual solution for the development of the Arctic shelf from seasonally flooded coastal areas. *Int. J. Min. Sci. Technol.* **2021**, in press. [CrossRef]

9. Filatova, I.; Nikolaichuk, L.; Zakaev, D.; Ilin, I. Public-Private Partnership as a Tool of Sustainable Development in the Oil-Refining Sector: Russian Case. *Sustainability* **2021**, *13*, 5153. [CrossRef]
10. Aramkitphotha, S.; Tanatavikorn, H.; Yenyuak, C.; Vitidsant, T. Low sulfur fuel oil from blends of microalgae pyrolysis oil and used lubricating oil: Properties and economic evaluation. *Sustain. Energy Technol. Assess.* **2019**, *31*, 339–346. [CrossRef]
11. Gulyaeva, L.A.; Lobashova, M.M.; Mitusova, T.N.; Shmel'kova, O.I.; Khavkin, V.A.; Nikul'shin, P.A. Production of Low-Sulfur Marine Fuel. *Chem. Technol. Fuels Oils* **2020**, *55*, 704–711. [CrossRef]
12. Stern, D.L.; Mauro, S.R.D.; Roccaro, A.; Bessonette, P.W. ExxonMobil Research and Engineering Co. Fuel Compositions and Methods for Making Same. US Patent 8999011 B2, 28 March 2012.
13. Droubi, D.F.; Branch, M.A.; Delaney-Kinsella, C.; Lipinsky, D.T.; Kraus, L.S.; Brumfield, T.L.; Bru, A.; Steernberg, K.; Tardif, P.; Boudreaux, S. Shell Internationale Research Maatschappij BV. Fuel Compositions. US Patent 8,987,537 B1, 23 March 2015.
14. Buchanan, K.D. Sunoco Partners Marketing & Terminals L.P. Low Sulfur Marine Fuel. US Patent 20150353851A1, 10 December 2015.
15. Wohaibi, M.; Pruitt, T.F. Mawetal LLC. Environment-Friendly Marine Fuel. US Patent 20190093026A1, 18 October 2016.
16. Zhang, Y.; Siskin, M.; Gray, M.R.; Walters, C.C.; Rodgers, R.P. Mechanisms of Asphaltene Aggregation: Puzzles and a New Hypothesis. *Energy Fuels* **2020**, *34*, 9094–9107. [CrossRef]
17. Speight, J.G. Petroleum asphaltenes—Part 2: The effect of asphaltenes and resin constituents on recovery and refining processes. *Oil Gas Sci. Technol.* **2004**, *59*, 479–488. [CrossRef]
18. Speight, J.G. *Refinery Feedstocks*; CRC Press: Boca Raton, FL, USA, 2021; 356p.
19. Sultanbekov, R.; Nazarova, M. The influence of total sediment of petroleum products on the corrosiveness of the metal of the tanks during storage. *E3S Web Conf.* **2019**, *121*, 01015. [CrossRef]
20. Sultanbekov, R.; Islamov, S.; Mardashov, D.; Beloglazov, I.; Hemmingsen, T. Research of the Influence of Marine Residual Fuel Composition on Sedimentation Due to Incompatibility. *J. Mar. Sci. Eng.* **2021**, *9*, 1067. [CrossRef]
21. Struchkov, I.A.; Rogachev, M.K.; Kalinin, E.S.; Roschin, P.V. Laboratory investigation of asphaltene-induced formation damage. *J. Pet. Explor. Prod. Technol.* **2019**, *9*, 1443–1455. [CrossRef]
22. Golubev, I.; Golubev, A.; Laptev, A. Practice of using the magnetic treatment devices to intensify the processes of primary oil treating. *J. Min. Inst.* **2020**, *245*, 554–560. [CrossRef]
23. Khormali, A.; Sharifov, A.R.; Torba, D.I. Experimental and modeling study of asphaltene adsorption onto the reservoir rocks. *Pet. Sci. Technol.* **2018**, *36*, 1482–1489. [CrossRef]
24. Rogachev, M.; Aleksandrov, A. Justification of a comprehensive technology for preventing the formation of asphalt-resin-paraffin deposits during the production of highlyparaffinic oil by electric submersible pumps from multiformation deposits. *J. Min. Inst.* **2021**, *250*, 596–605. [CrossRef]
25. Hosseini-Dastgerdi, Z.; Tabatabaei-Nejad, S.A.R.; Khodapanah, E.; Sahrae, E. A comprehensive study on mechanism of formation and techniques to diagnose asphaltene structure; molecular and aggregates: A review. *Asia-Pac. J. Chem. Eng.* **2014**, *17*, 743–753. [CrossRef]
26. Anchita, J. HYDRO-IMP technology for heavy oil refining. *J. Min. Inst.* **2017**, *224*, 229–234.
27. Ilyina, M.G.; Khamitov, E.M.; Galiakhmetov, R.N.; Mustafin, I.A.; Akhmetov, A.F.; Shayakhmetova, R.K.; Mustafin, A.G. Light gasoil of catalytic cracking: A quantitative description of the physical properties by joint use of chromato-mass-spectrometry and molecular dynamics. *J. Chin. Chem. Soc.* **2020**, *67*, 33–40. [CrossRef]
28. Badikova, A.D.; Muhamadeev, R.U.; Shiryaeva, R.N.; Mustafin, A.G.; Rullo, A.V.; Ibragimov, I.G. Spectral methods of analysis capabilities for investigation of the composition of oil sludges. *SOCAR Proc.* **2019**, *4*, 32–38. [CrossRef]
29. Speight, J.G. *Heavy Oil Recovery and Upgrading*; CRC Press Taylor & Francis Group: Boca Raton, FL, USA, 2019; 821p.
30. Luo, P.; Wang, X.; Gu, Y. Characterization of asphaltenes precipitated with three light alkanes under different experimental conditions. *Fluid Phase Equilib.* **2010**, *291*, 103–110. [CrossRef]
31. AlHumaidan, F.S.; Rana, M.S.; Tanoli, N.J.; Lababidi, H.M.; Al-Najdi, N.A. Changes in asphaltene surface topography with thermal treatment. *Arab. J. Chem.* **2020**, *13*, 5377–5389. [CrossRef]
32. AlHumaidan, F.S.; Hauser, A.; Rana, M.S.; Lababidi, H.M.; Behbehani, M. Changes in asphaltene structure during thermal cracking of residual oils: XRD study. *Fuel* **2015**, *150*, 558–564. [CrossRef]
33. Siddiqui, M.N.; Ali, M.F.; Shirokoff, J. Use of X-ray diffraction in assessing the aging pattern of asphalt fractions. *Fuel* **2002**, *81*, 51–58. [CrossRef]
34. Ancheyta, J.; Trejo, F.; Rana, M.S. *Asphaltenes Chemical Transformation during Hydro Processing of Heavy Oils*; CRC Press Taylor & Francis Group: Boca Raton, FL, USA, 2009; 441p.
35. Yen, T.F.; Erdman, J.G.; Pollack, S.S. Investigation of the Structure of Petroleum Asphaltenes by X-ray Diffraction. *Anal. Chem.* **1961**, *33*, 1587–1594. [CrossRef]
36. Bouhadda, Y.; Bormann, D.; Sheu, E.; Bendedouch, D.; Krallafa, A.; Daaou, M. Characterization of Algerian Hassi-Messaoud asphaltene structure using Raman spectrometry and X-ray diffraction. *Fuel* **2007**, *86*, 1855–1864. [CrossRef]
37. Jameel, A.G.; Alkhateeb, A.; Telalović, S.; Elbaz, A.M.; Roberts, W.L.; Sarathy, S.M. Environmental Challenges and Opportunities in Marine Engine Heavy Fuel Oil Combustion. In *Proceedings of the Fourth International Conference in Ocean Engineering (ICOE2018) 2019*; Springer: Singapore, 2019; pp. 1047–1055.
38. Chesnokov, V.V.; Chichkan, A.S.; Parmon, V.N. Effect of Nickel-Containing Catalyst on the Tar Coking Process. *Catal. Ind.* **2021**, *13*, 143–149. [CrossRef]

39. Shaltout, A.A.; Gomma, M.M.; Ali-Bik, M.W. Utilization of standardless analysis algorithms using WDXRF and XRD for Egyptian iron ore identification. *X-ray Spectrom.* **2012**, *41*, 355–362. [CrossRef]
40. SciPy.org. 2020. Available online: https://docs.scipy.org/doc/scipy/reference/generated/scipy.optimize.curve_fit.html (accessed on 5 November 2021).
41. Bragg, W.L. The Structure of Some Crystals as Indicated by Their Diffraction of X-rays. *Proc. R. Soc. A Math. Phys. Eng. Sci.* **1913**, *89*, 248–277.
42. Wulff, G. Über die Kristallröntgenogramme. *Phys. Zeitschrift.* **1913**, *14*, 217–220.
43. Shirokoff, J.W.; Siddiqui, M.N.; Ali, M.F. Characterization of the Structure of Saudi Crude Asphaltenes by X-ray Diffraction. *Energy Fuels* **1997**, *11*, 561–565. [CrossRef]
44. Scherrer, P. *Bestimmung der inneren Struktur und der Größe von Kolloidteilchen mittels Röntgenstrahlen. Kolloidchemie Ein Lehrbuch*; Springer: Berlin/Heidelberg, Germany, 1912; pp. 387–409.
45. Warren, B.E. X-ray diffraction in random layer lattices. *Phys. Rev.* **1941**, *59*, 693–698. [CrossRef]
46. Feret, F.R. Determination of the crystallinity of calcined and graphitic cokes by X-ray diffraction. *Analyst* **1998**, *123*, 595–600. [CrossRef]
47. Gibbs, J.W. The Scientific Papers of J. Willard Gibbs. *Nature* **1907**, *75*, 361–362.
48. Kurnakov, N.S. Introduction to Physicochemical Analysis. *Izv. Akad. Nauk SSSR.* **1940**.
49. Andersen, S.I.; Jensen, J.O.; Speight, J.G. X-ray Diffraction of Subfractions of Petroleum Asphaltenes. *Energy Fuels* **2005**, *19*, 2371–2377. [CrossRef]

Journal of
Marine Science and Engineering

MDPI

Article

Research of the Influence of Marine Residual Fuel Composition on Sedimentation Due to Incompatibility

Radel Sultanbekov [1], Shamil Islamov [2], Dmitry Mardashov [2], Ilia Beloglazov [3,*] and Tor Hemmingsen [4]

1 Department of Transport and Storage of Oil and Gas, Saint Petersburg Mining University, 199106 Saint Petersburg, Russia; Sultanbekov_RR@pers.spmi.ru
2 Department of Development and Operation of Oil and Gas Fields, Saint Petersburg Mining University, 199106 Saint Petersburg, Russia; Islamov_ShR@pers.spmi.ru (S.I.); Mardashov_DV@pers.spmi.ru (D.M.)
3 The Department of Automation of Technological Processes and Production, Saint Petersburg Mining University, 199106 Saint Petersburg, Russia
4 Faculty of Science and Technology, University of Stavanger, 4036 Stavanger, Norway; tor.hemmingsen@uis.no
* Correspondence: Beloglazov_II@pers.spmi.ru

Abstract: The article shows studies of the problem of active sediment formation during mixing of residual fuels, caused by the manifestation of incompatibility. To preserve the quality and reduce sediment formation during transshipment, storage, and transportation of marine residual fuels, a laboratory method for determining the compatibility and stability of fuels has been developed, which makes it possible to determine the quantitative characteristics of the sediment formation activity. According to the method developed, laboratory studies have been carried out to determine incompatible fuel components and the influence of composition on the sedimentation process. Tests were carried out to determine the quality indicators and the individual group composition of the fuel samples. Based on the results of the studies, the dependences of the influence of normal structure paraffins in the range from 55 to 70 wt. % and asphaltenes in the range from 0.5 to 3.5 wt. % in the fuel composition on the sedimentation activity due to incompatibility were obtained. To obtain a convenient tool that is applicable in practice, a nomogram has been developed on the basis of the dependences obtained experimentally. It was also determined that, after reaching the maximum values of sediment formation with a further increase in the content of n-paraffins, saturation is observed, and the value of the sediment content remains at the same level. Maximum total sediment values have been found to depend on asphaltene content and do not significantly exceed them within 10%. The results of the research presented in this article allow laboratory and calculation to determine the possibility of incompatibility and to preserve the quality of marine residual fuels.

Keywords: incompatibility; stability; total sediment; precipitation; marine residual fuels; mixing; storage; group composition

Citation: Sultanbekov, R.; Islamov, S.; Mardashov, D.; Beloglazov, I.; Hemmingsen, T. Research of the Influence of Marine Residual Fuel Composition on Sedimentation Due to Incompatibility. *J. Mar. Sci. Eng.* **2021**, *9*, 1067. https://doi.org/10.3390/jmse9101067

Academic Editors: María Dolores Esteban Pérez, José-Santos López-Gutiérrez, Vicente Negro and M. Graça Neves

Received: 6 September 2021
Accepted: 26 September 2021
Published: 29 September 2021

1. Introduction

Today, when providing civil ships and vessels with fuel, there are certain difficulties associated with the quality characteristics of marine residual fuels, namely in connection with the tightening of requirements for the sulfur content due to the introduction by the International Maritime Organization of a new convention in MARPOL 73/78 Annex VI [1]. Since 1 January 2020, according to this convention, it is regulated that, throughout the world's oceans, the sulfur content in marine fuels should not exceed 0.5% of the mass. As a result, the permissible sulfur content in marine fuels has decreased 7 times, from 3.5% to 0.5% of the mass. Oil refineries in Russia are currently not able to fully provide a new type of fuel with sulfur up to 0.5%, as this requires a significant restructuring of the refining process and requires significant capital investments [2–4]. Therefore, in order to meet the demand for a new type of marine fuels, bunkering companies are actively carrying out operations to mix fuels to obtain the required quality indicators [5,6]. That is why there is a sharp increase in the share of composite fuels for ship installations, which, in turn, increases the risks of active sedimentation, which is caused by the manifestation of incompatibility of fuels [7,8].

In this regard, during storage and transportation, especially marine, the problem of active sedimentation of residual and mixed fuels caused by loss of stability is relevant [9]. According to the international standard ISO 8217 for fuel quality, the content of the total sediment should not exceed 0.1% of the mass. Note that there are risks of incompatibility even when mixing the same brand of fuels due to differences in composition. The incompatibility of residual fuels is manifested due to the occurrence of strong intermolecular interactions, which are caused by a change in the structural group individual composition, as well as a change in the concentration ratio of high-molecular compounds of residual fuels [10,11]. All this contributes to the formation of associates of molecules, colloidal particles by volume of various shapes and structures [12,13].

There are several laboratory ways and methods to determine the compatibility and stability of residual fuels before actual mixing in tanks and storages: the spot method GOST 33365 and ASTM D 4740, determination of xylene and toluene equivalents in accordance with GOST 33288 and GOST 33296, patent RU 2305836 C1, patent RU 2444005 C1, and international application No. WO 2018185520 [14–17]. However, after their consideration and application, these methods and techniques have been found to have a high error rate and low precision, so that they have not been widely used in practice [18].

The first investigations of sediment formation due to loss of stability were carried out in order to study the compounding process of various types of heavy fuels. For the first time, in 1951, Martin used the term "incompatibility", to describe the sharp increase in precipitation observed when mixing "pure" fuel with a low content of total sediment [19].

Holmes et al. and other research groups have studied the manifestation of incompatibility when mixing fuels. These authors have developed basic theoretical provisions but have not formulated clear recommendations to preserve the quality of fuels during mixing to address this problem [20–23].

Other research groups have carried out studies to determine the stability and compatibility of mixtures of marine fuels. The main results of these studies are the acquisition of theoretical information and the development of a three-phase diagram to determine the stability of the mixture of marine residual fuels (Figure 1).

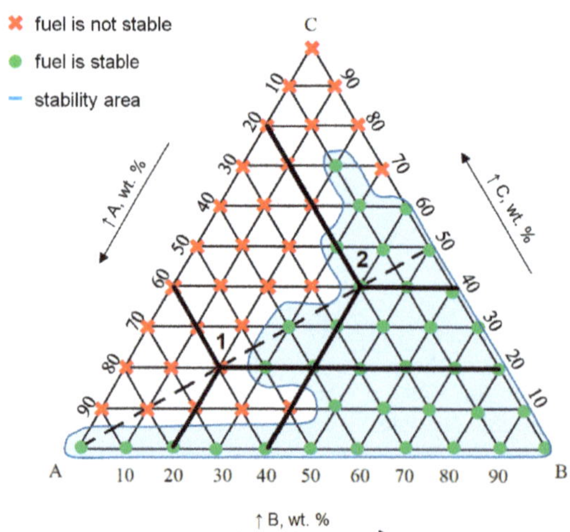

Figure 1. Three-phase stability diagram of marine residual fuel A–B–C [24–26].

The stability assessment method allows the establishment of areas of stability and the determination of the boundary content of each fuel component in a multicomponent mixture, taking into account the use of incompatible components, where the remaining quality requirements for marine fuels are also taken into account, such as density, viscosity, sulfur content, etc.

The compatibility evaluation criteria in this method were adopted by the authors according to ISO 10307-2:2009, which regulates the determination of fuel incompatibility if the total sediment value is greater than 0.1. Less than this value, the fuel composition is compatible and stable. Therefore, when mixing fuels with a total sediment value greater than 0.1%, this method will not allow assessing the stability of fuels.

The study of this problem was carried out in the specially created company "Soluble Solutions", which was directed by Wiehe [27,28]. Researchers have identified more than 100 incompatible oil components that are on the U.S. market. Unfortunately, due to commercial secrets, basic information (species, physicochemical properties, and qualitative characteristics) cannot be disclosed, and information is not publicly available.

However, exact dependences of the mutual influence of hydrocarbon composition on sedimentation due to the manifestation of incompatibility have not been obtained.

Analysis of existing research and patent documents showed that extensive laboratory studies were carried out to study the process of sedimentation. Most of the research focuses on sedimentation in oil [29]. It is also worth noting that precise, reliable, and practically applicable methods for predicting the potential incompatibility of mixtures that allow assessing the activity of sedimentation have not yet been developed. There is no dependency between the mutual influence of hydrocarbon composition on sediment formation, and many researchers limited themselves to only general recommendations for maintaining quality.

2. Materials and Methods

In the present work, the marine residual fuels of the KMS, TSU-80, TSU-380. and M-100 brands (according to the classification ISO 8217, corresponding to the brands RMD-80—sample No.1, RMD-80—sample No.2, RMG-380—sample No.3, and RMK 700—sample No.4) were used as research objects, which are shown in Figure 2. The samples were taken according to GOST 2517 from storage and railway tanks. The choice of the marine residual fuels presented is based on the different group composition, which allows the required fuel composition to be obtained when mixing to determine the influence of each group of residual fuels composition on the compatibility and stability of the fuel system. Isostructured paraffins predominate in the TSU-80 fuel and normal paraffins in the KMS fuel. The TSU-380 and M-100 fuel brands contain heavier components, including asphaltenes and resins. By combining the composition, it is possible to determine the influence of each component on the stability of the fuel system.

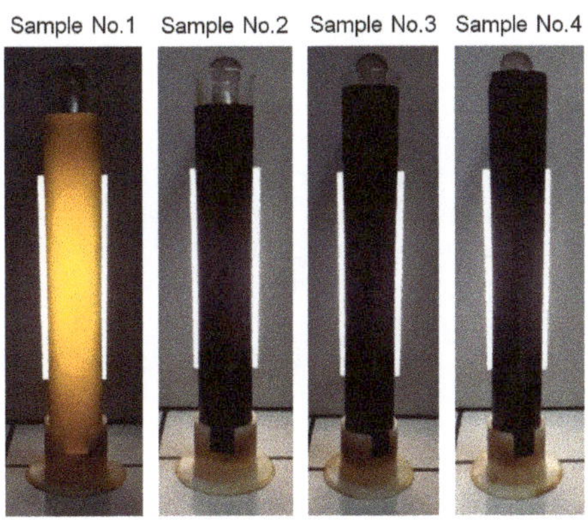

Figure 2. Samples of marine residual fuels.

Modern techniques and the corresponding laboratory equipment base of Scientific Center "Issues of Processing Mineral and Technogenic Resources" of Saint Petersburg Mining University were used to determine the main indicators of the quality of marine fuels, namely density, kinematic viscosity, PMCC flashpoint, sulfur and water content, pourpoint, total sediments accelerated (TSA), and total sediments potential (TSP) values.

In the work, for ease of labeling and conducting research, the components are distributed by numbers:

- Sample No.1—Compound oils grade A type 1 (KMC);
- Sample No.2—Fuel for TSU-80 marine installations (RMD-80);
- Sample No.3—Fuel for TSU-380 marine installations (RMG-380);
- Sample No.4—Residual oil fuel M-100, low ash (RK-700).

The group hydrocarbon composition of the samples was analyzed using actual methods, namely the SARA analysis, which is based on the separation of the petroleum products studied into saturated hydrocarbons, aromatic compounds, resins, and asphaltenes, according to their polarity and solubility (Figure 3).

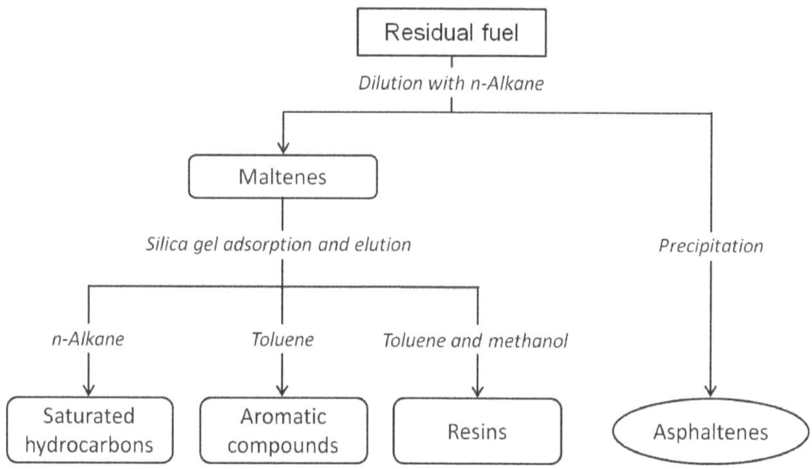

Figure 3. Residual fuel dilution scheme to determine the group composition of fuels.

Analysis of the group hydrocarbon composition of fuels was carried out on the laboratory gas chromatograph "Gradient-M" with a heat conductivity detector by liquid adsorption chromatography on glass columns with a height of 30 cm and a diameter of 1.2–1.4 mm [30,31].

According to the presented method, n-heptane was added for the separation of asphaltenes in a ratio of 1:40 by mass. The remaining maltenes are further separated by passing through a chromatographic column with adsorbent. The absorbent was a modified silica gel of the ASC brand, which is precalcined for 4 h at 150 °C. A complex mixture of solvents with gradient-displacement mode of supply was used as eluates. This method allows testing of samples for paraffin-naphthenic, aromatic hydrocarbons, resins, and asphaltenes [32,33].

Due to the newly developed method for determining fuel compatibility and stability, it became possible to determine the quantitative characteristics of sedimentation [34]. The essence of the method is as follows: initially, the contents of the total sediment of each component of the mixture are determined; further, the components are mixed in the required proportions, after which the value of the total sediment of the fuel mixture is determined, on the basis of which the compatibility index (CI) is calculated [35,36].

The CI of a fuel mixture is defined as the sum of the average total sediment of the fuel mixture components (S_{av}) and the convergence value (r). Calculation formula (1) is presented below [35]:

$$CI = S_{av} + r. \quad (1)$$

If the actual TSP values of the S_{av} mixture are greater than CI, then the fuel mixture is incompatible; if S_{av} is less than CI, then the fuel mixture is compatible and stable during long-term storage.

For example, consider the results of laboratory tests carried out according to the method developed. Samples No.1 and 3 are mixed in a 1:1 ratio. The result is a manifestation of fuel incompatibility, which contributes to a sharp increase in total sediment, the TSP value = 0.17 wt. %, where TSP is the total sediment after aging a sample of residual fuel for 24 h at 100 °C under prescribed conditions (Figure 4).

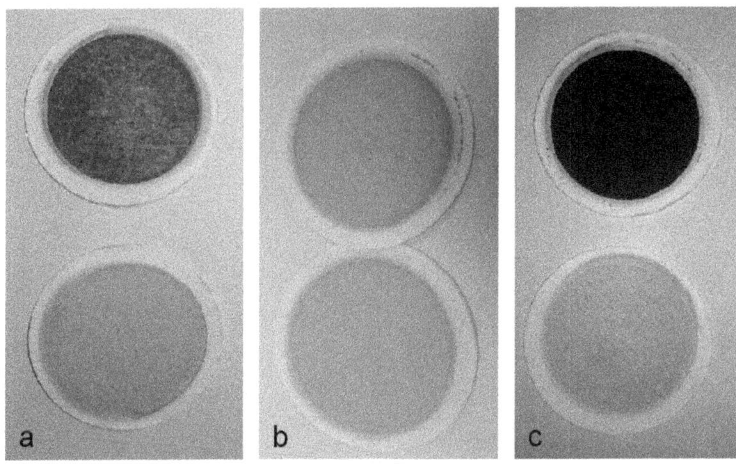

Figure 4. Filters after filtration and determination of total sediment content: (**a**)—sample No.1; (**b**)—sample No.2; (**c**)—mixture of samples No.1 and No.2.

Figure 4 shows as an example filter after thermostating and filtration of 10 g of fuel components (Figure 4a,b) and fuel mixture (Figure 4c) of samples No.1 and 3 with an asphaltene content of 1.0 wt. % and n-paraffins 63 wt. %. According to the developed method [35] on the basis of GOST 33360-2015, the samples are filtered through 2 filters and then washed with a mixture of heptane (85%) and toluene (15%). The test shall be considered valid if the lower filter is light-colored. Even visually, a high content of the total sediment in the fuel mixture is visible, since after mixing there is a loss of stability of the dispersed system and coagulation of molecules, which leads to the precipitation of asphaltene compounds. The arithmetic mean of two measurements to determine the amount of TSP value in the mixture is 0.17 wt. %. Hence, $r = 0.123\sqrt{X} = 0.05\%$, which means $CI = S_{av} + r = \left(\frac{0.02+0.01}{2}\right) + 0.05 = 0.065$. As a result, $0.17 \geq 0.065$, i.e., $S_{av} \geq CI$, it follows that this mixture of fuels is incompatible, which causes active sedimentation.

Theoretical studies of the influence of the composition of residual fuels on sediments were carried out for laboratory experiments. On the basis of known theoretical data, sediments in the fuel system are primarily influenced by changes in n-alkane concentration, as well as asphaltene content. Theoretical evidence of no influence of isoparaffins on the loss of stability of Asphaltenes was taken into account in the design of the experiment, as was the minimal influence of aromatic hydrocarbons with less than 10 wt. % in the fuel system [32]. The correlation-regression analysis performed confirmed these data. All these factors allowed us to optimize the number of laboratory tests required.

3. Results

Laboratory tests have been performed according to arbitration methods to determine the quality indicators of the fuel samples considered. The results of the tests are presented in Table 1.

Table 1. Quality indicators of residual fuels.

No.	Indicator Name	Unit of Measurement	Regulation of the Test Method	Result			
				Sample No.1	Sample No.2	Sample No.3	Sample No.4
1	Density at 15 °C	kg/m^3	ISO 12185	833.5	901.0	956.0	976.0
2	Kinematic viscosity at 50 °C	mm^2/s	GOST 33 ISO 3104	12.1	34.5	321.5	680.1
3	Flashpoint PMCC	°C	GOSTREN ISO 2719	181.0	110.0	98.0	110.0
4	Sulfur mass fraction	%	GOST R 51947 ISO 8754	0.004	0.046	1.276	2.668
5	Pourpoint	°C	ASTM D 6749	26.0	10.0	16.0	20.0
6	Mass fraction of water	%	GOST R 51946 ISO 3733	–	–	0.05	0.1
7	Total sediment accelerated (TSA)	%	GOST R 50837.6 ISO 10307-2	0.01	0.01	0.02	0.03
8	Total sediment potential (TSP)	%	GOST R 50837.6 ISO 10307-2	0.01	0.01	0.02	0.04

According to the results of laboratory tests of samples to determine quality indicators, the fuels submitted according to the classification ISO 8217, GOST R 54299-2010, and GOST 32510-2013 correspond to the brands RMD-80—sample No.1 and 2, RMG-380—sample No.3, and RMK 700—sample No.4.

Table 2 presents the results of laboratory tests to determine the group hydrocarbon composition of the fuels.

Table 2. Analysis of the hydrocarbon group composition of fuel samples.

Fuel Samples	Asphaltenes, wt. %	n-Alkanes, wt. %	Isoalkane, wt. %	Naphthenes, wt. %	Alkenes, wt. %	Aromatic Hydrocarbons, wt. %	Resins, wt. %
No.1	–	71.93	18.35	2.62	5.55	1.55	–
No.2	–	38.68	55.09	4.95	–	1.28	–
No.3	2.53	54.14	22.74	5.43	0.57	14.12	0.47
No.4	7.15	51.15	20.71	4.37	0.41	15.70	0.51

Based on the results of determining the group hydrocarbon composition of the studied samples of marine residual fuels, it can be concluded that samples No.1, 3, and 4 are secondary oil refinery products due to the content of alkenes in their composition. Sample No.1 has a high content of normal type alkanes, and sample No.2 contains isoparaffins. The samples are selected in such a way that the components can be mixed to select the compositions of the fuel mixture, with a subsequent determination of the compatibility and stability of the fuels [33].

According to the results obtained to determine the group composition of the study objects, namely samples No.1, 2, 3, and 4, 112 different fuel compositions were prepared with an asphaltene content from 0.5 to 3.5 wt. %, and simultaneously with an n-paraffin content from 55 to 70 wt. %, due to the developed computer program. On the basis of the results of the calculations, 7 series of single-factor experiments were planned and carried out using the newly developed method. In the first series of laboratory experiments, the asphaltene content in the fuel mixture corresponded to 0.5 wt. %, while the n-paraffin content varied from 55 to 70%. By analogy, a series of tests have been carried out with asphaltene content of 1.0, 1.5, 2.0, 2.5, 3.0, and 3.5 wt. %. The main indicators of the quality of the fuel compositions obtained and the TSP sediment content values for a series of tests with an asphaltene content of 0.5 wt. % are presented in Table 3.

Table 3. Group composition, main quality indicators, and TSP values at an asphaltene content of 0.5 wt. % in the fuel composition.

No.	Asphaltenes, %	n-Paraffins, %	Isoparaffins, %	Aromatic Hydrocarbons, %	Density at 15 °C, kg/m^3	Kinematic Viscosity at 50 °C, mm^2/s	Sulfur, %	TSP Values, %
1	0.5	54.91	34.05	4.00	885.7	30.74	0.29	0.02
2		55.92	32.93	4.01	883.6	29.63	0.29	0.02
3		56.94	31.82	4.01	881.5	28.57	0.29	0.02
4		57.95	30.71	4.01	879.4	27.56	0.28	0.02
5		58.95	29.61	4.01	877.3	26.60	0.28	0.02
6		59.95	28.51	4.01	875.2	25.69	0.28	0.02
7		60.94	27.42	4.02	873.2	24.82	0.28	0.02
8		61.93	26.33	4.02	871.1	24.00	0.27	0.05
9		62.92	25.25	4.02	869.1	23.20	0.27	0.13
10		64.06	24.00	4.02	866.7	22.33	0.27	0.20
11		65.04	22.93	4.02	864.7	21.61	0.26	0.25
12		66.01	21.86	4.02	862.7	20.93	0.26	0.29
13		66.98	20.80	4.02	860.7	20.28	0.26	0.33
14		67.94	19.74	4.02	858.7	19.65	0.25	0.39
15		69.03	19.02	3.59	853.5	17.75	0.23	0.47
16		70.05	18.66	2.85	846.4	15.40	0.20	0.58

For each fuel composition, parallel laboratory tests were performed, and the result was given as the arithmetic mean of the two tests.

Laboratory investigations carried out on the basis of the developed method for determining fuel compatibility and stability made it possible to determine the activity of precipitation formation. In the first series of experiments, the asphaltenes content in the composition of fuel mixtures was 0.5 wt. %; only the content of n-paraffins changed. According to the results obtained, presented in Table 3, the asphaltenes precipitation begins when the content of n-paraffins is 62 wt. %, and the most active precipitation is observed at 70 wt. % of the n-paraffin content.

When the paraffin content of normal structure is from 55 to 61%, the reaction is not observed. Due to the approximation, the dependences of the influence of n-paraffins on the deposition of agglomerates and heavy hydrocarbon components (1) were obtained with the approximation value $R^2 = 0.9805$ (Figure 5) with n-paraffin content in the fuel composition in fuels ranging from 62 to 70 wt. % and of asphaltene content 0.5 wt. %:

$$y = 0.06 \cdot x - 3.6611. \quad (2)$$

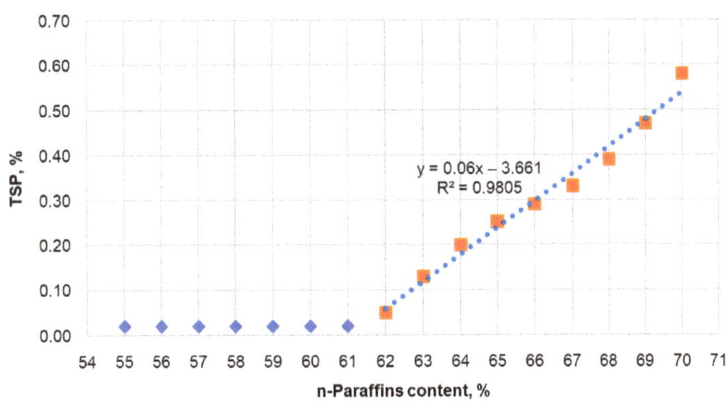

Figure 5. Influence of n-paraffins on the manifestation of incompatibility with an asphaltene content of 0.5 wt. %.

It should be noted that laboratory tests were also performed, which were not included in the experimental plan, namely, at a content of 71 wt. % normal-type paraffins, and the TSP value was 0.56%. The result shows that higher values of the TSP values with an increase in the proportion of n-paraffins are not observed, since saturation and maximum sedimentation occur.

Furthermore, the following experimental studies were carried out in a similar way at different asphaltene content, the results of which are shown in Figure 6.

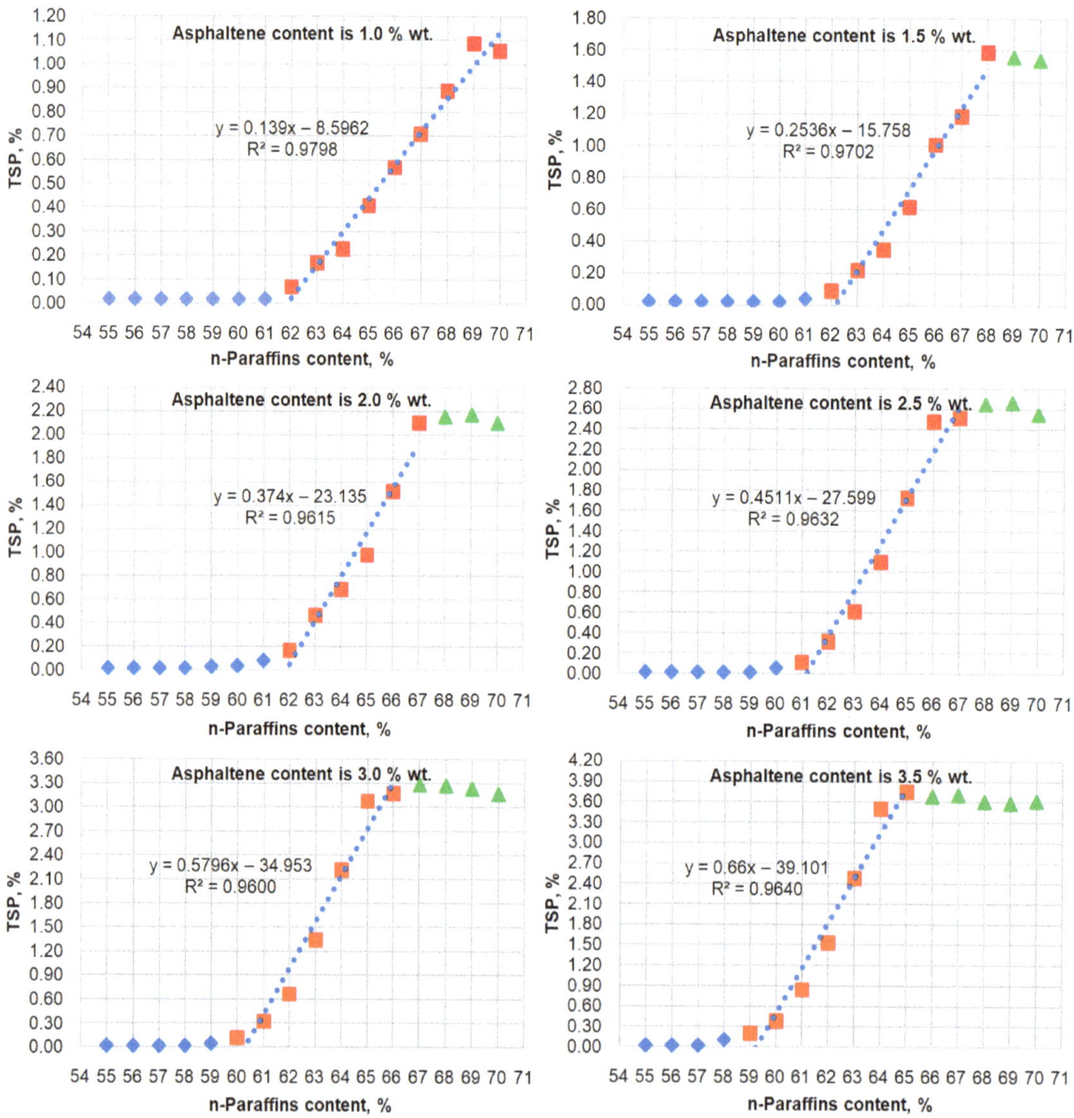

Figure 6. Dependences of the influence of n-paraffins on total sediments at an asphaltene content from 1.0 to 3.5 wt. %. Blue diamonds—no reaction, Red squares—coagulation process, Green triangles—saturation of the coagulation reaction.

As part of the planning of the experiment, 7 series of single-factor experiments were performed on the laboratory method developed to determine the compatibility and stability of fuel components. Based on the results of the studies carried out, the dependences of the influence of n-paraffins on sedimentation activity in fuel mixtures with asphaltene content from 0.5 to 3.5% with a step of 0.5% were obtained. A three-dimensional visualization of the research results is shown in Figure 7.

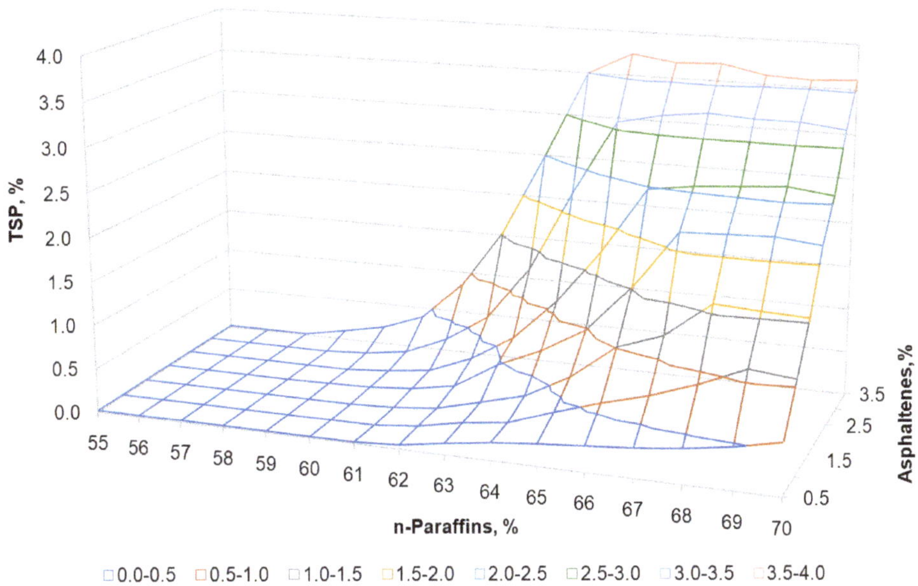

Figure 7. Three-dimensional visualization of the resulting experimental data on the influence of n-paraffins and asphaltenes on the manifestation of incompatibility.

The numerical interpretation of this three-dimensional model of the influence of asphaltenes and n-paraffins on sediment formation is presented in Table 4.

Table 4. Indicators of the influence of n-paraffins and asphaltenes on sedimentation.

n-Paraffins, %	Asphaltenes, %						
	0.5	1.0	1.5	2.0	2.5	3.0	3.5
55	0.02	0.02	0.02	0.02	0.02	0.02	0.02
56	0.02	0.02	0.02	0.02	0.02	0.02	0.02
57	0.02	0.02	0.02	0.02	0.02	0.02	0.02
58	0.02	0.02	0.02	0.02	0.02	0.02	0.11
59	0.02	0.02	0.02	0.03	0.02	0.05	0.21
60	0.02	0.02	0.02	0.04	0.06	0.12	0.39
61	0.02	0.02	0.04	0.08	0.12	0.33	0.85
62	0.05	0.07	0.09	0.17	0.32	0.65	1.54
63	0.13	0.17	0.22	0.47	0.62	1.28	2.49
64	0.20	0.23	0.35	0.69	1.10	2.22	3.50
65	0.25	0.41	0.62	0.98	1.73	3.08	3.75
66	0.29	0.57	1.01	1.52	2.48	3.19	3.68
67	0.33	0.71	1.19	2.18	2.55	3.25	3.70
68	0.39	0.89	1.59	2.20	2.59	3.22	3.60
69	0.47	1.11	1.56	2.22	2.63	3.22	3.58
70	0.58	1.06	1.54	2.17	2.57	3.17	3.61

On the basis of the experimental data obtained, it is possible to determine the dependencies of the influence of the composition of fuel mixtures on sediments. However, the data presented characterize dependences only for specific contents of the fuel composition, which is difficult to apply in practice, so it is necessary to develop a tool to measure sedimentation activity at different values of n-paraffins and asphaltenes in the ranges considered.

In addition, it is worth noting the resulting trend, namely that the maximum TSP values correlate with the asphaltene content that within 10% (Table 4). This can be explained by the fact that asphaltenes are fully precipitated, where heavy fuel components are also present. To numerically confirm this hypothesis, a correlation-regression analysis of the numerical data received is performed, which is presented in Table 5. Data for maximum TSP values were taken with an asphaltene content of 0.5–3.5 wt. %. The results show a high dependency, and the reliability R values are close to1.

Table 5. Correlation and regression analysis of the maximum content of TSP and asphaltene.

Multiple R	0.999471
R-square	0.998943
Normalized R-square	0.998731
Standard error	0.038477
Significance of F	1.24E-08
Student *t*-test	68.7256

To obtain a universal tool that will allow to determine the manifestation of incompatibility of the components of the fuel mixture at different values of n-alkanes (55–70 wt. %) and asphaltenes (0.5–3.5 wt. %) in the ranges considered, a nomogram based on experimental data has been developed and is presented in Figure 8.

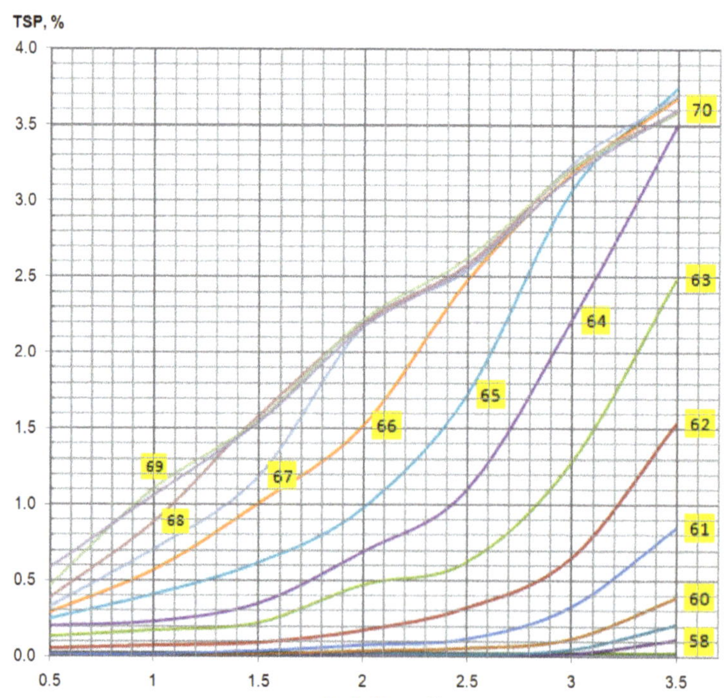

Figure 8. Nomogram of the influence of asphaltene and n-paraffin (marked with percentage) on the activity of precipitation formation.

The nomogram is built using MS Excel software complex, based on the experimental data obtained. On the X-axis, the asphaltene values are represented in wt. % and on the Y-axis, and the data on the sediment content TSP in %. The linear graphs represent n-paraffins in the fuel mixture. The line graphs represent n-paraffins in the composition of the fuel mixture. This makes it possible to determine intermediate values of the sediment content.

4. Discussion

It is important to note that, today, oil refineries in the EU and Russia are not able to fully provide the consumer with a new type of marine fuel, since this requires a significant revision of the approach to oil refining and spending significant capital investments. Since bunkering companies actively carry out fuel mixing operations, in order to obtain the required quality indicators, the article attempts to assess the viability of this approach and analyze possible solutions.

Based on the data obtained and laboratory tests, several conclusions can be drawn.

First, when the content of n-paraffins is less than 57 wt. % in the composition of fuel mixtures, sedimentation is not observed. The absence of coagulation in Figures 5 and 6 is shown with a blue diamond (rhombus).

Second, the higher the asphaltene content of the fuel mixture, the lower the n-paraffin content required for the manifestation of incompatibility. In asphaltene content 0.5 wt. %, incompatibility is observed at 62 wt. % of n-paraffins, and, in asphaltene content, 3.5 wt. %—n-paraffins 58 wt. %. The coagulation process in Figures 5 and 6 is shown in red squares.

Finally, the maximum TSP values depend on the asphaltene content. Similar values are observed with an increase in the proportion of n-paraffin. With further increases in n-alkanes, sedimentation remains at the same level. This is due to the fact that, at the critical concentration of n-paraffins, the asphaltenes are fully precipitated and saturated. This is due to the fact that, when the critical concentration of n-paraffins is reached, asphaltenes completely precipitate, and saturation is observed (shown by the green triangles in Figures 5 and 6).

Present results make it possible to calculate the possibility of incompatibility in the mixing of marine residual fuels and their applicability in practice. Due to the obtained dependences of the influence of n-paraffins and asphaltenes on the manifestation of incompatibility, it becomes possible to reduce risks in advance when building a logistics chain for the supply of fuels to marine fuel terminals and oil tank farms. In connection with the absence of asphaltene and n-paraffin content in the list of basic indicators in the international standard ISO 8217, authors will recommend to include these parameters in the standard. This will allow shipowners and buyers to assess in advance the risks of incompatibility and preserve the quality of fuels. Furthermore, the new method considered for determining the compatibility and stability of fuel components makes it possible, before the actual pumping, to obtain data on possible sedimentation and select the optimal component ratio without deterioration of quality. The developed method and the dependencies received are patented [35,36] and introduced into production, namely at the marine fuel terminal of KONTUR St. Petersburg LLC, which is a subsidiary of Baltic Fuel Company LLC.

5. Conclusions

(1) A method for determining the compatibility and stability of components of a fuel mixture has been developed and patented. The proposed method makes it possible to determine the stability and compatibility of fuels prior to actual fuel blending in tanks and storage facilities by means of a developed laboratory test algorithm and the calculation of a compatibility index, allowing for determination of quantitative characteristics of precipitation.

(2) The dependence of asphaltenes and n-paraffins in the fuel mixture on precipitation by experimental method is obtained, and correlation and regression analysis of the obtained data is performed. A nomogram has been developed for the dependence of sediment

formation in the fuel mixture on the concentration of n-paraffins and asphaltenes in the fuel composition, with the content of n-paraffins from 55 to 70 wt. %and asphaltenes from 0.5 to 3.5 wt. %.

(3) It has been determined that, once maximum TSP values have been reached with a further increase in the n-paraffin content, saturation is observed, and sedimentation values remain at the same level as a result of fully precipitation of asphaltenes in the residual fuel mixture. Maximum TSP values have been found to be correlated with asphaltene content and to not significantly exceed them within 10%.

6. Patents

Method of determining compatibility and stability of fuel mixture components. Available online: https://new.fips.ru/ofpstorage/Doc/IZPM/RUNWC1/000/000/002/733/748/%D0%98%D0%97-02733748-00001/document.pdf (accessed on 25 July 2021).

Program for calculating kinematic viscosity, density, sulfur and water content for a mixture of crude oil and petroleum products. Available online: https://new.fips.ru/ofpstorage/Doc/PrEVM/RUNWPR/000/002/020/613/357/2020613357-00001/document.pdf (accessed on 25 July 2021).

Author Contributions: Conceptualization, R.S., D.M., and I.B.; methodology, R.S.; software, S.I.; validation, D.M., I.B., and T.H.; formal analysis, I.B.; investigation, R.S. and S.I.; resources, S.I. and D.M.; data curation, R.S.; writing—original draft preparation, R.S. and S.I.; writing—review and editing, S.I., I.B. and T.H.; visualization, D.M.; supervision, I.B. and T.H.; project administration, R.S. All authors have read and agreed to the published version of the manuscript.

Funding: The research was performed at the expense of the subsidy for the state assignment in the field of scientific activity for 2021 No. FSRW-2020-0014.

Institutional Review Board Statement: Not applicable.

Informed Consent Statement: Not applicable.

Data Availability Statement: Not applicable.

Acknowledgments: The authors thank Saint Petersburg Mining University for enabling the laboratory experiments. The investigations were carried out using the equipment of the Scientific Centers "Arctic", "Issues of Processing Mineral and Technogenic Resources", and Center for Collective Use of Saint Petersburg Mining University.

Conflicts of Interest: The authors declare no conflict of interest.

Abbreviations

GOST	Russian government standard
CI	Compatibility index
ISO	International organization for standardization
SARA	Saturate, aromatic, resin, and asphaltene
TSA	Total sediment accelerated
TSP	Total sediment potential

References

1. Stratiev, D.; Dinkov, R.; Shishkova, I.; Yordanov, D. Can we manage the process of asphaltene precipitation during the production of IMO 2020 fuel oil? *Erdoel Erdgas Kohle/EKEP* **2020**, *12*, 32–39.
2. Litvinenko, V.S.; Tsvetkov, P.S.; Molodtsov, K.V. The social and market mechanism of sustainable development of public companies in the mineral resource sector. *Eurasian Min.* **2020**, *2020*, 36–41. [CrossRef]
3. Litvinenko, V.S. Digital economy as a factor in the technological development of the mineral sector. *Nat. Resour. Res.* **2020**, *29*, 1521–1541. [CrossRef]
4. Filatova, I.; Nikolaichuk, L.; Zakaev, D.; Ilin, I. Public-private partnership as a tool of sustainable development in the oil-refining sector: Russian case. *Sustainability* **2021**, *13*, 5153. [CrossRef]
5. Dubovoy, E.S.; Shmatkov, A.A.; Shtonda, N.V.; Lyapin, A.Y. On the approach to performance evaluation oil blending stations. *Sci. Technol. Oil Oil Prod. Pipeline Transp.* **2018**, *8*, 540–546. [CrossRef]

6. Demidova, N.P.; Marchenko, A.A.; Onishchenko, O.A. Evaluation of the compatibility of marine heavy fuels. *Bull. Kamchatka State Tech. Univ.* **2016**, *35*, 15–20. [CrossRef]
7. Stratiev, D.; Russell, C.; Sharpe, R.; Shishkova, I.; Dinkov, R.; Marinov, I.; Petkova, N.; Mitkova, M.; Botev, T.; Obryvalina, A.; et al. Investigation on sediment formation in residue thermal conversion based processes. *Fuel Process. Technol.* **2014**, *128*, 509–518. [CrossRef]
8. Abramova, E.A.; Shuvalov, G.V.; Yasyrova, O.A. Development of a method for assessing the stability and compatibility of marine fuels. *Geo-Sibir.* **2011**, *5*, 206–209.
9. Bykova, M.V.; Pashkevich, M.A.; Matveeva, V.A.; Sverchkov, I.P. Assessment and abatement of the soil oil-contamination level in industrial areas. In Proceedings of the International Forum-Contest of Young Researchers: Topical Issues of Rational Use of Natural Resources, St. Petersburg, Russia, 18–20 April 2018; pp. 347–359.
10. Karimov, R.M.; Mastobaev, B.N. Effect of paraffins, resins and asphaltenes on commodity quality of oil. *Bashkir Chem. J.* **2012**, *19*, 97–102.
11. Sergeenko, S.R. *High Molecular Weight Compounds*; Chemistry Press: Moscow, Russia, 1959.
12. Petrovich, N.I.; Ostroukhov, N.N.; Vasilyev, V.V.; Salamatova, E.V.; Strakhov, V.M. Comparison of light fuel-oil fractions produced by semicoking of kukersite shale in a gas generator and in a solid-fuel system. *Coke Chem.* **2019**, *62*, 249–257. [CrossRef]
13. Hemmingsen, T.; Hinnaland, D.; Saasen, A. The co-effect of asphaltene on wax precipitation and crystallisation temperature. In Proceedings of the 9th International Oil Field Chemical Symposium, Norway, Geilo, 22–25 March 1998; pp. 1–6.
14. Method for Determining Secondary Residual Products of Oil Refining in Mixed Fuels. Available online: https://new.fips.ru/Archive/PAT/2011FULL/2011.08.10/DOC/RUNWC1/000/000/002/426/116/DOCUMENT.PDF (accessed on 25 July 2021).
15. Method for Determining Stability Factor of Fuel Compositions Containing Residual Products of Oil Processing. Available online: https://new.fips.ru/Archive/PAT/2007FULL/2007.09.10/DOC/RUNWC1/000/000/002/305/836/DOCUMENT.PDF (accessed on 25 July 2021).
16. Method of Determining Compatibility of Components of Fuel Mixture. Available online: https://new.fips.ru/Archive/PAT/2012FULL/2012.02.27/DOC/RUNWC1/000/000/002/444/005/DOCUMENT.PDF (accessed on 25 July 2021).
17. Fuel Compatibility and Stability Analyzer. Available online: https://patentscope.wipo.int/search/en/detail.jsf?docId=WO2018185520 (accessed on 25 July 2021).
18. Pelenitsyna, O.A.; Litvinenko, N.A.; Gaile, A.A. Development of optimal formulations of ship low-viscosity and residual fuels on the basis of commodity oil products. *Bull. SPBSIT* **2018**, *45*, 30–36.
19. Martin, C.W.G. The stability and compatibility of fuel oils. In Proceedings of the 3rd World Petroleum Congress, The Hague, The Netherlands, 28 May 1951; pp. 66–75.
20. Holmes, J.W.; Bullin, J.A. Fuel oil compatibility probed. *Hydrocarb. Process.* **1983**, *62*, 101–103.
21. Murzakov, R.M.; Sabanenkov, S.A.; Syunyaev, Z.I. Influence of petroleum resins on colloidal stability of asphaltene-containing disperse systems. *Chem. Technol. Fuels Oils* **1980**, *16*, 674–677. [CrossRef]
22. Kassinger, R. Fuel blending—How to minimize risk of incompatibility. In Proceedings of the SPE 6th International Conference on Stability and Handling of Liquid Fuels, Vancouver, BC, Canada, 13–17 October 1997; pp. 489–502.
23. Evdokimov, I.N. *Problems of Incompatibility of Oils When They are Mixed*; National University of Oil and Gas "Gubkin University" Press: Moscow, Russia, 2008.
24. Mitusova, T.N.; Kondrasheva, N.K.; Lobashova, M.M.; Ershov, M.A.; Rudko, V.A. Influence of dispersing additives and blend composition on stability of marine high-viscosity fuels. *J. Min. Inst.* **2017**, *228*, 722–725.
25. Gulvaeva, L.A.; Khavkin, V.A.; Shmel'kova, O.I.; Mitusova, T.N.; Ershov, M.A.; Lobashova, M.M.; Nikul'shin, P.A. Production technology for low-sulfur high-viscosity marine fuels. *Chem. Technol. Fuels Oils* **2019**, *54*, 759–765. [CrossRef]
26. Kondrasheva, N.K.; Rudko, V.A.; Nazarenko, M.Y.; Gabdulkhakov, R.R. Influence of parameters of delayed asphalt coking process on yield and quality of liquid and solid-phase products. *J. Min. Inst.* **2020**, *241*, 97–104. [CrossRef]
27. Wiehe, I.A.; Kennedy, R.J. The oil compatibility model and crude oil incompatibility. *Energy Fuels* **2000**, *14*, 56–59. [CrossRef]
28. Wiehe, I.A.; Yarranton, H.W.; Akbarzadeh, K.; Rahimi, P.M.; Teclemariam, A. The paradox of asphaltene precipitation with normal paraffins. *Energy Fuels* **2005**, *19*, 1261–1267. [CrossRef]
29. Kopteva, A.V.; Koptev, V.Y. Automated monitoring system for asphaltene-resin-paraffin deposits in main oil pipelines. *Int. J. Appl. Eng. Res.* **2016**, *11*, 2191–2198.
30. Cheremisina, O.V.; Ponomareva, M.A.; Sagdiev, V.N. Thermodynamic characteristics of sorption extraction and chromatographic separation of anionic complexes of erbium and cerium with Trilon B on weakly basic anionite. *Russ. J. Phys. Chem.* **2016**, *90*, 664–670. [CrossRef]
31. Cheremisina, O.; Kulagin, V.; El-Saleem, S.; Nikulchev, E. Multisensory gas analysis system based on reconstruction attractors. *Symmetry* **2020**, *12*, 964. [CrossRef]
32. Korshunov, G.I.; Eremeeva, A.M.; Drebenstedt, C. Justification of the use of a vegetal additive to diesel fuel as a method of protecting underground personnel of coal mines from the impact of harmful emissions of diesel-hydraulic locomotives. *J. Min. Inst.* **2021**, *247*, 39–47. [CrossRef]
33. Prakoso, A.; Punase, A.; Klock, K.; Rogel, E.; Ovalles, C.; Hascakir, B. Determination of the stability of asphaltenes through physicochemical characterization of asphaltenes. In Proceedings of the SPE Western Regional Meeting, Anchorage, AK, USA, 23 May 2016; pp. 1–20.

34. Povarov, V.G.; Lopatnikov, A.I. A simple multisensor detector based on tin dioxide in capillary gas chromatography. *J. Anal. Chem.* **2016**, *71*, 901–906. [CrossRef]
35. Method of Determining Compatibility and Stability of Fuel Mixture Components. Available online: https://new.fips.ru/ofpstorage/Doc/IZPM/RUNWC1/000/000/002/733/748/%D0%98%D0%97-02733748-00001/document.pdf (accessed on 25 July 2021).
36. Program for Calculating Kinematic Viscosity, Density, Sulfur and Water Content for a Mixture of Crude Oil and Petroleum Products. Available online: https://new.fips.ru/ofpstorage/Doc/PrEVM/RUNWPR/000/002/020/613/357/2020613357-00001/document.pdf (accessed on 25 July 2021).

Journal of
Marine Science and Engineering

MDPI

Article

Machine Learning Based Moored Ship Movement Prediction

Alberto Alvarellos [1,*], Andrés Figuero [2], Humberto Carro [2], Raquel Costas [2], José Sande [2], Andrés Guerra [3], Enrique Peña [2] and Juan Rabuñal [4,*]

1 Software Engineering Laboratory (ISLA), Computer Science Department, Universidade da Coruña, Campus Elviña s/n, 15071 A Coruña, Spain
2 Water and Environmental Engineering Group (GEAMA), Universidade da Coruña, Campus Elviña s/n, 15071 A Coruña, Spain; andres.figuero@udc.es (A.F.); humberto.fidalgo@udc.es (H.C.); raquel.costas.gomez@udc.es (R.C.); jose.sande@udc.es (J.S.); enrique.penag@udc.es (E.P.)
3 Port Authority of A Coruña, Avenida de la Marina 3, 15001 A Coruña, Spain; aguerra@puertocoruna.com
4 Center for Technological Innovation in Construction and Civil Engineering (CITEEC), RNASA Group, Computer Science Department, Universidade da Coruña, Campus Elviña s/n, 15071 A Coruña, Spain
* Correspondence: alberto.alvarellos@udc.es (A.A.); juan.rabunal@udc.es (J.R.)

Abstract: Several port authorities are involved in the R+D+i projects for developing port management decision-making tools. We recorded the movements of 46 ships in the Outer Port of Punta Langosteira (A Coruña, Spain) from 2015 until 2020. Using this data, we created neural networks and gradient boosting models that predict the six degrees of freedom of a moored vessel from ocean-meteorological data and ship characteristics. The best models achieve, for the surge, sway, heave, roll, pitch and yaw movements, a 0.99, 0.99, 0.95, 0.99, 0.98 and 0.98 R^2 in training and have a 0.10 m, 0.11 m, 0.09 m, 0.9°, 0.11° and 0.15° RMSE in testing, all below 10% of the corresponding movement range. Using these models with forecast data for the weather conditions and sea state and the ship characteristics and berthing location, we can predict the ship movements several days in advance. These results are good enough to reliably compare the models' predictions with the limiting motion criteria for safe working conditions of ship (un) loading operations, helping us decide the best location for operation and when to stop operations more precisely, thus minimizing the economic impact of cargo ships unable to operate.

Keywords: machine learning; neural networks; deep learning; gradient boosting; decision trees; ship movement prediction; cargo ship; port management

Citation: Alvarellos, A.; Figuero, A.; Carro, H.; Costas, R.; Sande, J.; Guerra, A.; Peña, E.; Rabuñal, J. Machine Learning Based Moored Ship Movement Prediction. *J. Mar. Sci. Eng.* **2021**, *9*, 800. https://doi.org/10.3390/jmse9080800

Academic Editors: M. Dolores Esteban, Vicente Negro, M. Graça Neves and José-Santos López-Gutiérrez

Received: 29 June 2021
Accepted: 22 July 2021
Published: 24 July 2021

1. Introduction

Spain is the European Union country with the longest coastline, with a length of 8000 km. It is also the closest European country to the axis of one of the world's major maritime routes. Its geographical location positions it as a strategic element in international shipping and a logistics platform in Southern Europe.

The Spanish Port System includes 46 ports of general interest. The importance of ports in the Spanish economy, as links in the logistics and transport chains, is reflected in the following figures [1]:

- They handle nearly 60% of exports and 85% of imports, accounting for 53% of Spanish foreign trade with the European Union and 96% with third countries.
- The state port system's activity contributes to nearly 20% of the transport sector's GDP, accounting for 1.1% of the Spanish GDP.
- It employs more than 35,000 workers directly and around 110,000 indirectly.

These figures reveal that events that could disrupt the normal operations of a port and actions aimed to improve or optimize processes can have a significant economic impact. Port infrastructures are subject to different meteorological conditions such as waves, wind, and currents that can produce such disruptions. We can classify these conditions into extreme and normal ones. When these conditions are extreme (extreme range regime), the only way to minimize the negative impact is to build infrastructures that ensure the structural integrity of the port. Usually, regular ship operations (berthing, loading, and unloading) are not possible in this regime. However, in normal meteorological and ocean conditions, the port must minimize the effect of climate forcers on ship movements, ensuring that they can operate safely.

Port operability is usually quantified based on the movements of moored ships. When a vessel experiences large movements during its stay in port, these harm the operation, causing delays or even operational stoppages, with the corresponding economic cost. There are numerous regulations and recommendations that propose limits for the amplitudes of the movements compatible with the operation [2–5].

It is important to note that these criteria represent generic values applicable to port facilities worldwide, which in most cases have been established based on subjective observations of operators and shipmasters, without particularized studies to support them. The general nature of these criteria means that they are not adequately adjusted to the characteristics of each port facility, and it would therefore be advisable to make them more specific.

The variables involved in the moored vessel movement are mainly of two different types. In first place, there are the aforementioned ocean-meteorological conditions. Secondly, there are the variables related to the infrastructure, the ship and the port. The same weather conditions affect the vessel movement differently, depending on the size of the ship, the state of the cargo and the type of mooring, among others. Thus, the relationship between the variables is complex and, consequently, difficult to obtain.

In order to obtain the movement of a vessel, we can follow two approaches. One of them is using a physical or numerical model of a vessel and the port infrastructure and using physical or numerical simulations of the ocean-meteorological variables that influence the vessel movement and their interaction with the vessel and the infrastructure, obtaining the vessel movements. As mentioned in [6,7], this approach has a series of limitations. Physical modeling was one of the most widely used techniques to analyze the vessel behavior, the operational conditions [8,9] and to quantify the influence of mooring tension and fender stiffness on the operability of a berth [10,11]. However, it has several limitations:

- It is a simplified reproduction of reality.
- The construction and calibration processes must be highly meticulous, generating a high economic and time cost.
- This methodology is very rigid from modifications of test configuration (mooring configuration, fender stiffness, mooring tension and type vessel).

Numerical modeling of the moored vessel is another widely used methodology. In the past decades, its evolution has allowed it to be a viable alternative to physical modeling [12–15]. Cummins proposed the mathematical description of this phenomenon [16]. The main advantage of this tool is its high adaptability to changes in the geometry, the wave conditions or the vessel type. However, it has some limitations; firstly, the hypotheses adopted for solving the equations, generating certain deviations from reality. Furthermore, their validation is usually carried out using data from physical modeling [17] and not with real data. Finally, it does not reproduce the variations experienced by the ship during the operation.

The field campaign is the third alternative for the moored vessel's analysis. It is the least explored methodology despite providing very relevant information on the vessel's operation reality. The main reason is related to the high economic cost of the monitoring equipment and the personnel required to carry out extensive field campaigns containing a large number of ships. The most widespread measurement technology currently used is the GPS-RTK system [18–20]. Another frequently used technology is laser equipment [21]. There are only two examples where an extensive monitoring campaign was carried out with different vessel typologies during operation [22,23]. During the past decades, this type of work has focused mainly on the study of the long wave [21,24–26] and to validate physical or numerical models [13,27].

Currently, several ports of authority throughout Europe, such as Algeciras, A Coruña, or Rotterdam, are involved in the R+D+i projects for the analysis of the interaction between the physical environment, the behavior of the ship and the berthing infrastructure. Their main objective is to develop advanced management tools that help identify and predict possible risk situations and, therefore, replant operations and the allocation of resources to offer the desired level of service or to reinforce port safety and security measures. This work followed the latter approach to obtain a dataset large enough (with at least several hundred hours of recorded ships movements) and use it to create several machine learning-based prediction models that predict the different vessel movements. The purpose of these models is to predict how a vessel will move before it is operating. When the models predict that the vessel movements are excessive for loading and unloading cargo, port operators can decide to take actions that reduce the vessel movements. For instance, they could change the mooring arrangement (the geometric arrangement of mooring lines between the ship and the berth) or choose a better mooring location, i.e., the one where the models predict movements more favorable for the vessel operations.

We selected a subset of all variables that could influence the ship movement for the models' parameters: we used some vessel characteristics, the docking location, the weather conditions and the sea state.

The information regarding the vessel characteristics is always available [28,29]. Using this information allows the models to create different internal representations for different vessel types, thus allowing them to be more precise.

If we want to predict future movements, we have to use a weather and sea state forecast as input, i.e., the models provide the ship's movement given its characteristics, the sea state and the weather conditions. Therefore, if we want to predict how the movement will be, we need to use forecasting data for the sea state and weather conditions. This necessity of using meteorological forecasting limits the variables that can be used as inputs for the models, i.e., we can only use the same variables that the forecast systems provide.

Several European and international institutions have proposed limiting criteria regarding vessel movement during operations [3–5]. The state-owned Spanish port system compiles and publishes its criterion in a document named "*Recomendaciones para el Proyecto y ejecución en Obras de Atraque y Amarre*" ("Recommendations for the Design and Execution of Berthing and Mooring Works", from now on referred to as ROM). The models created in this work provide the ship's movement in meters or degrees (depending on the type of movement). This value can be directly compared with the ROM ship movement limits to check whether the ship's movement will exceed the limits imposed by ROM.

Once we collected data and created the dataset, we trained and tested several machine learning models. Machine learning has been successfully used to solve knowledge-based prediction and classification problems in civil engineering, both in coastal engineering [30–32] and other subdisciplines [33–35]. These applications show that machine learning is a powerful tool for producing results applicable in a real environment.

The data used to train and test the machine learning based models was collected at the Outer Port of Punta Langosteira (A Coruña), Spain (Figure 1). Its geographical location entails that the meteorological conditions it is subject to are more severe than other inland ports. This port is located 10 km southeast of A Coruña, Spain, and is protected by a 3360 m long main breakwater and a 1320 m long spur breakwater. The current berthing line is 914 m long with an average depth of 22 m. The dock's orientation is N62.7W and reaches a crest height of 6.5 m above the zero datum of the port. A set of bollards spaced 31 m apart with a load capacity of 200 T is situated 0.75 m from the dock. In addition, the vessel operations are streamlined by a double-fender system protected by a shield.

Figure 1. Aerial view of the outer port of Punta Langosteira, A Coruña, Spain.

The section Materials and Methods explains the technical details of the dataset creation, which machine learning models were used, and how the models' training and testing processes were carried out. The Results section presents the performance of the models.

2. Materials and Methods

The data used for training and testing the machine learning models was collected in five field campaigns, carried out in the outer port of Punta Langosteira (A Coruña), from October of 2015 until February of 2020. During these campaigns, we recorded the movement of 46 moored vessels for 1609 h total. We chose which vessels to monitor, from all available ships that operate in said port, so they reflect the typical fleet of the port. We also collected the environmental conditions (weather and sea state) the moored vessels were subject to during their port stay.

2.1. Dataset

The variables used to create the prediction models can be classified into meteorological conditions (weather conditions and sea state), ship characteristics, berthing location and ship movements. The models use the first two as inputs and output for the latter.

We gathered the weather conditions and sea state data using three measurement systems provided by the port technological infrastructure: a directional wave buoy, a weather station and a tide gauge.

The directional wave buoy is located 1.8 km off the main breakwater of the port (43°21′00″ N 8°33′36″ W). This buoy belongs to the Coastal Buoy Network of Puertos del Estado (REDCOS [36]) and has the code 1239 in said network. Due to the sensibility of the buoy data to noise, the system provides its data aggregated in 1 h intervals. This quantization allows calculating some statistical parameters that reflect the sea state while mitigating the noise's effects.

The weather station is located on the main port's breakwater. It provides the data aggregated in 10 min intervals.

The tide gauge is also placed at the end of the main port's breakwater. This tide gauge belongs to the Coastal Tide Gauges Network of Puertos del Estado (REDMAR [37]) and has the code 3214 on said network. It provides the data in 1 min intervals.

From all the variables these data sources provide, we chose the variables also available in the port's forecast system. This work's objective is to create vessel movement models that, when used with weather and sea state forecasts as inputs, can predict future movements of a moored ship (as is further explained later in this section). In order to fulfill this objective, the models' inputs regarding ocean-meteorological variables have to be those also available in the port's forecast system. In summary, these are the meteorological variables we used as input variables to the model:

- $\mathbf{H_s}$ (m): significant wave height, i.e., the mean of the highest third of the waves in a time-series of waves representing a certain sea state.
- $\mathbf{T_p}$ (s): peak wave period, i.e., the period of the waves with the highest energy, extracted from the spectral analysis of the wave energy.
- $\boldsymbol{\theta_m}$ (deg): mean wave direction, i.e., the mean of all the individual wave directions in a time-series representing a certain sea state.
- $\mathbf{W_s}$ (km/h): mean wind speed.
- $\mathbf{W_d}$ (deg): mean wind direction.
- $\mathbf{H_0}$ (m): sea level with respect to the zero of the port.
- $\mathbf{H_{sm}}$ (m): significant wave height measured by the tide gauge, i.e., the mean of the highest third of the waves in a time-series of waves representing a certain sea state.

The buoy provides H_s, T_p and θ_m, the weather station provides W_s and W_d and the tide gauge provides H_0 and H_{sm}.

The other category of variables used as inputs to the models is the ship characteristics and the berthing location. The ship's geometric characteristics have high relevance in its dynamic behavior. Within the existing variables for the physical characteristics of a vessel, we used those whose acquisition was straightforward and always available. We also used the berthing zone as input to allow the model to capture the characteristics of the different locations of the port. The berthing zone would also allow creating a tool that loops over all available berthing zones and calculates the output of the models for each berthing zone, allowing the port operator to choose the best berthing location. The variables used are:

- **L** (m): ship length.
- **B** (m): ship breadth.
- **DWT** (tonnes): deadweight tonnage, a measure of how much weight a ship can carry.
- **BZ**: berthing zone. The port operator provides it. The port is divided into 12 berthing zones.

The port operator provided all the values for the previous variables.

The last category of data, the ship's movements, was used as outputs of the model. A moored vessel has six degrees of freedom (see Figure 2): three displacements (surge, sway and heave) and three rotations (roll, pitch and yaw). The variables used for the vessel movement are:

- **Surge** (m): linear longitudinal motion (bow-stern).
- **Sway** (m): linear lateral motion (port-starboard).
- **Heave** (m): linear vertical motion.
- **Roll** (deg): tilting rotation of the vessel about its longitudinal axis (bow-stern).
- **Pitch** (deg): up/down rotation of the vessel about its lateral axis (port-starboard).
- **Yaw** (deg): turning rotation of the vessel about its vertical axis.

The methodology used to measure the ships' movement, in its six degrees of freedom, is described in [6,7], where we used three complementary and fully synchronized measurement systems: an inertial measuring unit (IMU, a device that can measure and record the angular rate and specific gravity of the object to which it is attached to) (Figure 3a), two laser distance meters (Figure 3b) and two cameras (Figure 3c). The combination of these devices allowed the recording of the movements at a minimum frequency of 1 Hz, which is enough to detect the movements of a moored vessel. The IMU provided the pitch and roll, two laser distance meters provided the sway and yaw and two cameras provided the surge and heave.

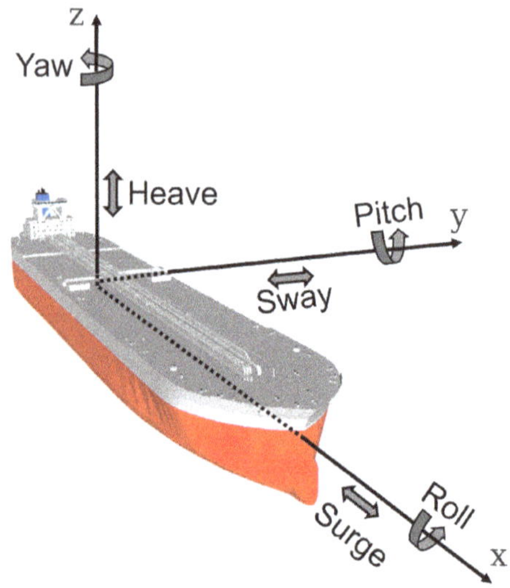

Figure 2. Vessels motions along the three axes.

(a) (b) (c)

Figure 3. The ship movement measurement system comprises an IMU (**a**), two laser distance meters (**b**) and two cameras (**c**).

This work aimed to create vessel movement models that, when used with weather and sea state forecasts as inputs, can predict future movements of a moored ship. We created different machine learning models (artificial neural network, from now on referred to as ANN and gradient boosting with decision trees) and compared them to check which one provides the best results. In order to compare different machine learning models, we need them to use the same dataset so that we can compare them appropriately.

The first step in creating the dataset was joining the data provided by the several data sources (buoy, weather station, tide gauge, ship's characteristics and ship's movement monitoring). These data sources have a different frequency, so we had to aggregate the data to have the same frequency as the frequency of the data with the largest period, the sea state data, which is the same frequency as the data we will use as the models' inputs in production. The variables that the forecast systems provided had a period of 1 h, so we needed to aggregate the rest of the variables and provide them with a 1 h period. When aggregating the data for the ship's movements, we calculated the mean, maximum and significant (mean of the highest third) value for the amplitude of each movement for every hour.

A moored vessel may experience a maximum point motion much greater than its significant or average motion under the action of certain ocean-meteorological conditions. This value abandons the primary trend of the movement. It could occasionally be caused by external agents such as the waves generated by the passage of other ships or the punctual modification of the mooring lines' tension to adapt them to the variations of the tidal range, making them difficult to predict. Therefore, we decided to select the significant value of each movement as the output of the models.

The wave parameters (significant wave height and peak period) were obtained similarly. These ship's motion forcers are among those that most affect the motion of a moored ship. Therefore, using the same statistic for the inputs and outputs of the models could help the models to obtain a better relationship between these motion forcers and the ship's movement. Moreover, other works used these values when studying moored vessels' behavior [14,21], which demonstrates its feasibility as a motion descriptor.

Once we integrated and aggregated all data sources, we separated the dataset in two, creating a training and a testing dataset that we used to train and test the different models. Usually, the test dataset is created by randomly choosing data from the whole dataset but, had we had done this, the test results would have been inaccurate (overly optimistic). This is because the movement data of a ship for a specific stay is highly correlated, as it is temporal data: the ship's movement for a given hour has a high correlation with the movement of the previous hour (they are close in the data distribution). Had we chosen random data from the whole dataset to create the testing dataset, the models would have obtained excellent results in testing since they would have been trained with data very similar to that used for testing (belonging to the same ship, similar environmental conditions and close temporal events). To avoid this, we created the test set by separating the data corresponding to two monitored vessels, so we did not train the models with data for those vessels. Thus, the test results will provide a more realistic estimation of how the models will perform in the production environment. In production, the port operators will use the models to predict a vessel's movements in a real scenario (with vessels the model was not trained on).

The two selected vessels had characteristics in the same data distribution used for training: we chose a small general cargo ship (Dominica: 13,000 dwt and 127 m length) and a large bulk carrier (Fu Da: 71,500 dwt and 225 m length). We chose these vessels considering that they should provide enough information to conclude the performance of the models in a production environment, but without reducing too much the training dataset size. Thus, the testing dataset is around 10% of the whole dataset when using these two vessels.

To summarize, the dataset characteristics are the following:

- We had two separate files for each movement: one for training and one for testing.
- In each file, we had the previously explained variables: L, B, DWT, BZ, H_s, T_p, θ_m, W_s, W_d, H_0 and H_{sm} and the corresponding mean movement, maximum movement and significant movement.
- Each row in each file represents one hour of the ship's movement (the data had a period of 1 h).
- Table 1 shows the number of rows in each file.

Table 1. Dataset: available data for each movement and purpose (training or testing).

Movement	Training (h)	Testing (h)
Surge	364	16
Sway	1451	158
Heave	364	18
Roll	1348	51
Pitch	1348	51
Yaw	1248	158

The dataset is publicly available in [38], where we provide a summarized description of the dataset and explain how to interpret it.

The dataset also stores variables not intended to be used as predictors, such as the ship's name, the date of the ship's stay and the ship's type (general cargo or bulk carrier).

2.2. The Vessel Movement Predictors Creation

The vessel movement method used to create the dataset relies on the computer vision to measure several vessel movements [6,7]. A port's infrastructure is not a controlled environment as a laboratory could be, so there are many technical limitations when using a camera to measure a vessel's movement, occlusion being the most limiting. Furthermore, the vessels that operate in the outer port of Punta Langosteira are cargo ships. When a vessel is loading/unloading, there are many types of machinery such as cranes and trucks that occlude the camera field of view, making it impossible to measure the movements that rely on the computer vision. Given this limitation, not all movements are available for all the vessels at all times. To surpass this limitation and create precise prediction models, we decided to create one prediction model for each vessel movement, resulting in six models (one for each of the six degrees of freedom). Thus, each model uses the 11 input variables explained in the previous section and outputs one value for the corresponding movement.

For creating the prediction models, we chose two well-known and tested machine learning models for regression problems: gradient boosting [39] and neural networks [40].

Gradient boosting creates a prediction model via an ensemble of weak prediction models. In our case, the weak prediction models used are decision trees. When a decision tree is used as the weak learner, the boosting algorithm is usually called gradient boosted trees. Gradient boosted trees usually outperforms the random forest and have been used successfully to solve civil engineering problems [41,42].

An artificial neural network is an interconnected group of nodes, where each interconnection has an associated weight. A node computes its output by applying a non-linear function to the sum of its inputs multiplied by the weight associated with each input. The variables used as inputs determine the ANN's input nodes and they form the input layer. The variables used as outputs determine the ANN's output nodes and they form the output layer. The layers of nodes between the input and output layers are called the hidden layer.

There is no proven methodology for choosing how many hidden layers and how many nodes per hidden layer are necessary to solve a given problem. Usually, one creates many different ANN architectures (with different layers and different nodes per layer), trains them and compares them using cross-validation. We followed this approach in this work.

The training process consisted of finding the appropriate weights for each interconnection, so the error for the training dataset was minimized (the output of the ANN was as close as possible to the desired output when we gave the corresponding input data to the ANN).

Figure 4 shows a representation of the overall architecture of the ANNs created in this work. The hidden layers and the dropout ratio varied for each of the six models obtained, but the overall structure remained the same.

The objective of an ANN is to approximate a function f that provides the desired output given the input. While a sufficiently complex ANN can approximate any function f, an ANN is a black box. Analyzing the ANN structure will not give insights into the function being approximated: there is no simple link between the weights and the function being approximated. Despite this lack of interpretability, ANN has successfully solved coastal engineering problems [30–32] and other subdisciplines of civil engineering [33–35].

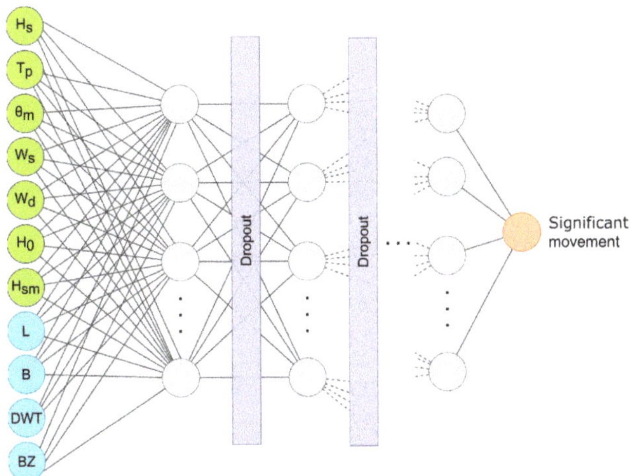

Figure 4. Overall structure of the ANNs created.

We trained and tested both types of models using the same training and testing dataset to check, which one gives the best results. The training process, in each case, consisted of creating multiple models with different hyperparameters and comparing them using 10-fold cross-validation. Once the best model, the one that performed the best using cross-validation, was selected, we assessed its performance by using the testing datasets. We explain the overall process in Algorithm 1.

Algorithm 1 Pseudocode for the iterative grid-based training process algorithm.

```
do
  manually define grid of model hyperparameters
  for each grid cell (model) do
      for each resampling iteration do
          hold-out specific samples
          fit model on the remainder
          predict the hold-out samples
      end
    calculate average performance across hold-out predictions
   end
  determine best hyperparameter
  fit final model (best hyperparameter) to all training data
while error >= ε
evaluate final model performance in test set
```

The hyperparameters' value for a specific model is chosen from a grid created as the outer product of all the values manually specified used for each hyperparameter.

The hyperparameters we need to define and the values used for the gradient boosting models are the following:

- **N.trees**: this hyperparameter represents the number of gradient boosting iterations (number of trees). Increasing N reduces the error on the training set, but setting it too high may lead to overfitting. We tried the values 1000, 2000, 3250, 5000, 7000, 8500, 10,000 and 15,000 for this hyperparameter.
- **Shrinkage**: this hyperparameter represents the learning rate. We tried the values 0.1, 0.01 and 0.001 for this hyperparameter.
- **Interaction.depth**: this parameter represents the depth of each tree. We tried the values 1, 5 and 9 for this hyperparameter.

- **N.minobsinnode**: this hyperparameter represents the minimum number of observations required in the tree terminal nodes (leaves). We tried the values 3, 5 and 10 for this hyperparameter.

ANNs are a more powerful model than gradient boosting, but they require more data and time to train and there are more hyperparameters we need to tune. The hyperparameters we need to define and the values we used for the ANN models are the following:

- **Network architecture**: we tested networks with 1, 2, 3 and 4 layers. We tried with 8, 16, 32, 64, 128 and 256 neurons per layer in each case. The layers were fully connected in all cases. Although the largest networks could seem too complex for the problem at hand, we used them because we planned to use dropout regularization and one is likely to get better performance when dropout is used on a larger network (dropout is going to prune the network resulting in a smaller network). Using a small network with dropout could result in a model too simple to be a good predictor.
- **Weights (kernel) initialization**: we used several weights initialization, both the uniform and the normal variations, for random, He [43] and Glorot (a.k.a. Xavier) [44] initializations.
- **Optimizer**: we used 3 different optimizers, stochastic gradient descent (SGD) [45], RMSprop [46] and Adam [47].
- **Training iterations**: we tested the optimizer with 2000, 5000, 7000 and 10,000 iterations. The results did not improve from 5000 iterations up, so we chose 5000 for all the following iterations.
- **Activation function**: we tested several activation functions for the neurons in the hidden layers: ReLU, SeLu, eLu and tanh.
- **Training iterations**: we tested each optimizer with 500, 1000, 2000, 2500 and 5000 training iterations.
- **Regularization**: we used dropout regularization [48], with one dropout layer for each hidden layer (except the output layer). We tested several dropout rates: none, 0.0625, 0.125, 0.25 and 0.375, where none indicates that no dropout was used.
- **Learning rate**: we tested several values for the learning rate: [0.0001, 0.001, 0.01, 0.1]. Although 0.01 and 0.1 are usually considered too large learning rates, large values are suggested in the original dropout paper [48].

Once we trained all the models using cross-validation (as described in Algorithm 1), we selected the best 12 models, i.e., we had two models for each of the six degrees of freedom of a vessel movement. One is an ANN model and another a gradient boosting model. We tested both models with the test data for the corresponding movement to select the best one for each movement. We present all the training and testing results in the next section.

3. Results and Discussion

A machine learning model was fully specified by its inputs, outputs and hyperparameters. The hyperparameters for the best gradient boosting model for each movement, obtained using the method previously described, are shown in Table 2.

Table 2. Gradient boosting: best models' hyperparameters.

Movement	Hyperparameter			
	N.Trees	Shrinkage	Interaction.Depth	N.Minobsinnode
Surge	8500	0.001	5	3
Sway	4000	0.01	9	5
Heave	8500	0.001	9	10
Roll	3250	0.01	9	3
Pitch	3250	0.01	9	3
Yaw	3250	0.01	9	3

We can see in Table 2 that in all movements, but surge, the best models were the ones with more complexity in terms of tree depth, but only the models for surge and heave required to create a higher number of trees to provide the best results. None of the models required more than 8500 trees, although the values 10,000 and 15,000 were also used.

We show the hyperparameters for the best ANN based models for each movement in Table 3.

Table 3. Neural networks: best models' hyperparameters.

Movement	Neurons Per Layer	Activation Function	Optimizer	Hyperparameter Kernel Initialization	Learning Rate	Iterations	Dropout Ratio
Surge	[128, 128, 128]	ReLU	Adam	He uniform	0.001	5000	0.25
Sway	[128, 128]	ReLU	Adam	He uniform	0.001	5000	0.0625
Heave	[128, 128, 128]	ReLU	Adam	He uniform	0.001	5000	0.375
Roll	[256, 256]	ReLU	Adam	He uniform	0.001	5000	0.125
Pitch	[128, 128]	ReLU	Adam	He uniform	0.001	5000	0.25
Yaw	[128, 128]	ReLU	Adam	He uniform	0.001	5000	0.125

We can see in Table 3 that in every case, using ReLU units provided the best results. Additionally, in every case, the He uniform initialization gave the best results. These results were consistent with the state-of-the-art recommendation of using He uniform initialization when using ReLU units.

We can see that the results we show in Tables 2 and 3 for the model of the surge movement and the model of the heave movement were consistent: they required the most complex models, both in the ANN and the gradient boosting cases. The gradient boosting models required more trees for modeling the movement, and the ANN models required a larger network. This need for more complex models was caused by the situation previously explained: we had less data for these movements, as they rely on computer vision techniques to measure them.

Table 4 shows the comparison between the training error, in terms of the root mean squared error (RMSE), of both the ANN and the gradient boosting model for each movement. The RMSE alone does not measure how good a model is if we do not compare it with the range of the movement that the model is predicting. We considered that a model has good performance if its RMSE is less than or equal to 10% of the range for the movement it is predicting. We show this limit as the "0.1 range" value in Table 4 to quickly compare the RMSE with this limit. We can see that all models for every movement had a lower RMSE than 10% of the movement range, so we considered that they achieved the goal we initially set.

Table 4. Training error for neural networks and gradient boosting.

Movement	0.1 Range	Training RMSE		Training R^2	
		Neural Networks	Gradient Boosting	Neural Networks	Gradient Boosting
Surge	0.23 m	0.03 m	0.11 m	0.99	0.86
Sway	0.25 m	0.02 m	0.04 m	0.99	0.92
Heave	0.09 m	0.05 m	0.03 m	0.95	0.92
Roll	0.43°	0.06°	0.11°	0.99	0.91
Pitch	0.28°	0.04°	0.02°	0.98	0.98
Yaw	0.48°	0.01°	0.06°	0.98	0.95

Table 4 also shows the coefficient of determination (R^2) that each model achieved in the training dataset. We can see that the ANN models achieved a higher R^2 than the gradient boosting models for every movement, being higher than 0.95 in every case. The worst R^2

for a gradient boosting model was 0.86, achieved in the case of the surge movement. This result was consistent with having less data for this movement, as previously explained.

To analyze where the models failed the most, we could plot the true values for each movement against the corresponding model predictions. A perfect prediction would fall on the line y = x.

Figure 5 shows the plots of the true outputs versus the predictions of the gradient boosting models for the training dataset. In these plots, we show the y = x line in blue.

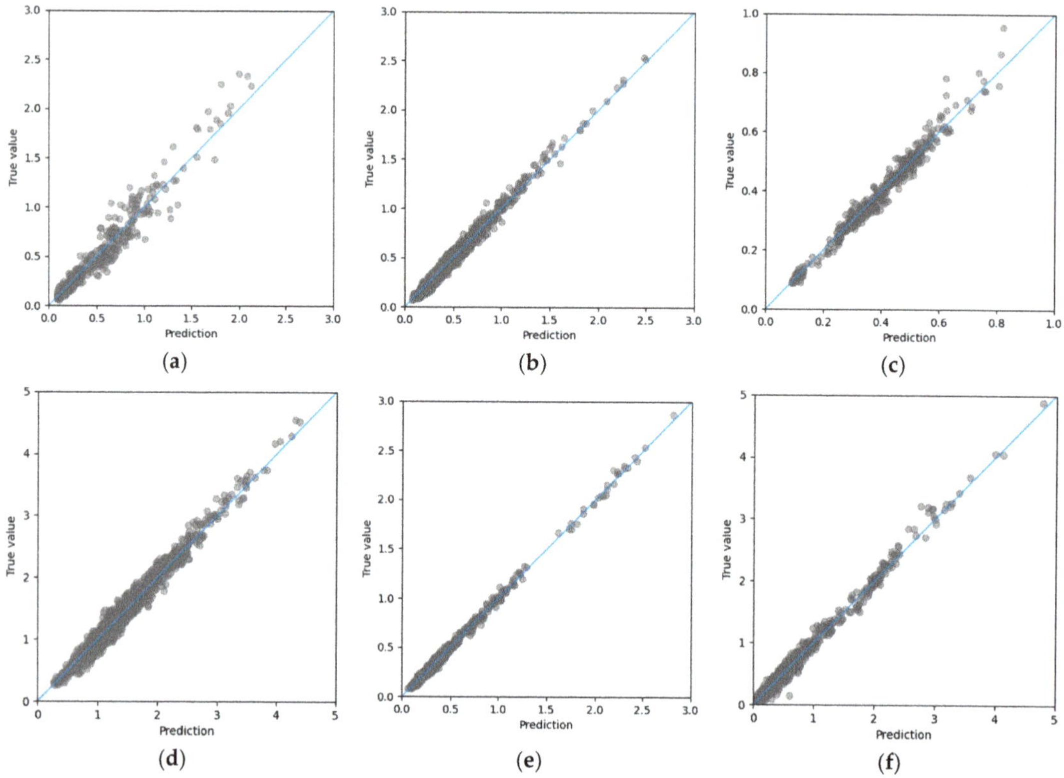

Figure 5. Training error for (**a**) surge, (**b**) sway, (**c**) heave, (**d**) roll, (**e**) pitch and (**f**) yaw gradient boosting models.

We can see in Figure 5 that the surge prediction model error increased as the amplitude of the motion increased, and the error for the heave movement was primarily due to a worse prediction of the larger movements. We can also see that the error for the roll movement was committed during all the movement ranges. Figure 6 shows the plots of the true outputs versus the predictions of the ANN models for the training dataset. In these plots, we show the y = x line in blue.

We can see in Figure 6 that the predictions were overall good. The model for the sway movements underestimated the amplitudes of the larger movements a bit. The model for the heave movement underestimated the shorter movements and overestimated the larger movements. In the model for the pitch movement, the mid and large movements predictions were more dispersed and the same happened for the larger movements in the yaw model.

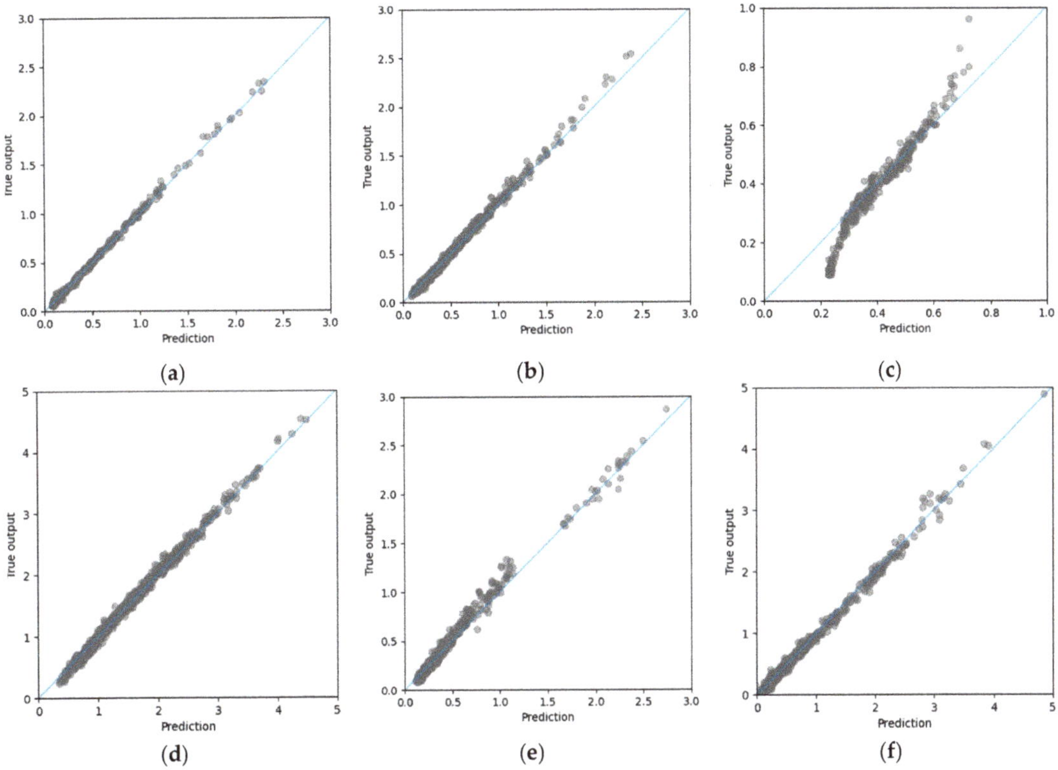

Figure 6. Training error for (**a**) surge, (**b**) sway, (**c**) heave, (**d**) roll, (**e**) pitch and (**f**) yaw ANN models.

After the training phase, where we obtained the best models for each movement, both using the ANN and the gradient boosting models, we tested them using the testing dataset. The results, in terms of the RMSE obtained for each movement and type of model are shown in Table 5. We also show the "0.1 range" value to quickly compare the testing RMSE with this limit.

Table 5. Testing errors for each artificial neural networks and gradient boosting ship movement models.

Movement	0.1 Range	Testing RMSE	
		ANN	**Boosting**
Surge	0.23 m	0.10 m	0.39 m
Sway	0.25 m	0.15 m	0.11 m
Heave	0.09 m	0.12 m	0.09 m
Roll	0.43°	0.90°	1.04°
Pitch	0.28°	0.11°	0.24°
Yaw	0.48°	0.15°	0.28°

In testing, we also considered that a model had good performance if its RMSE was less than or equal to 10% of the range for the movement it was predicting. We can see in Table 5 that all ANN models but the one for heave and roll were below the 10% error margin. In the gradient boosting models case, all the models but the one for surge and roll were below the limit and the model for the heave movement was right on the limit. As previously explained, we had less training data for the heave and surge movements, so we could expect that the corresponding model predictions in the testing dataset were in the error limit or above it.

The two models for the roll motion showed the least satisfactory results, probably because the ships used were out of range for this motion. The roll movement of a vessel shows a greater variability for similar ocean-meteorological conditions than the other movements, increasing the difficulty for a model to represent the relationship between inputs and outputs when not enough data is available.

4. Conclusions

In this work, we developed prediction models for the movement of a moored ship using data from 46 ships obtained between October 2015 and February 2020 in the outer port of Punta Langosteira, Spain. Using the gathered data, we created prediction models based on two well-known machine learning models that have been used successfully to solve civil engineering problems: gradient boosted trees [21,22] and artificial neural networks [9–14].

We set the limit for deciding if the models performed satisfactorily as having an RMSE lower than 10% of the corresponding movement range.

The best models achieved satisfactory results in the training dataset, having an RMSE lower than the limit in every case. The coefficient of determination (R^2) was 0.99, 0.99, 0.95, 0.99, 0.98 and 0.98 for the surge, sway, heave, roll, pitch and yaw movements, respectively, for the best models. These R^2 values were achieved by the ANN models, which showed to be more powerful than the gradient boosting models (they had lower RMSE and higher R^2) as they proved to have less bias. The counterpart is that ANN requires tuning more hyperparameters, more time to train and more data than the gradient boosting models to achieve better results.

The surge, sway, pitch and yaw models performed satisfactorily in the testing dataset, having an RMSE lower than 10% of the corresponding movement range. Although we created the heave and surge movements models, we did not have enough data to feel confident about the corresponding models' performance. However, the ANN heave model performed satisfactorily both in the training and testing.

In any case, increasing the volume of data available for training the models will improve their performance, especially in the ranges of motion where we did not have enough data to be confident in the model predictions. In addition, increasing the dataset is another way to avoid overfitting, so the need for regularization will be reduced and the training process could be accelerated.

The main limitations of the approach used in this work, and in any machine learning-based solution, were the lack of (good) data and the interpretability.

Many machine learning models require large amounts of data to provide useful results. For instance, in the case of neural networks, the larger the model's architecture, the more data we need to use for training it to produce good results. Some techniques, such as data augmentation, are useful to some extent, but the preferred solution is always obtaining more data, when possible.

In future works, we plan to increase the dataset via an intensive monitoring campaign focusing on the surge and heave movements to improve the models results. Unfortunately, in the last two campaigns (2020 and 2021), due to the situation generated by COVID-19, access to the interior of moored ships in the outer port of Punta Langosteira was prohibited. Hence, we could not use the IMU to record the ships' movements, so the roll and pitch dataset did not increase as much as intended.

Another data-related limitation is when data with potential bias finds its way into the data set or the models (via the training process). This circumstance could lead to situations where models that are seemingly performing well may actually be picking up noise in the data. This problem is more prominent in data with high variability than in slow-changing or bounded data, i.e., in data with high variability, we cannot accurately detect when there is noise in the data (in slow-changing data or data with values restricted to a small range, the noise is more evident).

The ideal solution to this problem is collecting data from multiple sources. Having multiple sources limits the effect that bias has on the models, resulting in higher quality models. To mitigate this problem, in future works, we plan to study and search for theoretical new variables that could improve predicting movements with high variability, such as the roll movement.

Interpretability is one of the primary problems with machine learning, as any user related problem in computer science.

The machine learning models created in this work are intended to replace traditional statistical methods used in port management tools. These traditional models are interpretable, a characteristic that port operators are accustomed to.

Using machine learning models we cannot explain to port operators how and why the model came to the decision it did. The only tool a machine learning practitioner has in this situation is to convince through results, i.e., the models need to prove that they are superior to traditional statistical methods by providing more accurate predictions than traditional models.

During the development of this work, and after obtaining the results, we maintained several meetings with the A Coruña port authority and the port operators. After reviewing the result obtained in this work, they are convinced that machine learning methods provide better results than traditional methods, despite the lack of interpretability. They intend to use the models created in this work in their decision-making pipeline.

To facilitate this, we intend to create a decision-aiding tool that uses the models created in this work. Currently, port operators do not have a tool such as this one that allows predicting whether a ship will exceed the recommended limits for the cargo ship movements during loading and unloading operations (determined by the Spanish Port Authority). Instead, they take a conservative approach, sometimes stopping the operations unnecessarily. Having this tool will improve the decision-making pipeline and optimize the ship's operations.

To summarize all future works, after gathering more data and new variables, we intend to optimize the models, retrain them and incorporate them in a web tool that will allow easy access to the model predictions to the port operators. The tool will use forecast data for the weather conditions and sea state as inputs for the models (and the ship characteristics and berthing location) to predict the ship movements several days in advance. These predictions can aid in deciding the best location for the operation and in deciding when to stop the ship operations more precisely, thus minimizing the economic impact the hydrodynamic phenomena have due to cargo ships being unable to operate.

Author Contributions: Conceptualization, J.R., E.P., A.F., A.A., J.S. and A.G.; methodology, J.R., A.A., A.F. and J.S.; software, A.A. and H.C.; validation, A.A., A.F. and H.C.; formal analysis, E.P., A.F., A.A. and J.S.; investigation, A.A., A.F., H.C. and R.C.; resources, J.R., E.P. and A.G.; data curation, A.A., A.F., H.C. and R.C.; writing—original draft preparation, A.A. and A.F.; writing—review and editing, A.A. and A.F.; visualization, A.F., E.P., J.S. and A.G.; supervision, J.R. and E.P.; project administration, E.P. and J.R.; funding acquisition, A.F., E.P. and J.R. All authors have read and agreed to the published version of the manuscript.

Funding: This research was funded by the Spanish Ministry of Economy, Industry, and Competitiveness, R&D National Plan, within the project BIA2017-86738-R, the FPI predoctoral grant from the Spanish Ministry of Science, Innovation, and Universities (PRE2018-083777) and the Spanish Ministry of Science and Innovation, Retos Call, within the project PID2020-112794RB-I00.

Institutional Review Board Statement: Not applicable.

Informed Consent Statement: Not applicable.

Data Availability Statement: The dataset used to provide the results shown in this work can be accessed through the repository information provided in http://doi.org/10.5281/zenodo.5113707 (accessed on 20 July 2021) or directly in https://github.com/aalvarell/ship-movement-dataset/tree/v1.0.0 (accessed on 20 July 2021).

Acknowledgments: The authors wish to thank the staff of Port Authority of A Coruña. Moreover, the authors are grateful to the owners and crew of the vessels for their collaboration on board.

Conflicts of Interest: The authors declare no conflict of interest.

References

1. Puertos del Estado (Spanish Port System). Available online: http://www.puertos.es/en-us (accessed on 28 June 2021).
2. *Mooring of Ships to Piers and Wharves*; Coasts, Oceans, Ports and Rivers Institute (American Society of Civil Engineers); Gaythwaite, J. (Eds.) ASCE manuals and reports on engineering practice; American Society of Civil Engineers: Reston, VA, USA, 2014; ISBN 978-0-7844-1355-5.
3. PIANC General Secretariat; Maritime Navigation Commission; World Association for Waterborne Transport Infrastructure. *Criteria for the (Un) Loading of Container Vessels*; PIANC Secrétariat Général: Bruxelles, Belgique, 2012; ISBN 978-2-87223-195-9.
4. *Criteria for Movements of Moored Ships in Harbours: A Practical Guide*; Supplement to Bulletin; Permanent International Association of Navigation Congresses: Bruxelles, Belgique, 1995; ISBN 978-2-87223-070-9.
5. Llorca, J.; González Herrero, J.M.; Ametller, S.; Ente Público Puertos del Estado (España). *ROM 2.0-11: Recomendaciones Para el Proyecto y Ejecución en Obras de Atraque y Amarre*; Puertos del Estado: Madrid, Spain, 2012; ISBN 978-84-88975-78-2.
6. Figuero, A.; Rodriguez, A.; Sande, J.; Peña, E.; Rabuñal, J.R. Dynamical Study of a Moored Vessel Using Computer Vision. *J. Mar. Sci. Technol.* **2018**, *26*, 240–250. [CrossRef]
7. Figuero, A.; Rodriguez, A.; Sande, J.; Peña, E.; Rabuñal, J.R. Field Measurements of Angular Motions of a Vessel at Berth: Inertial Device Application. *J. Control. Eng. Appl. Inform.* **2018**, *20*, 79–88.
8. Baker, S.; Frank, G.; Cornett, A.; Williamson, D.; Kingery, D. Physical Modelling and Design Optimizations for a New Container Terminal at the Port of Moin, Costa Rica. In *Ports 2016*; American Society of Civil Engineers: New Orleans, LA, USA, 2016; pp. 560–569.
9. Cornett, A.; Wijdeven, B.; Boeijinga, J.; Ostrovsky, O. 3-D Physical Model Studies of Wave Agitation and Moored Ship Motions at Ashdod Port. In Proceedings of the 8th International Conference on Coastal and Port Engineering in Developing Countries, Chennai, India, 20–24 February 2012.
10. Jorge Rosa-Santos, P.; Taveira-Pinto, F. Experimental Study of Solutions to Reduce Downtime Problems in Ocean Facing Ports: The Port of Leixões, Portugal, Case Study. *J. Appl. Water Eng. Res.* **2013**, *1*, 80–90. [CrossRef]
11. Rosa-Santos, P.; Taveira-Pinto, F.; Veloso-Gomes, F. Experimental Evaluation of the Tension Mooring Effect on the Response of Moored Ships. *Coast. Eng.* **2014**, *85*, 60–71. [CrossRef]
12. Kumar, P.; Batra, G.; Kim, K.I. A Moored Ship Motion Analysis in Realistic Pohang New Harbor and Modified PNH. In *Modern Mathematical Methods and High Performance Computing in Science and Technology*; Singh, V.K., Srivastava, H.M., Venturino, E., Resch, M., Gupta, V., Eds.; Springer Proceedings in Mathematics & Statistics; Springer: Singapore, 2016; Volume 171, pp. 207–214. ISBN 978-981-10-1453-6.
13. Howle, M.; Bont, J.D.; Molen, W.; Lem, J.; Ligteringen, H.; Mühlenstein, D. Calculations of the Motions of a Ship Moored with MoorMaster Units. In Proceedings of the 32nd PIANC International Navigation Congress, Liverpool, UK, 10–14 May; The World Association for Waterborne Transport Infrastructure (PIANC): Liverpool, UK, 2010; pp. 622–635.
14. Van der Molen, W.; Scott, D.; Taylor, D.; Elliott, T. Improvement of Mooring Configurations in Geraldton Harbour. *J. Mar. Sci. Eng.* **2016**, *4*, 3. [CrossRef]
15. Kwak, M. Numerical Simulation of Moored Ship Motion Considering Harbor Resonance. In *Handbook of Coastal and Ocean Engineering*; WORLD SCIENTIFIC: Singapore, 2018; pp. 1081–1110. ISBN 978-981-320-401-0.
16. Cummins, W.E. *The Impulse Response Function and Ship Motions*; Navy Dept., David Taylor Model Basin: Carderock, MD, USA, 1962.
17. Van der Molen, W.; Ligteringen, H.; Van der Lem, J.C.; De Waal, J.C.M. Behavior of a Moored LNG Ship in Swell Waves. *J. Waterw. Port Coast. Ocean. Eng.* **2003**, *129*, 15–21. [CrossRef]
18. Van Deyzen, A.F.J.; Beimers, P.B.; Van der Lem, J.C.; Messiter, D.; De Bont, J.A.M. To Improve the Efficiency of Ports Exposed to Swell. In Proceedings of the Australasian Coasts & Ports Conference 2015: 22nd Australasian Coastal and Ocean Engineering Conference and the 15th Australasian Port and Harbour Conference, Auckland, New Zealand, 15–18 September 2015; pp. 919–926.
19. Trejo, I.; Pérez, J.; Guerra, A.; Iribarren, J.R.; Gómez, J.; Sopelana, J.; Peña González, E. Onsite Measurement of Moored Ships Behaviour (RTK GPS), Waves and Long Waves in the Outer Port of A Coruña (Spain). In Proceedings of the 33rd PIANC World Congress, San Francisco, CA, USA, 1–5 June 2014.
20. Van Zwijnsvoorde, T.; Vantorre, M.; Ides, S. Container Ships Moored at the Port of Antwerp: Modelling Response to Passing Vessels. In Proceedings of the 34th PIANC World Congress, Panama City, Panama, 7–11 May 2018; pp. 1–18.
21. López, M.; Iglesias, G. Long Wave Effects on a Vessel at Berth. *Appl. Ocean. Res.* **2014**, *47*, 63–72. [CrossRef]
22. Jensen, O.J.; Viggosson, G.; Thomsen, J.; Bjordal, S.; Lundgren, J. Criteria for Ship Movements in Harbours. In Proceedings of the Coastal Engineering 1990, Delft, The Netherlands, 20 May 1990; American Society of Civil Engineers: Delft, The Netherlands, 1991; pp. 3074–3087.
23. Figuero, A.; Sande, J.; Peña, E.; Alvarellos, A.; Rabuñal, J.R.; Maciñeira, E. Operational Thresholds of Moored Ships at the Oil Terminal of Inner Port of A Coruña (Spain). *Ocean. Eng.* **2019**, *172*, 599–613. [CrossRef]

24. Hiraishi, T.; Atsumi, Y.; Kunita, A.; Sekiguchi, S.; Kawaguchi, T. Observation of Long Period Wave and Ship Motion in Tomakomai-Port. In Proceedings of the 7th International Offshore and Polar Engineering Conference, Honolulu, HI, USA, 25–30 May 1997.
25. Uzaki, K.; Matsunaga, N.; Nishii, Y.; Ikehata, Y. Cause and Countermeasure of Long-Period Oscillations of Moored Ships and the Quantification of Surge and Heave Amplitudes. *Ocean. Eng.* **2010**, *37*, 155–163. [CrossRef]
26. Sakakibara, S.; Saito, K.; Kubo, M.; Shiraishi, S.; Nagai, T.; Shimanoe, S. A Study on Long-Period Moored Ship Motions in a Harbor Induced by a Resonant Large Roll Motion during Long Waves. *J. Jpn. Inst. Navig.* **2001**, *104*, 187–196. [CrossRef]
27. Kwak, M.; Pyun, C. Computer Simulation of Moored Ship Motion Considering Harbor Resonance in Pohang New Harbor. In *Ports 2013*; American Society of Civil Engineers: Seattle, WA, USA, 2013; pp. 1415–1424.
28. MarineTraffic: Global Ship Tracking Intelligence | AIS Marine Traffic. Available online: https://www.marinetraffic.com/en/ais/home/centerx:-12.1/centery:25.0/zoom:4 (accessed on 28 June 2021).
29. Free AIS Ship Tracking of Marine Traffic-VesselFinder. Available online: https://www.vesselfinder.com/ (accessed on 28 June 2021).
30. Yin, J.; Zou, Z.; Xu, F. On-Line Prediction of Ship Roll Motion during Maneuvering Using Sequential Learning RBF Neuralnetworks. *Ocean. Eng.* **2013**, *61*, 139–147. [CrossRef]
31. Kumar, N.K.; Savitha, R.; Mamun, A.A. Regional Ocean Wave Height Prediction Using Sequential Learning Neural Networks. *Ocean. Eng.* **2017**, *129*, 605–612. [CrossRef]
32. Rigos, A.; Tsekouras, G.E.; Vousdoukas, M.I.; Chatzipavlis, A.; Velegrakis, A.F. A Chebyshev Polynomial Radial Basis Function Neural Network for Automated Shoreline Extraction from Coastal Imagery. *Comput. Aided Eng.* **2016**, *23*, 141–160. [CrossRef]
33. Cha, Y.-J.; Choi, W.; Büyüköztürk, O. Deep Learning-Based Crack Damage Detection Using Convolutional Neural Networks. *Comput. Aided Civ. Infrastruct. Eng.* **2017**, *32*, 361–378. [CrossRef]
34. Lin, Y.; Nie, Z.; Ma, H. Structural Damage Detection with Automatic Feature-Extraction through Deep Learning. *Comput. Aided Civ. Infrastruct. Eng.* **2017**, *32*, 1025–1046. [CrossRef]
35. Xue, Y.; Li, Y. A Fast Detection Method via Region-Based Fully Convolutional Neural Networks for Shield Tunnel Lining Defects. *Comput. Aided Civ. Infrastruct. Eng.* **2018**, *33*, 638–654. [CrossRef]
36. Red Costera de Boyas de Oleaje de Puertos Del Estado (REDCOS). Available online: https://bancodatos.puertos.es/BD/informes/INT_1.pdf (accessed on 28 June 2021).
37. Red de Medida Del Nivel Del Mar y Agitación de Puertos Del Estado (REDMAR). Available online: https://bancodatos.puertos.es/BD/informes/INT_3.pdf (accessed on 28 June 2021).
38. Alvarellos, A. aalvarell/ship-movement-dataset: Outer Port of Punta Langosteira ship movement dataset (Version v1.0.0). *Github* **2021**. [CrossRef]
39. Hastie, T.; Tibshirani, R.; Friedman, J. Neural Networks. In *The Elements of Statistical Learning: Data Mining, Inference, and Prediction*; Hastie, T., Tibshirani, R., Friedman, J., Eds.; Springer Series in Statistics; Springer: New York, NY, USA, 2009; pp. 389–416. ISBN 978-0-387-84858-7.
40. Hastie, T.; Tibshirani, R.; Friedman, J. Boosting and Additive Trees. In *The Elements of Statistical Learning: Data Mining, Inference, and Prediction*; Hastie, T., Tibshirani, R., Friedman, J., Eds.; Springer Series in Statistics; Springer: New York, NY, USA, 2009; pp. 337–387. ISBN 978-0-387-84858-7.
41. Piryonesi, S.M.; El-Diraby, T.E. Data Analytics in Asset Management: Cost-Effective Prediction of the Pavement Condition Index. *J. Infrastruct. Syst.* **2020**, *26*, 04019036. [CrossRef]
42. Madeh Piryonesi, S.; El-Diraby, T.E. Using Machine Learning to Examine Impact of Type of Performance Indicator on Flexible Pavement Deterioration Modeling. *J. Infrastruct. Syst.* **2021**, *27*, 04021005. [CrossRef]
43. He, K.; Zhang, X.; Ren, S.; Sun, J. Delving Deep into Rectifiers: Surpassing Human-Level Performance on ImageNet Classification. *arXiv* **2015**, arXiv:1502.01852.
44. Glorot, X.; Bengio, Y. Understanding the Difficulty of Training Deep Feedforward Neural Networks. In Proceedings of the Thirteenth International Conference on Artificial Intelligence and Statistics, Sardinia, Italy, 13–15 May 2010; Volume 9, pp. 249–256.
45. Kiefer, J.; Wolfowitz, J. Stochastic Estimation of the Maximum of a Regression Function. *Ann. Math. Statist.* **1952**, *23*, 462–466. [CrossRef]
46. Hinton, G.E. RMSPROP. 2014. Available online: http://www.cs.toronto.edu/~{}tijmen/csc321/slides/lecture_slides_lec6.pdf (accessed on 28 June 2021).
47. Kingma, D.P.; Ba, J. Adam: A Method for Stochastic Optimization. *arXiv* **2017**, arXiv:1412.6980.
48. Srivastava, N.; Hinton, G.; Krizhevsky, A.; Sutskever, I.; Salakhutdinov, R. Dropout: A Simple Way to Prevent Neural Networks from Overfitting. *J. Mach. Learn. Res.* **2014**, *15*, 1929–1958.

Journal of
Marine Science and Engineering

MDPI

Article

Soil Classification and Site Variability Analysis Based on CPT—A Case Study in the Yellow River Subaqueatic Delta, China

Zhongnian Yang [1], Xuesen Liu [1,2], Lei Guo [2,*], Yuxue Cui [1], Xiuting Su [3] and Xianzhang Ling [1,4]

[1] School of Civil Engineering, Qingdao University of Technology, Qingdao 266033, China; yzhnqd@qut.edu.cn (Z.Y.); liuxuesen1996@gmail.com (X.L.); cyxsnow9656@gmail.com (Y.C.); Lingxianzhang@qut.edu.cn (X.L.)
[2] Institute of Marine Science and Technology, Shandong University, Qingdao 266237, China
[3] College of Environmental Science and Engineering, Ocean University of China, Qingdao 266100, China; suxiuting@stu.ouc.edu.cn
[4] School of Civil Engineering, Harbin Institute of Technology, Harbin 150001, China
* Correspondence: 201894900036@sdu.edu.cn

Abstract: The Yellow River Delta is located at the junction of the Yellow River and Bohai. The impact function from the river and the dynamics of the ocean tides make the soil composition and distribution in this area substantially complicated. In order to test the distribution and variation of the soil layers in the Yellow River Delta, the soil layers in the test area were classified and the variation was calculated using the cone penetration test (CPT). The following conclusions were drawn: (1) the soil in the measured area is mainly composed of sensitive fine-grained soil, accounting for about 70% of all soil types, and the content of sensitive fine-grained soil in the far-sea position is higher than that in the offshore position in the direction perpendicular to the coastline. (2) It has a high vertical variability index (VVI) at the near-shore location, above 45%, and a low vertical variability at the far coast, generally below 20%. (3) The horizontal variability index (HVI) changes significantly near the coast, and it remains below 45% in the test area.

Keywords: Yellow River Delta; cone penetration test (CPT); soil behavior type (SBT); classification; vertical variability index (VVI); horizontal variability index (HVI)

Citation: Yang, Z.; Liu, X.; Guo, L.; Cui, Y.; Su, X.; Ling, X. Soil Classification and Site Variability Analysis Based on CPT—A Case Study in the Yellow River Subaquatic Delta, China. *J. Mar. Sci. Eng.* **2021**, *9*, 431. https://doi.org/10.3390/jmse9040431

Academic Editors: M. Dolores Esteban, José-Santos López-Gutiérrez, Vicente Negro and M. Graça Neves

Received: 2 March 2021
Accepted: 14 April 2021
Published: 16 April 2021

1. Introduction

A large number of cables and pipelines are distributed on the bottom of the Yellow River Delta, which is an important area for offshore oil and gas production [1]. Many international researchers have paid full attention to the investigations of the geological disasters in the Yellow River Delta [2–5], finding that the study area is characterized by complex topography and geological disasters. The shoreline boundary on either side of the estuary moves towards the sea at the average speed of 100 m/month [6]. The engineering characteristics of the sensitive fine-grained soil are relatively complicated, and may produce fluid mud to cause potential marine geological disasters [7,8]. The wave actions will directly lead to changes in the seabed topography at the mouth of the Yellow River, making the construction of the area, the engineering survey of the soil, and the prediction of geological disasters on the seabed difficult [9,10]. The Yellow River Delta is located at the intersection of the Yellow River and Bohai, which will cause sediment accumulation, and the effect of waves will cause the erosion of silt and sediment [11]. Changes in the erosion and deposition of the Yellow River Delta will seriously affect the development of ports [12]. Besides this, the deposition of the seabed results in frequent instability accidents, making it challenging to develop offshore projects on a large scale [13]. Therefore, the geological conditions in the Yellow River Delta are relatively complex, with an uneven distribution of soil layers, diverse soil composition, the influence of ocean tide dynamics, river erosion and accumulation on the coast, etc. Various conditions make the assessment of the strata of the Yellow River Delta region and exploration work very complicated. As a result, it is very important to investigate the variability of the entire test area. The variability is composed of the value and change of horizontal and vertical variability.

The cone penetration test (CPT) and its enhanced version (piezocone, CPTu) have seen extensive applications in many kinds of soils due to their continuity and repeatability [14]. Compared with traditional laboratory tests, CPT—with the advantages of fast, continuous, repeatable and reliable measurements—was often the preferred method for in-situ marine geological investigation, and abundant experience has been gained in the interpretation of CPT results for soil types [15], meaning that CPT is used in engineering geological monitoring widely. Due to the effect of ocean dynamics, sedimentary minerals will be transported and accumulated again [16], and the mineral composition of surface sediments in the modern Yellow River Delta is rather complicated [17]. These factors made the stratigraphic variability of the delta area relatively large. The soil classification information obtained by CPT can provide essential and useful information to geotechnical engineers [18–20]. The proposed soil classification charts by Robertson [21] have been popular and widely used [22,23]. Using the data obtained by the cone penetration test, it is convenient to investigate the changes of the soil layer and calculate the properties and parameters of various soil bodies. It can meet the simple operation and protect the nature of the original soil. Using the measured data to comprehensively investigate the variability values can draw more accurate results.

Due to the complex composition of the seabed soil in the Yellow River Delta, the study of the quantity and composition of the soil layers cannot provide the characteristics of the changes in the soil layers in detail. Therefore, it is necessary to continue to quantify the changes and calculate the changes in variability. Around the location of the area promoted by the Yellow River Delta, we established several points that can be used to investigate the horizontal and vertical changes, that is, the changes in the direction of the vertical coast of the Yellow River Delta near the coast and parallel to the coast.

Then, we used the variability calculation method summarized by previous scholars to investigate the characteristics of the soil variability in this area. Finally, we preliminarily summarized the change characteristics of the soil layer. Combined with the advantages of the cone penetration test, the composition of the soil layers in the Yellow River Delta region can be explored and the variation of the seabed soil can be quantified. The classification of the soil layers and the analysis of the site variability provide support for the Yellow River Delta underwater engineering construction, and make a preliminary exploration of the composition and variation of the underwater seabed soil.

2. In-Situ Survey

In order to determine accurate geological conditions and soil classification in the north of the Yellow River Subaquatic Delta, ten stations were arranged, of which one is perpendicular to the coastline, two others are parallel to it, as shown in Figure 1. The independently-developed Submarine Engineering Geological Environment Investigation System (SEGEIS) carried out a one-month CPT survey—as shown in Figure 2.

The surface layer of the area consisted of yellow-brown silt with a thickness of 4.7 m and a medium density, of which parts were interspersed in the silty clay. The standard test shot number was 12–31, and the median diameter of the topsoil particles was about 0.03 mm, of which the saturation and density were 27.4–32.3%, and 19.4–20.1 kN/m^3, respectively. The in-situ investigation system of the seabed engineering geological environment had many functions, such as in-situ observation, in-situ detection, static penetration, penetration, the extraction of sediment static sampling tubes, seabed environmental lighting, video shooting, and photography. The underwater CPT equipment, developed independently, was an in-situ investigation system for the subsea engineering geological environment. A high-quality pressure strain sensor was used to measure the cone tip resistance (range 0–50 MPa) and the sidewall friction resistance (range 0–1 MPa). The pore water pressure was measured by a piezoelectric element (range 0–2 MPa). The tilt angle was measured by an accelerometer (inclination angle in both the X and Y directions, range 0–15°). Finally, the sampling and penetration methods were static and hydraulic. The parameters of the underwater cone sounding test equipment are shown in Table 1.

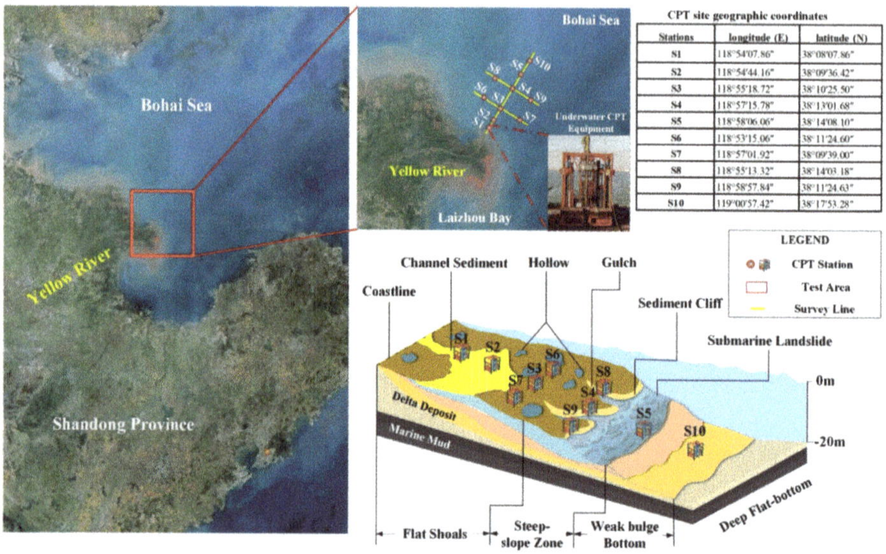

Figure 1. The stations and survey lines.

Figure 2. Underwater CPT system, developed independently.

Table 1. The main parameters of the independently-developed underwater CPT system.

Specifications	Parameters
Water depth of working	100 m
Probe parameters	Φ36 mm
Sampling tube parameters	Φ75 mm/Φ110 mm (drive pipe)
Penetration rate	1.2 m/min
Penetration force	2.5 t
Hydraulic pressure	10 MPa (Penetration)/12 MPa (Extraction)
Cone tip resistance range	50 Mpa
Side friction force	1 Mpa
Cone tip cross-sectional area	10 cm^2
Sample diameter	75 mm
Sampling depth	5 m

3. Soil Classification

Soil classification charts have been used widely to assess soil behavior based on CPT results [22,23]. The soil classification chart [21] is divided into twelve zones in order to represent the different grain size distributions, as well as the different soil states or behaviors coded by numbers. For example, Zone 1 indicates a soil type of sensitive fine grains, while the CPT data points of classical stiff fine-grained soil will fall in Zone 11, according to Figure 3, as given by Robertson [21].

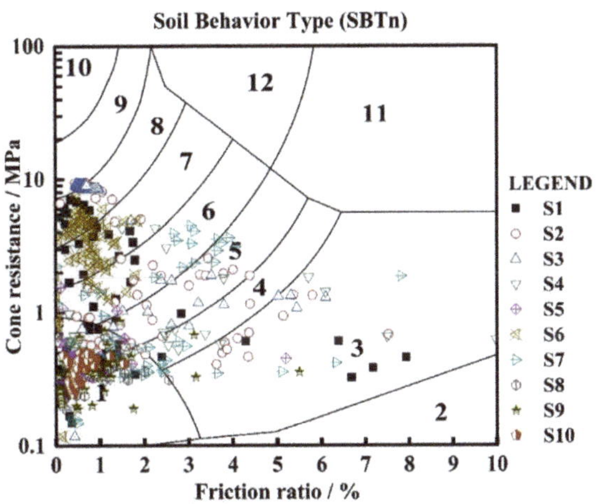

Figure 3. Soil behavior type chart.

Fine-grained soil is widely distributed in the Yellow River Delta region, and some sandy soil with larger particle sizes also exists. As shown in Figure 4, there is an increase by 10 MPa sharply at depths of 1.5 m and 2 m, etc. The continuous accumulation of pore water pressure under the action of tides and waves in the sediments of the Yellow River Delta results in the continuous accumulation of the strength of the surface soil [24]. Typically, there is high cone resistance in sands but low resistance in clays, while the friction ratio is contrary, according to Robertson's research [25].

The pore water pressure is distributed in 150 kPa–200 kPa, and the different sites vary significantly in depth. The pore water pressure will drop steeply at a certain depth. As such, the excess pore pressure shows a decreasing trend near 150 kPa–200 kPa in 0.5 m, at 3.4 m depth below from the surface for S9 and 2.1 m depth below for S4. At 4.5 m, it drops by about 140 kPa in S5. There is a high negative pore pressure in the case of high-density fine sands and dilative silts according Cai et al. [23]. Affected by the over-consolidation state, the pore water pressure may also become small or even negative [26].

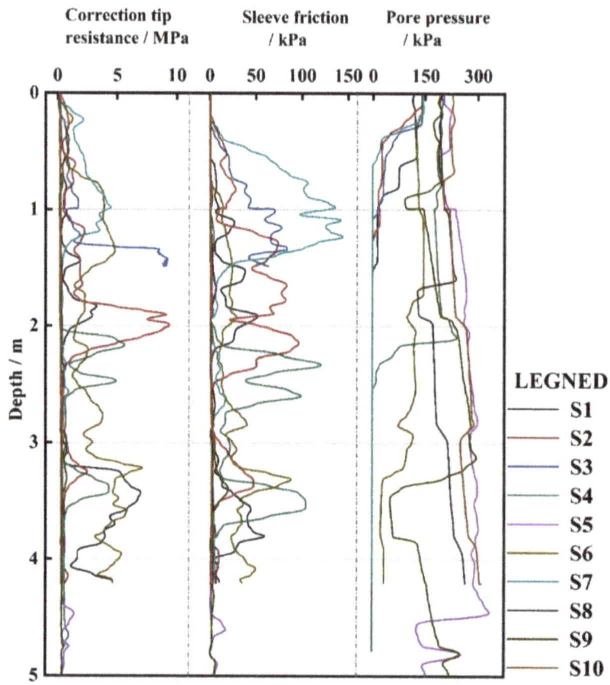

Figure 4. The in-situ CPTu data.

The data for the locations of the 10 stations are arranged in the chart [21]. There is a clear distribution of 10 measuring points in area 1, and there no points in areas 11 and 12, revealing preliminarily that the soil composition in the area is mainly sensitive fine-grained soil, as shown in Figure 3. 50% of the points are in zone 1, which is mainly related to the sediment accumulation in the Yellow River Delta, according to the relevant research [27]. The sediment of the Yellow River is mainly from the Loess Plateau, and the sensitive fine-grained soil is mainly in the depth of 40 cm below the surface, while the deep layer is dominated by clay [28] where there is an obvious effect from the wave actions [27]. The method of soil layer classification in this area is shown in Table 2. By referring to the classification results in Figure 3 and Table 2, the soil layer results are obtained in the test area. As shown in Figure 5, there is very fine sand on the delta front, which is consistent with previous studies [4]. Sedimentation in the sedimentary area is slow when it is near the land in the west, and steep when it is far away from the land in the east. According to previous research [29], from the coast to the deep sea, the sediment composition shows a coarse-fine-coarse variation.

Table 2. Non-normalized CPT soil behavior type (SBT) chart [21].

SBT Zone Robertson et al. (1986)	Proposed Common SBT Description
1	Sensitive fine-grained
2	Clay—organic soil
3	Clays: clay to silty clay
4 & 5	Silt mixtures: clayey silt & silty clay
6 & 7	Sand mixtures: silty sand to sandy silt
8	Sands: clean sands to silty sands
9 & 10	Dense sand to gravelly sand
12	Stiff sand to clayey sand
11	Stiff fine-grained

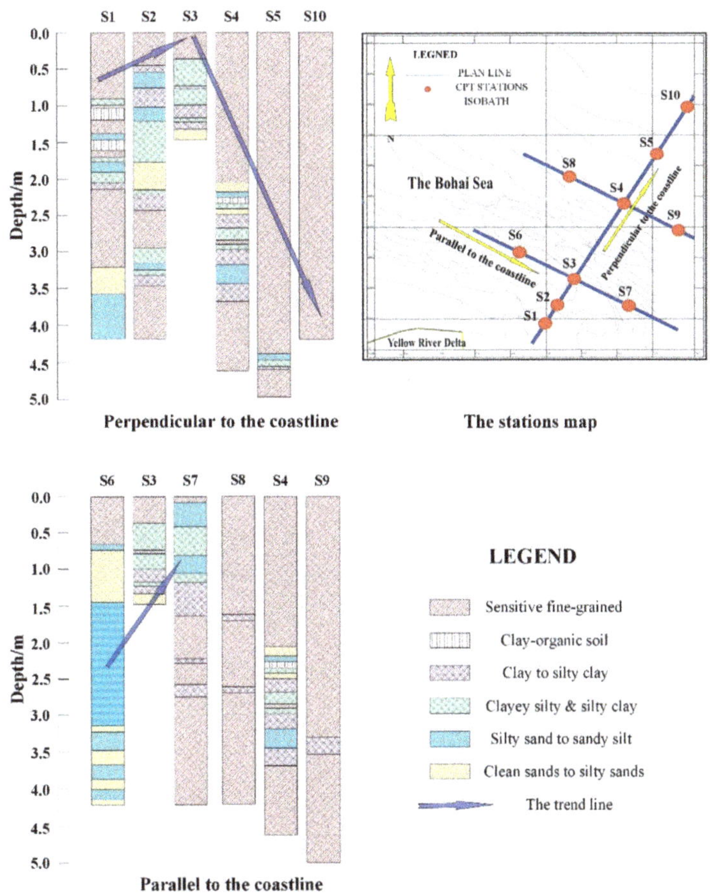

Figure 5. The soil layers profile.

4. Site Variability Analysis Based on CPT

By calculating the site variability, the geological conditions of the measured area can be evaluated, and the civil engineering construction of the site can be guided. There are three main parts about the analysis on the site variability of the static penetration test. The first part is the calculation of the vertical variability, including the intra-layer variation index (VVI_{IL}), log vertical variability index (VVI_{log}), and cone resistance vertical variability index (VVI_{qc}). The second part is the calculation of the horizontal variability. The final part is a comprehensive analysis of the obtained data of the horizontal variability and vertical variability, as well as the variability of the site considering the geology and marine factors of the measured area.

When processing the CPT data, all of the abnormal data caused by instrument error, improper operation, and special geological phenomena should be correctly identified and interpreted reasonably.

4.1. Vertical Variability Index (VVI)

First, the depth and q_c values measured by CPT are detrended. It is necessary to fit the data with functions. The polynomial function is used to fit the q_c data and depth for the data detrending processing in this paper. The polynomial function used is denoted as $f(x)$. After the fitting of the function is completed, the function is detrended, and the function used is as shown in the following Equation (1):

$$\omega = \frac{y - f(x)}{f(x)} \quad (1)$$

where $f(x)$ is the q_c value obtained after the polynomial function processing, y is the q_c value in the original data, and ω is the q_c value after the detrending. The relationship between the q_c data and the depth after the detrend processing of S8 is shown as an example in Figure 6.

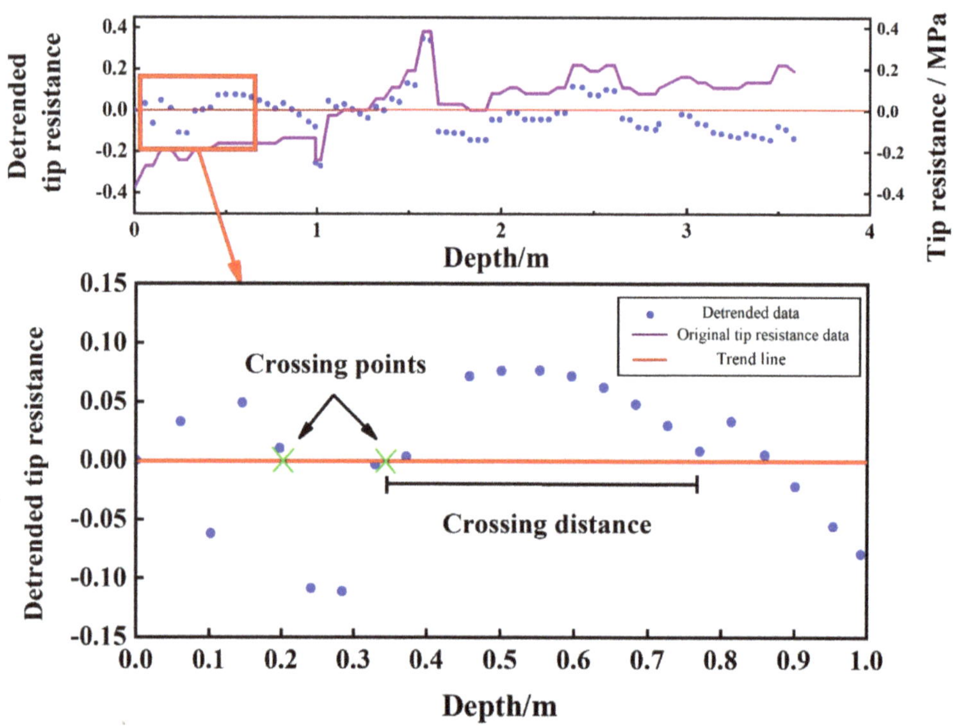

Figure 6. CPT detrended data with the fitted trendline.

The value after detrending ranges from −1 to 1. The crossing point and crossing distance are shown in Figure 6; the scale of the fluctuation-normalized coefficient of coefficient variation (COV) is defined as SNC, which will indicate the variety of the soil layers. Furthermore, the area of one side of the trendline will indicate the value of the scale of fluctuation (SF), which represents the variety of the soil layers. Although the cone penetration test method suffers a measurement error [30] when the soil layer thickness is small, the influence of the layer thickness on the variability is not significant. Therefore, the error of the cone penetration test in the case of a minimal-thickness soil layer is ignored in this paper.

4.2. Intra-Layer Variation Index (VVI_{IL})

By considering that the intra-layer variability index is affected by COV and SF, a new parameter—SNC—is defined. The variation index of intra-layer size has a significant connection with the each-layer soil thickness and the number of soil layers in the entire soil section. First of all, the data obtained from the CPT is used in calculating each soil

COV with soil layer maps and different soil type layer classification. Then, the relationship between the data after the detrending process with crossing points is established. The SF value used in this paper has been introduced [31], which relates to the model and the soil type used in the calculation. According to the research, many differences vary from 0.07 m to 4 m in marine or clay soil types regarding the SF value [32]. Finally, the SNC value is found with the formula proposed [33]. Figure 7 clearly illustrates the calculation process.

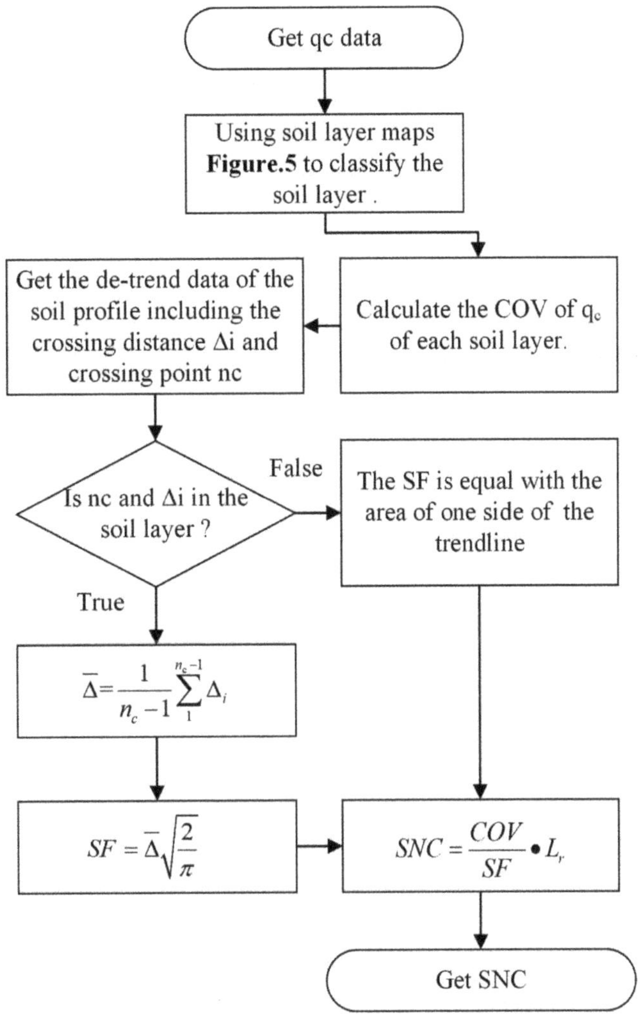

Figure 7. The SNC calculation method (adopted by Salgado et al. 2019) [33].

Here, n_c is the number of the crossing point, Δ_i is the crossing distance, and the value of L_r is 1. ACF means the Autocorrelation Function, which is used to determine the distance of the correlation, which changes with the function type [34]. The twice-positive area formed by the fitting function and the trend line is regarded as the SF value in this paper. The main factor affecting the variability in the soil layer is q_c; thus, only the SNC of q_c needs calculating [33]. The SNC results are shown in Table 3.

Table 3. The SNC.

Station	S1	S2	S3	S4	S5	S6	S7	S8	S9	S10
SNC (%)	4	9	34	15	13	22	3	11	89	85

The results are quite different, where S9 and S10 approach 90%; S3 and S6 are near 30%; S2, S5, and S8 are about 10%; and S1 and S7 are just about 3%. The soil stratification, SNC data, and the variability index of the intra-layer size are related to the thickness of the soil layer, the number of normalized q_c crossing points in each layer, and the COV of each layer of the variation coefficient. Therefore, the SNC at S9 and S10 is bigger than those of the other stations. Furthermore, the thickness of a single layer and the COV is large, resulting in a large SNC. There is a prime feature within S4 and S5, given that both of them have a dominant thickness; even the thickness of S5 is over 4 m in a single layer between S4 and S5. However, there is clay or silty clay type in both S4 and S5, making the VVI_{IL} in S4 and S5 small. All in all, the thickness of a single layer has a significant effect on the SNC.

4.3. Log Vertical Variability Index (VVI_{log})

The logarithmic vertical variability index mainly calculates the diversity factors (DF) within a layer. Firstly, the soil layer is classified according to Table 4. The numbers of the sand, clay, and mixed soil obtained are respectively recorded as N_S, N_C, and N_{MS}. Then, the proportions of the D_S, D_C, and D_{MS}—of which the sum is 1—are calculated. The D* shown in Figure 8 represents the distance between the soil of any content of the three, and the three of them are 1/3, which is the degree of variation of the three. Theoretically, it fluctuates between 0 and 1, such that the DF ranges from 0 to 100. The lower the DF, the higher the dominance of the soil; therefore, the vertical variability will be smaller. Eventually, the final VVI_{log} value is obtained. The main steps in the DF and VVI_{log} are in Figure 9.

The NDL is the number of layers per unit length. In this test, various values of NDL were calculated, with a range of 1 to 5.7, revealing that there are complex and diverse soil types and layers in the Yellow River Delta. The calculated VVI_{log} value is normalized with the $\log_{10}(x)$ function in order to keep the value obtained in the range of 0 to 1. As illustrated in Table 5, S1, S5, S8, S9, and S10 are below 90%, indicating that there is a dominant soil layer, or that the number of soil layers is small. Taking S4 as an example, there is a large proportion of sensitive fine-grained soil, but the number of soil layers at this point is 14, which is exceptionally numerous. Therefore, the VVI_{log} value of S4 is relatively large, resulting in a large vertical variability, which means that the value of VVI_{log} can intuitively show the influence of the number of soil layers and the thickness variation of the different soil layers.

Table 4. Soil group classification for the modified Robertson chart used to group soils for the calculation of dominance factor D_i and diversity factor DF.

Soil Group	Soil Types
Sand	Gravelly sand to sand (very loose to very silty sand) Clean sand to silty sand (very loose to very dense)
Clay	Organic clay Sensitive fine grained
Mixed soil	Silty sand to sand silty Clayey silty to silty clay Clay to silty clay

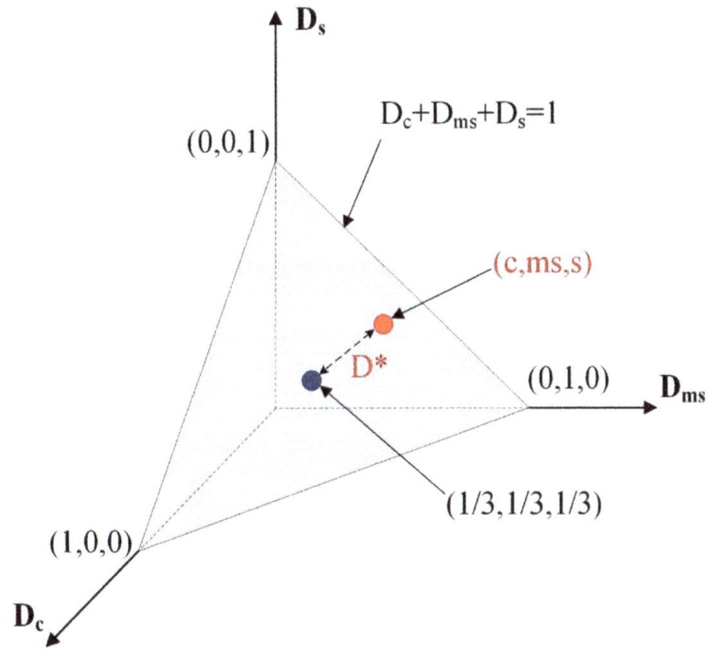

Figure 8. D* schematic diagram (modified from Salgado et al., 2019) [33].

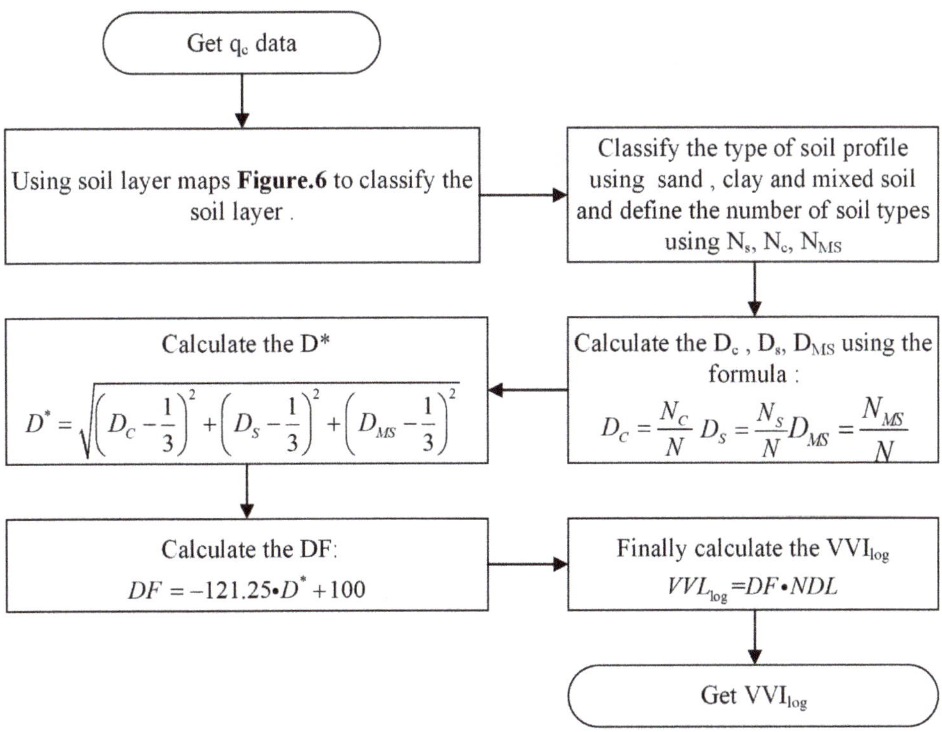

Figure 9. The VVI_{log} calculation method (modified from Salgado et al., 2019) [33].

Table 5. The VVI_{log}.

Station	S1	S2	S3	S4	S5	S6	S7	S8	S9	S10
VVI_{log} (%)	82	94	96	99	79	98	99	83	77	46

4.4. Cone Resistance Vertical Variability Index (VVI_{qc})

VVI_{qc} is a COV of q_c, and the standardized COV of q_c as the VVI_{qc} value of the soil layer, because the VVI_{qc} value is related to the soil layer profile and the composition of the soil type. We can calculate the COV from the obtained qc value, and we should normalize the result if it is between 0–1. In a soil layer comprising a single soil, especially a single clay or silty clay layer, the VVI_{qc} value is below 10%, such as for S5, S8, S9 and S10, as listed in Table 6. On the contrary, if the profile consists of multiple soil types, the number of soil layers is plentiful, and the physical characteristics of the soil types are quite different, then the VVI_{qc} is relatively large, just as for S1, S2, S4, and S7. Comparing S8 with S7, the number of layers is similar, and the main components of the soil are clay and silty clay, but the number of soil layers in S7 is five times as many as for S8, so the VVI_{qc} of S7 is higher than that of S8.

Table 6. The VVI_{qc}.

Station	S1	S2	S3	S4	S5	S6	S7	S8	S9	S10
VVI_{qc} (%)	58	66	29	48	10	33	42	3	6	3

Eventually, the vertical variability index can be calculated by Equation (2), and the VVI value is listed in Table 7.

$$VVI = 0.2VVI_{IL} + 0.2VVI_{\log} + 0.6VVI_{qc} \quad (2)$$

Table 7. The VVI.

Station	S1	S2	S3	S4	S5	S6	S7	S8	S9	S10
VVI (%)	52	60	43	52	10	24	46	21	37	28

The VVI is related to the soil layer content, number, and thickness. When the composition of the soil layer is single within three layers, the vertical variability is relatively single, which makes the value of VVI small, i.e., under about 40%. The vertical variability is various; meanwhile, when a single soil layer is thicker where there are five or more layers in one meter, and the number of soil layers is larger in one soil profile where the totally different layer numbers exceed ten, this will also cause more considerable vertical variability, and the value of VVI will even reach 50%. Therefore, the number of layers has a significant effect on the vertical variability.

4.5. Horizontal Variability Index (HVI)

When calculating the coefficient of the horizontal variation index, the size of the cone tip resistance and the depth of the sounding need to be considered as two factors affecting the horizontal variability, because they have a deficient level of variability to some extent if the changes in the two groups of qc data have a high correlation.

First, the influence factors of q_c should be indicated. The differences between q_c and the average of q_c are calculated. The f_0 is defined by dividing the previous maximum value q_c with the absolute q_c value $|q_c|$. Secondly, the correlation coefficient ρ_{xy} between q_c and the depth is calculated under the consideration of the depth effect. The parameters of f_1 and f_2 are figured with the Equations (3) and (4). The eventual HVI value can be calculated using Equation (6). The main calculation process is shown in Figure 10, where ρ_{xy} is the correlation coefficient, and s is the vertical distance between the two CPT data points [33].

$$f_1 = \frac{\rho_{xy} + 1}{2} \tag{3}$$

$$f_2 = 1 - e^{\frac{s}{4}} \tag{4}$$

$$f = [0.8(1 - f_0) + 0.2f_1]f_2 \tag{5}$$

$$HVI = 1 - \frac{\sum_1^n f}{n} \tag{6}$$

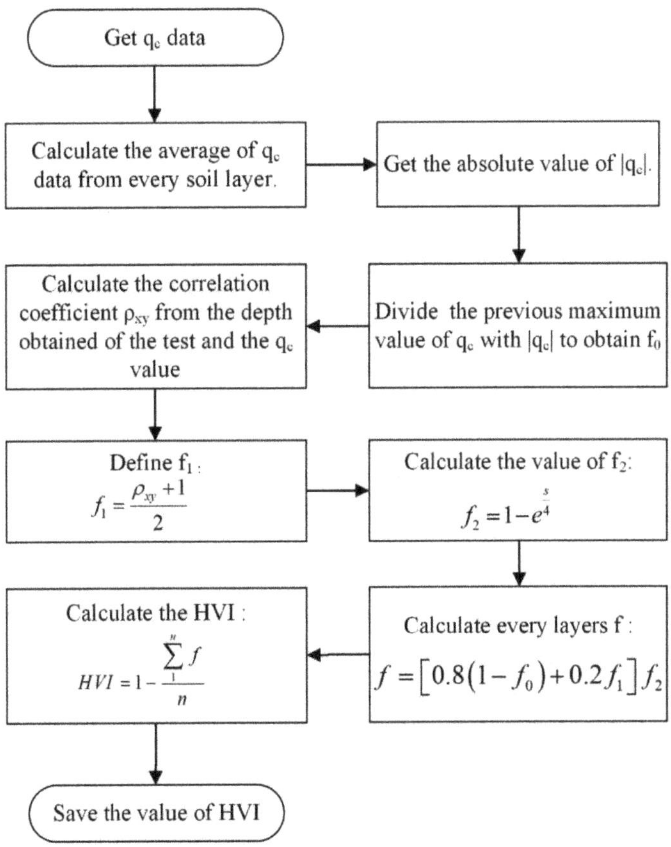

Figure 10. The HVI calculation method (modified from Salgado et al., 2019) [33].

The result of HVI is shown in Table 8. The horizontal variability index is related to the content of hard clay. According to the previous research [33], the high content of hard clay or silt will make the horizontal variability index large; otherwise, the horizontal variability index will be small. Besides this, due to changes in the carrying capacity of the Yellow River and the carrying capacity when entering the sea, the particle size of the Yellow River Delta varies from the estuary to the sea. Therefore, the horizontal variability has a major change with the direction of the vertical coastline [35]. In terms of the soil stratification in the overall soil profile, the content of hard clay in S6 is higher. Therefore, the horizontal coefficient of variation of S6 is the largest, while the rest are medium.

Table 8. The HVI.

Station	S1	S2	S3	S4	S5	S6	S7	S8	S9	S10
HVI	44	45	38	59	56	66	45	43	42	35

5. Site Variability Rating

The site variability evaluation and analysis are performed after obtaining the above horizontal variability index and the vertical variability index. The values of the HVI and VVI are divided into three grades of L, M, and H depending on the three criteria of (0–33%), (33–66%), and (66–100%). For the analysis of the horizontal and vertical variability of the entire measurement station, the overall horizontal and vertical variability is high throughout the exploration area.

As shown in Figures 11 and 12, the overall regularity of the vertical variability gradually decreases from the shore to the distance. As can be seen from Figure 13, the seabed soil in the Yellow River Delta region has complex geological conditions [36]. The maximum value of the near coast is about twice the minimum value of the far coast. Furthermore, the local law of vertical variability is roughly equivalent to the decrease of undulation. In a direction parallel to the coastline, the vertical variability of the measured area in the northwest direction is smaller than that in the southeast direction, as shown in Figure 12a. The horizontal variability in the direction perpendicular to the coastline gradually escalates, with an increase of about 20% according to Figure 12b. In the direction parallel to the coastline, the horizontal variability gradually decreases from northwest to southeast near the coast. Near the shore, the bottom of the stratum is mainly composed of relatively dense silt, and the fine-grained sand layer on the upper part of the soil layer far from the shore is relatively thick, which is related to the suspension of sand particles caused by the convergence of seawater and Yellow River water when the Yellow River enters the sea. These results are consistent with the previous studies [37,38]. The silty soil and cohesive soil of the Yellow River sediments were found on the sticky bed of the Yellow River Delta. Furthermore, because the Yellow River Delta belongs to the strong accumulation type estuary with a weak and sandy tide, the combination of runoff and tidal current causes the saltwater and freshwater to meet, leading a large amount of suspended sediment to flocculate and settle, which results to the development of the sand in the Yellow River Mouth. S1, S2, and S4 contain sand, clay, and silty soil, such that the vertical variability index is approximately similar, around 50%. However, compared with S6, the vertical variability index of S1, S2, and S4 are 25% higher than S6, although their number of soil layers are approximately equal. This is because S6 mainly contains stiff soil, while S1, S2, and S4 are more comprehensive, resulting in a smaller VVI of S6. It has been revealed that the entire soil section is dominant sensitive to fine-grained soil in the Yellow River Delta [39]. Therefore, the vertical variability is generally medium, around 40–50%. Considering all of the above factors and Figure 11a,b, in the direction perpendicular to the coastline, the VVI values are smaller than 33% far from the coastline, and generally above 50% near the coastline due to various components of soil layers. The HVI values hold at a medium level. Moreover, in the direction parallel to the coastline, the values of VVI far from the coast are smaller and more varied than those near the coast. On the contrary, the HVI values near the coast are more varied than those far from the coast. As shown in Figure 13, there are many complex geological hazards, such as gas-bearing soils, fault zones, eroded coasts, and slippery bodies in the Yellow River Delta. Besides this, the front geological body of the delta is also divided into smooth delta front and disturbing delta front. Combined with the calculated values of the horizontal and vertical variability of the measured points, the above comprehensive geological distribution and hazards also affect the distribution of the horizontal and vertical variability. It can be seen from Figures 12 and 13, in the direction perpendicular to the coastline, that the vertical variability gradually decreases with the distance of the Yellow River Delta into the sea. In other words, the water depth has a substantial influence on the vertical variability but little effect on the horizontal variability.

In the direction parallel to the coastline, the horizontal and vertical variability mainly change along the contour line. Far away from the coastline, the impact of the marine hydrodynamic environment on the seabed is weakened as the water depth increases, so that the position far away from the coastline has less horizontal and vertical variability. In conclusion, the VVI value and the HVI value have something to do with the number of soil layers in a profile, the soil thickness, and its composition, which is mainly determined by the characteristics of the sand carried by the Yellow River into the sea and the sedimentation characteristics of the estuary.

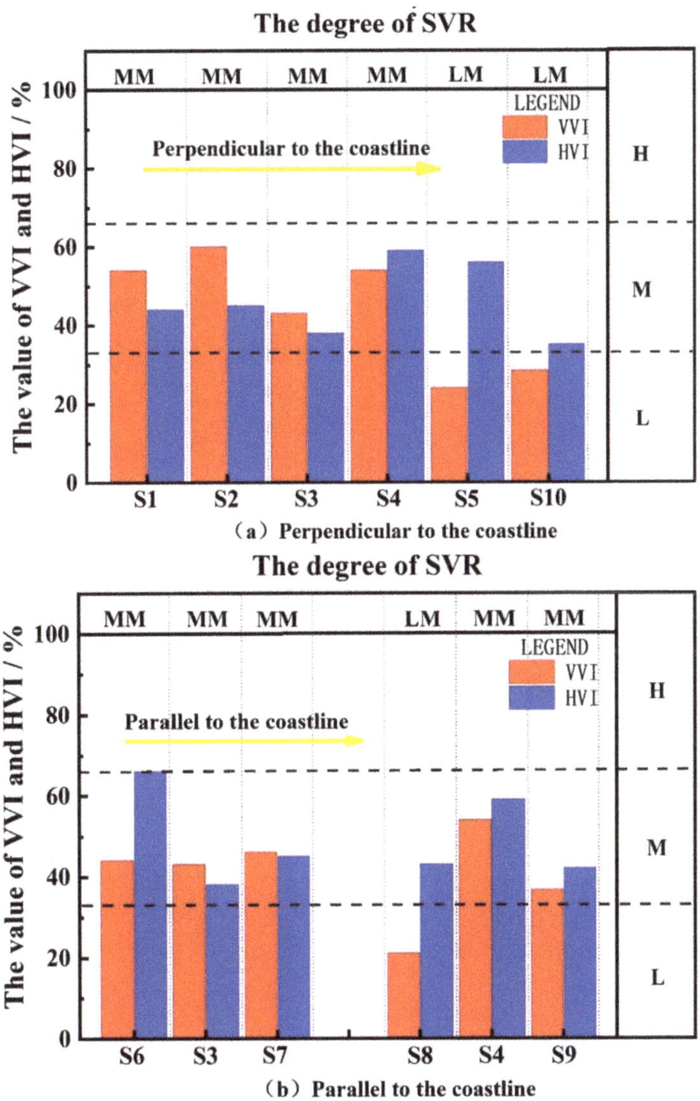

Figure 11. The SVR data. (**a**) Perpendicular to the coastline; (**b**) Parallel to the coastline.

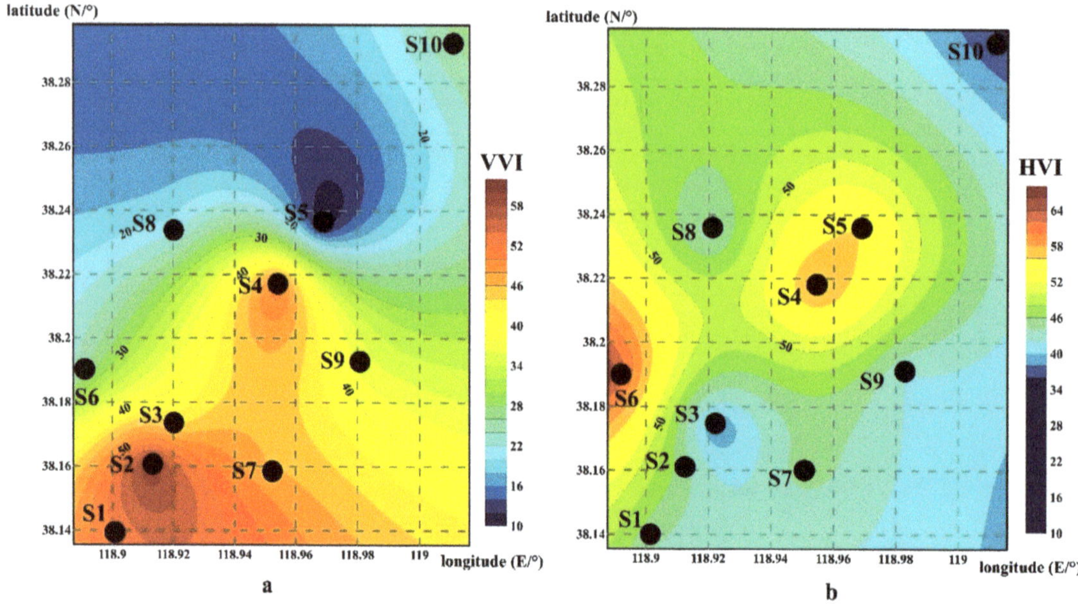

Figure 12. VVI (a) and HVI (b) contour sketch maps.

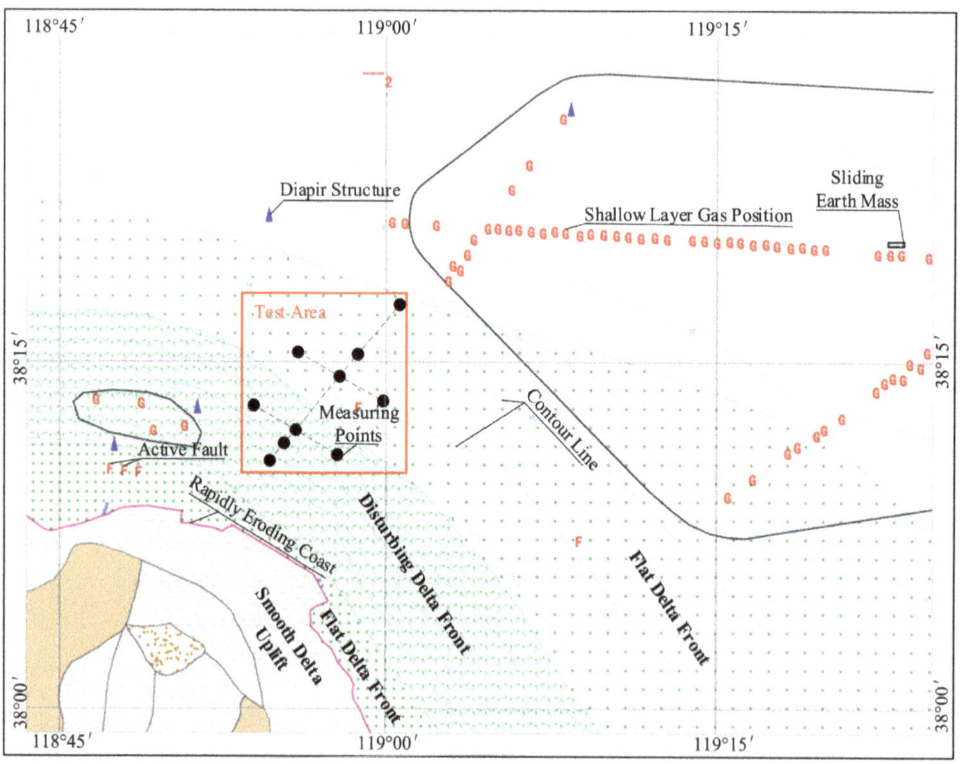

Figure 13. Measuring points and the surrounding geological hazards map (modified from Zhou et al. 2003) [36].

6. Conclusions

In-situ tests were carried out on the Yellow River Delta using independently-designed submarine CPT equipment. The classification of the soil layers and the analysis of the site variability were performed based on the obtained data. The main conclusions are as follows:

(1) Sensitive fine-grained soil is widely distributed in the Yellow River Delta. The classification results revealed that 70% of the soil layers in the test area are mainly sensitive fine-grained soil, and the soil composition type is distributed variously. There is an obvious change in the direction perpendicular to the coastline where the content of sensitive fine-grained soil in the near-coast is less than that in the far-coast.

(2) Vertical to the coast, the water depth changes complexly and is obviously affected by hydrodynamic forces. The vertical variability index varies obviously in the direction perpendicular to the coastline. The vertical variability in the offshore areas is basically maintained in the range of more than 45%, which is much larger than the value in distant sea areas, which is less than 20%.

(3) The soil layer changes are complex because of the shallow water depth and the obvious effect of the wave load near the shore. The diversification of the horizontal variability mainly occurs near the coast, and gradually decreases from the northwest direction to the southeast direction. There is no significant change of the horizontal variability distribution in the measured area, and its value is maintained at a moderate level of about 45% or less.

(4) Based on the previously-available geological composition of the Yellow River Delta, the vertical variability and soil distribution are significantly different in the disturbing delta front and the flat delta front. In the distribution delta front, the soil composition is diverse, with high vertical variability basically exceeding 40%; in the flat delta front, the soil composition is homogeneous, with low variability, and is below 20%.

Author Contributions: Conceptualization, Z.Y. and L.G.; methodology, Z.Y.; software, X.L. (Xuesen Liu); validation, Y.C. and X.S.; formal analysis, X.L. (Xuesen Liu); investigation, L.G.; resources, Z.Y. and L.G.; data curation, Z.Y. and L.G.; writing—original draft preparation, X.L. (Xuesen Liu); writing—review and editing, X.L. (Xuesen Liu) and Z.Y.; visualization, X.L. (Xuesen Liu); supervision, Z.Y.; project administration, Z.Y.; funding acquisition, Z.Y., L.G. and X.L. (Xianzhang Ling). All authors have read and agreed to the published version of the manuscript.

Funding: This research was funded by Shandong Province Focused Research and Development Program, grant number 2019GHY112075; the Special Project Fund of Taishan Scholars of Shandong Province, grant number 2015-212; the National Major Scientific Research Instrument Development Project, grant number 41627801; the Ministry of Science and Technology of the People's Republic of China, Key R&D Program, grant number 2018YFB2000901.

Institutional Review Board Statement: Not applicable.

Informed Consent Statement: Not applicable.

Data Availability Statement: Not applicable.

Conflicts of Interest: The authors declare no conflict of interest.

References

1. Song, B.; Sun, Y.; Song, Y.; Dong, L.; Du, X.; Zhou, Q.; Zhao, X. Post-liquefaction re-compaction effect on the cyclic behavior of natural marine silty soil in the Yellow River delta. *Ocean. Eng.* **2020**, *195*, 106753. [CrossRef]
2. Prior, D.B.; Yang, Z.-S.; Bornhold, B.D.; Keller, G.H.; Lu, N.Z.; Wiseman, W.J.; Wright, L.D.; Zhang, J. Active slope failure, sediment collapse, and silt flows on the modern subaqueous Huanghe (Yellow River) delta. *Geo-Mar. Lett.* **1986**, *6*, 85–95. [CrossRef]
3. Xiuli, F.; Hongshuai, Q.; Teng, W.; Anlong, L.; Lin, L. Geomorphological evolution and geological disasters analysis in Chengdao sea area of the Yellow River Delt. *Rock Soil Mech.* **2004**, *25*, 17–20. [CrossRef]
4. Yuling, Z.; Xiuli, F.; Sheng, S.; Donghui, T. Geochemical partition of surface sediments in the seas near the modern Yellow River Delta. *Mar. Sci.* **2016**, *40*, 98–104.

5. Liangyong, Z.; Jian, L.; Xiqing, L.; Guangxue, L.; Zhengxin, C. Hazardous Geology in the Shallow Sea Area of Modern Yellow River Delta. *Mar. Geol. Quat. Geol.* **2004**, *24*, 19–27.
6. Haoran, H.; Zhili, Z.; Chengshu, C. Numerical Simulation of Morphological Evolution Process of the Yellow River Mouth. *Coast. Eng.* **2018**, *37*, 1–14.
7. Yang, Z.; Zhu, Y.; Liu, T.; Sun, Z.; Ling, X.; Yang, J. Pumping effect of wave-induced pore pressure on the development of fluid mud layer. *Ocean. Eng.* **2019**, *189*, 106391. [CrossRef]
8. Yang, Z.; Zhu, Y.; Liu, T.; Sun, Z.; Ling, X.; Zheng, Y. Contribution of Pumping Action of Wave-Induced Pore-Pressure Response to Development of Fluid Mud Layer. *J. Mar. Sci. Eng.* **2019**, *7*, 317. [CrossRef]
9. Jia, Y.; Zhang, L.; Zheng, J.; Liu, X.; Jeng, D.-S.; Shan, H. Effects of wave-induced seabed liquefaction on sediment re-suspension in the Yellow River Delta. *Ocean. Eng.* **2014**, *89*, 146–156. [CrossRef]
10. Ren, Y.; Xu, G.; Xu, X.; Zhao, T.; Wang, X. The initial wave induced failure of silty seabed: Liquefaction or shear failure. *Ocean. Eng.* **2020**, *200*, 106990. [CrossRef]
11. Zheng, J.-W.; Jia, Y.-G.; Liu, X.-L.; Shan, H.-X.; Zhang, M.-S. Experimental study of the variation of sediment erodibility under wave-loading conditions. *Ocean. Eng.* **2013**, *68*, 14–26. [CrossRef]
12. Xiushan, G.; Qinian, Z.; Mingzuo, F.; Peiying, L.; Xiaoshi, X. Quantitative analysis and evaluation of scouring and silting in the Yellow River Estuary. *Acta Oceanol. Sin.* **1988**, *10*, 712–719.
13. Anlong, L.; Guangxue, L.; Lihua, C.; Qingde, Z.; Shenggui, D. Coastal erosion and shoreline evolution of abandoned leaf petals in the Yellow River Delta. *Acta Geogr. Sin.* **2004**, *59*, 731–737. [CrossRef]
14. Robertson, P.K. Evaluating Soil Liquefaction and Post-earthquake deformations using the CPT. In Proceedings of the Geotechnical and Geophysical Site Characterization, Porto, Portugal, 19–22 September 2004.
15. Lakusic, S.; Librić, L.; Jurić-Kaćunić, D.; Kovačević, M.S. Application of cone penetration test (CPT) results for soil classification. *J. Croat. Assoc. Civ. Eng.* **2017**, *69*, 11–20. [CrossRef]
16. Dong, D.; Wan, D. Preliminary Study on the Erosion of the Modern Yellow River Delta. *Mar. Geol. Quat. Geol.* **1988**, *8*, 53–60.
17. Liuqi, W.; Zaixing, J. Mineralogical characteristics and sedimentary genesis of land surface sediments in the Yellow River Delta. *J. Univ. Pet. China* **1993**, *17*, 1–6.
18. Eslami, A.; Fellenius, B.H. Pile capacity by direct CPT and CPTu methods applied to 102 case histories. *Can. Geotech. J.* **1997**, *34*, 886–904. [CrossRef]
19. Guo, L.; Liu, X.; Yang, Z.; Jia, C.; Shi, W.; Ling, X. CPT-based analysis of structured soil characteristics and liquefaction failure of the Yellow River Subaquatic Delta. *Mar. Georesour. Geotechnol.* **2021**, 1–20. [CrossRef]
20. Cetin, K.O.; Ozan, C. CPT-Based Probabilistic Soil Characterization and Classification. *J. Geotech. Geoenviron. Eng.* **2009**, *135*, 84–107. [CrossRef]
21. Robertson, P.K.; Campanella, R.G.; Gillespie, D.; Greig, J. Use of piezometer cone data. In *Use of In Situ Tests in Geotechnical Engineering*; ASCE: Blacksburg, VA, USA, 1986.
22. Long, M. Key Note Lecture: Design parameters from in situ tests in soft ground—Recent developments. In Proceedings of the 3rd International Conference on Geotechnical and Geophysical Site Characterisation (ISC'3), Taipei, Taiwan, 1–4 April 2008.
23. Cai, G.; Liu, S.; Puppala, A.J. Comparison of CPT charts for soil classification using PCPT data: Example from clay deposits in Jiangsu Province, China. *Eng. Geol.* **2011**, *121*, 89–96. [CrossRef]
24. Robertson, P.K. Soil behavior type from the CPT: An updata. In Proceedings of the 2nd International Symposium on Cone penetration Testing, Huntington Beach, CA, USA, 9–11 May 2010.
25. Jianmin, Z.; Hongxian, S.; Yonggang, J.; Hongjun, L.; Guohui, X. Experimental study on non-uniform consolidation of soil using fast-deposited seafloor in the Yellow River and estuary under wave tide. *Rock Soil Mech.* **2007**, *7*, 88–94. [CrossRef]
26. Hongxian, S.; Jianmin, Z.; Yonggang, J.; Hongjun, L.; Guohui, X. Study on the Consolidation Process of Rapidly Deposited Seabed Soil in the Yellow River Estuary. *Chin. J. Rock Mech. Eng.* **2006**, *25*, 1676. [CrossRef]
27. Xiaohua, W.; Hongjun, L.; Yonggang, J. Study on Mineral Composition Characteristics of Silty Soil in the Yellow River Estuary and Its Response to Hydrodynamic Conditions. *Mar. Geol. Lett.* **2004**, *5*, 34–39.
28. Yonggang, J.; Hongxian, S.; Xiujuan, Y.; Xiangmei, M.; Fangqiang, C.; Jiewen, Z. *Sediment Dynamics and Geologic Hazards in the Estuary of Yellow River, China*; Science Press: Beijing, China, 2011.
29. Xiaoying, C.; Shenliang, C.; Yongsheng, L. Characteristics and regularity of sediments in the coastal area of the Yellow River Delta. In Proceedings of the 9th National Symposium on Estuary and Coast, Xiamen, China, 6–9 September 2006.
30. Ganju, E.; Prezzi, M.; Salgado, R. Algorithm for generation of stratigraphic profiles using cone penetration test data. *Comput. Geotech.* **2017**, *90*, 73–84. [CrossRef]
31. Vanmarcke, E.H. Probabilistic Modeling of Soil Profiles. *J. Geotech. Eng. Div.* **1977**, *103*, 1227–1246. [CrossRef]
32. Bombasaro, E.; Kasper, T. Evaluation of spatial soil variability in the Pearl River Estuary using CPTU data. *Soils Found.* **2016**, *56*, 496–505. [CrossRef]
33. Salgado, R.; Ganju, E.; Prezzi, M. Site variability analysis using cone penetration test data. *Comput. Geotech.* **2019**, *105*, 37–50. [CrossRef]
34. Jaksa, M.B.; Kaggwa, W.S.; Brooker, P.I. Experimental evaluation of the scale of fluctuation of a stiff clay. In Proceedings of the 8th International Conference on the Application of Statistics and Probability, Sydney, Australia, 12–15 December 1999.

35. Ping, Y.; Naishuang, B.; Xiao, W.; Yong, Z.; Houjie, W. Spatial Distribution of Surface Sediments in the Modern Yellow River Delta. *Mar. Geol. Quat. Geol.* **2016**, *36*, 49–55.
36. Liangyong, Z.; Jian, L.; Xiqing, L. Geo Cloud. 2003. Available online: http://geocloud.cgs.gov.cn/#/portal/geologicalDatabase/DetailsPage?child_id=v_cpgl_dzt_33302020520183866214&tableCode=v_cpgl_dzt&jddm=333 (accessed on 2 March 2021).
37. Jianmin, Z. *Super-Consolidation Characteristics of Sediments in Modern Yellow River Delta and Their Genesis*; Ocean University of China: Qingdao, China, 2006.
38. Guohui, X.; Ruifang, C.; Anlong, L.; Guanghai, H. Experimental study on the behavioral changes of sticky silty bed under wave action. *Adv. Mar. Sci.* **2020**, *18*, 19–26.
39. Xiujuan, Y.; Yonggang, J.; Jie, L. Analysis of engineering geological characteristics of sediments in the Yellow River Delta. *Yellow River* **2010**, *32*, 201–202.

Journal of
Marine Science and Engineering

MDPI

Article

Comparison of Existing Equations for the Design of Crown Walls: Application to the Case Study of Ericeira Breakwater (Portugal)

Fábio Pereira [1], Maria Graça Neves [1,2,*], José-Santos López-Gutiérrez [3,*], María Dolores Esteban [3] and Vicente Negro [3]

1 Faculdade de Ciências e Tecnologia, Universidade Nova de Lisboa, 2829-516 Monte de Caparica, Portugal; fmm.pereira@campus.fct.unl.pt
2 Harbours and Maritime Structures Division (NPE), Hydraulics and Environment Department (DHA), National Laboratory for Civil Engineering (LNEC), 1700-066 Lisbon, Portugal
3 Grupo de Investigación de Medio Marino, Costero y Portuario, y Otras Áreas Sensibles, Universidad, Politécnica de Madrid, 28040 Madrid, Spain; mariadolores.esteban@upm.es (M.D.E.); vicente.negro@upm.es (V.N.)
* Correspondence: gneves@lnec.pt (M.G.N.); josesantos.lopez@upm.es (J.-S.L.-G.)

Citation: Pereira, F.; Neves, M.G.; López-Gutiérrez, J.-S.; Esteban, M.D.; Negro, V. Comparison of Existing Equations for the Design of Crown Walls: Application to the Case Study of Ericeira Breakwater (Portugal). *J. Mar. Sci. Eng.* **2021**, *9*, 285. https://doi.org/10.3390/jmse9030285

Academic Editor: Alessandro Antonini

Received: 26 January 2021
Accepted: 26 February 2021
Published: 5 March 2021

Abstract: The correct calculation of forces and moments caused by wave action over crown wall structures is critical for their design. There are several existing equations for this, some of which are sanctioned in practice as it is the case for Jensen (1984) and Bradbury et al. (1998), Günback and Gökce (1984), Martin et al. (1999), Berenguer and Baonza (2006), and Pedersen (1996) and Nørgaard et al. (2013). These equations are the main tool for the design of breakwater crown walls and their accuracy is crucial to ensure the stability of the crown wall, especially when considering the sea level rise due to climate change and the possible damage of the armor, since both aspects are not usually considered in most original design studies. In a scenario of climate change, it is very important to estimate the possible changes in security factors due to both these aspects, comparing the results with the original design ones. This paper has as main objective to analyze it for the case study of Ericeira rubble mound breakwater in Portugal. For this, a comparison of the results using those equations and different scenarios including the current, considering sea level rise and armor damage, were performed to extract some conclusions: the increase in the sea level in the case study was not significant and therefore its incidence is very small; and the damage to the main armor by losing pieces at the berm is much more important in this case study, so it is essential to carry out the proper maintenance of the design section. On the other hand, horizontal forces are more conservative using Pedersen and Nørgaard equations, obtaining the lowest value with Martin. Regarding uplift pressures, Martin gives the lowest value, while the most conservative values are given by Günbak and Gökce's for two scenarios, and Pedersen and Nørgaard for the other two scenarios. Furthermore, the sliding safety coefficient is more conditioning than overturning the safety coefficient in all the scenarios.

Keywords: crown wall; rubble mound breakwater; climate change; armor damage

1. Introduction

Rubble mound breakwaters are coastal structures that play a fundamental role in creating safety conditions in ports by reducing sea agitation in their sheltered areas. They enable favorable conditions for the permanence and berthing of ships and small boats and can also contribute to the non-accumulation of sediment, responsible for the conditioning of navigation. A rubble mound breakwater consists mainly of the core and the main and secondary armor, interior (leeward side) and exterior (seaward side), the armors being constituted by the material of significant weight to maintain the stability of the breakwater. The core consists of material of extensive granulometry, with the objective of controlling the transmission of energy resulting from the breaking of the waves and guaranteeing the stability of the armors. In case of need, filters are installed between the core and the armor to prevent core material to escape through the gaps in the armor pieces. Another important part in rubble mound breakwater is the crown wall, installed in its upper part. This superstructure (crown wall) can enable an important reduction in overtopping [1] and resistance to wave action, allowing an increase in the top level of the complete structure without compromising the financial viability of the construction process.

The superstructures of rubble mound breakwaters are usually made of concrete or masonry and are based directly on top of them. From a constructive point of view, in the case of superstructures with curtain walls, the great advantage is that the presence of a superstructure allows the height of the edge of the outer armor of the slope to be reduced, which results in less financial expense in armor rock or concrete pieces for the constitution of the armors, filters and core and consequently, less environmental impact due to the use of materials existing in the breakwater construction area [2].

From a structural point of view, due to their own weight and the friction generated between the base and the foundation material, the superstructures also guarantee the greater stability of the armor's constituent materials [3]. In matters of functionality, and in view of rubble mound breakwater solutions without a superstructure, the presence of a superstructure allows accessibility for maintenance purposes and the installation of elements for monitoring the behavior and conservation status of the constituent elements of breakwater [4].

In recent decades, the effects of climate change have become evident in the coastal environment throughout the world [5], a key condition being that of the sea level rise [6–8]. This sea level rise is not uniform in all regions of the planet so each particular case must be studied to determine the value of this rise and its possible impact on coastal structures. In view of the worsening conditions of maritime unrest in the context of climate change, it is becoming increasingly important to study the forces to which the superstructures of the slope breakwaters are and will be subjected, both to ensure their correct dimensioning and guarantee efficient rehabilitation processes, since it is intended to guarantee the conditions of stability of these structural elements throughout their useful life.

This paper summarizes the results of forces and moments in crown wall superstructures due to wave conditions considering some of the equations sanctioned by practice: (1) Jensen (1984) and Bradbury et al. (1998); (2) Günback and Gökce (1984); (3) Martin et al. (1999); (4) Berenguer and Baonza (2006); and (5) Pedersen (1996) and Nørgaard et al. (2013). The formulation proposed by Molines et al. [9] was not considered because in this case study, the percentage of breakwater overtopping is 12.3%, so the results are not comparable with those of the rest of the formulas used in the research. Anyway, the Molines et al. equation should be used in the case where the overtopping rates match their research test conditions.

For that, the state of the art of those equations are explained in the second section of the paper (Methods). Section 3 includes the description of the case study, the Ericeira rubble mound breakwater in Portugal, the meteocean data and the different scenarios considered in this study, including not only the current case, but also some of them associated with the sea level rise due to the climate change, and the possible damage of the armor layer losing the upper concrete piece.

The consideration of all these scenarios allows to obtain some important conclusions comparing the forces, moments, and security factors of the current scenario, similarly to the original design of the crown wall, the ones due to the sea level rise caused by climate change, and the ones due to the damage of the upper part of the armor layer. Section 4 exposes the results of the application of the five equations to the case study and for the different scenarios. Section 5 includes some comparisons between all the results exposed in Section 4, and conclusions are summarized in Section 6.

2. Methods

This manuscript aimed to compare the results of the application of the different existing equations for the design of crown walls in a specific location. In addition, to determine which of the expressions best fits the original state of the design, it will also determine if the design is adequate for the change in metocean conditions in different damage scenarios.

The actions on the superstructures of the rubble mound breakwaters are essentially due to the interaction between the sea waves and the breakwater, that is, the interaction of waves with the outer armor and the superstructure resulting in pressures and consequently, forces and moments. As the wave approaches a breakwater, the depth decreases and its steepness increases [10].

If the Iribarren number in deep waters and considering peak period (ξ_{0p}) value is greater than 3, the breaking due to the depth effect occurs, which means that the wave has already broken at the foot of the breakwater. This scenario is favorable from the functional point of view as it allows to a priori disregard a situation of wave breaking directly on the breakwater superstructure. If there is no depth breaking, it is still possible to determine empirically whether the breaking occurs directly on the superstructure or on the outer armor through the relation between the relative width of the berm of the outer main armor layer, B_e/H, and its relative height, A_c/H as it is indicated in [11].

On the other hand, it is important to verify if waves break before reaching the slope of the breakwater. One of the most used equations for that verification is:

$$H_r = 0.78\ d \tag{1}$$

This parameter represents the maximum wave height compatible with the depth without any breaking due to the seabed. Considering H_s as the significant wave height measured at the foot of the breakwater, if $H_s > H_r$ it is assumed that the break occurs before the wave reaches the breakwater, thus allowing the possibility of a break to be excluded on the superstructure.

Run-up, R_u, is the phenomenon that occurs after the wave breaks and translates, in the case of slope breakwater, as the rise of the water mass associated with the wave over the main outer armor. R_u is thus defined as the distance measured vertically, between the maximum level reached by the wave on the armor layer and the sea level [12]. This is one of the most important parameters in the design of a rubble mound breakwater, since it is based on its value that the crowning height, W_c, is defined, depending on whether the occurrence of overtopping (the occurrence of overtopping is verified if $R_u > F_c$), where Fc is the distance from the superstructure foundation in relation to the sea level [13,14]. Ru depends on the height and slope of the wave, the slope of the armor, and the permeability, roughness and porosity of the materials that make up the breakwater.

This research considers different equations for the design of the crown wall: (1) Jensen (1984) and Bradbury et al. (1998); (2) Günback and Gökce (1984); (3) Martin et al. (1999); (4) Berenguer and Baonza (2006); and (5) Pedersen (1996) and Nørgaard et al. (2013).

2.1. *Jensen (1984) and Bradbury et al. (1988)*

The formulation proposed by Bradbury, Allsop and Stephens (1988) was based on the results of the experimental work carried out by Jensen (1984) [15]. It is thus an empirical formulation, resulting from the performance and analysis of test results on different types of sections and geometries of slope breakwater superstructures. By performing the tests on the different cross sections, Jensen verified the existence of a linearity relationship between the horizontal force induced by the wave in the superstructure crown wall, F_h, the significant wave height value, H_s, and the distance measured vertically between the sea level and the top of the outer main layer (A_c) [4].

Bradbury, Allsop and Stephens (1988) [16] proposed an equation that makes it possible to determine the horizontal force induced by the wave in the crown wall which is exceeded by only 0.1% of the waves of the sea states considered, $F_{h,0.1\%}$ [4], considering the relationship of a line with two calibration parameters, a and b:

For the different standard section geometries considered, called A to E, the values of the empirical coefficients a and b were defined in Table 1.

Table 1. Empirical coefficients a and b for each type of cross section.

Cross Section	a	b
A	0.054	0.032
B	0.025	0.015
C	0.043	0.038
D	0.036	0.031
E	0.013	0.011

The vertical force induced by the wave at the base of the superstructure that is exceeded by only 0.1% of the waves of the sea states considered, $F_{b,0.1\%}$, is obtained by assuming a linear and constant pressure distribution in the superstructure's crown wall, and linear and triangular at the base of crown wall, with the continuity of pressures at the seaward side and zero pressure at the lee-ward side.

Given the pressure distributions assumed by the authors, the horizontal pressure acting on the crown wall, P_h, is given by the quotient between $F_{h,0.1\%}$ and h_w. The vertical pressure acting on the base of the seaward side, P_v, is equal to P_h.

The validity limits to be considered in the application of the formulation were suggested by the authors according to the type of cross section tested. The limits of validity, as well as the type of geometry, parameter, and author to which they are associated, are presented in Table 2.

Table 2. Limits of validity for the application of the formulation of Jensen, 1984, and Bradbury et al., 1988.

Cross Section	A_c (m)	S_{0p}	H_s/A_c	ξ_{0p}
A	5.6–10.6	0.016–0.036	0.76–2.5	2.6–4
B	1.5–3.0	0.005–0.011	4.8–7.1	4.8–7.1
C	0.1	0.023–0.07	0.9–2.1	1.9–3.3
D	0.14	0.04–0.05	1.43	2.2–2.5
E	0.18	0.04–0.05	1.11	2.2–2.5

2.2. *Günbak and Gökce (1984)*

Günbak and Gökce's (1984) semi-empirical formulation [3] was proposed considering the incidence of regular waves and the effect of run-up on the superstructure, thus allowing the vertical and horizontal pressures acting on the superstructure to be obtained.

In terms of pressure distribution in the superstructure, these authors assumed that the pressure acting on the crown wall corresponds to the sum of the dynamic pressure, P_d, with the hydrostatic pressure, P_{hd}, both in the area protected by the curtain edge and in the unprotected area. It was also assumed by the authors the linear variation of P_{hd}, occurring the maximum value at the level of the base of the superstructure. In turn, the dynamic pressure in the protected zone of the superstructure, $P_{d\,(p)}$, corresponds to half of the dynamic pressure in the unprotected zone, $P_{d\,(np)}$. The pressure at the base of the superstructure, P_b, has a triangular distribution and is zero at the lee-ward side ($P_{b\,(s)}$) [3].

In this case, the different pressures on the superstructures are related to the run-up [17,18], R_u, and to the height of the armor berm, A_c. The authors also take into account the angle of the armor slope by subtracting the angle formed between the wave screen and the slope of the armor layer, whose value is 15°, assumed by the authors.

For the present formulation, R_u is obtained through:

$$R_u = \begin{cases} 0.4\xi H & if\ \xi < 2.5 \\ H & if\ \xi > 2.5 \end{cases} \tag{2}$$

The authors assumed the height of run-up exceeded by only 0.1% of the waves of the sea states considered, $R_{u,0.1\%}$.

2.3. Martín et al. (1999)

The formulation proposed by Martín et al. (1999) [14] is based on the concept of run-up suggested earlier by Günbak and Gökce (1984) and is only valid in situations where the wave breaks before the superstructure.

Based on tests performed on small-scale models, the present authors concluded that, for each breaker wave, the two pressure peaks in the superstructure, which occurred at different times, should be analyzed separately:

- First pressure peak: dynamic pressure (P_d);
- Second pressure peak: pseudo-hydrostatic or reflection pressure (P_{ph}).

Based on the experimentally verified pressure distributions, the authors assumed the pressure diagrams presented as an assumption for the application of their suggested formulation, considering a triangular pressure distribution at the base of the superstructure of maximum value at seaward side at the first pressure peak and trapezoidal at the second.

As the separate analysis of the two pressure peaks is suggested, when applying the present formulation for sizing purposes, the authors suggest that the largest of the pressure valleys obtained between the first and second pressure peaks should be considered.

Pd values in the protected zone are lower in comparison to the values in the non-protected zone of the superstructure. The dynamic pressure in the protected zone of the superstructure, Pd (p) as a function of the z coordinate, being the vertical distance measured from the sea level, and λ, that is the adimensional parameter representative of the decrease in the dynamic pressure in the protected zone of the superstructure due to the constituent materials of the main outer armor. The range of values obtained experimentally for this coefficient varied between 0.25 and 0.65, being in competition with the decrease in dynamic pressure of 50% in the protected zone compared to the non-protected zone assumed by Jensen (1984) and Günback and Gökce (1984).

2.4. Berenguer and Baonza (2006)

The semi-empirical formulation of Berenguer and Baonza (2006) [2] was developed based on the results of tests on small scale models with irregular wave action, obtained in 2001. Through the relationships established by the authors for the different geometric characteristics of the models tested and their agitation parameters, it was possible to obtain a time-dependent graphical representation of the horizontal pressures acting on the crown wall and vertical pressures acting on the base of the superstructure. By integrating the maximum pressure value into each diagram, it was possible to obtain the values of the horizontal force, F_x, and the vertical force, F_y.

The authors concluded, through the analysis of the pressure diagrams obtained, that the maximum horizontal force acting on the crown wall of the superstructure, $F_{x,\,max}$, occurred in a phased shift and later in time in relation to the maximum vertical force acting on the base of the superstructure, $F_{y,\,max}$. Even if this has been verified, the authors in the elaboration of the present formulation have assumed, in a conservative way, the action of both forces simultaneously.

Although the results of the tests have allowed the variation of the pressure in the superstructure as a function of time, the method developed only provides an approximation for the acting forces and moments.

2.5. Pedersen (1996) and Nørgaard et al. (2013)

Of a semi-empirical nature, the formulation proposed by Perdersen (1996) [19,20] was developed on the basis of the theory of Günbak and Gökce (1988) [3] and the results of tests on small-scale models for different agitation conditions and mantle and superstructure geometries.

By carrying out different parametric analyses based on the results obtained in the tests, Pedersen concluded that the maximum forces generated in the crown wall of the superstructure were mainly due to the impact forces of the incident waves. Therefore, he proposed that the dynamic pressure, P_d, should be calculated as a kinetic pressure, resulting from the ascending velocity of the water mass at the time of run-up, v_0, and disregarding the loss of energy by friction between the run-up height and the crest of the armor.

Based on the above conclusions, the author suggested a superstructure pressure distribution resulting only from dynamic pressure, assuming that the contribution of pseudo-hydrostatic pressure, P_{ph}, was small and could be disregarded.

According to Pedersen (1996), $R_{u,0.1\%}$, is calculated using the method suggested by Van der Meer and Stam (1992) [21] for breakwater armors composed of permeable or non-permeable core rockfill and regular or irregular wave action. The linear pressure distribution at the base of the superstructure, exceeded only by 0.1% of the waves of the sea states, has a linear relationship with the dynamic pressure of the unprotected zone with the parameter V being the wave impact reduction parameter:

The parameter V results from the ratio between the volume of the crest of the outer main armor, V_1, and the volume of the hypothetical run-up, V_2. In cases of reduced run-up, the parameter V assumes a value lower than unit, reflecting the effect of the size of the top of the armor on the reduction in the impact imposed by the waves. In the opposite situation, the volume of water associated with run-up is large enough to make the consideration of the effect of reducing the impact of the waves negligible, in which case V is considered equal to the unit. In 2013, Nørgaard, Lykke-Andersen and Burcharth [22] proposed a modification to the formulation of Pedersen (1996) with the aim of validating it for small depths, since it was originally limited to large and intermediate depths. The proposed modifications were based on the results of 162 tests on small scale models and irregular wave action, considering agitation conditions at large and small depths and using the same experimental equipment used by Pedersen (1996).

The first modification to the original formulation of Pedersen (1996) was made with respect to the $R_{u,0.1\%}$ level. For the $P_{d\ (np),0.1\%}$ and $P_{b\ (b),0.1\%}$ calculation expressions, no changes were suggested. For the calculation of $F_{h,0.1\%}$ the expression suggested by Pedersen (1996) was considered, but changing the empirical coefficient b, to b = 1. This modification was intended to bridge the differences of the new results obtained compared to those obtained by Pedersen (1996), which occurred in the readings carried out in the unprotected area of the superstructure curtain wall. These differences were due to the fact that the transducer used was different from that previously used by Pedersen (1996).

3. Case Study Description

In the following paragraphs, the section of the Ericeira rubble mound breakwater will be described, in which the design formulas stated in the previous section will be applied. The metocean data to be used have also been defined, as have the scenarios considered in the study.

3.1. *Ericeira Rubble Mound Breakwater Description*

Ericeira rubble mound breakwater was built in the 1970s, on the west coast of mainland Portugal. Ericeira port has as main objective to provide shelter for a small artisanal fishing fleet. The breakwater of the port is 440 m long and has a slipway (Figures 1 and 2). The breakwater suffered several damages during its useful life, the last intervention being in 2008, going through its total rehabilitation and requalification, since some 200 m of its length were in complete ruin and the remaining 240 m were in high condition of degradation. During the reparation phase, the breakwater was damaged again, being repaired again in March 2016.

For the current research, the P–P cross section was selected (Figure 3). The selection of this profile was because it is one of the most critical parts for the operation of the port since it has an attached quay right on the side of that P–P profile protected against storms, and any failure would affect the use of the quay infrastructure.

Figure 1. Aerial view of Ericeira breakwater (Google Earth).

(**a**) (**b**)

Figure 2. Some views of Ericeira breakwater: (**a**) panoramic view; and (**b**) the view from the base.

(**a**) (**b**)

Figure 3. Cross section of the breakwater tested: (**a**) Location of the cross section on breakwater; (**b**) diagram of the cross section.

The cross section is composed of three layers: outer armor of 300 kN tetrapods, filter of 20–40 kN riprap and a core of sand and gravel. On top of the breakwater, there is a crown wall which is made up by mass concrete.

The main parameters and symbols of the geometry of the main outer armor of the rubble mound breakwater are included in Figure 4:

- Vertically measured distance between the sea level and the top of the crest of the outer main armor (A_c);
- Crest width of the outer main armor layer (B_e);
- Slope of the outer main armor (α);
- Depth at the toe of the breakwater (h).

The parameter A_c depends on the reference considered, that is, it corresponds to the distance, measured vertically, between the crowning edge and the hydrographic zero (ZH), or between this and the resting sea level/sea level (NR or NM). In terms of notation, the designation of A_c (zero) will be used for the first case since the level considered is the hydrographic zero (ZH). For the second case, the designation A_c will be used. The depth at the foot of the breakwater (h) is a measured distance from the ZH. The depth at the bottom of the breakwater measured in relation to sea level (NM) is called d.

In the case of superstructures where it is possible to make a clear distinction between the base and the curtain wall, the height of the superstructure base can also be defined. The geometric parameters of the superstructure must, therefore, be defined according to the geometry of the main outer armor, the set of which consists of (Figure 4):

- Height of the superstructure (h_w);
- Height of the base of the superstructure (h_{wl});
- Crown wall width (B′);
- Width of the base of the superstructure (B);
- Freeboard of the superstructure (W_c);
- Superstructure foundation level (W_f);

The top and foundation levels identified above are measured in relation to the hydrographic zero (ZH). Bearing in mind that the sea level (NM) is also measured in relation to ZH, subtracting the value of NM from the levels mentioned, it is possible to obtain the following distances using the NM as a reference level:

- The distance from the crown of the superstructure in relation to the sea level (R_c);
- Distance from the superstructure foundation in relation to sea level (F_c).

The height of the superstructure can be totally or partially protected by the edge of the outer main armor. In cases where it is partially protected, it is possible to define the measures fc and h′, which are, respectively, the height of the superstructure not protected by the shoulder of the outer main armor and the height of the superstructure protected by the shoulder of the outer main armor.

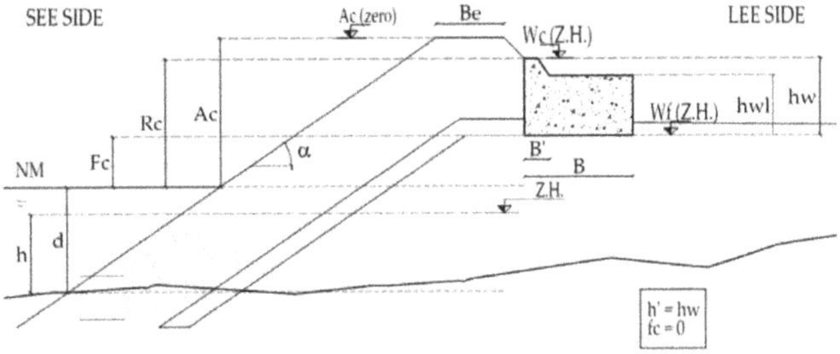

Figure 4. Geometrical parameters of the crown wall and main outer armor of a rubble mound breakwater with a monolithic crown wall.

The main values of different parameters in the Ericeira P–P cross section are shown in Tables 3 and 4.

Table 3. Geometric and material parameters of the main outer armor.

Parameter	Value	Units
$A_{c\,(zero)}$	+10.2 (ZH)	(m)
B_e	4.08	(m)
α	0.588	(rad)
Outer armor	Tetrapods	(-)
W_r	300	(kN)
Inner armor	Quarry stone	(-)
W_i	20–40	(kN)
h	−4.5 (ZH)	(m)
R_c	9.0	(m)

Table 4. Geometric parameters of the monolithic crown wall.

Parameter	Value	Units
h_w	4.5	(m)
h_{wl}	3.5	(m)
B′	1.5	(m)
B	6.5	(m)
W_c	+9.0 (ZH)	(m)
W_f	+4.5 (ZH)	(m)
f_c	0.0	(m)
h′	4.5	(m)

3.2. Metocean Data

For calculation purposes, a data series on agitation and sea level [23] at the toe of the breakwater corresponding to the period between 1 January 2009 and 31 December 2019 was considered. The approach selected for this research is to consider the real values of the waves over the breakwater and their effects over it during this period of time. The length of the data series has been chosen due to the fact that a breakwater repair was carried out in 2008, so the focus of the research is from the beginning of 2009.

Wave data were obtained using the results of a hindcast model, achieving spectral values of significant wave height (H_s), peak period (T_p), and wave directions (dir_w). The data series corresponds to the sum of four daily measurements, every 6 h, during the period indicated above, making a total of 16,068 values for each of the parameters before mentioned (H_s, T_p, dir_w).

Given the 11 years of wave series hindcast data, it was found that the maximum value of H_s is 5.98 m, being the highest percentage of occurrences in the interval between 1 and 2 m with a total number of occurrences of 5198, which corresponds to approximately 33% of the total series (Figure 5); here were 426 occurrences with H_s values greater than 4.5 m, a value normally used as a minimum for storm conditions, which correspond to 2.7%. Considering temporal cases in which $H_s > 4.5$ m during at least 1 day and considering those that occur with less than 0.75 as the same storm days apart, in these 11 years there have been 18 storms, with durations between 1 and 2.50 days. The maximum value of T_p is 20.4 s with the highest number of occurrences being between 12 and 13 s, with a total of 2919 occurrences, that is, approximately 18% (Figure 5).

Relating the data of H_s and T_p, it was verified that as the values of H_s increase, there is also an increase in T_p (Figure 6), which means that for the highest values of H_s, longer wavelengths are associated and consequently so are greater efforts on the superstructure [19], since the wavelength depends on T_p by airy or linear wave theory.

On the other hand, to obtain the sea level corresponding to each wave data, the astronomical tide levels obtained by the XTide program [24] were added to the respective values of the storm surge [23], thus obtaining the sea levels considered for the purpose calculation. Storm surge values correspond to a drop or rise in astronomical tide levels according to the increase or decrease in atmospheric pressure, respectively.

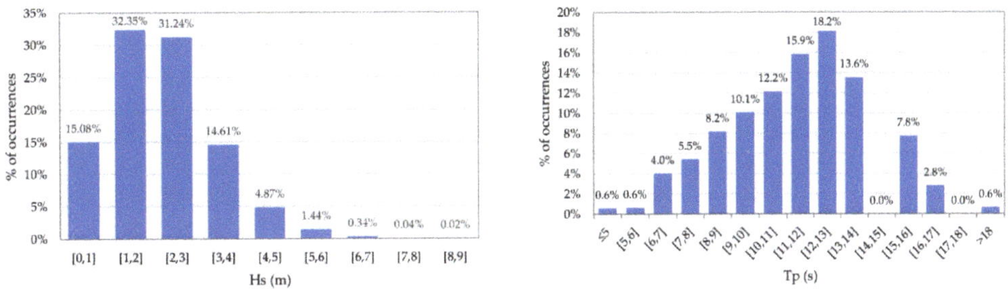

Figure 5. Intervals of H_s **(left)** and T_p **(right)** and the percentage of occurrences.

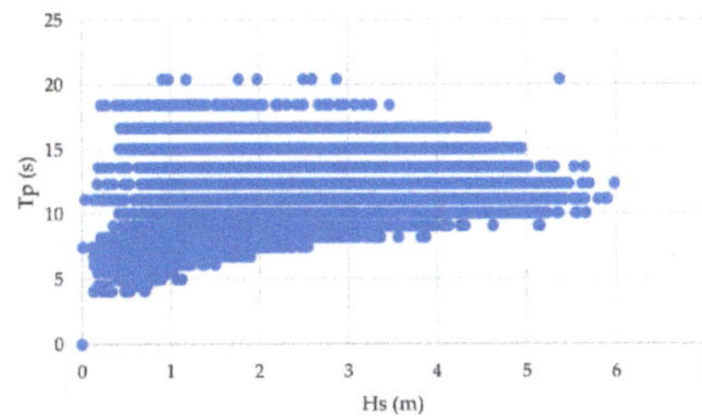

Figure 6. Relation between H_s and T_p.

Meteorological tide levels for the period relating to the wave data series ranged between 0.66 and 3.71 m, with the highest number of occurrences associated with the interval between 2.8 and 3.0 m with a total of 1725 occurrences (approximately 11% of the total data in the series) (Figure 7).

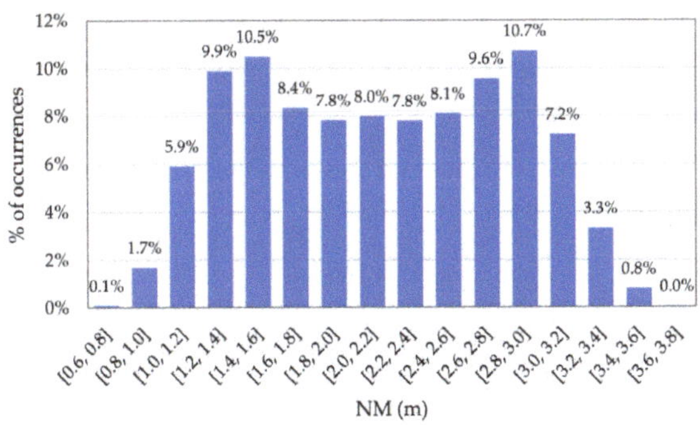

Figure 7. Intervals of sea level and percentage of occurrences.

3.3. Scenarios Considered for the Case Study

Different scenarios were going to be analyzed for Ericeira breakwater P–P section, using for all of them the different existing equations fr. The scenarios are:

- First scenario, considering wave series and sea level data previously described, representing the current situation;
- Second scenario, considering a sea level rise of 5 mm/year (expected value in the area close to Ericeira port due to the Climate Change) [25];
- Third scenario, considering a lowering of the armor berm. A tetrapod is removed at the armor berm leaving the crown wall partially unprotected;
- Fourth scenario, considering the second and third scenarios simultaneously.

The sea level rise included in the second and fourth scenarios enables analyzing the impact of climate change in the short term (10 years are considered in this research, the total sea level rise in the research being 50 mm) in rubble mound breakwater structures. The lowering of the armor berms represented in the third and fourth scenario simulates the partial failure of the rubble mound breakwater, losing some concrete pieces. All these scenarios allow the comparison of the loads in different cases: the real and current ones, the ones due to the sea level rise, the ones due to the failure on the top of the armor, and a combination of the sea level rise and the failure on the top of the armor. So, some conclusions can be obtained for being prepared for different possible scenarios.

4. Results

Formula application and validity are here presented together with the results obtained for the different formulas and for the different scenarios, which are shown below according to the different equations previously exposed: (1) Jensen (1984) and Bradbury et al. (1988); (2) Günbak and Gökce (1984); (3) Martín et al. (1999); (4) Berenguer and Baonza (2006); and (5) Pedersen (1996) and Nørgaard et al. (2013). The results are presented in terms of forces due to the wave action (F_h: horizontal; and F_b: uplift pressures), and corresponding destabilizing moments (M_h: horizontal forces moment; and M_b: uplift pressure moment) and safety coefficients (C_{des}: sliding; and C_{der}: overturning).

4.1. *Jensen (1984) and Bradbury et al. (1988)*

For the application of Jensen and Bradbury equations to the case study, it was necessary to associate the geometry of the P-P profile to the geometry of one of the type sections tested by the authors. As mentioned in previous sections, each theoretical section is characterized by a validity interval associated with the parameters A_c, S_{0p}, H_s/A_c and ξ_{op} (or I_{r0p}). To determine the most suitable section, the intervals of the validity parameters of each type of section (Table 2) were compared with the intervals obtained for the same parameters for the series of agitation data and sea levels, and for each study scenario (Table 5). Looking to the validity range of A_c and S_{op}, cross section A is the only one where the values are within the range of the validity of the formulation. Looking to H_s/A_c and ξ_{op}, cross section A is again the most appropriate for the parameter range of this case study. Consequently, the empirical coefficients of cross section A were adapted hereafter. Table 6 presents the valid number of cases for formula validity parameters. It should be pointed out that this formula does not include any reduction in horizontal pressures because of the existence of an armor layer facing the wave wall and the crown wall analyzed here is totally protected by the armor layer in scenarios 1 and 2.

Table 5. Parameter range for all the studied cases.

Scenario	A_c (m)		S_{0p}		H_s/A_c		ξ_{0p}	
	Min	Max	Min	Max	Min	Max	Min	Max
Scenario 1	6.49	9.54	0.00012	0.0398	0.0013	0.86	3.34	32.7
Scenario 2	6.48	9.51			0.0013	0.86		
Scenario 3	4.19	7.24			0.0018	1.28		
Scenario 4	4.18	7.21			0.0018	1.29		

Table 6. Number of cases valid for the formula validity parameters 1.

Scenario	A_c	S_{0p}	H_s/A_c	ξ_{0p}
Scenario 1	16,068	2531	12	164
Scenario 2	16,068		12	
Scenario 3	9047		439	
Scenario 4	8885		453	

As it can be seen, in none of the four scenarios are the formulas within the limits of parameters for all cases. H_s/A_c and ξ_{op} are the most restricted parameters for all scenarios, even for $H_s > 4.5$ m, i.e., temporal conditions.

The next figures include relevant information about horizontal and vertical forces for the first scenario compared with the second to fourth scenarios (Figure 8), as well as sliding safety coefficients for the first scenario compared with the second to fourth scenarios (Figure 9).

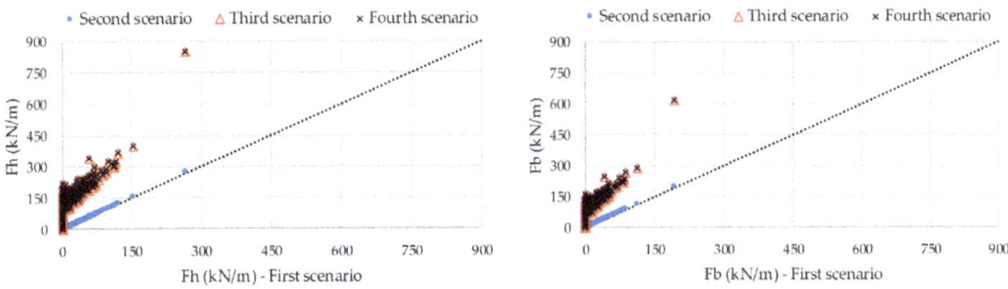

Figure 8. Horizontal and vertical forces for the first scenario compared with the second to fourth scenarios.

A small response is observed in the horizontal and vertical forces to the sea level variation, i.e., between scenarios 1 to 2 and 3 to 4, since the maximum sea level variation expected in 10 years was also small—50 mm. Differences in the maximum values are less than 2% with the increase in sea level considered. The maximum F_h varies from 267.6 (scenario 1) to 272.7 kN/m (scenario 2) and from 852.8 (scenario 3) to 864.2 kN/m (scenario 4). The same happens for Fb, with the maximum values being lower than F_h and varying from 193.2 kNm/m for scenario 1 to 197.0 kNm/m for scenario 2 but being 615.9 and 624.1 kNm/m for scenarios 3 and 4, respectively. However, if there is a change in the protection of the crown wall, with an area being unprotected, the differences deeply increase, especially for the higher values of forces that become more than two times the value for a design protected crown wall. The same trend was found for momentum.

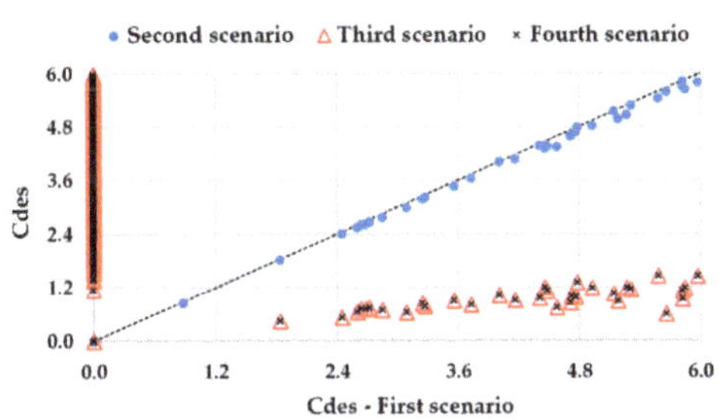

Figure 9. Sliding safety coefficients for the first scenario compared with the second to fourth scenarios.

These results lead to differences in safety coefficients. For both scenarios 1 and 2, both safety coefficients are always higher than 1.2. When the protection of the crown wall is reduced, part of the structure suffers from the direct wave impact; for 43 wave conditions C_{der} becomes less than 1.2, corresponding to Hs higher than 4.3 m and with sea levels higher than 2.15. C_{des} remains higher than 1.2. In other words, the crown wall becomes unstable for sliding for some of the storm conditions.

4.2. *Günbak and Gökce (1984)*

Although it does not impose validity limits of the equation, it is important to emphasize the run-up values, since the equation depends on this parameter, with the highest percentage between 2 and 3 m, followed by the intervals of 3 and 4 m, 4 and 5 m, and 1 and 2 m for the four scenarios (Figure 10).

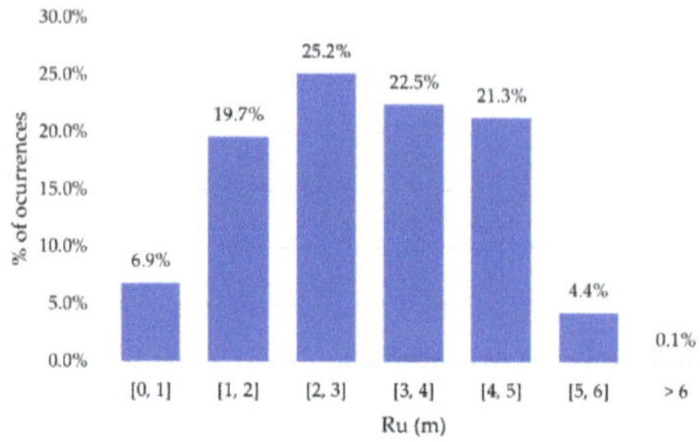

Figure 10. Intervals of run-up for the four scenarios.

The next figures include relevant information about these values: horizontal and vertical forces for the first scenario compared with the second to fourth scenarios (Figure 11); and sliding and overturning safety coefficients for the first scenario compared with the second to fourth scenarios (Figure 12).

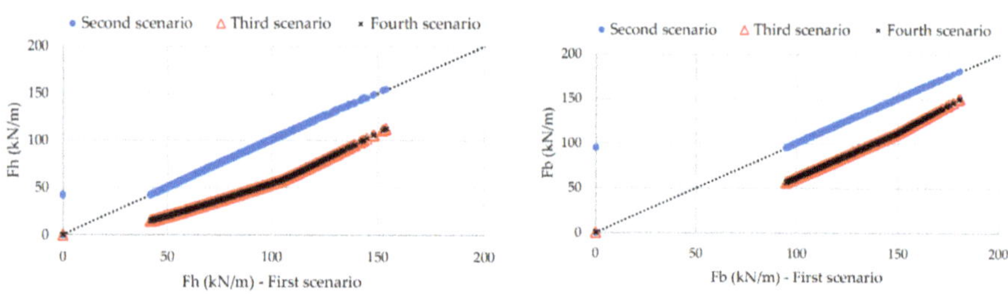

Figure 11. Horizontal and vertical forces for the first scenario compared with the second to fourth scenarios.

As expected, minor changes are observed in horizontal and vertical forces to the sea level variation, while the maximum forces increase by less than 2% from scenarios 1 to 2 and from scenarios 3 to 4. The same happens with the momentum. With a change in the protection of the crown wall, and an area being unprotected, the differences are not as expected. In fact, the value of forces reduces as well as the correspondent momentum and the maximum of forces reduces by more than 25%. F_h goes from a maximum of 153.8 kN/m (scenario 1) and 154.9 kN/m (scenario 2) to 112.8 kN/m (scenario 3) and 114.1 kN/m (scenario 4). This happens because the reduction in A_c in scenarios 3 and 4 does not correspond a reduction in the value used for the authors to calculate dynamic pressure. This value depends on the difference between R_u and A_c. R_u is only a function of the wave height and ξ, not depending on A_c, making a reduction in A_c lead to an increase in dynamic pressures and consequently, on forces and momentum.

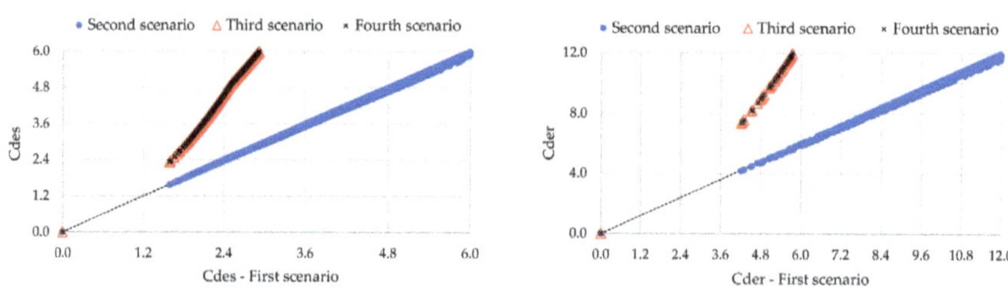

Figure 12. Safety coefficients for the first scenario compared with the second to fourth scenarios.

These results lead to a surprising increase in the safety coefficient values. For all the scenarios, both safety coefficients are always higher than 1.2, even when the protection of the crown wall is reduced, and part of the structure suffers from the direct wave impact. It should be pointed out that Günbak and Gökce (1984) claimed that their method was designed to be applied to Mediterranean ports, where the wave characteristics are clearly different to the one from Ericeira, on the Portuguese west Atlantic coast.

4.3. Martín et al. (1999)

The validity for the application of the Martín et al. equation depends on the validity limits of expressions for calculating μ and λ parameters, and the surf, only being valid in cases where the surf occurs before the wave reaches the superstructure. Furthermore, it is considered that the wave breaks before the superstructure if the bottom breaks ($\xi_{0p,\ max} > 3$) or if the relationship between B_e/H_{max} and A_c/H_{max} will respect the non-impact zone. The number of cases in the data series in which the formulation was valid according to the surf is shown in Table 7 for each study scenario.

Table 7. Number of cases valid for the formula validity parameters 2.

Scenario	$\xi_{0p,\ max}$
Scenario 1	378
Scenario 2	385
Scenario 3	378
Scenario 4	385

The number of cases valid for the formulas is rather small, since for most of the cases, $\xi_{0p,\ max}$ is less than 3. Hs is less than 3.2 m for the valid cases, showing that the formulas is applied out of its validity range for storm conditions.

The next figure presents information about horizontal and vertical forces for the first scenario compared with the second to fourth scenarios (Figure 13). The forces of the first and second peaks was calculated separately, as indicated by the authors, and the addition the first and second peaks were made in order to obtain the total force.

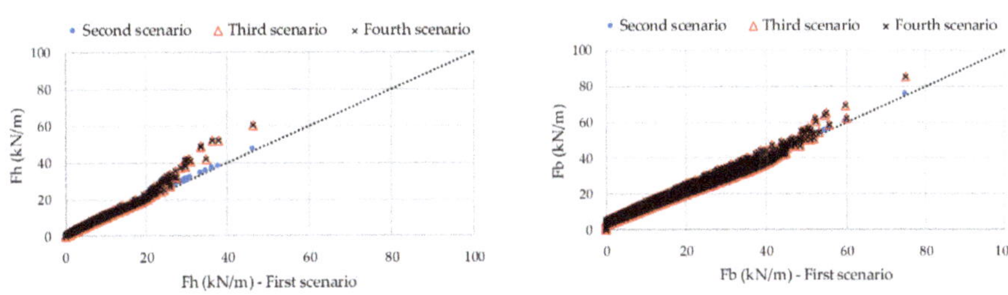

Figure 13. Horizontal and vertical forces for the first scenario compared with the second to fourth scenarios.

Horizontal and vertical forces reduce very little to the sea level variation, i.e., between scenarios 1 to 2 and 3 to 4, with the maximum forces reducing by less than 4%. With a change in the protection of the crown wall, between scenarios 1 to 3 and 2 to 4, with an area being unprotected, the differences in forces are small, increasing for the higher values of forces: difference in the maximum reaches 30% for F_h and 14% for F_b. F_h present a maximum value of 46.5 kN/m for scenario 1, and a value of 47.7 Kn/m, 60.3 kN/m and 62.4 Kn/m for scenarios 2, 3 and 4. Vertical forces are a little higher than F_h, with a maximum of 75.0 kN/m (scenario 1) to 75.9 (scenario 2) and from 84.9 kN/M (scenario 3) to 86.5 kN/m (scenario 4).

Both safety coefficients remain higher than 1.2 for all scenarios and wave conditions tested. In other words, the stability of the crown wall is considered not affected by the increase in sea level or by the crown wall being partially unprotected to the directly wave action.

4.4. Berenguer and Baonza (2006)

The application of this equation was carried out assuming the presence of concrete blocks in the outer main armor and Ericeira is composed of tetrapods. Table 8 shows the values of R_u for which there is no limit on validity. In view of this, the only validity condition for applying the formulation lies in calculating the moment due to the vertical forces, M_b, including the minimum and maximum figures obtained in Table 8, and the number of valid cases in Table 9.

Table 8. Parameter range.

Scenario	M_b (kN/m) Min	M_b (kN/m) Max	R_u (m) Min	R_u (m) Max
Scenario 1	0.0	533.7	0.0	8.5
Scenario 2		534.8		
Scenario 3		542.9		
Scenario 4		544.2		

Table 9. Number of cases valid for formula validity parameters.

Scenario	M_b (kN/m)	R_u (m)
Scenario 1	2972	16,068
Scenario 2	2934	
Scenario 3	2972	
Scenario 4	2934	

As can be seen, all R_u values are within the validity limits of the formulation. However, less than 20% of the data have M_b values within their range of validity.

Differences in forces due to increasing sea level are not relevant, with the maximum values increasing by less than 2% for F_h and less than 0.5% for F_b between scenarios 1 (maximum F_h and F_b of 229.0 and 158.7 kN/m, respectively) and 3 (maximum F_h and F_b of 232.4 and 159.3 kN/m, respectively), and between scenarios 2 (maximum F_h and F_b of 232.4 and 161.4 kN/m, respectively) and 4 (maximum F_h and F_b of 330.6 and 162.1 kN/m, respectively). The change in the protection of the crown wall, scenarios 1 to 3 and 2 to 4, with an area being unprotected, leads to larger differences on F_h than on F_b, with a difference in the maximum value reaching 42% and less than 2%, respectively (Figure 14). Differences in momentum follow the same trend as on forces.

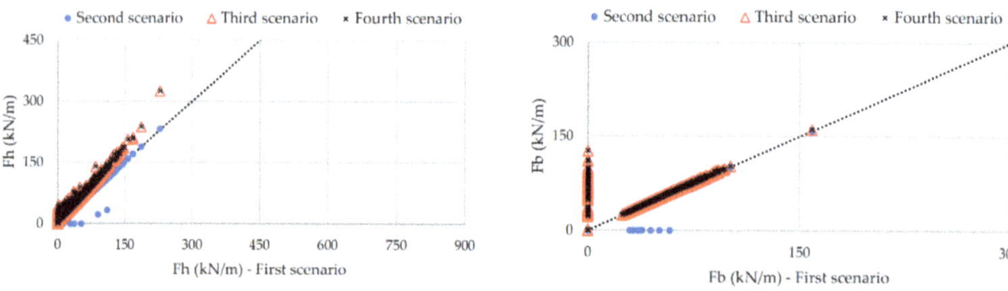

Figure 14. Horizontal and vertical forces for the first scenario compared with the second to fourth scenarios.

For both scenarios 1 and 2, both safety coefficients are always higher than 1.2. When the protection of the crown wall is reduced and part of the structure suffers from the direct wave impact, only for the highest H_s does the C_{des} become less than 1.2. C_{des} remains higher than 1.2. In other words, the crown wall becomes unstable for sliding for the worst storm condition tested (Figure 15).

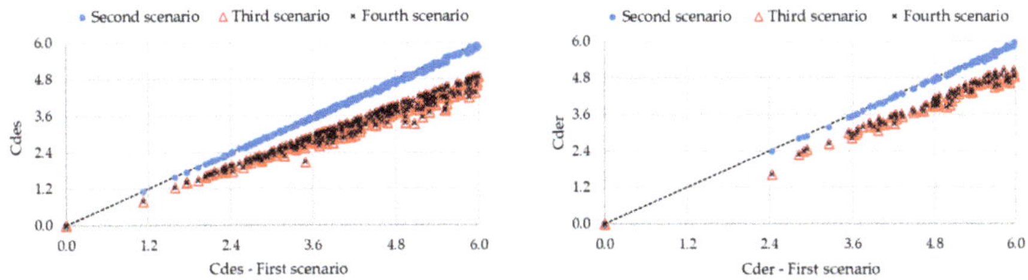

Figure 15. Safety coefficients for the first scenario compared with the second to fourth scenarios.

4.5. Pedersen (1996) and Nørgaard et al. (2013)

Table 10 shows the values of different parameters essential to detect the validity for the application of the formula, and Tables 11 and 12 include the number of valid cases.

Table 10. Parameter range for Pedersen et al. (1996) and Nørgaard et al. (2013).

<table>
<tr><th rowspan="2" colspan="2">Scenario</th><th colspan="4">Pedersen et al. (1996) and Nørgaard et al. (2013)</th><th colspan="2">Nørgaard et al. (2013)</th><th>Pedersen et al. (1996)</th></tr>
<tr><th>ξ_0</th><th>H_s/A_c</th><th>R_c/A_c</th><th>A_c/B_e</th><th>H_s/D</th><th>H_s/L_0</th><th>cot α</th></tr>
<tr><td rowspan="2">Scenario 1</td><td>min</td><td>0.00</td><td>0.00</td><td>0.82</td><td>1.59</td><td>0.00</td><td>0.00</td><td rowspan="8">1.5</td></tr>
<tr><td>max</td><td>56.34</td><td>0.86</td><td>0.87</td><td>2.34</td><td>0.77</td><td>0.05</td></tr>
<tr><td rowspan="2">Scenario 2</td><td>min</td><td>0.00</td><td>0.00</td><td>0.81</td><td>1.59</td><td>0.00</td><td>0.00</td></tr>
<tr><td>max</td><td>56.34</td><td>0.86</td><td>0.87</td><td>2.33</td><td>0.77</td><td>0.05</td></tr>
<tr><td rowspan="2">Scenario 3</td><td>min</td><td>0.00</td><td>0.00</td><td>1.15</td><td>0.65</td><td>0.00</td><td>0.00</td></tr>
<tr><td>max</td><td>56.34</td><td>1.28</td><td>1.26</td><td>1.13</td><td>0.77</td><td>0.05</td></tr>
<tr><td rowspan="2">Scenario 4</td><td>min</td><td>0.00</td><td>0.00</td><td>1.15</td><td>0.65</td><td>0.00</td><td>0.00</td></tr>
<tr><td>max</td><td>56.34</td><td>1.29</td><td>1.26</td><td>1.12</td><td>0.77</td><td>0.05</td></tr>
</table>

Table 11. Number of cases valid for the formula validity parameters for Pedersen et al. (1996).

<table>
<tr><th>Scenario</th><th>ξ_0</th><th>H_s/A_c</th><th>R_c/A_c</th><th>A_c/B_e</th><th>cot α</th></tr>
<tr><td>Scenario 1</td><td rowspan="4">810</td><td>810</td><td rowspan="2">0</td><td rowspan="2">0</td><td rowspan="4">16,068</td></tr>
<tr><td>Scenario 2</td><td>830</td></tr>
<tr><td>Scenario 3</td><td>3986</td><td rowspan="2">16,068</td><td>12,592</td></tr>
<tr><td>Scenario 4</td><td>3532</td><td>12,824</td></tr>
</table>

Table 12. Number of cases valid for the formula validity parameters for Nørgaard et al. (2013).

<table>
<tr><th>Scenario</th><th>ξ_0</th><th>H_s/A_c</th><th>R_c/A_c</th><th>A_c/B_e</th><th>H_s/D</th><th>H_s/L_0</th></tr>
<tr><td>Scenario 1</td><td rowspan="2">3046</td><td>810</td><td rowspan="4">16,068</td><td rowspan="2">0</td><td>10,363</td><td rowspan="2">3373</td></tr>
<tr><td>Scenario 2</td><td>830</td><td>10,374</td></tr>
<tr><td>Scenario 3</td><td rowspan="2">1096</td><td>3986</td><td rowspan="2">16,068</td><td>10,363</td><td rowspan="2">2239</td></tr>
<tr><td>Scenario 4</td><td>3532</td><td>10,374</td></tr>
</table>

As can be seen, some parameters are within the formulation validity range, but not all of them and for some parameters—R_c/A_c and A_c/B_e for Pedersen et al. (1996) and A_c/B_e for Nørgaard et al. (2013)—all data are out of limits for formula applications.

The next figures include relevant information about horizontal and vertical forces for the first scenario compared with the second to fourth scenarios (Figure 16); and sliding and overturning safety coefficients for the first scenario compared with the second to fourth scenarios (Figure 17) for Pedersen et al. (1996). Figures 18 and 19 show the same aspects for Nørgaard et al. (2013).

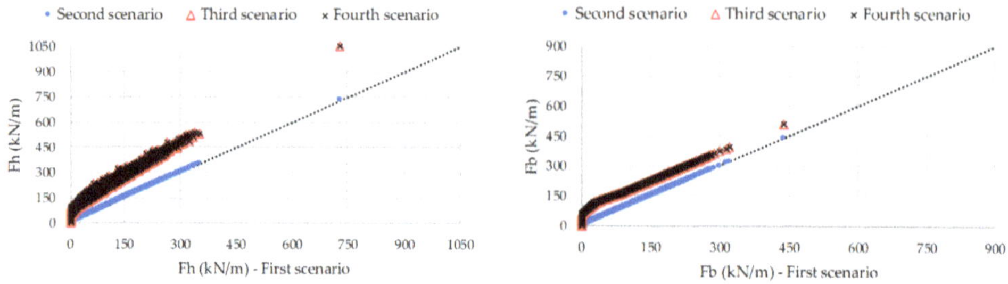

Figure 16. Horizontal and vertical forces for the first scenario compared with the second to fourth scenarios—Pedersen et al. (1996).

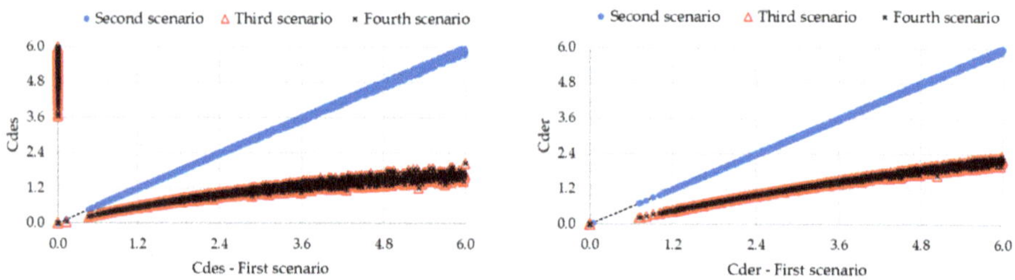

Figure 17. Safety coefficients for the first scenario compared with the second to fourth scenarios—Pedersen et al. (1996).

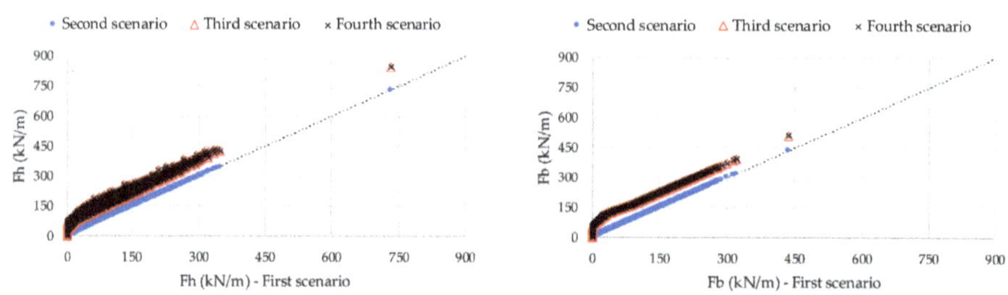

Figure 18. Horizontal and vertical forces for the first scenario compared with the second to fourth scenarios—Nørgaard et al. (2013)

The response of the horizontal and vertical forces to the sea level variation, i.e., between scenarios 1 to 2 and 3 to 4, is not relevant for both formula, as expected, with differences in the maximum values of F_h and F_b less than 0.5%. The maximum F_h for Pedersen (1996) varies from 730.6 kN/m (scenario 1) to 732.3 (scenario 2) and from 1049.6 (scenario 3) to 1051.6 kN/m (scenario 4). Vertical force is smaller than F_h, with the maximum varying from 437.4 kNm/m for scenario 1 to 512.6 kNm/m for scenario 2 but being 512.6 and 513.5 kNm/m for scenarios 3 and 4, respectively. With F_h for scenarios 1 and 2 and F_b for all scenarios calculated with Nørgaard et al. (2013), the formula gives similar results as Pederson (1996). Larger differences are found in F_h and scenarios 3 and 4, with maximum values of F_h being, for Nørgaard et al. (2013), almost 24% higher than for Pederson (1996): 1049.6 and 1051.6 kN/m, respectively.

The impact of the crown wall will become partially unprotected by the armor layer due to the wave action, which is especially important for the higher values of forces. Differences in maximum values of F_h reach 44% for Pederson (1996) and 16% for Nørgaard et al. (2013) and on the maximum values of F_b, are 17% for both formulas.

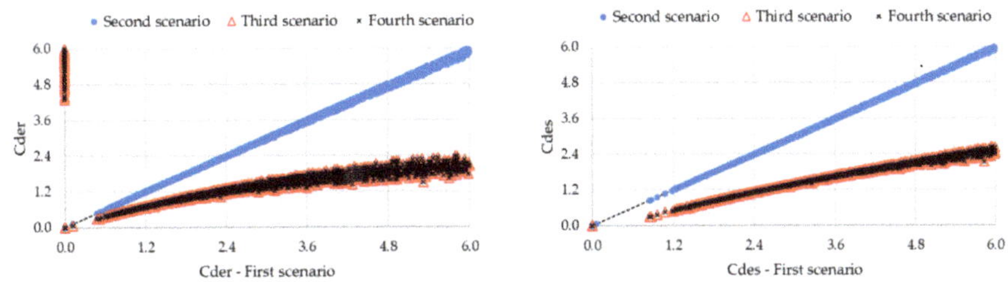

Figure 19. Safety coefficients for the first scenario compared with the second to fourth scenarios—Nørgaard et al. (2013).

In what concerns safety coefficients, values lower than 1.2 for scenarios 1 and 2 are presented in less than 3% of cases (with H_s higher than 2.6 m) and in 0.1% of cases (with H_s higher than 4.5 m) for C_{des} and C_{der}, respectively. However, the crown wall did not suffer any damage during these 11 years (scenario 1), these results being considered conservative for these types of structures/wave conditions.

When the protection of the crown wall is reduced and part of the structure suffers from the direct wave impact, the number of cases with C_{des} and C_{der} less than 1.2 increases to 16% (with H_s higher than 1.9 m) and 4% (H_s higher than 2.5 m), respectively, for Pederson (1996). For Nørgaard et al. (2013), these values are lower: 11 and 2% for C_{des} and C_{der}, respectively. In other words, the crown wall is unstable for sliding and overturning in a small number of wave conditions tested for scenarios 1 and 2, but these values increase for scenarios 3 and 4, with wave conditions becoming less energetic.

5. Discussion

To compare the results presented previously, firstly values of run-up, a very important parameter, are compared for each equation and scenario for the case study, Figure 20. As it can be observed in the figure, run-up values are highest in the case of Pedersen and Nørgaard et al., followed by Berenguer and Baonza, Günbak and Gökce, and Martín et al. Pederson and Nørgaard et al., calculate $R_{u0.1\%}$ as well as Günbak and Gökce and Martín et al. However, Pederson and Nørgaard et al. consider the Irribaren number in deep water, leading to higher run-up values and a wider range of values. Berenguer and Baonza calculate $R_{u2\%}$ but take the wave direction into account in the run-up whereas the other formulations have not. Jensen and Bradbury forces does not depend on R_u. This happens in the four scenarios, reaching similar values in all of them. The run-up value does not considerably change with the water level because the difference is only in the ξ_0 value, and ξ_0 does not change significantly. Run-up formulas depend on the average porosity, H or H_{max}, and ξ_{m0}. Since changes in A_c have no influence on R_u, the value of R_u in the four scenarios is the same.

Figure 21 allows to compare the values of horizontal pressures for the four equations analyzed in this paper where pressure is calculated (Jensen and Bradbury and Berenguer and Baonza calculate forces directly) and for the four scenarios. As it can be observed, the values for the first and second scenarios are very similar. This is because the only difference between both scenarios is the sea level, and the difference is very small to imply high variation in the forces. Differences among the third and fourth scenarios with the first scenario values are expected because of the damage in the armor. In fact, the comparison between the third and fourth scenario gives similar values except in the case of Günback and Gökce. On the other hand, horizontal pressure values are the highest in the case of Günbak and Gökce, followed by Pedersen and Nørgaard, and later by Martín et al. The higher maximum value and the wider range of values obtained by Pederson and Nørgaard et al. follow the trend presented in R_u, presented before. The lower values of Martin's formulation could be because in this formulation, unlike the others analyzed here, the effect of B_e and the material of the armor in the energy dissipate into the structure.

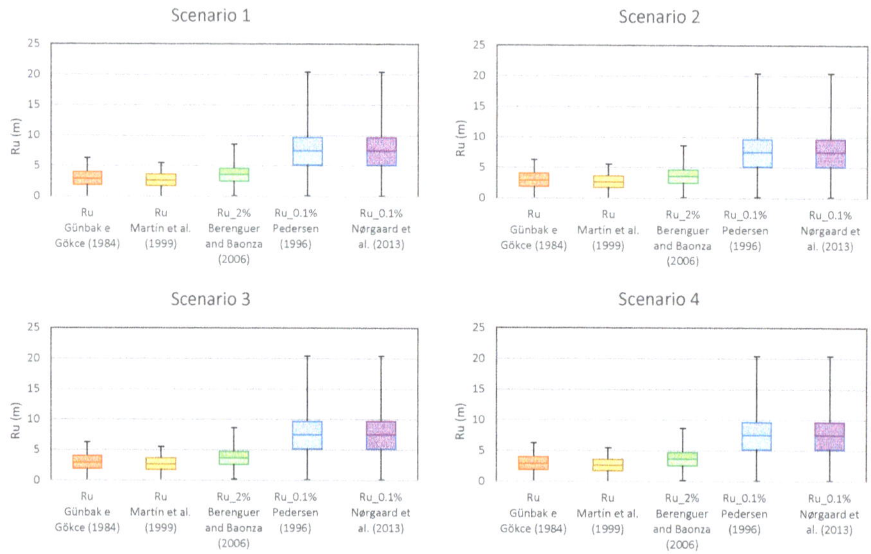

Figure 20. Results of the run-up for the different formulas for scenarios 1–4.

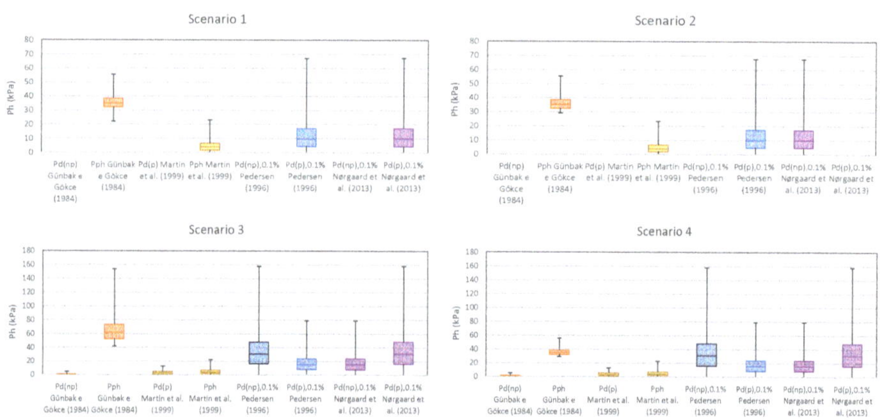

Figure 21. Results of P_h for the different formulas for scenarios 1–4.

The same aspects observed in the previous figure can be concluded for horizontal forces analyzing Figure 22: scarce differences between scenarios 1 and 2, and the same happens when comparing scenarios 3 and 4. The justification is the same as previously exposed: the low sea level variation in the area due to the climate change in the lifetime studied. Horizontal forces differences between scenario 1 and scenario 3 are clear, based on the damage of the armor. Günbak and Gökce is the only formulation were the reduction in A_c leads to a reduction in horizontal forces. In fact, the A_c reduction leads to a reduction in the pseudo-hydrostatic pressure and consequently in horizontal forces. On the other hand, medium horizontal forces from highest to lowest values are according to the following equations: Pedersen and Nørgaard, followed by Günbak and Gökce, Berenguer and Baonza, Jensen and Bradbury, and Martín. As referred to previously, Pedersen and Nørgaard consider $H_{0.1\%}$ as the wave height parameter, leading to the higher maximum values of R_u and P_d, as referred to before and consequently, higher maximum values of horizontal forces. Martin and Berenguer and Baonza formulation have the lower values and this could be due to the fact that these formulations, unlike the others analyzed here, have taken into account Be and the material of the armor on the energy dissipation. Moreover, Berenguer and Baonza only allow considering blocks or rip-rap, while Martín et al. allow considering tetrapods that have a higher porosity value leading to lower R_u values.

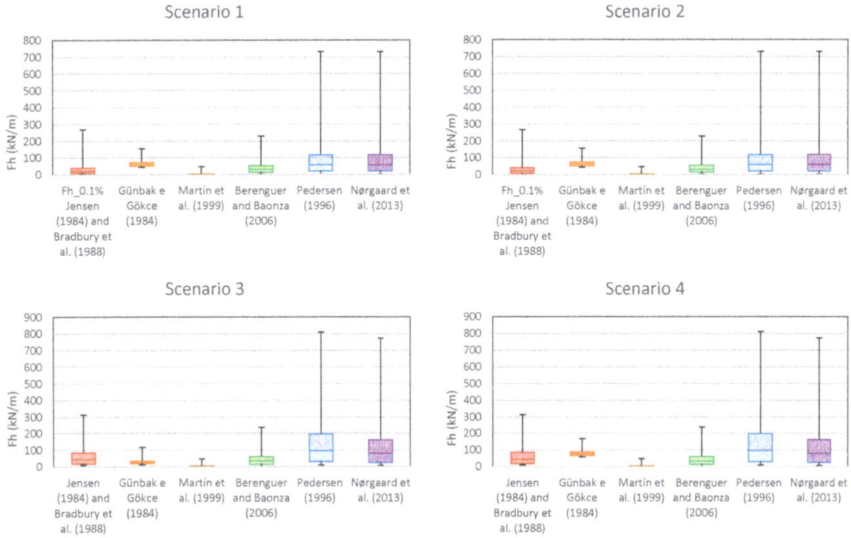

Figure 22. Results of F_h for the different formulas for scenarios 1–4.

Uplift pressure action values considerably change depending on the scenario in the case of Günbak and Gökce, but it is similar in the rest of the equations (Figure 23). On the other hand, uplift pressure action values from highest to lowest values are according to the next equations: Günbak and Gökce, followed by Pedersen and Nørgaard, and Martín. The trend is the same as in the horizontal pressure, for the same reasons already explained since the uplift pressure depends on the horizontal pressure in the base of the superstructure.

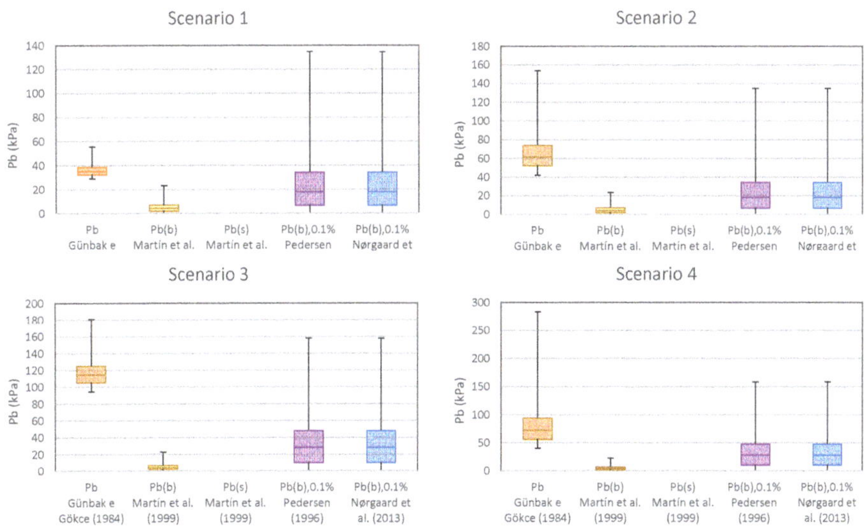

Figure 23. Results of Pb for the different formulas for scenarios 1–4.

Uplift forces follow the pressure trend, with the change depending on the scenario in the case of using the Günbak and Gökce equation and the Pedersen and Nørgaard (Figure 24). On the other hand, sub-pressure action values from highest to lowest values are according to the next equations: Günbak and Gökce, followed by Pedersen and Nørgaard, Berenguer and Baonza, Jensen and Bradbury, and Martín.

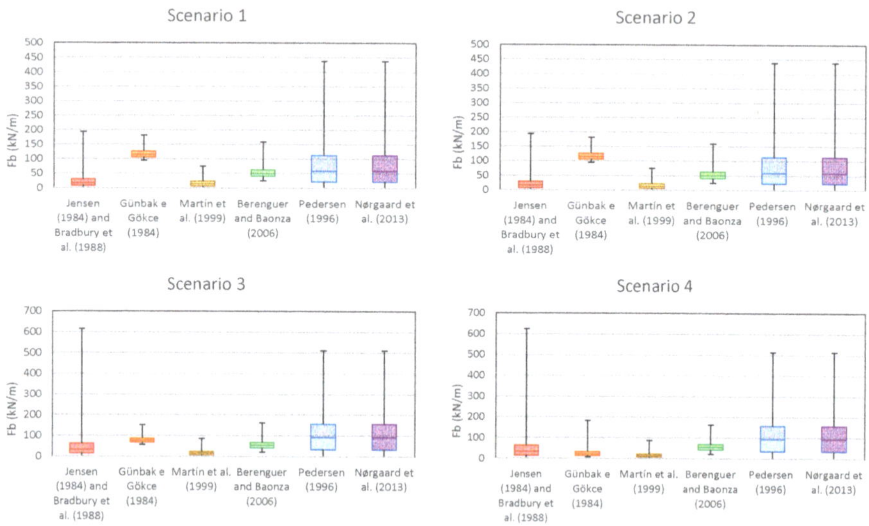

Figure 24. Results of Fb for the different formulas for scenarios 1 to 4.

Moments present the same trend as forces, for the reasons presented before. Safety coefficients C_{des} (sliding) and C_{der} (overtuning) present a small reduction between scenarios 1 and 2. This was expected, since the sea level rise considered is small and leads to small differences in forces and momenta. However, this decrease was more accentuated in scenarios 3 and 4 for all formulations since the reduction in A_c leads to a considerable increase in the forces and momenta calculated by the formulations. The only exception is that of the results of the Günbak and Gökce equation, where the opposite occurred, for the reasons explained before. Nørgaard and Pedersen is the only formulation with values of C_{der} less than 1.2, as the forces and momenta calculated by this formulation lead to higher values of forces and momenta, as referred to previously.

6. Conclusions

Calculating the forces on the crown walls is not a simple task. There are several equations that have been sanctioned by practice, so it is necessary to know the calculation hypotheses of each of them and check that the section under study fits those hypotheses. The methods of calculation of the crown walls respond to very different criteria in terms of schemes (forces or pressures), slopes and shape of the crests. To cite some examples, in the case of schemes, Martin et al. (1999) provides pressures or Pedersen and Burcharth (1992) and Berenguer and Baonza (2006) provide forces. In the case of slopes, Bradbury et al. (1988) for slopes 2/1, Martín et al. (1999) with slope 3/2 and Günbak and Gökce (1984) with slope 2/5. In addition, the geometry considered in all of them is with the mantle berm below the crest of the crown wall. Therefore, great caution should be exercised in using any crown wall calculation expression because of its dispersion in results and scenarios or because of the uniqueness in the geometry that does not exactly fit the application ranges of the known formulas.

Ericeira breakwater has a different geometry. It is a crown wall completely protected by the outer armor and the artificial blocks that form the armor are interlocking pieces (tetrapods). With all this, it can be observed that energy transmission takes place before reaching the parapet. It should also be noted that wave periods of high value can be recorded which, together with a macro-tidal tidal range, can lead to the presence of combinations of broken and unbroken waves on the armor. For the Ericeira crown wall, in the 11 years of data analyzed, only few or no wave conditions follow within the range of all parameters where the analyzed formulas are valid. Thus, for this structure, the formula should be used out of the limits of application proposed by the authors.

For this case, the values of the horizontal forces are more conservative in the case of the Pedersen and Nørgaard equations, the lowest value being that of Martin. With respect to the uplift pressures, Martin's is still the least value, while for scenarios 1 and 2 the most conservative is Günbak and Gökce's, and for scenarios 3 and 4, Pedersen and Nørgaard's as in the case of the horizontal forces. Günbak and Gökce's is the only formulation where damage to the main armor by losing pieces at the berm led to a reduction in forces, which was not expected. Furthermore, it can be observed that notable difference between the values of the forces when each of the formulas was used, which forces further research in this field since these variations imply important differences in the design of the crown walls.

For all scenarios, a sliding safety coefficient was more a conditioning than overturning safety coefficient. The increase in the sea level for the case study was not significant to the stability of the crown wall and therefore its incidence is very small. This does not mean that, in other regions of the world, where the rate of sea level rise per year is more important, the influence of sea level rise is not a determining factor in the design. On the other hand, damage to the main armor by losing pieces at the berm is much more important in this case study, so it is essential to carry out the proper maintenance of the design section.

Author Contributions: Conceptualization, M.G.N. and F.P.; methodology, M.G.N. and F.P.; software, F.P.; validation, M.G.N., J.-S.L.-G., M.D.E. and V.N.; formal analysis, F.P., M.G.N., J.-S.L.-G., M.D.E. and V.N.; investigation, F.P., M.G.N., J.-S.L.-G., M.D.E. and V.N.; writing—original draft preparation, J.-S.L.-G. and M.D.E.; writing—review and editing, F.P., M.G.N., J.-S.L.-G., M.D.E. and V.N.; supervision, M.G.N., J.-S.L.-G. and M.D.E. All authors have read and agreed to the published version of the manuscript.

Funding: This research was partially funded by the Portuguese Foundation for Science and Technology (FCT) for the project BSAFE4SEA-Breakwaters Safety Control through a Forecast and Decision Support System Analysis (PTDC/ECI-EGC/31090/2017).

Institutional Review Board Statement: Not applicable.

Informed Consent Statement: Not applicable.

Data Availability Statement: Data available on request due to privacy restrictions. The data presented in this study are available on request from the corresponding author. The data are not publicly available due to confidential reasons.

Acknowledgments: J.-S.L.-G. and M.D.E. are very grateful to LNEC for giving the opportunity to perform research in the Hydraulics and Environment Department.

Conflicts of Interest: The authors declare no conflict of interest. The funders had no role in the design of the study; in the collection, analyses, or interpretation of data; in the writing of the manuscript, or in the decision to publish the results.

References

1. Lacasa, M.C.; Esteban, M.D.; López-Gutiérrez, J.S.; Negro, V.; Zang, Z. Feasibility study of the installation of wave energy converters in existing breakwaters in the north of Spain. *Appl. Sci.* **2019**, *9*, 5225. [CrossRef]
2. Berenguer, J.M.; Baonza, A. On the design of crown walls of rubble mound breakwaters. In Proceedings of the 31st PIANC Congress, Estoril, Portugal, 14–18 May 2006; pp. 608–624.
3. Günbak, A.R.; Gökce, T. Wave screen stability of rubble-mound breakwaters. In Proceedings of the International Symposium of Maritime Structures in the Mediterranean Sea, Athens, Greece, 22–25 November 1984; pp. 2.99–2.112.
4. Negro, V.; López-Gutiérrez, J.S.; Polvorinos Flors, J.I. Comparative study of breakwater crown wall—Calculation methods. *Proc. Inst. Civ. Eng. Marit. Eng.* **2013**, *166*, 25–41.
5. Webpage of Intergovernmental Panel on Climate Change (IPCC). Available online: www.ipcc.ch (accessed on 3 January 2021).

6. Diez, J.J.; Esteban, M.D.; Paz, R.; López-Gutiérrez, J.S.; Negro, V.; Monnot, J.V. Urban coastal flooding and climate change. *J. Coast. Res.* **2011**, *64*, 205–209.
7. Negro, V.; López-Gutiérrez, J.S.; Esteban, M.D.; del Campo, J.M.; Luengo, J. Action strategy for studying marine and coastal works with climate change on the horizon. *J. Coast. Res.* **2018**, *85*, 506–510. [CrossRef]
8. Negro, V.; López-Gutiérrez, J.S.; Esteban, M.D.; Matutano, C. An analysis of recent changes in Spanish Coastal Law. *J. Coast. Res.* **2014**, *70*, 448–453. [CrossRef]
9. Molines, J.; Herrera, M.P.; Medina, J.R. Estimation of wave forces on crown walls based on wave overtopping rates. *Coast. Eng.* **2018**, *132*, 50–62. [CrossRef]
10. CIRIA/CUR/CERMEF Construction Industry Research, Information Association, Civieltechnisch Centrum Uitvoering Research en Regelgeving (Netherlands); Centre d'études Maritimes et Fluviales (France). Physical processes and design tools: Crown walls. In *The Rock Manual: The Use of Rock in Hydraulic Engineering*; CIRIA/CUR/CERMEF: Amsterdam, The Netherlands, 2007.
11. Martín, F.L.; Vidal, C.; Losada, M.Á.; Medina, R. Un método para el cálculo de las acciones del oleaje sobre los espaldones de los diques rompeolas. *Ing. Agua* **1995**, *2*, 37–52. [CrossRef]
12. Díaz-Sánchez, R.; López-Gutiérrez, J.S.; Lechuga, A.; Negro, V.; Esteban, M.D. Direct estimation wave setup as a medium level in swash. *J. Coast. Res.* **2013**, *65*, 201–206. [CrossRef]
13. United States Army Corps of Engineers (USACE). *Coastal Engineering Manual*; USACE: Washington, DC, USA, 2008.
14. Martín, F.L.; Losada, M.A.; Medina, R. Wave loads on rubble mound breakwater crown walls. *Coast. Eng.* **1999**, *37*, 149–174. [CrossRef]
15. Jensen, O.J. *A Monograph on Rubble Mound Breakwaters*; Danish Hydraulic Institute: Hørsholm, Denmark, 1984.
16. Bradbury, A.P.; Allsop, N.W.H.; Stephens, R.V. *Hydraulic Performance of Breakwater Crown Walls*; Report SR146; HR Wallingford: Wallingford, UK, 1988.
17. Yamamoto, Y.; Horikawa, K. New methods to evaluate wave run-up height and wave overtopping rate. In Proceedings of the 23rd International Conference on Coastal Engineering, ASCE, Venice, Italy, 4–9 October 1992; pp. 1734–1747.
18. Losada, M.A. Recent developments in the design of mound breakwaters. In *Handbook of Coast and Ocean Engrg*; Gulf Publishing: Houston, TX, USA, 1990; pp. 939–1050.
19. Pedersen, J. Wave Forces and Overtopping on Crown Walls of Rubble Mound Breakwaters: An Experimental Study. Ph.D. Thesis, Aalborg University, Aalborg, Denmark, 1996.
20. Pedersen, J.; Burcharth, H.F. Wave force on crown walls. In Proceedings of the 23rd International Conference on Coastal Engineering, Venice, Italy, 4–9 October 1992; ASCE: Reston, VA, USA, 1993; pp. 1489–1502.
21. Van der Meer, J.W.; Stam, C.J.M. Wave runup on smooth and rock slopes of coastal structures. *J. Waterway Port. Coast. Ocean Eng.* **1992**, *118*, 534–550. [CrossRef]
22. Nørgaard, J.Q.H.; Andersen, T.L.; Burcharth, H.F. Wave loads on rubble mound breakwater crown walls in deep and shallow water wave conditions. *Coast. Eng.* **2013**, *80*, 137–147. [CrossRef]
23. Zózimo, A.C.; Fortes, C.J.M.E.; Reis, M.T.; Pinheiro, L.; Smithers, N. Desenvolvimento do sistema To-SEAlert para aplicação ao porto da Ericeira. In Proceedings of the 15° Congresso da Água, Lisboa, Portugal, 22–26 March 2020.
24. XTide: Harmonic Tide Clock and Tide Predictor. Available online: https://flaterco.com/xtide/ (accessed on 1 September 2020).
25. Andrade, C.; Freitas, M.C.; Taborda, R.; Carvalho, M.R.; Santos, F.D. Sector de zonas costeiras. In *Estratégia Nacional de Adaptação aos Impactos das Alterações Climáticas Relacionadas com os Recursos Hídricos*; Agência Portuguesa do Ambiente: Amadora, Portugal, 2008.

Journal of
Marine Science and Engineering

MDPI

Article

Evaluation of Groundwater in the Coastal Portion of Guasave, Sinaloa for White Shrimp Farming (*Penaeus vannamei*) through VES, Chemical Composition, and Survival Tests

Mauro Espinoza Ortiz [1], Juan Pablo Apún Molina [2,*], Héctor José Peinado Guevara [3,*], Jaime Herrera Barrientos [4,*], Salvador Isidro Belmonte Jiménez [5], María de los Ángeles Ladrón de Guevara Torres [5] and Omar Delgado Rodríguez [6]

1 Instituto Politécnico Nacional, CIIDIR Unidad Sinaloa, Programa de Doctorado en Red en Ciencias en Conservación del Patrimonio Paisajístico, Guasave C.P. 81100, Sinaloa, Mexico; mespinozao1700@alumno.ipn.mx
2 Instituto Politécnico Nacional, CIIDIR Unidad Sinaloa, Guasave C.P. 81100, Sinaloa, Mexico
3 Escuela de Ciencias Económicas y Administrativas, Universidad Autónoma de Sinaloa, Guasave C.P. 81100, Sinaloa, Mexico
4 Centro de Investigación Científica y de Educación Superior de Ensenada, Baja California (CICESE), Ensenada C.P. 22860, Baja California, Mexico
5 Instituto Politécnico Nacional, CIIDIR Unidad Oaxaca, Santa Cruz Xoxocotlán, Oaxaca C.P. 71230, Oaxaca, Mexico; salvador.belmonte.j@gmail.com (S.I.B.J.); maladron@ipn.mx (M.d.l.Á.L.d.G.T.)
6 División de Geociencias Aplicadas, Instituto Potosino de Investigación Científica y Tecnológica (IPICyT), San Luis Potosi C.P. 78216, San Luis Potosí, Mexico; omar.delgado@ipicyt.edu.mx
* Correspondence: japun@ipn.mx (J.P.A.M.); hpeinado75@hotmail.com (H.J.P.G.); jherrera@cicese.mx (J.H.B.); Tel.: +52-687-104-7869 (J.P.A.M.); +52-687-122-1862 (H.J.P.G.); +52-646-126-0074 (J.H.B.)

Citation: Espinoza Ortiz, M.; Apún Molina, J.P.; Peinado Guevara, H.J.; Herrera Barrientos, J.; Belmonte Jiménez, S.I.; Ladrón de Guevara Torres, M.d.l.Á.; Delgado Rodríguez, O. Evaluation of Groundwater in the Coastal Portion of Guasave, Sinaloa for White Shrimp Farming (*Penaeus vannamei*) through VES, Chemical Composition, and Survival Tests. *J. Mar. Sci. Eng.* **2021**, *9*, 276. https://doi.org/10.3390/jmse9030276

Academic Editors: M. Dolores Esteban, José-Santos López-Gutiérrez, Vicente Negro and M. Graça Neves

Received: 14 January 2021
Accepted: 25 February 2021
Published: 4 March 2021

Abstract: We studied the potential of white shrimp farming with groundwater from the coastal aquifer of Guasave, Sinaloa. The use of water from the aquifer in aquaculture represents challenges due to variability in the quality of groundwater. We did twenty-three vertical electrical soundings (VES), performed to guide the search for continental groundwater, obtaining the resistivity of the saturated formation (Ro) and a relationship with resistivity of aquifer water (Rw) = 0.4478 Ro + 0.8371. We obtained ionic content and nutrients from shrimp farming water. Also, a positive correlation was found between the electrical conductivity of the aquifer water (inverse of Rw) with chlorine, sodium, magnesium, and calcium ions in 34 water samples. The analysis of ions and ammonia-nitrogen, nitrates, phosphates, potassium, manganese, and calcium were used to select suitable sites to perform in two shorts bioassays: natural aquifer water and adding KCl and Mg_2Cl to simulate diluted seawater. In most natural waters, survival of larvae was higher than 60%, and in simulated seawater survival improved only in two sites. Building the Rw-Ro relation allowed us to infer the quality and suitability of water and the positioning of the most suitable place for drilling. Finally, Rw relation with dissolved ions allowed us to estimate aquifer water quality and reduce uncertainty.

Keywords: continental water; white shrimp; groundwater; ionic composition; resistivity

1. Introduction

Aquaculture is a primary sector activity with high growth potential. This, in addition to being a food source of great nutritional value, has provided socioeconomic benefits [1]. Mexico is one of the largest aquaculture producing countries in Latin America and the Caribbean [2]. The main species considered as targets are snapper, mollusk, white shrimp, and tilapia [3].

Aquaculture is a source of jobs in Sinaloa. It contributes to food security and poverty alleviation [4]. Shrimp farming in continental waters is a viable alternative for the development of small and medium producers in rural communities [5].

The quality of the water that supplies a crop is one of the predominant factors in aquaculture. In Mexico, the water sources for aquaculture developments come to a greater extent from lagoons, rivers and the sea [6]. Recent research shows the potential for the development of systems for aquaculture organisms using groundwater in tilapia [7] and white shrimp [5,8], obtaining considerable productive returns. So, groundwater is a viable alternative for continental aquaculture [9], and according to Valenzuela et al. [10] and Valenzuela et al. [8] the ionic composition of groundwater is more important in shrimp farming at low salinities.

Extensive aquaculture farms show a gradual tendency to promote configurations for more intensive production [11]. This due to the pathologies observed in shrimp farming.

In the production processes with groundwater it is fundamental to preserve its quality and quantity, so it is a priority to determine the appropriate places for its use in aquaculture [12,13], as well as its availability and quality so that the physicochemical parameters are in optimal ranges for the growth of aquatic organisms prior to the construction of an aquaculture farm [14].

To directly evaluate the water characteristics of an aquifer is a costly process, since it requires drilling that makes large-scale evaluation impossible [15]. Given this, the use of indirect methods (VES) in conjunction with direct information is an alternative to characterize the water quality of an aquifer. This is achieved through the study of the electrical properties of the water of the aquifer formation using vertical electrical sounding (VES) [16–21]. This is a method of geophysical prospection by direct current [22] which allows knowing the electrical resistivity of the Ro aquifer formation [23]. On the other hand, Peinado et al. [24] has characterized the water quality of an aquifer by establishing relations between the electrical resistivity of the water (Rw) and electrical resistivity of the aquifer formation (Ro). It is proposed here to find the Rw–Ro relation for the study area, as well as the relation between Rw and the concentration of groundwater ions necessary for the survival of shrimp. So, once the aforementioned relations have been established, in a place in the study area where there is no drilling through the Ro value obtained by VES, the quality of the groundwater for shrimp farming can be inferred.

2. Materials and Methods

2.1. Description of Study Area

The study area is within the aquifer of the Río Sinaloa. It is located at coordinates 25° 16′ 30.9″ to 25° 41′ 31.5″ north latitude and from 108° 21′ 35.9″ to 108° 40′ 54″ west longitude (Figure 1). Based on studies carried out of geological, geophysical, and hydrogeological evidence, the Río Sinaloa aquifer has been defined as of free type, heterogeneous, and anisotropic, with alluvial and fluvial sediments of varied granulometry. The thickness of the sediments in the central portion of the coastal plain in Guasave, Sinaloa is of several hundred meters [25].

The climate is dry, warm, and very hot. For the period of 1986 to 2016, the average annual temperature is 24 °C [26]. Average annual rainfall is 429.2 mm. There are rains in two periods, one in summer with abundant rainfall from July to October and another in winter with rains from November to January with precipitation less than 5% of the annual average [25].

The aquifer structure in terms of aquifer material availability, boundaries, and geometry is determined by correlating six lithologic columns of wells (3, 29, 7, 10, 16, and 19) with depths ranging from 120 m to 150 m, provided by the National Water Comission (profile K-K′, Figure 2).

2.2. Groundwater Sampling

In the study area, water samples were collected in 34 wells (Figure 1); 28 of them were active and six were drilled for the extraction of water samples. The samples were labeled and stored in 1 L plastic bottles. At the moment of the extraction, the depth of the water table was measured with a tape. Regarding pH, electrical conductivity (EC), Total Dissolved Solids (TDS), and temperature, the measurement was carried out with a Hanna

HI98130 equipment. The content of the NH_4^+ samples was measured with a portable photometer brand YSI 9500. Each sampling point was georeferenced with a GPS Garmin eTrex brand.

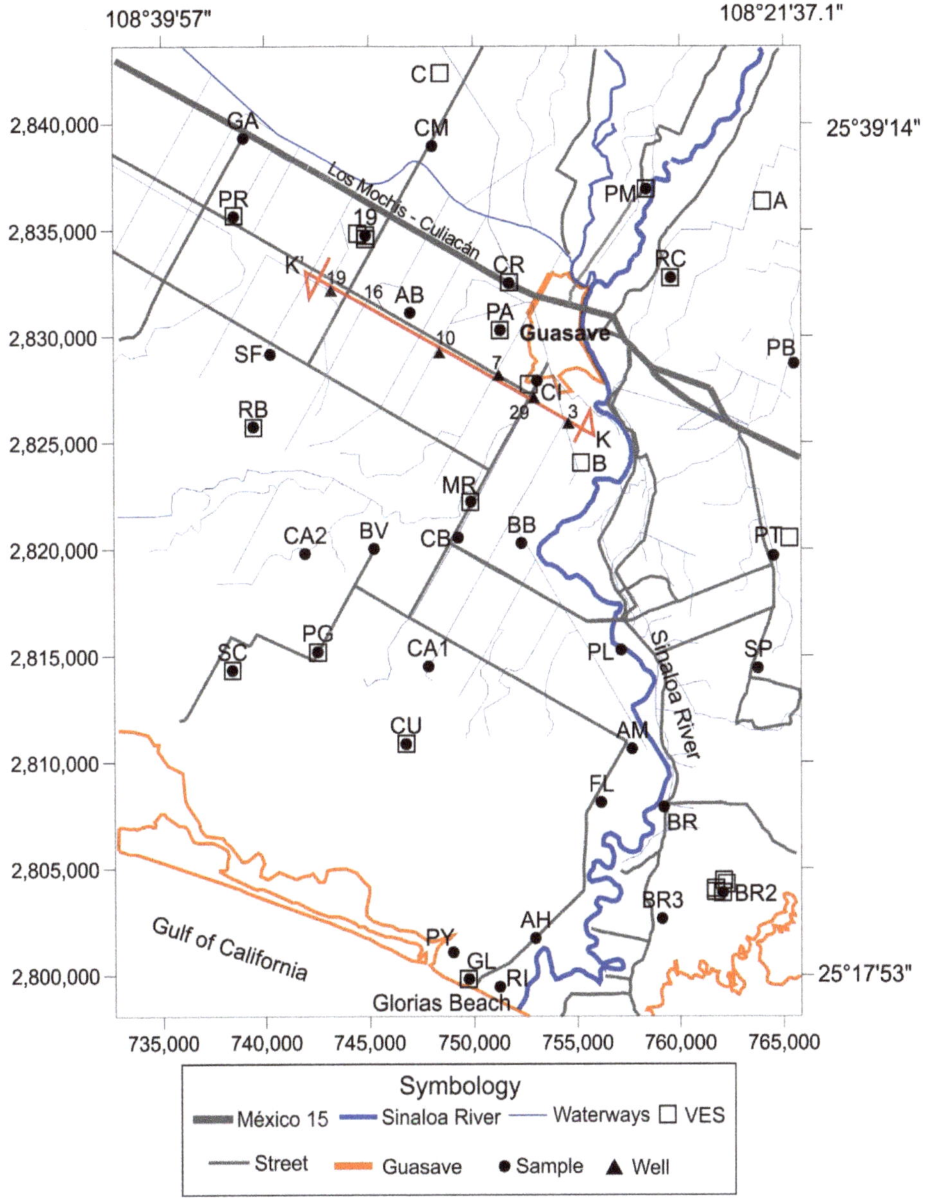

Figure 1. Study area. Locations of the water collections sites disignated with acronyms.

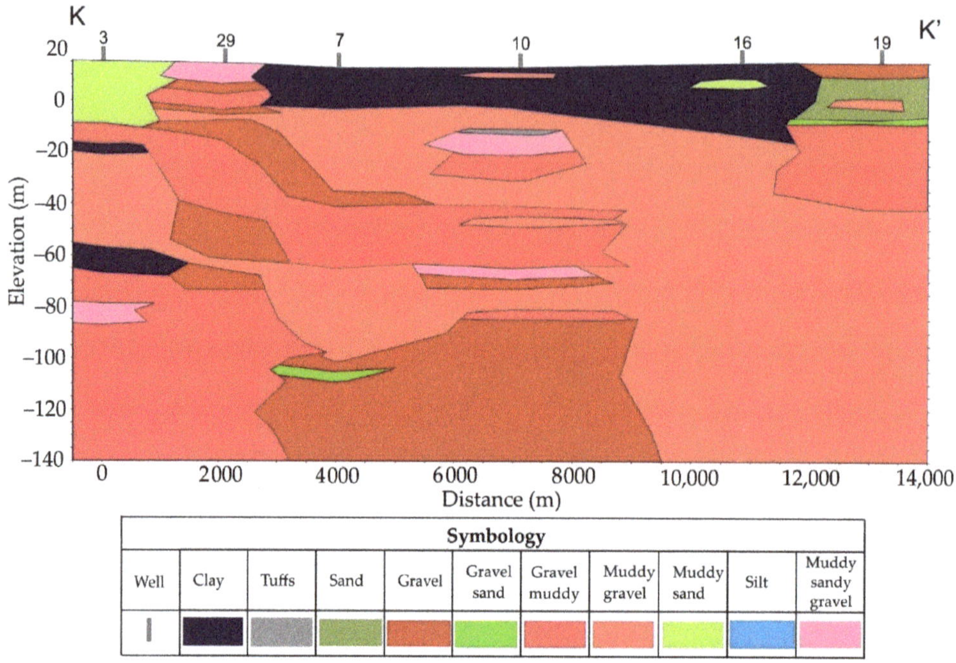

Figure 2. Hidrogeologycal cross section (K-K′) in six litological columns on the study area.

Wells are designated with acronyms that refer to the place where they are located: AB, AH, AM, BB, BR, BR2, BR3, BV, CA1, CA2, CB, CI, CM, CR, CU, FL, GA, GL, MR, PA, PB, PG, PL, PM, PR, PT, PY, RB, RC, RI, SC, SF, SP y 19.

The samples were placed on ice and transported to the Interdisciplinary Research Center for Integral Regional Development Sinaloa Unit (CIIDIR Sinaloa) fish aquaculture laboratory, Sinaloa, where the presence of nutrients (NO_3^-, NO_2^-, K^+, NH_4^+, PO_4^{3-}) was determined through colorimetric techniques using a YSI 9500 photometer (Yellow Springs, OH, USA). Metals Mn and Fe were determined using a HANNA 82,300 photometer (Hanna Instruments, Limena, Italy). Table 1 indicates the method to measure each parameter, as well as the range, wavelength, accuracy, and resolution of these instruments.

To measure the concentration of the parameters, 10 mL volume glass cuvettes were used for each sample following Table 1. Samples that exceeded the measurement concentration range of the spectrophotometer were diluted.

Table 1. Parameter description, method, range, wavelength, accuracy, and resolution of the YSI 9500 and HANNA 83,200 photometers.

Parameter	Method	Range (mg/L)	Wavelength	Accuracy	Resolution (mg/L)
Manganese	Standard methods for the examination of water and wasterwater	0.0 to 20.0	Tungsten lamp @525 nm	±0.2%/L to 25 °C	0.1
Iron	EPA Phenantroline method 315B, for natural and treated waters	0.00 to 5.00	Tungsten lamp @525 nm	±0.04 mg/L to 25 °C	0.01
Ammonium	Idophenol method	0.00 to 1.00	445 ± 5 nm, 495 ± 5 nm, 555 ± 5 nm, 570 ± 5 nm, 605 ± 5 nm, 655 ± 5 nm, automatic wavelength selection	±0.1 mg/L to 20 °C	0.01
Nitrites	YSI Nitricol method	0.00 to 0.5		±0.1 mg/L to 20 °C	0.001
Nitrates	YSI Nitratest method	0.00 to 1.00		±0.1 mg/L to 20 °C	0.001
Phosphates	YSI Phosphate LR method	0.00 to 4.00		±0.1 mg/L to 20 °C	0.01
Potassium	YSI Potassium test based on reagent of sodium tetraphenylboron	0.0 to 12.0		±0.1 mg/L to 20 °C	0.1

The ionic composition (Na^+, Cl^-, Ca^{+2}) of the samples was determined in a laboratory of the 075 Irrigation District belonging to the National Water Comission (CNA).

2.3. Vertical Electrical Soundings

To carry out the vertical electrical soundings (VES) a tetraelectrodic device was used in its Schlumberger mode (Figure 3). The center of symmetry of the device is O so that the distances OA = OB = L y OM= ON = a, in addition, it maintains the relation:

$$MN \leq (AB/5), \tag{1}$$

A continuous electric current of intensity I is injected through electrodes A and B, which will create an electric field in the subsoil. This field produces a potential difference ΔV between the electrodes M and N. Then, with these measurements, the apparent resistivity of the ground is obtained:

$$\rho a = K\,\Delta V/I, \tag{2}$$

where

$$K = \pi\, L\hat{}2/a, \tag{3}$$

Keeping the center of the device at O and with the purpose that the injected electric current penetrate the subsoil deeper, the separation between the electrodes is gradually increased, taking care that the relation given by equation 1 is fulfilled, thus obtaining the value of ρa as a function of each of the AB/2, separations, obtaining the apparent resistivity curve of the subsoil (AB/2 vs ρa). From this, by direct modeling considering that the terrain is formed by horizontal homogeneous and isotropic layers, the electrical resistivity of the aquifer formation, Ro, is obtained.

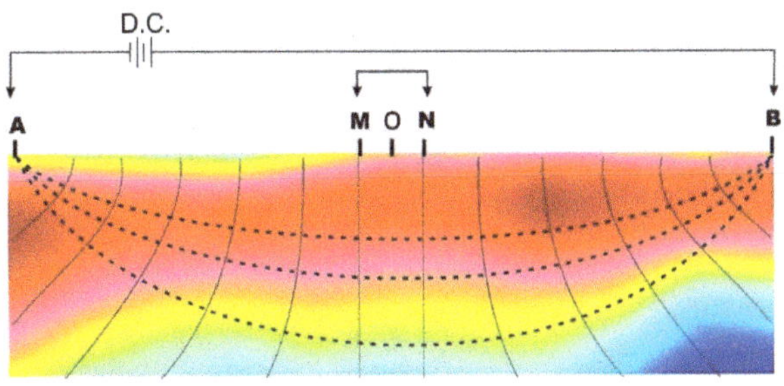

Figure 3. Tetraelectrodic device with Schlumberger setting diagram used.

The number of VES were 23, each one was georeferenced in Universal Transverse Mercator coordinates (UTM). AB/2 separations of: 5, 10, 20, 30, 40, 50, 60, 80, and 100 m were used. Of the VES carried out, 13 were made at sites where there are wells and groundwater EC measurements. VES were modeled with the software RESIXP. The product of the modeling, the saturated thickness and the electrical resistivity of the aquifer formation (Ro) were obtained.

This was related to the electrical resistivity of the water (Rw) in the aquifer. The latter was obtained from the following expression:

$$Rw = 10/EC, \tag{4}$$

where EC is the electrical conductivity of the aquifer water in mS/cm.

2.4. Selection of Sites for Aquaculture Farming

Table 2 was prepared from a bibliographic review. It contains the range of variation, maximum values and concentrations of various physical and chemical parameters of groundwater for shrimp farming. Considering the values of these parameters, water samples from wells that met these values were selected.

Table 2. Recommended concentration for culture of *P. vannamei* and number of samples that are under limits for culture.

No.	Parameter	Recommended Concentration	Results		Samples with Adequate Concentration		Reference
			Minimum	Maximum	Cant.	%	
1	EC (mS/cm)	>1	0.62	128.4	30	90.91	[5]
2	ppt	>0.5	0.31	64	29	87.88	-
3	T° (°C)	28–32 °C	22.6	32.3	34	100.00	[27]
4	pH	6.00–9.00	7.48	8.83	34	100.00	[5]
5	NH_4^+ (mg/L)	<0.5	0	28	28	84.85	[28]
6	PO_4^{3-} (mg/L)	-	0.41	2.69	-	-	-
7	NO_2^- (mg/L)	<1.45	0	2.685	32	96.97	[29]
8	NO_3^- (mg/L)	< 60	0.31	71.86	31	93.94	[27]
9	K^+ (mg/L)	-	0	2440	-	-	-
10	Fe (mg/L)	<0.3	0	2.46	29	87.88	[27]
11	Mn (mg/L)	<1	0	8.3	28	84.85	[30]
12	Mg^{2+} (mg/L)	-	5	2000	-	-	-

The parameters are: electrical conductivity (EC), pH, total dissolved solids, ionic composition (Na^+, Mg^{+2}, Ca^{+2}, Cl^-), heavy metals present (Mn, Fe) and nutrients (NO_3^-, NO_2^-, K^+, NH_4^+, PO_4^{3-}).

2.5. *Forty Eight Hours Survival Evaluation Test of P. vannamei Test*

With the purpose of evaluating the survival of shrimp culture in groundwater, 70 L of groundwater were collected from each of the following seven sites: GL, CU, RB, SC, PA, PR and La 19. The samples were transported to the Department of Aquaculture of CIIDIR, Sinaloa and there they were kept in aeration with diffuser stones from the beginning of the cultures until the end of the same. The shrimp larvae culture was developed over a 48 h period in two phases: (a) with direct transfer; and (b) with direct transfer adding KCl and Mg_2Cl to the water, in a similar way to that proposed by Valencia et al. [5].

2.5.1. Forty Eight Hours Survival with Groundwater

Groups of 10 postlarvae (initial weight 20.48 $\pm$ 2.08 mg) gradually acclimatized at a rate of (2 g/L/h) until a salinity of 4 g/L were transferred to containers with 2 L of water from each well for 48 h. At the end of this period, a count was made and a survival rate greater than or equal to 60% was considered as a criterion to select water samples from the wells for possible field cultivation.

As control treatments: C1 sea water was used at 30 g/L and C2 sea water diluted at 4 g/L. The cultures were performed in triplicate and fed twice a day (09:00 a.m. and 17:00 p.m.) *ad libitum.*

2.5.2. Forty Eight Hours Survival with Addition of Salts to Groundwater

Groups of 10 postlarvae (30.42 $\pm$ 2.42 mg) acclimated to 4 g/L were transferred to containers with 2 L of water from each well. The water samples where short-term survival (48 h) was $\geq$ 60% were selected.

As control treatments: C1 sea water was used at 30 g/L and C2 sea water diluted at 4 g/L. For the addition of salts in each well, the concentration of potassium chloride and magnesium chloride was calculated following the equation proposed by Boyd and Thunjai [31].

It was supplemented with potassium and magnesium salts using potassium chloride (KCl 99.0% Faga Lab®, Mocorito, Sinaloa, Mexico) and magnesium chloride ($MgCl_2$ 99.0% Faga Lab®, Mexico) for cultivation at theoretical salinity according to Boyd and Thunjai [31]. The cultures were performed in triplicate and fed twice a day (09:00 a.m. and 17:00 p.m.) *ad libitum.*

2.6. Statistic Analysis

The survival percentage of the bioassays were transformed with arcsine$\sqrt{p}$ [5,32] due to the lack of normality of the percentages from 0 to 100% with binomial distribution [33]. A one-way analysis of variance (ANOVA) ($p < 0.05$) and a Tukey test ($p < 0.05$) was performed to identify differences between groups.

3. Results

3.1. Aquifer Structure, Rw and Ro Relation

Table 3 shows 13 wells with their coordinates, static level depth (SL), electrical conductivity (EC) of the water in the well, resistivity of the aquifer formation (Ro), and resistivity of the aquifer water (Rw). VES was performed in each well and, from the apparent resistivity curve obtained and interpreted with the program RESIXP, the value of Ro indicated in Table 3 was obtained. With the Ro and Rw values, a linear regression analysis was performed resulting in a correlation coefficient of 0.933 (Figure 4). The equation that describes the relation between Ro and Rw is: Rw = 0.447Ro + 0.8371. That is, the higher the resistivity of the saturated formation, the higher the resistivity of the aquifer water.

Table 3. Data obtained from static level depth (SL) and electrical conductivity (EC) of the aquifer water, resistivity of the saturated formation (Ro) and electrical resistivity of the aquifer water (Rw).

Well	Site	Coordinate X	Coordinate Y	SL (m)	EC (mS/cm)	Ro (ohm-m)	Rw (ohm-m)
17	Culebras(CU)	746,764	2,810,846	5.2	5.87	0.54	1.70357751
30	La Brecha 2 (BR2)	762,024	2,803,857	1	61	0.39	0.16393443
1	Las Glorias (GL)	749,750	2,799,817	3.5	6.68	0.66	1.49700599
5	Roberto Barrios (RB)	739,441	2,825,759	2.45	5.69	3.32	1.75746924
26	Las Parritas (PR)	738,519	2,835,627	2	3.71	2.99	2.69541779
9	Sacrificio (SC)	738,408	2,814,322	2.7	3.15	4.98	3.17460317
32	La 19	744,854	2,834,744	1.1	1.95	6.1	5.12820513
6	El Progreso (PR)	745,516	2,825,179	2.2	2.98	6.39	3.37837838
33	El Pato (PA)	751,308	2,830,279	-	2.61	6.85	3.83141762
10	Cruces (CR)	751,738	2,832499	3.4	1.73	11	5.78034682
22	Ranchito de Castro (RC)	759,584	2,832,749	2.7	2.06	11.95	4.85436893
20	Marcol (MR)	749,883	2,822,226	5.2	1.35	14.59	7.40740741
31	CIIDIR (CI)	753,078	2,827,903	4.1	1.09	17.2	9.17431193

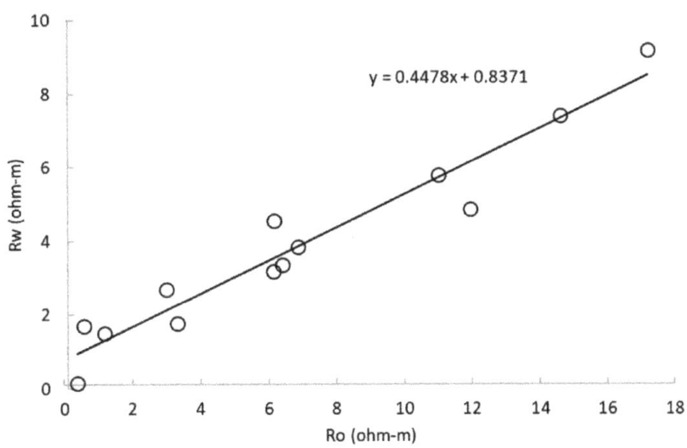

Figure 4. Linear regression of resistivity of the saturated formation Ro and electrical resistivity of the aquifer water Rw obtained from 13 wells in the communities of the study area.

The structure of the aquifer was determined by the lithologic sequences of wells 3, 29, 7, 10, 16, and 19 (Figure 2). The lithologic column of these six wells was used to construct the section of the aquifer environment corresponding to the K-K′ profile (Figure 1). The columns were topographically located and correlated, showing that the aquifer environment behaves as free in wells 3, 29, and 19 and as confined in wells 7 and 16. The distribution of the aquifer materials is mainly due to the eolian and fluvial deposits of the Sinaloa River. These detrital sediments were transported and deposited by the migratory dynamics of the Sinaloa River, arranged so that coarse materials such as gravels were deposited in the deeper portions and fine materials in the shallower part. This is due to variations in the historical intensity of the flow velocity.

The coarse materials are characterized by a high hydraulic transmissivity. On the other hand, the lithological columns in this section did not penetrate the geological and hydrogeological basement, which indicates that the wells partially penetrate the aquifer.

Regarding the upper and lower limits of the aquifer, the upper limit (between wells 3, 29, and 19) is the atmospheric pressure and the pressure of an impermeable medium (between wells 7, 10 and 16). As a lower boundary, the aforementioned wells did not cut impermeable medium.

3.2. Analysis of Collected Groundwater Samples

It was indicated the behavior of the physical-chemical parameters whose values are included in the intervals of Table 2. The static level depth average of 34 water samples was about 2.98 $\pm$ 1.85 m under natural terrain and the bottom of the well was not greater than 15 m in all wells analyzed in the present study.

The pH of the water extracted from the wells was measured in situ. This showed an average of 8.02 $\pm$ 0.32 with a maximum value of 8.83 at the La 19 site and a minimum of 7.48 at the Las Glorias (GL) site. In areas outside the city, there were areas registered with groundwater with a neutral trend.

The behavior of the electrical conductivity of the groundwater predominated as saline in the coastal zone and in the localities near the municipal seat. The average was 11.06 mS/cm and the maximum values were in El Pitahayal (PY), AH, and La Brecha 2 (BR2) with 128.4 mS/cm, 95 mS/cm, and 61 mS/cm, respectively. Figure 5 (upper left) shows the variation in electrical conductivity, higher on the coastline, decreasing as it moves towards the continent.

The ammonia nitrogen concentration (see Figure 5, upper right) at the BV, CA2, PT, BR, CB, BB, MR, and PA sites was zero with the equipment YSI 9500, which has a minimum detection resolution of 0.01 mg/L, these values are mainly found in communities near the municipal seat. In contrast to this value, the La Brecha 2 (BR2) site stands out, which had the maximum concentration of 28 mg/L. The behavior of ammonia nitrogen can be observed in Figure 5, with a predominant central strip of low value, flanked to the north and south by higher levels.

Nitrites increase from south to north with lateral variations (Figure 5, lower left). The 34 samples in this study showed a mean of 0.18 $\pm$ 0.46 mg/L with a maximum of 2.68 mg/L at the site El Gallo (GA), near a lined irrigation canal. Only in La Bebelama (BB) site, no concentration was detected with the detection resolution 0.001 mg/L. This variable did not show an increasing trend in the coastal zone.

Nitrates (Figure 5, lower right) increase from south to north. Their average is 16.30 mg/L and they present a variation from 0.31 to 71.86 mg/L. The maximum value was presented in the northeast area at the San Fernando (SF) site, to the northwest in Palos Blancos (PB) with a value of 62.46 mg/L and a minimum value of 0.31 mg/L was presented in El Sacrificio (SC).

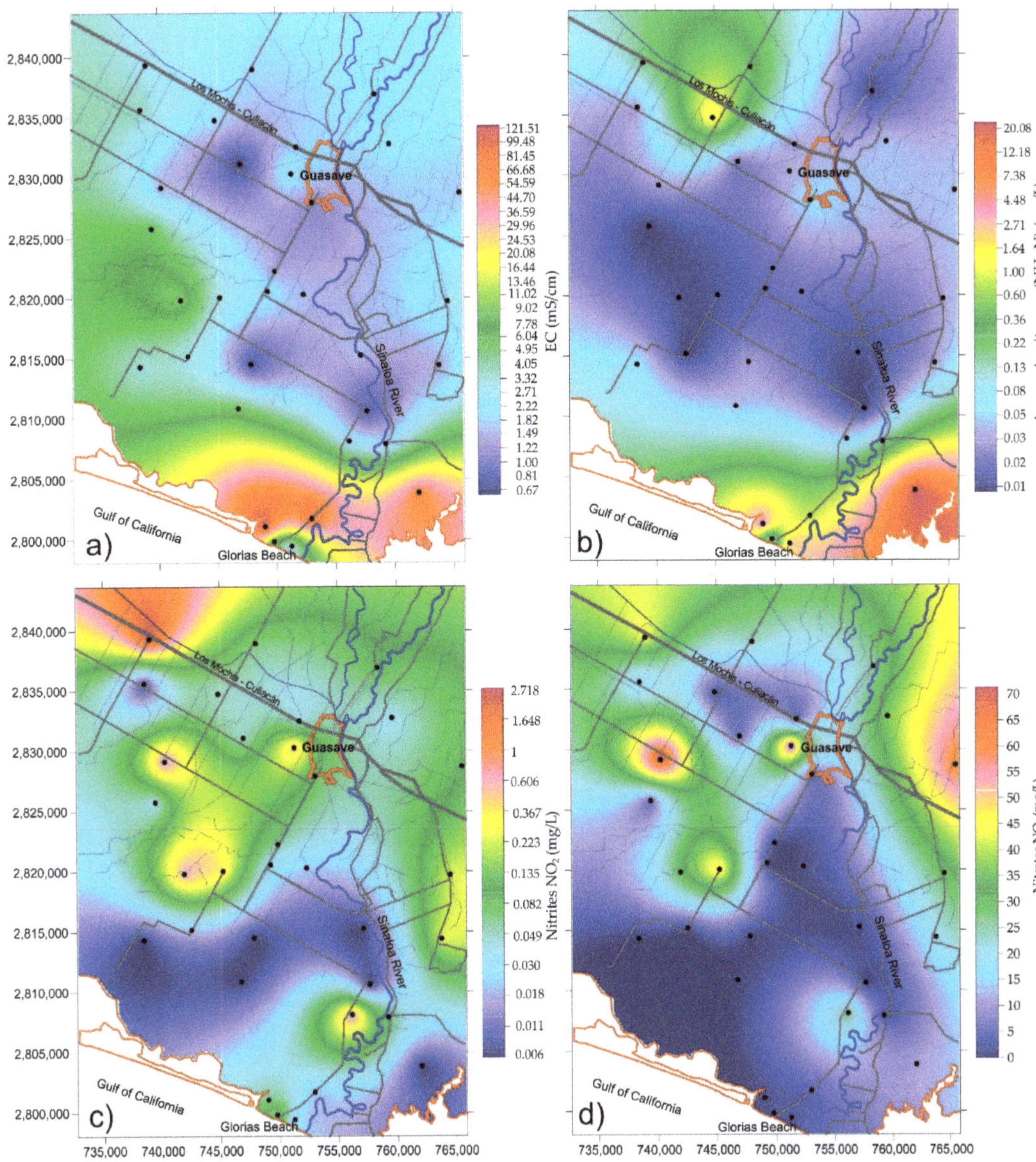

Figure 5. Behavior of (**a**) electrical conductivity (EC), (**b**) ammoniacal nitrogen, (**c**) nitrites and (**d**) nitrates present in groundwater in the study area.

The behavior of phosphates in groundwater in the study area is shown in Figure 6, upper left. This presents alternations of maximums without defined behavior. The data showed an average of 1.46 ± 0.62 mg/L with a maximum value of 2.69 mg/L in Las Flores (FL) and 2.44 mg/L in San Fernando (SF). The minimum value was presented in Boca del Rio (RI) with a concentration of 0.41 mg/L.

Potassium (Figure 6, upper right) varies from the coast to the mainland. On the coast it acquires its highest value and in the continent the lowest. Considering the resolution of the measuring equipment, the potassium ion concentration was not detected at four sites (RB, PL, CB, and CI). The other sites presented concentrations in the range of 0.1 to 70 mg/L in the area near the municipal seat. Elevated values of 780 mg/L were found in coastal areas and 2440 mg/L at the HA site.

Figure 6, bottom left and right, shows the variation of iron and manganese, respectively. Both present a variation of high in the south and low in the north. In the case of iron, the low concentration is distributed to the NW and, for Mn, to the NE. The highest and lowest values were found in localities PY and BR2, respectively.

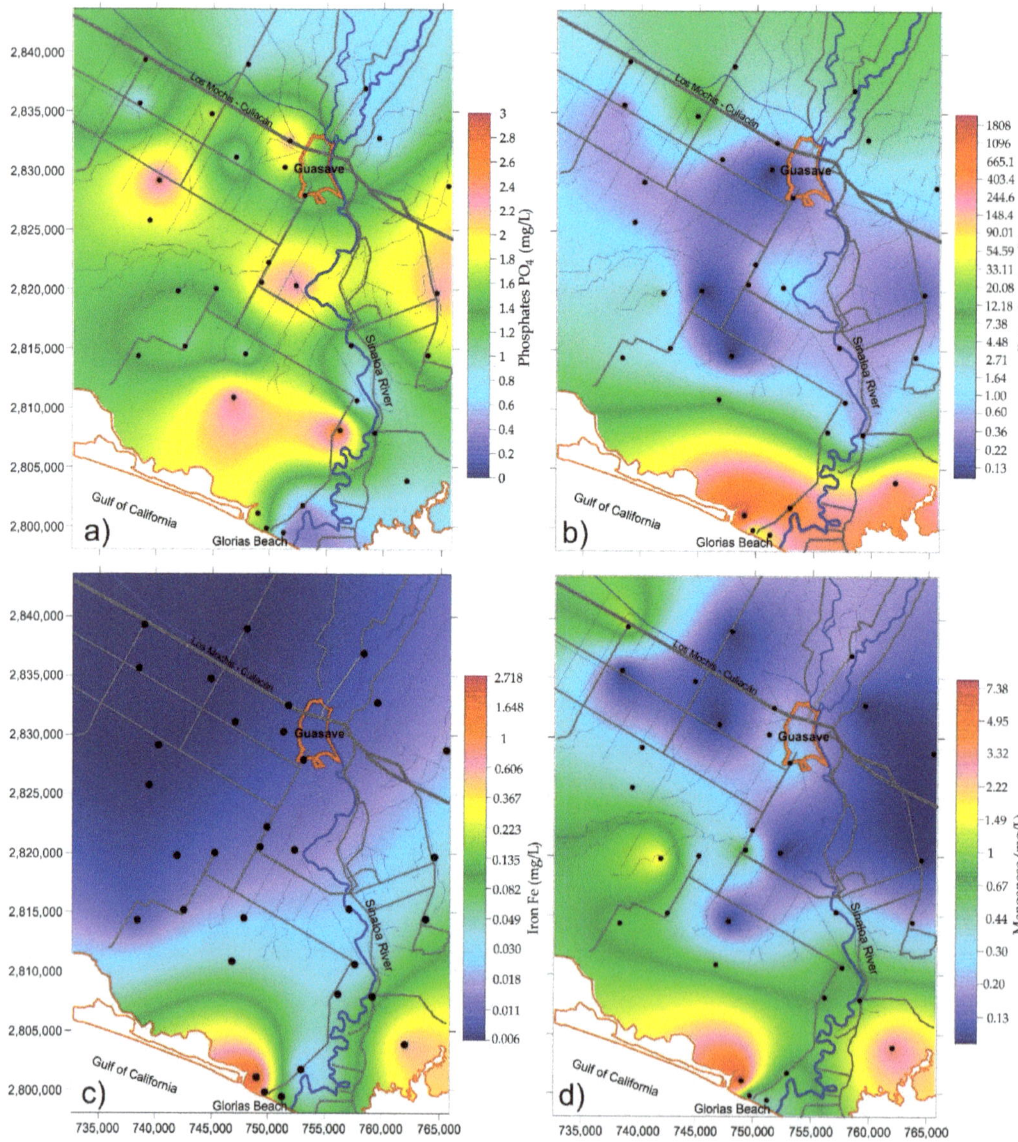

Figure 6. Behavior of the concentrations of (**a**) phosphate, (**b**) potassium, (**c**) iron and (**d**) manganese in the groundwater in the study area.

3.3. Correlation Matrix Matrix of Physicochemical Parameters

Table 4 shows the correlation matrix between the parameters measured in groundwater, observing high correlations between the ions of potassium, magnesium, calcium, chlorine, sodium, and some metals such as manganese and iron with electrical conductivity with correlation coefficients of 0.92, 0.99, 0.99, 0.98, 0.99, 0.82, and 0.80 respectively, this indicates that by increasing the electrical conductivity the concentration of these ions or metals increases in this study area. That is, the electrical conductivity of the water is a parameter that is related to parameters that intervene in the survival of shrimp and, in turn, the said conductivity is related to the resistivity of the water in the Ro formation, obtained by means of VES. Hence the importance of VES as a guide in the search for water for shrimp farming.

Table 4. Pearson correlations between the water parameters: EC, pH, NH_4^+, PO_4^{3-}, NO_2^-, NO_3^-, K^+, Fe, Mn, Mg^{+2}, Ca^{+2}, Na^+, and Cl^-. Data in red correspond to $p < 0.05$.

	EC	pH	NH_4^+	PO_4^{3-}	NO_2^-	NO_3^-	K^+	Fe	Mn	Mg^{+2}	Ca^{+2}	Na^+	Cl^-
EC	1.00												
pH	−0.32	1.00											
NH_4^+	0.47	−0.23	1.00										
PO_4^{3-}	−0.16	0.22	−0.17	1.00									
NO_2^-	−0.10	0.00	−0.08	0.16	1.00								
NO_3^-	−0.19	0.05	−0.10	0.23	0.47	1.00							
K^+	0.92	−0.25	0.29	−0.06	−0.08	−0.18	1.00						
Fe	0.80	−0.25	0.38	0.01	−0.07	−0.15	0.95	1.00					
Mn	0.82	−0.35	0.50	0.02	0.04	−0.14	0.90	0.96	1.00				
Mg^{+2}	0.99	−0.30	0.37	−0.16	−0.10	−0.20	0.94	0.80	0.80	1.00			
Ca^{+2}	0.99	−0.37	0.51	−0.13	−0.10	−0.20	0.93	0.85	0.87	0.98	1.00		
Na^+	0.99	−0.32	0.52	−0.15	−0.08	−0.17	0.94	0.86	0.86	0.98	0.99	1.00	
Cl^-	0.98	−0.37	0.60	−0.16	−0.10	−0.19	0.89	0.81	0.85	0.96	0.99	0.98	1.00

3.4. Ion Concentration Versus Total Salinity of Diluted Seawater

For shrimp farming in low salinity water or, where appropriate, groundwater, Boyd and Thunjai [31] and Roy et al. [34] emphasize that ion concentrations for an adequate shrimp culture should remain in a range equivalent to diluted seawater, however, in this study area, ideal concentrations were not present in some ions such as potassium and magnesium.

Figure 7 shows the behavior of the concentration of magnesium, calcium, sodium, and chlorine ions at sites with salinity between 0 and 6.5 g/L and their respective equivalent concentration of diluted seawater. The calcium ion concentration of the samples was above the theoretical dilute seawater. In the PY locality it reached a maximum of 9000 mg/L with a salinity of 64 g/L and, in CI, a minimum value of 30 mg/L. The sodium concentration generally showed a lower value than the equivalent concentration of diluted seawater, these in a range of 11.5 to 13,800 mg/L. Chlorine presented a minimum concentration of 17.75 mg/L in the CI site and a maximum value of 26,625 mg/L in PY. The magnesium ion maintained values above the equivalent concentration of seawater, with a maximum value in the PY locality of 3600 mg/L and a minimum of 6 mg/L in CI.

3.5. Bioassays

3.5.1. Bioassay 1 (48 h)

Figure 8A shows the result of the survival percentages of the shrimp postlarvae of the bioassay without adding salts. Preliminary 48 h cultures were performed with water from seven locations to observe the survival of the organisms. From the statistical analysis it was observed that there is no significant difference ($p < 0.05$) between the two controls (C1 and C2) and in the Las Glorias (GL) and El Sacrificio (SC) sites, but a difference was observed with the Las Culebras (CU), Roberto Barrios (RB), Progreso (PG), and Las Parritas (PR) sites. In the 48-h test it was observed that at only in two sites, La 19 (19) and La Progreso (PG), survival was not greater than 60%.

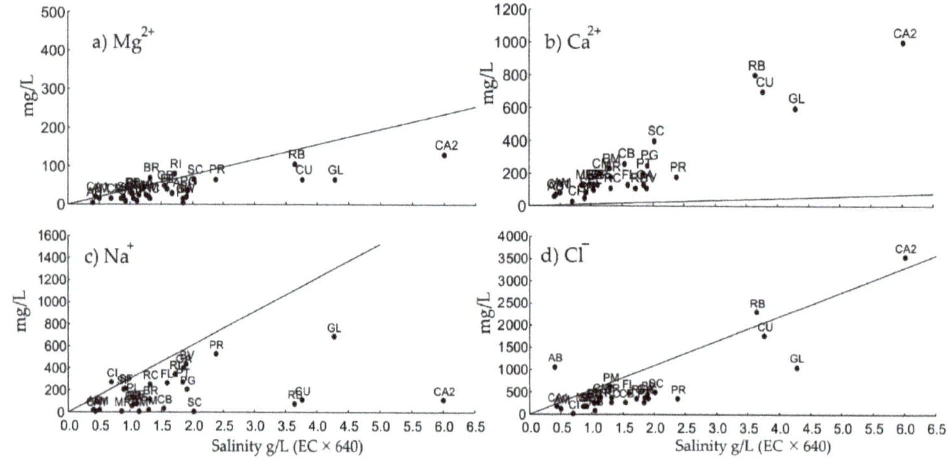

Figure 7. Concentrations (mg/L) of (a) Mg^{+2}, (b) Ca^{+2}, (c) Na^{+}, and (d) Cl^{-} versus salinity in groundwater obtained by multiply EC by 640. The solid lines represent the concentrations of the respective ion in normal seawater diluted to different salinities.

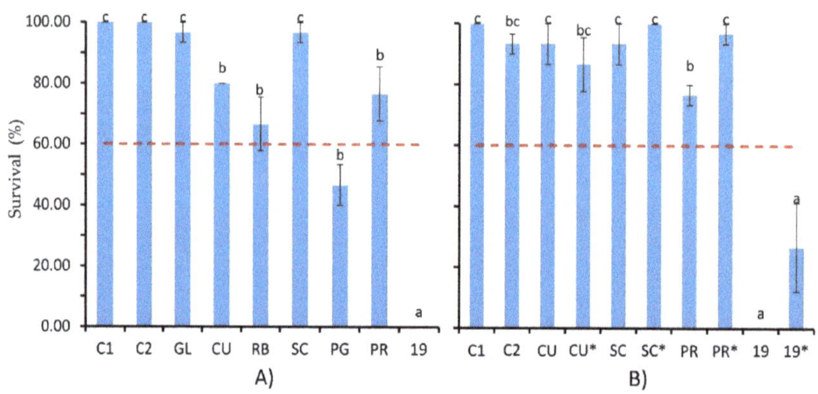

Figure 8. Survival of *P. vannamei* during the first 48 h, the letters indicate a significant difference $p < 0.05$ (* indicates that salts were added) (A) Survival of *P. vannamei* PL_{20} (B) Survival of *P. vannamei* (30.40 + 2.42 mg).

From the results of the first bioassay, four sites (CU, SC, PR, and 19) were taken to perform the second bioassay with the addition of salts of potassium chloride and magnesium chloride at the concentration proposed by Boyd and Thunjai [31] at the respective salinity for a groundwater culture.

3.5.2. Bioassay 2 (Addition of Salts)

Figure 8B shows the results of bioassay 2. Potassium chloride and magnesium chloride were added to the CU, SC, PR, and 19 sites. The first three obtained survival rates greater than 60%, with the exception of site La 19, which had had a survival rate of 0% in bioassay 1. The results did not show significant differences with respect to the controls (C1 and C2) and the study sites, with the exception of the PR site and La 19. Site 19 observed without the addition of salts a survival of 0% and with the addition of salts the survival was 26 ± 14%; that is, the addition of potassium and magnesium ions improved the survival in organisms cultured with La 19 water, however, it was not significant to be selected as a potential site for shrimp culture.

The result observed in the organisms cultured with the Las Parritas water source (significant difference) indicates that it can potentially be used for shrimp cultures, since an increase in survival was observed in the first 48 h. It is important to highlight that it was observed that in the cultures where salts were added as in the control bioassays, that the shrimp showed more activity (less lethargy) and energy, unlike the organisms grown in the bioassays where no salts were added.

4. Discussion

4.1. Variations of EC and Its Relation to Various Ions

In coastal areas, the electrical conductivity (EC) of water varies from moderately saline to saline, according to the Heath [35] classification. This variation is due to the interface of groundwater and seawater. Chidambaram et al. [36] reported similar behavior in a coastal aquifer, commonly influenced by the interaction of groundwater and sea. In this study the groundwater EC ranged from 2.68 mS/cm (RI site) to 95.6 mS/cm (AH site). Similar results have been reported by Peinado et al. [21] who characterized the coastal zone of Guasave in Las Glorias beach with 24 water samples from drillings, finding a variation in the EC that ranges from 1.96 to 88.60 mS/cm.

At the Guasave seat, Sinaloa Peinado et al. [37] found from 66 water samples that the EC varies from 0.18 to 8.08 mS/cm. These values are similar to those obtained in this work, ranging from 0.62 to 9.38 mS/cm.

The correlation of EC with Na^+, Mg^{+2}, Ca^{+2}, Cl^-, Mn, Fe, y K^+ (Table 4) is positive and similar to that reported by Shroff et al. [38]. These authors showed good correlation between EC and the variables Cl^-, Na^+, Mg^{+2} y Ca^{+2} of the groundwater of an aquifer in India. Other authors have also identified relations between EC and other ions [37,39,40].

Regarding the Ca^{+2} concentration of groundwater, this turned out to be higher than that of diluted seawater. Boyd and Thunjai [31], as well as Valencia et al. [5], indicate that this is explained by the genesis of the aquifer.

At a couple of sites, the iron concentration is 0.56 mg/L and 2.56 mg/L, not suitable for shrimp farming or human health, since they are higher than the 0.3 mg/L established by NOM-127-SSA1-1994. Saoud et al. [41] and Rodríguez et al. [42] carried out sampling in the municipality of Guasave without reporting iron concentrations higher than 0.3 mg/L.

Near the coastal zone (site BR2) in a well at a depth between 6 and 15 m, ammonia nitrogen was high at 28 mg/L. This is explained by the presence of organic matter. Similarly, Mastrocicco et al. [43] reported elevated concentrations of $NH_4{}^+$ in a coastal aquifer mainly due to the presence of muddy sediments with fine material. Valencia et al. [5] found a maximum concentration of nitrates of 108 mg/L and 2.88 mg/L of nitrites. In this work the maximum values for these two variables were 71.86 mg/L and 2.68 mg/L, respectively.

4.2. Application of Ro–Rw and Rw–Ion Concentration Relationships

The resistivity of the aquifer water Rw and the resistivity of the aquifer medium, Ro (obtained by vertical electrical sounding) are related in a linear way. So that through this relation, Ro–Rw in connection with the strong Person correlations that exists between Rw and the ions Na^+, Mg^{+2}, Ca^{+2}, and Cl^{-1} the aquifer is characterized and the places with groundwater are determined most suitable for growing shrimp *P. vannamei*.

The relation between Ro–Rw is linear and is given by Rw = 0.449Ro + 0.74 with a correlation coefficient $R^2 = 0.933$. This result is similar to that obtained by Peinado et al. [44] of Rw = 0.427Ro + 1.987.

The Ro–Rw relation under a linear scheme presented a good adjustment like the one found by Archie [45] from data of records in wells, so the Archie relation is applicable in aquifers like the one studied here. The Ro resistivity at the BR2 site presented values from 0.19 to 1.139 ohm-m characterized as saline water, similar to that corresponding to classifications of the Ro-based saline groundwater proposed by Llanes et al. [46].

The following is an example of the application of Ro–Rw and Rw relations and ionic relations. At sites A, B and C (Figure 1) there is no well, only VES. Thus, from the interpretation of VES, values for Ro of 11.12, 4.61, and 7.74 ohm-m were found, respectively. Values for water salinity of 6.0, 3.7, and 4 ohm-m respectively can be obtained by extrapolation from Figure 4. With these values in Figure 7 the expected concentrations of calcium, magnesium, sodium, and chlorine ions are obtained from the aforementioned places. The concentrations of these sites would vary from 100 to 800, from 175 to 200, from 1400 to 1600, and from 2000 to 3500 mg/L, respectively. These values in relation to those found for the GL, CU, RB, SC, and PR sites where the water allowed a survival greater than 60% of the shrimp larvae are similar for the calcium and magnesium ions. The values of sodium and chlorine ions are higher than those of the five places mentioned above, however, given that it is sodium and chlorine ions and the water is slightly brackish, it would be expected to be useful for aquaculture.

4.3. *P. vannamei* Survival Bioassay

From seven water samples (salinity from 1 to 4.5 g/L) used in the bioassay from 1 to 48 h, in five of them the survival of the shrimp larvae was greater than 60%. The recommended values of the parameters in Table 2 were followed, since at low salinities (0.5 to 10 g/L) the survival of the shrimp culture depends to a large extent on the ionic composition, rather than on the salinity [41].

At the El Progreso (PG) and La (19) sites, survival was less than 60%. At the La 19 site, despite having been adequate in all the analyzed physicochemical variables, it showed null survival; this can be attributed to the fact that this site corresponds to an agricultural property where fertilizers and insecticides have been used for agricultural practice. The species *P. vannamei* is susceptible to insecticides [27].

In the 2 to 48 h bioassay, potassium and magnesium salts were added, increasing the survival of the PR and La 19 sites. The addition of these salts was made considering that Roy et al. [34] and Valencia et al. [5] found that adding these salts until reaching the theoretical concentration equivalent to diluted seawater increases survival. Mariscal et al. [47] added KCl and $MgNO_3$ to the salinity groundwater of 0.65 mg/L used in the *P. vannamei* bioassay, obtaining a survival of $56.3 \pm 1.1\%$. The survival obtained after 48 h of acclimatization with water (from 2 to 4.5 g/L) of five sites presented survivals greater than 60%, that is, with a mean of $83.3 \pm 4.86\%$ sown. This is comparable to that obtained by Baldi et al. [48], who carried out acclimatization work with groundwater and obtained survival of $80 \pm 10\%$ in PL19 larvae, similar to that reported by McGraw et al. [49] with a survival of 87% after 48 h of acclimatization for *P. vannamei* PL_{20} at a salinity of 2 g/L.

Bioassays to check the viability of groundwater are necessary to decide whether it is potentially suitable for aquaculture [27].

Short periods of time, such as 48 h, are not enough to indicate productivity, due to the fact that on aquaculture in a commercial-size intervene variables that depend of the aquacultor such as feed rate, density, and type of culture, among others [14]. However, the 48 h survival results suggest the possibility of carrying out a complete fattening cycle with different strategic techniques.

Valencia et al. [5] reported that 11 out of 21 groundwater samples showed greater survival at 60%. The poor survival is attributed to the low quality ionic composition. The number of sites with survival greater than 60% in their work is possibly due to the fact that they did not perform a previous selection for culture, since they performed a bioassay on all the samples from the study sites. On the other hand, in this work the selection of the cultivation sites was made after the analysis of the ionic composition of the groundwater.

Laramore et al. [50] report that *P. vannamei* can be cultivated in low salinities without causing differences in survival and growth compared to the culture at 30 g/L, therefore in Guasave the results show that the cultivation of shrimp with groundwater is presented as a potential alternative. However, Valencia et al. [5] highlight that, to carry out low-salinity cultivation with groundwater for more than 28 days under laboratory conditions, the water exchange is necessary to avoid quality deterioration and thereby improve survival and growth. In this way, the evaluation of the aquifer in relation to its water quality is important to estimate the possible success of a commercial aquaculture project.

5. Conclusions

Aquaculture of white shrimp through the use of continental water represents a challenge, since it departs from the traditional use of seawater that is homogeneous in relation to the continental water.

Of the 13 parameters (EC, pH, NH_4^+, PO_4^{3-}, NO_2^-, NO_3^-, K^+, Fe, Mn, Mg^{+2}, Ca^{+2}, Na^+, and Cl^-) of groundwater from a coastal aquifer analyzed, it was found by Pearson correlation that only four (Na^+, Mg^{+2}, Ca^{+2}, and Cl^{-1}) of them present a strong correlation (greater than 0.98) with the electrical conductivity of the water. It was also found that these four parameters are important in the survival of white shrimp larvae.

When planning the cultivation of white shrimp, as in the case of other species, it is necessary to know a priori whether the water quality is adequate for this purpose. Thus, given the dependence of the electrical conductivity of groundwater (inverse of Rw) with its concentration of Na^+, Mg^{+2}, Ca^{+2}, and Cl^{-1} ions, as well as the geophysical relationship between the resistivity of the water of the Ro formation and that of the Rw groundwater, it is possible to couple them and use them jointly in places where there are no wells and it is possible to perform VES to obtain Ro.

So, the use of VES has been useful in the characterization of groundwater quality in the obtention of the Ro–Rw relation. This relation, in connection with the Rw-ionic concentrations relations (Na^+, Mg^{+2}, Ca^{+2}, Cl^-), allow for characterization and evaluation of the quality of groundwater for shrimp farming. These relations are helpful in selecting the ideal groundwater extraction sites for white shrimp aquaculture.

Laboratory tests of survival of white shrimp larvae at 48 h with water from the coastal water of the aquifer of Guasave, Sinaloa indicate that, in five places (GL, CU, RB, SC, and PR) a survival of 60% was observed, which can be increased by adding KCl and Mg_2Cl so a fattening cycle can be put into practice, to go from the laboratory phase to the field phase.

An analysis such as that carried out for shrimp can be done for other species reared in aquaculture (fish, mollusks, cetaceans, and plants), considering the specific parameters of water quality for each species.

Author Contributions: Conceptualization, M.E.O., H.J.P.G. and J.P.A.M.; methodology, M.E.O., H.J.P.G. and J.P.A.M.; software, M.E.O. and J.H.B.; validation, H.J.P.G., J.H.B. and S.I.B.J.; formal analysis, M.E.O., H.J.P.G. and O.D.R.; investigation, M.E.O. and H.J.P.G.; resources, J.P.A.M., H.J.P.G. and S.I.B.J.; data curation, J.P.A.M., H.J.P.G., S.I.B.J. and M.d.l.Á.L.d.G.T.; writing—original draft preparation, M.E.O., H.J.P.G. and J.P.A.M.; writing—review and editing, J.H.B., S.I.B.J., M.d.l.Á.L.d.G.T. and O.D.R.; visualization, J.H.B., S.I.B.J., M.d.l.Á.L.d.G.T. and O.D.R.; supervision, J.H.B., S.I.B.J., M.d.l.Á.L.d.G.T., O.D.R.; project administration, H.J.P.G. and J.P.A.M.; funding acquisition, H.J.P.G. and J.P.A.M. All authors have read and agreed to the published version of the manuscript.

Funding: This research was funded by Instituto Politécnico Nacional [grant number SIP20196805].

Institutional Review Board Statement: Not applicable.

Informed Consent Statement: Not applicable.

Data Availability Statement: MDPI Research Data Policies.

Acknowledgments: Aquaculture department of the IPN CIIDIR Sinaloa for the infrastructure support in the development of the bioassays with shrimp larvae. To CONACYT for the fellowship (grant number 762021) received by the student Mauro Espinoza Ortiz. Juan Pablo Apún Molina, Salvador Isidro Belmonte Jiménez and María de los Ángeles Ladrón de Guevara Torres received financial support from COFAA and E.D.I. scholarships from the INSTITUTO POLITECNICO NACIONAL.

Conflicts of Interest: The authors declare no conflict of interest. The funders had no role in the design of the study; in the collection, analyses, or interpretation of data; in the writing of the manuscript, or in the decision to publish the results.

References

1. Alvarez Torres, P.; Soto, F.; Aviles Quevedo, S.; Díaz Luna, C.; Treviño Carrillo, L.M. Panorama de la investigación y su repercusión sobre la produccion acuicola en Mexico. *Av. Nutr. Acuicola III* **1996**, *3*, 1–18.
2. Oldepesca, La Acuícultura y Sus Desafíos. *LA Acuicultura y sus Desafíos*; Organización Latinoamericana de Desarrollo Pesquero: Mexico City, DF, Mexico, 2009; pp. 1–26.
3. SAGARPA, Carta Nacional Acuícola. 2012. Available online: http://www.sagarpa.gob.mx/saladeprensa/2012/Paginas/2013B536.aspx (accessed on 11 December 2017).
4. Food and Agriculture Organization. *El Estado Mundial de la Pesca y Acuicultura Organización de las Naciones Unidad para la Alimnetación y la Agricultura*; FAO: Roma, Italy, 2016; p. 224.
5. Valencia Castañeda, G.; Millán Almaraz, M.I.; Fierro Sañudo, J.F.; Fregos López, M.G.; Páez Osuna, F. Monitoring of inland waters for culturing shrimp *Litopenaeus vannamei*: Application of a method based on survival and chemical composition. *Environ. Monit. Assess.* **2017**, *189*, 395. [CrossRef]
6. Sosa Villalobos, C.; Castaneda Chavez, M.; Amaro Espejo, I.; Galaviz Villa, I.; Lango Reynoso, F. Diagnosis of the current state of aquaculture production systems with regard to the environment in Mexico. *Lat. Am. J. Aquat. Res.* **2016**, *44*, 193–201. [CrossRef]
7. De Silva, S.S.; Nguyen, T.T.; Abery, N.W.; Amarasinghe, U.S. An evaluation of the role and impacts of alien finfish in Asian inland aquaculture. *Aquac. Res.* **2006**, *37*, 1–17. [CrossRef]
8. Valenzuela Quiñonez, W.; Rodrígues Quiroz, G.; Esparza Leal, H.M. Cultivo intensivo de camarón blanco *Litopenaeus vannamei* (boone) en agua de pozo de baja salinidad como alternativa acuícola para zonas de alta marginación. *Ximhai Rev. Soc. Cult. Desarro. Sustentable* **2010**, *6*, 1–8.
9. Summerfelt, R.C. Water Quality Considerations for Aquaculture. *Aquac. Netw. Inf. Cent.* **1998**, *50*, 11–21.
10. Valenzuela, M.; Suárez, J.; Sánchez, A.; Rosas, C. Cultivo de camarón blanco del golfo *Litopenaeus setiferus* en estanques de manto freático. In Proceedings of the II Congreso Nacional de la Asociación Mexicana de Limnología, Mexico City, DF, Mexico, 23–25 October 2002; pp. 1–9.
11. Gleeson, T.; Alley, W.M.; Allen, D.M.; Sophocleous, M.A.; Zhou, Y.; Taniguchi, M.; VanderSteen, J. Towards Sustainable Groundwater Use: Setting Long-Term Goals, Backcasting, and Managing Adaptively. *Ground Water* **2012**, *50*, 19–26. [CrossRef]
12. Chávez Sánchez, M.C.; Montoya Rodríguez, L. *Buenas Prácticas y Medidas de Bioseguridad en Granjas Camaronícolas*; Centro de Investigación en Alimentación y Desarrollo: Mazatlán, Sinaloa, Mexico, 2006; p. 95.
13. Jang, C.S.; Chen, C.F.; Liang, C.P.; Chen, J.S. Combining groundwater quality analysis and a numerical flow simulation for spatially establishing utilization strategies for groundwater and surface water in the Pingtung Plain. *J. Hydrol.* **2016**, *533*, 541–556. [CrossRef]
14. Cuéllar, J.; Lara, C.; Morales, V.; De Gracia, A.; García, O. *Manual de Buenas Prácticas de Manejo para el Cultivo del Camarón Blanco Penaeus Vannamei*; OIRSA-OSPESCA: Merida, Yucatán, Mexico, 2010; p. 132. ISBN 978-9962-661-05-4.
15. *Conagua, Manual de Agua Potable, Alcantarillado y Saneamiento. Prospección Geoeléctrica y Registros Geofisicos de Pozos*; Secretaría de Medio Ambiente y Recursos Naturales: Mexico City, DF, Mexico, 2007; p. 204. ISBN 978-968-817-880-5.
16. Muchingami, I.; Hlatywayo, D.J.; Nel, J.M.; Chuma, C. Electrical resistivity survey for groundwater investigations and shallow subsurface evaluation of the basaltic-greenstone formation of the urban Bulawayo aquifer. *Phys. Chem. Earth* **2012**, *50–52*, 44–51. [CrossRef]
17. Urish, D.W.; Frohlich, R.K. Surface electrical resistivity in coastal groundwater exploration. *Geoexploration* **1990**, *26*, 267–289. [CrossRef]
18. Frohlich, R.K.; Urish, D.W.; Fuller, J.; O'Reilly, M. Use of geoelectrical methods in groundwater pollution surveys in a coastal environment. *J. Appl. Geophys.* **1994**, *32*, 139–154. [CrossRef]
19. Worthington, F. Hydrogeophysical equivalence of water salinity, porosity and matrix conduction in arenaceous aquifers. *Ground Water* **1976**, *14*, 224–232. [CrossRef]
20. Ladrón de Guevara-Torres, M.L.; Peinado-Guevara, H.J.; Delgado-Rodríguez, O.; Shevnin, V.; Herrera-Barrientos, J.; Belmonte-Jiménez, S.I.; Peinado-Guevara, V.M. Geoelectrical and geochemical characterization of groundwater in a shallow coastal aquifer. *Polish J. Environ. Stud.* **2017**, *26*, 1511–1519. [CrossRef]
21. Peinado-Guevara, H.J.; Delgado-Rodríguez, O.; Herrera-Barrientos, J.; Ladrón de Guevara, M.D.L.A.; Peinado-Guevara, V.M.; Herrera-Barrientos, F.; Campista Leon, S. Determining water salinity in a shallow aquifer and its vulnerability to coastline erosion. *Polish J. Environ. Stud.* **2017**, *26*, 2001–2011. [CrossRef]
22. Orellana, E. *Prospección Geoeléctrica en Corriente Continua*; Paraninfo: Madrid, Spain, 1982; p. 578.
23. CNA. Lineamientos para el Otorgamiento de Concesiones o Asignaciones de Agua Subterránea salada Proveniente de Captaciones Ubicadas en la Proximidad del Litoral. D. Of. la Fed. 2017. Available online: https://www.dof.gob.mx/nota_detalle.php?codigo=5482204&fecha=11/05/2017 (accessed on 20 December 2020).
24. Peinado-guevara, H.J.; Herrera-barrientos, J.; Ladron-de-Guevara, M.A. Determinación de la conductividad hidráulica mediante medidas de resistividad eléctrica. *Ing. Hidraul. Mex.* **2009**, *24*, 123–134.

25. Conagua. *Actualización de la Disponibilidad Media Anual Agua en el Acuífero río Sinaloa (2502)*; Secretariat of the Interior: Mexico City, DF, Mexico, 2020; pp. 2–29.
26. INEGI. *Anuario Estadístico y Geográfico de Sinaloa*; Instituto Nacional Estadística Geografia: Aguascalientes City, Aguascalientes, Mexico, 2017; p. 475.
27. Van Wyk, P.; Scarpa, J. Water quality requirements and management in farming marine shrimp in recirculating freshwater systems. In *Farming Marine Shrimp in Recirculating Freshwater Systems*; Harbor Branch Oceanographic Institution, Florida Atlantic University: Boca Raton, FL, USA, 1999; pp. 141–161.
28. Li, E.; Chen, L.; Zeng, C.; Chen, X.; Yu, N.; Lai, Q.; Qin, J.G. Growth, body composition, respiration and ambient ammonia nitrogen tolerance of the juvenile white shrimp, *Litopenaeus vannamei*, at different salinities. *Aquaculture* **2007**, *265*, 385–390. [CrossRef]
29. Gross, A.; Abutbul, S.; Zilberg, D. Acute and Chronic Effects of Nitrite on White Shrimp, *Litopenaeus vannamei*, Cultured in Low-Salinity Brackish Water. *J. World Aquac. Soc.* **2004**, *35*, 7. [CrossRef]
30. Boyd, C.E. Trace Metals Toxic at High Concentrations. *Goblal Aquac. Adv.* **2009**, 24–26. Available online: https://www.aquaculturealliance.org/advocate/trace-metals-toxic-at-high-concentrations/?headlessPrint=AAAAAPIA9c8r7gs82oWZBA (accessed on 17 April 2019).
31. Boyd, C.E.; Thunjai, T. Concentrations of major ions in waters of inland shrimp farms in China, Ecuador, Thailand, and the United States. *J. World Aquac. Soc.* **2003**, *34*, 524–532. [CrossRef]
32. Daniel, W.W. *Bioestadística*; Limusa Wiley: Mexico City, DF, Mexico, 2002; pp. 335–337. ISBN 968-18-0178-4.
33. Zar, J.H. *Biostatistical Analysis*, 5th ed.; Pearson: London, UK, 2010; pp. 291–300. ISBN 978-0-13-100846-5.
34. Roy, L.A.; Davis, D.A.; Saoud, I.P.; Boyd, C.A.; Pine, H.J.; Boyd, C.E. Shrimp culture in inland low salinity waters. *Rev. Aquac.* **2010**, *2*, 191–208. [CrossRef]
35. Heath, R.C. *Basic Ground-Water Hydrology*; U.S. Geological Survey: Reston, VA, USA, 1983; p. 86. ISBN 0-607-68973-0.
36. Chidambaram, S.; Senthil Kumar, G.; Prasanna, M.V.; John Peter, A.; Ramanthan, A.; Srinivasamoorthy, K. A study on the hydrogeology and hydrogeochemistry of groundwater from different depths in a coastal aquifer: Annamalai Nagar, Tamilnadu, India. *Environ. Geol.* **2009**, *57*, 59–73. [CrossRef]
37. Peinado Guevara, H.; Green-Ruíz, C.; Herrera-Barrientos, J.; Escolero-Fuentes, O.; Delgado-Rodríguez, O.; Belmonte-Jiménez, S.; Ladrón de Guevara, M. Relationship between chloride concentration and electrical conductivity in groundwater and its estimation from vertical electrical soundings (VESs) in Guasave, Sinaloa, Mexico. *Cienc. e Investig. Agrar.* **2012**, *39*, 229–239. [CrossRef]
38. Shroff, P.; Vashi, R.T.; Champaneri, V.A.; Patel, K.K. Correlation study among water quality parameters of groundwater of valsad district of south gujarat (India). *J. Fundam. Appl. Sci.* **2015**, *7*, 340–349. [CrossRef]
39. Bhasin, S.K.; Singh, A.; Kaushal, J. Statistical Analysis of Water Falling Into Western Yamuna Canal From Yamunanagar Industrial Belt and Its Effect on Agriculture. *J. Environ. Res. Dev.* **2008**, *2*, 393–401.
40. Abyaneh, H.Z.; Nazemi, A.H.; Neyshabori, M.R.; Mohammadi, K.; Majzoobi, G.H. Chloride estimation in Ground water from electrical conductivity measurement. *Tarim Bilim. Derg.* **2005**, *11*, 110–114.
41. Saoud, I.P.; Davis, D.A.; Rouse, D.B. Suitability studies of inland well waters for *Litopenaeus vannamei* culture. *Aquaculture* **2003**, *217*, 373–383. [CrossRef]
42. Rodríguez Meza, D.; Sapozhnikov, D.; Vargas-Ramírez, C.; Vallejo Soto, A.; Verdugo-Quiñonez, G.; Michel-Rubio, A. Monitoreo de la Calidad del agua del Acuífero de Guasave, Sinaloa. 2008. Available online: http://sappi.ipn.mx/cgpi/archivos_anexo/20080587_6101.pdf (accessed on 10 January 2021).
43. Mastrocicco, M.; Giambastiani, B.M.S.; Colombani, N. Ammonium occurrence in a salinized lowland coastal aquifer (Ferrara, Italy). *Hydrol. Process.* **2013**, *27*, 3495–3501. [CrossRef]
44. Peinado-Guevara, H.J.; Green-Ruiz, C.R.; Herrera-Barrientos, J.; Escolero-Fuentes, O.A.; Delgado-Rodríguez, O.; Belmonte-Jiménez, S.I.; Ladrón de Guevara, M.Á. Quality and suitability of the agricultural and domestic water use of the Sinaloa river aquifer, coastal zone. *Hidrobiologica* **2011**, *21*, 63–76.
45. Archie, G.E. The electrical resistivity logs as an aid in determining some reservoir characteristics. *SPE-AIME Trans.* **1942**, *146*, 54–62. [CrossRef]
46. Llanes Cárdenas, O.; Norzagaray Campos, M.; Muñoz Sevilla, P.; Ruiz Guerrero, R. Agua subterránea: Alternativa acuícola en el Noroeste de Mexico. *Aquat. Rev. electrónica Acuic.* **2013**, *38*, 10–20.
47. Mariscal Lagarda, M.M.; Páez Osuna, F.; Esquer Méndez, J.L.; Guerrero Monroy, I.; del Vivar, A.R.; FélixGastelum, R. Integrated culture of white shrimp (*Litopenaeus vannamei*) and tomato (*Lycopersicon esculentum* Mill) with low salinity groundwater: Management and production. *Aquaculture* **2012**, *366–367*, 76–84. [CrossRef]
48. Baldi, F.; Rosas, J.; Velásquez, A.; Cabrea, T.; Maneiro, C. Aclimatación a baja salinidad de postlarvas del camarón marino *Litopenaeus vannamei* (Bonne, 1931) provenientes de dos criaderos comerciales. *Rev. Biol. Mar. Oceanogr.* **2005**, *40*, 105–109.
49. Mcgraw, W.J.; Davis, D.A.; Teichert-Coddington, D.B.; Rouse, D.B. Acclimation of *Litopenaeus vannamei* Postlarvae to Low Salinity: Influence of Age, Salinity Endpoint, and Rate of Salinity Reduction. *J. World Aquac. Soc.* **2002**, *33*, 78–84. [CrossRef]
50. Laramore, S.; Laramore, C.R.; Scarpa, J. Effect of Low Salinity on Growth and Survival of Postlarvae and Juvenile *Litopenaeus vannamei*. *J. World Aquac. Soc.* **2001**, *32*, 385–392. [CrossRef]

Journal of
Marine Science and Engineering

MDPI

Article

Application of the BIM Method in the Management of the Maintenance in Port Infrastructures

Paola Valdepeñas [1], María Dolores Esteban Pérez [1,*], Carlos Henche [1], Raúl Rodríguez-Escribano [1], Gonzalo Fernández [1] and José-Santos López-Gutiérrez [2]

1 Department of Civil Engineering, Universidad Europea de Madrid, 28670 Madrid, Spain; pao.vfuentes@gmail.com (P.V.); carlos.henche@ext.universidadeuropea.es (C.H.); raulruben.rodriguez@universidadeuropea.es (R.R.-E.); gonzalo.fernandez@ext.universidadeuropea.es (G.F.)

2 Grupo de Investigación de Medio Marino, Costero y Portuario, y Otras Áreas Sensibles, Universidad Politécnica de Madrid, 28040 Madrid, Spain; josesantos.lopez@upm.es

* Correspondence: mariadolores.esteban@universidadeuropea.es

Received: 10 November 2020; Accepted: 30 November 2020; Published: 2 December 2020

Abstract: This work represents a paradigm shift in the approach to manage port infrastructure maintenance where the use of the building information modeling (BIM) method is very scarce, and it is now beginning. The new infrastructure conservation management method proposed in this manuscript combines traditional methods with new technology, covering the gap of the implementation of BIM in port maintenance. This new technology is the BIM method, and its use is already mandatory in public infrastructure projects in some countries. In this work, the BIM method is applied to the management of the conservation of port infrastructures to show the benefits that can be obtained using it. After the analysis of the state of the art in the conservation of port infrastructures, two of the main maritime works of the Outer Port of A Coruña were modeled in Revit. This as-built model serves as an inventory model and forms a large database of all the information regarding the conservation of the structures. In this way, all the data on the conservation of the port's infrastructures are concentrated in a single source of information in which, in addition, there is a three-dimensional and georeferenced representation of the port in question. The main finding of this work is the development of a step by step method to implement BIM philosophy in port maintenance, highlighting the positive and negative points detected during the validation of the method in a real case, detecting future improvement research lines. Although BIM has a lot of advantages and possibilities just now, some improvements are still necessary.

Keywords: facility management; BIM; port infrastructure; maintenance and operation management

1. Introduction

Ports are considered as strategic infrastructures within the international market and, in most cases, they are presented as the main economic source of the cities to which they belong, increasing the potential of being part of shipping routes [1]. This fact highlights the importance of having a continuous service in the ports, which must always be operational. To do so, it is necessary to establish maintenance management systems appropriate to port requirements, as these facilities are exposed to numerous external agents that will hinder the correct state of conservation of the infrastructures that make up the port. Maintenance has more importance in a life cycle approach and, moreover, considering climate change [2,3]. Most port facilities are exposed to a very aggressive environment that makes them especially susceptible to the development of certain pathologies that will need to be repaired. A port's facilities represent a necessary link in the chain of port operations, and any disruption can lead to poor port efficiency. In general, each year, about 11% of the operating expenses of ports are allocated to the maintenance of the facilities [4]. Furthermore, in the cost life cycle, maintenance activities are around 25% of initial investment [5]. In other studies, the maintenance contribution is around 60% [6].

Materials used in the construction stage have a direct impact on the life cycle of ports in a sustainable and carbon footprint approach [7] but also on operational activities during conservation and the durability of the infrastructure [8–10].

In recent years, a digital transformation is being experienced in the field of civil engineering through the implementation of the building information modeling (BIM) method in construction projects, like in the case of bridges [11–20].

The BIM method allows for reducing costs, building in a more efficient way, and the possibility for all the agents involved in a project to carry out collaborative work. The advantages offered by this method applied to new construction projects are numerous, but not in the case of the maintenance phase. Therefore, it is necessary to implement BIM considering the entire project lifetime: design, construction, and operation.

Until now, the use of the BIM method in the management of conservation and maintenance has only been carried out in the building sector for the maintenance of facilities (e.g., [21]). However, in the infrastructure sector, it is an issue that today is it still very undeveloped, specifically in seaports, as already noted [22], with a critical review in relation to BIM and facility management in this kind of infrastructure. BIM and its application to port infrastructure in its life cycle (construction and maintenance) has not much literature, but there are some approaches related to BIM development for port facility operations. Some of these approaches [23] regard the application of BIM technology in the operation of port marine structures, Refs [24,25] are related to the use of BIM technology for the design of the operation of ports, Ref [26] show the application of BIM technology for the design of power pipes for liquid chemical berth, Ref [27] relate to electrical design using BIM technology, and Ref [28] show some software for the use of BIM in port infrastructures. All of them conclude with the lack of documentation of existing infrastructures, the great interconnections, and the continuous operation and effects on the infrastructure.

This research paper focuses on the development of a new conservation management method in which the BIM method is applied as the main tool to serve as a database for the maintenance of port infrastructures. The maintenance and operation manager systems linked with BIM is the near future for facility management in such infrastructures as ports, where the information and database are essential for the efficiency of the whole process [29].

The main objective of this research is the creation of a new port infrastructure conservation management method that allows for combining the traditional method used so far with new technologies, specifically, the BIM method. This is very important in the current moment, when BIM is starting to be a reality, so it is essential to develop new practical methods to be able to use it based on the previous knowledge. The use of BIM for port maintenance has several advantages, one of them being the control of all the generated information in an ordered way. For most public infrastructures, the use of BIM in developed countries is compulsory, so it is a complete need.

To integrate traditional and BIM methods, the maintenance protocols based on the "Standard Practice for Underwater Investigations" have been considered [30–32]. Following the literature related to asset management, facility management, and BIM (e.g., [33]), the problem and main limitation in this kind of study is not the lack of information, but just the opposite (a large collection of data), so to propose processes and protocols will be necessary for achieving success in such a purpose.

Some aspects and steps to be considered to achieve the main objective are: to know how the management of the conservation of port infrastructures is currently carried out and to study possible improvements through the application of the BIM method in that process; to create a three-dimensional (3D) BIM model of a port infrastructure and expose its optimal capacity to extract information at any time during its life cycle; to give continuity to the application of the BIM Guide developed by Puertos del Estado and published in July 2019, it being the objective of the research project exposed in this paper to expand certain aspects of the maintenance and conservation management; to check the feasibility and effectiveness of the implementation of the BIM method in the management of structure maintenance, applying it to a practical case to exemplify the process; to study up to what stage of maintenance management the use of the BIM method is viable; to establish the procedures and requirements necessary for the application of the new method for the management of the maintenance of a port; to present the benefits that can be obtained from the use of BIM in maintenance management and that compensate for the great effort required to integrate the traditional method in the BIM model.

The structure of the paper is: (1) introduction of the motivation, the objectives, and the structure of the manuscript; (2) methods used in the research, including some details about the software applied and the port selected as the case study; (3) results of the definition of the proposal created in this research for the application of the BIM method to the maintenance of port infrastructures; (4) discussion of the application of the proposal exposed in the results section to the case study to validate its feasibility; (5) a conclusion, highlighting the key aspects of the research.

The significance of the paper is key in the current state of BIM development for port infrastructures. Although there are some general studies related to the use of BIM in ports, there is not a clear method for BIM use in port maintenance. This manuscript shows a new proposed method to implement BIM in the maintenance phase of ports, and its validation with a real case.

2. Research Methods

The method starts with the analysis of the existing documentation regarding how the management of the conservation of infrastructures in port environments is currently carried out, including not only the assets management method but also the main failure to be detected in main port infrastructures, like breakwaters. The aim of that analysis is achieving a technical and global vision of the fundamental aspects that will be addressed in the research. Furthermore, the state of the art of BIM technology applied to port infrastructures and even to other civil infrastructures has been reviewed. The knowledge of BIM and associated software has been essential in this study.

Based on the knowledge of how the maintenance of ports is carried out up to now, and the advantages of the BIM method for maintenance operations, a new method for assets management in ports has been created using BIM philosophy. The main steps to follow and the necessary requirements for the application of this new method have been established.

To validate the proposed method, two port infrastructure models were developed in the Revit Autodesk application, including the data referring to the maintenance and conservation of the structures, managing to create a new work method for the improvement of the management of the conservation of port infrastructures.

2.1. Building Information Modeling (BIM)

Every project prepared using the BIM method must have a document known as the BIM execution plan (BEP). A BEP is a tool that provides a standardized workflow and consists of a general guide for the implementation of BIM in a project or group of projects [34]. The BEP document should provide an overview of the project, in addition to defining the uses of BIM in detail. It serves as a record of agreements between the interested parties about their functions, responsibilities, and the way in which communications and information exchanges should be established.

The BEP must contain at least the following information [35]: general information about the project and the scope and objectives of the project, BIM uses and processes, organization of the models (coordinates, development levels, element classification system, BIM software), common data environment (CDE), quality control (QA), deliverables, and reference documents and standards.

To manage the information and the exchange of data and documents in an organized way, it is essential to establish a common data environment (CDE). A CDE consists of a computer tool that is used to manage, archive, and share project data and documents (regardless of their size) among the different multidisciplinary teams that make it up. In addition, it allows the process to be auditable, transparent, and controllable [36].

In this way, it is ensured that the information is generated only once and can be used by all the agents involved in the project whenever necessary. This tool should allow: (a) collecting all the information on the project, both documents and models, as well as communications between stakeholders, (b) managing access, that is, not all the agents involved in the project can access all the information; (c) sharing information in the form of links; (d) controlling the versions of documents; (e) viewing files and models in the cloud; (f) having tools for planning the BIM project.

The BEP must establish the CDE to be used, as well as the organization of the directories in which each type of information and documentation will be published, the format in which the files must be published and the standard for their naming. In this case, it was decided to use Google Drive. This application allows online storage, as well as its installation on the computer itself, synchronizing local files automatically.

To take full advantage of the benefits offered by having a common data environment, it is also necessary to establish which will be the standard formats recognizable by any BIM tool to be used in the project. Generally, BIM models will be exchanged in industry foundation classes (IFC) format. It is a common standard that enables data exchange and ensures that information can be shared regardless of the application being used and its size. This format allows for sharing the tangible elements of the modeled structures (graphic information) and all those parameters that are associated with the elements that form it (metadata). The main objective is to facilitate the interoperability and collaborative work within the construction sector.

It is important to highlight the most important current disadvantages of BIM: resistance to change, lack of proven method in most civil infrastructures, high implementation cost and need for software training, and problems of software interoperability. In fact, it is essential to constantly review the state of the art of the BIM method since it is a work method that continues to be updated and renewed day by day. So, this research has some limitations due to the current state of the art, and it is certain the application of BIM is going to be improved in the future because it is a prevailing topic not only in R&D projects but also in the implementation in public and private companies.

2.2. Software

To carry out the model, the Revit Autodesk tool (Version 2019, US) was chosen. This is one of the current existing pieces of software, and it was selected for this research because it is one of the most mature ones. There are other options for using other software programs, but the method is the same as the one used in this research. This choice is because Revit meets the characteristics required of a BIM method program [37], explained in the following paragraphs.

The project carried out with the BIM method must have a single model as the source of all the information regarding the project, giving accessibility to all interested agents to said information. Since all the information is stored in a single database, it is easy to consult or modify data at any time during the useful life of the work. In this particular project, a 3D model was developed in Revit that will basically be a large database that will serve as an "inventory model" and will contain the information of the inspections, graphic documentation and all the history of repairs carried out at the facilities.

This type of design consists of the modeling of a geometry based on a series of parameters and the correlations between them. Thanks to this parameterization of the model, when any type of modification is made to the model, the model is automatically updated in all the project views linked to these parameters, avoiding possible geometric errors or manual changes of different data. In addition, the models developed with parameterized elements are capable of quantifying non-formal parameters, such as measurements of materials, volumes, surfaces, energy consumption, etc.

Collaborative work is one of the greatest advantages of the BIM method in general, and it is partly achieved thanks to the interoperability of the different programs and applications used, guaranteeing communication and information exchange between them. Nowadays, when carrying out a project, one works with different types of programs, calculation, graphics, planning, etc. This fact makes necessary the existence of a universal format that allows communication between all platforms. This format is the industry foundation classes (IFC) format. Thanks to it, all BIM applications can read the information from the files and exchange it between different software models.

It is also important to understand how information is managed in Revit. At a graphical level, the model data can be managed by: (a) the organization of the project browser itself; (b) the creation of new views; (c) through display filters. However, the application has another very useful information management tool, known as planning tables. These tables consist of a tabulated presentation of the information relating the elements of the model to their properties and parameters. Following the general operation of the program, the tables will be automatically updated with the changes introduced in the model [38].

This software also has some disadvantages, and it is certain that it is going to improve in the future. One of the limitations that can be found in this application is that it is intended mainly for building projects and, therefore, in civil works projects, certain additional processes and the use of BIM objects are required to be able to perform a correct modeling as true to reality as possible. The new elements to be incorporated are incorporated into the model from some components called "families". Each family contains the graphic and non-graphic information of the element. Today, there are already many families created by different organizations, which facilitate the modeling and design process.

It is important to bear in mind that, like many other applications, Revit cannot open files created in higher versions so the entire workflow must be carried out with a pre-established version from the beginning, and this information must be reflected in the BEP "Building Execution Plan". Finally, it should be noted that a tool, in this case Revit, cannot create a BIM as a product by itself. Do not lose sight of the fact that BIM is a method, not software. In other words, this method requires managing various workflows, not just creating a model.

On the other hand, it is important to emphasize that the use of the Revit program in this process does not avoid the need to continue using CMMS (computerized maintenance management system), asset management software. In this new method, a model is developed that will serve as a database, but the handling of this data must be carried out, for the moment, with the software used to date for this purpose.

3. Case Study

The scope of this practical case is to put into practice and validate the new BIM application method in the management of port infrastructure maintenance. For this, the Outer Port of A Coruña "Langosteira" (Galicia) was selected to model a 180 m section of the main breakwater and another 85.5 m section of the spur breakwater. Both sections were modeled with the data included in the constructive project. The hypothesis was made that the modeled infrastructures present a series of pathologies with the aim of achieving the objectives of this research. These pathologies do not actually exist in the infrastructures of the Outer Port of A Coruña, and their incorporation into the model is intended to validate the BIM method created in this document.

To carry out the practical case, the Outer Port of A Coruña, located in Punta Langosteira (Galicia), was selected. The reasons for this choice are based on the fact that, in addition to being one of the ports on the Spanish coast that registers the highest maximum wave heights in the nearby buoys (with the repercussions this has on the maintenance of the infrastructures), it is a port which houses one of the largest maritime engineering works carried out in Spain: the main breakwater of 3.360 m in length.

The Outer Port of A Coruña arose from the need to have an alternative Galician port that could accommodate ships in emergency situations, due to the disasters that had taken place in the Inner Port (port with a very limited space and that is very close to the urban center). The new Punta Langosteira project was proposed at the Council of Ministers held in A Coruña in January 2003, after the Prestige oil slick in 2002, presenting itself as a leading project for the economic development of the area. Finally, the works began in April 2005 with an execution period of six and a half years [39].

The port is in the municipality of Arteixo, about 7 km southwest of the center of A Coruña, between Punta Langosteira and Punta do Pelón [40]. Currently, after several extensions, the port is sheltered by the following works (Figure 1) [41]: (a) a main breakwater 3360 m long in three alignments, protected with 150 t concrete blocks and a crown wall at an elevation of +25 m. It is one of the largest breakwaters in the world; (b) a 1350 m long breakwater, in two alignments, protected by 15 t and 20 t cubipod concrete pieces; (c) a 391 m long spur breakwater with a vertical section, which also serves as a berth for oil tankers on the inside part.

Figure 1. Outer Port of A Coruña "Langosteira" (Google Earth).

It should be noted that the port and especially the main breakwater are subject to strong waves and storms due to their location. These waves usually cause overtopping in the shelter area. The Langosteira Point buoy, in 2018, registered a wave of 19.23 m in height.

4. New Proposed Method

Figure 2 shows the main steps that have been followed for the application of BIM in the management of infrastructure maintenance, differentiating between design, construction, and operation phases. This method proposal has been essential for this research, being the foundation of the study. For all the phases, it is essential to create a database from the design phase to be used during construction and operation phases. It must be considered that some of the defined phases already existed without the application of the BIM method, so the workflow will remain almost the same, but will be developed in a different way due to the different philosophies of both methods. The application of this method will result in a single 3D model of the infrastructure in which all the information necessary to carry out the conservation management of the infrastructure will be collected.

Table 1 describes in a simple way the main advantages and disadvantages of the BIM method in comparison to traditional methods, so decision makers can have the all the information to use one method or other.

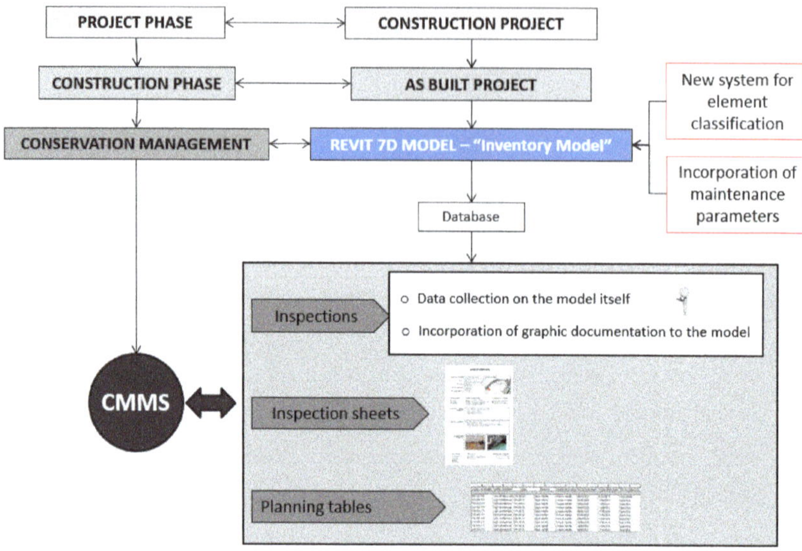

Figure 2. Proposal for the application of building information modeling (BIM) in the maintenance of infrastructures.

Table 1. Advantages and disadvantages of BIM method in comparison to traditional ones.

Advantages	Disadvantages
Collaborative work	New method: resistance to change
Information for all phases: design, construction, and operation	Lack of proven method in most civil infrastructures
One change in one document is directly implemented in all project documents	High implementation cost and need for software training
Easy detection of mistakes	Problems of software interoperability

Next, the phases to follow in the application of this new conservation management method in port infrastructures are briefly explained, this proposal being key in the future development of BIM for the operation and maintenance of port infrastructures:

- Phase 1: Construction project. Modeling in Revit of the construction project (project phase).
- Phase 2: "As-built" project. Updating of the construction project with the data of the project built in the construction phase. In this step, the "as-built" model is generated, which acts as an "inventory model" and is the point from which to start working in the infrastructure exploitation phase.
- Phase 3: Incorporation of new maintenance parameters into the model. In this phase, a series of parameters associated with each of the inventoried elements are incorporated into the model. These parameters basically consist of all the data that are recorded for each element in a conventional maintenance inspection.
- Phase 4: Inspections. One of the objectives that this new method aims to achieve is to be able to take data from inspections directly from the Revit model. Once in the port, the 3D model could be accessed from a tablet, and the data collected directly on the model, filling in all the parameters entered in Phase 3, in the same way as a conventional inspection sheet would be filled out. Graphical information could even be collected, taking photographs of the defects found, and associating them directly with the corresponding model element.

- Phase 5: Generation of inspection files. In this phase, the inspection sheets in PDF format are generated and entered into the model. It is proposed that these inspection sheets be incorporated into the Revit model as drawings, with a specific numbering.
- Phase 6: Planning tables. This phase consists of obtaining data from the model. To do this, schedule tables must be generated. The planning tables allow professionals to view in an orderly way, through filters, all the information required from the model.
- Phase 7: Data extraction from the model for CMMS. From the generation of the maintenance planning tables, the necessary information is extracted for its incorporation into the specific maintenance management programs currently used.

Next, some details of the previously described phases are given. The generation of the "as-built" model must begin with the search for the cartographic data of the environment in which the project is located (orthophotos, topography, bathymetry, etc.), to contextualize the infrastructure and be able to carry out a correct georeferencing of all the elements. With the cartographic information, an identical or very similar topo surface to the existing one is generated in Revit. There are different possibilities to model the existing topography: (a) import from other applications, such as ArcGIS; (b) from contour lines in DWG (DraWinG) format; (c) manual modeling using coordinate points; (d) from CAD to Earth using a surface taken from Google Earth; (e) from other data sources such as the National Geographic Institute or port authorities.

Once the topography has been modeled, the modeling process can be done in two different ways. The first one is to model it in two-dimensional projections. This type of modeling consists of creating a 3D model from 2D plans of the infrastructures (floors, elevations, sections, etc.), generally existing DWG plans. In this case, the model is created by incorporating these planes in Revit, which will serve as a drawing template. Each template is placed in the view and at the corresponding elevation (which will facilitate the georeferencing of the structure), filtering the drawing to display only the necessary information from the 2D plan and enabling a cleaner job. Once the templates have been entered, three-dimensional modeling is performed by extrusions of the two-dimensional drawing and by using BIM objects and families to achieve a more detailed model.

The second way is to model from three-dimensional data. This is modeling from point clouds or meshes that can be obtained using the devices described below in a very simplified way: (a) topographic devices, carrying out the survey of point clouds by means of a total station that carries out measurements from a fixed point. It is a simple job, but expensive in terms of time and budget; (b) laser scanner, a device whose operation is like that of a total station, but it moves automatically, collecting millions of points in each data collection, but the cost of the device is high; (c) photogrammetry, the operation of this technique being based on taking pictures of the element to be modeled, to create point clouds from them; to use this system, specialized personnel in photography are required, but they can obtain very good results.

It is important to bear in mind that the construction project model must be updated to obtain the model of the project actually executed ("as-built" model), which is the one with which it will be possible to carry out the management of infrastructure conservation. This model will serve as an "inventory model" for the management of port maintenance. For this, all the elements of the model must contain, in the form of parameters, a series of properties that allow for ordering and classifying all the relevant data. These properties must be at least: the name of the element, the identification code of the element, and the assembly code according to the Puertos del Estado Element Classification System.

A fundamental element in the information management of the models made with the BIM method is the element classification system. Revit elements have a built-in parameter known as the assembly code. This code consists of a method to organize the information of the model, based on functional elements of the model or parts that make up the infrastructure [42].

This classification system will be applied to the model to group the information of each of the elements in an organized and logical way. Currently, there are different systems for classifying elements. At the national Spanish level, the GuBIMClass Classification System is available. This system came to light in 2017 and was designed for application in the field of construction (equipment, infrastructures, and facilities) in Spain. It is a living document that is currently working on increasing the number of assembly codes [43]. At an international level, the most used element classification systems are Omniclaas, Uniclass 2015, and Uniformat.

However, none of them has a sufficient degree of development to classify the entire set of elements that may come into being in a civil work, much less include the elements that make up port and maritime works.

The BIM Guide of Puertos del Estado includes, in its Annex 2 "Element Classification System", a proposal for the alphanumeric classification of elements of port infrastructure [44]. What is proposed in this research is to carry out an extension of this classification system of elements of the BIM Guide of Puertos del Estado. In this extension, the new elements of this type of maintenance management model include "pathologies". The main and most representative pathologies that can occur in this type of infrastructure were analyzed in the first phases of this research.

These new parameters have been classified alphanumerically depending on the type of work or element to which they belong, following the coding established by Puertos del Estado. Table 2 shows the proposal to expand the classification system of the State Ports BIM Guide.

Table 2. Proposal to expand the Puertos del Estado Classification System.

Code	Description
PT.150	Pathologies
PT.150.010	Rubble mound breakwater
PT.150.010.010	Loss of armor units
PT.150.010.020	Armor units break
PT.150.010.030	Loss of filter material
PT.150.010.040	Toe destabilization
PT.150.010.050	Crown wall movement
PT.150.010.060	Cracking of the crown wall
PT.150.010.070	Slab cracking
PT.150.010.080	Seabed scour
PT.150.010.090	Settlements
PT.150.020	Vertical breakwater
PT.150.020.010	Toe destabilization
PT.150.020.020	Crown wall movement
PT.150.020.030	Cracking of the crown wall
PT.150.020.040	Slab cracking
PT.150.020.050	Caisson racking
PT.150.020.060	Seabed scour
PT.150.020.070	Corrosion of the caisson reinforcement
PT.150.020.080	Filler outlet through joints
PT.150.030	Pile Quay
PT.150.030.010	Corrosion of pile reinforcement
PT.150.030.020	Corrosion of upper slab reinforcement
PT.150.030.030	Backfill leak
PT.150.030.040	Seabed scour
PT.150.030.050	Sinking
PT.150.040	Block Quay
PT.150.040.010	Toe loss
PT.150.040.020	Seabed scour
PT.150.040.030	Filler outlet through joints
PT.150.040.040	Tilting or displacement

As recommended in the BIM Guide of Puertos del Estado, for all those elements that are not included in the classification or that are not exclusive to maritime work, the use of the GuBIMClass Classification System is recommended. Likewise, all these factors must be specified in the BEP.

Once the model has been generated, it is necessary to incorporate a series of parameters into the model to include the general information of the inventories. These parameters must be included in the BEP and added to each corresponding element of the model. The parameters are one of the most important parts of Revit objects, since they are the ones that contain all the information of each of the elements. The parameters assigned to each element can be seen in the element's properties tab. In this case, a series of parameters will be assigned to the elements of the model, to contain the information regarding their state of conservation. For this, the elements that are defined are, basically, those corresponding to the data to be filled in in a conventional maintenance inspection sheet. The choice of these parameters was made based on the recommendations established in the guide "Conservation Management in the Port Environment" [31,32] and in the "Standard Practice Manual for Underwater Investigations" [30], being the minimum amount of data that an inspection sheet must contain.

Next, each of the parameters included in the model are defined: (a) name of the element and associated identification code; (b) situation, with the indication of the relevant infrastructure to which it belongs (for example: spur breakwater, dock, etc.); (c) localization map, with the image that allows visualization of the approximate location of the element in a sketch of the port; (d) inspection sheet, with the image of the inspection sheet filled in with all the data; (e) date on which the item's inspection is being performed; (f) date the item was last inspected; (g) indication of the type of inspection to be carried out. The types of inspections are defined based on the provisions of the "Standard Practice Manual for Underwater Investigations" [30]; (h) the work team that will carry out the inspection; (i) the type of conservation work to be carried out in the port, which can be of three types; (j) the detailed description of the work to be carried out during the inspection and the aspects to which special importance will be given; (k) detailed description of the status of the item and the anomalies that occur in it; (l) photographs showing the state of the elements or the defects detected in the inspection; (m) state level represents the situation of a certain element at a given moment and is based on parameters that must be previously defined in the port's conservation management system (it aims to give an idea of the health status of the elements of a port through five predefined levels: very good, good, average, poor, ruined); (n) the service level, a coefficient that serves to measure the attention given to the user, as well as the way in which those responsible for port maintenance react, in terms of speed and efficiency, to the different problems that may occur in the installation (three levels can be established for its definition: unrestricted use, use with some restriction, and out of use); (o) image of the inspection file filled in with all the previously mentioned data; (p) specific surveillance service and scheduled routine work equipment. This is the team that carries out routine inspections at the foreman level; (q) the team of scheduled inspections (team made up of specialized personnel who are normally part of the port maintenance team in question); (r) specific inspections team (in the event of exceptional circumstances that require resorting to a specific inspection. Usually it is a team of outsourced personnel).

According to the type of conservation work, the following can be found: (a) help with port operations (all those activities necessary for the port to function at the established service levels); (b) ordinary conservation (all those conservation works designed to extend the useful life of the elements of the port); (c) extraordinary conservation (all those activities necessary to carry out the repair of an element of the installation).

Thanks to new technologies, the model can be viewed on portable devices and tablets and can even use virtual and augmented reality. One of the objectives of this new method is for the data collection in the inspections to be carried out on the model itself and not on paper, as has been done to date, saving time and resources. The idea is that the inspections are carried out with a tablet from which the edition of the Revit model can be accessed. The use of tablets to manage the Revit application can be somewhat uncomfortable when working with the graphical interface of the software to model or design. However, since the objective is to introduce alphanumeric data to the parameters of the selected elements, this can be considered a very advantageous solution.

Another possible option is to access the Revit model by remotely connecting the tablet to a computer that contains the program, making it possible for the remote computer to run the program and not the tablet itself. In this way, once in the port, the operator will take the data directly on the Revit model, filling in all the parameters mentioned in the previous section, in the same way that the inspection form would be filled in in the conventional way. With this, the data are recorded at that moment and associated with each element to be inspected. In addition, another advantage of this method is the ease of collecting graphic information on defects or breakdowns that appear in the structures, since with a tablet, workers can take photographs of them and link them directly to the element in question in the model.

Once the data are collected in the inspection, all the elements inspected will have associated the necessary information to be able to fill in the inspection sheets. All these files will be included in the model, generating a database of all the inspections carried out at the facility. To do this, it is proposed to create new plans, with a specific name, which include an image of the item's inspection sheet, as well as a view of the model in which the pathology or pathology-associated defect appears.

The defects and breakdowns that appear in the infrastructures can be represented in the model, to have a global vision of the state of the installation. With this representation, the inspection work is facilitated in the next actions, being able to visualize the pathologies prior to the inspection and knowing what its development has been. This allows for reducing redundancies of information between one inspection and another. The most comfortable way to view all the pathologies added to the model is to create new specific views of them. These views must be named in a certain way to differentiate them from other views in the model. Considering that the program, by default, orders the views in alphabetical order, the following name is proposed: "PATHOLOGY_Identifying code of the element_Type of pathology"; for example: PATHOLOGY_LDM-08-20_Cracking slab.

Each inspection file will be included in the model as an image within a plan, with the particularity that these will not be used as plans per se, but as a tool to archive the inspection files within the model. The reason for having to include the inspection cards as images is that Revit, in its 2019 version, does not support the import of PDF files. In the same way as with the views, the plans must be numbered in a specific way, to have control of them and facilitate the data search process. The following is proposed: "FI_Identifying code of the element_Date of inspection"; for example: FI_LDM-08-20_06082020. In this plane, on the left side, the image of the inspection sheet will be inserted, and on the right side, the view of the previously generated pathology. The inspection type sheet used in the application of the method is included in Figure 3.

At this point, it will have a 7D "as-built" model with a large amount of information, which will allow professionals to carry out the conservation management of the port by extracting and processing the data that are always necessary. For this, they have the planning tables. The planning tables can be considered one more view of the model and as such, they allow for visualizing and extracting all the data that are necessary, for later export to the asset management software used today.

It is important to bear in mind that the information that is modified in the planning tables is automatically updated in the rest of the model and vice versa. Therefore, the tables will also serve to control and edit the model. Schedules are created from the elements of the model and their associated parameters. In this case, the parameters of interest are those related to the maintenance and upkeep of the infrastructures.

Therefore, "maintenance planning tables" are to be obtained. To do this, within Revit, from the View tab, in the Create group, the schedule tables appear. From all the masses of the model, the parameters corresponding to maintenance will be selected, as well as the parameter "Assembly code", which will be the one that will allow us to filter the masses, to keep only the information that is needed.

INSPECTION SHEET

ELEMENT NAME [ ]

CODE [ ]

LOCATION [ ]

Inspection date [ ]

Last inspection date [ ]

Location map

Inspection type:
- ❑ Routine
- ❑ Scheduled
- ❑ Specific

Work team:
- ❑ Specific surveillance service
- ❑ Routine scheduled work team
- ❑ Scheduled Inspections team
- ❑ Specific inspections team

Conservation work type:
- ❑ Support to port operation
- ❑ Ordinary conservation
- ❑ Extraordinary conservation

DESCRIPTION OF THE WORK TO BE DONE

OBSERVED ASPECTS

GRAPHIC DOCUMENTATION

State level
- ❑ Very Good, V
- ❑ Good, G
- ❑ Average, A
- ❑ Poor, P
- ❑ Ruinous, R

Service level:
- ❑ Unrestricted use, 1
- ❑ Use with some restriction, 2
- ❑ Out of use, 3

Estimate cost of the repair: [€]

Figure 3. Inspection file.

In addition to these, more specific parameters can also be added, such as gross volume of items or area measurements. These data can be exported to a database directly coordinated with a CMMS program. In case of the modification of the model data, the results and the analysis are automatically updated.

5. Discussion

5.1. Examples of Typical Pathologies

The detail of the port sections selected for modeling in Revit (Version 2019) were entered. The works chosen were the main breakwater (rubble mound section) and the spur breakwater (vertical section), as they are considered to represent the two main and more usual types of breakwater executed in all maritime works, in addition to being those for which data were available for modeling.

A 180 m section of the main breakwater (Section type "C") was modeled in Revit 2019. Figure 4 shows the plan of the main breakwater (Section type "C").

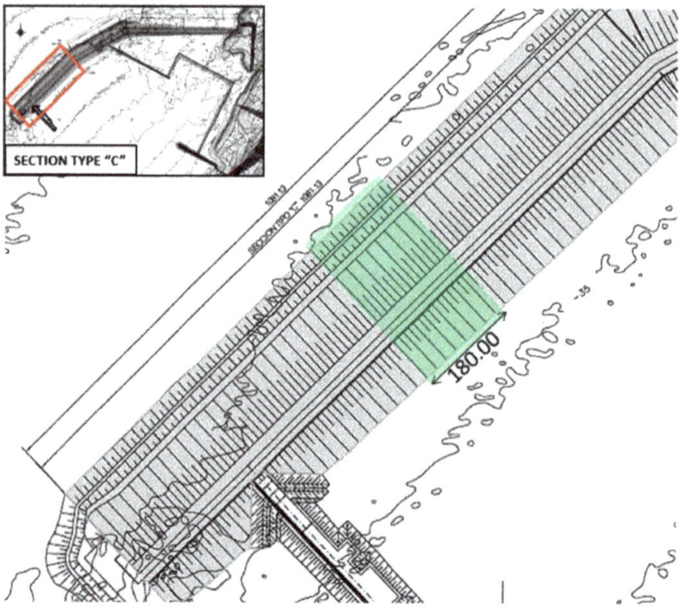

Figure 4. Plan of Section C of the main breakwater of the Outer Port of A Coruña (Langosteira).

The modeling was carried out by incorporating an Autocad template of the section of the main breakwater in the Revit program. The elements were created by means of masses in situ and the 3D section was created by means of extrusions of these elements. Likewise, with the aim of creating a more elaborate model that is more faithful to reality, new families of elements were created to represent the 15 t, 50 t, and 150 t concrete blocks that are part of the breakwater's armor. All elements of the model were classified with the Puertos del Estado Element Classification System. The result of the modeling in Revit can be seen in Figure 5.

In the same way, an 85.5 m section of the spur breakwater was modeled in Revit 2019. Figure 6 shows the hammer plan, and Figure 7 shows the Revit model.

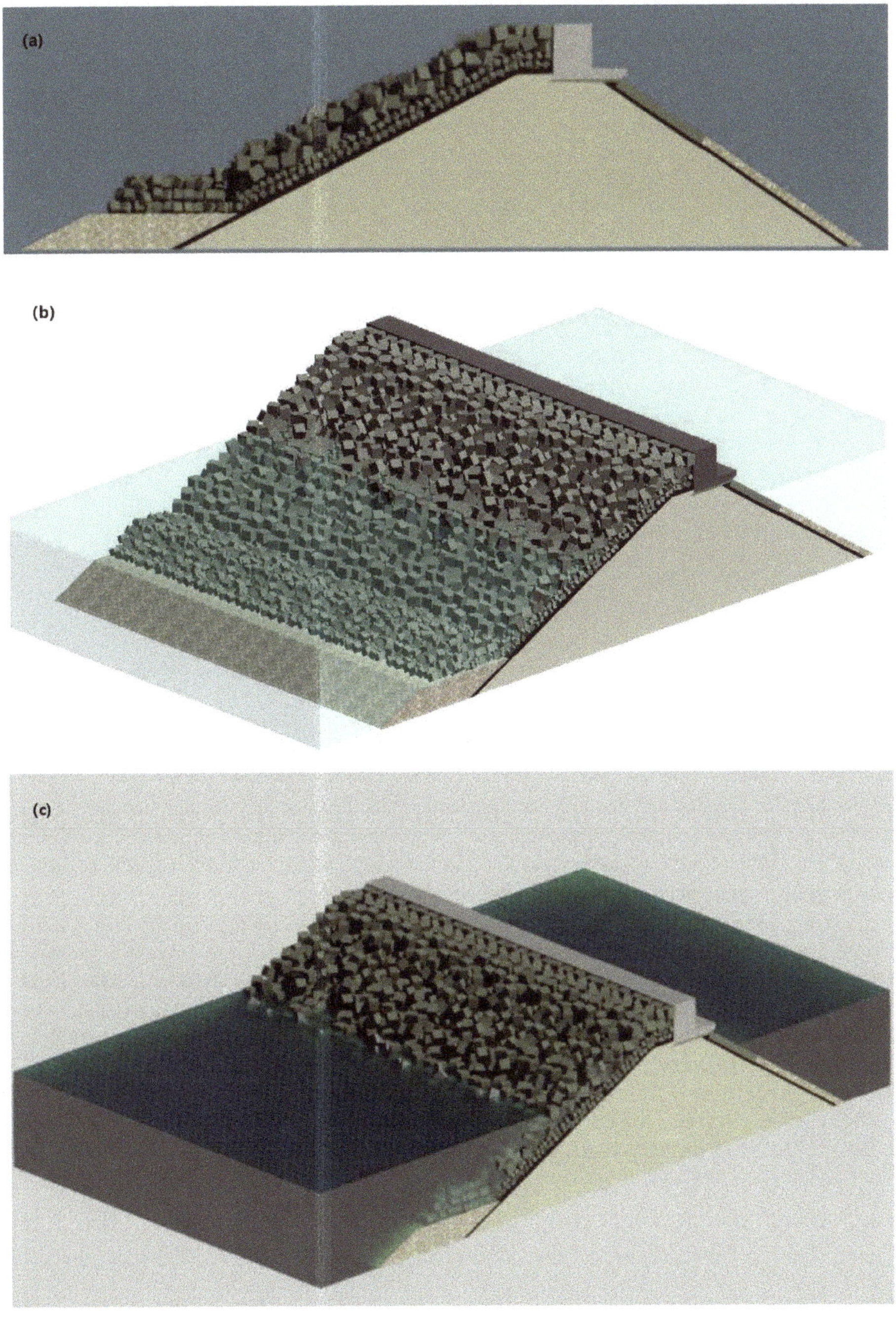

Figure 5. Revit model of the main breakwater of the Outer Port of A Coruña (Langosteira); (**a**) front view; (**b**) diagonal view with transparent sea level; (**c**) diagonal view with non-transparent sea level.

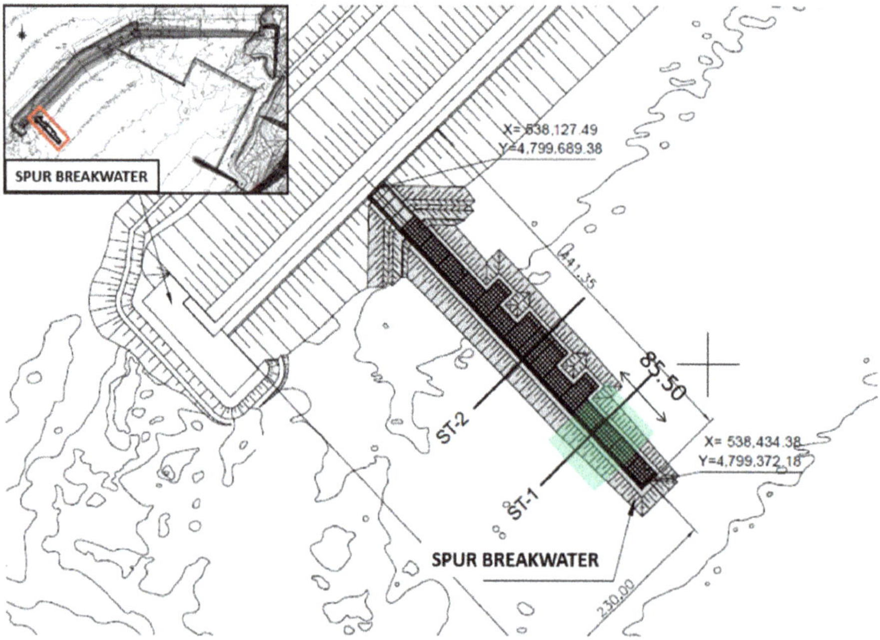

Figure 6. Plan of the spur breakwater of the Outer Port of A Coruña (Langosteira).

One of the main phases of this new method is the incorporation of a series of parameters, referring to the maintenance and upkeep of the infrastructures, into the elements of the model. All the established parameters were included in the model, in such a way that, when selecting any element of the infrastructure, these parameters and their associated data can be seen. With this step, the "as-built" model is converted into a "facility management", which includes all the information regarding the management of infrastructure conservation. The information regarding each parameter can be included in the model from the port itself during inspection work.

To carry out this practical case, the hypothesis was made that the infrastructures have a series of defects or pathologies (not real) in certain elements thereof, and the information on the parameters referring to the maintenance of said elements has been filled in, as if filling out an inspection sheet.

Once the inspection data have been collected, the detected defects can be modeled as one more element. These new elements that represent the defects of the structures must be classified as pathologies in the element classification system, assigning them an assembly code from 150 onwards. The pathologies represented in the models are shown below as an example.

Breaking of 150 t concrete blocks of the armor of the main breakwater is shown in Figure 8. Assembly code: PT.150.010.020 Armor units breakage.

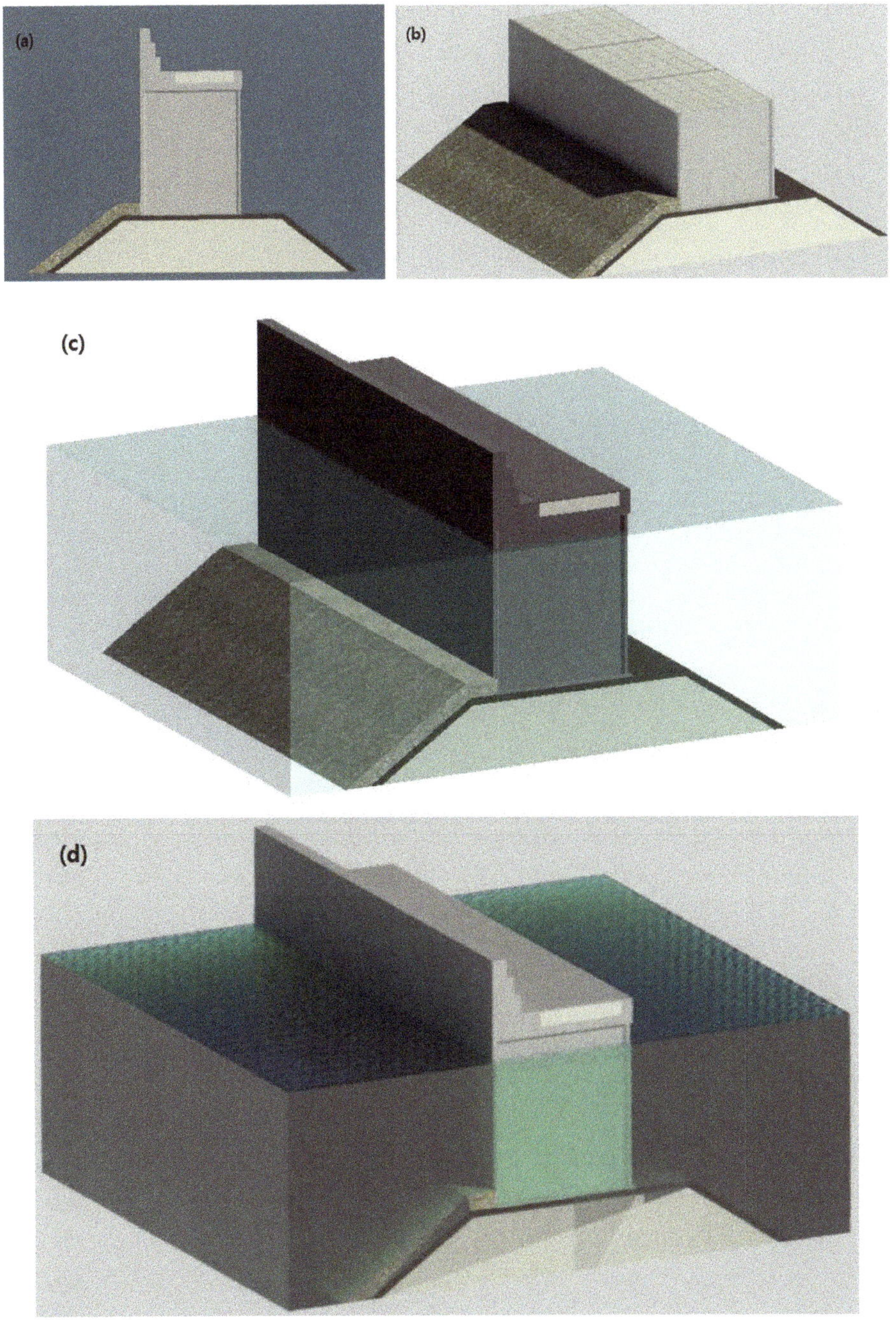

Figure 7. Revit model of the spur breakwater of the Outer Port of A Coruña (Langosteira): (**a**) front view; (**b**) diagonal view without sea level; (**c**) diagonal view with transparent sea level; (**d**) diagonal view with non-transparent sea level.

Figure 8. Breaking of 150 t concrete blocks of the armor of the main breakwater.

The presence of cracks and corrosion of the armor of the reinforcement of the caissons that make up the spur breakwater is shown in Figure 9. Assembly code: PT.150.020.070 Corrosion of the reinforcement of the caissons.

Figure 9. Presence of cracks and corrosion of the armor of the reinforcement of the caisson that makes up the spur breakwater.

Cracking of the spur breakwater slab is shown in Figure 10. Assembly code: PT.150.020.040 Cracking of the slab.

Figure 10. Cracking of the spur breakwater slab.

With all the information included in the model so far, the inspection sheets can be generated in PDF format for filing in the model. The inspection files must be imported into the model as an image (.jpg) since Revit 2019 does not allow the import of PDF documents. Their filing in the Revit model is done by generating planes, named in a certain way to differentiate them from the rest of the model's planes:

"FI_Identifying code of the element_Date of inspection"; for example: FI_LDM-08-20_06082020.

On the other hand, specific views of the pathologies modeled in the infrastructures were generated. These views, whose names are as follows, are also included in the plan: "PATHOLOGY_Identifying code of the element_Type of pathology"; for example: PATHOLOGY_LDM-08-20_Cracking slab.

At this point, it has been possible to generate a model with a significant amount of information and that serves as a database for the conservation and maintenance of infrastructures. What is intended at this point is to extract said information, and then use it as appropriate. Generally, the information will be exported to a database to be managed with the CMMS software. To extract the information from the model, the maintenance planning tables were generated.

To generate the tables, all the masses of the model were selected with their maintenance parameters and their assembly code. The program generated a table that contained all the masses that make up the structure of the model, therefore, by applying a filter, only those masses that represent the pathologies are selected, that is, the masses that have an assembly code numbering 150 or higher, leaving only the elements affected by some pathology in the table. The maintenance planning tables extracted from both models and an example of their interpretation are included below.

Main breakwater maintenance planning table: the armor of the main breakwater is in an average state, requiring the replacement of 17 pieces of the armor in the 180 m studied. This would involve, approximately, the removal of a total of 623 m^3 of concrete from the exterior armor, for replacement.

Spur breakwater maintenance planning table: the spur breakwater is in an average state, requiring the repair of an approximate area of 430 m^2, due to the appearance of cracks and subsidence. The studied caissons are in a poor condition. Approximately 60 m^2 of the exterior wall of the spur breakwater have cracks that expose the reinforcement of the caissons.

Ultimately, what is intended to be demonstrated with these examples is the large amount of data that can be extracted from the model through these planning tables. These data must be managed by another type of software since, now, the Revit application is not prepared for such exhaustive data processing.

5.2. Challenges and Lessons Learned

These results are focused on the maintenance stage of a port infrastructure. From a cost life cycle approach, maintenance is acquiring more and more importance, with costs between 25 and 60% [5,6]. Considering the aggressive environment of this kind of project, the quality of materials is essential for conservation and maintenance activities [7–10]. In this sense, some of the main pathologies caused by such environment are used in the case study (cracks, corrosion, and breaking of concrete blocks).

This proposal and results focus on the maintenance stage, proposing a new method for facility management in ports using BIM that allows for handling the great interconnection and continuous operation of a big amount of information. This problem is marked in the literature as the main issue in facility management in ports [23–28]. In this sense, a 3D model with its 7 dimension related to facility management serves as a visual tool plus a database that support decision making during the conservation of any port. This theoretical and practical proposal is a first step, where transport infrastructures are somewhat delayed in their integration of BIM compared with other kinds of infrastructures and buildings [11,12,21].

The BIM method, including facility management as an integrated system as proposed here, allows for the coordination of information and visual issues between all different stakeholders of a port infrastructure with clear operability and connections, solving some problems identified by some authors [24]. There were not clear classification or codes for the existing elements of marine ports, so an expansion is proposed, with inspection sheet templates and scheduling maintenance activities during the conservation stage. In this paper, the case study is already projected and built. However, with the know-how from the results presented here, BIM technology can be implemented with a greater impact from the planning and design stage to the whole life cycle, as some authors highlight [45], so the facility management system can have a greater impact from the beginning. Additionally, the facility management system based on BIM allows for the automation of planning the maintenance costs and scheduling the interventions, and serves as a database of all correctives observed during the operations, solving some problems already identified in the literature [46].

6. Conclusions

While in the architecture sector the use of the BIM method is in a much more advanced state internationally, even in the field of facility maintenance, the civil engineering sector is lagging in this regard, and much more so in the sector of port infrastructures and their maintenance and conservation. In the state of the art, there are some general studies about the use of BIM in ports, but there is not a method to use BIM philosophy in the maintenance of ports, as there is in this case study. Up to now, traditional methods (not BIM) have been used for characterizing the assets, evaluating their state, carrying out the registering of inspection and repairs, etc. The new proposed method created and validated in this manuscript is the main result of this research; this can be used as a tool for its implementation in the maintenance of port infrastructures.

With this research, the numerous studies that exist today on this subject have taken ground, and it is presented as a paradigm shift in the approach to infrastructure maintenance management.

Once this new conservation management method is put into practice, it can be concluded that a correct application of the BIM method in projects can greatly optimize the maintenance and conservation management work of port infrastructures, thanks to the collection of all the information in a single model. Although the case study is a Spanish port, this new method can be useful for every seaport authority worldwide; so, there is not a limitation on its application in any country. It has been proven that this is possible from the creation of a three-dimensional model with the Revit program, in which all the information regarding the maintenance and upkeep of the modeled infrastructures is incorporated. It should be noted that the application of this method requires workers to be trained in this type of tool.

Likewise, the advantages of applying this method have been disclosed. The main benefits it offers are: (a) the three-dimensional representation of the infrastructures in an integrated data environment; (b) the storage of the history of inspections and repairs carried out on the infrastructures; (c) faster and more efficient work processes, thanks to the parameterization of all the elements of the model, allowing the automatic updating of all the data contained in it; (d) collaborative work is promoted, facilitating the exchange of information between the different agents interested in the project, in all its phases; (e) the ease of locating elements and assets of the infrastructures, which allows for greater control and management of all of them; (f) the connection and information exchange between the BIM and CMMS tools, which allows for better planning of the maintenance tasks to be carried out, making it possible to visualize the changes in the 3D model, as well as to consult all the necessary information of an intuitive and simple way; (g) it allows for the abandonment of the information silos of the works of conservation and maintenance of existing structures; (h) although the BIM method is designed for application to new construction projects, all the benefits of its incorporation in the exploitation phase of a project must be considered, even if the effort for its application is greater than if it had been implemented in the design phase.

During the development of this project, certain aspects have been detected that for the moment are not developed and that are beyond the established scope. Therefore, the following lines of work are proposed to be developed in future research: (a) creation of new parametric families that allow for a simpler and more comfortable modeling of port infrastructures; (b) creation of new families that allow for a modeling more adjusted to the reality of the most representative pathologies of this type of infrastructure; (c) to reduce the current disadvantages of BIM philosophy: resistance to change, lack of proven method in most of civil infrastructures, high implementation cost, need for software training, and problems of software interoperability. Constant review of the state of the art of the BIM method is needed since it is a work method that continues to be updated and renewed day by day; (d) integration of the agents responsible for managing the maintenance of port infrastructures in the process of applying BIM to maintenance. In this way, the information that needs to be obtained from BIM will be specifically defined.

Author Contributions: Conceptualization, P.V., M.D.E.P., C.H., and R.R.-E.; method, P.V., M.D.E.P., C.H., and R.R.-E.; software, P.V. and C.H.; validation, P.V., M.D.E.P., C.H., and R.R.-E.; formal analysis, P.V., M.D.E.P., C.H., and R.R.-E.; investigation, G.F. and J.-S.L.-G.; writing—original draft preparation, G.F. and J.-S.L.-G.; writing—review and editing, G.F. and J.-S.L.-G.; supervision, G.F. and J.-S.L.-G. All authors have read and agreed to the published version of the manuscript.

Funding: This research received no external funding.

Conflicts of Interest: The authors declare no conflict of interest.

References

1. Wang, D.; Li, D.; Gong, Y.; Wang, R.; Wang, J.; Huang, X. Development situation and future demand for the ports along the Northern Sea Rute. *Res. Transp. Bus. Manag.* **2019**, *33*, 100465. [CrossRef]
2. Diez, J.J.; Paz, R.; Esteban, M.D.; López-Gutiérrez, J.S.; Negro, V. Urban Coastal Flooding and Climate Change. *J. Coast. Res.* **2011**, *SI 64*, 205–209.
3. Negro, V.; Lopez-Gutierrez, J.S.; Esteban, M.D.; del Campo, J.M.; Luengo, J. Action Strategy for Studying Marine and Coastal Works with Climate Change on the Horizon. *J. Coast. Res.* **2018**, *85*, 506–510. [CrossRef]
4. Lin, D.; Jin, B.; Chang, D. A PSO approach for the integrated maintenance model. *Reliab. Eng. Syst. Saf.* **2020**, *193*, 106625. [CrossRef]
5. PIANC. *Life Cycle Management of Port Structures Recommended Practice for Implementation*; PIANC: Brussels, Belgium, 2008.
6. Guillen, A.J.; Crespo, A.; Gómez, J.; Kobbacy, K.; Shariff, S. Building Information Modeling as Asset Management Tool. *IFAC Pap.* **2018**, *49*, 191–196. [CrossRef]

7. de los Reyes, R.S.; Fernández-Sánchez, G.; Esteban, M.D.; Rodríguez, R.R. Carbon Footprint of a Port Infrastructure from a Life Cycle Approach. *Int. J. Environ. Res. Public Health* **2020**, *17*, 7414. [CrossRef]
8. Cejuela, E.; Negro, V.; Esteban, M.D.; López-Gutiérrez, J.S.; Ortega, J.M. From Julius Caesar to Sustainable Composite Materials: A Passage through Port Caisson Technology. *Sustainability* **2018**, *10*, 1225. [CrossRef]
9. Cejuela, E.; Negro, V.; Del Campo, J.M.; Martín-Antón, M.; Esteban, M.D.; López-Gutiérrez, J.S. Recent History, Types, and Future of Modern Caisson Technology: The Way to More Sustainable Practices. *Sustainability* **2018**, *10*, 3839. [CrossRef]
10. Ortega, J.M.; Esteban, M.D.; Sánchez, I.; Climent, M.A. Performance of Sustainable Fly Ash and Slag Cement Mortars Exposed to Simulated and Real in Situ Mediterranean Conditions along 90 Warm Season Days. *Materials* **2017**, *10*, 1254. [CrossRef]
11. Wan, C.; Zhou, Z.; Li, S.; Ding, Y.; Xu, Z.; Yang, Z.; Xia, Y.; Yin, F. Development of a Bridge Management System Based on the Building Information Modeling Technology. *Sustainability* **2019**, *11*, 4583. [CrossRef]
12. Shin, H.; Lee, H.; Oh, S.; Chen, J. Analysis and Design of Reinforced Concrete Bridge Column Based on BIM. *Procedia Eng.* **2011**, *14*, 2160–2163. [CrossRef]
13. Liu, W.; Guo, H.; Li, H.; Li, Y. Retracted: Using BIM to Improve the Design and Construction of Bridge Projects: A Case Study of a Long-Span Steel-Box Arch Bridge Project. *Int. J. Adv. Robot. Syst.* **2014**, *11*, 125. [CrossRef]
14. Lee, K.M.; Lee, Y.B.; Shim, C.S.; Park, K.L.; Shim, C. Bridge information models for construction of a concrete box-girder bridge. *Struct. Infrastruct. Eng.* **2012**, *8*, 687–703. [CrossRef]
15. McGuire, B.; Atadero, R.; Clevenger, C.; Ozbek, M. Bridge Information Modeling for Inspection and Evaluation. Transportation and roads. *J. Bridg. Eng.* **2016**, *21*, 04015076. [CrossRef]
16. Costin, A.; Adibfar, A.; Hu, H.; Chen, S.S. Building Information Modeling (BIM) for transportation Infrastructure—Literature review, applications, challenges, and recommendations. *Autom. Constr.* **2018**, *94*, 257–281. [CrossRef]
17. Biancardo, S.A.; Viscione, N.; Cerbone, A.; Dessi, E.J. BIM-Based Design for Road Infrastructure: A Critical Focus on Modeling Guardrails and Retaining Walls. *Infrastructures* **2020**, *5*, 59. [CrossRef]
18. Cho, H.; Lee, K.H.; Lee, S.H.; Lee, T.; Cho, H.J.; Kim, S.H.; Nam, S.H. Introduction of Construction Management Integrated System Using BIM in the Honam High-Speed Railway Lot No.4-2. In Proceedings of the 28th ISARC, Seoul, Korea, 29 June–2 July 2011.
19. Biancardo, S.A.; Capano, A.; Oliveira, S.G.; Tibaut, A. Integration of BIM and Procedural Modeling Tools for Road Design. *Infrastructures* **2020**, *5*, 37. [CrossRef]
20. Dell'Acqua, G.; De Oliveira, S.G.; Biancardo, S.A. Railway-BIM: Analytical review, data standard and overall perspective. *Ing. Ferrov.* **2018**, *73*, 901–923.
21. Aziz, N.D.; Nawawi, A.H.; Ariff, N.R.M. Building Information Modelling (BIM) in Facilities Management: Opportunities to be Considered by Facility Managers. *Procedia Soc. Behav. Sci.* **2016**, *234*, 353–362. [CrossRef]
22. Edirisinghe, R.; London, K.; Kalutara, P.; Aranda-Mena, G. Building information modelling for facility management: Are we there yet? *Eng. Constr. Archit. Manag.* **2017**, *24*, 1119–1154. [CrossRef]
23. Garibin, P.; Ol'khovik, E. Application of BIM Technology in operation of port marine structures. *E3s Web. Conf.* **2019**, *135*, 01022. [CrossRef]
24. Zhang, P.; Chen, Q. Application research of BIM technology used in port handling system design. *IOP Conf. Ser. Mater. Sci. Eng.* **2018**, *452*, 042012. [CrossRef]
25. Zunxiang, L.H. Application and Exploration of BIM Technology in Port Operation Design. *Port Oper.* **2016**, *2*, 32–34.
26. Puhao, H.; Yaoyu, X. The application of BIM in the power pipe design for liquid chemical berth. *Wuhan Investig. Des.* **2016**, *3*, 45–48.
27. Yang, W.; Jian, X. The Application of BIM in the Harbor Electrical Design. *Constr. Des. Proj.* **2015**, *12*, 81–84.
28. Q Web Page. Available online: https://www.qps.nl/qps-and-vrt-cooperate-with-bim-software-solution/ (accessed on 1 June 2020).
29. Akcamete, A.; Akinci, B.; Garrett, J.H. Potential Utilization of Building Information Models for Planning Maintenance Activities. In Proceedings of the EG-ICE 2010—17th International Workshop on Intelligent Computing in Engineering, Nottingham, UK, 30 June–2 July 2010.
30. ASCE. *Standard Practice Manual for Underwater Investigations*; American Society of Civil Engineers: Reston, VA, USA, 2001.

31. ATPyC. *Gestión de la Conservación en el Entorno Portuario*; Asociación Técnica de Puertos y Costas: Madrid, Spain, 2012.
32. ATPyC. *Guía de Reparación de Estructuras de Hormigón en Ambiente Marino*; Asociación Técnica de Puertos y Costas: Madrid, Spain, 2018.
33. Munir, M.; Kiviniemi, A.; Finnegan, S.; Jones, S.W. BIM business value for asset owners through effective asset information management. *Facilities* **2019**, *38*, 181–200. [CrossRef]
34. McArthur, J.J.; Sun, X. Best practices for BIM Execution Plan development for a Public–Private Partnership Design-Build-Finance-Operate-Maintain project. Building Information Modelling (BIM) in Design, Construction and Operations. *WIT Transit. Built Environ.* **2015**, *149*, 119–130.
35. Puertos del Estado. Guia BIM del Sistema Portuario de Titularidad Estatal. 2019. Available online: http://www.puertos.es/es-es/BibliotecaV2/Guia%20BIM%2009.pdf (accessed on 15 May 2020).
36. BuildingSMART. (n.d.). Available online: https://www.buildingsmart.es/bim/ (accessed on 3 May 2020).
37. Pitarch, C.M.; Vidal, F.J. *Impacto del BIM en la Gestión del Proyecto y la Obra de Arquitectura*; Un proyecto con REVIT; Universidad Politécnica de Valencia: Valencia, Spain, 2014.
38. Autodesk Knowledge Network. (n.d.). Available online: https://knowledge.autodesk.com/es (accessed on 20 May 2020).
39. Ricondo Cueva, A. *Estudio del Rebase en la Zona del Nuevo Pantalán de REPSOL en el Puerto de Langosteira (A Coruña) y Análisis de las Cargas Ejercidas Sobre el Mismo*; Universidad de Cantabria: Cantabria, Spain, 2018.
40. Autoridad Portuaria de A Coruña. (n.d.). Available online: http://www.puertocoruna.com/es/index.html (accessed on 7 May 2020).
41. Santos, M.; Corredor, A.; Maciñeira, E.; Bajo, V.; Gómez-Martín, M.E.; Medina, J.R. Aspectos innovadores en el diseño y construcción del dique Oeste en el puerto exterior de A Coruña en Punta Langosteira (España). *Ribagua* **2016**, *3*, 89–100. [CrossRef]
42. BIM Chapters: Revit Assembly Code Basics. (n.d.). Retrieved 2 September 2020. Available online: https://bimchapters.blogspot.com/2019/02/revit-assembly-code-basics.html (accessed on 10 June 2020).
43. GuBIMClass. (n.d.). Available online: https://gubimclass.org/es/ (accessed on 15 May 2020).
44. Puertos del Estado. *Guía de Buenas Prácticas Para la Ejecución de Obras Marítimas*; Puertos del Estado: Madrid, Spain, 2008.
45. Cao, M.; Duan, X.; Zhao, X. Construction of integrated evaluation system for green port construction based on BIM technology. *J. Coast. Res.* **2019**, *93*, 1073–1079. [CrossRef]
46. Garibin, P.; Ol'khovik, E. Development of requirements for the BIM model of the Arctic port facilities exploitation. MATEC Web Conference. International Science Conference SPbWOSCE-2017. *Bus. Technol. Sustain. Urban Dev.* **2018**, *170*, 03027. [CrossRef]

Journal of
*Marine Science
and Engineering*

MDPI

Article

Research on the Bearing Capacity of a Damaged Jacket Repaired by a Grouting Clamp Based on a Type of Wedge Gripping

Bo Zhang [1,*], Qing Rui Zhang [1], Tao Wang [2] and Zhuo Wang [1]

[1] College of Mechanical and Electrical Engineering, Harbin Engineering University, Harbin 150001, China; chenyin@hrbeu.edu.cn (Q.R.Z.); wangzhuo_heu@hrbeu.edu.cn (Z.W.)
[2] School of Mechanical Engineering, Hebei University of Technology, Tianjin 300130, China; 18846166436@hrbeu.edu.cn
* Correspondence: zhangbo_heu@hrbeu.edu.cn

Received: 26 October 2020; Accepted: 23 November 2020; Published: 30 November 2020

Abstract: In the process of repairing a damaged jacket with a grouting clamp, it is necessary to analyze the bearing capacity of the damaged structure to determine whether it meets requirements for maintenance and reinforcement. On the basis of previous research results, this paper proposes a design scheme of a grouting clamp based on wedge gripping, which can effectively improve the local bearing capacity of the jacket by more than 35%. The elastic–plastic theory was used to analyze the bearing capacities of steel pipe piles under three different conditions (ideal steel pipe, bending steel pipe and steel pipe with a defective section) and with reinforcement by a grouting hoop. The bearing capacity of the mechanical grouting clamp was analyzed, and the resulting theoretical design was verified by a bearing performance experiment. The results support the conclusion that the wedge mechanical grouting clamp can effectively strengthen local jacket members.

Keywords: grouting clamp; bearing capacity analysis; damaged jacket; wedge gripping

1. Introduction

Over the 21st century, oil and gas reserves on land have been rapidly depleted, and marine environments are thus increasingly explored to obtain these resources. Many of the world's offshore platforms are approaching or have exceeded their design life. At present, more than 50% of the offshore facilities on the Norwegian continental shelf (NCS), the British continental shelf (UKCS) and the Gulf of Mexico continental shelf have exceeded their original service lives [1]. Currently, the most commonly repaired component of offshore platforms is the grouting clamp because this strategy is characterized by effectiveness, convenient construction and low cost [2]. Therefore, this type of maintenance technology has received extensive attention from experts around the world.

In the past two decades, countries such as the United States, the United Kingdom, Norway and Japan have invested considerable research efforts into the repair and reinforcement of damaged components of offshore platforms, and several breakthroughs have been made [3]. From 1985 to 1987, the Elnashai team of the Imperial College of Technology completed an experimental mechanical study on the pressure grouting clamp for the first time. In 1989, Shuttleworth F P and Billington C J et al. in Britain analyzed the prestressed grouting connection of an offshore structural steel pipe. A small model was used to test the shrinkage of grouting, the initial creep in the grouting ring and the long-term shrinkage creep [4]. In 2002, Grundy Paul and Yee Teck Lee of Monash University in Australia carried out a study on the bearing capacity of grouting clamps with a shear key structure under dynamic load [5]. In 2003, the team of Jin Weiliang and Gong Shunfeng of Zhejiang University confirmed through research that the grouting repair and reinforcement scheme could improve the bearing capacity of the damaged components of the studied platform [6]. In 2006, Zhao Xiaolin's team from Monash University in Australia analyzed and studied the bearing capacity of grouting clamps under a harsh environment [7]. In 2007, a K-type grouting clamp was designed for an ocean environment. The annular space between the clamp and the damaged parts was filled with cement. In this case, both ends of the clamp needed to be sealed to avoid the corrosion of cement by seawater [8]. In 2010, the team of Shi Xiang and Yang Bin from the Ocean University of China analyzed the traditional self-stress grouting clamp and proposed a new type of expansion pressure self-stress grouting clamp [9]. In 2011, Jiang Shouchao of Tongji University and colleagues summarized the latest developments in grouting pile-to-casing connections and characterized their mechanical properties under axial, combined and cyclic loads [10]. In 2013, Samarakoon, S.M.S.M.K and colleagues of the Singapore University of Technology highlighted the wide use of grouting repair for offshore platforms and explained the latest developments in grouting connections and their advantages and disadvantages relative to the welding or mechanical connection [11]. In 2014, Li Cong and others from the Ocean University of China designed a shorter bolt-type expansion grouting clamp with a simpler structure, aiming to address the complicated structure and installation of the long bolt-type expansion grouting clamp. In 2015, Advanced Composite Structures Australia Co., Ltd., developed the composite clamp with Petronas [12]. In 2016, researchers from Petronas, such as Sum W S, Leong K H and colleagues, conducted a further study on the use of composite clamps to repair damaged pipelines. Based on the optimization of the finite element model, a method of designing clamps for different pipe diameters was developed [13]. In 2017, Jeong Hwa Lee's team from Korea University conducted concentric and eccentric load tests on the grouting connection of piles and casings of an offshore oil platform and natural gas platform [14]. In the same year, Shi Xiang, Jiao Guoyang and colleagues from the Ocean University of China tested and analyzed the load-bearing performance of a large-size expansion clamp model [15]. In 2018, Norwegian researchers Gunnar Soland and AtlE Johansen et al. summarized the current design rules of the grouting pile sleeve connection, carried out compression and tension axial load experiments on the grouting clamp and proposed a new design formula [16]. In 2018, Norwegians Johansen, A., Solland, G. and colleagues conducted a study on the grouting connection with shear keys of an offshore sheathed platform and determined its axial bearing capacity under cyclic load [17]. In 2019, Chellappan, N. Vignesh and colleagues from the Indian Institute of Technology conducted a comprehensive study on the load transfer between the two parts of a tubular member connected by a fractured grouting joint. The axial tension and moment loads were simulated [18].

In view of the previous achievements of different research institutions, this paper proposes a new type of wedge gripping for the grouting clamp. First, the bearing capacities of an ideal steel pipe, bending steel pipe, steel pipe with a defective section and a steel pipe pile reinforced by a grouting clamp were analyzed. Second, the bearing capacity of the mechanical grouting clamp was evaluated, and the ability of the grouting clamp to repair the damaged jacket was established. Finally, the effectiveness of the mechanical grouting clamp in strengthening the jacket members was verified by experiments.

2. Analysis of the Bearing Capacity of a Damaged Steel Pipe Pile

2.1. Buckling Analysis of an Ideal Steel Tubular Pile Under Axial Compression

Assuming that the jacket steel pipe is subjected to uniform axial pressure, its buckling deformation is as shown in Figure 1: uniform axial pressure is applied to the steel pipe, and the Karman–Donnell nonlinear buckling equation is used to obtain the solution [19].

$$U\nabla^8\omega - \nabla^4\left(P_x\frac{\partial^2\omega}{\partial x^2} + P_y\frac{\partial^2\omega}{\partial y^2} + 2P_{xy}\frac{\partial^2\omega}{\partial x\partial y}\right) + \frac{Et}{R^2}\frac{\partial^4\omega}{\partial x^4} = 0 \quad (1)$$

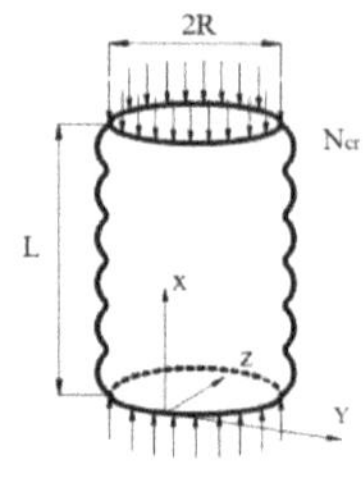

Figure 1. Shell-shaped tube.

Here, R is the shell radius, t is the shell thickness, L is the shell length, E is the elastic modulus, U is the bending rigidity, $U = \left(Et^3\right)/\left[12\left(1-\mu^2\right)\right]$, μ is Poisson's ratio, P_x is the axial load, P_y is the transverse load, and ω is the lateral displacement, as shown in Figure 2.

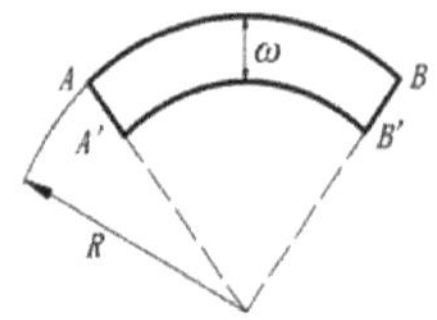

Figure 2. Shell lateral displacement.

∇ is the Hami operator, the 4th and 8th orders of which are calculated as follows:

$$\nabla^4 = \left(\frac{\partial^4}{\partial x^4} + 2\frac{\partial^4}{\partial x^2 \partial y^2} + \frac{\partial^4}{\partial y^4}\right),$$

$$\nabla^8 = \left(\frac{\partial^8}{\partial x^8} + 4\frac{\partial^8}{\partial x^6 \partial y^2} + 6\frac{\partial^8}{\partial x^2 \partial y^6} + \frac{\partial^8}{\partial y^8}\right).$$

Because the steel tube is only subjected to axial load (only P_x), $P_x = -\sigma_x t$. Equation (1) can be written as follow:

$$U\nabla^8\omega + \sigma_x t \nabla^4 \frac{\partial^2 \omega}{\partial x^2} + \frac{Et}{R^2}\frac{\partial^4 \omega}{\partial x^4} = 0 \tag{2}$$

Before calculating the critical stress, the mode of lateral displacement ω is assumed to be as follow:

$$\omega = \omega_0 \sin\frac{m\pi x}{L} \sin\frac{ny}{R},$$

where m is the half-wave number in the x-direction, n is the half-wave number in the y-direction, and ω_0 is the maximum deflection value.

In order to simplify the calculation, let $\beta = m\pi R/L$; then,

$$\sigma_x = \frac{U\beta^2}{tR^2} + \frac{E}{\beta^2} \tag{3}$$

For the analysis in this paper, the critical stress of the steel pipe is required. Equation (3) is derived from β, and the result is zero. The calculated σ_x is the minimum value, which is the critical stress; the result is as follow:

$$\beta^4 = \frac{EtR^2}{U} \tag{4}$$

When Equation (4) is introduced into Equation (3), the critical stress of the thin cylindrical shell under axial compression can be obtained as follows:

$$\sigma_{cr} = \frac{1}{\sqrt{3(1-\mu^2)}}\frac{Et}{R} \tag{5}$$

In order to avoid local buckling of the thin-walled steel pipe before yielding, the yield stress of the steel must be $f_y \leq \sigma_{cr}$, $D = 2R$, and Equation (5) can be obtained as follows:

$$D/t \leq 1072\frac{235}{f_y} \tag{6}$$

2.2. Buckling Analysis and Horizontal Bearing Capacity Analysis of a Bending Steel Tubular Pile under Axial Compression

When the axial compression bar is subjected to a force with a pressure load of P, it is assumed that the deformation of the bar is a small deflection bending deformation, and the size and direction of the axial compression P remain unchanged, as shown in Figure 3a,b. The upper end is restrained by an anti-rotation spring and horizontal tension-compression spring, with stiffnesses of $k_{\theta 1}$ and k_{v1}, respectively; the lower end of the rod has an anti-rotation spring with a stiffness of $k_{\theta 2}$, a horizontal tension-compression spring with a stiffness of k_{v2} and a vertical rigid hinge.

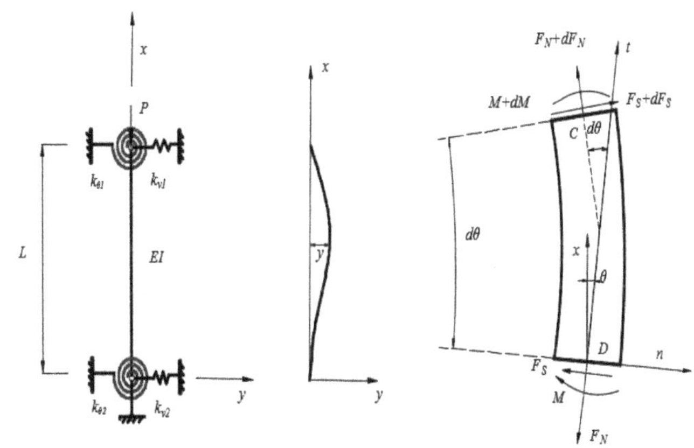

Figure 3. Jacket piles under axial compression. (**a**) Steel pipe compression diagram. (**b**) Simplified bending deformation of the small deflection. (**c**) the forces balance diagram.

The equilibrium equation is established according to Figure 3c: $\sum M_C = 0$.

$$dM + (F_S \sin\theta - F_N \cos\theta)dy - (F_N \sin\theta + F_S \cos\theta)dx = 0 \tag{7}$$

According to the equivalence of the force system, the force is equivalent to H parallel to the x-axis and P parallel to the y-axis, as shown in Figure 4.

$$\begin{cases} P = F_N \cos\theta - F_s \sin\theta \\ H = F_N \sin\theta + F_S \cos\theta \end{cases} \tag{8}$$

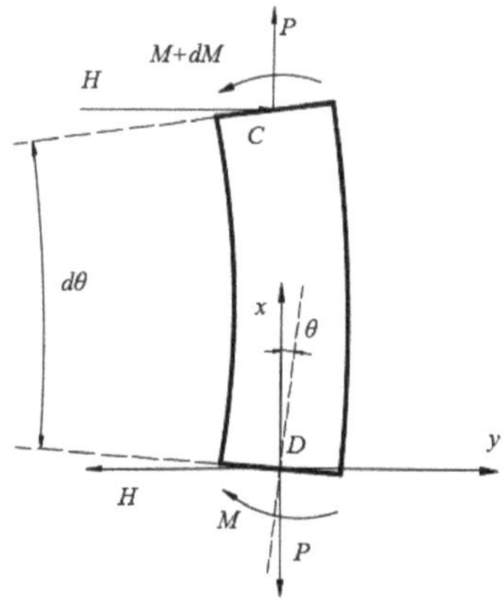

Figure 4. Equal-effectiveness analysis.

By applying the bending moment formula and the bending equation $EIy'' = -M$ under a small disturbance to the above formula, three differential equations of the steel pipe pile can be obtained:

$$EIy''' + Py' + H = 0 \tag{9}$$

According to Equation (9) and the bending equation, the boundary conditions of the end of the jacket pile with elastic constraints corresponding to $x = 0$ and $x = l$ are established:

$$\text{When } x = 0, \begin{cases} EIy''' + Py' + k_{v2}y = 0 \\ EIy'' - k_{\theta 2}y' = 0 \end{cases} \tag{10}$$

$$\text{When } x = l, \begin{cases} EIy''' + Py' - k_{v1}y = 0 \\ EIy'' - k_{\theta 1}y' = 0 \end{cases} \tag{11}$$

with the order $k_{v1}/EI = \alpha_1$, $k_{\theta1}/EI = \beta_1$, $k_{v2}/EI = \alpha_2$, $k_{\theta2}/EI = \beta_2$, $P/EI = k^2$, and $u = kl$. According to its general solution, the following equation can be obtained:

$$\begin{aligned} &\left\{(\alpha_1+\alpha_2)\left(\tfrac{u}{l}\right)^6 + [\beta_1\beta_2(\alpha_1+\alpha_2) - \alpha_1\alpha_2 l]\left(\tfrac{u}{l}\right)^4 - \alpha_1\alpha_2(\beta_2+\beta_1\beta_2 l+\beta_1)\left(\tfrac{u}{l}\right)^2\right\}\sin u + \\ &\left[(\alpha_1+\alpha_2)(\beta_1-\beta_2)\left(\tfrac{u}{l}\right)^5 + \alpha_1\alpha_2 l(\beta_2-\beta_1)\left(\tfrac{u}{l}\right)^3 - 2\alpha_1\alpha_2\beta_1\beta_2\tfrac{u}{l}\right]\cos u + 2\alpha_1\alpha_2\beta_1\beta_2\tfrac{u}{l} = 0 \end{aligned}$$

In order to obtain the solution of the above equation, the coefficients before each variable function must be determined. By changing the stiffness of the support at one end of the jacket pile, different constraint conditions at both ends of the jacket pile can be obtained.

When $\alpha_1 = \alpha_2 = \infty$, there is no lateral displacement at the end of the jacket pile; when $\alpha_1 = \alpha_2 = 0$, there is no lateral constraint at the end of the jacket pile; when $\beta_1 = \beta_2 = \infty$, there is no change in the angle of the jacket pile; when $\beta_1 = \beta_2 = 0$, there is no rotation constraint at the end of the jacket pile.

The lateral bearing capacity of the steel pipe pile is mainly determined by the bending bearing capacity of the plastic hinge area of the steel pipe pile section. The diameter and material of the steel pipe pile determine the bending capacity of the plastic hinge area. When the plastic hinge is present in the structure, all parts except for the plastic hinge area can be regarded as rigid bodies [20,21].

As shown in Figure 5, it is assumed that the force and moment conservation equations of the steel tube column are applied under the condition of full-section yield:

$$\begin{cases} f_{ys}(2\pi - 4\Phi)r_p\delta = N \\ 2\Phi r_p\delta f_{ys}(x_{2c} + x_1) + 2(\pi - \Phi)r_p\delta f_{ys}(x_{1c} - x_1) = M \end{cases} \tag{12}$$

where f_{ys} is the yield strength of steel with defects, which generally refer to corrosion defects in this analysis; Φ is the angle between the intersection line of the neutral axis and the steel pipe and the loading direction; r_p is the inner radius of the steel pipe; δ is the thickness of the steel pipe; N is the axial force; x_1 is the offset of the neutral axis; x_{1c} is the displacement of the centroid axis in the area under tension; and x_{2c} is the displacement of the centroid axis in the compression area. The horizontal bearing capacity of the bending steel pipe pile can be obtained from the data.

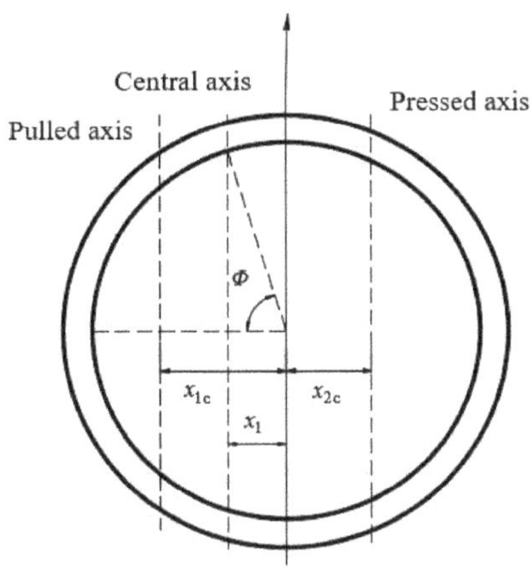

Figure 5. Calculation diagram of the steel pipe pile.

2.3. *Ultimate Bending Moment and Axial Bearing Capacity of a Steel Pipe Pile with a Defective Section*

When the jacket is damaged, defects are produced. It is assumed that the defect is a circular arc and that the rest is the steel pipe with a constant radius [6,20]. The defect leads to a change in the inertia of the section and the movement of the neutral axis, which results in the loss of strength. It is assumed that the axial length of the defective section is the same as that of the ideal section, and the defect does not change the equal area axis of the section. As shown in Figure 6, the defect depth δ is as follow:

$$\frac{\delta}{R_p} = 1 + \frac{\pi\left[\cos\alpha - \cos\left(\frac{\alpha + \sin\alpha}{2}\right)\right]}{\alpha + \sin\alpha} \quad (13)$$

where δ is the defect depth, R_p is the outer radius of the steel pipe, and α is the angle from the center of the deformed circle to the defect and the Z-axis.

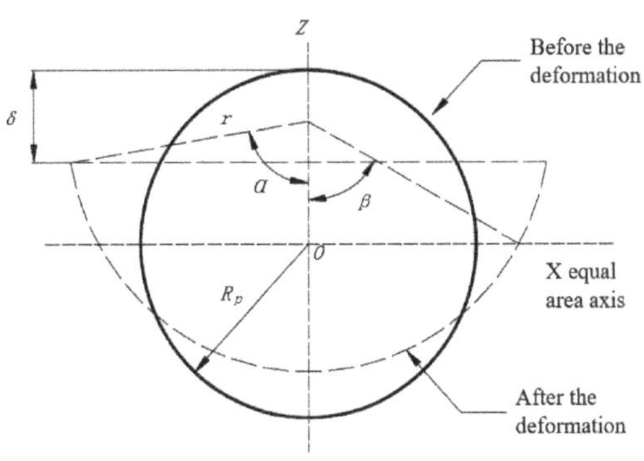

Figure 6. The form of local defects.

Considering the center of the deformed circle as the origin, the ultimate bending moment of the defective section is as follow:

$$M = 2\sigma_y t_p r^2 [2\sin\beta - \sin\alpha(1+\cos\alpha)] \tag{14}$$

where σ_y is the yield limit, R_p is the outer radius of the steel pipe, t_p is the wall thickness, and r is the radius after deformation.

When calculating the axial bearing capacity of the defective section, the area converted under the impact load is determined first:

$$A_1 = A\frac{\pi - \theta}{\pi} \tag{15}$$

The tensile capacity of the defective steel pipe is as follow:

$$F_p = F_t A_1 \tag{16}$$

The compressive bearing capacity of the defective steel pipe is as follow:

$$F_{cp} = F_{ct} A_1 \tag{17}$$

F_{ct} in the above formula is as follow:

$$F_{ct} = \frac{\left[1 - \frac{\left(Kl/r_p\right)^2}{2C_c}^2\right]\sigma_y}{\frac{5}{3} + \frac{3\left(Kl/r_p\right)}{8C_c} - \frac{\left(Kl/r_p\right)^3}{8C_c}} \quad Kl/r_p < C_c$$

$$F_{ct} = \frac{12\pi^2 E}{23\left(Kl/r_p\right)^2} \quad Kl/r_p \geq C_c$$

$$C_c = \left(\frac{12\pi^2 E}{\sigma_y}\right)^{\frac{1}{2}}$$

where F_t is the allowable tensile strength, F_{ct} is the allowable compressive strength, K is the effective length coefficient, l is the unsupported length, r_p is the inner diameter of the steel pipe, and E is the modulus of elasticity and equal to 206 GPa.

As shown in Figure 7, the compressive bearing capacity changes with the unsupported length l and the converted area A_1. As shown in Figure 7, the unsupported length l has little effect on the change in the tensile bearing capacity because the value of l is too small compared with C_c to have an effect. As the converted area A_1 increases, the defect area decreases, and the tensile bearing capacity of the steel pipe pile increases in strength, this is consistent with the actual application.

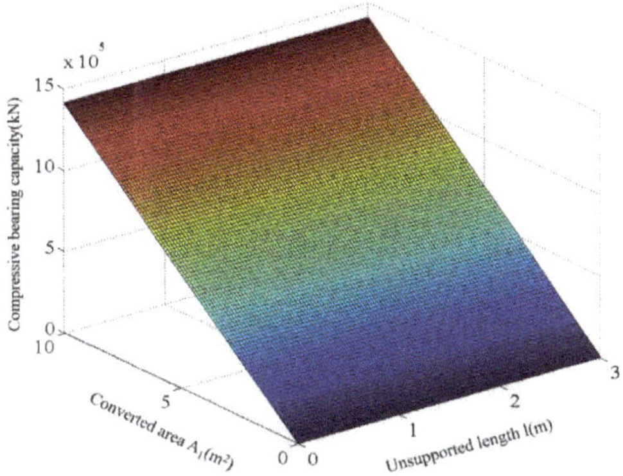

Figure 7. Compressive capacity.

3. Analysis of the Bearing Capacity of a Grouting Clamp Based on a Type of Wedge Gripping

3.1. Analysis of the Axial Bearing Capacity of the Grouting Clamp

In this analysis, a grouting hoop without a shear key was used to reinforce the defective steel pipe. The clamps are connected with high-strength bolts, and the annular space between the light pipe and the outer clamp is filled with cement. The structural section after installation is shown in Figure 8, and the shaded part is the filler cement.

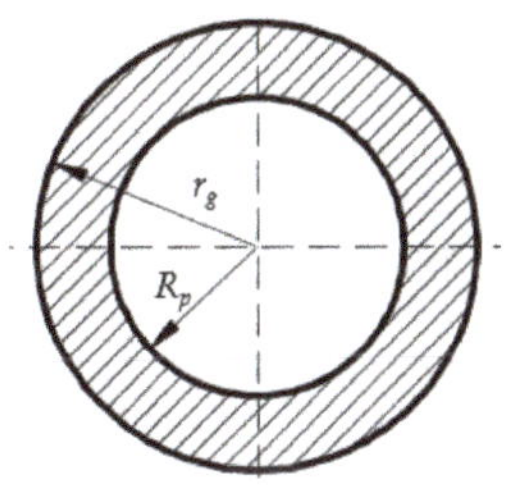

Figure 8. A sectional view of the grouting clamp.

Under the action of axial force, the bearing capacity F_1 of the damaged steel pipe with the grouting clamp is composed of two parts: the bearing capacity F_p or F_{cp} of the defective steel pipe itself and the force F_2 between the concrete and the steel pipe.

$$F_2 = f_{ba}\left(2\pi R_p L_s - A_d\right) + 2\pi r_g L_s f_{ba} \tag{18}$$

Here, f_{ba} is the axial load transfer stress value and has a value of 0.184 MPa, R_p is the outer radius of the defective steel pipe, r_g is the inner radius of the grouting clamp, L_s is the length of the clamp, and A_d is the area of the defect in the steel pipe, with $A = A_1 + A_d$.

3.2. Analysis of the Moment Bearing Capacity of the Grouting Clamp

After the grouting clamp is installed on the defective steel pipe, its bending capacity M_t consists of three parts: the bending moment capacity of the defective steel pipe itself, the bending moment capacity of the grouting material and the bending moment capacity of the clamp itself. The bending moment capacities of the clamp itself and the undamaged steel pipe can be calculated according to Equation (19):

$$M_1 = \sigma_{\max}\omega_z \tag{19}$$

The value of $\sigma_{\max}$ in the above formula is as follow:

$$
\begin{aligned}
&\sigma_{\max} = 0.75\sigma_y & \frac{D}{t} \le \frac{1500}{\sigma_y} \\
&\sigma_{\max} = \left(0.84 - 1.74\frac{\sigma_y D}{Et}\right)\sigma_y & \frac{1500}{\sigma_y} < \frac{D}{t} < \frac{3000}{\sigma_y} \\
&\sigma_{\max} = \left(0.72 - 0.58\frac{\sigma_y D}{Et}\right)\sigma_y & \frac{3000}{\sigma_y} < \frac{D}{t} < 300 \\
&\omega_Z = \frac{\pi}{64}\left(D^4 - d^4\right) &
\end{aligned}
$$

where D is the outer diameter, d is the inner diameter, and $\sigma_{\max}$ is the maximum normal stress.

Because the local structure of the jacket is damaged, its bending capacity and the safety performance of the offshore platform are reduced, so the jacket needs to be strengthened.

Figure 9a is a three-dimensional model diagram of the grouting clamp; Figure 9b is the force analysis diagram of the grouting clamp and shows the strengthened part of the jacket. In Figure 9a, the grouting clamp is composed of the outer flap, self-locking mechanism and bolted connection. In the self-locking mechanism, the wedge clamping mode is adopted to realize unidirectional locking in the bearing capacity direction, so the clamping force can be improved, and the axial bearing capacity can be increased accordingly. The part of the steel pipe that is not covered by the clamp is subject to tangential and normal forces from the external environment. The axial shear force produced by the grouting material is oriented in different directions at different positions. Because the normal force is a certain distance from the centroid, a bending moment is produced.

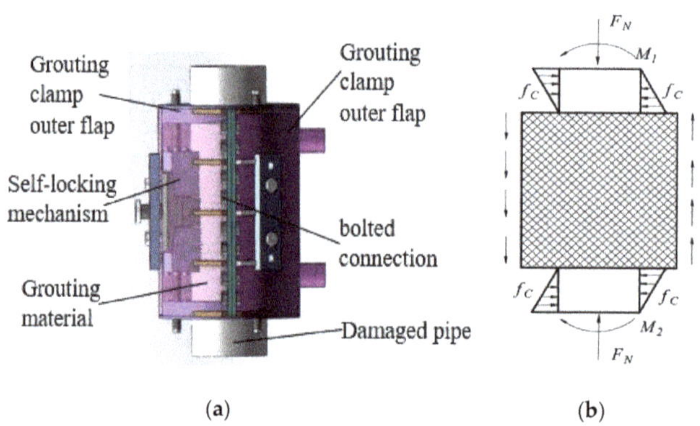

Figure 9. Grouting clamp reinforcement. (**a**) Clamp structure diagram. (**b**) Simplified stresses diagram.

Compared with the whole steel pipe, the length of the grouting clamp is small, so the bending moment M_1 can be regarded as equal to M_2. Thus, the bending moment of the section is taken as M_2.

According to the analysis of the bottom section of the grouting clamp, which is derived from the bottom of the overall device (corresponding to Figure 10a), the bending moment of the reinforcement area is M_2, and the bending moment of the steel pipe is M_s, as shown in Figure 10b.

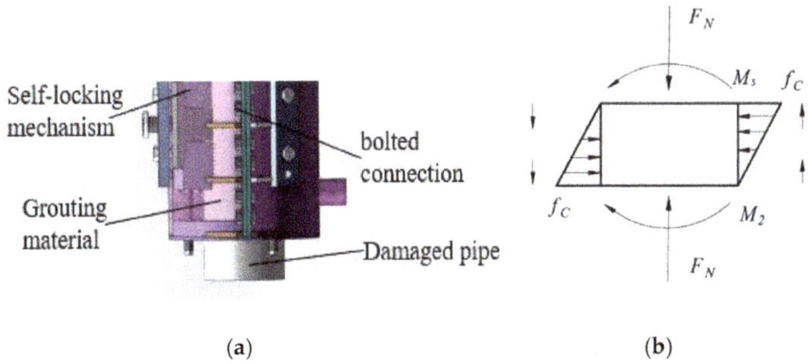

Figure 10. Analysis diagram of the grouting clamp bottom. (**a**) Clamp structure of lower part. (**b**) Simplified stress diagram of lower part.

Thus, $M_2 = M_s + M_g$.

One of the functions of grouting material is to provide a degree of compressive bearing capacity for clamps and steel pipes; thus, there is sufficient compressive bearing capacity in the grouting itself, as shown in Figure 11.

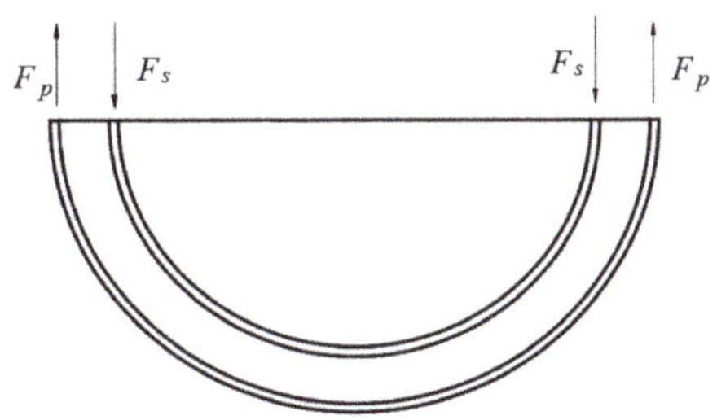

Figure 11. Calculation diagram of the steel pipe and clamp.

It is assumed that the grouting material is in an elastic state and that the following conditions are met:

$$\begin{cases} F_g = \dfrac{f_c \cdot R_s}{2t_g} \\ F_p = \dfrac{f_c \cdot R_s}{2t_p} \end{cases} \tag{20}$$

When Equation (20) is satisfied, the maximum radial bending moment provided by the grouting material is as follow:

$$M_g = \frac{1}{2} f_c R_p L_s^2 \tag{21}$$

According to the bending capacity of the defective steel pipe itself and that of the concrete between the damaged steel pipe and the grouting clamp, the bending moment capacity of the clamp itself can be obtained as follows:

$$M_t = 2\sigma_y t_p r^2 [2\sin\beta - \sin\alpha(1+\cos\alpha)] + \frac{1}{2} f_c R_p L_s^2 + \sigma_{\max}\omega_z \tag{22}$$

It can be seen in Figure 12 that with the increase in α, the bending moment bearing capacity of the defective steel pipe after the grouting clamp is installed decreases. When 0.2 rad < α < 1.5 rad, the bending moment decreases rapidly with α; when 1.5 rad < α, the reduction of the bending moment slows down.

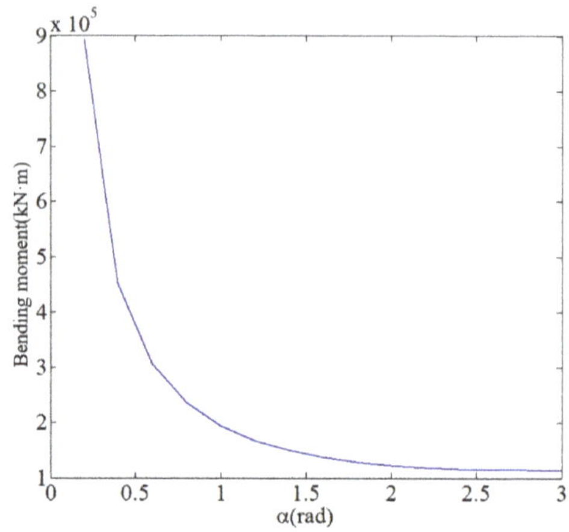

Figure 12. Bending moment diagram.

4. Mechanical Experiment Verification of the Grouting Clamp Based on a Type of Wedge Gripping

According to the principle of similarity [22] (geometric similarity, kinematic similarity and dynamic similarity), in engineering applications, the grouting hoop is reduced in equal proportion. The inner diameter of the steel pipe is 84 mm, the outer diameter is 89 mm, and the length is 300 mm. The inner diameter of the mechanical grouting clamp is 159 mm, the outer diameter is 168 mm, and the length is 221 mm. In this experiment, a grouting clamp prototype was used to obtain the data. The composition of the experimental device is shown in Figures 13 and 14. The loading tools in this experiment included a screw jack with a range of 10 t, a vehicle jack, a strain gauge, and a PLC. The screw jack simulates the axial force on the steel pipe and the slippage force on the clamp, and the vehicle jack simulates the radial force on the steel pipe and the clamp. While force is exerted on the steel pipe and clamp, the current signal of the pressure sensor enters the transmitter and is sent to the PLC, which communicates with the computer that ultimately reads and saves the collected data.

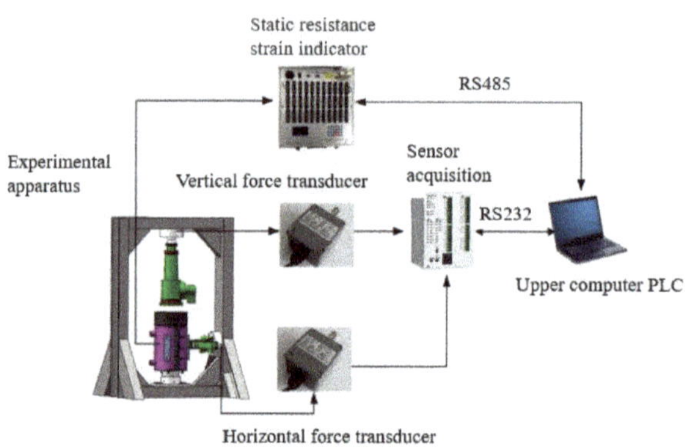

Figure 13. Schematic diagram of the experimental device.

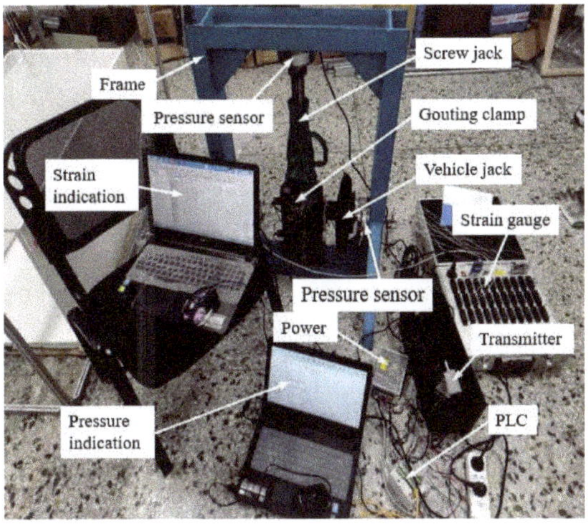

Figure 14. Experimental device.

The jack force range used in the experiment was 0~10t. A large number of wheels spoke load cells were used to measure and record the force exerted in the horizontal and vertical directions.

The primary goal of the experiment was to determine the deformation size by measuring the strain value so as to establish the reinforcement effect. If the strain value of the defective steel pipe reinforced by the clamp is less than or equal to the strain value of the ideal steel pipe under the same conditions when the load is applied, then the design of the mechanical grouting clamp is successful.

(1) First, axial force was applied to three types of steel pipes (no defect, defect with a diameter of 20 mm and defect with a diameter of 35 mm), and the axial stress distribution was measured while applying increasing force. Then, each strain measurement was obtained from the strain gauge. The strain value of the sheet was determined, and then the car jack was used to apply radial force to the clamp to determine the strain value for the entire pipe.

(2) Then, the reinforcement effect of the clamp was verified. To install the clamp on the defective steel pipe, the screw jack was first used to apply axial force to the clamp, and each strain value was obtained from the strain gauge. Next, the vehicle jack was used to apply radial force to the clamp to obtain the strain value.

(3) The strain values obtained for the defective section were compared with the corresponding values of the non-defective steel pipes to verify the reinforcement performance.

Figures 15 and 16 show the three steel pipes and grouting mechanical clamp.

Figure 15. Steel tubes with three different damage holes.

Figure 16. Fully assembled clamp for experiments.

In the jacket axial compression experiment, the strain values of all measurement points of the ideal steel pipe under axial compression were used as a control group to test the reinforcement effect.

Since the clamp needs to be subsequently loaded to prevent the clamp and its internal metal parts from contacting and crushing the strain gauge, the sticking position of the non-defective steel pipe strain gauge was placed as shown in Figure 17.

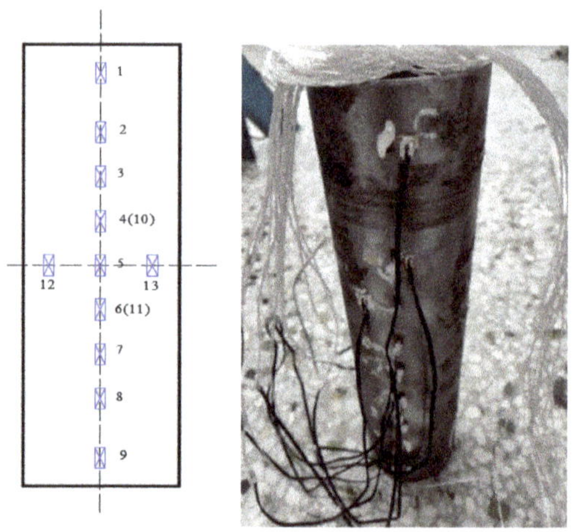

(a) (b)

Figure 17. Ideal position of the steel tube strain gauge. (a) Mark the position of steel pipe by sticking stress piece. (b) Pasted the stress piece on the steel pipe.

The sticking position of the ideal steel pipe strain gauge, as shown in Figure 17, needs to be compared with that of the defective steel pipe strain gauge. For this analysis, the values measured at strain gauges 10, 11, 12 and 13 were used. Because the clamp cannot contact the strain gauge, the gaps between No. 1 and No. 2 and between No. 8 and No. 9 of the strain gauge are large.

The mechanical grouting hoop is subjected not only to axial load but also to current and wave loads. Therefore, the radial loading experiment was carried out for the mechanical grouting hoop. The loading direction was in the defect direction of the steel pipe, and the jack was loaded onto the pipe through two clips.

As shown in Figure 18, in the experiment to verify the influence of radial load on the steel pipe, the pipe was subjected to a certain axial force (ensuring that the pre-applied axial force is the same) to prevent it from deviating when the radial load is applied.

Figure 18. Axial loading of the steel pipe.

The installation mode and loading scheme of the mechanical grouting clamp are the same as those of the steel pipe, both of which are axial and radial loadings under certain axial pressure. In this experiment, strain gauge No. 13 was selected as the data recording point.

According to the formula $\sigma = e \cdot \varepsilon$, the obtained strain values were converted into stress values, and the stress diagrams are depicted in Figures 19 and 20. As shown in Figure 19, as the axial load increases, the maximum stress of the steel pipe with defects increases and is greater than that of the ideal steel pipe, and the larger the defect diameter, the greater the stress. The stress of the steel pipe strengthened by the mechanical grouting clamp is notably less than that of the ideal steel pipe: the stress is reduced by nearly 28% compared with the unreinforced steel pipe. As shown in Figure 20, the change trend of stress under axial and radial loads is the same as that under only the axial load. The stress reduction is about 35%, which is equivalent to a 35% increase in bearing capacity, indicating that the mechanical grouting clamp can effectively reinforce the defect. Although the clamping block can cause damage to the jacket surface, this damage is local plastic deformation, which will not affect the overall strength of the jacket. Furthermore, the effective bearing capacity can be improved as a result of the plastic deformation caused by the block embedded in the jacket surface. Therefore, it is feasible to improve the overall strength of the jacket by local plastic deformation damage.

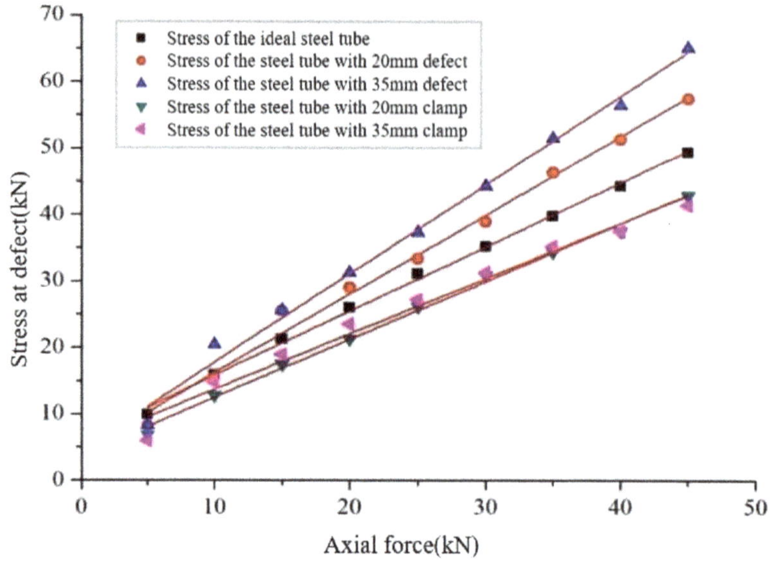

Figure 19. Stress at the measuring point of axial load.

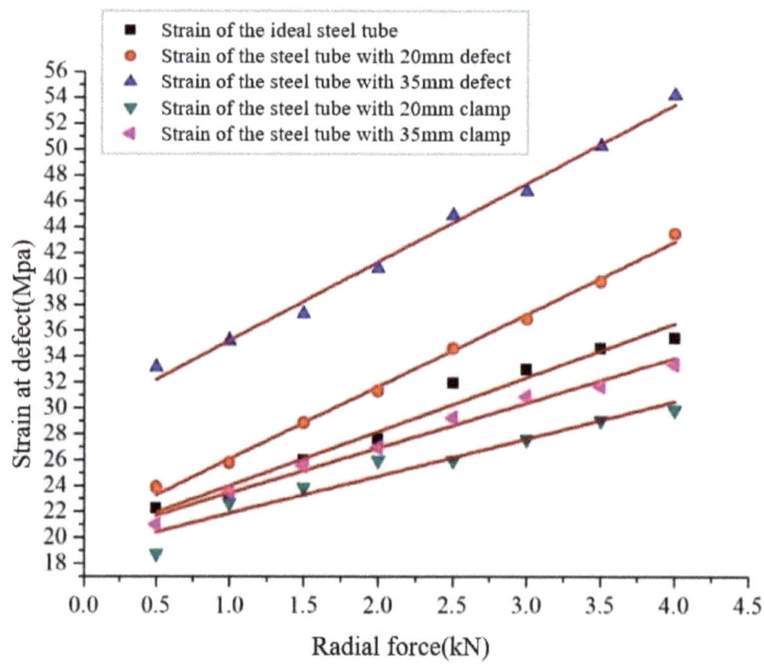

Figure 20. Stress at the measuring point of radial loading.

5. Conclusions

In this study, the bearing capacity of a jacket before and after damage was analyzed, and then the bearing capacity of the damaged jacket strengthened by a mechanical grouting clamp was assessed. These analyses were performed under axial and radial loads, and the results obtained were more representative of the actual working conditions than those under the single-direction load. Then, the bearing capacity of the damaged jacket and ideal jacket before and after hoop reinforcement were compared. It is concluded that the mechanical clamp can increase the bearing capacity by 35%. Finally, a comparison of results obtained from the experiment reveals that the surface defects will have a negative impact on the safety performance of the steel pipe, and the mechanical grouting clamp can effectively strengthen the damaged jacket and enhance the strength.

Author Contributions: The first author, B.Z., conceived the framework of the article and wrote the article; the second author, Q.R.Z., analyzed the bearing capacity of damaged steel pipe pile; the third author, T.W., analyzed the capacity of the grouting clamp; the fourth author, Z.W., was responsible for the mechanical experiment verification of the grouting clamp. All authors have read and agreed to the published version of the manuscript.

Funding: This paper was funded by NSFC (Contract name: Research on ultimate bearing capacity and parametric design for the grouted clamps strengthening the partially damaged structure of jacket pipes, Grant number: 51879063; Contract name: Research on analysis and experiments of gripping and bearing mechanism for large-scale holding and lifting tools on ocean foundation piles, Grant number: 51479043).

Conflicts of Interest: The authors declare no conflict of interest.

References

1. Ashish, A.; Sudath, C.S.; Ove, M.; Ivar, L. A framework to assess structural integrity of ageing offshore jacket structures for life extension. *Mar. Struct.* **2017**, *562*, 37–259.
2. Gong, S.-F.; Jin, W.; Wang, Q.-Z. Repair and assessment of damaged components of offshore fixed platforms. *China Ocean Platf.* **2001**, *16*, 37–41.
3. Gong, S.-F.; Shen, X.-W.; Li, F.; Wang, Q.-Z. Research on grouting clamp technology for offshore platforms. *Ocean Eng.* **2001**, *19*, 32–37.
4. David, L.; Giovanni, S.; Paul, D. Design of novel drilled-and grouted pile in sand for offshore oil & gas structures. *Mar. Struct.* **2014**, *39*, 39–49.
5. Zhao, X.L.; Grundy, P.; Lee, Y.T. Grout sleeve connections under large deformation cyclic loading. In Proceedings of the Twelfth International Society of Offshore and Polar Engineers, ISOPE, Kitakyushu, Japan, 26–31 May 2002; Volume IV, pp. 53–59.
6. Jin, W.-L.; Song, J.; Gong, S.-F.; Bai, B.-R. Bearing capacity and reinforcement analysis of damaged components of offshore platform. *Eng. Mech.* **2003**, *20*, 37–42.
7. Zhao, X.L.; Ghojel, J.; Grundy, P.; Han, L.H. Behavior of grouted sleeve connections at elevated temperatures. *Thin-Walled Struct.* **2006**, *447*, 751–758. [CrossRef]
8. Jan, D.; Michael, M. Advanced representation of tubular joints in jacket models for offshore wind turbine simulation. *Energy Procedia* **2013**, *35*, 234–243.
9. Yang, B.; Shi, X.; Li, H.-J. Pressure measurement method of expansion self-stress grouting clamp. *Offshore Eng.* **2010**, *28*, 80–85.
10. Jiang, S.C.; Wang, Z.; Zhao, X.L. Structural Performance of Prestressed Grouted Pile-to-Sleeve Connectionsin. In Proceedings of the Twelfth East Asia-Pacific Conference on Structural Engineering and Construction, Hong Kong, China, 26–28 January 2011; Volume 14.
11. Samarakoon, S.M.S.M.K.; Ratnayake, R.M.C.; Siriwardane, S.A.S.C. Structural integrity control of ageing offshore structures: Repairing and strengthening with grouted connections. In Proceedings of the Asme 32nd International Conference on Ocean, Offshore and Arctic Engineering, Nantes, France, 9–14 June 2013; Volume 3.
12. Djukic, L.P.; Sum, W.S.; Leong, K.H. Development of a fibre reinforced polymer composite clamp for metallic pipeline repair. *Mater. Des.* **2015**, *706*, 8–80. [CrossRef]
13. Sum, W.S.; Leong, K.H.; Djukic, L.P. Design, testing and field deployment of a composite clamp for pipeline repairs. *Plast. Rubber Compos.* **2016**, *45*, 81–94. [CrossRef]
14. Lee, J.H. Interfacial shear behavior of a high-strength pile to sleeve grouted connection. *Eng. Struct.* **2017**, *1517*, 704–723. [CrossRef]
15. Jiao, G.; Zhou, L.; Shi, X.; Fang, K. Load-bearing performance test of large-size model of expansion self-stress grouting clamp. *J. Ocean Univ. China* **2017**, *47*, 111–118.
16. Gunnar, S.; Atle, J. Design recommendations for grouted pile sleeve connections. *Mar. Struct.* **2018**, *601*, 1–14.

17. Johansen, A.; Solland, G.; Lervik, A.; Strande, M.; Nybo, T. Testing of jacket pile sleeve grouted connections exposed to variable axial loads. *Mar. Struct.* **2018**, *582*, 254–277. [CrossRef]
18. Chellappan, N.V.; Nallayarasu, S. Axial and moment carrying capacity of split sleeve grouted connections for repair of tubular member. In Proceedings of the Asme 38th International Conference on Ocean, Offshore and Arctic Engineering, Glasgow, Scotland, 9–14 June 2019; Volume 3.
19. Wu, X.; An, W. Effect of distributed axisymmetric defects on buckling reliability of thin-walled cylindrical shells. *J. Harbin Eng. Univ.* **2009**, *30*, 635–638.
20. Yang, X. Experimental study on flexural behavior of corroded reinforced concrete beams. *J. Nat. Disasters* **2018**, *27*, 70–78.
21. Yuan, Y.; Jia, F.; Cai, Y. Degradation of mechanical properties of corroded steel bars. *Ind. Archit.* **2000**, *30*, 43–46.
22. Wang, X.Q.; Zhang, S.M.; Liao, J.; Huang, Y.S. Research and development on similarity principle of model pile. In *Applied Mechanics and Materials;* Trans Tech Publications Ltd.: Stafa-Zurich, Switzerland, 2014; Volume 580–583.

Journal of
Marine Science and Engineering

MDPI

Article

Framework for Development of an Economic Analysis Tool for Floating Concrete Offshore Wind Platforms

David Cordal-Iglesias [1], Almudena Filgueira-Vizoso [2,*], Eugenio Baita-Saavedra [3], Manuel Ángel Graña-López [4] and Laura Castro-Santos [1]

[1] Department of Naval and Industrial Engineer, University of A Coruña, Escola Politécnica Superior, Esteiro, 15471 Ferrol, Spain; david.cordal@udc.es (D.C.-I.); laura.castro.santos@udc.es (L.C.-S.)

[2] Department of Chemist, University of A Coruña, Escola Politécnica Superior, Esteiro, 15471 Ferrol, Spain

[3] Saitec Offshore Technologies, Business Park Ibarrabarri, A-2 Building, Leioa, 48940 Bilbao, Spain; eugenio.baita@udc.es

[4] Department of Industrial Engineer, University of A Coruña, Escuela Universitaria Politécnica, 15405 Ferrol, Spain; manuel.grana@udc.es

* Correspondence: almudena.filgueira.vizoso@udc.es; Tel.: +34-881013289

Received: 20 October 2020; Accepted: 19 November 2020; Published: 24 November 2020

Abstract: The objective of this work is to establish a framework for the development of an economic analysis tool for floating offshore wind platforms built in concrete. The operation and usefulness of the beta version of a software, called Arcwind, is explained. It calculates the main economic aspects of offshore wind platforms built in concrete considering different locations in the European Atlantic Arc. This software allows the user to select different input parameters such as: the type of platform, the installation area of the farm, its specific location and in this way create different analysis scenarios. This paper analyzes the case study to install TELWIND® offshore concrete floating platforms in the Canary Islands, in Spain. The software provides data on the main economic parameters of the farm, such as Levelized Cost Of Energy (LCOE), Net Present Value (NPV), Internal Rate of Return (IRR), Pay-Back Period (PBP), as well as the main costs: cost of conception and definition, cost of design and development, cost of manufacturing, cost of installation, cost of operation and cost of dismantling. Using these parameters, a first analysis of the viability of this type of floating technology built on concrete is shown.

Keywords: TELWIND; floating offshore wind; concrete; feasibility; economics

1. Introduction

The global economy is committed to reducing greenhouse gas emissions [1]. With this purpose, policies to reduce them are being promoted. The 2015 Paris agreement [2] highlights the importance of cross-border cooperation in environmental matters between member states, including Spain and Portugal. The priority objective is to reduce greenhouse gases by 20% compared to 1990 and reach 80% by 2050 [3]. In recent years, the European Union has made considerable efforts to increase electricity generation through renewable sources. The percentage of renewable energies in the final electrical consumption has doubled from 8.5% in 2004 to 17% in 2016 [4,5].

According to the WEO 2018 report [6] of the IEA (International Energy Agency), two thirds of the world's electricity generation will come from renewable sources in 2040, and solar and wind energy must produce seven times more than coal in 2040 if the objective of the Paris agreement; to achieve this it is necessary to enhance the use of this type of energy. Within wind energy, offshore wind power has been experiencing a great increase in recent years due to other factors because its visual impact is lower [7–9], there are more possible locations [10–12], and the powers of the turbines are greater.

Numerous changes in technology are being made within offshore wind power in order to make them more economically viable [13]. One of the main changes has been to modify the materials with which the platform is made [14], moving from traditional steel platforms to concrete platforms [15].

The platform to be studied in this article is the TELWIND® platform, developed by the Esteyco company [16]. Esteyco is an independent Spanish civil engineering consultancy and architecture firm founded in 1970 which is dedicated, among other civil sectors, to the development of structures both for onshore and offshore wind power. In this field, this company revolutionized the offshore wind energy sector by building concrete structures replacing steel structures, a predominant material in the construction of subtructures for wind turbines. The steel is light and resistant, but due to restrictions in road transport, it can only be used to manufacture towers up to 80 or 90 m high; this would imply increasing the diameter of the base above 4.5 m which is prohibited in many countries [17]. In addition to this, the problem associated with the corrosion [18] of steel in marine environments that causes maintenance costs to be high is reduced [19], thus reducing the economic viability of the park. Taking these problems into account, Esteyco developed concrete structures as an alternative to steel structures. Although concrete is heavier than steel, it is a cheaper material, less susceptible to temporary price fluctuations and is intensive in terms of both local work-force and raw materials, thus promoting local content.

To this end, Esteyco has participated in several research projects partially funded by the European Commission under the Horizon 2020 program that have facilitated the development of its different concrete wind solutions both for bottom-fixed platforms and floating platforms. Examples of these projects are the ELICAN [20] and TELWIND projects. In the case of ELISA technology for bottom-fixed platforms, it consists of a self-floating gravity foundation (Gravity Base System-GBS) in its maritime transport, together with a self-installing telescopic tower, both made of concrete. The buoyancy of the joint structure, and the telescopic configuration of the tower, allow each unit to be fully mounted on land, including the turbine, and then be conventionally towed to its final position at sea. Once there, it is ballasted to the seabed and is self-hoisted using hydraulic jacks to its final height. The ELICAN project is the construction of a real-scale prototype with ELISA technology. This company subsequently developed the TELWIND® technology which consist of an evolved spar with an integrated telescopic tower [21] both made of concrete, also using the concepts developed in ELISA gravity.

The aim of the present paper is to establish a framework for the development of an economic analysis tool for floating offshore wind platforms built in concrete. The operation and usefulness of the beta version of a software, called Arcwind, is explained. It calculates the main economic aspects of offshore wind platforms built in concrete considering different locations of the European Atlantic Arc. It allows the user to select the type of platform, the area and location of the farm and the alternative to be studied. The tool establishes the beginning to create maps of the locations for concrete platforms in future researches. Floating concrete platforms for offshore wind represent a revolutionary new concept of offshore wind. Therefore, the calculation of their economic aspects is very important in order to carry out the future of offshore wind technology. The research gap of this paper is to calculate them in order to compare them with other substructures. The case study considered takes into account a farm with uses the floating offshore evolved spar platform called TELWIND®, whose technology is designed by the enterprise Esteyco® and located in the waters of Gran Canaria Island (Spain, Europe). The software gives information about the main economic parameters of the farm, such as its costs and economic feasibility indicators (Internal Rate of Return, Net Present Value or Pay-Back Period).

2. Materials and Methods

2.1. Software Description

The objective of the software created is to simplify the way to calculate the main economic aspects of offshore wind platforms built in concrete considering different locations of the European Atlantic

Arc. Castro-Santos et al. previously developed a software to calculate the main economic aspects of floating offshore wind energies, which have been built using steel [22].

This software is a very useful tool for quickly calculating the economic parameters that allow us to know the viability of an offshore wind farm. By changing the location, the program immediately calculates the costs. This program can be useful for companies and entrepreneurs related to renewable energy offshore port which greatly facilitates calculations and saves time.

The program is developed with Matlab to simplify the mathematical formulas that the software contains. In the future it will be shared with the project partners, and although it is developed with Windows, it will be available for Windows and MacOS.

The software can be installed as another application on the computer, without the need to have Matlab installed, using the MATLAB Runtime installer. If the computer has a modern version of Matlab, it would not be necessary to install any additional software.

Figure 1 describes the main operations of the software and Figure A1 (in Appendix A) is an extract of the code that calculates the cost of electrical wiring.

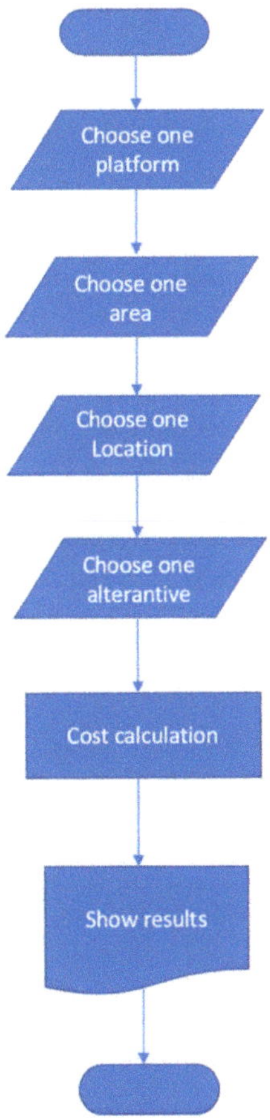

Figure 1. Flowchart.

The welcome window (Figure 2) allows the user to select the different options (steps) that they will see again in the main window. This window gives a quick access to the user.

Figure 2. Welcome window.

The software comprises four steps, as Figure 3 is shown on the top:

Step 1: the user should select the platform between the alternatives: TELWIND®. The next versions of the software will allow the selection of more platforms.

Step 2: he user should select the area of the farm. The software was designed for areas included in the Atlantic Arc of the European continent. Therefore, these locations are: Spain (Peninsula), Spain (Canary Islands), Portugal (Peninsula), Portugal (Madeira), Portugal (Azores), France, UK (North Scotland), Ireland (West Ireland) and Ireland (South Ireland). However, it is important to notice that these locations can be increased in the future in order to extend the study to the whole world.

Step 3: the user should select the specific location of the area selected.

Step 4: the user should choose an alternative of study. From the main menu they can create new alternatives to calculate new scenarios. In this beta version of the software, three alternatives have been considered: electricity rate, insurance price and dismantling cost.

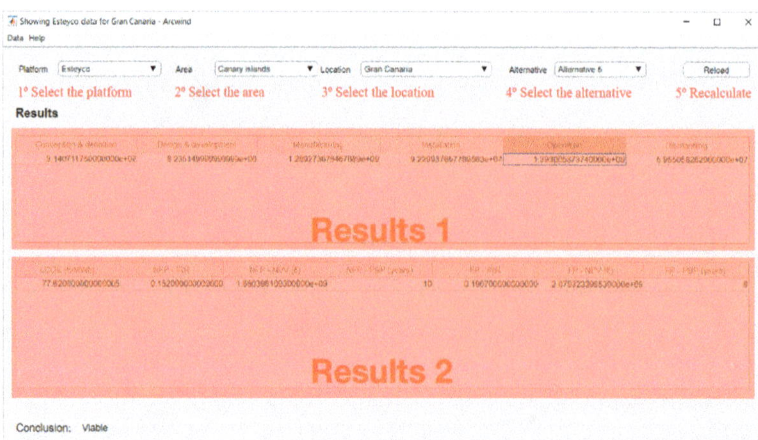

Figure 3. Main window of the software and types of the results depending on the scenario selected.

2.2. Input Values

The input values for each of the steps previously defined are:

Step 1: Platforms

Inputs related to the platform:

- Dimensions of the platform.
- Mass of the main material of the platform without the wind turbine (kg).
- Mass of the auxiliary systems of the platform without the wind turbine (kg).
- Material cost of building the platform (€/platform).
- Labour cost of building the platform (€/platform).
- Indirect cost of building the platform (€/platform).
- Total cost of building the platform (€/platform).
- Inputs related to the mooring lines:
- Material of the mooring lines.
- Type of mooring lines (tensioned, non-tensioned).
- Number of mooring lines.
- Mass of the mooring lines (kg).
- Cost of the mooring lines (€/kg).
- Inputs related to the anchoring:
- Material of the anchoring.
- Type of anchoring.
- Number of anchors.
- Mass of the anchors (kg).
- Cost of the anchors (€/kg).
- Inputs of installing:
- Description of the installation process.
- Description of the decommissioning process.
- Cost of installing one platform (€/platform).
- Inputs related to the maintenance procedure.
- Description of the maintenance process.
- Cost of the maintenance of one platform over one year (€/platform year).
- Cost of the maintenance of one mooring line over one year (€/platform year).
- Cost of the maintenance of one anchoring system over one year (€/platform year).

Step 2: Area & **Step 3:** Location

- Medium wind speed.
- Depth location.
- Height of waves.
- Period of waves.
- Scale parameter.
- Shape parameter.
- Distance from farm to shore.
- Onshore distance of the electric cable.
- Distance from shipyard to port.
- Distance from farm to shipyard.

Step 4: Alternatives

- Assurance cost.
- Dismantling cost.
- Electric tariff.

In the alternatives window (Figure 4) the user can verify the different alternative values and create new ones. By changing the chosen alternative, we can verify whether the wind farm is viable for the scenario selected.

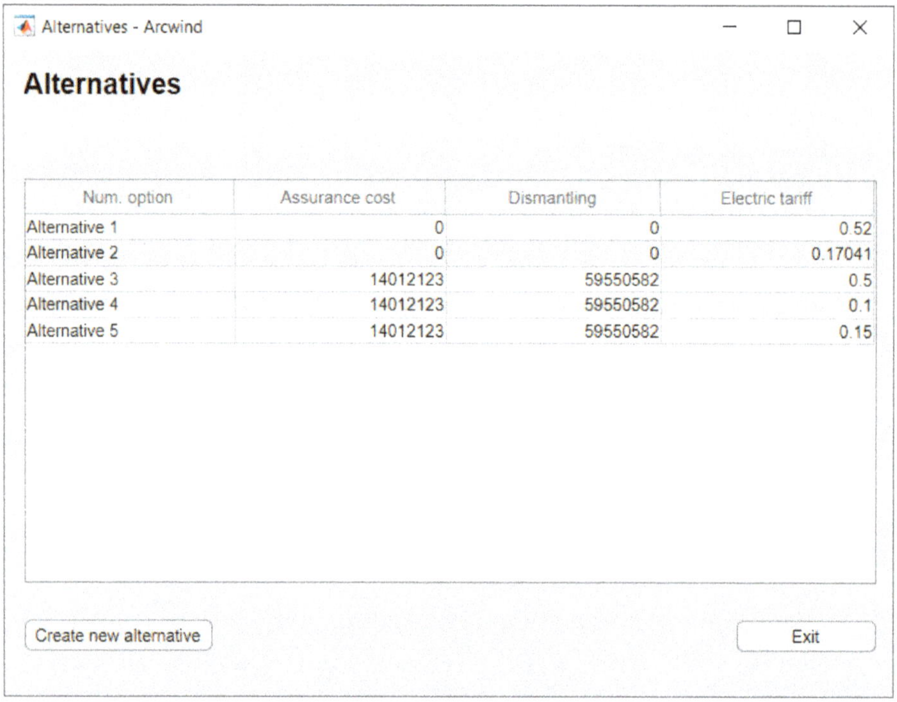

Figure 4. Alternatives window.

2.3. *Output Values*

The bottom of the Figure 3 shows the two types of results (output values):

(a) Results 1: the costs of the farm are shown [23–25]:

- C1: cost of conception and definition.
- C2: cost of design and development.
- C3: cost of manufacturing.
- C4: cost of installation.
- C5: cost of operation.
- C6: cost of dismantling.

The cost of dismantling is eligible from the alternative menu because as it is developed in the last year of the project it could or could not be considered in the feasibility study.

(b) Results 2: the economic values of the feasibility of the farm are shown:

- LCOE (€/MWh): Levelized Cost Of E Energy.
- NPV (€): Net Present Value of the Non Funded Project (NFP) and the Funded Project (FP).
- IRR (%): Internal Rate of Return of the Non Funded Project (NFP) and the Funded Project (FP).
- PBP (years): Pay-Back Period of the Non Funded Project (NFP) and the Funded Project (FP).

2.4. Functioning

Once the platform and location have been chosen, the software begins to internally calculate the costs of conception and definition, design and development, manufacturing, installation, operation and dismantling. Each process is divided into its own function, where the location and platform values are given and the total cost of each process is returned. Internally, the value of each value can be known, for example: cost of electric cables, cost of anchors, cost of installation of the platform, cost of installation of anchors, etc. For simplicity, the values shown are only the totals. With these total results, the rate of return, the net present value and the payback period are calculated.

To calculate the maintenance, the scheduled checks of the turbine, the cables or the platform are taken into account. The rental price of transport to the park is also calculated based on the distance to the port, the number of people needed and the type of ship.

For example, some of the cost calculations are internally separated into: annual maintenance, crew, maintenance every 5 years, preventive maintenance of the turbine, corrective maintenance of the turbine, cost of spare parts for the turbine, maintenance and inspection of electric cables, maintenance and inspection of the anchors, equipment for inspections, cost of foundation repair, ...

All these calculations are reflected in the operating cost of the park (C5).

3. Case Study

The costs depend on many variables, such as bathymetry, which determines the length of the submarine cables and the mooring and anchor lines dimensions. However, there are many variables that depend on the characteristics and design of the floating structures selected for each location.

The case study will be focused on the TELWIND® platform (see Figure 5), which is an evolved spar platform designed by the Spanish Enterprise Esteyco®, which is a partner of the ARCWIND Project, that financed the present research. It is an evolved spar floating platform consisting of two bodies connected by steel tendons together with a self-erecting telescopic tower.

Regarding the turbine that this floating structure must support, the project has chosen to use the DTU 10 MW Reference Wind Turbine. The characteristics of the turbine, such as power and height, influence costs and energy production, two very important factors in the economic analysis of the project.

The farm will have a total power of 500 MW. Moreover, the area selected for the case study is the Canary Islands (Spain) and the particular location is the Gran Canaria Island.

The alternative studied has the following values:

- Assurance cost: 14,012,124.74 €
- Dismantling: 59,550,582 €
- Electric tariff: 0.150 €/kWh

Figure 5. TELWIND® platform. Source: Figure courtesy of Esteyco (Reproduced from [16], with permission from Esteyco, 2020).

4. Results

Figure 6 shows the main components of costs of the TELWIND® platform. Manufacturing and Operation costs are the most representative costs in the chart.

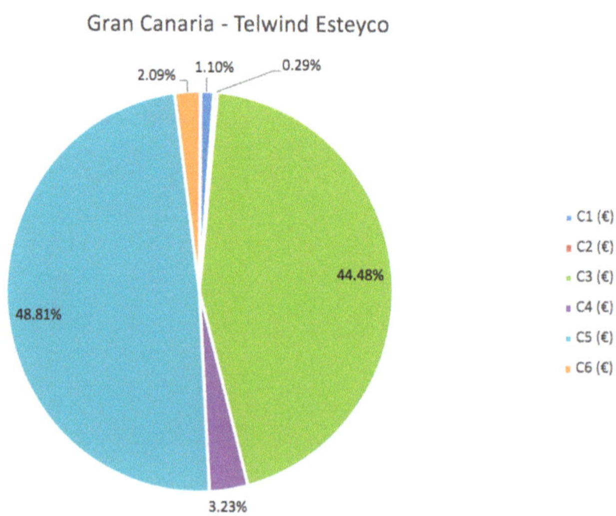

Figure 6. Main components of costs for the TELWIND® platform.

Table 1 shows the results of costs and subcosts for the case study considered: TELWIND® platform in the Gran Canaria location. Offshore logistics is the most important sub-phase for exploitation, reaching 800 M€, due to the high tariff of the means necessary for offshore maintenance tasks, such as special marine cranes. Regarding manufacturing costs, the generators (540 M€) and the floating platforms (473 M€) are the most important sub-phases.

Table 1. Cost estimation for the TELWIND® platform.

Costs	Value
C1 (€)	31,407,117.50 €
C12—Legal aspects	5,000,000.00 €
C13—Wind resource study, sea states study, seabed study	26,407,117.50 €
C2 (€)	8,235,150.00 €
C3 (€)	1,263,873,385.65 €
C31—Generator manufacturing cost	540,771,740.60 €
C32—Floating platform manufacturing cost	473,837,908.79 €
C33—Mooring manufacturing cost	183,531,825.6
C34—Anchoring manufacturing cost	9,983,040.00 €
C35—Electric systems manufacturing cost	55,748,870.66 €
C4 (€)	92,299,376.58 €
C41—Generator installation cost	6,875,000.00 €
C42—Floating platform installation cost	33,560,240.84 €
C43—Mooring and anchoring installation cost	31,334,135.73 €
C44—Electric systems installation cost	20,530,000.00 €
C5 (€)	1,392,951,900.30 €
C51—Assurance cost	13,958,150.30 €
C52—Administration and Operations cost	50,468,750.00 €
C53—Maintenance cost	444,150,000.00 €
C54—Onshore logistics	17,187,500.00 €
C55—Offshore logistics	867,187,500.00 €
C6 (€)	59,550,582.62 €
C61—Generator dismantling cost	28,304,668.59 €
C62—Mooring and anchoring dismantling cost	28,200,722.16 €
C63—Electric system dismantling cost	3,045,191.86 €

Regarding the installation costs, the turbine does not involve so many costs as compared to the floating platform and its anchorage to the seabed due to its ability to be commissioned onshore and under controlled weather conditions.

Regarding the economic feasibility results, considering the alternative selected, the Internal Rate of Return (IRR) is 19.07%, the Net Present Value (NPV) is 2,079,323,398 € and the Discounted Pay-Back Period (DPBP) is 8 years. These results indicate the economic feasibility of the project because the IRR is higher than the capital cost considered (6%), the NPV is higher than zero and the DPBP is less than the life-cycle of the project (20 years). Obviously, these results are very good because the alternative selected is considered to be a good electric tariff (0.150 €/kWh) considering the Spanish tariff. However, this tariff is closed to other European countries where the offshore wind is greatly developed.

5. Discussion

The goal of the present work has been to explain the beta version of a software that calculates the main economic aspects of offshore wind platforms built in concrete considering different locations of the European Atlantic Arc.

The software allows the user to select the type of structure, the location where the farm will be installed and the alternative of study (predefined or created by the user).

The case study analyzed studied a farm which uses the floating offshore spar platform called TELWIND®, whose technology is designed by the enterprise Esteyco®. The location selected has been the Gran Canaria Island (Spain, Europe), a region close to the location of the 5 MW prototype that was installed in 2017 by the same enterprise. In addition, it is important to notice that this type of technology can be very useful for islands, whose isolation and low onshore surface causes other types of renewable technologies to become less effective in the future.

The software gives information about the main economic parameters of the farm, such as its costs and economic feasibility indicators (Internal Rate of Return, Net Present Value or Pay-Back Period).

The most important costs are the manufacturing costs of the platforms and the wind turbine, and Operation and Maintenance costs. The costs of offshore operations represent an important cost to the life cycle of an offshore wind farm, due to the need to use marine cranes that work at high altitude to repair the rotor and the blades of the wind turbine (the turbine hub is located at more than 100 m high). O and M costs can be reduced by optimizing the maintenance strategy.

Another influential factor in costs is the manufacturing of the floating structure, which should develop the concept innovation and subsequently use the economy of scale to reduce costs and reach commercial level.

Generally, the lower the costs in the cycle life of an offshore wind farm, the greater the probability that this project will reach an acceptable economic viability. However, it is important to keep in mind that there are many other factors, such as the energy produced.

The energy generated by the wind turbines at each location must be taken into account. Energy production depends on the characteristics of the location, the distance to the coast, the meteorological conditions and the metaoceanic data. Therefore, the costs of the life cycle of a project are not definitively indicative of feasibility.

It is also important to mention that this software allows us to identify the phases of the life cycle that have the greatest impact, and therefore where measures should be taken to reduce costs and improve the competitiveness of floating technology in the future.

Author Contributions: Conceptualization, A.F.-V. and E.B.-S.; Investigation, D.C.-I., A.F.-V., E.B.-S., M.Á.G.-L. and L.C.-S.; Methodology, D.C.-I. and L.C.-S.; Resources, M.Á.G.-L.; Software, D.C.-I. and A.F.-V.; Supervision, L.C.-S.; Writing—original draft, E.B.-S.; Writing—review & editing, A.F.-V. and L.C.-S. All authors have read and agreed to the published version of the manuscript.

Funding: This research was funded by EAPA_344/2016 ARCWIND project, co-financed by the European Regional Development Fund though the Interreg Atlantic Area Programme.

Acknowledgments: The authors would like to thank the enterprise Esteyco for the information provided.

Conflicts of Interest: The authors declare no conflict of interest.

Appendix A

```
% section
section_11 = 1;
auxCable(2) = num2cell(section_11);
% power
power_11 = P * section_11;
auxCable(3) = num2cell(power_11);
% Intensity
I_11 = (power_11 * 10^6) / (cos_phi * (3^0.5) * V_inter * 10^3);
auxCable(4) = num2cell(I_11);
% Max intensity

interCable = sortrows(Cable, 8, 'ascend');
validCable = cell(1, 4);
valid = false;
i = 1;
while valid == false
    if str2double(interCable(i,8)) > I_11
        validCable(1) = interCable(i, 8);
        validCable(2) = interCable(i, 9);
        validCable(3) = interCable(i, 10);
        validCable(4) = interCable(i, 11);
        valid = true;
    end
    i = i + 1;
end
I_max_11 = str2double(validCable(1));
auxCable(5) = num2cell(I_max_11);
% Section (mm^2)
section_11 = str2double(validCable(2));
auxCable(6) = num2cell(section_11);
% line cost
valid = false;
i = 1;
aux_line_cost_11 = 0;
while valid == false
    if cell2mat(triCost(i,1)) == section_11
        aux_line_cost_11 = cell2mat(triCost(i, 2));
        valid = true;
    end
    i = i + 1;
end
line_cost_11 = aux_line_cost_11;
auxCable(7) = num2cell(line_cost_11);
% length (m)
length_11 = L_turbine;
auxCable(8) = num2cell(length_11);
% Total cost (€)
total_cost_11 = line_cost_11 * (length_11 / 1000);
auxCable(9) = num2cell(total_cost_11);
% resistence
resistence_11 = Cu_resistivity * length_11 / (section_11 / 10^6);
auxCable(10) = num2cell(resistence_11);
% Resistence 90°
resistence_90_11 = (resistence_11 * (1 + (coefficient_var_tmp * temp_diff))) / (length_11 / 1000);
auxCable(11) = num2cell(resistence_90_11);
```

Figure A1. Example of the software code.

References

1. Enkvist, P.-A.; Nauclér, T.; Rosander, J. *A Cost Curve for Greenhouse Gas Reduction*; McKinsey and Company: New York, NY, USA, 2016; pp. 35–45.
2. European Union. Acuerdo de París|Acción por el Clima. Available online: https://ec.europa.eu/clima/policies/international/negotiations/paris_es (accessed on 31 July 2019).
3. 2009/28/EC, D. *Directive 2009/28/ec of the European Parliament and of the Council of 23 April 2009 on the Promotion of the Use of Energy from Renewable Sources and Amending and Subsequently Repealing Directives 2001/77/EC and 2003/30/EC (Text with EEA Relevance)*; Oficial Journal of the European Union: Brussels, Belgium, 2009; pp. 16–62.

4. Estadísticas de Energía Renovable. Eurostat Stat. Explain. 2018. Available online: http://ec.europa.eu/eurostat/statistics-explained/index.php (accessed on 6 November 2020).
5. Heesterman, A. The pace and practicality of decarbonization. *Clean Technol. Environ. Policy* **2017**, *19*, 295–310. [CrossRef]
6. IEA. *World Energy Outlook 2018*; IEA: Paris, France, 2018; Available online: https://www.iea.org/reports/world-energy-outlook-2018 (accessed on 6 November 2020).
7. Real, H.-P.B.S. Global renewable energies: A dynamic study of implementation time, greenhouse gas emissions and financial needs. *Clean Technol. Environ. Policy* **2006**, *8*, 159–173.
8. Molina-Ruiz, J.; Martínez-Sánchez, M.J.; Pérez-Sirvent, C.; Tudela-Serrano, M.L.; Lorenzo, M.L.G. Developing and applying a GIS-assisted approach to evaluate visual impact in wind farms. *Renew. Energy* **2011**, *36*, 1125–1132. [CrossRef]
9. Ray, P. Renewable energy and sustainability. *Clean Technol. Environ. Policy* **2019**, *21*, 1517–1533. [CrossRef]
10. Bidwell, D. Ocean beliefs and support for an offshore wind energy project. *Ocean Coast. Manag.* **2017**, *146*, 99–108. [CrossRef]
11. Ou, L.; Xu, W.; Yue, Q.; Ma, C.L.; Teng, X.; Dong, Y.-E. Offshore wind zoning in China: Method and experience. *Ocean Coast. Manag.* **2018**, *151*, 99–108. [CrossRef]
12. Wang, J.; Zou, X.; Yu, W.; Zhang, D.; Wang, T. Effects of established offshore wind farms on energy flow of coastal ecosystems: A case study of the Rudong offshore wind farms in China. *Ocean Coast. Manag.* **2019**, *171*, 111–118. [CrossRef]
13. Castro-Santos, L.; Filgueira-Vizoso, A.; Álvarez-Feal, C.; Carral, L. Influence of size on the economic feasibility of floating offshore wind farms. *Sustainability* **2018**, *10*, 4484. [CrossRef]
14. Borrell, C.M.; Hortigüela, A.C. WindCrete. *Wind. Int.* **2016**, *12*, 1–4.
15. Quilligan, A.; O'Connor, A.; Pakrashi, V. Fragility analysis of steel and concrete wind turbine towers. *Eng. Struct.* **2012**, *36*, 270–282. [CrossRef]
16. Esteyco. Available online: https://www.esteyco.com (accessed on 27 June 2010).
17. Cinco Días. Available online: https://cincodias.elpais.com (accessed on 6 November 2020).
18. Momber, A.W. Quantitative performance assessment of corrosion protection systems for offshore wind power transmission platforms. *Renew. Energy* **2016**, *94*, 314–327. [CrossRef]
19. Castro-Santos, L.; Filgueira-Vizoso, A.; Carral-Couce, L.; Formoso, J.Á.F. Economic feasibility of floating offshore wind farms. *Energy* **2016**, *112*, 868–882. [CrossRef]
20. Serna, J.; Nieto, L.C.J. ELISA Technology. Concrete Offshore Wind Tower, Self-Floating and Telescopic. *Ing. Civil.* **2017**, *188*, 8.
21. Dankelmann, S.; Visser, B.; Gupta, N.; Serna, J.; Couñago, B.; Urruchi, Á.; Fernández, J.L.; Cortés, C.; Guanche García, R.; Jurado, A. TELWIND-Integrated Telescopic tower Combined with an Evolved Spar Floating Substructure for Low-Cost Deep Water Offshore Wind and Next Generation of 10 MW+ Wind Turbines. Available online: https://windeurope.org (accessed on 6 November 2020).
22. Castro-Santos, L.; Filgueira-Vizoso, A. A software for calculating the economic aspects of floating offshore renewable energies. *Int. J. Environ. Res. Public Health* **2020**, *17*, 218. [CrossRef] [PubMed]
23. Castro-Santos, L.; Diaz-Casas, V. Life-cycle cost analysis of floating offshore wind farms. *Renew. Energy* **2014**, *66*, 41–48.
24. Cartelle-Barros, J.J.; Cordal-Iglesias, D.; Baita-Saavedra, E.; Filgueira-Vizoso, A.; Couñago-Lorenzo, B.; Vigara, F.; Cortés, C.; Cerdán, L.; Nieto, J.; Serna, J.; et al. Pathways to bring the costs down of floating offshore wind farms in the Atlantic Area. *Wind Energy Sci.* **2019**. [CrossRef]
25. Baita-Saavedra, E.; Cordal-Iglesias, D.; Filgueira-Vizoso, A.; Castro-Santos, L. Economic aspects of a concrete floating offshore wind platform in the Atlantic Arc of Europe. *Int. J. Environ. Res. Public Health* **2019**, *16*, 4122. [CrossRef] [PubMed]

Publisher's Note: MDPI stays neutral with regard to jurisdictional claims in published maps and institutional affiliations.

MDPI
St. Alban-Anlage 66
4052 Basel
Switzerland
Tel. +41 61 683 77 34
Fax +41 61 302 89 18
www.mdpi.com

Journal of Marine Science and Engineering Editorial Office
E-mail: jmse@mdpi.com
www.mdpi.com/journal/jmse

www.ingramcontent.com/pod-product-compliance
Lightning Source LLC
Chambersburg PA
CBHW061455220926
51850CB00004B/98
9783036585918